**elements of water supply
and wastewater disposal**

## Gordon Maskew Fair
*Late Abbott and James Lawrence Professor
of Engineering and Gordon McKay Professor
of Sanitary Engineering, Emeritus
Harvard University*

## John Charles Geyer
*Professor of Environmental Engineering
The Johns Hopkins University*

## Daniel Alexander Okun
*Professor of Environmental Engineering
The University of North Carolina at Chapel Hill*

Wiley International Edition

# elements of water supply and wastewater disposal

**second edition**

*With a chapter on groundwater by Dr. Jabbar K. Sherwani,
Associate Professor of Hydrology, The University of North Carolina at Chapel Hill*

**John Wiley & Sons, Inc.**
New York, London, Sydney, Toronto

**Toppan Company, Ltd.**
Tokyo, Japan

# preface

During the final preparation of this edition, the senior author, Professor Gordon Maskew Fair, died. He devoted most of his time and effort during the last two years of his life to the preparation of this volume. As in our earlier books, the style and content reflect his mastery of the language and his dedication to improving the scientific basis for engineering practice. In the normal course of events Professor Fair would have prepared this preface. Much of his thought and language from the original edition appears in this edition.

Today, effective design and efficient operation of engineering works ask, above all, for a fuller understanding and application of scientific principles. Thus, in water engineering as in other fields, the results of scientific research are being incorporated with remarkable success in new designs and new operating procedures.

The study of scientific principles is best accomplished in the classroom; the application of these principles follows as a matter of practice. To bridge the way from principle to practice, we suggest that the study of this textbook be supplemented by (1) visits to water and wastewater works, (2) examination of plans and specifications of existing plants, (3) readings in the periodicals listed in the appendix to this book, (4) study of the data and handbook editions of trade journals, and (5) examination of the catalogues and bulletins of equipment manufacturers.

This volume shares much of its purpose, content, and statement with the two-volume work *Water and Wastewater Engineering* (Wiley, 1966), in which water-quality management is considered as a unified undertaking. However, in order to direct this book more specifically to the needs of undergraduate students and to the needs of others seeking fundamental principles in this field, some of the subject matter has been simplified, some of it has been recast, and some has been shortened.

The chapter on groundwater has been revised and expanded by Jabbar K. Sherwani, whose assistance is gratefully acknowledged. We thank Mrs. William Hutchinson for preparing the bulk of the typescript.

*November 1970*

John C. Geyer
Daniel A. Okun

# contents

| | | |
|---|---|---|
| 1 | Water and Wastewater Systems | 1 |
| 2 | Systems Capacities | 17 |
| 3 | Statistical Hydrology | 41 |
| 4 | Surfacewater Collection | 69 |
| 5 | Groundwater Development | 112 |
| 6 | Water Transmission | 163 |
| 7 | Water Distribution | 190 |
| 8 | Flow in Sewers and Their Appurtenances | 224 |
| 9 | Wastewater Removal | 247 |
| 10 | Water-Quality Management | 280 |
| 11 | Unit Operations and Treatment Kinetics | 310 |
| 12 | Aeration and Gas Transfer | 343 |
| 13 | Screening, Sedimentation, and Flotation | 362 |
| 14 | Filtration | 397 |
| 15 | Flocculation, Adsorption, Desalination, and Ion Exchange | 437 |
| 16 | Chemical Coagulation, Precipitation, and Stabilization, Including the Mitigation of Corrosion | 464 |
| 17 | Disinfection | 500 |
| 18 | Biological Transfer Processes | 525 |
| 19 | Biological Treatment Systems | 558 |
| 20 | Treatment and Disposal of Waste Solids | 592 |
| 21 | Ecology and Management of Natural and Receiving Waters | 636 |
| 22 | Engineering Projects | 678 |

| | | |
|---|---|---|
| *Appendix* | | 688 |
| *Bibliography* | | 719 |
| *Index* | | 727 |

**elements of water supply
and wastewater disposal**

one

# water and
# wastewater systems

## 1-1 Water for Towns and Cities

Water is led into communities for many purposes: (1) for drinking and culinary uses; (2) for washing, bathing, and laundering; (3) for cleaning windows, walls, and floors; (4) for heating and air conditioning; (5) for watering lawns and gardens; (6) for washing cars and sprinkling streets; (7) for filling swimming and wading pools; (8) for display in fountains and cascades; (9) for producing hydraulic and steam power; (10) for carrying on numerous and varied industrial processes; (11) for protecting life and property against fire; and (12) for removing offensive and possibly dangerous wastes from household (sewage) and industry (industrial wastewaters). To provide for these various uses, which average about 100 gallons per capita daily (gpcd) in average United States *residential* communities and 150 gpcd in large *industrial* cities, the supply of water must be satisfactory in quality and adequate in quantity, on tap day and night, readily available to the user, relatively cheap, and easily disposed of after it has served its purpose (Chap. 2). Necessary engineering works are waterworks, or water-supply systems, and wastewater works, or wastewater disposal systems.[1]

[1] The terms *sewage*, *sewage works*, and *sewerage systems* are more limited in concept than the terms *wastewater*, *wastewater works*, and *wastewater-disposal systems*. They find less use today than they once did.

1

Waterworks tap natural sources of supply, treat and purify the water collected, and deliver it to the premises of the consumer. Wastewater works remove the *spent water*—about 70% of the water supplied—together with such groundwater and surface runoff as may enter or be admitted to the collecting system. Street gutters, open channels, and covered structures for the removal of runoff from rainstorms and melting snow and ice are among the earliest public works of urban communities.[2] They kept the low-lying and often highly developed portions of the community from being flooded by converging and areally increasing overland flows. The collecting conduits emptied into nearby streams and other bodies of water that composed the natural drainage channels of the region. Not until the industrial revolution of the nineteenth century and the associated rapid rise of cities were these *storm sewers* called into service as everyday conveyors also of wastewaters and water-carried waste matters from dwellings and working places.[3] Originally single-purpose systems were thus converted into *combined* systems (Sec. 1-4). Although still in duty in the older communities of the world, few, if any, combined systems have been built in the United States since the turn of the last century. Health departments began to object to them at that time, and coming decades should see their total displacement by individualized or *separate* systems of *sanitary* sewers and *stormwater* sewers or by surrogate constructions. The objective is to reduce and eventually prevent the pollution of surface water sources by overflows from combined systems.

The connecting link between water supply and wastewater disposal is the system of water and wastewater piping within dwellings, mercantile and commercial establishments, and industries. More often than not, the collection and disposal of solid wastes is an independent undertaking. Exceptions are (1) the grinding of garbage and its discharge to sewers and (2) the operation of refuse incinerators in conjunction with wastewater-treatment works. Figure 1-1 illustrates, from the householder's point of view, the progress from the individualistic practices of rural populations to the communal services provided for the urban dweller. Associated problems of water-quality management are indicated.

## 1-2 Water-Supply Systems

Municipal water systems generally comprise (1) collection or intake works, (2) purification or treatment works, (3) transmission works, and (4) distri-

---

[2] The great sewer of Rome, known as the *cloaca maxima* and constructed to drain the Forum, is an example. It is still in operation.

[3] That storms presumably had long before washed street and market refuse into existing storm drains is suggested in Jonathan Swift's description of a London (England) shower in October 1710: "Now from all parts the swelling kennels flow, and bear their trophies with them as they go: Filth of all hues and odour, seem to tell what street they sailed from, by their sight and smell."

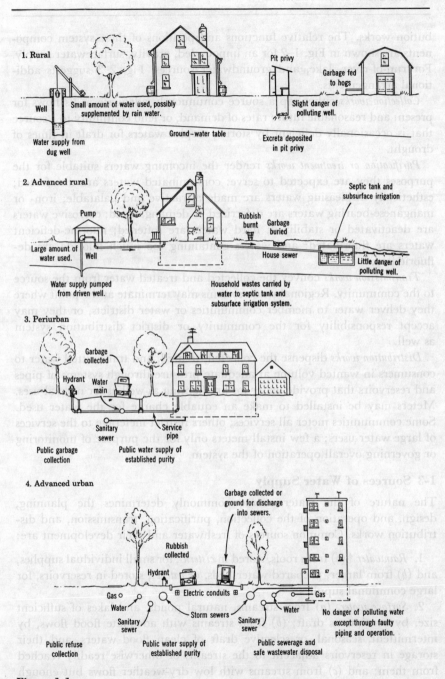

Figure 1-1.
Rural and urban water supply and wastewater disposal.

bution works. The relative functions and positions of these system components are shown in Fig. 1-2 for an impounded, gravity, surfacewater supply. For run-of-river, lake, and groundwater sources, Fig. 2-3 suggests additional arrangements.

*Collection works* either tap a source continuously adequate in its flows for present and reasonable future rates of demand, or lend continuity to a source that is occasionally deficient by storing surplus waters for draft in times of drought.

*Purification or treatment works* render the incoming waters suitable for the purposes they are expected to serve: contaminated waters are disinfected; esthetically displeasing waters are made attractive and palatable; iron- or manganese-bearing waters are deferrized or demanganized; corrosive waters are deactivated or stabilized; hard waters are softened; fluorine-deficient waters are fluoridated; and waters containing too much fluoride are defluoridated.

*Transmission works* convey the collected and treated water from the source to the community. Regional water systems may terminate at the point where they deliver water to member communities or water districts, or they may accept responsibility for the community or district distribution system as well.

*Distribution works* dispense the collected, treated, and transmitted water to consumers in wanted volume at adequate pressure through systems of pipes and reservoirs that provide water for fire-fighting as well as for normal uses. Meters may be installed to make an equable charge for the water used. Some communities meter all services; others restrict meterage to the services of large water users; a few install meters only for the purpose of monitoring or governing overall operation of the system.

## 1-3 Sources of Water Supply

The nature of the water source commonly determines the planning, design, and operation of the collection, purification, transmission, and distribution works. Common sources of freshwater and their development are:

1. *Rainwater.*[4] (a) from roofs, stored in *cisterns*, for small individual supplies, and (b) from larger, prepared watersheds, or *catches*, stored in reservoirs, for large communal supplies.

2. *Surface water.* (a) from streams, natural ponds, and lakes of sufficient size, by continuous draft; (b) from streams with adequate flood flows, by intermittent, seasonal, or selective draft of clean flood waters, and their storage in reservoirs adjacent to the streams or otherwise readily reached from them; and (c) from streams with low dry-weather flows but enough

[4] Strictly speaking, rainwater is usually collected as surface runoff.

Sources of Water Supply/5

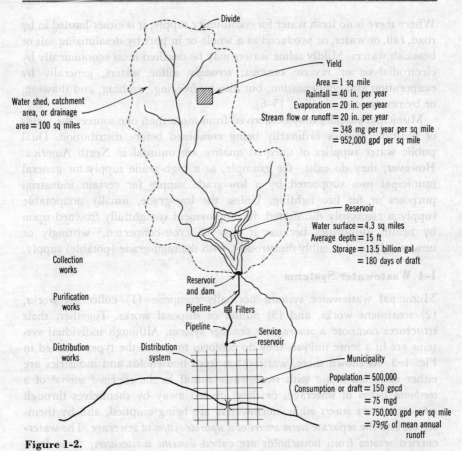

**Figure 1-2.**
**Rainfall, runoff, storage, and draft in the development of surfacewater supplies.**

annual discharge, by storage of wet-weather flows in reservoirs impounded by dams thrown across stream valleys.

3. *Groundwater.* (a) from natural springs; (b) from wells; (c) from infiltration galleries, basins, or cribs; (d) from wells, galleries and, possibly, springs, with flows augmented from some other source (1) spread on the surface of the gathering ground, (2) carried into charging basins or ditches, or (3) led into diffusion galleries or wells; and (e) from wells or galleries with flows maintained by returning to the ground water previously withdrawn from the same aquifer for cooling or similar purposes.

On board ship and in arid lands, salt or brackish water may have to be supplied for all but drinking and culinary uses. Ships usually carry drinking water in tanks, but they may produce fresh water also by desalting seawater.

Where there is no fresh water for community supply, it is either hauled in by road, rail, or water, or produced as a whole or in part by desalinizing salt or brackish waters. Mildly saline waters may be desalted most economically by electrodialysis or reverse osmosis; strongly saline waters, generally by evaporation and condensation, but also by freezing, washing, and thawing, or by reverse osmosis (Sec. 15-6).

Municipal supplies may be derived from more than one source, the yields of multiple sources ordinarily being combined before distribution. Dual public water supplies of unequal quality are unusual in North America. However, they do exist—for example, as a high-grade supply for general municipal uses supported by a low-grade supply for certain industrial purposes or for fire fighting. Unless the low-grade, usually nonpotable supply is rigorously disinfected, its employment is rightfully frowned upon by health authorities because it may be cross-connected,[5] wittingly or unwittingly, and possibly disastrously, with the high-grade (potable) supply.

## 1-4 Wastewater Systems

Municipal wastewater systems normally comprise (1) collection works, (2) treatment works, and (3) outfall or disposal works. Together, their structures compose a *sewerage* or *drainage* system. Although individual systems are in a sense unique, they do conform to one of the types outlined in Fig. 1-3. As shown there, wastewaters from households and industries are either collected along with stormwater runoff in the *combined sewers*[6] of a *combined system* of sewerage, or they are led away by themselves through *separate sanitary sewers* while stormwaters are being emptied, also by themselves, into the separate *storm sewers* of a *separate system* of sewerage. The water-carried wastes from households are called *domestic wastewaters;* those from manufacturing establishments are referred to as *industrial* or *trade wastes; municipal wastewaters* include both kinds. Combined sewerage systems are common to the older cities of the world,[7] where they generally evolved from existing systems of storm drains (Sec. 1-1).

The converging conduits of wastewater-collection works remove wastewaters from households and industries or stormwater in *free* flow as if the waters were traveling along branch or tributary streams into the trunk or main stem of an underground river system. Sometimes the main collector of combined systems has, in fact, been a brook eventually covered over when

---

[5] A *cross-connection* is a junction between water supply systems through which water from a doubtful or unsafe source may enter an otherwise safe supply.

[6] The word *sewer* is derived ultimately from the Latin *ex*, out and *aqua*, water. In the eighteenth and early nineteenth centuries, the common form of the word was *shor*.

[7] The sewerage systems of London, England, Paris, France, New York City, and Boston, Mass., are examples of this evolution.

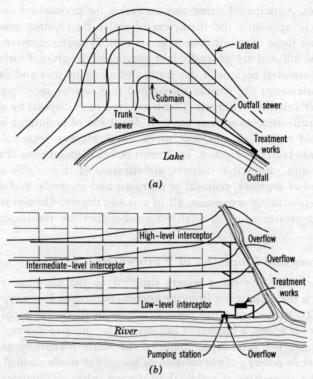

**Figure 1-3.**
**Plans of sanitary and combined sewerage systems.** (*a*) Sanitary system; (*b*) combined system.

pollution made its waters too unsightly, malodorous, and otherwise objectionable. For free or gravity flow, sewers and drains must head continuously downhill, except where pumping stations lift flows through force mains into higher-lying conduits. Pumping either avoids the costly construction of deep conduits in flat country or bad ground, or it transfers wastewaters from low-lying subareas to higher-lying main drainage schemes. Gravity sewers are not intended to flow under pressure. However, some systems designs are looking to the future provision of comminuters or grinders and pumps in individual buildings for the purpose of discharging nonclogging wastewaters from individual plumbing and drainage systems through pressurized pipes.

Hydraulically, gravity sewers are designed as *open channels* flowing partly full or, at most, just filled. Vitrified-clay and asbestos-cement pipes are generally the material of choice for small gravity sewers, and concrete pipes or conduits for larger ones. Plastic tubes reinforced with glass fibers are on the market. Anticipated sizing and operation for pressurized networks are expected to approach the diameters of water distribution systems more nearly than those of gravity sewerage systems. Pressurized sewers and force mains flow full and are generally laid parallel to the ground surface.

In well-watered regions of the earth and along its seas and oceans, collected wastewaters are usually discharged into nearby receiving bodies of water after suitable treatment. This is referred to as disposal by *dilution*, but natural purification as well as physical dispersion of pollutants is involved. In semiarid regions or otherwise useful circumstances, terminal discharge may be onto land by *irrigation*. Treatment before disposal aims at removal of unsightly and putrescible matters, stabilization of degradable substances, elimination of organics, removal of nutrients and minerals, and destruction of disease-producing organisms, all in suitable degree. Conservation of the water and protection of its quality for other uses are the important considerations.

## 1-5 Sources and Properties of Wastewaters

Wastewaters are the spent waters supplied to the community. *Domestic wastewater* is the spent water from the kitchen, bathroom, lavatory, toilet, and laundry. To the mineral and organic matter in the water supplied to the community is added a burden of human excrement, paper, soap, dirt, food wastes, and other substances. Some of the waste matters remain in suspension, while others go into solution or become so finely divided that they acquire the properties of colloidal (dispersed, ultramicroscopic) particles. Many of the waste substances are organic and serve as food for saprophytic microorganisms, that is, organisms living on dead organic matter. Because of this, domestic wastewater is unstable, biodegradable, or putrescible.

Here and there, and from time to time, intestinal pathogens reach domestic and other wastewaters. Accordingly, it is prudent to consider such wastewaters suspect at all times.

The carbon, nitrogen, and phosphorus in most wastewaters are good plant nutrients. They, as well as the nutrients in natural runoff, add to the eutrophication[8] of receiving waters and may produce massive algal blooms, especially in lakes. As pollution continues, the depth of heavily eutrophic lakes may be reduced by the benthal buildup of dead cells and other plant debris. In this way, lakes may be turned into bogs in the course of time and eventually completely obliterated.

*Groundwater* may enter gravity sewers through pipe joints.[9] In combined systems and stormwater drains, *runoff from rainfall and melting ice and snow* adds

---

[8] Greek *eu*, well and *trophein*, to nourish.

[9] Vitrified-clay sewer pipe, 4 to 36 in. in diameter, is ordinarily 3 to 6 ft long. Unreinforced-concrete sewer pipe, 4 to 24 in. in diameter, is generally 2 to 4 ft long. Preformed joints made of resilient plastic materials greatly increase the tightness of the system.

the washings from streets, roofs, gardens, parks, and yards. Entering dirt, dust, sand, gravel, and other gritty substances are heavy and inert and form the bed load. Leaves and organic debris are light and degradable and float on or near the water surface. Waters from street flushing, fire fighting, and water-main scouring, as well as wastewaters from fountains, wading and swimming pools, and drainage waters from excavations and construction sites swell the tide.

*Domestic wastewaters* flow through house or building drains directly to the public sewer. Runoff from roofs and paved areas may be directed first to the street gutter or immediately to the storm sewer. In combined systems, water from roof and yard areas may be led into the house drain. Other storm runoff travels over the ground until it reaches a street gutter along which it flows until it enters a stormwater inlet or catch basin and is piped to a manhole and to the drainage system. In separate systems, connections to the wrong sewer are commonly in violation of sewer regulations. The dry-weather flows of combined sewers are primarily domestic and industrial wastewaters; the wet-weather flows are predominantly storm runoff. The first flush of stormwater may scour away deposited and stranded solids and increase the discharge of putrescent organic matter through stormwater overflows.

## 1-6 Systems Planning and Management

The planning, design, and construction of water and wastewater systems for metropolitan areas usually bring together sizable groups of engineering practitioners and their consultants, for months and even for years. Under proper leadership, task forces perpetuate themselves in order to attack new problems or deal with old ones in new ways. The science and practice of water supply and wastewater disposal are preserved and promoted in these ways.

For publicly owned water and wastewater systems, studies, plans, specifications, and construction contracts are prepared by engineers normally engaged by the cities and towns or by the water or wastewater districts to be served. In the United States, private water companies are not increasing in number, and private sewerage corporations have always been rare institutions. Water and wastewater engineers may therefore belong to the professional staff of municipal or metropolitan governmental agencies responsible for designing and managing public works, or they may be attached to private works or to firms of consulting engineers. To accomplish large new tasks, permanent staffs may be temporarily expanded. For smaller works, consultant groups may be given most and possibly all of the responsibility. Engineers for manufacturers of water and wastewater equipment also play a part in systems development. The engineers of construction companies bring the designs into being.

## 10/Water and Wastewater Systems

Construction of new water and wastewater systems, or the improvement and extension of existing ones, progresses from preliminary investigations or planning through financing, design, and construction to operation, maintenance, and repair. Political and financial procedures are involved, as well as engineering.

**Cost of Municipal Water Supplies.**[10] For North American communities in excess of 10,000 people, replacement costs of water supplies run to about $300 per capita, with much of the investment in small communities chargeable to fire protection. Of the systems components, collection and transportation works cost about a fourth, distribution works slightly less than a half, purification and pumping works about a tenth, and service lines and meters nearly a sixth of the total. Including interest and depreciation as well as charges against operation and maintenance, water can be delivered at the tap for costs of $50 to $500 per mg and is charged for accordingly. As one of man's most prized commodities, water is nevertheless remarkably cheap—as low as 2 cents a ton delivered to the premises of large consumers and as little as 4 cents a ton to the taps of small consumers. See Fig. 11-1 for types of treatment works.

---

*Example 1-1.*

Roughly, what is the replacement cost of the waterworks of a city of 10,000 people?

1. Assuming a per capita cost of $300, the total first cost is 300 × 10,000 = $3,000,000.

2. Assuming that 30% of this amount is invested in the collection works, 10% in the purification works, and 60% in the distribution works, the breakdown is as follows: (*a*) collection works, 0.3 × 3,000,000 = $900,000; (*b*) purification works, 0.10 × 3,000,000 = $300,000; and (*c*) distribution works, 0.60 × 3,000,000 = $1,800,000.

---

**Cost of Municipal Wastewater Systems.**[10] The first cost of sanitary sewers lies between $30 and $100 per capita in the United States. Storm drains and combined sewers, depending on local conditions, cost about three times as much. The first cost of wastewater-treatment works varies with the degree of treatment they provide. Depending on plant size which, for wastewater, is more clearly a function of the population load than of the volume of water treated (Sec. 18-14), the per capita cost of conventional

[10] For 1965 price levels. For other years, multiply by the ratio of applicable *Engineering News-Record* or other pertinent construction indexes. The January 1971 indices stood at about 150% of the 1965 indices.

wastewater treatment works as of 1967, for communities of 10,000 inhabitants,[11] ranged from $20 to $100 depending on the process used and the degree of treatment provided. See Fig. 11-2 for types of treatment works.

For communities of 10,000 people,[11] the annual per capita cost for treatment-plant operation and maintenance ranged from $1 to $5.

Including interest and depreciation, as well as charges against operation and management, the removal of domestic wastewaters and their safe disposal cost from $50 to $100 per mg. In comparison with water-purification plants, wastewater-treatment works are relatively twice as expensive; in comparison with water-distribution systems, collection systems for domestic sewage are about half as expensive. Sewer use-charges, sometimes called sewer rentals, like charges for water, can place the cost of sewerage on a *value-received* basis. Use charges may cover part or all of the cost of the service rendered and are generally related to the water bill as a matter of equity.

No general costs can be assigned to separate treatment of industrial wastewaters. When they are discharged into municipal sewerage systems, treatment costs can be prorated according to the loads imposed on the municipal works in terms of flow, suspended solids, putrescible matter, or other pertinent and controlling characteristics, singly or in combination.[12]

## 1-7 History as Prologue

"Sanitation has its history, its archeology, its literature, and its science. Whoever, indeed, would study the subject with a knowledge worthy of its magnitude must consider it from all angles and with a . . . wealth of learning."[13] Applying this broad interpretation of requisite knowledge to the hydraulic component of sanitation removes the incentive for hinging the story of man's quest[14] for water merely on a statistic of municipal works whether or not they were constructed in antiquity, allowed to decay during the Dark Ages, or returned to urban design following the industrial revolution of the 19th century. Nor can adequate understanding of works past or present be gained from engineering descriptions alone. Even the majestic structures conceived and erected in the days of Imperial Rome, and often still extant though rarely in service, would fail to give a meaningful account of the supply

[11] Robert Smith, Cost of Conventional and Advanced Treatment of Wastewater, *J. Water Pol. Control Fed.*, **40**, 1546 (1968).

[12] Y. Maystre and J. C. Geyer, Charges for Treating Industrial Wastewaters in Municipal Plants, *J. Water Pol. Control Fed.* **42**, 1277 (1970).

[13] Reginald Reynolds, *Cleanliness and Godliness*, George Allen and Urwin, London, 1943, p. 4.

[14] M. N. Baker, *The Quest for Pure Water*, Am. Water Works Assoc., New York, N.Y., 1948.

without a suitable commentary such as that provided, for example, by Julius Sextus Frontinus,[15] Water Commissioner of Rome, A.D. 97.

During the first half of the 19th century, the provision of water within towns and cities of the United States shifted from (1) pumping relatively small volumes of water by hand from wells within the community to (2) bringing in piped water in abundance from external sources. Private "watering" companies led the way to new and better service. Well before the end of the century, however, public waterworks had overtaken them both in number of supplies and in volume of water delivered. New works were seldom authorized without hot debate over health as well as financial issues; printing and circulation of pamphlets for and against proposed schemes; politicking at municipal and state levels; and the occasional revelation of sharp practices and entrepreneurial dishonesty. In spite of the pressing water requirements of the growing cities of America, battles for pure and adequate supplies were often waged for years, sometimes indeed for decades. Great personages entered the fray, and names well known in national as well as local history were attached to the advocacy or condemnation of this or that proposal.[16] Thus, Benjamin Franklin, in 1789, added a codicil to his will leaving one thousand pounds each to Philadelphia and Boston to be compounded by 5% loans to "young married artificers" over the course of a hundred years, when each city would be free to expend £100,000 on useful public works, including waterworks. In New York, both Aaron Burr and Alexander Hamilton played a part in the incorporation in 1799 of the Manhattan Company "for supplying the city of New York with pure and wholesome water." The charter also gave the company the right to employ its "surplus capital . . . in the purchase of public or other stock, or in any other monied transactions. . . ." Aaron Burr's intention to create not so much a water company as a great banking business, the Bank of the Manhattan Company, was thereby fulfilled. In Boston, the "Aqueduct Corporation" was chartered in 1795 as a "Body Politic" for the purpose of bringing fresh water into the city by subterranean pipes. Jamaica Pond was chosen as the source of this supply. Before the middle of the 19th century, however, both New York and Boston were forced to go in search of new supplies. In New York, completion of the Croton dam and aqueduct was celebrated in

---

[15] As reported by Frontinus, Rome was supplied with water by nine aqueducts varying in length from 10 to over 50 miles and in cross-section from 7 to over 50 sq. ft. Clemens Herschel (1842–1930), inventor of the venturi meter, and an engineer endowed with a sense of history, estimated the aggregate capacity of the aqueducts at 84 mgd in his translation *Frontinus and the Water Supply of the City of Rome* (Longmans, Green and Co., New York, 1913).

[16] N. M. Blake, *Water for the Cities*, a history of the urban water-supply problem in the United States, Syracuse University Press, 1956.

1840. John B. Jervis was the engineer for these works. In Boston, an aqueduct brought in the waters of Lake Cochituate at the suggestion of Loammi Baldwin, the leading engineer of his day. At the ground-breaking ceremonies in 1846, Mayor Josiah Quincy took out the first sod and John Quincy Adams, the venerable former President of the United States, proposed the toast[16]: "The waters of Lake Cochituate—May they prove to the citizens of after ages, as inspiring as ever the water of Helicon to the citizens of ancient Greece," Mount Helicon being the residence of Apollo and the Muses and containing the fountains of Aganippe and Hippocrene.

Whoever the responsible persons might be, when at last, the supplies had been completed and were turned on, there was general rejoicing and an upwelling of civic pride great enough to wipe out the bad feelings often engendered before general agreement could be reached on the source to be tapped, the line to be taken by the aqueduct, the location of the distribution reservoir, and the kind of pipe to be built into the reticulated system.

Among the principal water-connected perils to life and limb in the cities of the United States during the 18th and 19th centuries were disease and fire. Examples of catastrophic happenings were the repeated sweeps of cholera across the Atlantic, the multitudinous outbreaks of typhoid fever in cities along the great rivers and lakes of the country, and the conflagrations that broke out in communities with inadequate water systems and none but volunteer fire departments with hand-operated pumps. To mention just one of each kind: a cholera epidemic in New York took the lives of 3500 persons in 1832; an outbreak of typhoid fever in Chicago in 1891 caused 2000 deaths; and the great fire of New York in 1776 wiped out a quarter of the city.

**The Great Sanitary Awakening.** Among the early critics of the quality of living in the industrializing cities and towns of the 19th century was Edwin Chadwick (1800–1890).[17] The public-health profession is indebted to him for the *Report of the Poor Law Commissioners on an Inquiry into the Sanitary Condition of the Labouring Population of Great Britain*, 1842. Engineers should have been but were not widely grateful to him for his advocacy of small pipe sewers and separate systems for wastewater removal in 1850. He epitomized separation in the slogan "the rain to the river and the sewage to the soil." His counterpart in the United States was Lemuel Shattuck (1793–1859),[18] schoolteacher and Boston bookseller who, almost single-handed, wrote the *Report of the Massachusetts Sanitary Commission*, 1850, and through its publication eventually provided the stimulus for the *Report of the Council of Hygiene and Public Health of the Citizens Association of New York upon the Sanitary Condition of the City*, 1865. The revelations contained in this

[17] S. E. Finer, *The Life and Times of Edwin Chadwick*, Methuen, London, 1952.
[18] G. C. Whipple, *State Sanitation*, Harvard University Press, 1917.

document by Dr. Stephen Smith in turn forced the passage of the Metropolitan Health Law of 1866.

Some of the historical contributions of British and American engineers in this era of the *Great Sanitary Awakening* can be summarized as follows:

**1829.** James Simpson built the first sizable water filter for the Chelsea Water Company, which drew water for London from the Thames.

**1848.** Robert Rawlinson studied the water and wastewater needs of English cities and towns and became the first chief engineer of the National Board of Health.

**1850.** John Bazalgette constructed the main drainage of London.

**1857.** Julius W. Adams designed the first comprehensive system of sewerage for Brooklyn, N.Y.

**1871.** James P. Kirkwood built the first sizable water filters for the Hudson River Supply of Poughkeepsie, N.Y.

**1886.** Hiram F. Mills became the engineer-member of the reorganized Massachusetts State Board of Health, supported the creation of the Engineering Department of the Board, and made the Lawrence Experiment Station available for experimentation on water and wastewater treatment.

## 1-8 History as Presence and Prospect

In 1965, more than 20,000 waterworks in the United States supplied in excess of 20 billion gallons of water daily to communities with almost 160 million inhabitants. In the decade from 1956 to 1965, more than $10 billion were expended for new construction and addition to plant. This is almost three times the total during the previous history of the industry.[19] Nevertheless, capital requirements for the decade beginning with 1966 were expected to total $24 billion.

The annual value of the product water places waterworks in the roster of the top ten largest industries, and the tonnage of water delivered (more than 80 million tons daily) is many times the weight of all other industrial products. About 100,000 people are employed directly by the industry.

In 1965, more than 13,000 public sewerage systems in the United States served almost 133 million people. In the decade from 1956 to 1965 more than $7.5 billion were expended in new construction and additions to plants. Capital requirements for the decade beginning with the year 1966 were expected to total $15 billion.

The prospect for increases, improvements, and replacements in water systems must continue to be based on the accepted national objective of providing safe, acceptable, and economically useful water supplies. In similar

---

[19] Staff Report, The Water Industry in the United States, *J. Am. Water Works Assoc.*, **58**, 767 (1966).

fashion, the prospect for increases, improvements, and replacements in wastewater systems must continue to be based on the accepted national objective of removing wastewaters economically, safely, and without nuisance or adverse side effects. The national scope of wastewater removal, treatment, and disposal is being enlarged and further refined to meet public demands for ever more exacting quality standards for receiving bodies of water. Acceptable quality criteria have reference (1) to the esthetic enjoyment of natural waters and their furtherance of aquatic recreation such as boating, bathing, camping, and fishing, and (2) to the protection of aquatic and related wild life in terms of sport fishing, commercial fisheries, shellfish culture, birds, and game. To meet these criteria, the required investment in water pollution control, both public and private, can presently be estimated as of the order of $100 billion.

Advances in water supply and wastewater removal include, in addition to economic benefits, an enhancement of human comfort and well-being that cannot readily be measured in monetary terms but is nevertheless very real. At the same time, great strides have been made in the protection of the public against waterborne diseases. Through the purification and protection of their waters, cities once supplied from heavily polluted streams have seen their annual death rate from typhoid fever drop from well over 100 per 100,000 to well below 1 per 100,000. At such a level there is little reason to suspect that typhoid fever and other intesinal infections will be acquired from drinking the water of American towns and cities in the normal course of events. In unusual circumstances, acts of nature and human failure may nevertheless lower the barriers against infection that were otherwise so carefully erected and maintained.

## 1-9 Resource Planning and Management

"All things are sprung from water! . . . All things are sustained by water!"[20] This pantheistic concept of life is expressed with much felicity in the evolution of human communities also.

The waters of surface streams and underground aquifers normally seek paths of least resistance from upland to the sea. To take advantage of the associated minimum grades, roads and rails normally follow river banks. Waters made available in the valleys of the world slake the thirst of man and his domestic animals, advance his health and cleanliness, and provide moisture for his crops, fish for his table, power for his mills, and transport for himself and his goods. Because of this, great cities have normally risen on the banks of streams, lakes, and tidal estuaries. Clearly, therefore, the responsi-

[20] Goethe's *Faust*, Part II, Act 2, "Alles ist aus dem Wasser entsprungen!! . . . Alles wird durch das Wasser erhalten!"

bility of engineers and government lies in the optimal planning and management of water quality within the river basins of the world.

Concern for the fitness of water and for its safety, attractiveness, and economic usefulness distinguishes water and wastewater engineering from hydraulic operations such as navigation and water power. "An awareness of quality inheres in underlying hydrological studies, in hydraulic and structural designs giving form to necessary engineering works, and in the operation of completed systems."[21] The greater the urbanization and industrialization and the more effective the response of urban and industrial leadership to human welfare, the more pronounced must become the promotion and preservation of water quality in nature and in use.

---

[21] G. M. Fair, J. C. Geyer, and D. A. Okun, *Water and Wastewater Engineering*, John Wiley & Sons, New York, 1966, Vol. 1, pp. 1–2.

# two

# systems
# capacities

## 2-1 Performance and Required Capacities

A knowledge of the performance and required capacities of water and wastewater systems is fundamental to systems design and management. In the United States the volumes of water supplied to cities and towns or removed from them are expressed in U.S. gallons per year, month, day, or minute. The U.S. gallon (gal) occupies a volume of 231 cubic inches (cu in.) or 0.1337 cubic feet (cu ft) and weighs 8.344 pounds (lb); the (British) Imperial gallon weighs 10 lb, and a cubic foot of water 62.43 lb (Table a2 in the Appendix). The fundamental metric unit in engineering work is the cubic meter (cu m or m$^3$), weighing $10^3$ kilograms (kg) and equaling $10^3$ liters (l). In the United States, annual water or wastewater volumes are conveniently recorded in million gallons (mg) if under 100 mg and in billion gallons (bg) if over 100 mg. Daily volumes are generally expressed in mgd if over 100,000 gpd; per capita daily volumes are stated in gpcd. Connected or tributary populations and numbers of services (active or total) or dwelling units may take the place of total populations.

Per capita and related figures generalize the experience. They permit comparison of the experience and practices of different communities and are helpful in estimating future requirements of specific communities. Fluctuations in flow are usefully expressed as ratios of

*17*

*maximum* or *minimum* annual, seasonal, monthly, weekly, daily, hourly, and peak rates of flow to corresponding *average* rates of flow.

## 2-2 Design Period

New water and wastewater works are normally made large enought to meet the needs and wants of growing communities for an economically justifiable number of years in the future. Choice of a relevant design period is generally based on (1) the useful life of component structures and equipment, taking into account obsolescence as well as wear and tear; (2) the ease or difficulty of enlarging contemplated works, including consideration of their location;

*Table 2-1*
Design Periods for Water and Wastewater Structures

| Type of Structure | Special Characteristics | Design Period, Years |
|---|---|---|
| *Water supply* | | |
| Large dams and conduits | Hard and costly to enlarge | 25–50 |
| Wells, distribution systems, and filter plants | Easy to extend | |
| | When growth and interest rates are low[a] | 20–25 |
| | When growth and interest rates are high[a] | 10–15 |
| Pipes more than 12 in. in diameter | Replacement of smaller pipes is more costly in long run | 20–25 |
| Laterals and secondary mains less than 12 in. in diameter | Requirements may change fast in limited areas | Full development |
| *Sewerage* | | |
| Laterals and submains less than 15 in. in diameter | Requirements may change fast in limited areas | Full development |
| Main sewers, outfalls, and intercepters | Hard and costly to enlarge | 40–50 |
| Treatment works | When growth and interest rates are low[a] | 20–25 |
| | When growth and interest rates are high[a] | 10–15 |

[a] The dividing line is in the vicinity of 3% per annum.

(3) the anticipated rate of population growth and water use by the community and its industries; (4) the going rate of interest on bonded indebtedness; and (5) the performance of contemplated works during their early years when they are expected to be under minimum load. Design periods often employed in practice are shown in Table 2-1.

## 2-3 Population Data

For information on the population of given communities or regions at a given time, engineers turn to the records of official censuses or enumerations. The government of the United States has made a decennial census since 1790. Some state and local enumerations provide additional information, usually for years ending in 5, and results of special surveys sponsored by public authorities or private agencies for political, social, or commercial purposes may also be available. United States census dates and intervals between censuses are listed in Table 2-2.

The information obtained in the decennial censuses is published by the U.S. Bureau of the Census, Department of Commerce. Political or geographic subdivisions for which population data are collated vary downward in size from the country as a whole, to its coterminous portion only, individual states and counties, metropolitan districts, cities and wards, townships and towns, and—in large communities—census tracts. The tracts are areas of substantially the same size and large enough to house 3000 to 6000 people.

## 2-4 Population Growth

Populations increase by births, decrease by deaths, and change with migration. Communities also grow by annexation. Urbanization and industrialization bring about social and economic changes as well as growth. Educational and employment opportunities and medical care are among the desirable changes. Among unwanted changes are the creation of slums and

*Table 2-2*

*U.S. Census Dates and Intervals between Censuses*

| Year | Date | Census Interval, Years |
| --- | --- | --- |
| 1790–1820 | First Monday in August | Approximately 10 |
| 1830–1900 | June 1 | Exactly 10, except between 1820 and 1830 |
| 1910 | April 15 | 9.875 |
| 1920 | January 1 | 9.708 |
| 1930 | April 1 | 10.250 |
| 1940–1970 | April 1 | Exactly 10 |

the pollution of air, water, and soil. Least predictable of the effects on growth are changes in commercial and industrial activity. Examples are furnished in Table 2-3, (1) for Detroit, Mich., where the automobile industry was responsible for a rapid rise in population between 1910 and 1950; (2) for Providence, R. I., where competition of southern textile mills was reflected in low rates of population growth after 1910; and (3) for Miami, Fla., where recreation added a new and important element to prosperity from 1910 onward.

### Table 2-3
*Rounded Census Populations of Detroit, Mich., Providence, R.I., and Miami, Fla., 1910–1960*

| Census Year | City | | |
| --- | --- | --- | --- |
| | Detroit | Providence | Miami |
| 1910 | 466,000 | 224,000 | 5,500 |
| 1920 | 994,000 | 235,000 | 30,000 |
| 1930 | 1,569,000 | 253,000 | 111,000 |
| 1940 | 1,623,000 | 254,000 | 172,000 |
| 1950 | 1,850,000 | 249,000 | 249,000 |
| 1960 | 1,670,000 | 207,000 | 292,000 |
| 1970 | 1,493,000 | 177,000 | 332,000 |

Were it not for industrial vagaries of the Providence type, human population kinetics would trace an S-shaped growth curve in much the same way as spatially constrained microbic populations. As shown in Fig. 2-1, the trend of *seed* populations is progressively faster at the beginning and progressively slower towards the end as a saturation value or upper limit is approached. What the future holds for a given community, therefore, is seen to depend on where on the growth curve the community happens to be at a given time.

The growth of cities and towns and characteristic portions of their growth curves can be approximated by relatively simple equations that derive historically from chemical kinetics (Sec. 11-12). The equation of a first-order chemical reaction, possibly catalyzed by its own reaction products, is a recurring example. It identifies also the kinetics of biological growth and other biological reactions including population growth, kinetics, or dynamics. This widely useful equation may be written

$$dy/dt = ky(L - y) \qquad (2\text{-}1)$$

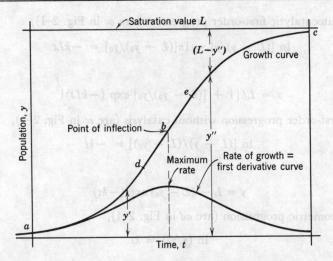

**Figure 2-1.**
**Population growth idealized. Note geometric increase from *a* to *d*; straight-line increase from *d* to *e* (approximately); and first-order increase from *e* to *c*.**

where $y$ is the population at time $t$, $L$ is the saturation or maximum population, and $k$ is a growth or rate constant with the dimension $[t^{-1}]$. It is pictured in Fig. 2-1 together with its integral, Eq. 2-6.

Three related equations apply closely to characteristic portions of this growth curve: (1) a first-order progression for the terminal arc *ec* of Fig. 2-1; (2) a logarithmic or geometric progression for the initial arc *ad*; and (3) an arithmetic progression for the transitional intercept, *de*, or:

For arc *ec*
$$dy/dt = k(L - y) \qquad (2\text{-}2)$$

For arc *ad*
$$dy/dt = ky \qquad (2\text{-}3)$$

For arc *de*
$$dy/dt = k \qquad (2\text{-}4)$$

If it is assumed that the initial value of $k$, namely $k_0$, decreases in magnitude with time or population growth rather than remaining constant, $k$ can be assigned the following value:

$$k = k_0/(1 + nk_0 t) \qquad (2\text{-}5)$$

in which $n$, as a coefficient of *retardance*, adds a useful concept to Eqs. 2-2 to 4.

On integrating Eqs. 2-1 to 2-4 between the limits $y = y_0$ at $t = 0$ and $y = y$ at $t = t$ for unchanging $k$ values, they become:

For autocatalytic first-order progression (arc $ac$ in Fig. 2-1),

$$\ln[(L-y)/y] - \ln[(L-y_0)/y_0] = -kLt$$

or

$$y = L/\{1 + [(L-y_0)/y_0]\exp(-kLt)\} \quad (2\text{-}6)$$

For first-order progression without catalysis (arc $ec$ in Fig. 2-1),

$$\ln[(L-y)/(L-y_0)] = -kt$$

or

$$y = L - (L-y_0)\exp(-kt) \quad (2\text{-}7)$$

For geometric progression (arc $ad$ in Fig. 2-1),

$$\ln(y/y_0) = kt$$

or

$$y = y_0 \exp(kt) \quad (2\text{-}8)$$

For arithmetic progression (arc $de$ in Fig. 2-1),

$$y - y_0 = kt \quad (2\text{-}9)$$

Substituting Eq. 2-5 in Eqs. 2-2 to 4 yields the following retardant expressions:

For retardant first-order progression,

$$y = L - (L-y_0)(1 + nk_0 t)^{-1/n} \quad (2\text{-}10)$$

For retardant, geometric progression,

$$\ln(y/y_0) = (1/n)\ln(1 + nk_0 t) \quad \text{or} \quad y = y_0(1 + nk_0 t)^{-1/n} \quad (2\text{-}11)$$

For retardant, arithmetic progression,

$$y - y_0 = (1/n)\ln(1 + nk_0 t) \quad (2\text{-}12)$$

These and similar equations are useful in water and wastewater practice, especially in water and wastewater treatment kinetics (Chap. 11).

## 2-5 Short-Term Population Estimates

Estimates of midyear populations for current years and the recent past are normally derived by arithmetic from census data. They are needed perhaps most often for (1) computing per capita water consumption and wastewater release, and (2) for calculating the annual birth and *general* death rates per 1000 inhabitants, or *specific* disease and death rates per 100,000 inhabitants.

Understandably, morbidity and mortality rates from waterborne and otherwise water-related diseases are of deep concern to sanitary engineers.

For years *between censuses* or *after the last census*, estimates are usually interpolated or extrapolated as *arithmetic* or *geometric* progressions. If $t_i$ and $t_j$ are the dates of two sequent censuses and $t_m$ is the midyear date of the year for which a population estimate is wanted, the rate of arithmetic growth is given by Eq. 2–9 as $k_{\text{arithmetic}} = (y_j - y_i)/(t_j - t_i)$, and the midyear populations, $y_m$, of intercensal and postcensal years are respectively:

(Intercensal)
$$y_m = y_i + (t_m - t_i)(y_j - y_i)/(t_j - t_i) \qquad (2\text{-}13)$$

(Postcensal)
$$y_m = y_j + (t_m - t_j)(y_j - y_i)/(t_j - t_i) \qquad (2\text{-}14)$$

In similar fashion, Eq. 2–8 states that $k_{\text{geometric}} = (\log y_j - \log y_i)/(t_j - t_i)$, and the logarithms of the midyear populations, $\log y_m$, for intercensal and postcensal years are respectively:

(Intercensal)
$$\log y_m = \log y_i + (t_m - t_i)(\log y_j - \log y_i)/(t_j - t_i) \qquad (2\text{-}15)$$

(Postcensal)
$$\log y_m = \log y_j + (t_m - t_j)(\log y_j - \log y_i)/(t_j - t_i) \qquad (2\text{-}16)$$

Geometric estimates, therefore, use the logarithms of the population parameters in the same way as the population parameters themselves are employed in arithmetic estimates; moreover, arithmetic increase corresponds to capital growth by simple interest, and geometric increase to capital growth by compound interest. Graphically, arithmetic progression is characterized by a straight-line plot against arithmetic scales for both population and time on double-arithmetic coordinate paper, and thus, geometric as well as first-order progression by a straight-line plot against a geometric (logarithmic) population scale and an arithmetic time scale on semilogarithmic paper. The suitable equation and method of plotting is best determined by inspection from a basic arithmetic plot of available historic population information.

---

*Example 2–1.*

As shown in Table 2–3, the rounded census population of Miami, Fla., was 249,000 in 1950 and 292,000 in 1960. Estimate the midyear population (1) for the fifth intercensal year and (2) for the ninth postcensal year by (a) arithmetic and (b) geometric progression. The two census dates were both April 1.

1. Intercensal estimates for 1955: $t_m - t_i = 5.25$ yr; $t_j - t_i = 10.00$ yr; and $(t_m - t_i)/(t_j - t_i) = 0.525$.

| | (a) Arithmetic | (b) Geometric |
|---|---|---|
| 1960 | $y_j = 292{,}000$ | $\log y_j = 5.4654$ |
| 1950 | $y_i = 249{,}000$ | $\log y_i = 5.3962$ |
| | $y_j - y_i = 43{,}000$ | $\log y_j - \log y_i = 0.0692$ |
| | $0.525(y_j - y_i) = 23{,}000$ | $0.525(\log y_j - \log y_i) = 0.03633$ |
| 1955 | $y_m = 272{,}000$ | $y_m = 268{,}000$ |

2. Postcensal estimate for 1969. $t_m - t_j = 9.25$ yr; $t_j - t_i = 10.00$ yr; $(t_m - t_j)/(t_j - t_i) = 0.925$.

| (a) Arithmetic | (b) Geometric |
|---|---|
| From (1a) $y_j - y_i = 43{,}000$ | From (1b) $\log y_j - \log y_i = 0.0692$ |
| $0.925(y_j - y_i) = 40{,}000$ | $0.925(\log y_j - \log y_i) = 0.0620$ |
| $y_m = 332{,}000$ | $y_m = 337{,}000$ |

Geometric estimates are seen to be lower than arithmetic estimates for intercensal years and higher for postcensal years. By recent census the 1970 population of Miami was 332,000.

---

The U.S. Bureau of the Census estimates the current population of the whole nation by adding to the last census population the intervening differences (1) between births and deaths, that is, the *natural increases*, and (2) between immigration and emigration. For states and other large population groups, postcensal estimates can be based on the *apportionment method*, which postulates that local increases will equal the national increase times the ratio of the local to the national intercensal population increase. Intercensal losses in population are normally disregarded in postcensal estimates; the last census figures are used instead.

Supporting data for short-term estimates can be derived from sources that reflect population growth in ways different from, yet related to, population enumeration. Examples are records of school enrollments; house connections for water, electricity, gas, and telephones; commercial transactions; building permits; and health and welfare services. These are translated into population values by ratios derived for the recent past. The following ratios are not uncommon: population:school enrollment = 5:1; population:number of water, gas, or electricity services = 3:1; and population:number of telephone services = 4:1.

## 2-6 Long-Range Population Forecasts

Long-range forecasts, covering design periods of 10 to 50 years, make use of available and pertinent records of population growth. Again dependence is placed on mathematical curve fitting and graphical studies. The *logistic growth curve* championed by P. F. Verhulst[1] is an example.

Verhulst's equation is derived from the autocatalytic, first-order equation (Eq. 2-6) by letting $p = (L - y_0)/y_0$ and $q = kL$, or

$$y = L/[1 + p \exp(-qt)] = L/[1 + \exp(\ln p - qt)] \quad (2\text{-}17)$$

and equating the first derivative of Eq. 2-1 to zero, or

$$d(dy/dt)/dt = kL - 2ky = 0$$

It follows that the maximum rate of growth $dy/dt$ obtains when $y = \frac{1}{2}L$, and $t = (-\ln p)/q = (-2.303 \log p)/q$.

It is possible to develop a *logistic* scale for fitting a straight line to pairs of observations as in Fig. 2-2. For general use of this scale, populations are

[1] See Raymond Pearl, *Medical Biometry and Statistics*, W. B. Saunders Co., Philadelphia, Chapter 18, 1940; and J. C. McLean, *Civ. Eng.*, **22**, 133 and (particularly), 886 (1952).

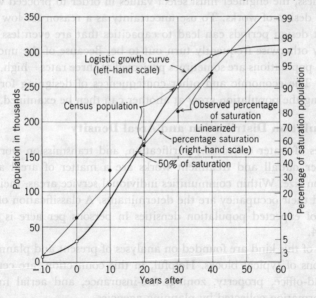

**Figure 2-2.**

**Logistic growth of a city. Calculated saturation population, confirmed by graphical good straight-line fit, is 313,000. Right-hand scale is plotted as log $[(100 - P)/P]$ about 50% at the center.**

expressed in terms of successive saturation estimates $L$, which are eventually verified graphically by lying closely in a straight line on a logistic-arithmetic plot. The percentage saturation $P = 100\,y/L = 100/[1 + p\exp(-qt)]$ and $\ln[(100 - P)/P] = \ln p - qt$. The straight line of best fit by eye has an ordinate intercept $\ln p$ and a slope $-q$ when $\ln[(100 - P)/P]$ is plotted against $t$ or values of $n[(100 - P)/P]$ are scaled in either direction from a 50-percentile or middle ordinate.[2]

Graphical forecasts offer a means of escape from mathematical forecasting. However, even when mathematical forecasting appears to give meaningful results, most engineers seek support for their estimates from plots of experienced and projected population growth on arithmetic or semilogarithmic scales. Trends in rates of growth rather than growth itself may be examined arithmetically, geometrically, or graphically with fair promise of success. Estimates of arithmetic and semilogarithmic straight-line growth of populations and population trends can be developed analytically by applying *least-squares procedures*, including the determination of the coefficient of correlation and its standard error.[3]

At best, since forecasts of population involve great uncertainties, the probability that the estimated values turn out to be correct can be quite low. Nevertheless, the engineer must select values in order to proceed with planning and design of works. To use uncertainty as a reason for low estimates and short design periods can lead to capacities that are even less adequate than they otherwise frequently turn out to be. Because of the uncertainties involved, populations are sometimes projected at three rates—high, medium, and low. The economic and other consequences of designing for one rate and having the population grow at another can then be examined.

## 2-7 Population Distribution and Areal Density

Capacities of water collection, purification, and transmission works and of wastewater outfall and treatment works are a matter of areal as well as population size. Within communities individual service areas, their populations, and their occupancy are the determinants. A classification of areas by use and of expected population densities in persons per acre is shown in Table 2-4.

Values of this kind are founded on analyses of present and planned future subdivisions of typical blocks. Helpful, in this connection, are census tract data; land-office, property, zoning, fire-insurance, and aerial maps; and other information collected by planning agencies.

[2] G. M. Fair, J. C. Geyer, and D. A. Okun, *Water and Wastewater Engineering*, John Wiley & Sons, New York, 1966, Vol. 1, pp. 5–8 and 5–9.

[3] See pp. 4–23 to 25 in reference 2 (footnote 2).

## Table 2-4
*Common Population Densities*

| | Persons per Acre |
|---|---|
| 1. Residential areas | |
|    (a) Single-family dwellings, large lots | 5–15 |
|    (b) Single-family dwellings, small lots | 15–35 |
|    (c) Multiple-family dwellings, small lots | 35–100 |
|    (d) Apartment or tenement houses | 100–1000 or more |
| 2. Mercantile and commercial areas | 15–30 |
| 3. Industrial areas | 5–15 |
| 4. Total, exclusive of parks, playgrounds, and cemeteries | 10–50 |

## 2-8 Water Consumption

Although the draft of water from distribution systems is commonly referred to as water *consumption*, little of it is, strictly speaking, *consumed;* most of it is discharged as *spent* or *waste* water. *Use* of water is a more exact term. True *consumptive use* refers to the volume of water evaporated or transpired in the course of use—principally in sprinkling lawns and gardens, in raising and condensing steam, and in bottling, canning, and other industrial operations.

Service pipes introduce water into dwellings, mercantile and commercial properties, industrial complexes, and public buildings (Fig. 7–3). The water delivered is classified accordingly. Table 2–5 shows approximate per capita daily uses in the United States. Wide variations in these figures must be expected because of differences in (1) climate, (2) standards of living, (3) extent of sewerage, (4) type of mercantile, commercial, and industrial activity, (5) water pricing, (6) resort to private supplies, (7) water quality for domestic and industrial purposes, (8) distribution-system pressure, (9) completeness of meterage, and (10) systems management.

**Domestic Consumption.** Although domestic water use is about 50% of the water drawn in urban areas, 90% of the consumers are domestic.[4] A breakdown of household flows apportions the various uses as follows:[5] 41% to flushing toilets; 37% to washing and bathing; 6% to kitchen use; 5% to drinking water; 4% to washing clothes; 3% to general household cleaning; 3% to watering lawns and gardens; and 1% to washing family

---
[4] F. P. Linaweaver, Jr., J. C. Geyer, and J. B. Wolff, Summary Report on the Residential Water Use Research Project, *J. Am. Water Works Assoc.*, **59**, 267 (1967).

[5] C. N. Dufor and Edith Becker, Public Water Supplies of the 100 Largest Cities in the United States, *U.S. Geol. Survey, Water Supply Paper* 1812, 5 (1964).

## Table 2-5
Normal Water Consumption

| Class of Consumption | Quantity, gpcd | |
|---|---|---|
| | Normal Range | Average |
| Domestic or residential | 20–90 | 55 |
| Commercial | 10–130 | 20 |
| Industrial | 20–80 | 50 |
| Public | 5–20 | 10 |
| Water unaccounted for | 5–30 | 15 |
| Total | 60–250 | 150 |

cars. Although domestic use is commonly expressed in gpcd, the draft per dwelling unit, gpud, may offer more meaningful information.

Extremes of heat and cold increase water consumption: hot and arid climates by frequent bathing, air conditioning, and heavy sprinkling; and cold climates by bleeding water through faucets to keep service pipes and internal water piping from freezing during cold spells. In metered and sewered residential areas, the observed average daily use of water for lawns and gardens, $Q_{sprinkling}$, in gpd during the growing season is about 60% of the estimated average potential evapotranspiration $E$ (Sec. 3-6), reduced by the average daily precipitation, $P$, effective in satisfying evapotranspiration during the period,[4] or

$$Q_{sprinkling} = 1.63 \times 10^4 \, A(E - P) \qquad (2\text{-}18)$$

Here $1.63 \times 10^4 = 0.6 \times 2.72 \times 10^4$, the number of gallons in an acre-inch; $A$ is the average lawn and garden acreage per dwelling unit, and $E$ and $P$ are expressed in inches. The average lawn and garden area is given by the observational relationship

$$A = 0.803 \, D^{-1.26} \qquad (2\text{-}19)$$

where $D$ is the gross housing density in dwelling units per acre.[4]

High standards of cleanliness, large numbers of water-connected appliances, oversized plumbing fixtures, and frequent lawn and garden sprinkling, all associated with wealth, result in heavy drafts. For sewered properties, the average domestic use of water $Q_{domestic}$ in gpd for each dwelling unit is related to the average market value $M$ of the units in thousands of dollars by the following observational equation:[4]

$$Q_{domestic} = 157 + 3.46 \, M \qquad (2\text{-}20)$$

**General Urban Water Demands.** Some commercial enterprises—hotels and restaurants, for instance—draw much water; so do industries such as breweries, canneries, laundries, paper mills, and steel mills. Industries, in particular, draw larger volumes of water when it is cheap than when it is dear. Industrial draft varies roughly inversely as the *manufacturing rate*[6] and is likely to drop by about half the percentage increase in cost when rates are raised. Hospitals, too, have high demands. Although the rate of draft in fire fighting is high (Table 7–1), the time and annual volume of water consumed in extinguishing fires are small and seldom identified separately for this reason.

Water of poor quality may drive consumers to resort to uncontrolled, sometimes dangerous, sources, but the public supply remains the preferred source when the product water is clean, palatable, and of unquestioned safety; soft for washing and cool for drinking; and generally useful to industry. The availability of groundwater and nearby surface sources may persuade large industries and commercial enterprises to develop their own process and cooling waters.

Hydraulically, leaks from mains and plumbing systems and flows from faucets and other regulated openings behave like orifices. Their rate of flow varies as the square root of the pressure head, and high distribution pressures raise the rate of discharge and with it the waste of water from fixtures and leaks. Ordinarily, systems pressures are not raised above 60 psig (lb per sq in. gage) in American practice, even though it is impossible to employ direct hydrant streams in fire fighting when hydrant pressures are below 75 psig (Sec. 7–3).

*Meterage* encourages thrift and normalizes the demand. The cost of metering and the running expense of reading and repairing meters are substantial. They may be justified in part by accompanying reductions in waste and possible postponement of otherwise needed extensions. Under study and on trial here and there is the encouragement of *off-peak-hour draft* of water by large users. To this purpose, rates charged for water drawn during off-peak hours are lowered preferentially. The objective is to reap the economic benefits of a relatively steady flow of water within the system and the resulting proportionately reduced capacity requirements of systems components (Fig. 2–3 and Sec. 2–9). The water drawn during off-peak hours is generally stored by the user at ground level even when this entails repumping.

Distribution networks are seldom perfectly tight. Mains, valves, hydrants, and services of well-managed systems are therefore regularly checked for

---

[6] Defined as the charge for more than 3 mg annually, compared with a *domestic rate* set at 0.3 mg annually.

leaks. Superficial signs of *controllable leakage* are (1) high night flows in mains, (2) water running in street gutters, (3) moist pavements, (4) persistent seepage, (5) excessive flows in sewers, (6) abnormal pressure drops, and (7) unusually green vegetation (in dry climates). Leakage is detected by (1) driving sounding rods into the ground to test for moist earth; (2) applying listening devices that amplify the sound of running water; and (3) inspecting premises for leaky plumbing and fixtures. Leakage detection of well-managed waterworks may be complemented by periodic and intensive but, preferably, routine and extensive water-waste surveys. Generally involved is the isolation of comparatively small sections of the distribution system by closing valves on most or all feeder mains and measuring the water entering the section at night through one or more open valves or added piping on fire hoses. Common means of measurement are pitot tubes, bypass meters around controlling valves, or meters on one or more hose lines between hydrants that straddle closed valves.

**Industrial Water Consumption.** The amounts of water used by industry vary widely. Some manufactories draw in excess of 50 mgd; others, no more than comparably sized mercantile establishments. On an average, U.S. industry satisfies more than 60% of its water requirements by internal reuse, and less than 40% by draft through plant intakes from its own water sources or through service connections from public water systems. Only about 7% of the water taken in is *consumed;* 93% is returned to open waterways or to the ground, whence it may be removed again by downstream users. On balance, industry's consumptive use is kept down to 2% of the

*Table 2-6*
*Water Consumed by Industry; Percentage of Total Water Intake*[a]

| Industry | Percent of Intake | Industry | Percent of Intake |
| --- | --- | --- | --- |
| Automobile | 6.2 | Meat | 3.2 |
| Beet sugar | 10.5 | Petroleum | 7.2 |
| Chemicals | 5.9 | Poultry processing | 5.3 |
| Coal preparation | 18.2 | Pulp and paper | 4.3 |
| Corn and wheat milling | 20.6 | Salt | 27.6 |
| Distillation | 10.4 | Soap and detergents | 8.5 |
| Food processing | 33.6 | Steel | 7.3 |
| Machinery | 21.4 | Sugar, cane | 15.9 |
|  |  | Textiles | 6.7 |

[a] *Source.* National Association of Manufacturers.

draft of all water users in the United States. Table 2-6 shows the relative amounts of water consumed by different industries. Not brought out is the fact that once-through cooling, particularly by the power industry, is by far the biggest use component and the principal contributor to the thermal pollution of receiving waters. For industrial fire supplies, see Sec. 7-9.

To draw comparisons between the water uses of different industries and of plants within the same industrial category, it is customary to express plant or process use in volumes of water—gallons, for instance—per unit of production (Table 2-7). For the chemical industry, however, this may not be meaningful, because of the diversity of chemicals produced.

**Table 2-7**
*Water Requirements of Selected Industries*

| Industry | Unit of Production | Gallons per Unit |
|---|---|---|
| Food products | | |
| Beet sugar | Ton of beets | 7,000[a] |
| Beverage alcohol | Proof gallon | 125– 170 |
| Meat | 1000 lb live weight | 600– 3,500[b] |
| Vegetables, canned | Case | 3– 250 |
| Manufactured products | | |
| Automobiles | Vehicle | 10,000 |
| Cotton goods | 1000 lb | 20,000–100,000 |
| Leather | 1000 sq ft of hide | 200– 64,000 |
| Paper | Ton | 2,000–100,000 |
| Paper pulp | Ton | 4,000– 60,000 |
| Mineral products | | |
| Aluminum (electrolytic smelting) | Ton | 56,000 (max) |
| Copper | | |
| Smelting | Ton | 10,000[c] |
| Refining | Ton | 4,000 |
| Fabricating | Ton | 200– 1,000 |
| Petroleum | Barrel of crude oil | 800– 3,000[d] |
| Steel | Ton | 1,500– 50,000 |

[a] Includes 2600 gal of flume water and 2000 gal of barometric condenser water.
[b] Lower values for slaughterhouses; higher for slaughtering and packing.
[c] Total, including recycled water; water consumed is 1400 gal.
[d] Total, including recycled water; water consumed is 30 to 60 gal.

Rising water use can be arrested by conserving plant supplies and introducing efficient processes and operations. Most important, perhaps, are the economies of multiple reuse through countercurrent rinsing of products, recirculation of cooling and condensing waters, and reuse of otherwise spent waters for secondary purposes after their partial purification or repurification.

About two thirds of the total water intake of U.S. manufacturing plants is put to use for cooling. In electric-power generation, the proportion is nearly 100%; in manufacturing industries, it ranges from 10% in textile mills to 95% in beet-sugar refineries. It averages 66% in industries covered in the 1965 report of the National Association of Manufacturers.[7]

Industry often develops its own supply. Chemical plants, petroleum refineries, and steel mills, for example, draw on public or private utilities for less than 10% of their needs. Food processors, by contrast, purchase about half their water from public supplies, largely because the bacterial quality of drinking water makes it *de facto* acceptable.

About 90% of the industrial draft is taken from surface sources. Groundwaters may be called into use in the summer because their temperature is then seasonally low. They may be prized, too, for their clarity and their freedom from color, odor, and taste.

Available sources may be drawn on selectively: municipal water for drinking, sanitary purposes, and delicate processes, for example, and river water for rugged processes and cooling, and for emergency uses such as fire protection. Treatment costs as well as economic benefits are the determinants.

Private water supplies are not inexpensive; the replacement cost of water-treatment facilities for approximately 3000 U.S. plants covering 90% of the total water intake by manufacturing industries was set[7] at about $400 million in 1965.

**Rural Water Consumption.** The minimum use of piped water in rural dwellings is about 20 gpcd; the average about 50 gpcd. Approximate drafts of rural schools, overnight camps, and rural factories (exclusive of manufacturing uses) are 25 gpcd; of wayside restaurants, 10 gpcd on a patronage basis; and of work or construction camps, 45 gpcd. Resort hotels need about 100 gpcd and rual hospitals and the like, nearly twice this amount.

Farm animals have the following approximate requirements: dairy cows, 20 gpcd; horses, mules, and steers, 12 gpcd; hogs, 4 gpcd; sheep, 2 gpcd; turkeys, 0.07 gpcd, and chickens, 0.04 gpcd. Cleansing and cooling water add about 15 pgd per cow to the water budget of dairies. Greenhouses may use as much as 70 gpd per 1000 sq ft and garden crops about half this amount.

Military requirements vary from an absolute minimum of 0.5 gpcd for troops in combat through 2 to 5 gpcd for men on the march or in bivouac,

[7] *Water in Industry*, National Association of Manufacturers, 1965.

and 15 gpcd for temporary camps, up to 50 gpcd or more for permanent military installations.

## 2-9 Variations in Water Demand

Water consumption changes with the seasons, the days of the week, and the hours of the day. Fluctuations are greater (1) in small than in large communities, and (2) during short rather than during long periods of time. Variations are usually expressed as ratios to the average demand. Estimates for the United States are as follows:

| Ratio of Rates | Normal Range | Average |
|---|---|---|
| Maximum day : average day | (1.5 to 3.5):1 | 2.0:1 |
| Maximum hour : average day | (2.0 to 7.0):1 | 4.5:1 |

**Domestic Variations.** Observations in a Johns Hopkins University study and Standards of the Federal Housing Administration for domestic drafts are brought together[4] in Table 2-8. Damping effects produced by network size and phasing of commercial, industrial, and domestic drafts explain the differences between communitywide and domestic demands. Observational values for peak hourly demands in gpud are given by the regression equation[4]

$$Q_{\text{peak-hr}} = 334 + 2.02\, Q_{\text{max day}} \qquad (2\text{-}21)$$

### Table 2-8
*Variations in Average Daily Rates of Water Drawn by Dwelling Units*[4]

| | gpd per Dwelling unit and (Ratios to Aver. Day in paren.) | | |
|---|---|---|---|
| | Average Day | Maximum Day | Peak hour |
| Federal Housing Administration Standards | 400 (1.0) | 800 (2.0) | 2000 (5.0) |
| *Observed Drafts—Metered Dwellings* | | | |
| National average | 400 (1.0) | 870 (2.2) | 2120 (5.3) |
| West | 460 (1.0) | 980 (2.1) | 2480 (5.2) |
| East | 310 (1.0) | 790 (2.5) | 1830 (5.9) |
| *Unmetered Dwellings* | | | |
| National average | 690 (1.0) | 2350 (3.4) | 5170 (7.6) |
| *Unsewered Dwellings* | | | |
| National average | 250 (1.0) | 730 (2.9) | 1840 (7.5) |

Calculation of the confidence limit of expected demands[8] makes it possible to attach suitably higher limits to these values. For design this may be the 95% confidence limit, which lies above the expected rate of demand by twice the variance, that is, by $2\sigma$,[2] where $\sigma$ is the standard deviation of the observed demands. Approximate values of this variance are shown in Table 2-9 for gross housing densities of 1, 3, and 10 dwelling units per acre and a potential daily evapotranspiration of 0.28 in. of water.[4]

Design demands determine the size of hydraulic components of the system; expected demands determine the rate structure and operation of the system.

*Fire Demands.* Height, bulk, congestion, fire resistance, and nature and cost of the contents of buildings determine the rate at which water should be made available at neighboring hydrants either as hydrant or engine streams numerous enough to extinguish localized fires and prevent their spread into areal or citywide conflagrations. In the United States, standards and recommendations for the prevention and control of fires are issued by the American Insurance Association (AIA),[9] the national fact-finding organization for fire-insurance companies. Among its activities is the assessment of the fire readiness of individual communities and their water-supply systems. Analysis by AIA of water demands actually experienced during fires in communities of different size underlies the formulation of general standards from which the designer should depart only for good and sufficient reasons (Sec. 7-2).

## Table 2-9

*Increase in Design Demands Above Expected Maximum Day and Peak Hour, gpud, Drafts by Dwelling Units.*

| Number of Dwelling Units | Maximum Demand (1000 gpud) for Stated Housing Density | | | Peak Demand (1000 gpud) for Stated Housing Density | | |
|---|---|---|---|---|---|---|
| | 1 | 3 | 10 | 1 | 3 | 10 |
| 1 | 5.0 | 4.8 | 4.6 | 7.0 | 7.0 | 7.0 |
| 10 | 1.6 | 1.5 | 1.5 | 3.5 | 2.3 | 2.2 |
| $10^2$ | 0.7 | 0.5 | 0.5 | 1.5 | 0.8 | 0.7 |
| $10^3$ | 0.6 | 0.3 | 0.2 | 1.5 | 0.4 | 0.3 |
| $10^4$ | 0.5 | 0.2 | 0.1 | 1.4 | 0.3 | 0.2 |

[8] See footnotes 2 and 4.

[9] Originally established as the National Board of Fire Underwriters, New York, N.Y., this organization publishes a *Standard Schedule for Grading Cities and Towns of the United States.*

## Variations in Water Demand / 35

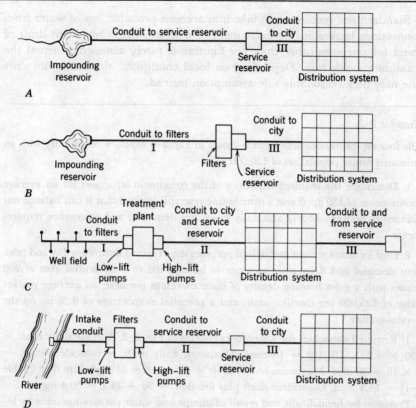

**Figure 2-3.**
**Required capacities of four typical waterworks systems. The service reservoir is assumed to compensate for fluctuations in draft and fire drafts, and to hold an emergency reserve.**

| | | Capacity of System, mgd | | | |
|---|---|---|---|---|---|
| Structure | Required Capacity | A | B | C | D |
| 1. River or well field | Maximum day | | | 36.0 | 36.0 |
| 2. Conduit I | Maximum day | 36.0 | 36.0 | 36.0 | 36.0 |
| 3. Conduit II | Maximum day | ... | ... | } 81.0 | 36.0 |
| 4. Conduit III | Maximum hour | 81.0 | 81.0 |  | 81.0 |
| 5. Low-life pumps | Maximum day plus reserve | ... | ... | 48.0 | 48.0 |
| 6. High-life pumps | Maximum hour plus reserve | ... | ... | 108.0 | 108.0 |
| 7. Treatment plant | Maximum day plus reserve | ... | 48.0 | 48.0 | 48.0 |
| 8. Distribution system high-value district | Maximum hour | 81.0 | 81.0 | 81.0 | 81.0 |

## 36/Systems Capacities

Standard fire requirements take into account probable loss of water from connections broken in the excitement of a serious fire. Coincident draft of water for purposes other than fire fighting is rarely assumed to equal the maximum *hourly* rate. Depending on local conditions, the maximum *daily* rate may be a reasonably safe assumption instead.

*Example 2-2.*

The four typical waterworks systems shown in Fig. 2-3 supply a community with an estimated future population of 120,000.

1. Determine the required capacities of the constituent structures for an average consumption of 150 gpcd and a distributing reservoir so sized that it can balance out differences between hourly and daily flows, fire demands, and emergency requirements.

2. Find for management and design purposes the expected maximum day and peak hour demand and daily rates of water to be supplied to a residential area of 200 houses with a gross housing density of three dwellings per acre, an average market value of $25,000 per dwelling unit, and a potential evaporation of 0.28 in. on the maximum day.[4]

1. Required capacities for waterworks systems of Fig. 2-3. Average daily draft = $150 \times 1.2 \times 10^5/10^6 = 18$ mgd. Maximum daily draft = coincident draft = $2 \times 18 = 36$ mgd. Maximum hourly draft = $4.5 \times 18 = 81$ mgd. Fire flow (Table 7-1) = 14.4 mgd. Coincident draft plus fire flow = $36 + 14.4 = 50.4$ mgd.

Provision for breakdowns and repair of pumps and water purification units by installing one reserve unit. Low-lift pumps: $4/3 \times$ maximum daily draft = $4/3 \times 36 = 48$ mgd. High-lift pumps: $4/3 \times$ maximum hourly draft = $4/3 \times 81 = 108$ mgd. Treatment works: $4/3 \times$ maximum daily draft = 48 mgd.

The resultant capacities of systems components are summarized in the table below Fig. 2-3.

2. Average maximum and peak daily domestic demands per dwelling unit:[4] By Eq. 2-20: $Q_{\text{domestic}} = 157 + 3.46 \times 25 = 244$ gpud. By Eq. 2-19: $A = 0.803/3^{1.26} = 0.20$ acre per dwelling unit. By Eq. 2-18: $Q_{\text{sprinkling}} = 1.63 \times 10^4 \times 0.20 \times 0.28 = 913$ gpud excluding precipitation.

*For management:* $Q_{\text{max day}} = 244 + 913 = 1160$ gpud. By Eq. 2-21: $Q_{\text{peak hr}} = 334 + 2.02 \times 1160 = 2680$ gpud.

*For design:* From Table 2-9: $Q_{\text{max day}} = 1160 + 400 = 1560$ gpud. From Table 2-9: $Q_{\text{peak hr}} = 2680 + 700 = 3380$ gpud.

Rates of rural water use and wastewater production are generally functions of the water requirements and discharge capacities of existing fixtures.

## 2-10 Wastewater Flows

Public sewers receive and transport one or more of the following liquids: spent water, groundwater seepage or infiltration, and runoff from rainfall.

***Spent Water.*** Spent waters are primarily portions of the public water supply discharged into sewers through the drain pipes of buildings, and secondarily waters drawn from private or secondary sources for air conditioning, industrial processing, and similar uses. Of the water introduced into dwellings and similar buildings, 60 to 70% becomes wastewater. The remainder is used consumptively (Sec. 2–8). Commercial areas discharge about 20,000 gpd per acre.

***Groundwater Seepage.*** Well-laid street sewers equipped with modern preformed joints and tight manholes carry little groundwater. This cannot be said of house sewers unless they are constructed of cast iron or other materials normally laid with tight joints.

The following seepage may be expected when the sewers are laid above the groundwater table: 500 to 5000 gpd per acre, average 2000; 5000 to 100,000 gpd per mile of sewer and house connection, average 20,000; and 500 to 5000 gpd per mile and inch diameter of sewer and house connection, average 2500 plus 500 gpd per manhole. These ranges in seepage were made so broad because of the great uncertainty regarding the magnitude of groundwater flows in leaky sewers. Where poor construction permits entrance of groundwater, the flows may be expected to vary with rainfall, thawing of frozen ground, changes in groundwater levels and nature of the soil drained. As suggested by the diurnal flow variations shown in Fig. 2–4, seepage makes up most of the early-morning flows. Tightness of sewers may be measured by blocking off a portion of the system, usually with balloons designed for the purpose, then subjecting the tested portion to an appropriate pressure and observing the rate of water loss. Various other tests have been devised. If

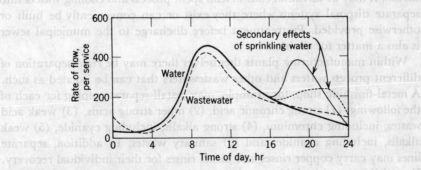

**Figure 2-4.**
**Flow variations of water and wastewater.**

contracts call for such tests and they are made, even occasionally, the estimates of the rate of groundwater seepage into street sewers can be greatly reduced and will have validity.

**Stormwater.** In well-watered regions, runoff from rainfall, snow, and ice normally outstrips spent water in intensity and annual volume. Spent water reaching combined sewers is indeed so small a fraction of the design capacity that it may be omitted from calculations for combined systems. As of 1960 about half the 130 million sewered persons in the United States were still connected to combined sewers. Relatively large volumes of raw wastewaters and large amounts of pollutional solids are, therefore, discharged in times of heavy rain into receiving waters through overflows from combined sewers (Sec. 21-14).

Although stormwater runoff is seldom allowed to be introduced into spent-water systems, it is difficult to enforce the necessary ordinances. Even when connections to the wastewater system can be made only by *licensed drain layers*, illicit connections to sanitary sewers must be expected to add to them some of the runoff from roof, yard, basement entrance, and foundation drains. Poorly sealed manhole covers permit further entrance of runoff. Total amounts vary with the effectiveness of enforcing regulations and conducting countermeasures. Allowances for illicit stormwater flow are as high as 70 gpcd and average 30 gpcd. A rainfall of 1 in. per hour may shed water at a rate of 12.5 gpm from 1200 sq ft of roof area, or 1.008 cfs from an acre of impervious surfaces. Leaky manhole covers may admit 20 to 70 gpm when streets are under an inch of water. The volume of illicit stormwater approximates the difference between normal dry-weather flows (DWF) and flows during intense rainfalls

***Industrial Wastewaters.*** Industrial wastewaters may be discharged into municipal wastewater systems at convenient points, provided they do not overload them or damage the collecting and treatment works. Nevertheless, it may be advantageous to lead spent process and cooling waters into separate disposal systems where they exist or can conveniently be built or otherwise provided. Pretreatment before discharge to the municipal sewer is also a matter for decision.

Within manufacturing plants themselves there may be rigid separation of different process waters and other wastewaters that can be isolated as such. A metal-finishing shop, for example, may install separate piping for each of the following: (1) strong chromic acid, (2) other strong acids, (3) weak acid wastes, including chromium, (4) strong alkalis, including cyanide, (5) weak alkalis, including cyanide, and (6) sanitary wastes. In addition, separate lines may carry copper rinses and nickel rinses for their individual recovery. Not all lines need be laid as underground gravity-flow conduits. Relatively small volumes of wastewaters may be collected in sumps and pumped through overhead lines instead.

When there is good promise of reasonable recovery of water or waste matters or of treatment simplification, spent industrial waters may be segregated even if collecting lines must be duplicated. Examples are (1) the pretreatment of stong wastewaters before admixture with similar dilute wastewaters and (2) the separation of cyanide wastewaters for destruction by chlorine before mixing them with wastewaters containing reaction-inhibiting nickel. However, it may also pay to blend wastewaters in order to (1) dilute strong wastes, (2) equalize wastewater flows and composition, (3) permit self-neutralization to take place, (4) foster other beneficial reactions, and (5) improve the overall economy. As a rule, it pays to separate wastewaters while significant benefits can still accrue. After that they may well be blended to advantage into a single waste stream.

Wash waters may require special collection and treatment when they differ from other process waters. Thus, most wastewaters from food processing contain nutrients that are amenable as such to biological treatment. However, they may no longer be so after strong alkalis, soaps, or synthetic detergents, sanitizers, and germicides have been added to them along with the wash waters of the industry.

## 2-11 Variations in Wastewater Flows

Imprinted on flows in storm and combined sewers is the pattern of rainfall and snow and ice melt. Fluctuations may be sharp and high for the storm rainfall itself and as protracted and low as the melting of snow and ice without the benefit of spring thaws. As shown in Fig. 2–4, (1) flows of spent water normally lie below and lag behind the flows of supplied water; and (2) some of the water sprayed onto lawns and gardens is bound to escape into yard and street drains.

*Damping Effects.* The *open-channel* hydraulics of sewers allow their levels to rise and fall with the volume rate of entrant waters. Rising levels store flows; falling levels release them. The damping effect of storage is reinforced by the compositing of flows from successive upstream areas for which shape as well as size are governing factors. Low flows edge upward and high flows move downward. Indeed, wastewater and stormwater flows are quite like stream and flood flows and hence subject also to analysis by flood-routing procedures (Sec. 4–9).

H. M. Gifft[10] has evaluated damping effects by the following observational relationships:

$$Q_{max}/Q_{avg} = 5.0\ P^{-1/6} \qquad (2\text{-}22)$$

[10] H. M. Gifft, Estimating Variations in Domestic Sewage Flows, *Waterworks & Sewerage*, **92**, 175 (1945). Eight comonly used sets of ratios, including Gifft's, are compared in Fig. 3, p. 33, *Manual of Practice*, ASCE-No. 37, WPCF-No. 9 (1969). A somewhat different approach is recommended by J. C. Geyer and J. J. Lentz, An Evaluation of Problems of Sanitary Sewer System Design, *Jour. WPCF*, **38**, 1138 (1966).

40/Systems Capacities

$$Q_{min}/Q_{avg} = 0.2\ P^{1/6} \quad (2\text{-}23)$$

$$Q_{max}/Q_{min} = 25.0\ P^{-1/3} \quad (2\text{-}24)$$

Here $Q_{max}$, $Q_{avg}$, and $Q_{min}$ are respectively, the maximum, average, and minimum daily rates of flow of spent water and $P$ is the population in thousands.

Expected flow rates from sewered areas of moderate size (tens of sq miles) are as follows:

$$Q_{max} = 2\ Q_{avg} \quad \text{and} \quad Q_{min} = \tfrac{2}{3}\ Q_{avg}$$

$$Q_{max\ peak} = {}^3\!/_2\ Q_{max} = 3\ Q_{avg} \quad \text{and} \quad Q_{min\ peak} = \tfrac{1}{2}\ Q_{min} = \tfrac{1}{3}\ Q_{avg}$$

*Example 2-3.*

Estimate the average, peak, and low rates of flow in a spent-water sewer serving 9000 people, an area of 600 acres, and a community of 45,000 inhabitants with an average rate of water consumption of 150 gpcd.

1. Spent water: $0.7 \times 150 = 105$ gpcd.

2. Maximum hour: $3 \times 105 = 315$ gpcd compared with Eq. 2-22: $Q_{max} = 105 \times 5/(9)^{1/6} = 360$ gpcd.

3. Minimum hour: $^1\!/_3 \times 105 = 35$ gpcd compared with Eq. 2-23: $Q_{min} = 105 \times 0.2(9)^{1/6} = 30$ gpcd.

Expected low flows in spent-water and combined sewers are as meaningful as expected high flows because suspended solids are deposited or stranded as flows and velocities decline. Flow may be obstructed and malodorous, and dangerous gases may be released from the accumulating detritus.

**Industrial and Rural Wastewaters.** Variations in the flow of *industrial wastewaters* are dictated by working hours and process schedules. Continuous flow normally contributes to the economy of industrial operations. However, batch processes are still in use. Their presence is reflected in heavy intermittent drafts on water systems and sudden releases of large quantities of process waters during short periods of time, often toward the end of each working shift.

Spent-water flows from *rural dwellings* are usually relatively small and unsteady; comparatively fresh and concentrated; and quite warm and greasy or soapy. There may be little, if any, flow at night.

# three

# statistical
# hydrology

## 3-1 Definitions

Hydrology[1] is the science of water in nature: its properties, distribution, and behavior. *Statistical hydrology* is the application of statistical methods of analysis to measurable hydrological events for the purpose of arriving at engineering decisions. By the introduction of suitable statistical techniques, "enormous amounts of quantitative information can often be reduced to a handful of parameters that convey, clearly and incisively, the underlying structure of the original or *raw* data."[2]

***Global Water Resources.*** The total water resource of the earth is approximately $330 \times 10^6$ cu miles or $363 \times 10^9$ billion gallons (bg)—about 95% in the oceans and seas and 2% in the polar ice caps.[3] However, the 35 grams of salt in a liter of sea water and the remoteness as well as fundamentally ephemeral nature of the polar ice caps interfere with their use. This leaves as the potential fresh-water resource of the earth no more than $10.9 \times 10^9$ bg in lakes, streams, permeable soils, and the atmosphere; only

---

[1] From the Greek *hydor*, water, and *logos*, science.

[2] G. M. Fair, J. C. Geyer, and D. A. Okun, *Water and Wastewater Engineering*, John Wiley & Sons, 1966, Vol. 1, pp. 4–2, 4–26, 4–32.

[3] R. Colas, Producing Fresh Water from Sea Water, *L'Eau*, **49**, 205 (1962). 1 cu mile = $1.10 \times 10^3$ billion gallons.

*41*

about 3% is in the atmosphere, the remainder being split almost equally between surface and ground.

Fortunately, the hydrosphere is not static; its waters circulate (Fig. 3–1). Between $110 \times 10^6$ and $130 \times 10^6$ bg of water fall annually from the skies, about a quarter onto the continents and islands and the remainder onto the seas; $10 \times 10^6$ to $11 \times 10^6$ bg return to the oceans as annual runoff. In overall estimates about a third of the land mass of the earth is classified as well watered, the remainder as semiarid and arid.

Within the coterminous United States, the 100th meridian is the general dividing point between annual rainfalls of 20 in. or more to the East and, except for the Pacific slopes, less than 20 in. to the West. The total length of surface streams is about $3 \times 10^6$ miles and the five Great Lakes, four of them shared with Canada, hold $7.8 \times 10^6$ bg, and constitute geographically the largest surface storage of freshwater on earth.

***The Water Cycle.*** As shown in Fig. 3–1, water is transferred to the earth's atmosphere (1) through the *evaporation* of moisture from land and water surfaces and (2) through the *transpiration* of water from terrestrial and emergent aquatic plants. Solar radiation provides the required energy.

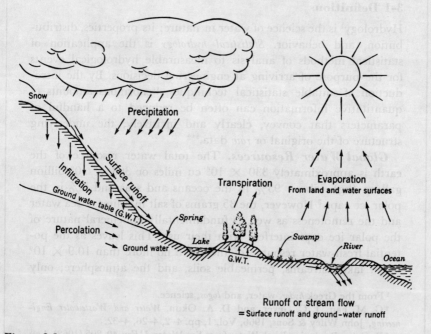

**Figure 3–1.**
**The water cycle.**

Internal or dynamic cooling of rising, moisture-laden air and its exposure to cold at high altitudes eventually lower the temperature of ascending air masses to the dewpoint,[4] condense the mositure, and precipitate it on land and sea. Overland and subsurface flows complete the hydrological cycle.

## 3-2 Collection of Hydrological Data

Without adequate quantitative information on the earth's water resource, its use and development become an economic uncertainty and an engineering gamble. The collection of pertinent data is, therefore, an imperative social responsibility that is generally assumed by government. When and to what extent the government of the United States has accepted this responsibility is summarized in Table 3-1.

*Table 3-1*
*Nation-wide Collection of Hydrological Data for the United States*

| Measurement and Apparatus | Beginning Date | Number of Stations | Areal Density per 1000 sq miles |
|---|---|---|---|
| Rain and snow | 1870 | $> 10^5$ | 4 |
| Nonrecording rain gages | | ⅔ of total | 2.7 |
| Recording rain gages | | ⅓ of total | 1.3 |
| Storm-tracking radar | 1945 | — | — |
| Snow courses | 1910 | $> 10^3$ annually | — |
| Stream flow | | | |
| Stream gages | 1890 | $< 10^4$ | 0.5–11 |
| Groundwater | 1895 | — | — |
| Evaporation | | Several hundred | |

Of the different hydrological parameters shown in this table, annual precipitation is a measure of the maximum annual renewal of the water resource of a given region. About one fourth to one third of the water falling on continental areas reaches the oceans as runoff. The balance is returned to the atmosphere by evaporation and transpiration. Where melting of the winter's snows produces a major part of the annual runoff, or where spring thaws cause serious floods, accurate methods for determining the annual snowfall on a watershed are essential requirements. Otherwise, storm rain-

[4] The temperature at which the air becomes saturated with water vapor and below which dew is formed.

falls monitored by recording gages form the basis for predicting the flood stages of river systems and for estimating the expected rate of runoff from areas yet to be provided with drains or sewers.

Much larger volumes of water are returned to the atmosphere over continents by transpiration than by evaporation. In the well-watered regions of the United States, annual evaporation from free water surfaces more or less equals annual precipitation; in the arid and semiarid portions, it exceeds precipitation manyfold.

About half the annual rainfall normally enters the earth's crust and about half the water stored in its interstices lies within half a mile of the surface and can normally be drawn upon for water supply.

## 3-3 Hydrological Frequency Functions

Arrays of many types of hydrological as well as other measurements trace bell-shaped curves like the *normal frequency distribution* shown above Table 5 in the Appendix to this book. Arrays are numerical observations arranged in order of magnitude from the smallest to the largest, or in reverse order. They are transformed into frequency distributions by subdividing an abscissal scale of magnitude into a series of usually equal intervals of size—called class intervals—and counting the number of observations that lie within them. The resulting numbers of observations per class interval, by themselves or in relation to the total number of observations, are called the frequencies of the observations in the individual sequent intervals. To simplify subsequent calculations, frequencies are referenced to the central magnitudes of the sequent class intervals.

*Averages.* The tendency of a bell-shaped curve to cluster about a central magnitude of an array, that is, its central tendency, is a measure of its average magnitude. The *arithmetic mean* magnitude of an array generally offers the most helpful concept of central tendency. Unlike the *median* magnitude of an array, which is positionally central irrespective of the magnitudes of the individual observations composing the array, and unlike the *mode* of the array, which is the magnitude having the highest frequency of the observations composing the array, the arithmetic mean is a collective function of the magnitudes of all component observations.

*Variability and Skewness.* The tendency of a bell-shaped curve to deviate by larger and larger amounts from the central magnitude of an array with less and less frequency is a measure of its variability, variation, deviation, dispersion, or scatter. The tendency of a bell-shaped curve to be asymmetrical is a measure of its skewness.

Mathematically, the symmetrical bell-shaped *normal* frequency curve of Table 5 in the Appendix is described fully by two parameters: (1) the arithmetic mean, $\mu$, as a measure of central tendency, and (2) the *standard deviation*, $\sigma$, as a measure of dispersion, deviation, or variation. The equation of the curve, which is referred to as the *Gaussian*[5] or *normal probability curve*, is

$$y/n = F(x)/n = f(x) = [1/(\sigma\sqrt{2\pi})] \exp\{-\tfrac{1}{2}[(x-\mu)/\sigma]^2\} \quad (3\text{-}1)$$

Here $y$ is the number or frequency of observations of magnitude $x$ deviating from the mean magnitude by $x - \mu$, $n$ is the total number of observations, $\sigma$ is the standard deviation from the mean, $F(x)$ symbolizes a function of $x$, and $\pi$ is the mathematical constant 3.1416.

As shown in Table 5 of the Appendix and calculated from Eq. 3-1, the origin of the coordinate system of the Gaussian curve lies at $x - \mu = 0$ where the frequency is $y = n/(\sigma\sqrt{2\pi})$, and the distance from the origin to the points of inflection of the curve is the standard deviation $\sigma = \sqrt{\Sigma(x-\mu)^2/n}$. This presupposes that deviations are referenced to the true mean $\mu$, whereas calculations must be based on the observed mean $\bar{x} = \Sigma x/n$, which is expected to deviate from the true mean by a measurable although small amount. However, it can be shown that when the observed standard deviation is calculated as $s = \sqrt{\Sigma(x-\mu)^2/(n-1)}$, it closely approximates $\sigma$. As $n$ becomes large, there is indeed little difference between $s$ and $\sigma$ as well as between $\bar{x}$ and $\mu$, and the Greek letters are often used even when, strictly speaking, the Latin ones should be. The ratios $y/n$ and $(x-\mu)/\sigma$ generalize the equation of the *normal frequency function*.

Engineers are usually interested in the expected frequencies of observations below or above a given value of $x$ or falling between two given values of $x$. The magnitudes of areas under the normal curve provide this useful information better than do point frequencies. Table 5 in the Appendix is a generalized table of these areas.[6] It is called the probability integral table because a given area is a measure of the expected fraction of the total area (as unity) corresponding to different departures from the mean $(x - \mu)$ in terms of the standard deviation $(\sigma)$, that is, $(x-\mu)/\sigma = t$.

[5] After its formulator, the German astronomer and mathematician, Karl Friedrich Gauss (1777–1855).

[6] Measuring the area from the center out, and replacing exp $\tfrac{1}{2}[(x-\mu)/\sigma]^2$ by the convergent series

$$1 - \left(\frac{x-\mu}{\sigma\sqrt{2}}\right)^2 + \frac{1}{2!}\left(\frac{x-\mu}{\sigma\sqrt{2}}\right)^4 - \frac{1}{3!}\left(\frac{x-\mu}{\sigma\sqrt{2}}\right)^6 + \cdots$$

the ratio of the wanted area to the total number of observations $n$ becomes

$$\frac{1}{\sqrt{\pi}}\left[\left(\frac{x-\mu}{\sigma\sqrt{2}}\right) - \frac{1}{3}\left(\frac{x-\mu}{\sigma\sqrt{2}}\right)^3 + \frac{1}{5\times 2!}\left(\frac{x-\mu}{\sigma\sqrt{2}}\right)^5 - \cdots\right]$$

## 3-4 Moments of Distributions

The moments of continuous distributions provide useful information on their properties. They are analogous to moments of structural shapes such as I-beams. Thus if $y = F(x)$ is a continuous distribution, such as the normal frequency distribution, the $r$th moment *about the origin* is written

$$\mu_r' = \int_{-\infty}^{+\infty} x^r f(x)\, dx \qquad (3-2)$$

and it can be shown that the set of moments for $r = 1, 2, 3, \ldots, \infty$ completely defines the function $F(x)$. For $r = 1$, the first moment about the origin is

$$\mu' = \int_{-\infty}^{+\infty} x f(x)\, dx \qquad (3-3)$$

and equals the *arithmetic mean*. When the origin of moments is at the mean, $\mu = 0$, and the $r$th moment, now written without a prime, has the equation

$$\mu_r = \int_{-\infty}^{+\infty} (x - \mu')^r f(x)\, dx \qquad (3-4)$$

and again the set of moments for $r = 1, 2, 3, \ldots, \infty$ completely defines the function.

The sets of equations in $\mu_r'$ and $\mu_r$ can be derived from each other. For example, if $r = 2$, the second moments about the mean and the origin are related as follows:[7]

$$\mu_2 = \mu_2' - \mu^2 \qquad (3-5)$$

The second moment is called the *variance* and obviously equals $\sigma^2$. Because $y = F(x)$ and $(x - \mu)^2$ are never negatives, the variance is never negative; its positive square root is the *standard deviation* $\sigma$. Both $\mu$ and $\sigma$ have the same dimensions as $x$. The first moment measures the magnitude of the mean or central tendency of the distribution; the second moment the magnitude of the variance or dispersion of the distribution; and the third moment the *skewness* or asymmetry of the distribution. The third moment may be positive or negative. Its magnitude is a measure of the distance to the right ($+$) or the left ($-$) of the arithmetic mean and median from the mode. Higher moments measure other properties of frequency curves, among them their *kurtosis* or tendency to be excessively tall or excessively squat, in the region of peak frequencies.

---

[7] Because $\mu_2'$ from Eq. 3-2 is $\int x^2 F(x)dx$ and $\mu_2$ from Eq. 3-4 is $\int x^2 F(x)dx - 2\mu \int xF(x)dx + \mu^2 \int F(x)dx$ and equal to $\mu_2' - 2\mu(\mu) + \mu^2(1) = \mu_2' - \mu^2$.

**Probability Paper.** As suggested by Allen Hazen,[8] the probability integral (Appendix, Table 5) can be used to develop a system of coordinates on which normal frequency distributions plot as straight lines. The companion to the probability scale can be (1) arithmetic for true Gaussian normality, (2) logarithmic for geometric normality, or (3) some other function of the variable for some other functionally Gaussian normality. Observations that have a lower limit at or near zero may be geometrically normal ($\log 0 = -\infty$).

Helpful and useful associations in arithmetic and geometric probability plots can be listed as follows:

| Observed or Derived Frequency | Observed or Derived Magnitude: Arithmetic | Geometric |
|---|---|---|
| 50% | $\mu$ | $\mu_g$ |
| 84.1% | $\mu + \sigma$ | $\mu_g \times \sigma_g$ |
| 15.9% | $\mu - \sigma$ | $\mu_g / \sigma_g$ |

As shown in Fig. 3-2, two or more series of observations can be readily compared by plotting them on the same scales. The ratio of $\sigma$ to $\mu$ is called the coefficient of variation, $c_v$. It is a useful, dimensionless, analytical measure of the relative variability of different series.

Examination of a series of equally good arrays of information shows that their statistical parameters, their means and standard deviations for example, themselves form bell-shaped distributions. Their variability, called their *reliability* in such instances, is intuitively a function of the size of the sample. Expressed as a standard deviation, the reliability of the common parameters for normal distributions is shown in the following list:

| Parameter | Computation | Standard Deviation |
|---|---|---|
| Arithmetic mean | $\bar{x} = \Sigma x_i / n$ | $\sigma / \sqrt{n}$ |
| Median | midmost observation | $1.25\sigma / \sqrt{n}$ |
| Arithmetic standard deviation | $s = \sqrt{\Sigma(x_i - \bar{x})^2 / (n-1)}$ | $\sigma / \sqrt{2n} = 0.707 \sigma / \sqrt{n}$ |
| Coefficient of variation | $c_v = s / \bar{x}$ | $\sqrt{1 + 2c_v^2} / \sqrt{2n}$ |
| Geometric mean | $\log \mu_g = (\Sigma \log x_i) / n$ | $(\log \sigma_g) / \sqrt{n}$ |
| Geometric standard deviation | $\log \sigma_g = \sqrt{\Sigma \log^2(x_i / \bar{x}_g) / (n-1)}$ | $\log \sigma_g / \sqrt{2n}$ |

[8] Allen Hazen, Storage to be Provided in Impounding Reservoirs, *Trans. Am. Soc. Civil Engrs.*, 77, 1539 (1914).

## 48 / Statistical Hydrology

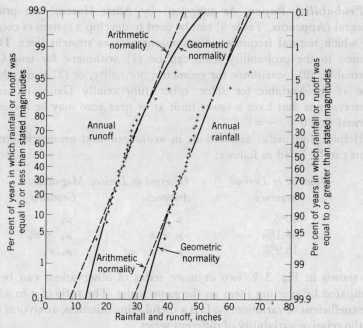

**Figure 3-2.**
**Frequency distribution of annual rainfall and runoff plotted on arithmetic-probability paper.**

### 3-5 Least Squares and Regression Analysis

Engineers are called upon again and again to fit an analytical function to observed data as well as to evaluate the parameters of some prescribed functional representation such as arithmetically or geometrically normal frequency distributions.

*Least Squares.* The most useful method of finding the coefficients $\alpha$ and $\beta$ in the linear relationship $y = \alpha + \beta(x - \mu_x)$, where $y$ is the dependent and $x$ is the independent variable, minimizes the sum of the squares of the residuals $R$ or

$$\sum_{i=1}^{n} R_i^2 = \sum [\alpha + \beta(x_i - \mu_x) - y_i]^2 = \text{a minimum}$$

A minimum is obtained when the first derivatives of $\Sigma R_i^2$ with respect to $\alpha$ and $\beta$ are set equal to zero; thus

$$\partial(\sum R_i^2)/\partial \alpha = 2 \sum [R_i/\partial \alpha)] = 2 \sum [\alpha + \beta(x_i - \mu_x) - y_i] = 0$$
$$\partial(\sum R_i^2)/\partial \beta = 2 \sum [R_i(R_i/\partial \beta)] =$$
$$2 \sum \{[\alpha + \beta(x_i - \mu_x) - y_i](x_i - \mu_x)\} = 0$$

Summing over $i$ and simplifying, the following simultaneous equations or normal equations of the bivariate array $(x_i, y_i)$ are found:

$$n\alpha + \beta \sum x_i - \beta n \mu_x - \sum y_i = 0$$

$$\alpha \sum x_i + \beta \sum x_i^2 - \beta \mu_x \sum x_i - \sum x_i y_i = 0 \qquad (3\text{-}6)$$

Because $\beta \sum x_i = \beta n \mu_x$,

$$\alpha = \sum y_i / n = \mu_y \qquad (3\text{-}7)$$

Hence $\beta(\sum x_i^2 - n\mu_x^2) = \sum x_i y_i - n\mu_x \mu_y$
or

$$\beta = (\sum x_i y_i - n\mu_x \mu_y)/(\sum x_i^2 - n\mu_x^2) \qquad (3\text{-}8)$$

and

$$y = \alpha + \beta(x - \mu_x) \qquad (3\text{-}9)$$

is the best estimate of $y$ in terms of $x$.

**Regression.** In some circumstances $y$ and $x$ can each be both dependent and independent variables. Hydrological examples are comparisons of rainfalls at nearby stations or stream flows in neighboring catchment areas. Such comparisons are called correlations or regressions,[9] and can be generalized in terms of the coefficient $\sigma$, which is called the coefficient of correlation and is the root mean square of the slopes of the lines of best fit of ($y$ out of $x$) and ($x$ out of $y$) respectively. In accordance with Eq. 3-8, therefore,

$$\rho = \sqrt{\beta_y \beta_x} = (\sum x_i y_i - n\mu_x \mu_y)/(\sigma_x \sigma_y) \qquad (3\text{-}10)$$

and it follows that

$$\beta_y = \rho \sigma_y / \sigma_x \quad \text{and} \quad \beta_x = \rho \sigma_x / \sigma_y \qquad (3\text{-}11)$$

respectively.

The standard deviation $\sigma_\rho = \sigma_y \sqrt{1 - \rho^2}$ is called the standard error of estimate of $y$ and is a measure of the unexplained variance. When $x_i$ and $y_i$ are colinear, $\rho = \pm 1$ and $\sigma_\rho$ is zero. When $x_i$ and $y_i$ are unrelated, $\rho = 0$, and the standard error, or unexplained variation, is not reduced by virtue of the regression of $y$ or $x$.

Correlation of hydrological records is useful in filling in missing information by *cross correlation* and in increasing the length of available records by *serial correlation*.[2]

---

[9] They were so named by Sir Francis Galton (1822–1911), English meteorologist, statistician, and biologist, in comparisons of body measurements of fathers and sons that showed evidence of a *regression* towards mean measurements.

## 3-6 Rainfall and Runoff Analysis

Two types of rainfall and runoff records are analyzed most frequently in the design of water and wastewater works: (1) records of the amounts of water collected by given watersheds in fixed calendar periods such as days, months, years, and (2) records of the intensities and durations of specific rainstorms and flood flows in given drainage areas. Statistical studies of *annual water yields* provide information on the safe and economic development of surfacewater supplies by direct draft and by storage, cast some light on the possible production of groundwater, and are needed in estimates of the pollutional loads that can be tolerated by bodies of water into which wastewaters are discharged. Statistical studies of *rainfall intensities* and *flood runoff* are starting points in the design of stormwater drainage systems, the dimensioning of spillways and diversion conduits for dams and related structures, the location and protection of water and wastewater works that lie in the flood plains of given streams, and the sizing of rainwater collection works.

## 3-7 Annual Rainfall and Runoff

The presumptive presence of a lower limit of annual rainfall and runoff skews annual rainfall and runoff frequency distributions to the right. The magnitude of this lower limit is generally smaller than the recorded minimum but greater than zero. Although it stands to reason that there must also be an upper limit, its value is less circumscribed and, from the standpoint of water supply, also less crucial. In spite of acknowledging these constraints, most records of annual rainfall and runoff are generalized with fair success as arithmetically normal series and somewhat better as geometrically normal series. Therefore, reasonably accurate comparisons are made in terms of the observed arithmetic or geometric means and the arithmetic or geometric standard deviations. For ordinary purposes, mean annual values and coefficients of variation are employed to indicate the comparative safe yields of water supplies that are developed with and without storage. For comparative purposes, drafts are expressed best as ratios to the mean annual rainfall or runoff, whatever the basis of measurement is.

*Rainfall.* In those portions of the North American continent in which municipalities have flourished, mean annual rainfalls generally exceed 10 in. and range thence to almost 80 in. In areas with less than 20 in. of annual rainfall and without irrigation, agriculture can be a marginal economic pursuit. For the well-watered regions, the associated coefficients of variation, $c_v$, are as low as 0.1; for the arid regions they are as high as 0.5, thus implying that a deficiency as great as half the mean annual rainfall is expected to occur in the *arid* regions as often as a deficiency as great as one-tenth the mean annual rainfall, or less, in the *well-watered* regions.[10] Therefore, high values of $c_v$ are warning signals of low maintainable drafts or high storage requirements.

**Runoff.** Evaporation and transpiration, together with unrecovered infiltration into the ground, reduce annual runoff below annual rainfall. Seasonally, however, the distribution of rainfall and runoff may vary so widely that it is impossible to establish a direct and meaningful relationship between the two. On the North American continent, the mean annual runoff from catchment areas for water supplies ranges from about 5 to 40 in. and the coefficient of variation of runoff lies between 0.75 and 0.15, respectively. The fact that the mean annual runoff is usually less than half the mean annual rainfall and the variation in stream flow is about half again as great as the variation in precipitation militates against the establishment of direct runoff-rainfall ratios. Storage of winter snows and resulting summer snow melts offer an important example of conflict between seasonal precipitation and runoff.

---

*Example 3–1.*

Analyze the 26-year record of a stream[11] in the northeastern United States and of a rain gage[11] situated in a neighboring valley and covering the identical period of observation (Table 3–2).

To plot the data, use the following information: the length of each record is $n = 26$ years; therefore, each year of record spans $100/26 = 3.85\%$ of the experience. However, the arrays are plotted on probability paper in Fig. 3–2 at $100\,k/(n+1) = (100\,k/27)\%$ in order to locate identical points for the left-hand and right-hand probability scales. The resulting plotting observations are shown in Table 3–3.

---

Necessary calculations are exemplified in Table 3–3 for rainfall and arithmetic normality alone. Calculations for runoff and arithmetic normality would substitute the array of runoff values for that of rainfall values; and assumption of geometric normality would require substitution of the loga-

---

[10] Reference to the probability integral, Appendix Table 5, will show that deficiencies, or negative deviations from the mean, equal to or greater than $c_v\,\mu = \sigma$ are to be expected $50.0 - 34.1 = 15.9\%$ of the time or $1/0.159 =$ once in 6.3 years, because $x/\sigma = 1.0$. However, these calculations yield only approximate results because normality is assumed where skewness may well exist.

[11] The Westfield Little River, which supplies water to Springfield, Mass., the rain gage being situated at the West Parish filtration plant. The years of record are 1906 to 1931.

rithms of the observations for the observations themselves. The calculated statistical parameters can be summarized as follows:

|  | Rainfall | Runoff | Runoff-Rainfall Ratio % |
|---|---|---|---|
| Length of record, $n$, years | 26 | 26 | — |
| Arithmetic mean, $\mu$, in. | 46.8 $\pm$ 1.2 | 26.6 $\pm$ 1.1 | 57 |
| Median, at 50% frequency, in. | 46.3 $\pm$ 1.5 | 25.5 $\pm$ 1.5 | 55 |
| Geometric mean, $\mu_g$, in. | 46.5 $\overset{\times}{\div}$ 1.02 | 26.1 $\overset{\times}{\div}$ 1.04 | 56 |
| Arithmetic standard deviation, $\sigma$, in. | 5.8 $\pm$ 0.8 | 5.8 $\pm$ 0.8 | 100 |
| Coefficient of variation, $c_v$ | 12.9 $\pm$ 1.8 | 21.8 $\pm$ 3.2 | 169 |
| Geometric standard deviation, $\sigma_g$ | 1.13 $\overset{\times}{\div}$ 1.02 | 1.24 $\overset{\times}{\div}$ 1.03 | 110 |

Examination of the results and the plots shows that (1) both annual rainfall and annual runoff can be fitted approximately by arithmetically normal distributions and somewhat better by geometrically normal distributions; (2) a little more than half the annual rainfall appears as streamflow; (3) runoff is about 1.7 times as variable as rainfall when measured by $c_v$; (4) the probable lower limits of rainfall and runoff are 30 in. and 10 in. respectively, as judged by a negative deviation from the mean of $3\sigma$, or a probability of occurrence of $1/(0.5 - 0.4987) =$ once in 700 years; and (5) for geometric normality, the magnitudes of the minimum yields expected once in 2, 5, 10, 20, 50, and 100 years, that is, 50, 20, 10, 5, 2, and 1% of the years, are 47, 42, 39, 37, 35, and 33 in. for rainfall and 27, 22, 20, 19, 17, and 15 in. for runoff.

## 3-8 Storm Rainfall

Storms sweeping over the country precipitate their moisture in fluctuating amounts during given intervals of time and over given areas. For a particular storm, recording rain gages measure the *point rainfalls* or quantities of precipitation collected during specified intervals of time at the points at which the gages are situated. Depending on the size of the area of interest, the sweep of the storm, and the number and location of the gages, the information obtained is generally far from complete. Statistical averaging of experience must then be adduced to counter individual departures from the norm. Given the records of one or more gages within or reasonably near the area of interest, the rainfall is generally found to vary in intensity (1) during the time of passage or duration of individual storms (time-intensity or intensity-duration); (2) throughout the area covered by individual storms

## Table 3-2
*Record of Annual Rainfall and Runoff (Example 3–1)*

| Order of Occurrence | Rainfall, in. | Runoff, in. |
|---|---|---|
| 1  | 43.6         | 26.5 |
| 2  | 53.8         | 35.5 |
| 3  | 40.6         | 28.3 |
| 4  | 45.3         | 25.5 |
| 5  | 38.9 (min.)  | 21.4 |
| 6  | 46.6         | 25.3 |
| 7  | 46.6         | 30.1 |
| 8  | 46.1         | 22.7 |
| 9  | 41.8         | 20.4 |
| 10 | 51.0         | 27.6 |
| 11 | 47.1         | 27.5 |
| 12 | 49.4         | 21.9 |
| 13 | 40.2         | 20.1 |
| 14 | 48.9         | 25.4 |
| 15 | 66.3 (max.)  | 39.9 |
| 16 | 42.5         | 23.3 |
| 17 | 47.0         | 26.4 |
| 18 | 48.0         | 29.4 |
| 19 | 41.3         | 25.5 |
| 20 | 48.0         | 23.7 |
| 21 | 45.5         | 23.7 |
| 22 | 59.8         | 41.9 (max.) |
| 23 | 48.7         | 32.9 |
| 24 | 43.3         | 27.7 |
| 25 | 41.8         | 16.5 (min.) |
| 26 | 45.7         | 23.7 |

(areal distribution); and (3) from storm to storm (frequency-intensity-duration or distribution in time).

**Intensity of Storms.** The intensity, or rate, of rainfall is conveniently expressed in inches per hour; and it happens that an inch of water falling on an acre in an hour closely equals a cubic foot per second (1 in./hr = 1.008 cfs per acre). By convention, storm intensities are expressed as the maximum arithmetic mean rates for intervals of specified length, that is, as progressive means for lengthening periods of time. Within each storm, they are highest

## Table 3-3
Calculation of Arithmetic Parameters of Annual Rainfall Frequency (Example 3-1)

| Magnitude of Observation (1) | Plotting Position % (2) | Deviation from Mean $(x - \mu)$ (3) | $(x - \mu)^2$ (4) |
|---|---|---|---|
| 38.9 | 3.7 | −7.9 | 62.41 |
| 40.2 | 7.4 | −6.6 | 43.56 |
| 40.6 | 11.1 | −6.2 | 38.44 |
| 41.3 | 14.8 | −5.5 | 30.25 |
| 41.8 | 18.5 | −5.0 | 25.00 |
| 41.8 | 22.2 | −5.0 | 25.00 |
| 42.5 | 25.9 | −4.3 | 18.49 |
| 43.3 | 29.6 | −3.5 | 12.25 |
| 43.6 | 33.3 | −3.2 | 10.24 |
| 45.3 | 37.0 | −1.5 | 2.25 |
| 45.5 | 40.7 | −1.3 | 1.69 |
| 45.7 | 44.5 | −1.1 | 1.21 |
| 46.1 | 48.2 | −0.7 | 0.49 |
| 46.6 | 51.9 | −0.2 | 0.04 |
| 46.6 | 55.5 | −0.2 | 0.04 |
| 47.0 | 59.3 | +0.2 | 0.04 |
| 47.1 | 63.0 | +0.3 | 0.09 |
| 48.0 | 66.7 | +1.2 | 1.44 |
| 48.0 | 70.4 | +1.2 | 1.44 |
| 48.7 | 74.1 | +1.9 | 3.61 |
| 48.9 | 77.8 | +2.1 | 4.41 |
| 49.4 | 81.5 | +2.6 | 6.76 |
| 51.0 | 85.2 | +4.2 | 17.64 |
| 53.8 | 88.9 | −7.0 | 49.00 |
| 59.8 | 92.6 | +13.0 | 169.00 |
| 66.3 | 96.3 | +19.5 | 380.25 |
| Sum, 1218.8 | n = 26 | 0.0 | 916.04 |

Mean, $\mu = 46.8$  (Standard deviation)$^2$, $\sigma^2 = 33.93$
Median by interpolation or from plot, 46.3  $\sigma = 5.8$
Coefficient of variation,  $c_v = 12.4\%$

during short time intervals and decline steadily with the length of interval. Time-intensity calculations are illustrated in Example 3–2.

*Example 3–2.*

Given the record[12] of an automatic rain gage, find the progressive arithmetic mean rates, or intensities, of precipitation for various durations. The record is shown in Cols. 1 and 2 of Table 3–4, converted into rates in Cols. 3 and 4, and assembled in order of magnitude for increasing lengths of time in Cols. 5 to 7. The maximum rainfall of 5 min. duration (0.54 in. = 6.48 in./hr) was experienced between the 30th and 35th min.

## 3-9 Frequency of Intense Storms

The higher the intensity of storms, the rarer is their occurrence or the lower their frequency. Roughly the highest intensity of specified duration in a station record of $n$ years has a frequency of once in $n$ years and is called the $n$-year storm. The next highest intensity of the same duration has a frequency of once in $n/2$ years and is called the $n/2$-year storm.

The recurrence interval $I$ is the calendar period, normally in years, in which the $k$th highest or lowest values in an array covering $n$ calendar periods is expected to be exceeded statistically. The associated period of time is $100/I$. The fifth value in a 30-year series would, therefore, be calculated to have a recurrence interval $I = n/k = 30/5 = 6$ years, or a chance of $100/6 = 16.7\%$ of being exceeded in any year. If, as in Example 3–1, the frequency of occurrence is calculated as $k/(n + 1)$, the recurrence interval is $(n + 1)/k = 31/5 = 6.2$ years. Other plotting positions could be chosen instead.

By pooling all observations irrespective of their association with individual storm records, a generalized intensity-duration-frequency relationship is obtained. The following empirical equations are of assistance in weeding out storms of low intensity from the analysis of North American station records:

$$i = 0.6 + 12/t \text{ for the northern United States} \quad (3\text{-}12)$$

$$i = 1.2 + 18/t \text{ for the southern United States} \quad (3\text{-}13)$$

Here $i$ is the rainfall intensity in inches per hour and $t$ is the duration in minutes. For a duration of 10 min, for example, intensities below 3 in./hr in

[12] Storm of October 27–28, 1908, at Jupiter, Florida.

## Table 3-4
*Time and Intensity of a Storm Rainfall (Example 3-2)*[a]

| Rain-Gage Record | | | | Time-Intensity Relationship | | |
|---|---|---|---|---|---|---|
| Time from Beginning of Storm, min (1) | Cumulative Rainfall, in. (2) | Time Interval, min (3) | Rainfall During Interval, in. (4) | Duration of Rainfall, min (5) | Maximum Total Rainfall, in. (6)[b] | Arithmetic Mean Intensity, in. per hr (7) |
| 5 | 0.31 | 5 | 0.31 | 5 | 0.54 | 6.48 |
| 10 | 0.62 | 5 | 0.31 | 10 | 1.07 | 6.42 |
| 15 | 0.88 | 5 | 0.26 | 15 | 1.54 | 6.16 |
| 20 | 1.35 | 5 | 0.47 | 20 | 1.82 | 5.46 |
| 25 | 1.63 | 5 | 0.28 | 30 | 2.55 | 5.10 |
| 30 | 2.10 | 5 | 0.47 | 45 | 3.40 | 4.53 |
| 35 | 2.64 | 5 | 0.54 | 60 | 3.83 | 3.83 |
| 40 | 3.17 | 5 | 0.53 | 80 | 4.15 | 3.11 |
| 45 | 3.40 | 5 | 0.23 | 100 | 4.41 | 2.65 |
| 50 | 3.66 | 5 | 0.26 | 120 | 4.59 | 2.30 |
| 60 | 3.83 | 10 | 0.17 | | | |
| 80 | 4.15 | 20 | 0.32 | | | |
| 100 | 4.41 | 20 | 0.26 | | | |
| 120 | 4.59 | 20 | 0.18 | | | |

[a] Storm of October 27–28, 1908, at Jupiter, Fla.

[b] Column 6 records maximum rainfall in consecutive periods. It proceeds out of Col. 4 by finding the value, or combination of consecutive values, that produces the largest rainfall for the indicated period. Col. 7 = 60 × Col. 6/Col. 5.

the Southern states need not receive attention. The storm recorded in Example 3-2 exhibits double this intensity. Storm rainfall can be analyzed in many different ways. However, all procedures normally start from a summary of experience such as that shown in Example 3-3. The results may be used directly, or after smoothing (generally graphical) operations that generalize the experience. The developed intensity-duration-frequency relationships may be left in tabular form, presented graphically, or fitted by equations.

## Example 3-3.

The number of storms of varying intensity and duration recorded by a rain gage[13] in 45 years is listed in Table 3-5. Determine the time-intensity values for the 5-year storm.

If it is assumed that the 5-year storm is equaled or exceeded in intensity $45/5 = 9$ times in 45 years, the generalized time-intensity values may be interpolated from Table 3-5 by finding (1) for each specified duration the intensity equaled or exceeded by 9 storms, and (2) for each specified intensity the duration equaled or exceeded by 9 storms. The results are as follows and are used in constructing Fig. 3-3.

| Duration, min | 5 | 10 | 15 | 20 | 30 | 40 | 50 | 60 | 80 | 100 |
|---|---|---|---|---|---|---|---|---|---|---|
| Intensity, in. per hr | 6.50 | 4.75 | 4.14 | 3.50 | 2.46 | 2.17 | 1.88 | 1.66 | 1.36 | 1.11 |

| Intensity, in. per hr | 1.0 | 1.25 | 1.5 | 1.75 | 2.0 | 2.5 | 3.0 | 4.0 | 5.0 | 6.0 |
|---|---|---|---|---|---|---|---|---|---|---|
| Duration, min | 116.0 | 89.9 | 70.0 | 52.5 | 46.7 | 29.0 | 25.7 | 16.0 | 9.3 | 7.5 |

Similar calculations for the 1-year, 2-year, and 10-year storms underlie the remaining members of the family of curves in Fig. 3-3.

## Table 3-5
Record of Intense Rainfalls (Example 3-3)[a]

| Duration, min | Number of Storms of Stated Intensity (inches per hour) or More | | | | | | | | | | | | |
|---|---|---|---|---|---|---|---|---|---|---|---|---|---|
| | 1.0 | 1.25 | 1.5 | 1.75 | 2.0 | 2.5 | 3.0 | 4.0 | 5.0 | 6.0 | 7.0 | 8.0 | 9.0 |
| 5 | | | | | | | 123 | 47 | 22 | 14 | 4 | 2 | 1 |
| 10 | | | | | 122 | 78 | 48 | 15 | 7 | 4 | 2 | 1 | |
| 15 | | | | 100 | 83 | 46 | 21 | 10 | 3 | 2 | 1 | | |
| 20 | | | 98 | 64 | 44 | 18 | 13 | 5 | 2 | 2 | | | |
| 30 | 99 | 72 | 51 | 30 | 21 | 8 | 6 | 3 | 2 | | | | |
| 40 | 69 | 50 | 27 | 14 | 11 | 5 | 3 | 1 | | | | | |
| 50 | 52 | 28 | 17 | 10 | 8 | 4 | 3 | | | | | | |
| 60 | 41 | 19 | 14 | 6 | 4 | 4 | 2 | | | | | | |
| 80 | 18 | 13 | 4 | 2 | 2 | 1 | | | | | | | |
| 100 | 13 | 4 | 1 | 1 | | | | | | | | | |
| 120 | 8 | 2 | | | | | | | | | | | |

[a] Record for New York City from 1869 to 1913.
[13] Recorded at New York City from 1869 to 1913.

# 58/Statistical Hydrology

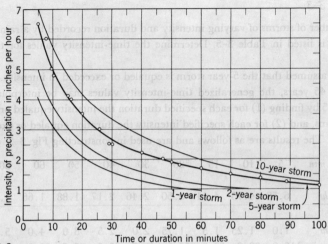

**Figure 3-3.**
**Intensity-duration-frequency of intense rainfalls.**

## 3-10 Intensity-Duration-Frequency Relationships

Time-intensity curves such as those in Fig. 3-3 are immediately useful in the design of storm-drainage systems and in flood-flow analyses. For purposes of comparison as well as further generalization, the curves can be formulated individually for specific frequencies or collectively for the range of frequencies studied.

Good fits are usually obtained by a collective equation of the form

$$i = cT^m/(t + d)^n \qquad (3\text{-}14)$$

where $i$ and $t$ stand for intensity and duration as before, $T$ is the frequency of occurrence in years, and $c$ and $d$ and $m$ and $n$ are pairs of regional coefficients and regional exponents respectively. Their order of magnitude is about as follows in North American experience: $c = 5$ to $50$ and $d = 0$ to $30$; $m = 0.1$ to $0.5$ and $n = 0.4$ to $1.0$.

Equation 3-14 can be fitted to a station record either graphically or by least squares. For storms of specified frequency, the equation reduces to

$$i = A(t + d)^{-n} \qquad (3\text{-}15)$$

where from Eq. 3-14,

$$A = cT^m \qquad (3\text{-}16)$$

**Graphical Fitting.** Equation 3-15 can be transformed to read $[\log i] = \log A - n[\log (t + d)]$ where the brackets identify the functional scales

$y = [\log i]$ and $x = [\log (t + d)]$ for direct plotting of $i$ against $t$ on double logarithmic paper for individual frequencies. Straight lines are obtained when suitable trial values of $d$ are added to the observed values of $t$. To meet the requirements of Eq. 3–14 in full, the values of $d$ and of $n$, the slope of the straight line of best fit, must be the same or averaged to become the same at all frequencies. Values of $A$ can then be read as ordinates at $(t + d) = 1$, if this point lies or can be brought within the plot. To determine $c$ and $m$, the derived values of $A$ are plotted on double logarithmic paper against $T$ for the frequencies studied. Because $[\log A] = \log c + m[\log T]$, the slope of the resulting straight line of best fit equals $m$, and the value of $c$ is read as the ordinate at $T = 1$.

---

*Example 3–4.*

Fit Eq. 3–14 to the 60-min record of intense rainfalls presented in Example 3–3.

Plot the values for the 5-year storm on double logarithmic paper as in Fig. 3–4. Because the high-intensity, short-duration values are seen to bend away from a straight line, bring them into line by adding 2 min to their duration periods, that is, $(t + d) = (t + 2)$. Derivation of the equation $i = A/(t + d)^n = 26/(t + 2)^{0.66}$ is noted on Fig. 3–4. Similar plots for the other storms of Fig. 3–3 would yield parallel lines of good fit. The intercepts $A$ of these lines on the $i$-axis at $(t + d) = 1$ themselves will plot as straight lines on double logarithmic paper against the recurrence interval $T$. Hence for $[\log A] = \log c + m[\log T]$, find the magnitudes $c = 16$ and $m = 0.31$ to complete the numerical evaluation of the coefficients and with them the equation $i = cT^m/(t + d)^n = 16\ T^{0.31}/(t + 2)^{0.66}$.

---

**Least-Squares Fitting.** Least-squares fitting of the equation $A = cT^m$ presents no difficulty when it is written in straight-line, in this case logarithmic, form. Fitting the equation $i = A(t + d)^{-n}$ is somewhat more taxing. The straight-line form of this equation is

$$[\log (-di/dt)] = \log n - (1/n) \log A + (1 + 1/n)[\log i]$$

If the storm intensities are recorded at uniform intervals of time, the slopes $(-di/dt)$ of the intensity-duration curves at $i_{k+1}$ are closely approximated by the relation $-(di/dt) = (i_k - i_{k+2})/(t_{k+2} - t_k)$ where the subscripts $k$, $k + 1$, and $k + 2$ denote sequences of pairs of observations in the series. Better fits are commonly obtained when the data for durations below and above 60 min are analyzed separately.

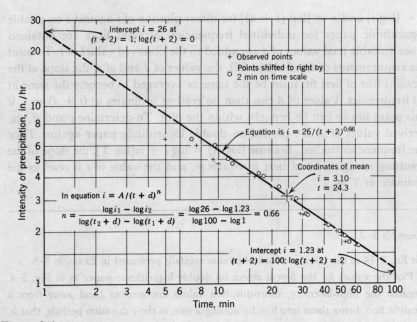

**Figure 3-4.**
**Intensity-duration of 5-year rainstorm.**

## 3-11 Storm Runoff and Flood Flows

The flood flows descending the arterial system of river basins or collecting in the storm drains or combined sewers of municipal drainage districts are derived from rains that fall upon the tributary watershed. The degree of their conversion into runoff is affected by many factors, especially in the varied environment of urban communities. Component effects and their relative importance must, therefore, be clearly recognized in the interpretation of storm runoff or flood experience in relation to intense rainfalls.

Flows normally reach flood crest at a given point on a stream or within a drainage scheme when runoff begins to pour in from distant parts of the tributary area. There are exceptions to this rule, but they are few. An important exception is a storm traveling upstream or sweeping across a catchment area so rapidly that runoff from distant points does not reach the point of concentration until long after the central storm has moved on. Diminution of effective area or *retardance* of this kind is rarely taken into consideration in American practice, but it should be in some circumstances. In a given storm the maximum average rate of rainfall is always highest for the shortest time interval or duration. Therefore, the shorter the elapsed time or *time of concentration* in which distant points are tributary to the *point of concentration*, the larger are the flows.

The time of concentration is shortest for small, broad, steep drainage areas with rapidly shedding surfaces. It is lengthened by dry soil, surface inequalities and indentations, and vegetal cover, and by storage in water courses, on flood plains, and in reservoirs. In short intense thunderstorms, peak urban flows often occur when only the impervious or paved areas are shedding water. The volume of runoff from a given storm is reduced by infiltration, freezing, and storage; it is swelled by snow and ice melt, seepage from bank storage, and release of water from impoundages either on purpose or by accident. Maximum rates obtain when storms move downstream at speeds that bring them to the point of discharge in about the time of concentration, making it possible for the runoff from the most intense rainfall to arrive at the point of discharge at nearly the same instant.

Among the ways devised for estimating storm runoff or flood flows for engineering designs are the following:

1. Statistical analyses based on observed records of adequate length. Obviously these can provide likely answers. Unfortunately, however, recorded information is seldom sufficiently extensive to identify critical magnitudes directly from experience. Information must be generalized to arrive at rational extrapolations for the frequency or recurrence interval of design flows or for the magnitude of flows of design frequency.

2. Statistical augmentation of available information through cross-correlation with recorded experience in one or more adjacent and similar basins for which more years of information are available; through correlation between rainfall and runoff when the rainfall record is longer than the runoff record; and through statistical generation of additional values.

3. Rational estimates of runoff from rainfall. This is a common procedure in the design of storm and combined sewers that are to drain existing built-up areas and satisfy anticipated change in the course of time, or areas about to be added to existing municipal drainage schemes.

4. Calculations based on empirical formulations not devised specifically from observations in the design area but reasonably applicable to existing watershed conditions. Formulations are varied in structure and must be selected with full understanding of the limitations of their derivation. At best, they should be applied only as checks of statistical or rational methods.

Where failure of important engineering structures is sure to entail loss of life or great damage, every bit of hydrological information should be adduced to arrive at economical but safe design values. Hydraulic models may also be helpful.

## 3-12 Analysis of Flood Flows

Records of maximum daily, weekly, monthly, or annual runoff trace frequency distributions that are skewed to the right, implying that their means lie to the right of their modes. Records of this kind can be generalized roughly as geometrically normal series, more closely as Pearsonian Type III and Gumbel distributions, or graphically as partial duration curves. Only the Gumbel distribution will be discussed here.[14]

**Gumbel's Distribution.** E. J. Gumbel[15] has concluded that extreme values of streamflow conform to the theoretical distribution of extreme values as follows:

$$F(x) = 1 - \exp\left[-\exp(-b)\right] \qquad (3\text{-}17)$$

where $b$ is the dimensionless variable $(x - \mu + 0.450\sigma)/0.780\sigma$ because the deviation of the modal flood from the mean annual flood is approximately $-0.577\sigma\sqrt{6}/\pi = -0.450\sigma$. For example, if the mean annual 24-hr flood of a stream is $288 \times 10^3$ cfs with a standard deviation of $113 \times 10^3$ cfs, $b = (700 - 288 + 0.450 \times 113)/(0.780 \times 113) = 5.25$ for a flood of $700 \times 10^3$ cfs, and $F(x) = 1 - \exp[-\exp(-5.25)] = 0.005$; or the recurrence interval $I = 1/(5 \times 10^{-3}) = 200$ years approximately.

## 3-13 Estimates of Storm Runoff

Among the various methods used to estimate storm runoff, two are of general interest: the rational method and the unit-hydrograph method.

**The Rational Method.** This is the method commonly used as a basis for the design of storm drains and combined sewers. Such facilities are expected to carry, without surcharge, the peak runoff expected to be equaled or exceeded on the average once in a period (a recurrence interval) of $T$ years. The interval selected is short, $T = 2$ to 5 years, when damage due to surcharge and street flooding is small, and is long, $T = 20$ to 100 years, when the damage is great as it sometimes can be when basements in residential and commercial districts are flooded.

The design peak flow is estimated using the equation

$$Q = cia \qquad (3\text{-}18)$$

where $Q$ is the peak rate of runoff at a specified place, in cu ft per sec, $a$ is the tributary area in acres, and $i$ the rainfall intensity in in. per hr (1 in. per hr on 1 acre = 1 cfs) for the selected values of $T$ in years and $t$ in minutes. The rainfall duration, $t$, is in fact a rainfall-intensity averaging time.

Of the three factors included in Eq. 3-18, $a$ is found from a regional map or survey, $i$ is determined for a storm of duration equal to the time of con-

---

[14] For a discussion of Pearson's Type III distribution, see H. A. Foster, Theoretical Frequency Curves, *Trans. Am. Soc. Civil Engrs.*, **87**, 142 (1924).

[15] E. J. Gumbel, Floods Estimated by the Probability Method, *Eng. News-Record*, **134**, 833 (1945); also see *Maximum Possible Precipitation*, Hydrometeorological Report No. 23, Dept. of Commerce, Washington, D.C., 1947.

centration (Secs. 8 to 12 of this chapter), and $c$ is estimated from the characteristics of the catchment area. The time of concentration is found (1) for flood discharge by estimating average velocities of flow in the principal channels of the tributary area; and (2) for runoff from sewered areas by estimating the inlet time, or time required for runoff to enter the sewerage system from adjacent surfaces, and adding to it the time of flow in the sewers or storm drains proper. Because rapid inflow from tributaries generates flood waves in the main stem of a river system, flood velocities are often assumed to be 30 to 50 percent higher than normal rates of flow (Sec. 8-8).

When Equation 3-18 is written in functional terms $Q(T) = ca\{i(t,T)\}$, it becomes evident that there is implicit in the rational method an assumption that the design peak runoff rate is expected to occur with the same frequency as the rainfall intensity used in the computations. As pointed out by Schaake et al.,[16] the value computed for $Q(T)$ does not correspond to the peak runoff rate that would be expected from any particular storm. If the peak runoff rates and rainfall intensities for a specified averaging time are analyzed statistically, coefficient $c$ is the ratio of peak unit runoff to average rainfall intensity for any value of $T$. Examination of the rational method shows that values of $c$, developed through experience, give reasonably good results even though the assumptions with regard to inlet times and contributing areas are often far from reality.

Peak flows at inlets occur during the very intense part of a storm and are made up of water that comes almost entirely from the paved areas. Lag times between peak rainfall rate and peak runoff rate at inlets are very short, often less than one minute. For areas up to 20 to 50 acres good values of peak runoff rates can be obtained by estimating the flow to each inlet and combining attenuated hydrographs to obtain design values.[17] Since in most intense storms very little runoff is contributed by the unpaved areas, good estimates can also be made by considering rainfall on impermeable surfaces only. These and other modifications of the rational method require the use of an appropriate set of coefficients. The selection of suitable values for $c$ in estimating runoff from sewered areas is discussed more fully in connection with the design of storm drains and combined sewers (Sec. 9-6).

**The Unit Hydrograph.** In dry weather, or when precipitation is frozen, the residual hydrograph or base flow of a river is determined by water released from storage in the ground or in ponds, lakes, reservoirs, and backwaters of the stream. Immediately after a rainstorm, the rate of discharge rises above base flow by the amount of surface runoff entering the drainage

[16] J. C. Schaake, Jr., J. C. Geyer, and J. W. Knapp, Experimental Examination of the Rational Method, *Proc. Am. Soc. of Civil Engrs.*, **93**, HY-6, 353 (Nov. 1967).

[17] A. B. Kaltenbach, Storm Sewer Design by the Inlet Method, *Public Works*, **94**, 1, 86 (1963).

system. That portion of the hydrograph lying above base flow can be isolated from it and is a measure of the true surface runoff (Fig. 3–5). The *unit hydrograph* method stems from studies of simple geometric properties of the surface-runoff portion of the hydrograph in their relation to an *effective rain* that has fallen during a *unit* of time, such as a day or an hour, and that, by definition, has produced surface runoff.

The important geometric properties of the unit hydrograph or surface runoff illustrated in Fig. 3–5 are (1) the abscissal length measuring time duration above base flow is substantially constant for all unit-time rains; (2) sequent ordinates, measuring rates of discharge above base flow at the end of each time unit, are proportional to the total runoff from unit-time rains irrespective of their individual magnitudes; (3) ratios of individual areas to the total area under the hydrograph, measuring the amount of water discharged in a given interval of time, are constant for all unit hydrographs of the same drainage area. These distribution ratios are generally referred to as the *distribution graph*, even when they are not presented in graphical form; and (4) rainstorms extending, with or without interruption, over several time units generate a hydrograph composed of a series of unit hydrographs superimposed in such manner as to distribute the runoff from each unit-time rain in accordance with the successive distribution ratios derived from unit-time rainfalls. This permits the construction of a hydrograph that might result from not-yet-experienced rainstorms.

These geometric properties do not apply when runoff originates in melting snow or ice, nor when the speed of flood waves in streams is changed appreciably as river stages are varied by fluctuating flows. Time is an important

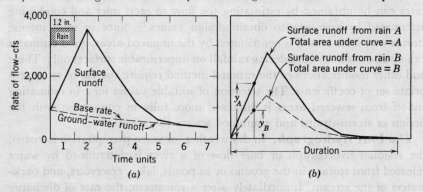

**Figure 3-5.**

**Origin and geometric properties of the unit hydrograph. (*a*) Hydrograph resulting from unit-time rain. See Example 3-6. (*b*) Distribution graph showing geometric properties of unit hydrograph; $y_A:A = y_B:B$; base duration is constant.**

element of this procedure, and rainfall data must be available for unit times shorter than the time of concentration of the drainage area. Unit times as long as a day can be employed successfully only for large watersheds (1000 sq miles or more). For sheds of 100 to 1000 sq miles, Sherman[18] has suggested values of 6 to 12 hr; for sheds of 20 sq miles, 2 hr; and for very small areas, one fourth to one third the time of concentration. The unit hydrograph method is illustrated in Example 3–5.

*Example 3–5.*

1. Given the rainfall and runoff records of a drainage area of 620 sq miles, determine the generalized distribution of runoff (the distribution graph) from isolated unit-time rainfalls. This involves first of all a search for records of isolated rainfalls and for records of the resulting surface runoff. The basic data for a typical storm are shown in Table 3–6, together with necessary calculations. Development of this table is straightforward, except for Col. 4, which records the estimated base flow and can be derived only from a study of the general hydrograph of the stream in combination with all related hydrological observations of the region.

2. Apply the average estimate of runoff distribution to the observed rainfall sequence presented in Table 3–7.

The calculations in Table 3–7 need little explanation except for Col. 3, the estimated loss of rainfall caused principally by infiltration. This estimate rests on all available information for the region. It is discussed in principle in this section and in Chap. 5. Column 5 is identical with Col. 7 of Table 3–6. Column 6 is the net rain of 0.5 in. during the first time unit multiplied by the distribution ratio of Col. 5. Columns 7, 8, and 9 are similarly derived for the net rains during the subsequent time units. Column 10 gives the sums of Cols. 6 to 9, and Col. 11 converts these sums from inches to cubic feet per second. If the base flow is estimated and added to the surface runoff shown in Col. 11, the hydrograph becomes complete.

---

The unit-hydrograph method is useful in estimating magnitudes of unusual flood flows, in forecasting flood crests during storms, and in the manipulation of storage on large river systems. It has the important property of (1) tracing the full hydrograph resulting from a storm rather than being confined to a determination of the peak flow alone, and (2) producing useful results from short records. For small drainage areas, the method depends on the readings of a recording rain gage. Refinements in procedure and aids to the rationalization of the various steps continue to be developed by hydrologists and engineers.

[18] O. E. Meinzer, *Hydrology*, McGraw-Hill Book Co., New York, p. 524, 1942.

## Table 3-6

Observations and Calculations for Unit Hydrograph (Example 3-5)

| Sequence of Time Units | Observed Rainfall, in. | Runoff, cfs Observed Total | Runoff, cfs Estimated Base Flow | Estimated Distribution of Surface Runoff cfs | Estimated Distribution of Surface Runoff % | Average Distribution Ratio for 10 Storms, % |
|---|---|---|---|---|---|---|
| (1) | (2) | (3) | (4) | (5) = (3) − (4) | (6) = 100(5)/6200 | (7) |
| 1 | 1.20 | 1,830 | 870 | 960 | 15.5 | 16 |
| 2 | 0.03 | 3,590 | 800 | 2,790 | 45.0 | 46 |
| 3 | 0.00 | 2,370 | 690 | 1,680 | 27.1 | 26 |
| 4 | 0.00 | 1,220 | 600 | 620 | 10.0 | 10 |
| 5 | 0.00 | 640 | 510 | 130 | 2.1 | 1 |
| 6 | 0.00 | 430 | 410 | 20 | 0.3 | 1 |
| 7 | 0.00 | 350 | 350 | 0 | 0.0 | 0 |
| Totals | — | ... | ... | 6,200 | 100.0 | 100 |

## 3-14 Flood-Flow Formulas

Flood-flow formulas derive from empirical evaluations of drainage-basin characteristics and hydrological factors falling rationally within the framework of the relation $Q = cia$. Frequency relations are implied even when they are not expressed in frequency terms. Time-intensity variations, likewise, are included, but (indirectly) as functions of the size of area drained. Equation 3-18 is thereby reduced to the expression $Q = Ca^m$, where $m$ is less than 1. This follows from the relative changes in $i$ and $a$ with $t$; namely, $i = A/(t+d)^n$ and $a = kt^2$ (Sec. 3-10). For $d$ close to zero, therefore, $i = \text{constant}/(a^{n/2})$ and, substituting $i$ in Eq. 3-18, $Q = \text{constant } a^{1-n/2} = Ca^m$, where $m = 1 - n/2$. Because $n$ varies from 0.5 to 1.0, $m$ must and does vary in different formulations from 0.8 to 0.5. The value of $C$ embraces the maximum rate of rainfall, the runoff-rainfall ratio of the watershed, and the frequency factor. An example is the Fanning formula listed in Table 3-8, together with other flood-flow formulas in which certain component variables or their influence on runoff are individualized.

## Table 3-7
*Application of Unit Hydrograph Method (Example 3–5)*

| Sequence of Time Units | Rainfall, in. | | | Average Runoff Distribution Ratio, % | Distributed Runoff for Stated Time Units, in. | | | | Compounded Runoff | |
|---|---|---|---|---|---|---|---|---|---|---|
| | Observed | Estimated Loss | Net | | 1st | 2nd | 3rd | 5th | in. | cfs[a] |
| | | | (4) = (2) − (3) | | | | | | | |
| (1) | (2) | (3) | (5) | (6) | (7) | (8) | (9) | (10) | (11) |
| 1 | 1.8 | 1.3 | 0.5 | 16 | 0.08 | ... | ... | ... | 0.08 | 1,300 |
| 2 | 2.7 | 1.6 | 1.1 | 46 | 0.23 | 0.18 | ... | ... | 0.41 | 6,900 |
| 3 | 1.6 | 1.1 | 0.5 | 26 | 0.13 | 0.50 | 0.08 | ... | 0.71 | 11,900 |
| 4 | 0.0 | 0.0 | 0.0 | 10 | 0.05 | 0.29 | 0.23 | ... | 0.57 | 9,500 |
| 5 | 1.1 | 0.2 | 0.9 | 1 | 0.01 | 0.11 | 0.13 | 0.14 | 0.39 | 6,500 |
| 6 | 0.0 | 0.0 | 0.0 | 1 | 0.00 | 0.01 | 0.05 | 0.42 | 0.48 | 8,000 |
| 7 | 0.0 | 0.0 | 0.0 | 0 | 0.00 | 0.01 | 0.01 | 0.23 | 0.25 | 4,200 |
| 8 | 0.0 | 0.0 | 0.0 | 0 | 0.00 | 0.00 | 0.00 | 0.09 | 0.09 | 1,500 |

[a] Rate of runoff in cubic feet per second = inches × 26.88 × 620 sq miles = 16,700 cfs if the time unit is a day. For other time units multiply by reciprocal ratio of length of time to length of day.

These examples are chosen only as illustrations of forms of flood-flow formulas. They are not necessarily the best forms, nor should they be applied outside the area for which they were derived.

The flood-flow characteristics of United States drainage basins have been compared by developing their envelope curve on a $Q$ versus $a$ plot as a function of $\sqrt{a}$ and identifying their Myers rating from the equation

$$Q = 100 \, p\sqrt{a} \tag{3-19}$$

where $Q$ is the extreme peak flow in cfs; $p$ is the percentage ratio of $Q$ to a postulated ultimate maximum flood flow of $Q_u = 10,000\sqrt{a}$; and $a$ is the drainage area, which must be 4 sq miles or more. In the Colorado River basin, the Myers rating is only 25%; in the northeastern United States it is seldom more than 50%; in the lower Mississippi basin, it is about 64%.

## Table 3-8
Examples of Flood-Flow Formulas

| Individualized Variable | Author and Region | Formula |
|---|---|---|
| None | Fanning, New England | $Q = Ca^{5/6}$, where $C = 200$ for $a$ in sq miles |
| Rainfall intensity and slope of watershed | McMath, St. Louis, Mo. | $Q = cia^{4/5}s^{1/5}$, where $s$ = slope in ‰ and $c = 0.75$ for $a$ in acres or 480 for $a$ in sq miles |
| Shape and slope of watershed | Potter, Cumberland Plateau | $Q = ca^{7/6}/(l/s^{1/2})$, where $l$ = length of principal waterway in miles, $s$ = slope of waterway in ft/mile, and $c = 1920d$ for $a$ in sq miles, the 10-year peak flood and a factor $d$ relating the basin to the base station at Columbus, Ohio |
| Shape, slope, and surface storage of watershed | Kinnison and Colby, New England | $Q = (0.000036h^{2.4} + 124)a^{0.85}/(rl^{0.7})$, where $h$ = median altitude of drainage basin in ft above the outlet; $r$ = % of lake, pond, and reservoir area; $l$ = average distance in miles to outlet; and $a$ = sq miles |
| Frequency of flood | Fuller, U.S.A. | $Q = Ca^{0.8}(1 + 0.8 \log T)(1 + 2a^{-0.3})$, where $T$ = number of years in the period considered, and $C$ varies from 25 to 200 for different drainage basins and $a$ in sq miles |

Fuller's formula[19] is of particular interest because it incorporates a frequency factor and is countrywide in scope.

---

[19] W. E. Fuller, Flood Flows, *Trans. Am. Soc. Civil Engrs.*, 77, 564 (1914). Weston Fuller (1879–1935) was a partner of Allen Hazen and subsequently professor of civil engineering at Swarthmore College.

**four**

# surface water collection

## 4-1 Sources of Surfacewater

In North America by far the largest volumes of municipal water are collected from surface sources. Possible yields vary directly with the size of the catchment area, or watershed, and with the difference between the amount of water falling on it and the amount lost by evapotranspiration.[1] The significance of these relations to water supply is illustrated in Fig. 1-2. Where surfacewater and groundwater sheds do not coincide, some groundwater may enter from neighboring catchment areas or escape to them.

*Continuous Draft.* Communities on or near streams, ponds, or lakes may withdraw their supplies by continuous draft if streamflow and pond or lake capacity are high enough at all seasons of the year to furnish requisite water volumes.[2] Collecting works include

[1] This awkward term has been accepted into the vocabulary of hydrology as including all water lost to the atmosphere, whether by evaporation, transpiration, or other processes.

[2] Examples of continuous draft from streams are the water supplies of Montreal, P.Q., St. Lawrence River; Philadelphia, Pa., Delaware and Schuylkill rivers; Pittsburgh, Pa., Allegheny River; Cincinnati, O., and Louisville, Ky., Ohio River; Kansas City, Mo., Missouri River; Minneapolis and St. Paul, Minn., Mississippi River; St. Louis, Mo., Missouri and Mississippi rivers; and New Orleans, La., Mississippi River. Examples of continuous draft from lakes are furnished by Burlington, Vt., Lake Champlain; Syracuse, N.Y., Lake

# 70/Surface Water Collection

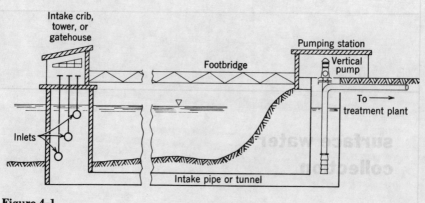

**Figure 4-1.**
**Continuous draft of water from large lakes and streams.**

ordinarily (1) an intake crib, gatehouse, or tower; (2) an intake conduit; and (3) in many places, a pumping station. On small streams serving communities of moderate size, intake or diversion dams can create a sufficient depth of water to submerge the intake pipe and protect it against ice. From intakes close to the community the water must generally be lifted to purification works and thence to the distribution system (Fig. 4-1).

Because most large streams are polluted by wastes from upstream communities and industries, their waters must be purified before use. Cities on large lakes must usually guard their supplies against their own and their neighbors' wastewaters and spent industrial-process waters by moving their intakes far away from shore and purifying both their water and their wastewater. Diversion of wastewaters and other plant nutrients from lakes will retard their eutrophication (Chap. 21).

*Selective Draft.* Low stream flows are left untouched when they are wanted for other valley purposes or are too highly polluted for reasonable use. Only clean flood waters are then diverted into reservoirs constructed in meadow lands adjacent to the stream or otherwise conveniently available.[3] The amount of water so stored must supply demands during seasons of

---

Skaneateles; Toronto, Ont., Lake Ontario; Buffalo, N.Y., and Cleveland, O., Lake Erie; Detroit, Mich., Lake St. Clair; Chicago, Ill., and Milwaukee, Wis., Lake Michigan; and Duluth, Minn., Lake Superior.

[3] London, England, meets part of its water needs from the Thames River by storing relatively clean floodwaters in large basins surrounded by dikes in the Thames Valley. The Boston, Mass., Metropolitan Water Supply diverts the freshets of the Ware River through a tunnel either to the previously constructed Wachusett Reservoir, which impounds the waters of a branch of the Nashua River, or to the subsequently completed Quabbin Reservoir, which impounds the Swift River.

unavailable streamflow. If draft is confined to a quarter year, for example, the reservoir must hold at least three fourths of the annual supply. In spite of its selection and long storage, the water may have to be purified.

**Impoundage.** In search of clean water and water that can be brought and distributed to the community by gravity, engineers have developed supplies from upland streams. Most of them are tapped near their source in high and sparsely settled regions. To be of use, their annual discharge must equal or exceed the demands of the community they serve for a reasonable number of years in the future. Because their dry-season flows generally fall short of concurrent municipal requirements, their floodwaters must usually be stored in sufficient volume to assure an adequate supply. Necessary reservoirs are impounded by throwing dams across the stream valley (Fig. 4–2). In this way, amounts up to about 70 or 80% of the mean annual flow can be utilized. The area draining to impoundages is known as the catchment area or watershed. Its economical development depends on the value of water in the region, but it is a function, too, of runoff and its variation, accessibility of catchment areas, interference with existing water rights, and costs of construction. Allowances must be made for evaporation from new water surfaces generated by the impoundage, and often, too, for release of agreed-on flows to the valley below the dam (compensating water). Increased ground storage in the flooded area and the gradual diminution of reservoir volumes by siltation must also be considered.

Intake structures are incorporated in impounding dams or kept separate. Other important components of impounding reservoirs are (1) spillways safely passing floods in excess of reservoir capacity and (2) diversion conduits safely carrying the stream past the construction site until the reservoir has been completed and its spillway can go into action. Analysis of flood records enters into the design of these ancillary structures.

Some impounded supplies[4] are sufficiently safe, attractive, and palatable to be used without treatment other than protective disinfection. However, it may be necessary to remove (1) high color imparted to the stored water by the decomposition of organic matter in swamps and on the flooded valley floor; (2) odors and tastes generated in the decomposition or growth of algae, especially during the first years after filling; and (3) turbidity (finely divided clay or silt) carried into streams or reservoirs by surface wash, wave action, or bank erosion. Recreational uses of watersheds and reservoirs may en-

[4] Examples of untreated, impounded, upland supplies are the Croton River, Catskill, and Delaware River supplies of New York, N.Y., and the Wachusett and Quabbin supplies of the Metropolitan District of Boston, Mass. Examples of treated impounded supplies are found at Baltimore, Md.; Providence, R.I.; Hartford, Conn.; Springfield, Mass.; and Springfield, Ill.

72/Surface Water Collection

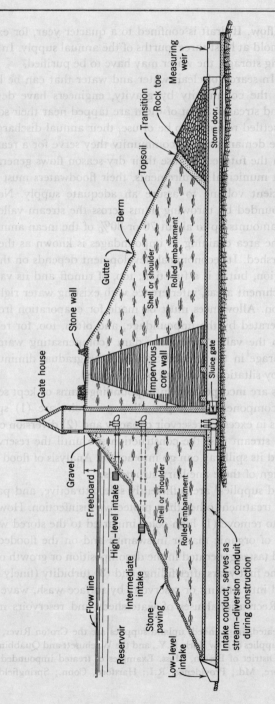

Figure 4-2.
Dam and intake towers for an impounded surface water supply.

danger the water's safety and call for treatment of the flows withdrawn from storage.

Much of the water entering streams, ponds, lakes, and reservoirs in times of drought, or when precipitation is frozen, is seepage from the soil. Nevertheless, it is classified as surface runoff rather than groundwater. Water seeps *from* the ground when surface streams are low, and *to* the ground when surface streams are high. Release of water from ground storage or from accumulations of snow in high mountains is a determining factor in the yield of some catchment areas. Although surfacewaters are derived ultimately from precipitation, the relations between precipitation, runoff, infiltration, evaporation, and transpiration are so complex that engineers rightly prefer to base calculations of yield on available stream gagings. For adequate information, gagings must extend over a considerable number of years.

## 4-2 Impounding Reservoirs

In the absence of adequate natural storage, engineers construct impounding reservoirs. More rarely they excavate storage basins in lowlands adjacent to streams. Natural storage, too, can be regulated. Control works (gates and weirs or sills) on outlets of lakes and ponds are examples.

Some storage works are designed to serve a single purpose only; others are planned to perform a number of different functions and to preserve the broader economy of natural resources. Common purposes include: (1) water supply for household, farm, community, and industry; (2) dilution and natural purification of sewage and other municipal and industrial wastewaters; (3) irrigation of arable lands; (4) harnessing water power; (5) low-water regulation for navigation; (6) preservation and cultivation of useful aquatic life; (7) recreation—fishing, boating, and bathing; and (8) control of destructive floods.

The greatest net benefit may accrue from a judicious combination of reservoir functions in multipurpose developments. The choice of single-purpose storage systems should indeed be justified fully.

Storage is provided when streamflow is inadequate or rendered unsatisfactory by heavy pollution. Release of stored waters then swells flows and dilutes pollution. Storage itself also affects the quality of the waters impounded. Both desirable and undesirable changes may take place. Their identification is the responsibility of *limnology*, the science of lakes or, more broadly, of inland waters.

If they must receive wastewaters, streamflows should be adjusted to the pollutional load imposed on them. Low-water regulation, as such, is made possible by headwater or upstream storage, but lowland reservoirs, too, may aid dilution and play an active part in the natural purification of river systems. Whether overall results are helpful depends on the volume and nature of wastewater flows and the chosen regimen of the stream.

## 4-3 Safe Yield of Streams

In the absence of storage, the safe yield of a river system is its lowest dry-weather flow; with full development of storage, the safe yield approaches the mean annual flow. The economical yield generally lies somewhere in between. The attainable yield is modified by (1) evaporation, (2) bank storage, (3) seepage out of the catchment area, and (4) silting.

Storage-yield relations are illustrated in this chapter by calculations of storage to be provided in impounding reservoirs for water supply. However, the principles demonstrated are also applicable to other purposes and uses of storage.

## 4-4 Storage as a Function of Draft and Runoff

A dam thrown across a river valley impounds the waters of the valley. Once the reservoir has filled, the water drawn from storage is eventually replenished by the stream, provided runoff, storage, and draft are kept in proper balance. The balance is struck graphically or analytically on the basis of historical records or replicates generated by suitable statistical procedures of operational hydrology.

Assuming, as in Fig. 4-3, that the reservoir is full at the beginning of a dry period, the maximum amount of water $S$ that must be withdrawn from storage to maintain a given average draft $D$ equals the maximum cumulative difference between the draft $D$ and the runoff $Q$ in a given dry period, or

$$S = \text{maximum value of } \Sigma(D - Q) \qquad (4\text{-}1)$$

To find $S$, $\Sigma(D - Q)$ is summed arithmetically or graphically. The mass diagram or Rippl[5] method illustrated in Fig. 4-3 is a most convincing and useful demonstration of finding $\Sigma(D - Q) = \Sigma D - \Sigma Q$. The shorter the interval of time for which runoff is recorded, the more exact is the result. As the maximum value is approached, therefore, it may be worthwhile to shift to short intervals of time—from monthly to daily values, for example. The additional storage identified by such a shift may be as much as 10 days of draft.

Assuming that inflow and drafts are repeated cyclically, in successive sets of $T$ years, Thomas and Fiering[6] have developed a *sequent peak* procedure for determining minimum storage for no shortage in draft based upon two

---

[5] W. Rippl, The Capacity of Storage Reservoirs for Water Supply, *Proc. Inst. Civil Engrs.*, **71**, 270 (1883).

[6] H. A. Thomas, Jr., and M. B. Fiering, personal communication.

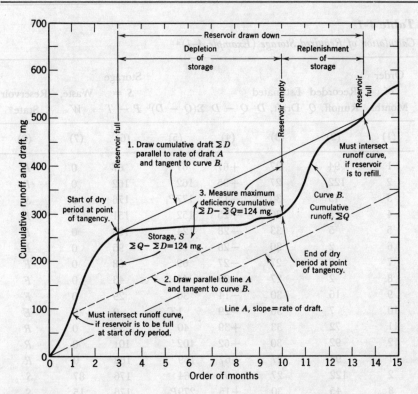

**Figure 4-3.**

**Mass-diagram or Rippl method for the determination of storage required in impounding reservoirs. Basic runoff values are for the Westfield Little River near Springfield, Mass. for March 1914 to March 1915. A constant draft of 750,000 gpd/sq mile = 23 mg/sq mile for a month of 30.4 days is assumed.**

needed cycles. Example 4–1 illustrates their procedure. Also recommended is the synthesis of runoff traces by operational methods.[7]

---

*Example 4–1.*

From the recorded monthly mean runoff values shown in Col. 2, Table 4-1, find the required storage for the estimated rates of draft listed in Col. 3, Table 4-1.

---

[7] G. M. Fair, J. C. Geyer, and D. A. Okun, *Water and Wastewater Engineering*, Vol. 1, *Water Supply and Wastewater Removal*, John Wiley & Sons, New York, 1966, Sec. 4–10, pp. 4–32 to 4–34.

## Table 4-1

*Calculation of Required Storage (Example 4-1)* [a]

| Order of Months | Recorded Runoff, $Q$ | Estimated Draft, $D$ | $Q - D$ | Storage $S = \Sigma(Q - D)$ [b] | $P - T$ | Waste, $W$ | Reservoir State [c] |
|---|---|---|---|---|---|---|---|
| (1) | (2) | (3) | (4) | (5) | (6) | (7) | (8) |
| 1 | 94 | 27 | +67 | 67 | 67 | 0 | R |
| 2 | 122 | 27 | +95 | 162 | 162 | 0 | R |
| 3 | 45 | 30 | +15 | $177P_1$ | 176 | 1 | S |
| 4 | 5 | 30 | −25 | 152 | 151 | 0 | F |
| 5 | 5 | 33 | −28 | 124 | 123 | 0 | F |
| 6 | 2 | 30 | −28 | 96 | 95 | 0 | F |
| 7 | 0 | 27 | −27 | 69 | 68 | 0 | F |
| 8 | 2 | 27 | −25 | 44 | 43 | 0 | F |
| 9 | 16 | 30 | −14 | 30 | 29 | 0 | F |
| 10 | 7 | 36 | −29 | $1T_1$ | 0 | 0 | E |
| 11 | 72 | 33 | +39 | 40 | 39 | 0 | R |
| 12 | 92 | 30 | +62 | 102 | 101 | 0 | R |
| 1 | 94 | 27 | +67 | 169 | 168 | 0 | R |
| 2 | 122 | 27 | +95 | 264 | 176 | 87 | S |
| 3 | 45 | 30 | +15 | $279P_2$ | 176 | 15 | S |
| 4 | 5 | 30 | −25 | 254 | 151 | 0 | F |
| 5 | 5 | 33 | −28 | 226 | 123 | 0 | F |
| 6 | 2 | 30 | −28 | 198 | 95 | 0 | F |
| 7 | 0 | 27 | −27 | 171 | 68 | 0 | F |
| 8 | 2 | 27 | −25 | 146 | 43 | 0 | F |
| 9 | 16 | 30 | −14 | 132 | 29 | 0 | F |
| 10 | 7 | 36 | −29 | $103T_2$ | 0 | 0 | E |
| 11 | 72 | 33 | +39 | 142 | 39 | 0 | R |
| 12 | 92 | 30 | +62 | $204P_3$ | 101 | 0 | R |

[a] Runoff, draft, and storage are expressed in mg/sq mile.
[b] $P$ = peak; $T$ = trough.
[c] $R$ = rising; $F$ = falling; $S$ = spilling; $E$ = empty.

Col. 2: These are observed flows for the Westfield Little River, near Springfield, Mass., for March 1914 to February 1915. Operational replicates might have been used instead (Sec. 4–10).

Col. 3: The values 27, 30, 33, and 36 mg/sq mile = 0.89, 0.11, 1.09, and 1.18 mgd/sq mile respectively for 30.4 days/month. For a total flow of 462 mg/sq mile in

12 months the average flow is $462/365 = 1.27$ mg/sq mile, and for a total draft of 360 mg the development is $100 \times 360/462 = 78\%$.

Col. 4: Positive values are surpluses, negative values deficiencies.

Col. 5: $P_1$ is the first peak, and $T_1$ is the first trough in the range $P_1P_2$, where $P_2$ is the second higher peak; similarly $T_2$ is the second trough in the range $P_2P_3$ presumably.

Col. 6: The required maximum storage $S_m = \max (P_j - T_j) = P_m - T_m = P_1 - T_1 = 177 - 1 = 176$ in this case. The fact that $P_2 - T_2 = 279 - 103 = 176$ also implies that there is seasonal rather than over-year storage. Storage at the end of month $i$ is $S_i = \min \{S_M, [S_{i-1} + (Q_i - D_i)]\}$; for example, in line 2, $S_M = 176$ and $[S_{i-1} + (Q_i - D_i)] = 67 + 95 = 162$, or $S_i = 162$; in line 3, however, $S_M = 176$ and $[S_{i-1} + (Q_i - D_i)] = 162 + 95 = 257$ or $S_i = S_M = 176$.

Col. 7: The flow wasted $W_i = \max \{0, [(Q_i - D_i) - (S_M - S_{i-1})]\}$; for example, line 3, $(Q_i - D_i) - (S_m - S_{i-1}) = 15 - (176 - 162) = 1$ or $W_i = 1$; in line 3 of the second series, however, $(Q_i - D_i) - (S_m - S_{i-1}) = 15 - (176 - 176) = 15$. There is no negative waste.

For variable drafts and inclusion of varying allowances for evaporation from the water surface created by the impoundage, the analytical method possesses distinct advantages over the graphical method. The principal value of the Rippl method, indeed, is not for the estimation of storage requirements, but for determining the yield of catchment areas upon which storage reservoirs are already established.

## 4-5 Design Storage

Except for occasional series of dry years and very high developments, seasonal storage generally suffices in the well-watered regions of North America. Water is plentiful, streamflows do not vary greatly from year to year, reservoirs generally refill within the annual hydrologic cycle, and it does not pay to go in for high or complete development of catchment areas. In semiarid regions, on the other hand, water is scarce, streamflows fluctuate widely from year to year, runoff of wet years must be conserved for use during dry years, and it pays to store for use a large proportion of the mean annual flow. In these circumstances, operational records of adequate length become important along with machine computation.[7]

Given a series of storage values for the flows observed or generated statistically, the engineer must decide which value he will use. Shall it be the highest on record, or the second, third, or fourth highest? Obviously, the choice depends on the degree of protection to be afforded against water shortage. This must be fitted into drought experience, which is a function of the length of record examined. To arrive at a reasonable answer and an economically justifiable design storage, the engineer may resort to (1) a statistical analysis of the arrayed storage values and (2) estimates of the difficulties and costs associated with shortage in supply. Storage values equaled or exceeded but once in 20, 50, or 100 years, that is, 5, 2, and 1% of the years, are often considered. For water supply, Hazen[8] suggested employing the 5% value in ordinary circumstances. In other words, design storage should be adequate to compensate for a drought of a severity not expected to occur oftener than once in 20 years. In still drier years, it may be necessary to curtail the use of water by limiting, or prohibiting, lawn sprinkling and car washing, for example.

Restricting water use is irksome to the public and a poor way to run a public utility. As a practical matter, moreover, use must be cut down well in advance of anticipated exhaustion of the supply. It would seem logical to consider not only the frequency of curtailment but also the depletion at which conservation should begin. In practice, the *iron ration* generally lies between 20 and 50% of the total water stored. Requiring a 25% reserve for the drought that occurs about once in 20 years is reasonable. An alternative is a storage allowance for the drought to be expected once in 100 years. This is slightly less in magnitude than the combination of a 25% reserve with a once-in-20-years risk.

In undeveloped areas, few records are even as long as 20 years. Thus, estimation of the 5, 2, and 1% frequencies, or of recurrence intervals of 20, 50, and 100 years, requires extrapolation from available data. Probability plots lend themselves well to this purpose. However, they must be used with discretion. Where severe droughts in the record extend over several years and require annual rather than seasonal storages, the resulting series of storage values becomes nonhomogeneous and is no longer strictly subject to ordinary statistical interpretations. They can be made reasonably homogeneous by including, besides all truly seasonal storages, not only all true annual storages, but also those seasonal storages that would have been identified within the periods of annual storage if the drought of the preceding year or years had not been measured. Plots of recurrence intervals should include minor storages as well as major ones. The results of these statistical analyses are then conveniently reduced to a set of draft-storage-frequency curves.

*Example 4–2.*

Examination of the 25-year record of runoff from an eastern stream[9] shows that the storages listed in Table 4–2 are needed in successive years to maintain a draft of 750,000 gpd per sq mile.

[8] Allen Hazen, Storage to Be Provided in Impounding Reservoirs, *Trans. Am. Soc. Civil Engrs.*, 77, 1539 (1914).

[9] The Westfield Little River near Springfield, Mass., for the years 1906 to 1930.

Estimate the design storage requirement probably reached or exceeded but once in 20, 50, and 100 years.

1. The 25 calculated storage values arrayed in order of magnitude are plotted on arithmetic-probability paper in Fig. 4–4 at $100\,k/26 = 3.8, 7.7, 11.5\%$, and so forth. A straight line of best fit is identified in this instance, but not necessarily others, the arithmetic mean storage being $\mu = 67$ mg and the standard deviation $\sigma = 33$ mg.

2. The storage requirements reached or exceeded once in 20, 50, and 100 years, or 5, 2, and 1% of the time, are read as 123, 137, and 146 mg respectively. Probability paper is used because it offers a rational basis for projecting the information beyond the period of experience. The once-in-20-years requirement with 25% reserve suggests a design storage of $123/0.75 = 164$ mg per sq mile of drainage area.

3. It should be noted that, in this instance, the coefficient of variation of the calculated storage $c_v = 100 \times 33/67 = 50\%$ is more than twice the variability of runoff ($c_v = 22\%$) for approximately the same period of observation (Example 3–1).

4. For comparison with other river records, draft and storage may be expressed in terms of the mean annual flow (MAF); storage may also be expressed in terms of daily draft. For a mean annual flow of 26.6 in., or $26.6 \times 0.0477 = 1.27$ mgd per sq mile (Example 3–1): (a) draft = 750,000 gpd/sq mile = $100 \times 0.750/1.27 = 59\%$ of MAF; (b) storage requirement equalled or exceeded once in 20 years = 123 mg/sq mile, or $(100 \times 123)/(1.27 \times 365) = 27\%$ of MAF; and (c) storage requirement = $123/0.750 = 164$ days of draft, or nearly half a year when 10 days are added to compensate for the use of monthly averages rather than daily stream flows.

When more than one reservoir is built on a stream, the overflow from each impoundage passes to the reservoir next below in the valley, together with the runoff from the intervening watershed. The amount of overflow is determined from the storage analysis for each year or for the critical year. If the reservoirs are operated jointly and those downstream are drawn on first, all reservoirs may be considered to be combined at the most-downstream location, provided the area tributary to each reservoir is large enough to fill its reservoir during the season of heavy runoff. The last-mentioned point requires special study.

**Table 4-2**
*Storage Requirements (Example 4–2)*

| Order of year | 1 | 2 | 3 | 4 | 5 | 6 | 7 | 8 | 9 | 10 | 11 | 12 | 13 |
|---|---|---|---|---|---|---|---|---|---|---|---|---|---|
| Calculated storage, mg | 47 | 39 | 104 | 110 | 115 | 35 | 74 | 81 | 124 | 29 | 37 | 82 | 78 |
| Order of year | 14 | 15 | 16 | 17 | 18 | 19 | 20 | 21 | 22 | 23 | 24 | 25 | |
| Calculated storage, mg | 72 | 10 | 117 | 51 | 61 | 8 | 102 | 65 | 73 | 20 | 53 | 88 | |

# 80/Surface Water Collection

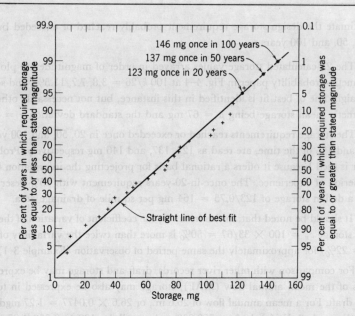

**Figure 4-4.**

**Frequency distribution of required storage plotted on arithmetic-probability paper.**

## 4-6 Generalized Storage Values

Hazen[10] has shown by an analysis of countrywide information that it is possible to base regional storage requirements on the mean annual flows of streams and their coefficients of variation. A partial summary of Hazen's generalized storage values is given in Table 4-3, and its use is illustrated in Example 4-3.

*Example 4-3.*

For the eastern stream dealt with in Fig. 4-3 and Example 4-2, find the generalized storage for a draft of 750,000 gpd per sq mile on the assumption that the coefficient of variation in annual flow is 0.22 and the mean annual flow 1.27 mgd (Example 3-1).

1. The draft is 59% of MAF as shown in Example 4-2.

2. For 59% and $c_v = 0.22$, Table 4-3 lists a storage of 0.30 MAF or 0.30 × 1.27 × 365 = 139 mg/sq mile.

[10] Allen Hazen as revised by Richard Hazen, in R. W. Abbott, Ed., *American Civil Engineering Practice*, John Wiley & Sons, New York, Vol. II, p. 18–09, 1956.

3. For 30 days' ground storage, deduct, according to Table 4–3, 0.048 from 0.30, making it 0.25 MAF or $0.25 \times 1.27 \times 365 = 116$ mg/sq mile.

The agreement between the results obtained by normal analytical procedures and by Hazen's generalized storage values is good in this case.

### Table 4-3
*Generalized Storage Values for Streams East of the Mississippi River, and in Oregon and Washington*[a]

| Draft | Storage for Stated Values of $c_v$ | | | | | | Deduction for 30 Days' Ground Storage |
|---|---|---|---|---|---|---|---|
| | 0.20 | 0.25 | 0.30 | 0.35 | 0.40 | 0.45 | |
| 0.9 | 0.85 | 1.05 | 1.31 | 1.60 | 1.88 | 2.20 | 0.074 |
| 0.8 | 0.54 | 0.64 | 0.78 | 0.97 | 1.19 | 1.39 | 0.066 |
| 0.7 | 0.39 | 0.43 | 0.50 | 0.62 | 0.76 | 0.92 | 0.058 |
| 0.6 | 0.31 | 0.32 | 0.34 | 0.40 | 0.49 | 0.60 | 0.049 |
| 0.5 | 0.23 | 0.23 | 0.24 | 0.26 | 0.32 | 0.39 | 0.041 |

[a] Both draft and storage are expressed in terms of the mean annual flow of the stream. The coefficient of variation in annual flows is designated $c_v$.

### 4-7 Loss by Evaporation, Seepage, and Silting

When an impounding reservoir is filled, the hydrology of the inundated area and its immediate surroundings is changed in a number of respects: (1) the reservoir loses water by evaporation to the atmosphere and gains water by direct reception of rainfall; (2) rising and falling water levels alter the pattern of groundwater storage and movement into and out of the surrounding reservoir banks; (3) at high stages, water may seep from the reservoir through permeable soils into neighboring catchment areas and so be lost to the area of origin; and (4) quiescence encourages subsidence of settleable suspended solids and silting of the reservoir.

*Water-Surface Response.* The response of the new water surface is to establish new hydrological equilibria (1) through loss of the runoff once coming from precipitation on the land area flooded by the reservoir $Qa$ (closely), where $Q$ is the areal rate of runoff of the original watershed, and $a$ is the water surface area of the reservoir; and through evaporation from the water surface $Ea$, where $E$ is the areal rate of evaporation; and (2) through gain of rainfall on the water surface $Ra$, where $R$ is the areal rate of rainfall. The net rate of loss or gain is $[R - (Q + E)]a$; a negative value records a net loss and a positive value a net gain.

Individual factors vary within the annual hydrologic cycle and from year to year. They can be measured. Exact calculations, however, are commonly handicapped by inadequate data on evaporation. Required hydrological information should come from local or nearby observation stations, areas of water surface being determined from contour maps of the reservoir site. The mean annual water surface, normally about 90% of the reservoir area at spillway level, is sometimes substituted to simplify calculations.

For convenience, the water-surface response is expressed in one of the following ways:

1. Revised runoff: $\quad\quad\quad Q_r = Q - (Q + E - R)(a/A) \quad\quad (4\text{-}2)$

2. Equivalent draft: $\quad\quad\quad D_e = (Q + E - R)(a/A) \quad\quad\quad\quad (4\text{-}3)$

3. Effective catchment area: $A_e = A - a[1 - (R - E)/Q] \quad (4\text{-}4)$

Here $A$ is the total catchment area and $a$ the reservoir surface area, and the values obtained are used as follows in recalculating storage requirements: $Q_r$ replaces $Q$; $D + D_e$ replaces $D$; and $A_e$ replaces $A$. A fourth allowance calls for raising the flow line of the reservoir by $Q + E - R$ expressed in units of length yearly. In rough approximation, the spillway level is raised by a foot or two in the eastern United States.

---

*Example 4–4.*

A mean draft of 30.0 mgd is to be developed from a catchment area of 40.0 sq miles. First calculations ask for a reservoir area of 1500 acres at flowline. The mean annual rainfall is 47.0 in., the mean annual runoff 27.0 in., and the mean annual evaporation 40.0 in. Find (1) the revised mean annual runoff, (2) the equivalent mean draft, (3) the equivalent land area, and (4) the adjusted flowline.

1. By Eq. 4–2, the revised annual runoff is $Q_r = 27.0 - (27.0 + 40.0 - 47.0)\,0.9 \times 1500/(640 \times 40.0) = 27.0 - 1.1 = 25.9$ in.

2. By Eq. 4–3, the equivalent mean draft is $D_e = 1.1$ in., or 52,000 gpd/sq mile, and the effective draft is $30.0 + 40.0 \times 0.052 = 32.1$ mgd.

3. By Eq. 4–4, the equivalent land area is $A_e = 40.0 - (0.9 \times 1500/640) \times [1 - (47.0 - 40.0)/27.0] = 40.0 - 1.6 = 38.4$ sq miles.

4. The adjusted flowline is $Q + E - R = 27.0 + 40.0 - 47.0 = 20$ in., equaling $20 \times 0.9 = 18$ in. at spillway level.

---

**Seepage.** If the valley enclosing a reservoir is underlain by porous strata, water may be lost by seepage. Subsurface exploration alone can foretell how much. Seepage is not necessarily confined to the dam site. It may occur wherever the sides and bottom of the reservoir are sufficiently permeable to permit water to escape through the surrounding hills.

**Silting.** Soil erosion on the watershed causes reservoir silting. Both are undesirable. Erosion destroys arable lands. Silting destroys useful storage. How bad conditions are in a given catchment area depends principally on soil and rock types, ground-surface slopes, vegetal cover, methods of cultivation, and storm-rainfall intensities.

Silt accumulations cannot be removed economically from reservoirs by any means so far devised. Dredging is expensive, and attempts to flush out deposited silt by opening scour valves in dams are fruitless. Scour only produces gullies in the silt. In favorable cirucmstances, however, much of the heaviest load of suspended silt can be steered through the reservoir by opening large sluices installed for this purpose. Flood flows are thereby selected for storage in accordance with their quality as well as their volume.

Reduction of soil erosion is generally a long-range undertaking. Involved are proper farming methods, such as contour plowing; terracing of hillsides; reforestation or afforestation; cultivation of permanent pastures; prevention of gully formation through construction of check dams or debris barriers; and revetment of stream banks.

In the design of impounding reservoirs for silt-bearing streams, suitable allowance must be made for loss of capacity by silting. Rates of deposition are especially high in impoundments on flashy streams draining easily eroded catchment areas. The proportion of sediment retained is called its *trap efficiency*. A simple calculation will show that 2000 mg per $l$ of suspended solids equals 8.3 tons per mg and that an acre-foot of silt weighs almost 1500 tons if its unit weight is 70 lb per cu ft. In some parts of the United States,[11] the volume of silt $V_s$ in acre-feet deposited annually can be approximated by the equation

$$V_s = cA^n \qquad (4\text{--}5)$$

where $A$ is the size of the drainage area in square miles, and $c$ and $n$ are coefficients with a value of $n = 0.77$ for southwestern streams and values of $c$ varying from 0.43 through 1.7 to 4.8 for low, average, and high deposition respectively, the corresponding values for southeastern streams being $c = 0.44$ only and $n = 1.0$. Understandably, the magnitudes of $c$ and $n$, here reported, apply only to the regions for which they were developed.

A plot of trap efficiency against the proportion of the mean annual flow stored in a reservoir traces curves quite similar to curves for the expected performance of settling basins of varying effectiveness. Close to 100% of the

---

[11] For basic information, see H. M. Eakin, Silting of Reservoirs, *U.S. Dept. Agr. Tech. Bull.*, 524, 1939, and J. E. Jenkins, C. E. Moak, and D. A. Okun, Sedimentation in Reservoirs in the Southeast, *Trans. Am. Soc. Civil Engrs.*, *III*, **68**, 3133 (1961).

sediment transported by influent streams may be retained in reservoirs storing a full year's tributary flow. Trap efficiency drops to a point between 65 and 85% when the storage ratio is reduced to 0.5 (half a year's inflow) and to 30 to 60% when the storage ratio is lowered to 0.1 (5 weeks' inflow). Silting is often fast when reservoirs are first placed in service and may be expected to drop off and reach a steady state as delta building goes on and shores become stabilized. An annual silting rate of 1.0 acre-ft per sq mile of watershed corresponds roughly to a yearly reduction in storage of 0.3 mg per sq mile, because an acre 3 ft deep is about 1 mg.

### 4-8 Area and Volume of Reservoirs

The surface areas and volumes of water at given horizons are found from a contour map of the reservoir site. Areas enclosed by each contour line are planimetered, and volumes between contour lines are calculated. The *average-end-area method* is generally good enough for the attainable precision of measurements.

For uniform contour intervals $h$ and successive contour areas $a_0, a_1, \ldots a_n$, the volume $V$ of water stored up to the $n$th contour is

$$V = \tfrac{1}{2}h[(a_0 + a_1) + (a_1 + a_2) + \ldots + (a_{n-1} + a_n)]$$
$$= \tfrac{1}{2}h\left(a_0 + a_n + 2\sum_{1}^{n-1} a\right) \quad (4\text{-}6)$$

For general use, surface areas and volumes are commonly plotted against contour elevations as in Fig. 4–5. It should be noted that volumes must be determined from the surface-area curve by planimetering the area enclosed between the curve and its ordinate.[12]

In reservoir operation, a small amount of water lies below the invert of the reservoir outlet. Constituting the dregs of the impoundage, this water is of poor quality. The associated reduction in *useful storage* is offset, in general, by bank storage released from the soil as the reservoir is drawn down. Moreover, the water below the outlet sill does form a conservation pool for fish and wildlife.

Surface areas and volumes enter not only into the solution of hydrological problems but also into the management of water quality, such as the control of algae by copper sulfate and destratification by pumping or aeration (Secs. 21–14, 15).

[12] The vertical scale implied by elevations generally leads engineers to plot elevations as ordinates. Fig. 4–5, therefore, is not in consonance with the injunction to plot as the ordinate the variable to be found.

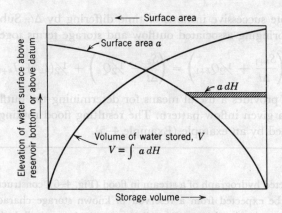

**Figure 4-5.**
**Surface area of a reservoir and volume of water stored.**

## 4-9 Spillway Capacity and Flood Routing

Impounding reservoirs must have spillways that can safely discharge the maximum peak flood the storage works are expected to pass. Unless flood storage is one of the planned purposes of the reservoir, entrant floods are assumed to occur when the reservoir is full. Before the maximum head on the spillway can be developed, however, flood waters will back up in the reservoir and fill the space between spillway level and flood crest. As a result, the flood peak is reduced, often by enough to lower the required discharge capacity of the spillway appreciably. However, if construction of the reservoir deprives the stream of significant amounts of valley storage within the reservoir site, studies of flood routing must make proper allowances for lost storage.

In other respects, retardation of floods by storage above spillway level is a function of reservoir inflow rates $I$, water storage $S$ above spillway level, and reservoir outflow rates $Q$. These variables are usually so irregular that they cannot be generalized mathematically. In the circumstances, engineers proceed to stepwise analyses of pertinent hydraulic sequences: varying inflow, changing water level, and varying outflow. For a specified time element, $\Delta t$,

$$Q\Delta t = I\Delta t - \Delta S \tag{4-7}$$

Assuming average rates of inflow and outflow closely equal to the arithmetic means of the rates obtaining at the beginning and end of time intervals $\Delta t$, mechanical integration proceeds to evaluate $Q\Delta t$ as $\Delta t(Q_k + Q_{k+1})/2$; $I\Delta t$ as $\Delta t(I_k + I_{k+1})/2$; and $\Delta S$ as $(S_{k+1} - S_k)$. Here the subscripts $k$ and

86/Surface Water Collection

$(k + 1)$ denote successive instants of time differing by $\Delta t$. Substituting in Eq. 4–7 and bringing associated outflow and storage terms together:[13]

$$\left(\frac{S_{k+1}}{\Delta t} + \tfrac{1}{2}Q_{k+1}\right) = \left(\frac{S_k}{\Delta t} - \tfrac{1}{2}Q_k\right) + \tfrac{1}{2}(I_k + I_{k+1}) \qquad (4-8)$$

Equation 4–8 provides a useful means for determining the outflow pattern produced by a given inflow pattern. The resulting flood routing procedure is best explained by an example (Example 4–5).

*Example 4–5.*

From the predicted hydrograph of a stream in flood (Fig. 4–6), construct the outflow hydrograph to be expected from a reservoir of known storage characteristics impounding the stream under the runoff conditions assumed to prevail during the flood.

The following assumptions are made: length of spillway tentatively selected $L = 250$ ft; appropriate weir coefficient $C = 3.8$; and time interval $\Delta t = 3$ hr = 10,800 sec.

1. Determine, for increasing heads $H$, the outflow $Q$ and storage $S$ above spillway level. To do this, the surface area $a$ of the reservoir must be found from a curve like that in Fig. 4–5.

2. Calculate the corresponding functional rates of storage $(S/\Delta t)$, $[(S/\Delta t) - \tfrac{1}{2} Q]$, and $[(S/\Delta t) + \tfrac{1}{2} Q]$ as in Table 4–4.

3. Plot the rates of discharge and storage against the heads on the spillway. This is done in Fig. 4–7. The resulting curves, known as *routing curves* and *discharge curves*, allow the stepwise graphical determination of spillway heads and outflows at the chosen time intervals, 3 hr in this instance.

4. In Fig. 4–7 add the average rate of inflow $\tfrac{1}{2}(I_k + I_{k+1})$ for each specified time interval to the corresponding value of $[(S/\Delta t) - \tfrac{1}{2} Q_k]$ in accordance with Eq. 4–8, and find at the resulting magnitude of $[(S_{k+1}/\Delta t - \tfrac{1}{2}Q_{k+1}]$ the spillway head and discharge that must obtain in order to satisfy these relationships. To provide a starting point, it is assumed that the reservoir is in equilibrium at the initial rate of inflow of 700 cfs shown in Fig. 4–6, the head on the spillway then being

$$H = [Q/(CL)]^{2/3} = [700/(3.8 \times 250)]^{2/3} = 0.82 \text{ ft}$$

Necessary calculations are shown in Table 4–5.

5. Plot the calculated outflow hydrograph against the observed inflow hydrograph. This is done in Fig. 4–7. It is seen that storage above spillway level lowers the peak flow from 34,000 cfs to 27,000 cfs, that is, to 80% of its uncontrolled magnitude. Accordingly, the head on the spillway is 9.2 ft (Fig. 4–7).

[13] Spillway discharge $Q = CL(H - H_0)^{3/2}$ by the common weir formula, while storage $S = V - V_0 = c(H^m - H_0^m)$. Here $H_0$ is the water depth at spillway level, $H - H_0$ the spillway head, $L$ the weir length, and $c$ and $m$ are coefficients for a given reservoir site.

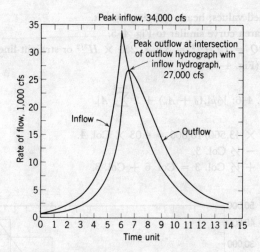

**Figure 4-6.**
**Flood routing or flood-flow modification by storage.**

*Table 4-4*
Calculation of Functional Rates of Storage (Example 4–5)

| Head on Spillway $H$, ft | Reservoir Area, $A$ acres | Calculated Outflow, $Q$ cfs | Calculated Storage $S$ Above Spillway Level, acre-ft | Functional Rates of Storage | | |
|---|---|---|---|---|---|---|
| | | | | $S/\Delta t$, cfs | $S/\Delta t - \tfrac{1}{2}Q$, cfs | $S/\Delta t + \tfrac{1}{2}Q$, cfs |
| (1) | (2) | (3) | (4) | (5) | (6) | (7) |
| 0 | 670 | 0 | 0 | 0 | 0 | 0 |
| 1 | 700 | 950 | 685 | 2,760 | 2,285 | 3,235 |
| 2 | 730 | 2,680 | 1,400 | 5,650 | 4,310 | 6,990 |
| 3 | 760 | 4,940 | 2,145 | 8,650 | 6,180 | 11,120 |
| 4 | 790 | 7,600 | 2,920 | 11,770 | 7,970 | 15,570 |
| 5 | 820 | 10,620 | 3,725 | 15,020 | 9,710 | 20,330 |
| 6 | 850 | 13,950 | 4,560 | 18,390 | 11,415 | 25,365 |
| 7 | 885 | 17,600 | 5,425 | 21,900 | 13,100 | 30,700 |
| 8 | 920 | 21,500 | 6,330 | 25,500 | 14,750 | 36,250 |
| 9 | 960 | 25,700 | 7,270 | 29,300 | 16,450 | 42,150 |
| 10 | 1,000 | 30,100 | 8,250 | 33,300 | 18,300 | 48,400 |

Col. 1: Assumed values; heads differing by 1 ft.
Col. 2: From area curve similar to Fig. 4–5.
Col. 3: From $Q = CLH^{3/2} = 3.8 \times 250 \times H^{3/2}$ or straight-line plot on log-log paper (Fig. 4–7).

Col. 4: By Eq. 4–6: $\frac{1}{2}h[A_0 + A_n) + 2\sum_{1}^{n-1} A]$.

Col. 5: Col. 4 $\times$ 43,560/10,800 = 4.03 $\times$ Col. 4.
Col. 6: Col. 5 $-$ ½ Col. 3.
Col. 7: Col. 5 $+$ ½ Col. 3 = Col. 6 + Col. 3.

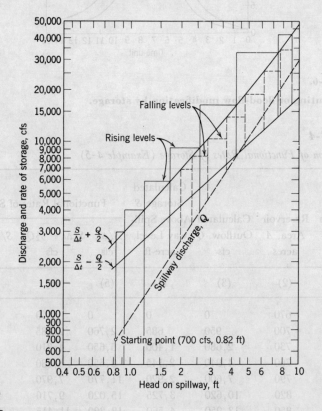

**Figure 4-7.**
**Stepwise graphical determination of head and discharge relationships in routing a flood through an impounding reservoir. See Eq. 4-8: $[(S_{k+1}/\Delta t) + \frac{1}{2}Q_{k+1}] = [(S_k/\Delta t) - \frac{1}{2}Q_k] + \frac{1}{2}(I_k + I_{k+1})$. Example: $(S_{k+1}/\Delta t + \frac{1}{2}Q_{k+1}) = 1900 + 1050 = 2950$; $H = 0.92$ and $(S_{k+1}/\Delta t + \frac{1}{2}Q_{k+1}) = 2100 + 1900 = 4000$; $H = 1.2$.**

### Table 4-5
*Calculation of Reservoir Outflows (Example 4–5)*

| Time Number | Observed Inflow $I$, cfs | $\frac{1}{2}(I_k + I_{k+1})$ Average Inflow, cfs | $S/\Delta t - Q/2$ At Beginning of Time Interval | $S/\Delta t + Q/2$ At End of Time Interval | Head on Spillway, ft | Outflow $Q$, cfs |
|---|---|---|---|---|---|---|
| (1) | (2) | (3) | (4) | (5) | (6) | (7) |
| 0 | 700 | — | ..... | ..... | 0.817 | 700 |
| 1 | 1,400 | 1,050 | 1,900 | 2,950 | 0.920 | 840 |
| 2 | 2,400 | 1,900 | 2,100 | 4,000 | 1.20 | 1,250 |
| 3 | 4,000 | 3,200 | 2,700 | 5,900 | 1.70 | 2,100 |
| 4 | 7,000 | 5,500 | 3,700 | 9,200 | 2.55 | 3,870 |
| 5 | 15,000 | 11,000 | 5,400 | 16,400 | 4.20 | 8,170 |
| 6 | 34,000 | 24,500 | 8,300 | 32,800 | 7.40 | 19,100 |
| 7 | 22,000 | 28,000 | 13,700 | 41,700 | 8.95 | 25,400 |
| 8 | 14,000 | 18,000 | 16,200 | 34,200 | 7.65 | 20,010 |
| 9 | 9,000 | 11,500 | 14,200 | 25,700 | 6.10 | 14,300 |
| 10 | 6,000 | 7,500 | 11,500 | 19,000 | 4.70 | 9,680 |
| 11 | 3,400 | 4,700 | 9,300 | 14,000 | 3.70 | 6,760 |
| 12 | 2,500 | 2,950 | 7,400 | 10,350 | 2.85 | 4,570 |
| 13 | 2,200 | 2,350 | 5,900 | 8,250 | 2.30 | 3,310 |
| 14 | 2,000 | 2,100 | 4,850 | 6,950 | 1.95 | 2,590 |

Col. 1: Each time interval is 3 hours.

Col. 2: The observed inflow is taken from the chosen flood hydrograph of the stream before impoundage. See Fig. 4–6.

Col. 3: Average of successive values in Col. 2.

Col. 4: Value of $[(S/\Delta t) - \frac{1}{2}Q]$ at beginning of time interval read during construction of Fig. 4–7.

Col. 5: Value of $[(S/\Delta t) + \frac{1}{2}Q]$ at end of time interval = Col. 4 + Col. 3 in accordance with Eq. 4–8.

Cols. 6 and 7. Read from Fig. 4–7 with exception of the initial values: 0.82 ft and 700 cfs. These identify the starting point of the step integration.

---

The principles involved in this as well as numerous other methods of flood routing can be put to use also in studies of the effect of channel storage, detention or retardation basins, and other types of storage upon flood flows and flows in general.

A rough determination of whether it will pay to perform calculations such as these can be made from generalized estimates suggested by Fuller[14] (Table 4-6). If the outflow is reduced to 90% or less of the inflow, more accurate calculations are normally justified.

**Table 4-6**
*Generalized Estimates of Reservoir Outflows*

| Ratio of storage above spillway level to flood flow in 24 hours, % | 5 | 10 | 20 | 30 | 40 | 50 | 60 | 70 |
|---|---|---|---|---|---|---|---|---|
| Ratio of peak outflow to peak inflow, % | 99 | 97 | 93 | 86 | 77 | 65 | 53 | 40 |

*Example 4-6.*

The maximum 24-hour flood flow in Example 4–5 is 27,000 cfs. For a maximum allowable head on the spillway of 10 ft and a storage above spillway level of 8250 acre-ft, find the ratio of peak outflow to peak inflow from Fuller's values (Table 4–6): The ratio of storage above spillway level to the 24-hour flood is $100 \times 8{,}250/27{,}600 = 30\%$; and from Fuller's values, the ratio of peak outflow to peak inflow is 86%, whereas the value ascertained in Example 4–5 is 80%, the maximum head on the spillway being 9.2 ft. Because Fuller's outflow ratio is less than 90%, a more accurate determination is warranted.

The principles set forth in the preceding sections of this chapter are applicable also to the storage and regulation of stormwater runoff collected by combined municipal drainage schemes. Stormwater stand-by or holding tanks used for this purpose may be incorporated in the collecting system itself or become auxiliary units in wastewater-treatment works (Fig. 9–8).

## 4-10 Catchment Areas

The comparative advantage of developing surface rather than underground waters is offset, in large measure, by the unsteadiness of surface runoff, both in quantity and quality, and the recurrence of flow extremes. That hydrological factors enter strongly into the development of surfacewater supplies must, therefore, be kept clearly in mind in their design and operation, with special reference to (1) the principles of selecting, preparing, and controlling catchment areas; (2) the choice and treatment of reservoir areas and the management of natural ponds and lakes as well as impoundages; and (3) the

[4] W. E. Fuller, Flood Flows, *Trans. Am. Soc. Civil Engrs.*, **70**, 564 (1914).

siting, dimensioning, construction, and maintenance of necessary engineering works, including dams and dikes, intake structures, spillways, and diversion works. It should also be kept in mind that river systems may have to be developed for multiple purposes, not just for municipal uses.

The gathering grounds for public water supplies vary *in size* from a few hundred acres to thousands of square miles, and *in character* from sparsely inhabited uplands to densely populated river valleys. The less developed they are, the better, relatively, do they lend themselves to exploitation for steady yields and the production of water of high quality.

**Upland Areas.** Occasionally, yet seldom, a water utility can, with economic justification, acquire the entire watershed of its source and manage it solely for water-supply purposes, excluding habitations and factories to keep the water safe and attractive; letting arable lands lie fallow to prevent wasteful runoff and high turbidities; draining swamps to reduce evaporation and eliminate odors, tastes, and color; and cultivating woodlots to hold back winter snows and storm runoff and help to preserve the even tenor of streamflow. As competition for water and land increases, land holdings of water utilities are understandably confined to the marginal lands of water courses, especially those closest to water intakes themselves. Yet water-quality management need not be neglected. Scattered habitations can be equipped with acceptable sanitary facilities; wastewaters can be adequately treated or, possibly, diverted into neighboring drainage areas not used for water supply; swamps can be drained; and soil erosion can be controlled. Intelligent land management of this kind can normally be exercised most economically when water is drawn from upland sources where small streams traverse land of little value and small area. However, some upland watersheds are big enough to satisfy the demands of great cities.[15]

**Lowland Areas.** When water is drawn from large lakes and wide rivers that, without additional storage, yield an abundance of water, management of their catchments ordinarily becomes the concern of more than one community,[16] sometimes of more than a single state,[17] and even of a single country.[18] Regional, interstate, and international authorities must be set up to manage and protect land and water resources of this kind.

**Quality Control.** To safeguard their sources, water utilities can fence and post their lands, patrol watersheds, and obtain legislative authority for enforcing reasonable rules and regulations for the sanitary management of

---

[15] The water supplies of Boston, Mass.; New York, N.Y.; and San Francisco, Calif. are examples.

[16] Examples are many on the Ohio and Mississippi rivers.

[17] Allocation of the waters of the Delaware River is a notable example.

[18] This is true for the Great Lakes shared with Canada and the Colorado River shared with Mexico.

the catchment area. When the cost of policing the area outweighs the cost of purifying its waters in suitable treatment works, purification is often preferred. It is likewise preferred when lakes, reservoirs, and streams become important recreational assets and their enjoyment can be encouraged without endangering their quality. That recreation must be properly supervised and recreational areas suitably located and adequately equipped with sanitary facilities needs no special plea.

**Swamp Drainage.** Three types of swamps may occur on catchment areas: (1) rainwater swamps where precipitation accumulates on flat lands, or rivers overflow their banks in times of flood; (2) backwater swamps or reaches of shallow flowage in sluggish, often meandering,[19] streams and at bends or other obstructions to flow; and (3) seepage-outcrop swamps where hillside meets the plain or sand and gravel overlie clay or other impervious formations.

Rainwater swamps can be drained by ditches cut into the flood plain; backwater swamps by channel regulation; and seepage-outcrop swamps by marginal interception of seepage waters along hillsides sometimes supplemented by the construction of central surface and subsurface drains.

## 4-11 Reservoir Siting

In the absence of natural ponds and lakes, intensive development of upland waters requires the construction of impounding reservoirs. Suitable siting is governed by interrelated considerations of adequacy, economy, safety, and palatability of the supply. Desirable factors include:

1. Surface topography that generates a low ratio of dam volume to volume of water stored; for example, a narrow gorge for the dam, opening into a broad and branching upstream valley for the reservoir. In addition, a favorable site for a stream-diversion conduit and a spillway, and a suitable route for an aqueduct or pipeline to the city are desired.

2. Subsurface geology that ensures (a) safe foundations for the dam and other structures, (b) tightness against seepage through abutments and beneath the dam, and (c) materials, such as sand, gravel, and clay, for construction of the dam and appurtenant structures.

3. A reservoir valley that is sparsely inhabited, neither marshy nor heavily wooded, and not traversed by important roads or railroads; the valley being *so shaped* that waters pouring into the reservoir are not short-circuited to the outlet, and *so sloped* that there is little shallow flowage around the margins. Natural purification by storage can be an important asset. Narrow reservoirs stretching in the direction of prevailing winds are easily short-circuited and

---

[19] After the river *Maiandros*, in Asia Minor, proverbial for its windings.

may be plagued by high waves. Areas of shallow flowage often support heavy growths of water plants while they are submerged, and of land plants while they are uncovered. Shore-line vegetation encourages mosquito breeding; decaying vegetation imparts odors, tastes, and color to the water.

4. Reservoir flowage that interferes as little as possible with established property rights, close proximity of the intake to the community served, and location at such elevation that supply can be by gravity.

***Site Preparation.*** Large reservoirs may inundate villages, including their dwellings, stores, and public buildings; mills and manufacturing establishments; farms and farmlands, stables, barns, and other outhouses; and gardens, playgrounds, and graveyards. Although such properties can be seized by the *right of eminent domain,* a wise water authority will proceed with patience and understanding. To be humane and foster good will, the authority will transport dwellings and other wanted and salvable buildings to favorable new sites, establish new cemeteries or remove remains and headstones to grounds chosen by surviving relatives, and assist in reconstituting civil administration and the regional economy.

When reservoir sites are flooded, land plants die and organic residues of all kinds begin to decompose below the rising waters; nutrients are released; algae and other microorganisms flourish in the eutrophying environment; and odors, tastes, and color are intensified. Ten to 15 years normally elapse before the biodegradable substances are minimized and the reservoir is more or less stabilized.

In modern practice, reservoir sites are cleared only in limited measure as follows:

1. Within the entire reservoir area: (a) dwellings and other structures are removed or razed; (b) barnyards, cesspools, and privies are cleaned, and ordure and manure piles are carted away; (c) trees and brush are cut close to the ground, usable timber is salvaged, and slash, weeds, and grass are burned; (d) swamp muck is dug out to reasonable depths, and residual muck is covered with clean gravel, the gravel, in turn, being covered with clean sand; and (e) channels are cut to pockets that would not drain when the water level of the reservoir is lowered.

2. Within a marginal strip between the high-water mark reached by waves and a contour line about 20 ft below reservoir level: (a) stumps, roots, and topsoil are removed; (b) marginal swamps are drained or filled; and (c) banks are steepened to produce shore-water depths close to 8 ft during much of the growing season of aquatic plants—to do this, upper reservoir reaches may have to be improved by excavation or fill or by building auxiliary dams across shallow arms of the impoundage.

*Soil stripping*, namely, the removal of all topsoil containing more than 1 or 2% organic matter from the entire reservoir area, is no longer economical.

In malarious regions, impounding reservoirs should be so constructed and managed that they will not breed dangerous numbers of anopheline mosquitoes. To this purpose, banks should be clean and reasonably steep. To keep them so, they may have to be protected by riprap.

## 4-12 Reservoir Management

The introduction of impounding reservoirs into a river system or the existence of natural lakes and ponds within it raise questions of quality control that are best discussed by themselves under the general title of limnology. This is done in Chap. 21 rather than here. What should be said here, however, is that some limnological factors are important not only in the management of ponds, lakes, and reservoirs but also in reservoir design.

***Quality Control.*** Of concern in the quality management of reservoirs is the control of water weeds and algal blooms; the bleaching of color; the settling of turbidity; destratification by mixing or aeration; and, in the absence of destratification, the selection of water of optimal quality and temperature by shifting intake depths in order to suit withdrawals to water uses or to downstream quality requirements.

***Evaporation Control.*** The thought that oil spread on water will suppress evaporation is not new. Yet the threshold of its realization has only just been crossed, thanks largely to Irving Langmuir's Nobel Prize-winning work.[20] What has been learned is that (1) certain chemicals spread spontaneously on water as layers no more than a molecule thick; (2) these substances include alcohol (hydroxyl) or fatty acid (carboxyl) groups attached to a saturated paraffin chain of carbon atoms; (3) the resulting *monolayers* consist of molecules oriented in the same direction and thereby offering more resistance to the passage of water molecules than do thick layers of oil composed of multilayers of haphazardly oriented molecules; and (4) the hydrophilic radicals (OH or COOH) at one end of the paraffin chain move down into the water phase while the hydrophobic paraffin chains, themselves, stretch up into the gaseous phase. Examples of suitable chemicals are alcohols and corresponding fatty acids. Cost and difficulty of maintaining adequate coverage of the water surface have operated against their use.

[20] V. K. La Mer, Ed., *Retardation of Evaporation by Monolayers: Transport Processes*, Academic Press, New York, 1962. Langmuir (1881–1957) received the Nobel Prize in chemistry in 1932 for his work on the structure of matter and surface chemistry. He is also known to hydrologists for his experiments on seeding clouds with silver iodide to start rainfall.

## 4-13 Dams and Dikes

Generally speaking, the great dams and barrages of the world are the most massive structures built by man. To block river channels carved through mountains in geological time periods, many of them are wedged between high valley walls and impound days and months of flow in deep reservoirs. Occasionally, impoundages reach such levels that their waters would spill over low saddles of the divide into neighboring watersheds if saddle dams or dikes were not built to complement the main structure. In other ways, too, surface topography and subsurface geology are of controlling influence. Hydraulically, they determine the siting of dams; volumes of storage, including subsurface storage in glacial and alluvial deposits; and spillway and diversion arrangements. Structurally, they identify the nature and usefulness of foundations and the location and economic availability of suitable construction materials. Soils and rock of many kinds can go into the building of dams and dikes. Timber and steel have found more limited application. Like most other civil-engineering constructions, therefore, dams and their reservoirs are derived largely from their own environment.

Structurally, dams resist the pressure of waters against their upstream face by gravity, arch action, or both. Hydraulically, they stem the tides of water by their tightness as a whole and the relative imperviousness of their foundations and abutments. Coordinately, they combine hydraulic and structural properties to keep seepage within tolerable limits and so channeled that the working structures are and remain safe.

Materials and methods of construction create dams of many types. The following are most common: (a) embankment dams of earth, rock, or both and (b) masonry dams (today largely concrete dams) built as gravity, arched, or buttressed structures.[21]

***Embankment Dams.*** Rock, sand, clay, and silt are the principal materials of construction for rock and earth embankments. Permeables provide weight, impermeables water tightness. Optimal excavation, handling, placement, distribution, and compaction with special reference to selective placement of available materials challenge the ingenuity of the designer and constructor. Permeables form the shells or shoulders, impermeables the core or blanket of the finished embankment. Depending in some measure on the abundance or scarcity of clays, relatively thick cores are centered in a substantially vertical position, or relatively thin cores are displaced towards the upstream face in an inclined position. Common features of an earth dam with a central clay core wall are illustrated in Fig. 4–2. Concrete walls can take the place of clay cores, but they do not adjust well to

---

[21] *Design and Construction of Dams*, Am. Soc. of Civil Engineers, 1967, 131 pp.

the movements of newly placed, consolidating embankments and foundations; by contrast, clay is plastic enough to do so. If materials are properly dispatched from borrow pits, earth shells can be ideally graded from fine at the watertight core to coarse and well-draining at the upstream and downstream faces. In rock fills, too, there must be effective transition from core to shell, the required change in particle size ranging from a fraction of a millimeter for fine sand through coarse sand (about 1 mm) and gravel (about 10 mm) to rock of large dimensions.

Within the range of destructive wave action, stone placed either as paving or as *riprap* wards off erosion of the upstream face. Concrete aprons are not so satisfactory, sharing as they do most of the disadvantages of concrete core walls. A wide *berm* at the foot of the protected slope helps to keep riprap in place. To save the downstream face from washing away, it is commonly seeded to grass or covering vines and provided with a system of surface and subsurface drains. Berms break up the face into manageable drainage areas and give access to slopes for mowing and maintenance. Although they are more or less horizontal, berms do slope inward to gutters; moreover, they are pitched lengthwise for the gutters to conduct runoff to surface or subsurface main drains and through them safely down the face or abutment of the dam, eventually into the stream channel.

Earth embankments are constructed either as *rolled fills* or *hydraulic fills;* rock embankments are built as *uncompacted* (dumped) or *compacted* fills. In *rolled earth fills*, successive layers of earth 4 to 12 in. thick are spread, rolled, and consolidated. Sheep's-foot rollers do the compacting, but they are helped in their work by heavy earth-moving vehicles bringing fill to the dam or bulldozing it into place. Portions of embankment that cannot be rolled in this way are compacted by hand or power tampers. Strips adjacent to concrete core walls, the walls of outlet structures, and the wingwalls of spillway sections are examples.

In *hydraulic fills* water-carried soil is deposited differentially to form an embankment graded from coarse at the two faces of the dam to fine in the central core.

Methods as well as materials of construction determine the strength, tightness, and stability of embankment dams. Whether their axis should be straight or curved depends largely on topographic conditions. Whether upstream curves are in fact useful is open to question. Intended is axial compression in the core and prevention of cracks as the dam settles. Spillways are incorporated into some embankment dams and divorced from others in separate constructions.

Where rock outcrops on canyon walls can be blasted into the stream bed or where spillways or stream-diversion tunnels are constructed in rock, rock embankment becomes particularly economical. In modern construction, rock fills are given internal clay cores or membranes in somewhat the same fashion as earth fills (Fig. 4–8). Concrete slabs or timber sheathing once much used on the upstream face can be dangerously stressed and fail as the fill itself, or its foundation, settles. They are no longer in favor.

**Masonry Dams.** In the construction of gravity dams, *cyclopean* masonry and mass concrete embedding great boulders have, in the course of time, given way to poured concrete; in the case of arched dams rubble has also ceded the field to concrete. Gravity dams are designed to be in compression under all conditions of loading. They will fit into almost any site with a suitable foundation. Some arched dams are designed to resist water pressures and other forces by acting as vertical cantilevers and horizontal arches simultaneously; for others, arch action alone is assumed, thrust being transmitted laterally to both sides of the valley, which must be strong enough to serve as abutments. In constant-radius dams, the upstream face is vertical or, at most, slanted steeply near the bottom; the downstream face is projected as a series of concentric, circular contours in plan. Dams of this kind fit well into U-shaped valleys, where cantilever action is expected to respond favorably to the high-intensity bottom loads. In constant-angle[21] dams, the upstream face bulges up-valley; the downstream face curves inward like the small of a man's back. Dams of this kind fit well into V-shaped valleys, where arch action becomes their main source of strength at all horizons.

Concrete buttresses are designed to support flat slabs or multiple arches in buttress dams. Here and there, wood and steel structures have taken the place of reinforced concrete. Their upstream face is normally sloped 1 on 1 and may terminate in a vertical cut-off wall.

All masonry dams must rest on solid rock. Foundation pressures are high in gravity dams; abutment pressures are intense in arched dams. Buttress dams are light on their foundations. Making foundations tight by sealing contained pockets or cavities and seams or faults with cement or cement-and-sand grout under pressure is an important responsibility. Low-pressure grouting (up to 40 psig) may be followed by high-pressure grouting (200 psig) from permanent galleries in the dam itself, and a curtain of grout may be forced into the foundation at the heel of gravity dams to obstruct seepage. Vertical drainage holes just downstream from the grout curtain help reduce uplift.

## 4-14 Common Dimensions of Dams

There is much to be said both for and against reporting a set of common dimensions or rules of thumb for the design of dams. What can be advanced in their favor is that they are generalizations of tested decisions in an otherwise uncertain area of design; what must be said against them is that they stultify the imagination of the designer and obstruct progress. Both state-

98/*Surface Water Collection*

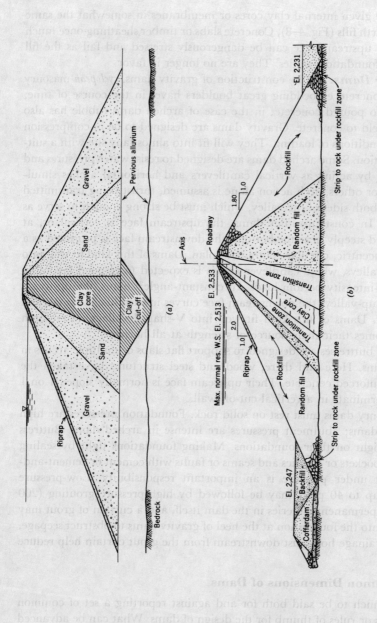

**Figure 4-8.**
Zoned earth-fill and rock-fill dams. Kindness of Arthur Casagrande. (*a*) Earthfill dam on pervious alluvium; (*b*) rock-fill dam on bedrock (Furnas Dam, Brazil).

ments are invalidated, however, if the fundamental reason for the citation of common dimensions is that they provide useful starting points—no more.

**Embankment Dams.** The detailed design of modern earth and rock fills has become the responsibility of specialists in soil mechanics. However, because detailed design is not the objective of this discussion, first estimates of the order of magnitude of needed structures should be enough. Only to this purpose are common dimensions shown in Fig. 4–9 and explained in the following.

Theoretically, the slope of a cohesionless material could be as steep as the angle of internal friction of its grains. In practice, the slope[22] is reduced for earth but not necessarily for rock fills by a factor of safety of 1.3 to 1.5. Maximum slopes of 1 on 2, for example, are flattened out to 1 on 2.6 to 3.0. The stability of cohesionless materials, such as sand, depends almost entirely on the friction between their grains. In a loose fill, voids are large, points of contact between grains few, and internal friction small. Accordingly, a loose fill is weak and easily disturbed by vibration or shock. If the material is dry, it will gradually shake down into a stabler, denser form and reach its *critical density* when its grains become so interlocked that there can be further movement within the material only if it expands in volume to allow particles to roll over one another. This makes the critical density the *maximum density* of cohesionless materials undergoing shear failure. However, if the material is full of water, the relative incompressibility of this fluid will force it out of the pores as the void space is reduced. Therefore, in the instant before the water escapes, because of the collapsing movement of the grains, the soil particles are actually suspended in the water, the material is deprived of its resistance to shear, and the entire mass can flow almost like a liquid. This explains the importance of compacting cohesionless materials in earth dams to more than critical density. In rolled fill, as has been shown, this is done by the travel of heavy equipment over successive layers of earth. In hydraulic fill, control is less certain. The fine-grained core substance is often quite loose, and its stability depends primarily on cohesion by molecular attraction. If the core does not drain, the shoulders must, indeed, be heavy enough to resist the fluid pressure not only of the stored water but also of the core, so long as it remains liquid. A fluid core weighs about 110 lb per cu ft—therefore, nearly twice as much as water.

The length $l$ and depth $y$ (Fig. 4–9) of a clay blanket beneath the upstream shell of an embankment dam are functions of permissible seepage and matters of practical construction. As shown in Fig. 4–9, $y$ is given a minimum thickness of $(2 + 0.02x)$ feet for practical reasons of construction, and the

---

[22] By convention, but also for ease of measurement in the field, the slope of embankments is reported as the tangent of the angle between the slope and its horizontal projection, namely, 1 (vertical) on $x$ (horizontal); or as the cotangent, namely, $x$:1.

100/Surface Water Collection

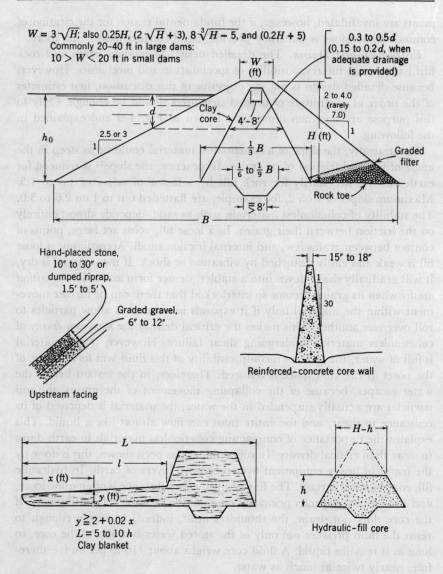

**Figure 4-9.**
**Common dimensions of earth dams.**
Upstream slope 2.5 on 1, downstream slope 2 to 4 (rarely 7) on 1 for homogeneous well-graded material; homogeneous silty clay, or clay when $H \lesssim 50$ ft; and sand, or sand and gravel, with reinforced concrete wall.
Upstream slope 3 on 1, downstream slope 2.5 on 1 for homogeneous coarse silt; homogeneous silty clay, or clay, when $H > 50$ ft; and sand, or sand and gravel, with clay core.

total length, $l$, of blanket and core material is made sufficient to ensure safety against piping.

In time, even well-constructed embankments shrink in volume or height, internally by consolidation of the embankment material, externally by consolidation of the foundation. Additional fill must be provided in compensation either then or, better, at the time of construction by sloping the crest of the embankment inward and upward from the abutments to the point of maximum height. Depending on soil type, long-time compression is generally measured at 0.2 to 0.4%. Understandably, dumped rock settles more than compacted rock.

Embankment volumes are computed in accordance with earthwork practice by average end areas, applying the prismoidal formula, or combining the two as follows:

$$V = \tfrac{1}{2}l[(a_1 + a_2) - \tfrac{1}{6}(c_1 - c_2)(b_1 - b_2)] \qquad (4\text{--}9)$$

where $V$ is the volume of earthwork between two parallel cross-sections of the dam, $l$ is the distance between these sections, and $a_1$ and $a_2$ are the end areas with center heights $c_1$ and $c_2$ and base width $b_1$ and $b_2$ as shown in Fig. 4-10. For computation of reservoir volumes, see Sec. 4-8.

The upstream shoulder of embankment dams is usually saturated with water—up to the flowline of the reservoir by entrant water and above it by capillary rise. This reduces the stability of the upstream shoulder and requires that the upstream slopes of earth dams be flatter than their downstream slopes. The slopes of embankments that rest on soft foundations or that include cohesive materials, clay for instance, must be flatter than normal. Embankments in narrow rock canyons can be given slopes that are steeper than normal.

The full weight of the central vertical core often used in earth dams is transmitted to its foundation. This minimizes leakage at contact with the foundation. Because the full weight of the sloping core common to rock-fill dams is not transmitted to the contact area with the foundation, more leakage must be expected. An advantage of rock-fill over earth-fill is that the bulky rock-fill can be placed in the wet as well as the dry season. This leaves only the core and upstream portion and the grouting of the foundation to be completed during the dry season.

Berms on the downstream face of embankment dams are about 30 ft apart vertically and wide enough to accommodate gutters and other drainage appurtenances as well as maintenance vehicles such as mowers. In practice, berms often cut into the dam as setbacks without increasing the volume of earthwork. To accomplish this, the slopes between berms must be somewhat steeper than they would be if there were no berms. A single berm on the

102/Surface Water Collection

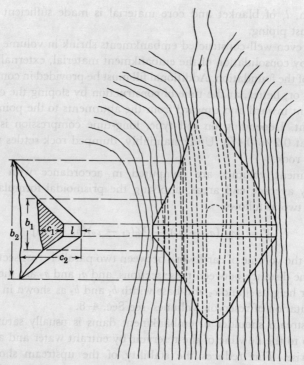

**Figure 4-10.**

**Plan and sections of earth dam for earthwork computations.**

upstream face supports needed riprap. The average depth and size of riprap may vary[23] from 12 in. of 10-in. rock for waves expected to be as high as 2 ft up to 30 in. of 18-in. rock for wave heights up to 8 ft. Transition from riprap to embankment material is by graded filters 6 to 12 in. deep.

*Masonry Dams.* For gravity dams, first estimates of the amount of masonry needed at a given site can be based on *practical profiles* developed for this purpose and in agreement with common assumptions. Wegmann's Practical Type No. 2 is shown in Fig. 4-11. This cross-section was designed by him[24] for zero uplift, masonry weighing 145.8 lb/cu ft, and zero ice pressure. However, the top was made 20 ft wide. To obtain the profile of a

---

[23] J. L. Sherard, R. J. Woodward, S. F. Gizienski, and W. A. Clevenger, *Earth and Earth-Rock Dams*, New York, John Wiley & Sons, 1963, p. 456.

[24] Edward Wegmann, *The Design and Construction of Dams*, 8th Ed., New York, John Wiley & Sons, 1927. Edward Wegmann (1850–1935) was the designer of the New Croton Dam, a lofty and lovely structure in the Croton Water Supply System of the City of New York, with which he was long associated.

*Common Dimensions of Dams/103*

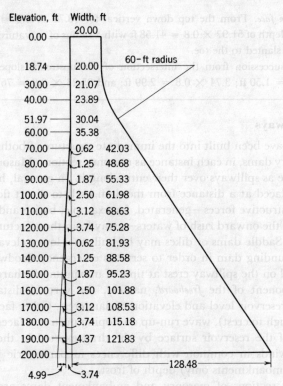

**Figure 4-11.**
**Dimensions of masonry dams; Wegmann's Practical Type 2.**

dam with this top width but less than 200 ft high, the unwanted lower portion is simply cut off. For a smaller top width, every dimension shown in Fig. 4–11 is reduced in the ratio of the desired width to 20 ft before the lower part of the modified structure is lopped off.

*Example 4-7.*

Estimate the principal dimensions of a dam 120 ft high with a top width of 16 ft.

Calculated from Wegmann's Practical Type No. 2 for a ratio of $16/20 = 0.8$, the dimensions for a 120-ft dam are read from the top $120/0.8 = 150$ ft of Fig. 4–11 and corrected as follows:

*Upstream face:* From the top down vertical for $60 \times 0.8 = 48$ ft; next sloping outward to a depth of $60 \times 0.8 = 48$ ft and a width of $3.74 \times 0.8 = 2.99$ ft; and finally vertical again for $120 - 2 \times 48 = 24$ ft.

*Downstream face:* From the top down vertical for 18.74 × 0.8 = 14.99 ft; next curved to a depth of 51.97 × 0.8 = 41.58 ft with a radius of curvature of 60 × 0.8 = 48 ft; finally slanted to the toe.

*Base:* In succession from the intersection of the upstream slope and the base 1.87 × 0.8 = 1.50 ft; 3.74 × 0.8 = 2.99 ft; and 95.23 × 0.8 = 76.18 ft.

## 4-15 Spillways

Spillways have been built into the immediate structure of both embankment and masonry dams, in each instance as masonry sections. Masonry dams may indeed serve as spillways over their entire length. In general, however, spillways are placed at a distance from the dam itself to divert flow and direct possible destructive forces—generated, for example, by ice and debris, wave action, and the onward rush of waters—away from the structure rather than towards it. Saddle dams or dikes may be built to a lower elevation than the main impounding dam in order to serve as emergency floodways.

The head on the spillway crest at time of maximum discharge is the principal component of the *freeboard*, namely, the vertical distance between maximum reservoir level and elevation of dam crest. Other factors are wave height (trough to crest), wave run-up on sloping upstream faces, wind set-up or tilting of the reservoir surface by the drag exerted in the direction of persistent winds in common with differences in barometric pressure, and (for earth embankments only) depth of frost.

Overflow sections of masonry and embankment dams are designed as masonry structures and separate spillways as *saddle, side channel*, and *drop inlet* or *shaft* structures. Spillways constructed through a *saddle* normally discharge into a natural floodway leading back to the stream below the dam. Usually they take the form of open channels and may include a relatively low overflow weir in the approach to the floodway proper. Overflow sections and overflow weirs must be calibrated if weir heads are to record flood discharges accurately, but their performance can be approximated from known calculations of similar structures. If their profile conforms to the ventilated lower nappe of a sharp-crested weir of the same relative height $d/h$ (Fig. 4–12), under the design head, $h$, the rate of discharge, $Q$, becomes

$$Q = \tfrac{2}{3}c\sqrt{2g}lh^{3/2} = Clh^{3/2} \qquad (4\text{--}10)$$

where $C = \tfrac{2}{3}c\sqrt{2g}$ is the coefficient of discharge, $g$ the gravity constant, and $l$ the unobstructed crest length of the weir. For a crest height $d$ above the channel bottom, the magnitude of $C$ is approximately[25] $4.15 + 0.65\ h/d$ for

---

[25] A. T. Ippen, Channel Transitions and Controls, in Hunter Rouse, Ed., *Engineering Hydraulics*, New York, John Wiley & Sons, 1950, pp. 534 and 535 (approximations of curves in Figs. 15 and 16 respectively).

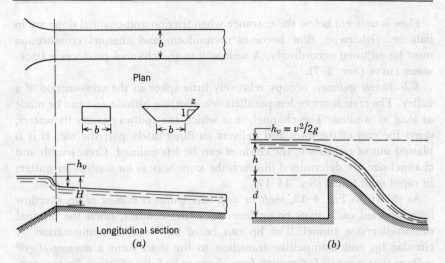

**Figure 4-12.**
(a) Channel spillway and (b) ogee spillway.

$h/d < 4$ or $C = 4.15$ to 6.75. Under heads $h'$ other than the design head $h$, $C$ approximates[25] $4.15(h'/h)^{7/5}$ up to a ratio of $h'/h = 3.0$.

If the entrance to the floodway is streamlined, little if any energy is lost—certainly no more than $0.05\ v^2/2g$. Otherwise, too, losses are presumably of the same relative magnitude as for entrances to pipes (Sec. 6–2). As suggested in Fig. 4–12, substantial quiescence within the reservoir must be translated into full channel velocity. Discharge is greatest when flow becomes critical. The velocity head $h_v$ then equals one-third the height $H$ of the reservoir surface above the entrance sill to a rectangular channel, and the rate of discharge $Q$ becomes

$$Q = \tfrac{2}{3}CbH\sqrt{2gH/3} = 3.087CbH^{3/2} \tag{4-11}$$

where $b$ is the width of the channel and $C$ is an entrance coefficient varying from 1.0 for a smooth entrance to 0.8 for an abrupt one. A trapezoidal channel with side slopes of 1:2 discharges

$$Q = 8.03\ Ch_v^{1/2}(H - h_v)[b + z(H - h_v)] \tag{4-12}$$

where

$$h_v = \frac{3(2zH + b) - (16z^2H^2 + 16zbH + 9b^2)^{1/2}}{10z} \tag{4-13}$$

Best hydraulic but not necessarily best economic efficiency is obtained when a semicircle can be inscribed in the cross-section (Sec. 8–1).

Flow is uniform below the entrance when friction and channel slope are in balance. Otherwise, flow becomes nonuniform and channel cross-section must be adjusted accordingly. A weir within the channel produces a backwater curve (Sec. 8-7).

*Side-channel spillways* occupy relatively little space in the cross-section of a valley. The crest more or less parallels one abutting hillside and can be made as long as wanted. The channel, into which the spillway pours its waters, skirts the end of the dam and delivers its flows safely past the toe. If it is blasted out of tight rock, the channel can be left unlined. Crest length and channel size are determined in much the same way as for washwater gutters in rapid sand filters (Sec. 14-17).

As shown in Fig. 4-13, *shaft* or *drop-inlet spillways* consist of an overflow lip supported on a shaft rising from an outlet conduit, often the original stream-diversion tunnel. The lip can be of any wanted configuration. A circular lip and trumpetlike transition to the shaft form a *morning-glory*[26] *spillway* that must lie far enough from shore to be fully effective. By contrast, a three-sided semicircular lip can be placed in direct contact with the shore; accessibility is its advantage. The capacity of shaft spillways is governed by their constituent parts and by flow conditions including air entrainment and hydraulic submersion. Hydraulic efficiency and capacity are greatest when the conduit flows full. Model studies are useful in arriving at suitable dimensions.

Flashboards or stop logs and gates of many kinds are added to spillways to take advantage of storage above crest level. They must be so designed and operated that the dam itself is not endangered in times of flood.

## 4-16 Intakes

Depending on the size and nature of the installation, water is drawn from rivers, lakes, and reservoirs through relatively simple submerged intake

[26] J. N. Bradley, W. E. Wagner, and A. J. Peterka, Morning-Glory Shaft Spillways—A Symposium, *Trans. Am. Soc. Civil Engrs.*, 121, 312 (1956).

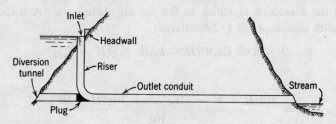

**Figure 4-13.**
**Shaft spillway and diversion tunnel.**

pipes, or through fairly elaborate towerlike structures that rise above the water surface and may house intake gates; openings controlled by stop logs; racks and screens, including mechanical screens, pumps, and compressors; chlorinators and other chemical feeders; venturi meters and other measuring devices; yes, even living quarters and shops for operating personnel (Fig. 4–1). Important in the design and operation of intakes is that the water they draw be as clean, palatable, and safe as the source of supply can provide.

**River Intakes.** Understandably, river intakes are constructed well upstream from points of discharge of sewage and industrial wastes. Optional location will take advantage of deep water, a stable bottom, and favorable water quality (if pollution hugs one shore of the stream, for example), all with proper reference to protection against floods, debris, ice, and river traffic (Fig. 4–14).[27] Small streams may have to be dammed up by *diversion* or *intake dams* to keep intake pipes submerged and preclude hydraulically wasteful air entrainment. The resulting intake pool will also work usefully as a settling basin for coarse silt and allow a protective sheet of ice to form in winter.

**Lake and Reservoir Intakes.** Lake intakes are sited with due reference to sources of pollution, prevailing winds, surface and subsurface currents, and shipping lanes. As shown in Fig. 4–15, shifting the depth of draft makes it possible to collect clean bottom water when the wind is offshore and, conversely, clean surfacewater when the wind is onshore. If the surrounding water is deep enough, bottom sediments will not be stirred up by wave action, and ice troubles will be few.

Reservoir intakes resemble lake intakes but generally lie closer to shore in the deepest part of the reservoir. They are often incorporated into the impounding structure itself (Fig. 4–2). Where a reservoir serves many purposes, the intake structure is equipped with gates, conduits, and machinery not only for water supply but also for regulation of low-water flows (including compensating water); generation of hydroelectric power; release of irrigation waters; and control of floods. Navigation locks and fish ladders or elevators complete the list of possible control works.

**Submerged and Exposed Intakes.** Submerged intakes are constructed as *cribs* or *screened bellmouths*. Cribs are built of heavy timber weighted down with rocks to protect the intake conduit against damage by waves and ice and to support a grating that will keep large objects out of the central intake pipe.

Exposed intake gatehouses, often still misnamed cribs, are towerlike structures built (1) into dams, (2) on banks of streams and lakes, (3) sufficiently near the shore to be connected to it by bridge or causeway, and (4) at

---

[27] For a full discussion of "Intakes on Variable Streams," see Lischer, V.C. and H. O. Hartung, *J. Am. Water Works Assoc.*, **33**, 873 (1952).

108/Surface Water Collection

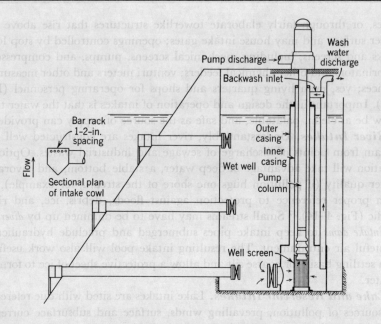

**Figure 4-14.**
**River or lake intake with vertical pump and backwashed well-type screen.**

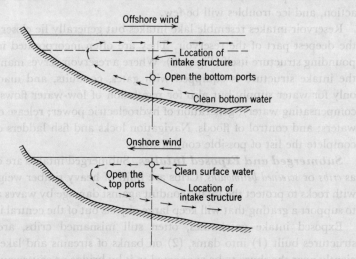

**Figure 4-15.**
**Effect of onshore and offshore winds on water quality at water intake. See Sec. 21-12 for effect on wastewater outlet.**

such distance from shore that they can be reached only by boat (Figs. 4-1 and 4-2). In *dry* intakes, ports in the outer wall admit water to gated pipes that bridge a circumferential dry well and open into a central wet well comprising the entrance to the intake conduit. In *wet* intakes, water fills both wells. Open ports lead to the outer well, whence needed flows are drawn through gated openings into the inner well.

**Intake Velocities and Depths.** In cold climates, ice troubles are reduced in frequency and intensity if intake ports lie as much as 25 ft below the water surface and entrance velocities are kept down to 3 or 4 in. a sec. At such low velocities, ice spicules, leaves, and debris are not entrained in the flowing water and fish are well able to escape from the intake current.

Of the three types of ice encountered in cold climates, sheet ice is seldom troublesome at intakes, but *frazil*[28] and *anchor* ice often are. Depending on the conditions of its formation, frazil ice takes the shape of needles, flakes, or formless slush. According to Barnes,[27] it is a surface-formed ice that does not freeze into a surface sheet. Carried to intakes or produced in them from supercooled water, it obstructs flow by attaching itself to metallic racks, screens, conduits, and pumps or being held back by them. Anchor ice behaves like frazil ice but is derived from ice crystals formed on the bottom of lakes and reservoirs and on submerged objects in much the same way that frost forms on vegetation during a clear night. Frazil and anchor ice are seldom encountered below sheet ice. Their generation can be prevented by heating metallic surfaces or raising the temperature of supercooled water by about 0.1°F, the common order of magnitude of maximum supercooling. Compressed air, backflushing, and light explosives (¼ lb of 60% dynamite) have successfully freed ice-clogged intakes. Telltale chains hung in front of ports will give warning of impending trouble.

Bottom sediments are kept out of intakes by raising entrance ports 4 to 6 ft above the lake or reservoir floor. Ports controlled at numerous horizons permit water-quality selection and optimization. A vertical interval of 15 ft is common. Submerged gratings are given openings of 2 to 3 in. Specifications for screens commonly call for 2 to 8 meshes to the inch and face (approach) velocities of 3 or 4 in. a sec (Sec. 13-2). Wet wells should contain blow-off gates for cleaning and repairs.

**Intake Conduits and Pumping Stations.** Intakes are connected to the shores of lakes and reservoirs (1) by pipelines (often laid with flexible joints) or (2) by tunnels blasted through rock beneath the lake or reservoir floor. Pipelines are generally laid in a trench on the floor and covered after completion. This protects them against disturbance by waves and ice.

---

[28] *Frazil* is a French-Canadian term derived from the French and applied to ice that resembles fine spicular forge cinders. See H. T. Barnes, *Ice Formation*, John Wiley & Sons, New York, 1906.

Except in rock, conduits passing through the foundations of dams are subjected to heavy loads and to stresses caused by consolidation of the foundation.

Intake conduits are designed to operate at self-cleansing velocities—at 3 to 4 fps, therefore. Flow may be by gravity or suction. Pump wells are generally located on shore. Suction lift, including friction, should not exceed 15 to 20 ft. Accordingly, pump wells or rooms are often quite deep. The determining factor is the elevation of the river, lake, or reservoir in times of drought. Placing pumping units in dry wells introduces problems of hydrostatic uplift and seepage in times of flood. Wet wells and deep-well pumps may be used instead.

## 4-17 Diversion Works

Depending on the geology and topography of the dam site and its immediate surroundings, streams are diverted from the construction area in two principal ways:

1. The entire flow is carried around the site in a diversion conduit or tunnel. An upstream cofferdam, and, if necessary, a downstream cofferdam lay the site dry. After fulfilling its duty of bypassing the stream and protecting the valley during construction, the diversion conduit is usually incorporated in the intake or regulatory system of the reservoir (Figs. 4–2 and 4–13).

2. The stream is diverted to one side of its valley, the other side being laid dry by a more or less semicircular cofferdam. After construction has progressed far enough in the protected zone, streamflow is rediverted through a sluiceway in the completed section of the dam, and a new cofferdam is built to pump out the remaining portion of the construction site.

Diversion conduits are built as grade aqueducts and tunnels, or as pressure conduits and tunnels. As a matter of safety, however, it should be impossible for any conduit passing through an earth embankment dam to be put under pressure; a leak might bring disaster. Accordingly, gates should be installed only at the inlet portal, never at the outlet portal. If a pipe must work under pressure, it should be laid within a larger access conduit. To discourage seepage along their outer walls, conduits passing through earth dams or earth foundations are often given projecting fins or collars that increase the length of path of seepage (by, say, 20% or more) and force flow in the direction of minimum as well as maximum permeability. At their terminus near the toe of the dam, moreover, emerging conduits should be surrounded by rock, through which residual seepage waters can escape safely.

The capacity of diversion conduits is determined by flood-flow requirements (Sec. 4–9). Variations in head and volume of floodwater impounded behind the rising dam are important factors in this connection. Rising heads normally increase the capacity of diversion conduits, and increasing storage reduces the intensity of floods. At the same time, however, dangers to the construction site and the valley below mount higher.

## 4-18 Collection of Rainwater

Rain is rarely the immediate source of municipal water supplies,[29] and the use of rain water is generally confined (1) to farms and towns in semiarid regions devoid of satisfactory groundwater or surfacewater supplies, and (2) to some hard-water communities in which, because of its softness, roof drainage is employed principally for household laundry work and general washing purposes, while the public supply satisfies all other requirements. In most hard-water communities, the installation and operation of municipal water-softening plants can ordinarily be justified economically. Their introduction is desirable and does away with the need for supplementary rainwater supplies and the associated objection of their possible cross-connection with the public supply.

For individual homesteads, rainwater running off the roof is led through gutters and downspouts to a rain barrel or cistern situated on the ground or below it (Fig. 1-1). Barrel or cistern storage converts the intermittent rainfall into a continuous supply. For municipal service, roof water may be combined with water collected from sheds or catches on the surface of ground that is naturally impervious or rendered so by grouting, cementing, paving, or similar means.

The gross yield of rainwater supplies is proportional to the receiving area and the amount of precipitation. Some rain, however, is blown off the roof by wind, evaporated, or lost in wetting the collecting area and conduits and in filling depressions or improperly pitched gutters. Also, the first flush of water contains most of the dust and other undesirable washings from the catchment surfaces and may have to be wasted. The combined loss is particularly great during the dry season of the year. A cutoff, switch, or deflector in the downspout permits selecting the quality of water to be stored. Sand filters are successfully employed to cleanse the water and prevents its deterioration (1) by growth of undesirable organisms and (2) by the bacterial decomposition of organic materials, both of which may give rise to tastes, odors, and other changes in the attractiveness and palatability of the water.

Storage to be provided in cisterns depends on seasonal rainfall characteristics and commonly approximates one third to one half the annual needs in accordance with the length of dry spells. If the water is to be filtered before storage, standby capacity in advance of filtration must be provided if rainfalls of high intensity are not to escape. Because of the relatively small catchment area available, roof drainage cannot be expected to yield an abundant supply of water, and a close analysis of storm rainfalls and seasonal variations in precipitation must be made if catchment areas, stand-by tanks, filters, and cisterns are to be proportioned and developed properly.

[29] A notable example is the water supply of the communities on the Islands of Bermuda, on which streams are lacking and groundwater is brackish.

five

# groundwater development

## 5-1 Introduction

Wells and springs have served as sources of domestic water supply since antiquity. Groundwater accounts for one sixth to one fifth of the total withdrawal requirements in the United States. Estimates of groundwater use in 1960 show that the total amount of groundwater withdrawn was 46.3 bgd (billion gallons per day) distributed as follows: irrigation, 30.2 bgd; industrial uses, 7.3 bgd; public water supplies, 6.3 bgd; and rural uses, 2.8 bgd.[1] Almost 80% of the rural domestic and stock supplies were provided by subsurface water. Of the total water withdrawn by public water supplies (20.3 bgd), about one third came from groundwater sources. Approximately one third (46 million) of the total population served by public water systems depended on groundwater. Groundwater works are many times more numerous than surfacewater installations; the average capacity of groundwater facilities is, however, much smaller. Contributions from groundwater also play a major role in the supplies depending on surface sources. It is the discharge of groundwater that sustains the dry-season flow of most streams.

[1] K. A. Mackichan and L. C. Kammerer, Estimated Use of Water in the United States, 1960, *U.S. Geological Survey Circular 456* (1961).

Groundwater is more widely distributed than surfacewater. Its nearly universal, albeit uneven, occurrence and other desirable characteristics make it an attractive source of water supply. Groundwater offers a naturally purer, cheaper and more satisfactory supply than do surfacewaters. It is generally available at the point of use and obviates the need of incurring substantial transmission costs. It occurs as an underground reservoir, thus eliminating the necessity of impoundment works. It is economical even when produced in small quantities.

To an increasing degree, engineers are being called upon to investigate the possibility of developing groundwater as a usable resource. The following factors need consideration:

1. The effective water content, that is, the maximum volume of water that can be withdrawn from a body of groundwater through engineering works. (Effective porosity and storage coefficient of the water-bearing material control the useful storage).

2. The ability of the aquifer to transmit water in requisite quantities to wells or other engineering installations (permeability and transmissivity are the indicators of this capability).

3. The suitability of the quality of water for the intended use, after treatment if necessary.

4. The reliability and permanence of the available supply with respect to both the quantity and the quality of water.

As a source of permanent and reliable water supply, only that portion of the subsurface water that is in the zone of saturation need be considered. In this zone almost all the interstices are completely filled with water under hydrostatic pressure (atmospheric pressure or greater) and is free to move in accordance with the laws of saturated flow from places where it enters the zone of saturation (recharge areas) to places where it is discharged. The main features of the groundwater phase of the hydrologic cycle are depicted in Figure 5-1.

## 5-2 Porosity and Effective Porosity

The amount of groundwater stored in saturated materials depends on the material's *porosity*, the ratio of the aggregate volume of interstices in a rock or soil to its total volume. It is usually expressed as a percentage. The concept of porosity involves all types of interstices, both primary (original) and secondary. Primary interstices were created at the time of the origin of the rock. In granular unconsolidated sediments, they coincide with intergranular spaces. In volcanic rocks, they include tubular and vesicular openings. Secondary interstices result from the action of geological, mechanical, and chemical forces on the original rock. They include joints, faults, fissures,

# 114/Groundwater Development

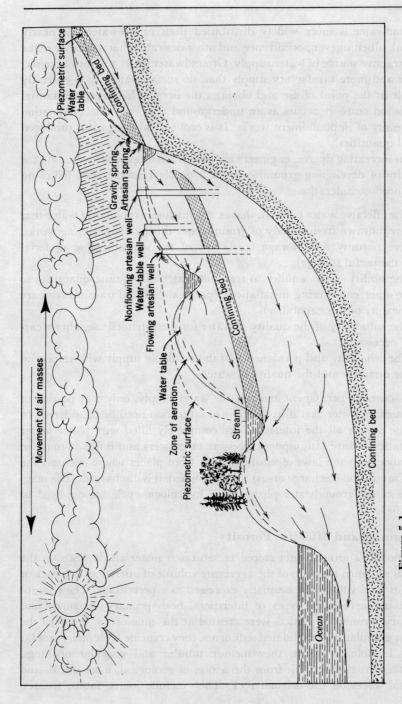

Figure 5-1. Groundwater features of the hydrologic cycle. (After McGuinness, footnote 8.)

solution channels, and bedding planes in hard rocks. The extent of fracturing and intensity of weathering exert a profound influence on the distribution of larger interstices. The importance of secondary porosity in determining the amount of water that can be obtained from a formation is often great in those hard rocks that lack intergrain porosity. This type of porosity is dependent on local conditions and gives water-bearing formations a heterogeneous character. The distribution of secondary porosity varies markedly with depth.

Porosity is a static quality of rocks and soils. It is not itself a measure of *perviousness* or *permeability*, which are dynamic quantities controlling the flow. Not all the water stored in a saturated material is available for movement. It is only the interconnected interstices that can participate in flow. Water in isolated openings is held immobile. Furthermore, water in a part of the interconnected pore space is held in place by molecular and surface-tension forces. This is the dead storage and is called *specific retention*. Thus not all the water stored in a geological formation can be withdrawn by normal engineering operations. Accordingly, there is a difference between total storage and useful storage. That portion of the pore space in which flow takes place is called *effective porosity*, or *specific yield* of the material, defined as the proportion of water in the pores that is free to drain away or be withdrawn under the influence of gravity. Specific yields vary from zero for plastic clays to 30% or more for uniform sands and gravels. Most aquifers have yields of 10 to 20%.

## 5-3 Permeability

The *permeability* or *perviousness* of a rock is its capacity for transmitting a fluid under the influence of a hydraulic gradient. Important factors affecting the permeability are the geometry of the pore spaces and of the rock particles. The nature of the system of pores, rather than their relative volume, determines the resistance to flow at given velocities. There is no simple and direct relationship between permeability and porosity. Clays with porosities of 50% or more have extremely low permeability; sandstones with porosities of 15% or less may be quite pervious.

A standard unit of intrinsic permeability, dependent only on the properties of the medium, is the *darcy*.[2] It is expressed as flow in cubic centimeters per second, of a fluid of one centipoise viscosity, through a cross-sectional area of one square centimeter of the porous medium under a pressure gradient of one atmosphere per centimeter. It is equivalent to a flow of 18.2 gallons of

[2] Named for Henri Philibert Gaspard Darcy (1803–1858), French engineer, member of the Corps des Ponts et Chaussees stationed at his native Dijon, *Les fontaines publiques de la Ville de Dijon* (The public wells of the city of Dijon), V. Dalmont, Paris, 1856; English translation by J. J. Fried, *Water Resources Bull.*, Am. Water Resources Assoc., **1**, 4 (1965).

water per day per square foot under a hydraulic gradient of 1 foot per foot at a temperature of 60°F.

The *homogeneity* and *isotropy* of a medium refer to the spatial distribution of permeability. A porous medium is isotropic if its permeability is the same in all directions. It is called *anisotropic* if the permeability varies with the direction. Anisotropy is common in sedimentary deposits where the peremability across the bedding plane may be only a fraction of that parallel to the bedding plane. The medium is *homogeneous* if the permeability is constant from point to point over the medium. It is *nonhomogeneous* if the permeability varies from point to point in the medium. Aquifers with secondary porosity are nonhomogeneous. Isotropy and homogeneity are often assumed in the analysis of groundwater problems. The effects of nonhomogeneity and anisotropy can, however, be incorporated into analysis under certain conditions.

**Classification of Rocks on the Basis of Permeability.** Rocks may be grouped into hydrologic units on the basis of their ability to store and transmit water. An *aquifer*[3] is a body of rock that acts as a hydrologic unit and is capable of transmitting significant quantities of water. An aquiclude[4] is a rock formation that contains water but is not capable of transmitting it in significant amounts. Aquicludes usually form the boundaries of aquifers, although they are seldom absolute barriers to groundwater movement. They often contain considerable water in storage, and there is frequently some interchange between the free groundwater above an aquiclude and the confined aquifer below. Materials that have permeabilities intermediate between those of aquifers and aquicludes have been termed aquitards.[5]

The boundaries of a geologic rock unit and the dimensions of an aquifer often do not correspond precisely. The latter are arrived at from the considerations of the degree of hydraulic continuity and from the position and character of hydrologic boundaries. An aquifer can thus be a geologic formation, a group of formations, or part of a formation.

## 5-4 Groundwater Geology

The geological framework of an area provides the most valuable guide to the occurrence and availability of groundwater. Rocks, the solid matter forming the earth's crust, are an assemblage of minerals. In the geologic sense the term rock includes both the hard, consolidated formations and loose, unconsolidated materials. With respect to their origin, they fall into three broad categories: *igneous*, *metamorphic*, and *sedimentary*.

The two classes of igneous rocks, *intrusive* and *extrusive*, differ appreciably in their hydrologic properties. Fresh intrusive rocks are compact and, in

---

[3] The word aquifer comes from Latin *aqua*, water, and *ferre*, to bear.
[4] The word aquiclude comes from the Latin *aqua*, and *cludere*, shut or close (out).
[5] From *aqua*, and *tardus*, slow.

general, not water-bearing. They have very low porosities (less than 1%) and are almost impermeable. When fractured and jointed they may develop appreciable porosity and permeability within a few hundred feet of the surface. Permeability produced by fracturing of unweathered rocks generally ranges from 0.001 to 10.0 darcy.[6] Extrusive or volcanic rocks can be good aquifers.

Metamorphic rocks are generally compact and highly crystalline. They are impervious and make poor aquifers.

Sedimentary formations include both consolidated, hard rocks (shale, sandstone, and limestone) and loose, unconsolidated materials (clay, gravel, and sand). Some sandstones may be almost impermeable, and others highly pervious. The degree of cementation plays a crucial role. Partially cemented or fractured sandstones have very high yields. Porosity of sandstones ranges from less than 5% to a maximum of about 30%. Permeability of medium-range sandstones generally varies from 1 to 500 millidarcy.

Limestones vary widely in density, porosity, and permeability. When not deformed, they are usually dense and impervious. From the standpoint of water yield, secondary porosity produced as a result of fracturing and solution is more important. The nonuniform distribution of interstices in limestones over even short distances results because of marked differences in secondary porosity, which depends on local conditions. They are second only to sandstones as a source of groundwater. Limestones are prolific producers under suitable conditions.

Although consolidated rocks are important sources of water, the areas served by them in the United States are relatively small. Most developments lie in granular, unconsolidated sediments. Unconsolidated, sedimentary aquifers include (a) marine deposits, (b) river valleys, (c) alluvial fans, (d) coastal plains, (e) glacial outwash, and, to a much smaller degree, (f) dune sand. Materials deposited in seas are often extensive; sediments deposited on land by streams, ice, and wind are less extensive and are usually discontinuous.

*Sands* and *gravels* are by far the best water-producing sediments. They have excellent water storage and transmission characteristics and are ordinarily so situated that replenishment is rapid, although extremely fine sands are of little value. Porosity, specific yield, and permeability depend on particle size, size distribution, packing configuration, and shape. Uniform or well-sorted sands and gravels are the most productive; mixed materials containing clay are least so. Boulder clay deposited beneath ice sheets is an example. Typical porosities lie between 25 and 65%. Gravel and coarse sands usually have specific yields greater than 20%.

[6] S. N. Davis and R-J. M. DeWiest, *Hydrogeology*, John Wiley and Sons, 1966, p. 320.

*Clays* and *silts* are poor aquifers. They are highly porous but have very low permeabilities. However, the permeability is seldom zero. They are significant only when they (a) confine or impede the movement of water through more pervious soils and (b) supply water to aquifers through leakage by consolidation.

## 5-5 Groundwater Situation in the United States

Geologic and hydrologic conditions vary greatly in various parts of the United States. To permit useful generalizations about the occurrence and availability of groundwater, Thomas[7] has divided the U.S. into ten major groundwater regions (Fig. 5-2). McGuiness[8] has provided an updated assessment of the groundwater situation in each of Thomas' regions and has also described the occurrence and development of groundwater in each of the states.

The Water Resources Division of the U.S. Geological Survey is the principal agency of the federal government engaged in groundwater investigations. The published reports and the unpublished data of the Division are indispensable to any groundwater investigation. In addition, many states have agencies responsible for activities in groundwater.

## 5-6 Types of Aquifers

Because of the differences in the mechanism of flow, three types of aquifers are distinguished: (1) *unconfined* or *water table*, (2) *confined* or *artesian*, and (3) *semiconfined* or *leaky*.

Unconfined aquifers (also known as water-table, *phreatic*, or *free* aquifers) are those in which the upper surface of the zone of saturation is under atmospheric pressure. This surface is free to rise and fall in response to the changes of storage in the saturated zone. The flow under such conditions is said to be unconfined. An imaginary surface connecting all rest or static levels in wells in an unconfined aquifer is its *water table* or *phreatic surface*. This defines the level in the zone of saturation, which is at atmospheric pressure. The water held by capillary attraction at less than atmospheric pressure may fully saturate the interstices to levels above those observed in wells. Thus the upper limit of the zone of saturation and water table are not coincident. The capillary fringe may be significant for sediments with small interstices and low permeability, such as clay.

---

[7] H. E. Thomas, Groundwater Regions of the United States—Their Storage Facilities, in U.S. 83rd Congress, House Interior and Insular Affairs Comm., *The Physical and Economic Foundation of Natural Resources*, Vol. 3 (1952).

[8] C. L. McGuiness, The Role of Groundwater in the National Water Situation, in *U.S. Geol. Survey Water Supply Paper 1800* (1963).

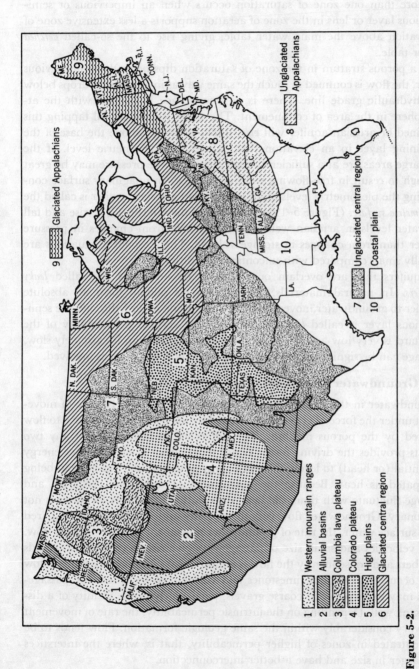

**Figure 5-2.**
Major groundwater regions of the United States, excepting Alaska and Hawaii. (After Thomas and McGuinness, footnotes 7 and 8, as illustrated in *Ground Water and Wells*, footnote 34.)

More than one zone of saturation occurs when an impervious or semipervious layer or lens in the zone of aeration supports a less extensive zone of saturation above the main water table, giving rise to the so-called *perched* water table.

If a porous stratum in the zone of saturation dips beneath an impervious layer, the flow is confined in much the same way as in a pipe that drops below the hydraulic grade line. There is no free surface in contact with the atmosphere in the area of confinement. The water level in a well tapping this confined or artesian aquifer will rise, under pressure, above the base of the confining layer to an elevation that defines the piezometric level. If the recharge areas are at a sufficiently high elevation, the pressure may be great enough to result in free-flowing wells or springs. An imaginary surface connecting the piezometric levels at all points in an artesian aquifer is called the *piezometric surface*. (Figure 5–1 depicts some of these terms.) The rise and fall of water levels in artesian wells result primarily from changes in pressure rather than from changes in storage volume. The seasonal fluctuations are usually small compared with unconfined conditions.

Aquifers that are overlain or underlain by aquitards are called *leaky aquifers*. In natural materials, confining layers seldom form an absolute barrier to groundwater movement. The magnitude of flow through the semipervious layer is called *leakage*. Although the vertical permeability of the aquitard is very low and the movement of water through it extremely slow, leakage can be significant because of the large horizontal areas involved.

## 5-7 Groundwater Movement

Groundwater in the natural state is constantly in motion. Its rate of movement under the force of gravity is governed by the frictional resistance to flow offered by the porous medium. The difference in head between any two points provides the driving force. Water moves from levels of higher energy potential (or head) to levels of lower energy potential, the loss in head being dissipated as heat. Because the magnitudes of discharge, recharge, and storage fluctuate with time, the head distribution at various locations is not stationary. Groundwater flow is both unsteady and nonuniform. Compared with surface water, the rate of groundwater movement is generally very slow. Low velocities and small size of passageways give rise to very low Reynolds numbers and consequently the flow is almost always laminar. Turbulent flow may occur in cavernous limestones and volcanic rocks, where the passageways may be large, or in coarse gravels, particularly in the vicinity of a discharging well. Depending on the intrinsic permeability, the rate of movement can vary considerably within the same geologic formation. Flow tends to be concentrated in zones of higher permeability, that is, where the interstices are larger in size and have a better interconnection.

In aquifers of high yield, velocities of 5 to 60 ft per day are associated with hydraulic gradients of 10 to 20 ft per mile. Underflow through gravel deposits may travel several hundred feet per day. Depending on requirements, flows as low as a few feet per year may also be economically useful.

In homogeneous, isotropic aquifers, the dominant movement is in the direction of greatest slope of water table or piezometric surface. Where there are marked nonhomogeneities and anisotropies in permeability, the direction of groundwater movement can be highly variable.

## 5-8 Darcy's Law

Although Hagen[9] and Poiseuille[10] were the first to propose that the velocity of flow of water and other liquids through capillary tubes is proportional to the first power of the hydraulic gradient, credit for verification of this observation and for its application to the flow of water through natural materials or, more specifically, its filtration through sand, must go to Darcy. The relationship known as Darcy's law may be written

$$v = K(dh/dl) = KI \qquad (5\text{-}1)$$

where $v$ is the hypothetical or face velocity through the gross cross-sectional area of the porous medium, $I = dh/dl$ is the hydraulic gradient, or the loss of head per unit length in the direction of flow, and $K$ is a constant of proportionality known as *hydraulic conductivity*, or the *coefficient of permeability*. The actual velocity, known as *effective velocity*, varies from point to point. The average velocity through pore space is given by

$$v_e = KI/\theta \qquad (5\text{-}2)$$

where $\theta$ is the effective porosity. Because $I$ is a dimensionless ratio, $K$ has the dimensions of velocity and is in fact the velocity of flow associated with a hydraulic gradient of unity.

The proportionality coefficient in Darcy's law, $K$, refers to the characteristics of both the porous medium and the fluid. By dimensional analysis,

$$K = Cd^2\gamma/\mu \qquad (5\text{-}3)$$

where $C$ is a dimensionless constant summarizing the geometric properties of the medium affecting flow, $d$ is a representative pore diameter, $\mu$ is the

---

[9] Gotthilf Heinrich Ludwig Hagen (1797–1874), Ueber die Bewegung des Wassers in engen cylindrischen Röhren (On the flow of water in narrow cylindrical tubes), *Ann. Physik und Chemie*, 46, 423 (1839).

[10] Jean Louis Poiseuille (1799–1869) (French physician interested in the flow of fluid through arteries and veins), Recherches experimentales sur le mouvement des liquides dans les tubes de tres petits diametres (Experimental Investigations of the Flow of Liquids in Tubes of Very Small Diameter), *Roy. Acad. Sci. Inst., France Math. Phys. Sci., Mem.* 9, 433 (1846). Translated by W. H. Herschel, Easton, Pa., 1840.

viscosity, and $\gamma$ is the specific weight of fluid. The product $Cd^2$ depends on the properties of the medium alone and is called the intrinsic or specific permeability of a water-bearing medium; $k = Cd^2$. It has the dimensions of area.

Hence,

$$K = k\gamma/\mu = k\rho g/\mu = kg/\nu \qquad (5-4)$$

where $\rho$ is the specific density and $\nu$, the kinematic viscosity.

The fluid properties that affect the flow are viscosity and specific weight. The value of $K$ varies inversely as the kinematic viscosity, $\nu$, of the flowing fluid (Table 3, Appendix). The ratio of specific weight to viscosity is affected by changes in temperature and salinity of groundwater. Measurements of $K$ are generally referred to a standard water temperature such as 60°F or 10°C. The necessary correction factor for field temperatures other than standard is provided by the relationship

$$K_1/K_2 = \nu_1/\nu_2 \qquad (5-5)$$

Most ground waters have relatively constant temperatures, and this correction is usually ignored in practice and $K$ is stated in terms of the prevailing water temperature. Special circumstances in which correction may be important include influent seepage into an aquifer from a surface water body where temperature varies seasonally.

Darcy's law is applicable only to laminar flow, and there is no perceptible lower limit to the validity of the law.

The volume rate of flow is the product of the velocity given by Darcy's law and the cross-sectional area $A$ normal to the direction of motion. Thus

$$Q = KA(dh/dl) \qquad (5-6)$$

and solving for $K$,

$$K = Q/[A(dh/dl)] \qquad (5-7)$$

Hydraulic conductivity may thus be defined as the volume of water per unit time flowing through a medium of unit cross-sectional area under a unit hydraulic gradient. In the standard coefficient used by the U.S. Geological Survey, the rate of flow is expressed in gpd per square foot under a hydraulic gradient of 1 ft per ft at a temperature of 60°F. This unit is called the meinzer.[11] For most natural aquifer materials, values of $K$ fall in the range of 10 to 5000 meinzers.

[11] After O. G. Meinzer, a noted American groundwater hydrologist.

## 5-9 Aquifer Characteristics

The ability of an aquifer to transmit water is characterized by its *coefficient of transmissivity*. It is the product of the saturated thickness of the aquifer, $b$, and the average value of the hydraulic conductivities in a vertical section of the aquifer, $K$. The transmissivity, $T = Kb$, gives the rate of flow of water through a vertical strip of the aquifer 1 ft wide extending the full saturated thickness of the aquifer under a unit hydraulic gradient. It has the dimensions of (length)$^2$/time, that is, ft$^2$/day, or gpd/ft. Equation 5–6 can be rewritten as

$$Q = TW(dh/dl) \qquad (5\text{–}8)$$

where $W$ is the width of flow.

The *coefficient of storage* is defined as the volume of water that a unit decline in head releases from storage in a vertical prism of the aquifer of unit cross-sectional area[12,13] (Fig. 5–4).

The physical processes involved when the water is released from (or taken into) storage in response to head changes are quite different in cases in which free surface is present from those in which it is not. A confined aquifer

---

[12] C. V. Theis, The Relation Between the Lowering of the Piezometric Surface and the Rate and Duration of Discharge of a Well Using Ground Storage, *Trans. Am. Geophys. Union*, **16**, 519 (1935).

[13] C. E. Jacob, On the Flow of Water in an Elastic Artesian Aquifer, *Trans. Am. Geophys. Union*, **21**, 574 (1940).

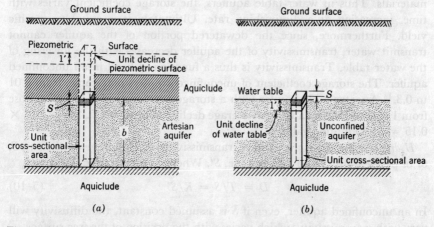

**Figure 5-3.**
Graphical representation of storage coefficient, the volume of water that a unit decline in head releases from storage in a vertical prism of the aquifer of unit cross-sectional area. (a) Confined aquifer; (b) unconfined aquifer.

remains saturated during the withdrawal of water. In the case of a confined aquifer the water is released from storage by virtue of two processes: (a) lowering of the water table in the recharge or intake area of the aquifer and (b) elastic response to pressure changes in the aquifer and its confining beds induced by the withdrawal of water. For this Jacob[14] expresses the storage coefficient as

$$S = \theta \gamma b [\beta + (\alpha/\theta)] \qquad (5\text{-}9)$$

in which $\theta$ is the average porosity of the aquifer; $\gamma$, the specific weight of water; $\beta$, compressibility of water; and $\alpha$, the vertical compressibility of aquifer material. In most confined aquifers, storage coefficient values lie in the range 0.00005 and 0.0005. These values are small and thus large pressure changes over extensive areas are required to develop substantial quantities of water.

A confined aquifer for which $S$ in Eq. 5-9 is $3 \times 10^{-4}$ will release from 1 square mile 64,125 gallons by lowering the piezometric surface by 1 ft.

A water-table aquifer also releases water from storage by two processes: (a) dewatering or drainage of material at the free surface as it moves downwards, and (b) elastic response of the material below the free surface. In general, the quantity released by elastic response is very small as compared to the dewatering of the saturated material at the water table. Thus the storage coefficient is virtually equal to the specific yield of the material. In unconfined aquifers, the full complement of storage is usually not released instantaneously. The speed of drainage depends on the types of aquifer materials. Thus in water-table aquifers, the storage coefficient varies with time, increasing at a diminishing rate. Ultimately it is equal to specific yield. Furthermore, since the dewatered portion of the aquifer cannot transmit water, transmissivity of the aquifer decreases with the lowering of the water table. Transmissivity is thus a function of head in an unconfined aquifer. The storage coefficient of unconfined aquifers may range from 0.01 to 0.3.[15] A water-table aquifer with a storage coefficient of 0.15 will release from 1 square mile area with an average decline in head of 1 ft, $209 \times 10^6 \times 0.15 = 31.35$ million gallons.

*Hydraulic diffusivity* is the ratio of transmissivity, $T$, to storage coefficient, $S$, or of permeability, $K$ to unit storage, $S'$. Where $D$ is hydraulic diffusivity,

$$D = T/S = K/S' \qquad (5\text{-}10)$$

In an unconfined aquifer, even if $S$ is assumed constant, the diffusivity will vary with transmissivity, which varies with the position of the free surface.

[14] C. E. Jacob, Radial Flow in a Leaky Artesian Aquifer, *Trans. Am. Geophys. Union*, **27**, 2, pp. 198–205 (1946).

[15] R. C. Heath and F. W. Trainer, *Introduction to Ground Water Hydrology*, John Wiley and Sons (1968).

The conductivity, the transmissivity, the storage coefficient, and the specific yield are usually referred to as *formation constants*, and provide measures of the hydraulic properties of aquifers.

**Measurement of Hydraulic Conductivity.** The capacity of an aquifer to transmit water can be measured by several methods: (1) laboratory tests of aquifer samples, (2) tracer techniques, (3) analysis of water-level maps, and (4) aquifer tests. The laboratory measurements of hydraulic conductivity are obtained by using samples of aquifer material in either a constant-head or a falling-head permeameter. Undisturbed core samples are used in the case of well-consolidated materials, and repacked samples in the case of unconsolidated materials. Observations are made of the time taken for a known quantity of water under a given head to pass through the sample. The application of Darcy's law enables hydraulic conductivity to be determined. The main disadvantage of this method arises from the fact that the values obtained are point measurements. Aquifers are seldom, if ever, truly homogeneous throughout their extent, and laboratory measurements are not representative of actual "in place" values. Most samples of the material are taken in a vertical direction, whereas the dominant movement of water in the aquifer is nearly horizontal, and horizontal and vertical permeabilities differ markedly. Also, some disturbance is inevitable when the sample is removed from its environment. This method cannot, therefore, be used to give a reliable quantitative measure of hydraulic conductivity.

The measurement of hydraulic conductivity in undisturbed natural materials can be made by measurement of hydraulic gradient and determination of the speed of groundwater movement through the use of tracers. A tracer (dye, electrolyte, or radioactive substance) is introduced into the groundwater through an injection well at an upstream location, and measurements are made of the time taken by the tracer to appear in one or more downstream wells. Uranin, a sodium salt of fluorescein, is an especially useful dye because it remains visible in dilutions of $1:(14 \times 10^7)$ without a fluoroscope and $1:10^{10}$ with one. Tritium has been used as a radioactive tracer. The time of arrival is determined by visual observation or colorimetry when dyes are added, by titration or electrical conductivity when salt solutions are injected, or by a Geiger or scintillation counter when radioactive tracers are used. The distance between the wells divided by the time required for half the recovered substance to appear is the median velocity. The observed velocity is the actual average rate of motion through the interstices of the aquifer material. The face velocity can be calculated, if effective porosity is known. The application of Darcy's law enables the hydraulic conductivity to be computed. The problems of direction of motion, dispersion and molecular diffusion, and the slow movement of groundwater limit the applicability of this method. The method is impractical for a heterogeneous aquifer that has large variations in horizontal and vertical hydraulic conductivity.

The drop in head between two equipotential lines in an aquifer divided by the distance traversed by a particle of water moving from a higher to a lower potential determines the hydraulic gradient. Changes in the hydraulic gradient may arise from either a change in flow rate, $Q$, hydraulic conductivity, $K$, or aquifer thickness, $b$ (Eq. 5-5). If no water is being added to or lost from an aquifer, the steepening of the gradient must be due to lower transmissivity, reflecting either a lower permeability, a reduction in thickness, or both (Eq. 5-8).

Of the presently available methods for the estimation of formation constants, aquifer tests (also called pumping tests) are the most reliable. The mechanics of a test involves the pumping of water from a well at a constant discharge and the observation of water levels in observation wells at various distances from the pumping well at different time intervals after pumping commences. The analysis of a pumping test comprises the graphical fitting of the various theoretical equations of groundwater flow to the observed data. The mathematical model giving the best fit is used for the estimation of formation constants. The main advantages of this method are that the sample used is large and remains undisturbed in its natural surroundings. The time and expense are reasonable. The main disadvantage of the method concerns the number of assumptions that must be made when applying the theory to the observed data. Despite the restrictive assumptions, pumping tests have been successfully applied under a wide range of conditions actually encountered. The theory of aquifer tests is given in Section 5-12.

## 5-10 Well Hydraulics

Well hydraulics deals with predicting yields from wells and in forecasting the effects of pumping on groundwater flow and on the distribution of potential in an aquifer. The response of an aquifer to pumping depends on the type of aquifer (confined, unconfined, or leaky), aquifer characteristics (transmissivity, storage coefficient, and leakage), aquifer boundaries, and well construction (size, type, whether fully or partially penetrating) and well operation (constant or variable discharge, continuous or intermittent pumping).

The first water pumped from a well is derived from aquifer storage in the immediate vicinity of the well. Water level (that is, piezometric surface or water table), is lowered and a *cone of depression* is created. The shape of the cone is determined by the hydraulic gradients required to transmit water through aquifer material toward the pumping well. The distance through which the water level is lowered is called the *drawdown*. The outer boundary of the drawdown curve defines the *area of influence* of the well. As pumping is continued, the shape of the cone changes as it travels outward from the well.

This is the dynamic phase, in which the flow is *time-dependent* (*nonsteady*), and both the velocities and water levels are changing. With continued withdrawals, the shape of the cone of depression stabilizes near the well and, with time, this condition progresses to greater distances. Thereafter the cone of depression moves parallel to itself in this area. This is the depletion phase. Eventually the drawdown curve may extend to the areas of natural discharge or recharge. A new state of equilibrium is reached if the natural discharge is decreased or the natural recharge is increased by an amount equal to the rate of withdrawal from the well. A *steady state* is then reached and the water level ceases to decline.

## 5-11 Nonsteady Radial Flow

Theis,[12] Jacob,[13,14] and Hantush[16,17,18] have developed solutions for nonsteady radial flow toward a discharging well. The methods of pumping test analysis for the determination of aquifer constants are based on solutions of unsteady radial flow equations.[19,20,21,22]

*Confined Aquifers.* In an effectively infinite artesian aquifer, the discharge of a well can only be supplied through a reduction of storage within the aquifer. The propagation of the area of influence and the rate of decline of head depend on the hydraulic diffusivity of the aquifer. The differential equation governing nonsteady radial flow to a well in a confined aquifer is given by

$$\partial^2 h/\partial r^2 + (1/r)(\partial h/\partial r) = (S/T)(\partial h/\partial t) \qquad (5\text{--}11)$$

Using an analogy to the flow of heat to a sink, Theis derived an expression for the drawdown in a confined aquifer due to the discharge of a well at a constant rate. His equation is really a solution of the Equation 5–11 based on the following assumptions: (a) the aquifer is homogeneous, isotropic, and of infinite areal extent; (b) transmissivity is constant with respect to time and

---

[16] M. S. Hantush and C. E. Jacob, Nonsteady Radial Flow in an Infinite Leaky Aquifer, *Trans. Am. Geophys. Union*, **36**, (1955).

[17] M. S. Hantush, Nonsteady Flow to Flowing Wells in Leaky Aquifers, *J. Geophys. Res.*, **64**, 1043 (1959).

[18] M. S. Hantush, Modification of the Theory of Leaky Aquifers, *J. Geophys. Res.*, **65**, 3713 (1960).

[19] J. G. Ferris, D. B. Knowles, R. H. Brown, and R. W. Stallman, Theory of Aquifer Tests, *U.S. Geol. Surv. Water Supply Paper 1536-E* (1962).

[20] W. C. Walton, Selected Analytical Methods for Well and Aquifer Evaluation, *Illinois State Water Survey, Bull.* **49** (1962).

[21] W. C. Walton, Leaky Artesian Aquifer Conditions in Illinois, *Illinois State Water Survey Report Invest.*, **39**, 1960.

[22] R. H. Brown, Selected Procedures for Analyzing Aquifer Test Data, *J. Am. Water Works Assn.*, **45** pp. 844–866 (1953).

space; (c) water is derived entirely from storage, being released instantaneously with the decline in head; (d) storage coefficient remains constant with time; and (e) the well penetrates, and receives water from, the entire thickness of the aquifer. The Theis equation may be written as

$$s = h_0 - h = \frac{Q}{4\pi T} \int_{r^2 S/4Tt}^{\infty} \frac{e^{-u}}{u} du \qquad (5\text{-}12)$$

where $h$ is the head at a distance $r$ from the well at a time $t$ after the start of pumping; $h_0$ is the initial head in the aquifer prior to pumping; $Q$ is the constant discharge of the well; $S$ is the storage coefficient of the aquifer; and $T$ is the transmissivity of the aquifer. The integral in the above expression is known as the exponential integral and is a function of its lower limit. In groundwater literature, it is written symbolically as $W(u)$, which is read "well function of $u$" where

$$u = (r^2 S)/(4Tt) \qquad (5\text{-}13)$$

Its value can be approximated by a convergent infinite series

$$W(u) = -0.5772 - \ln u + u - u^2/2 \times 2! + u^3/3 \times 3! \ldots \qquad (5\text{-}14)$$

Values of $W(u)$ for a given value of $u$ are tabulated in numerous publications. A partial listing is given in the Appendix, Table 7.

The drawdown $s$ (ft), at a distance $r$ (ft), at time, $t$ (days) after the start of pumping for a constant discharge $Q$ (gpm), is given by

$$s = h_0 - h = 1440 \, QW(u)/(4\pi T) = 114.6 \, QW(u)/T \qquad (5\text{-}15)$$

where

$$u = 1.87 \, (S/T)(r^2/t) \qquad (5\text{-}16)$$

and $T$ is transmissivity in gpd/ft.

The equation can be solved for any one of the quantities involved if other parameters are given. The solution for drawdown, discharge, distance from the well, or time is straightforward. The solution for transmissivity, $T$, is difficult, since it occurs both inside and outside the integral. Theis devised a graphical method of superposition to obtain a solution of the equation for $T$ and $S$.

If the discharge $Q$ is known, the formation constants of an aquifer can be obtained as follows: (1) plot the field or data curve with drawdown, $s$, as ordinate and $r^2/t$ as abscissa on logarithmic coordinates on a transparent paper; (2) plot a "type curve" of the well function, $W(u)$, as ordinate and its argument $u$ as abscissa on logarithmic coordinates and the same scale as field curve; (3) superimpose the curves shifting vertically and laterally, keeping the coordinate axes parallel until most of the plotted points of the observed data fall on a segment of the type curve; (4) select a convenient matching point anywhere on the overlapping portion of the sheets and record the coordinates of this common point on both graphs; (5) use the two ordinates, $s$ and $W(u)$ to obtain the solution for transmissivity, $T$, from Eq. 5-15; and (6) use the two abscissas, $r^2/t$ and $u$, together with the value of $T$, to obtain the solution for the storage coefficient $S$, from Eq. 5-16.

Values of $s$ are related to the corresponding values of $W(u)$ by the constant factor $114.6\, Q/T$, whereas values of $r^2/t$ are related to corresponding values of $u$ by the constant factor $T/(1.87S)$. Thus when the two curves are superimposed, corresponding vertical axes are separated by a constant distance proportional to $\log[114.6\,(Q/T)] = \log C_1$, whereas the corresponding horizontal axes are separated by a constant distance proportional to $\log[T/(1.87\,S)] = \log C_2$ as shown in Fig. 5-4.

*Example 5-1.*

The observed data from a pumping test are shown plotted in Fig. 5-4 along with a Theis-type curve, as if the transparency of the observed data had been moved into place over the type curve. The observation well represented by the data is 225 ft from a pumping well where rate of discharge is 350 gpm. The match-point coordinates are: $W(u) = 4.0$, $s = 5.0$ ft, $u = 10^{-2}$, and $r^2/t = 5 \times 10^6$. Compute the formation constants:

$$T = 114.6\, QW(u)/s = 114.6 \times 350 \times 4.0/5.0 = 3.2 \times 10^4 \text{ gpd/ft}$$
$$S = uT/(1.87\, r^2/t) = 10^{-2}(3.2 \times 10^4)/[1.87\,(5 \times 10^6)] = 3.4 \times 10^{-5}$$

**Semilogarithmic Approximation.** Jacob[23] recognized that when $u$ is small, the sum of the terms beyond $\ln u$ in the series expansion of $W(u)$, Eq. 5-14, is relatively insignificant. The Theis equation (Eq. 5-12) then reduces to

$$s = [Q/(4\pi T)]\{\ln[(4Tt)/(r^2S)] - 0.5772\}$$
$$= [Q/(4\pi T)]\{\ln[(2.25\, Tt)/(r^2S)]\} \qquad (5\text{-}17)$$

When $Q$ is in gpm, $T$ in gpd/ft, $t$ in days, and $r$ in ft, the equation becomes

$$s = [264(Q/T)]\{\log[(0.3\, Tt)/(r^2S)]\} \qquad (5\text{-}18)$$

[23] C. E. Jacob, Drawdown Test to Determine Effective Radius of Artesian Well, *Trans. Am. Soc. Civil Engineers*, 112, 1047 (1947).

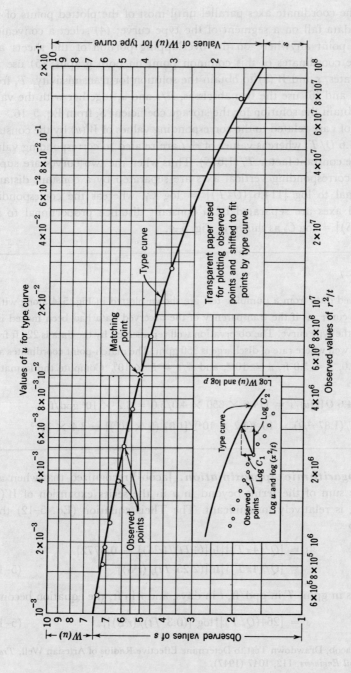

Figure 5-4. Theis type-curve determination of the formation constants of a well field. (Data by courtesy of the U.S. Geological Survey.)

Cooper and Jacob[24] proposed a graphical solution of this equation. If the drawdown is measured in a particular observation well (fixed $r$) at several values of $t$, the equation becomes

$$s = 264 \ (Q/T) \log (Ct)$$

where

$$C = 0.3 \ T/(r^2 S)$$

If on a semilogarithmic paper, the values of drawdown are plotted on the arithmetic scale and time on logarithmic scale, the resulting graph should be a straight line for higher values of $t$ where the approximation is valid. The graph is referred to as the time-drawdown curve. On this straight line an arbitrary choice of times $t_1$ and $t_2$ can be made and the corresponding values of $s_1$ and $s_2$ recorded. Inserting these values in Eq. 5–18, we obtain

$$s_2 - s_1 = 264 \ (Q/T) \log (t_2/t_1) \qquad (5\text{–}19)$$

Solving for $T$,

$$T = 264 \ Q \log (t_2/t_1)/(s_2 - s_1) \qquad (5\text{–}20)$$

Thus transmissivity is inversely proportional to the slope of the time-drawdown curve. For convenience, $t_1$ and $t_2$ are usually chosen one log cycle apart. The Eq. 5–20 then reduces to

$$T = 264 \ Q/\Delta s \qquad (5\text{–}21)$$

where $\Delta s$ is the change in drawdown, in feet, over one log cycle of time.

The coefficient of storage of the aquifer can be calculated from the intercept of the straight line on the time axis at zero drawdown, provided that time is converted to days. For zero drawdown, Eq. 5–18 gives

$$0 = 264 \ (Q/T) \log [0.3 \ Tt_0/(r^2 S)]$$

that is,

$$0.3 \ Tt_0/(r^2 S) = 1$$

which gives

$$S = 0.3 \ Tt_0/r^2 \qquad (5\text{–}22)$$

[24] H. H. Cooper and C. E. Jacob, A Generalized Graphical Method for Evaluating Formation Constants and Summarizing Well-Field History, *Trans. Am. Geophys. Union*, **27**, 526 (1946).

## Example 5-2.

A time-drawdown curve for an observation well at a distance of 225 ft from a pumping well discharging at a constant rate of 350 gpm is shown in Fig. 5-5. To determine the slope of the straight-line portion, select two points one log cycle apart, viz.

$$t_1 = 1 \text{ min} \qquad s_1 = 1.6 \text{ ft}$$
$$t_2 = 10 \text{ min} \qquad s_2 = 4.5 \text{ ft}$$

The slope of the line per log cycle, $\Delta s = 4.5 - 1.6 = 2.9$ ft. The line intersects the zero drawdown axis at $t_0 = 0.3$ min. The transmissivity and storage coefficient of the aquifers are

$$T = 264 \, Q/\Delta s = 264 \times 350/2.9 = 3.2 \times 10^4 \text{ gpd/ft}$$
$$S = 0.3 \, Tt_0/r^2 = 0.3(3.2 \times 10^4)(0.3/1440)/(225)^2 = 4.0 \times 10^{-5}$$

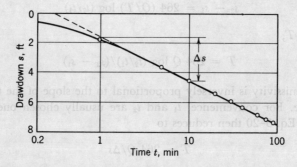

**Figure 5-5.**

**Time-drawdown curve. (Data by courtesy of the U.S. Geological Survey.)**

Equation 5-18 may also be used if the drawdown is measured at several observation wells at essentially the same time, that is, from the shape of the cone of depression. Drawdowns are plotted on the arithmetic scale and distance on the log scale and the resulting straight-line graph is called the distance-drawdown curve. It can be shown that the expressions for $T$ and $S$ in this case are

$$T = 528 \, Q/\Delta s \qquad (5\text{-}23)$$
$$S = 0.3 \, Tt/r_0^2 \qquad (5\text{-}24)$$

With the formation constants $T$ and $S$ known, Eq. 5-18 gives the drawdown for any desired value of $r$ and $t$, provided that $u$ (Eq. 5-13) is less than 0.01. The value of $u$ is directly proportional to the square of the distance and inversely proportional to time, $t$. The combination of time and distance at which $u$ passes the critical value is inversely proportional to the hydraulic diffusivity of the aquifer, $D = T/S$. The critical value of $u$ is reached much more quickly in confined aquifers than in unconfined aquifers.

**Recovery Method.** In the absence of an observation well, transmissivity can be determined more accurately by measuring the recovery of water levels in the well under test after pumping has stopped than by measuring the drawdown in the well during pumping. For this purpose, a well is pumped for a known period of time, long enough to be drawn down appreciably. The pump is then stopped, and the rise of water level within the well (or in a nearby observation well) is observed (Fig. 5-6). The drawdown after the shutdown will be the same as if the discharge had continued at the rate of pumping and a recharge well with the same flow had been superimposed on the discharge well at the instance the discharge was shut down. The residual drawdown, $s'$, can be found from Eq. 5-15 as

$$s' = (114.6\ Q/T)[W(u) - W(u')] \qquad (5\text{-}25)$$

where

$$u = 1.87\ r^2 S/4Tt \quad \text{and} \quad u' = 1.87\ r^2 S/(4Tt')$$

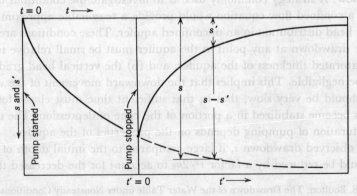

**Figure 5-6.**
**Water-level recovery after pumping has stopped.**

where $r$ is the effective radius of the well (or the distance to the observation well), $t$ is the time since pumping started, and $t'$ is the time since pumping stopped. For small values of $r$ and large values of $t'$, using Jacob's approximation, the residual drawdown may be obtained from Eq. 5-19 as

$$s' = (264\, Q/T)\, \log\, (t/t') \qquad (5\text{-}26)$$

Solving for $T$,

$$T = (264\, Q/s')\, \log\, (t/t') \qquad (5\text{-}27)$$

Plotting $s'$ on an arithmetic scale and $t/t'$ on a logarithmic scale, a straight line is drawn through the observations. The coefficient of transmissivity can be determined from the slope of the line, or for convenience, the change of residual drawdown over one log cycle can be used as

$$T = 264\, Q/\Delta s' \qquad (5\text{-}28)$$

Strictly speaking, the Theis equation and its approximations are applicable only to the situations that satisfy the assumptions used in their derivation. They undoubtedly also provide reasonable approximations in a much wider variety of conditions than their restrictive assumptions would suggest. Significant departures from the theoretical model will be reflected in the deviation of the test data from the type curves. Advances have recently been made in obtaining analytical solutions for anisotropic aquifers, for aquifers of variable thickness, and for partially penetrating wells.

**Unconfined Aquifers.** The partial differential equation governing nonsteady unconfined flow is nonlinear in $h$.[25] In many cases, it is difficult or impossible to obtain analytical solutions to the problems of unsteady unconfined flow. A strategy commonly used is to investigate the conditions under which a confined flow equation would provide a reasonable approximation for the head distribution in an unconfined aquifer. These conditions are that (a) the drawdown at any point in the aquifer must be small relative to the total saturated thickness of the aquifer, and (b) the vertical head gradients must be negligible. This implies that the downward movement of the water table should be very slow, that is, that sufficient time must elapse for the flow to become stabilized in a portion of the cone of depression. The minimum duration of pumping depends on the properties of the aquifer.

The observed drawdown $s$, if large compared to the initial depth of flow $h_0$, should be reduced by a factor $s^2/2h_0$ to account for the decreased thick-

---

[25] N. S. Boulton, The Drawdown of the Water Table under Nonsteady Conditions Near a Pumped Well in an Unconfined Formation, *Proc. Instn. Civil Engrs.*, London, Pt. 3, 564 (1954).

ness of flow due to dewatering before Equation 5-15 can be applied.[26] For an observation well at a distance greater than $0.2h_0$, the minimum duration of pumping beyond which the approximation is valid is given by Boulton as

$$t_{min} = 37.4 \, Sh_0/K$$

where $t$ is in days; $h_0$, saturated aquifer thickness in ft; and $K$, hydraulic conductivity in gpd/ft$^2$.

Based on electric analog studies, Stallman[27] describes the response of water-table aquifers during pumping tests.

**Leaky Aquifers.** The partial differential equation governing nonsteady radial flow toward a steadily discharging well in a leaky confined aquifer is[28]

$$\frac{\partial^2 s}{\partial r^2} + \frac{1}{r}\frac{\partial s}{\partial r} - \frac{s}{B^2} = \frac{S}{T}\frac{\partial s}{\partial t} \qquad (5\text{-}29)$$

where

$$B = \sqrt{T/(K'/b')} \qquad (5\text{-}30)$$

and $s$ is the drawdown at a distance $r$ from the pumping well, $T$ and $S$ are the transmissivity and storage coefficient of the lower aquifer, and $K'$ and $b'$ are the vertical permeability and thickness of the semipervious confining layer, respectively. Jacob and Hantush give a solution that can be written in an abbreviated form as

$$s = 114.6 \, Q/T[W(u, r/B)] \qquad (5\text{-}31)$$

where

$$W(u, r/B) = \int_u^\infty (1/y) \exp[-y - r^2/(4B^2 y)] dy$$

and

$$u = 1.87 \, r^2 S/(Tt) \qquad (5\text{-}32)$$

$W(u, r/B)$ is the well function of the leaky aquifer, $Q$ is the constant discharge of the well in gpm, $T$ is transmissivity in gpd/ft., and $t$ is the time in days.

[26] C. E. Jacob, Determining the Permeability of Water-Table Aquifers in Methods of Determining Permeability, Transmissibility and Drawdown, *U.S.G.S. Water Supply Paper 1536-I*, 245 (1963).

[27] R. W. Stallman, Effects of Water Table Conditions on Water Level Changes Near Pumping Wells, *Water Res. Research*, **1**, 295 (1965).

[28] M. S. Hantush, Hydraulics of Wells, in *Advances in Hydroscience*, Vol. 1, 281 (1964).

In the earlier phases of the transient state, that is, at very small values of time, the system acts like an ideal elastic artesian aquifer without leakage and the drawdown pattern closely follows the Theis type-curve. As time increases, the drawdown in the leaky aquifer begins to deviate from the Theis curve. At large values of time, the solution approaches the steady-state condition. With time, the fraction of well discharge derived from storage in the lower aquifer decreases and becomes negligible at large values of time as steady state is approached.

The solution to the above equation is obtained graphically by the match-point technique described for the Theis solution. On the field curve are plotted drawdown vs. time on logarithmic coordinates. On the type-curve are plotted the values of $W(u, r/B)$ vs. $1/u$ for various values or $r/B$ as shown in Fig. 5–7. The curve corresponding to the value of $r/B$ giving the best fit is selected. From the match-point coordinates $s$ and $W(u, r/B)$, $T$ can be calculated by substituting in Eq. 5–31. From the other two match-point coordinates and the value of $T$ computed above, $S$ is determined from Eq. 5–32. If $b'$ is known, the value of the vertical permeability of the aquitard can be computed from Eq. 5–30, knowing $r/B$ and $T$. Values of $W(u, r/B)$ for the practical range of $u$ and $r/B$ are given in Table 5–1.

## 5-12 Prediction of Drawdown

Predictions of drawdowns are useful when a new well field is to be established or where new wells are added to an existing field. To predict drawdowns, $T$, $S$ and proposed pumping rates must be known. Any of the several equations can be used. The Theis equation is of quite general applicability. Jacob's approximation does not accurately show drawdowns during the first few hours or first few days of withdrawals ($u > .01$). Because the equations governing flow are linear, the principle of superposition is valid.

**Constant Discharge.** Examples 5–3 and 5–4 illustrate the methods that can be used to evaluate the variation in drawdown with time and with distance when the pumping rate is constant.

---

*Example 5–3.*

In the aquifer represented by the pumping test in Example 5–1, a gravel-packed well with an effective diameter of 24 inches is to be constructed. The design flow of the well is 700 gpm. Calculate the drawdown at the well with total withdrawals from storage, (that is, with no recharge or leakages) after (a) 1 minute, (b) 1 hour, (c) 8 hours, (d) 24 hours, (e) 30 days, and (f) 6 months of continuous pumping, at design capacity.

## Table 5-1

Values of the Function $W(u, r/B) = \int_{u}^{\infty} (1/y) \exp[-y - r^2/(4B^2 y)]\, dy$

| u \ r/B | 0.005 | 0.01 | 0.025 | 0.05 | 0.075 | 0.10 | 0.15 | 0.2 | 0.3 | 0.4 | 0.5 | 0.6 | 0.7 | 0.8 | 0.9 | 1.0 | 1.5 | 2.0 |
|---|---|---|---|---|---|---|---|---|---|---|---|---|---|---|---|---|---|---|
| 0 | 10.8286 | 9.4425 | 7.6111 | 6.2285 | 5.4228 | 4.8541 | 4.0601 | 3.5054 | 2.7449 | 2.2291 | 1.8488 | 1.5550 | 1.3210 | 1.1307 | 0.9735 | 0.8420 | 0.4276 | 0.2278 |
| 0.000001 | 10.8283 | | | | | | | | | | | | | | | | | |
| 0.000005 | 10.6822 | 9.4413 | | | | | | | | | | | | | | | | |
| 0.00001 | 10.3963 | 9.4176 | | | | | | | | | | | | | | | | |
| 0.00005 | 9.2052 | 8.8827 | 7.6000 | | | | | | | | | | | | | | | |
| 0.0001 | 8.5717 | 8.3983 | 7.5199 | 6.2282 | 5.4228 | | | | | | | | | | | | | |
| 0.0005 | 7.0118 | 6.9750 | 6.7357 | 6.0821 | 5.4062 | 4.8530 | | | | | | | | | | | | |
| 0.001 | 6.3253 | 6.3069 | 6.1823 | 5.7965 | 5.3078 | 4.8292 | 4.0595 | 3.5054 | | | | | | | | | | |
| 0.005 | 4.7249 | 4.7212 | 4.6960 | 4.6084 | 4.4713 | 4.2960 | 3.8821 | 3.4567 | 2.7428 | 2.2290 | | | | | | | | |
| 0.01 | 4.0373 | 4.0356 | 4.0231 | 3.9795 | 3.9091 | 3.8150 | 3.5725 | 3.2875 | 2.7104 | 2.2253 | 1.8486 | 1.5550 | 1.3210 | 1.1307 | | | | |
| 0.05 | 2.4678 | 2.4675 | 2.4653 | 2.4576 | 2.4448 | 2.4271 | 2.3776 | 2.3110 | 2.1371 | 1.9283 | 1.7075 | 1.4927 | 1.2955 | 1.1210 | 0.9700 | 0.8409 | | |
| 0.1 | 1.8229 | 1.8227 | 1.8218 | 1.8184 | 1.8128 | 1.8050 | 1.7829 | 1.7527 | 1.6704 | 1.5644 | 1.4422 | 1.3115 | 1.1791 | 1.0505 | 0.9297 | 0.8190 | 0.4271 | 0.2278 |
| 0.5 | 0.5598 | 0.5598 | 0.5597 | 0.5594 | 0.5588 | 0.5581 | 0.5561 | 0.5532 | 0.5453 | 0.5344 | 0.5206 | 0.5044 | 0.4860 | 0.4658 | 0.4440 | 0.4210 | 0.3007 | 0.1944 |
| 1.0 | 0.2194 | 0.2194 | 0.2193 | 0.2193 | 0.2191 | 0.2190 | 0.2186 | 0.2179 | 0.2161 | 0.2135 | 0.2103 | 0.2065 | 0.2020 | 0.1970 | 0.1914 | 0.1855 | 0.1509 | 0.1139 |
| 5.0 | 0.0011 | 0.0011 | 0.0011 | 0.0011 | 0.0011 | 0.0011 | 0.0011 | 0.0011 | 0.0011 | 0.0011 | 0.0011 | 0.0011 | 0.0011 | 0.0011 | 0.0011 | 0.0011 | 0.0010 | 0.0010 |

Abstracted from Hantush, Analysis of Data from Pumping Tests in Leaky Aquifers, *Trans. Am. Geophys. Union*, **37**, 702 (1956).

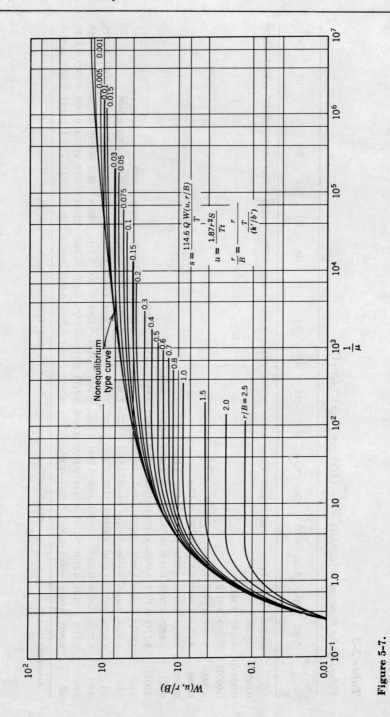

**Figure 5-7.**
Nonsteady-state leaky artesian-type curves. (After Walton, footnote 21.)

The drawdown constant, $114.6\ Q/T = 114.6 \times 700/(3.2 \times 10^4) = 2.51$ ft.

$u = 1.87\ r^2 S/Tt = 1.87 \times 1^2 \times 3.4 \times 10^{-5}/(3.2 \times 10^4) = (2.0 \times 10^{-9})/t$

The values of drawdown for various values of time are given in Table 5–2.

The Theis equation was used for illustrative purposes only. The values of $u$ are quite low; Jacob's approximation can be used with identical results.

## Example 5–4.

Determine the profile of a quasi-steady state cone of depression for a proposed 24-in. well pumping continuously at (a) 150 gpm, (b) 200 gpm, and (c) 250 gpm in an elastic artesian aquifer having a transmissivity of 10,000 gpd/ft and a storage coefficient of $6 \times 10^{-4}$. Assume that the discharge and recharge conditions are such that the drawdowns will be stabilized after 180 days.

The distance at which drawdown is approaching zero, that is, the radius of cone of depression, can be obtained from Eq. 5–24.

$r_0^2 = 0.3\ Tt/S = 0.3\ Dt$ where $D$ is the diffusivity of the aquifer
$= 0.3\ (1 \times 10^4)\ 180/(6 \times 10^{-4}) = 9 \times 10^8$
$r_0 = 0.3\ Dt = 3 \times 10^4$ ft

This is independent of $Q$ and depends only on the diffusivity of the aquifer.

The change in drawdown per log cycle from Eq. 5–23 is

$$\Delta s = 528\ Q/T$$

For 150 gpm, $\Delta s_1 = 528 \times 150/1 \times 10^4 = 7.9$ ft; for 200 gpm, $\Delta s_2 = 10.6$ ft; and for 250 gpm, $\Delta s_3 = 13.2$ ft.

Using the value of $r = 30{,}000$ ft as the starting point, straight lines having slopes of 7.9 ft, 10.6 ft, and 13.2 ft are drawn (Fig. 5–8).

The contours of the piezometric surface can be drawn by subtracting the drawdowns at several points from the initial piezometric surface.

## Table 5-2

*Variation of Drawdown with Time*

|     | Time, days | $u$ | $W(u)$ | Drawdown, $s$, ft. |
|-----|------------|-----|--------|---------------------|
| (a) | 1/1440     | $2.86 \times 10^{-6}$  | 12.19 | 30.6 |
| (b) | 1/24       | $4.8 \times 10^{-8}$   | 16.27 | 40.8 |
| (c) | 1/3        | $6.0 \times 10^{-9}$   | 18.35 | 46.0 |
| (d) | 1          | $2.0 \times 10^{-9}$   | 19.45 | 48.8 |
| (e) | 30         | $6.6 \times 10^{-11}$  | 22.86 | 57.3 |
| (f) | 180        | $1.1 \times 10^{-11}$  | 24.66 | 61.8 |

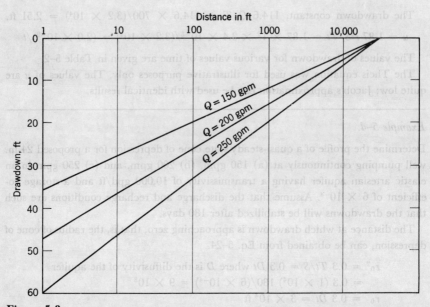

**Figure 5-8.**
**Distance-drawdown curves for various rates of pumping (Ex. 5-4).**

*Variable Discharge.* The rate at which water is pumped from a well field in a water-supply system will vary with time in response to changes in demand. The continuous rate of pumping curve can be approximated by a series of steps as shown in Fig. 5-9. Then each step can be analyzed, using one of the conventional equations. From the principle of superposition, the drawdown at any point at any specific time can be obtained as the sum of increments in drawdowns caused by the step increases up to that time.

$$s = \Delta s_1 + \Delta s_2 + \cdots + \Delta s_t \tag{5-33}$$

Using semilogarithmic approximation,

$$\Delta s_i = (264 \, \Delta Q_i / T) \log [(0.3 \, T t_i)/(r^2 S)] \tag{5-34}$$

where

$$t_i = t - t_{i-1}$$

Increments of drawdown $\Delta s_i$ are determined with respect to the extension of the preceding water-level curve.

*Intermittent Discharge.* In a water-supply system, a well (or a well field) may be operated on a regular daily cycle, pumping at a constant rate

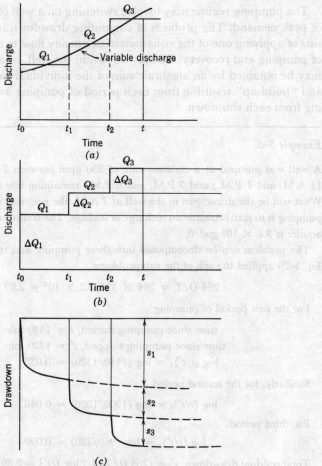

**Figure 5-9.**
**Step function approximation of variable discharge.**

for a given time interval, remaining idle for the rest of the period. Brown[29] gives the following expression for computing the drawdown in the pumped well after $n$ cycles of operation:

$$s_n = (264\ Q/T) \log [(1.2.3 \cdots n)/(1-p)(2-p) \cdots (n-p)] \quad (5\text{-}35)$$

where $p$ is the fractional part of the cycle during which the well is pumped, $Q$ is the discharge in gpm, and $T$ is the transmissivity in gpd/ft.

[29] R. H. Brown, Drawdowns Resulting from Cyclic Intervals of Discharge, in Methods of Determining Permeability, Transmissibility and Drawdown, *U.S. Geol. Surv. Water Supply Paper 1537-I* (1963).

*142/Groundwater Development*

The pumping regime may involve switching on a well only during periods of peak demand. The problem of computing drawdown in a well then consists of applying one of the equations of nonsteady flow to each of the periods of pumping and recovery. The drawdown in the well, or at any other point, may be obtained by an algebraic sum of the individual values of drawdown and "build up" resulting from each period of pumping and recovery resulting from each shutdown.

*Example 5-5.*

A well was pumped at a constant rate of 350 gpm between 7 A.M. and 9 A.M.; 11 A.M. and 1 P.M.; and 3 P.M. and 6 P.M., remaining idle the rest of the time. What will be the drawdown in the well at 7 A.M. the next day when a new cycle of pumping is to start? Assume no recharge or leakage. The transmissivity of the artesian aquifer is $3.2 \times 10^4$ gpd/ft.

The problem can be decomposed into three pumping and recovery periods and Eq. 5-26 applied to each of the sub-problems.

$$264\ Q/T = 264 \times 350/3.2 \times 10^4 = 2.89$$

For the first period of pumping,

time since pumping started, $t = 1440$ min
time since pumping stopped, $t' = 1320$ min
$\log (t/t')_1 = \log (1440/1320) = 0.038$

Similarly, for the second period,

$$\log (t/t')_2 = \log (1200/1080) = 0.046$$

For third period,

$$\log (t/t')_3 = \log (960/780) = 0.090$$

Total residual drawdown, $s' = (264\ Q/T)\ [\ \Sigma \log t/t'] = 2.89 \times 0.174 = 0.5$ ft.

## 5-13 Multiple Well Systems

As the equations governing the steady and unsteady flow are linear the drawdown at any point due to several wells is equal to the algebraic sum of the drawdowns caused by each individual well, that is, for $n$ wells in a well field

$$s = \sum_{i=1}^{n} s_i$$

where $s_i$ is the drawdown at the point due to the $i$th well. If the location of wells, their discharges, and their formation constants are known, the combined distribution of drawdown can be determined by calculating drawdown at several points in the area of influence and drawing contours.

*Example 5-6.*

Three 24-in. wells are located on a straight line 1000 ft apart in an artesian aquifer with $T = 3.2 \times 10^4$ gpd/ft and $S = 3 \times 10^{-5}$. Compute the drawdown at each well when (a) one of the outside wells is pumped at a rate of 700 gpm for 10 days, and (b) the three wells are pumped at 700 gpm for 10 days.

$$u_{1\text{ ft}} = (1.87 \times 1^2 \times 3.0 \times 15^5)/(3.2 \times 10^4 \times 10) = 1.75 \times 10^{-10},$$
$$W(u) = 21.89$$
$$u_{100\text{ ft}} = 10^6 u_1 = 1.75 \times 10^{-4}, \quad W(u) = 8.08$$
$$u_{2000\text{ ft}} = 4 \times 10^6 u_1 = 7 \times 10^{-4}, \quad W(u) = 6.69$$
$$(114.6\ Q)/T = (114.6 \times 700)/(3.2 \times 10^4) = 2.51 \text{ ft}$$

(a) Drawdown at the face of pumping well, $s_1 = 2.51 \times 21.89 = 54.9$ ft; drawdown in the central well, $s_2 = 2.51 \times 8.08 = 20.3$ ft; drawdown in the other outside well, $s_3 = 2.51 \times 6.69 = 16.8$ ft. (b) Drawdown in outside wells $= 54.9 + 20.3 + 16.8 = 92$ ft; drawdown in central well $= 54.9 + 20.3 + 20.3 = 95.5$ ft.

A problem of more practical interest is to determine the discharges of the wells when their drawdowns are given. This will involve simultaneous solution of linear equations, which can be undertaken by numerical methods or by trial and error.

*Example 5-7.*

Suppose it is desired to restrict the drawdown in each of the wells to 60 ft in the above problem. What will be the corresponding discharges for individual wells?

$$\text{well 1: } [114.6/T][Q_1 W(u_1) + Q_2 W(u_{1000}) + Q_3 W(u_{2000})] = 60$$
$$\text{well 2: } [114.6/T][Q_1 W(u_{1000}) + Q_2 W(u_1) + Q_3 W(u_{1000})] = 60$$
$$\text{well 3: } [114.6/T][Q_1 W(u_{2000}) + Q_2 W(u_{1000}) + Q_3 W(u_1)] = 60$$
$$Q_1 = Q_3 = 468 \text{ gpm} \quad \text{and} \quad Q_2 = 420 \text{ gpm}$$

When the areas of influence of two or more pumped wells overlap, the draft of one well affects the drawdown of all others. In closely spaced wells, interference may become so severe that a well group behaves like a single well producing a single large cone of depression. When this is the case, discharge-drawdown relationships can be studied by replacing the group of wells by an equivalent single well having the same drawdown distribution when producing water at a rate equal to the combined discharge of the group. The effective radius of a heavily-pumped well field could be a mile or more and have a circle of influence extending over many miles. By contrast, lightly pumped, shallow wells in unconfined aquifers may show no interference when placed 100 ft apart or even less. The number of wells, the geometry of the well field, and its location with respect to recharge and discharge areas and aquifer boundaries are important in determining the distribution of drawdown and well discharges. An analysis of the optimum location, spacing, and discharges should be carried out when designing a well field.

## 5-14 Aquifer Boundaries

Most methods of analysis assume that an aquifer is infinite in extent. In practice, all aquifers have boundaries. However, unless a well is located so close to a boundary that the radial flow pattern is significantly modified, the flow equations can be applied without appreciable error. Nevertheless, in many situations definite geologic and hydraulic boundaries limit aquifer dimensions and cause the response of an aquifer to deviate substantially from that predicted from equations based on extensive aquifers. This is especially true if the cone of depression reaches streams, outcrops, or groundwater divides; geologic boundaries, such as faults and folds; and valley fills of limited extent.

The effect of aquifer boundaries can be incorporated into analysis through the method of images. The *method of images* is an artifice employed to transform a bounded aquifer into one of an infinite extent having an equivalent hydraulic flow system. The effect of a known physical boundary (in the flow system) is simulated by introducing one or more hypothetical components, called *images*. The solution to a problem can then be obtained by using the equations of flow developed for extensive aquifers for this hypothetical system.

**Recharge Boundaries.** The conditions along a recharge boundary can be reproduced by assuming that the aquifer is infinite and by introducing a negative image well (for example, a recharge image well for a discharging real well) an equal distance on the opposite side of the boundary from the real well, the line joining the two being at right angles to the boundary (Fig. 5-10). The drawdown, $s$, at any distance, $r$, from the pumping well and $r_i$ from the image well is the algebraic sum of the drawdowns due to the real well, $s_r$, and build-up due to image well, $s_i$.

*Example 5–8.*

A gravel-packed well with an effective diameter of 24 in. pumps from an artesian aquifer having $T = 3.2 \times 10^4$ gpd/ft and $S = 3.4 \times 10^{-5}$. The well lies at a distance of 1000 ft from a stream that can supply water fast enough to maintain a constant head. Find the drawdown in the well after 10 days' pumping at 700 gpm. Determine the profile of the cone of depression with a vertical plane through the well normal to the stream.

In the region of interest, the use of semi-logarithmic approximation $s = (528\, Q/T) \log (r_i/r)$ is valid:

$$528Q/T = (528 \times 700)/(3.2 \times 10^4) = 11.55 \text{ ft}$$

Drawdown at the well, $s_w = 11.55 \log (2a/r_w) = 11.55 \log 2000 = 37.0$ ft

Drawdown at 500 ft from the stream $= 11.55 \log (1500/500) = 5.5$ ft
Drawdown at 1500 ft from the stream $= 11.55 \log (2500/500) = 8.1$ ft

Drawdowns at other points can be calculated in a similar manner. The results are shown in Fig. 5–10.

For a well located near a stream, the proportion of the discharge of the well diverted directly from the source of recharge depends on the distance of the well from the recharge boundary, the aquifer characteristics, and the duration of pumping. The contribution from a line source of recharge and distribution of drawdown in such a system can be evaluated[30] and are extremely useful in arriving at an optimal location of well fields.

The problem of recirculation between a recharge well and a discharge well pair is of great practical importance because of the use of wells (or other devices) for underground waste disposal (or artificial recharge) and for water supply in the same area. The recirculation can be minimized by locating the recharge well directly downstream from the discharge well. The critical value of discharge and optimum spacing for no recirculation can be evaluated.[31]

The permissible distance, $r_c$, between production and disposal wells in an isotropic, extensive aquifer to prevent recirculation is given by

$$r_c = 2\, Q/(\pi T I) \qquad (5\text{–}36)$$

where $Q$ is the equal pumping and disposal rate in gpd, $T$ is the transmissivity in gpd/ft, and $I$ is the hydraulic gradient of the water table or piezometric surface.

---

[30] C. V. Theis, The Effect of a Well on the Flow of a Nearby Stream, *Trans. Am. Geophys. Union*, **22**, 734 (1941); R. E. Glover and G. G. Balmer, River Depletion Resulting from Pumping a Well near a River, *Trans. Am. Geophys. Union*, **35**, 468 (1954); and C. V. Theis, Drawdowns Caused by a Well Discharging Under Equilibrium Conditions from an Aquifer Bounded on a Straight-Line Source, in Short Cuts and Special Problems in Aquifer Tests, *U.S. Geol. Surv., Water Supply Paper 1545-C* (1963).

[31] C. E. Jacob, Flow of Ground Water, in H. Rouse, Ed., *Engineering Hydraulics*, John Wiley & Sons, 1950, p. 321.

146/*Groundwater Development*

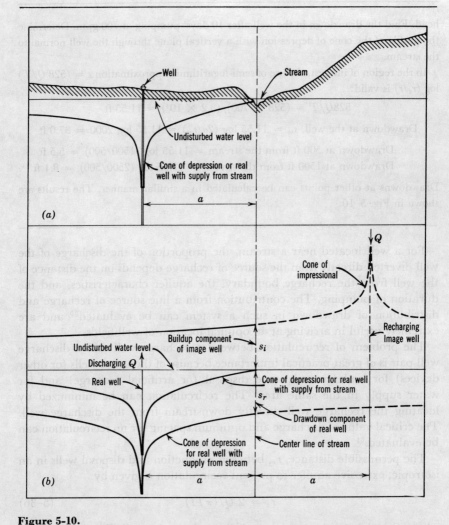

**Figure 5-10.**
**Application of the method of images to a well receiving water from a stream (idealized). (a) Real system. (b) Equivalent system in an infinite aquifer.**

*Location of Aquifer Boundaries.* In many instances, the location and nature of hydraulic boundaries of an aquifer can be inferred from the analysis of aquifer-test data. The effect of a boundary when it reaches an observation well causes the drawdowns to diverge from the Theis type-curve or Jacob method's straight line. The nature of the boundary, recharge, or barrier is given by the direction of departures. An observation well closer to the boundary shows evidence of boundary effect earlier than does an observation well at a greater distance. The theory of images can be used to estimate the distance to the boundary.[32] The analysis can be extended to locate multiple boundaries.

For the estimation of the formation constants, only those observations should be used that do not reflect boundary effects, that is, the earlier part of the time-drawdown curve. For the prediction of future drawdowns, the latter part of the curve incorporating the boundary effects is pertinent.

## 5-15 Characteristics of Wells

The drawdown in a well being pumped is the difference between the static water level and the pumping water level. The well drawdown consists of two components: (1) *Formation loss*, that is, the head expended in overcoming the frictional resistance of the medium from the outer boundary to the face of the well, and which is directly proportional to the velocity, if the flow is laminar; and (2) *well loss*, which includes (a) the entrance head loss caused by the flow through the screen and (b) the head loss due to the upward axial flow of water inside the screen and the casing up to the pump intake. This loss is associated with the turbulent flow and is approximately proportional to the square of the velocity. The well drawdown $D_w$ can be expressed as

$$D_w = BQ + CQ^2 \quad (5\text{-}37)$$

where $B$ summarizes the resistance characteristics of the formation and $C$ represents the characteristics of the well.

For unsteady flow in a confined aquifer, from Eq. 5–18,

$$B = (264/T) \log (0.3Tt/r_w^2 S) \quad (5\text{-}38)$$

This shows that the resistance of an extensive artesian aquifer increases with time as the area of influence of the well expands. For relatively low pumping rates, the well loss may be neglected, but for higher rates of discharge it can represent a sizable proportion of the well drawdown.

**Specific Capacity of a Well.** The productivity and efficiency of a well is generally expressed in terms of *specific capacity*, defined as the discharge per unit drawdown, that is, the ratio of discharge to well drawdown.

$$Q/D_w = 1/(B + CQ) \quad (5\text{-}39)$$

The specific capacity of a well depends on formation constants and hydrogeologic boundaries of the aquifer, on well construction and design, and on test conditions. It is sometimes useful to distinguish between *theoretical specific capacity*, which depends only on formation characteristics and ignores well

[32] S. M. Lang, Interpretation of Boundary Effects from Pumping Test Data, *J. Am. Water Works Assoc.*, **52**, 3 (1960).

losses, and actual specific capacity. The former is a measure of the productivity. The difference between the two, or their ratio, is a measure of the efficiency of the well.

For unsteady flow in a confined aquifer,

$$Q/D_w = 1/\{(264/T) \log [(0.3\ Tt)/(r_w{}^2 S)] + CQ\} \qquad (5\text{-}40)$$

Hence the specific capacity is not a fixed quantity, but decreases with both the period of pumping and the discharge. It is important to state not only the discharge at which a value of specific capacity is obtained but also the duration of pumping. Determination of specific capacity from a short-term acceptance test of a few hours duration can give misleading results, particularly in aquifers having low hydraulic diffusivity, that is, low transmissivity and high storage coefficients.

**Partial Penetration.** The specific capacity of a well is affected by partial penetration. A well that is screened only opposite a part of an aquifer will have a lower discharge for the same drawdown or larger drawdown for the same discharge, that is, a smaller specific capacity. The ratio of the specific capacity of a partially penetrating well to the specific capacity of a completely penetrating well in homogeneous artesian aquifers is given by the Kozeny formula valid for steady-state conditions.[33]

$$(Q/s_p)/(Q/s) = K_p\{1 + 7[r_w/(2K_p b)]^{1/2} \cos (\pi K_p/2)\} \qquad (5\text{-}41)$$

where $Q/s_p$ = specific capacity of a partially penetrating well, gpm/ft; $Q/s$ = specific capacity of a completely penetrating well, gpm/ft; $r_w$ = effective well radius, ft; $b$ = aquifer thickness, ft; and $K_p$ = ratio of length of screen to saturated thickness of the aquifer.

If the right-hand side of Eq. 5-41 is denoted by $F_p$, the equation may be written as

$$Q/s_p = (Q/s)F_p \qquad (5\text{-}42)$$

The formula is not valid for small $b$, large $K_p$, and large $r_w$.

A graph of $F_p$ vs. $K_p$ for various values of $b/r_w$ is given in Fig. 5-11[34] within the valid range of the formula.

**Effective Well Radius.** The effective radius of a well is seldom equal to its nominal radius. Effective radius is defined as that distance, measured radially from the axis of a well, at which the theoretical drawdown equals the actual drawdown just at the surface of the well. Depending on the method of construction and development, and the actual condition of the intake portion

---

[33] A. N. Turcan, Estimating the Specific Capacity of a Well, *U.S. Geol. Surv. Professional Paper 450-E* (1963).

[34] Edward E. Johnson, Inc., *Groundwater and Wells*, Saint Paul, Minn. (1966).

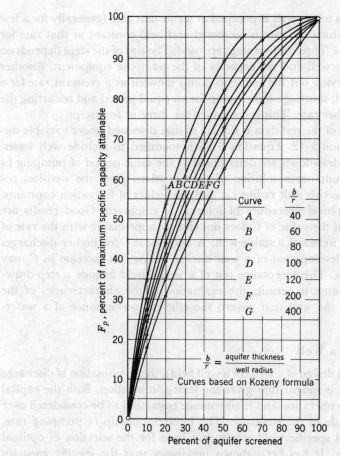

**Figure 5-11.**
**Relationship of partial penetration and specific capacity for wells in homogeneous artesian aquifers. (After *Ground Water and Wells*, footnote 34.)**

of the well, the effective radius may be greater than, equal to, or less than the nominal radius. The transmissivity of the material in the immediate vicinity of a well is the controlling factor. If the transmissivity of the material surrounding the well is higher than that of the aquifer, the effective well radius will be greater than the nominal radius. On the other hand, if the material around the well has a lower transmissivity due to caving or clogging because of faulty construction, the effective radius will be less than nominal radius.

***Measurement of Well Characteristics.*** The well-loss factor and the effective radius of a well can be determined by the multiple-step drawdown test.[23,35] In this test, a well is pumped at a constant rate (generally for a few hours), after which the rate is increased and held constant at that rate for the same period. Three or four steps are used. The size of the steps depends on the yield of the well and the capacity of the pumping equipment. Another method of carrying out the test is to pump the well at a constant rate for a specified period, shutting off the pump for an equal period and restarting the pump at a higher rate. This is continued for three or four steps.

The analysis of the test data is similar to that discussed under variable discharge in Section 5-12. Equation 5-34 is modified to include well losses. Increments of drawdown are determined at the same period of pumping in each step. Simultaneous solution of the equations gives the well-loss coefficient, $C$, effective well radius $r_w$, and the values of formation constants. Usually a graphical procedure of solution is employed. Good results are obtained only if the value of $C$ does not change appreciably with the rate of pumping. This indicates a stable well. A decrease in $C$ for higher discharges may indicate development of a well during testing; an increase in $C$ may denote clogging. The acceptance test of a well should include a step-drawdown test, because it permits the evaluation of the characteristics of the aquifer and of the well that govern the efficient performance of a water-supply system.

## 5-16 Yield of a Well

For the optimal design of a well (or a well field), the combination of discharge and drawdown is sought that gives the least-cost solution. Both the capital outlays and the operation and maintenance costs need to be considered over the economic life of the structure. The interrelationship of pumping rate, drawdown, and specific yield serves as a basis for the selection of optimal design capacity. It has been shown previously that the specific capacity decreases as the pumping rate is increased. Hence the earlier increments of drawdown are more effective in producing yields than the later ones; each additional unit of yield is more expensive than the previous one. Increasing the yield of a well by one unit is economically justified only if the cost of developing this unit from alternate sources, another well or surface supplies, is higher.

The yield obtainable from a well at any site depends on (a) hydraulic characteristics of the aquifer, which may be given in terms of a specific-capacity drawdown relationship, (b) the drawdown at the pumping well, (c) the length of the intake section of the well, (d) the effective diameter of the well, and (e) the number of aquifers penetrated by the well.

[35] M. I. Rorabaugh, Graphical and Theoretical Analysis of Step-Drawdown Test of Artesian Well, *Proc. Am. Soc. Civil Engrs.*, **79**, Separate No. 362 (1953).

**Maximum Available Drawdown.** The maximum available drawdown at a well site can be estimated by the difference in elevations between the static water level and a conservation level below which it is undesirable to let the water levels drop. The conservation level is controlled by hydrogeologic conditions (type and thickness of the aquifer and the location of the most permeable strata), maintenance of the efficiency of the well, preservation of water quality, and pumping costs. In an artesian aquifer, good design practice requires that the drawdown not result in the dewatering of any part of the aquifer. Hence the maximum allowable drawdown is the distance between the initial piezometric level and the top of the aquifer. In a water-table aquifer, the pumping level should be kept above the top of the screen. The yield-drawdown relationship of homogeneous water-table aquifers indicates that optimum yields are obtained by screening the lower one half to one third of the aquifer. A common practice is to limit the maximum available drawdown to $\frac{1}{2}$ to $\frac{2}{3}$ of the saturated thickness. In very thick aquifers, artesian or water table, the limiting factor in obtaining yields is not the drawdown but the cost of pumping. In some locations, the available drawdown may be controlled by the presence of poor quality water. The maximum drawdown should be such as to avoid drawing this poor quality water into the pumping well.

**Specific Capacity—Drawdown Curve.** A graph of specific capacity vs. drawdown is prepared from the data on existing wells in the formation if such data are available. Specific capacities should be adjusted for well losses and partial penetration and should be reduced to a common well radius and duration of pumping. If no data are available, a step-drawdown test is conducted on the production well. The curve is extended to cover the maximum available drawdown. For a well receiving water from more than one aquifer, the resultant specific capacity is the sum of the specific capacities of the aquifer penetrated reduced appropriately for partial penetration.

**Maximum Yield.** The following procedure is carried out to estimate the maximum yield of a well. (1) Calculate the specific capacity of the fully penetrating well having the proposed diameter from Eq. 5–15 or Eq. 5–18. (2) Reduce the specific capacity obtained above for partial penetration. This can be done by using Eq. 5–41. (3) Adjust the specific capacity for the desired duration of pumping from Eq. 5–40. (4) Calculate the maximum available drawdown. (5) Compute the maximum yield of the well by multiplying the specific capacity in (3) by the maximum available drawdown.

---

*Example 5–9.*

A well having an effective diameter of 12 in. is to be located in a relatively homogeneous artesian aquifer with a transmissivity of 10,000 gpd/ft and a storage coefficient of

$4 \times 10^{-4}$. The initial piezometric surface level is 20 ft below the land surface. The depth to the top of the aquifer is 150 ft and the thickness of the aquifer is 50 ft. The well is to be finished with a screen length of 20 ft. Compute the specific capacity of the well and its maximum yield after 10 days of pumping. Neglect well losses.

The specific capacity of a 12-in. fully penetrating well after 10 days of pumping can be calculated from Eq. 5–15, 16. $u = 1.87\, r^2 S/(Tt) = 1.87 \times 10^{-9}$, $W(u) = 19.52$, and $(Q/s) = T/114.6\, W(u) = 10,000/(114.6 \times 19.52) = 4.47$ gpm/ft. The percentage of aquifer screened is $K_p = 20/50 = 40\%$. The slenderness of the well factor is $b/r_w = 50/0.5 = 100$. The value of $F_p$ from Fig. 5–11 is 0.65. The expected specific capacity of the well $= 0.65 \times 4.47 = 2.9$ gpm/ft. The maximum available drawdown $= 150 - 20 = 130$ ft. The maximum yield of the well $= 130 \times 2.9 = 377$ gpm. In view of the approximations involved with the evaluation, the actual maximum yield will be between perhaps 300 and 450 gpm.

## 5-17 Well Design[36]

From the standpoint of well design, it is useful to think of a well as consisting of two parts: (1) the *conduit portion* of the well, which houses the pumping equipment and provides the passage for the upward flow of water to the pumping intake, and (2) the *intake portion*, where the water from the aquifer enters the well. In consolidated water-bearing materials, the conduit portion is usually cased from the surface to the top of the aquifer, and the intake portion is an uncased open hole. In unconsolidated aquifers, a perforated casing or a screen is required to hold back the water-bearing material and to allow water to flow into the well.

The depth of a well depends on the anticipated drawdown for the design yield, the vertical position of the more permeable strata, and the length of the intake portion of the well.

The well size affects the cost of construction substantially. The well need not be of the same size from top to bottom. The diameter of a well is governed by (a) the proposed yield of the well, (b) entrance velocity and loss, and (c) the method of construction. The controlling factor is usually the size of the pump that will be required to deliver the design yield. The diameter of the casing should be two nominal sizes larger than the size of the pump bowls, to prevent the pump shaft from binding and to reduce well losses. Table 5–3 gives the casing sizes recommended for various pumping rates.

The selection of the well size may depend on the size of the open area desired to keep entrance velocities and well losses at a reasonable value. In deep drilled wells, the minimum size of the hole may be controlled by the

---

[36] *Groundwater and Wells*, footnote 34, contains good coverage of all aspects of design and construction of wells.

## Table 5-3
*Recommended Well Diameters*

| Anticipated Well Yield, in gpm | Nominal Size of Pump Bowls, in inches | Optimum Size of Well Casing, in inches | Smallest Size of Well Casing, in inches |
|---|---|---|---|
| Less than 100 | 4  | 6 ID  | 5 ID  |
| 75 to 175     | 5  | 8 ID  | 6 ID  |
| 150 to 400    | 6  | 10 ID | 8 ID  |
| 350 to 650    | 8  | 12 ID | 10 ID |
| 600 to 900    | 10 | 14 OD | 12 ID |
| 850 to 1300   | 12 | 16 OD | 14 OD |
| 1200 to 1800  | 14 | 20 OD | 16 OD |
| 1600 to 3000  | 16 | 24 OD | 20 OD |

(From *Groundwater and Wells*, footnote 34).

equipment necessary to reach the required depth. Deep wells in consolidated formations are often telescoped in size to permit drilling to required depths.

The wells are generally lined or cased with mild steel pipe, which should be grouted in place in order to prevent caving and contamination by vertical circulation and to prevent undue deterioration of the well by corrosion. If conditions are such that corrosion is unusually severe, then asbestos cement, plastic, or glass fiber pipes can be used if practicable.

The intake portion of the well[37] should be as long as economically feasible to reduce the drawdown and the entrance velocities. In relatively homogeneous aquifers, it is not efficient to obtain more than 90% of the maximum yield. In nonhomogeneous aquifers, the best strategy is to locate the intake portion in one or more of the most permeable strata.

Perforated pipes or prefabricated screens are used in wells in unconsolidated aquifers. The width of the screen openings, called the *slot size*, depends on the critical particle size of the water-bearing material to be retained and on the grain-size distribution, and is chosen from a standard sieve analysis of the aquifer material. With a relatively coarse and graded material, slot sizes are selected that will permit the fine and medium-sized particles to wash into the well during development and to retain a specified portion of the aquifer material around the screen. A graded filter is thereby generated around the well, which has higher permeability than the undisturbed aquifer material.

[37] An excellent exposition of the available literature is provided in A. H. Blair, Well Screens and Gravel Packs, *Water Research Assoc.*, England, TP64 (1968).

Perforated casings are generally used in uncemented wells when relatively large openings are permissible. If the casing is slotted in place after installation, the smallest practical opening is ⅛ in. Machine-perforated casings are also available. Fabricated well screens are available in a wide variety of sizes, designs, and materials. The choice of material is governed by water quality and cost.

For maximum efficiency, the frictional loss of the screen must be small. The head loss through a screen depends on screen length, $L$; diameter, $D$; percentage open area, $A_p$; coefficient of contraction of openings, $C_c$; velocity in the screen, $v$; and the total flow into the screen, $Q$ (ft$^3$/sec). It has been shown that for minimum screen loss, $CL/D > 6$ where $C = 11.31\ C_c A_p$.[37] The value of $CL/D$ may be increased by increasing $C_c$, $A_p$, or $L$, or by decreasing $D$. Thus for the screen loss to be a minimum, the percentage of open area depends on the length and diameter of the screen. The screen length is usually fixed by the considerations of hydrogeology and cost.

The screen length and diameter can be selected from the slot size and the requirement that the entrance velocity be less than that needed to move the unwanted sand particle sizes into the well. Experience has shown that, in general, a velocity of 0.1 ft/sec gives negligible friction losses and the least incrustation and corrosion.

Where the natural aquifer material is fine and uniform (effective size less than 0.01 in. and uniformity coefficient less than 3.0), it is necessary to replace it by a coarser gravel envelope next to the screen. The slot size is selected to fit this gravel pack. The gravel pack increases the effective well radius and acts as a filter and a stabilizer for the finer aquifer material. A gravel pack well is shown in Fig. 5-12. There are no universally accepted rules for the selection of slot sizes or for the design of a gravel pack. A correctly designed well should provide a virtually sand-free operation (less than 3 ppm). The thickness of the gravel pack should not be less than 3 in. or more than 9 in. and the particle size distribution curve of the pack should approximately parallel that of the aquifer.

## 5-18 Well Construction

There is no one optimum method of well construction. The size and depth of the hole, the rocks to be penetrated, and the equipment and experience of local drillers control the method of well sinking and determine the cost of construction. Well sinking is a specialized art that has evolved along a number of more or less regional lines. In the United States, well drillers are generally given much latitude in the choice of a suitable method. What they undertake to do is to sink a well of specified size at a fixed price per foot. Ordinarily, therefore, the engineer gives his attention not so much to drilling

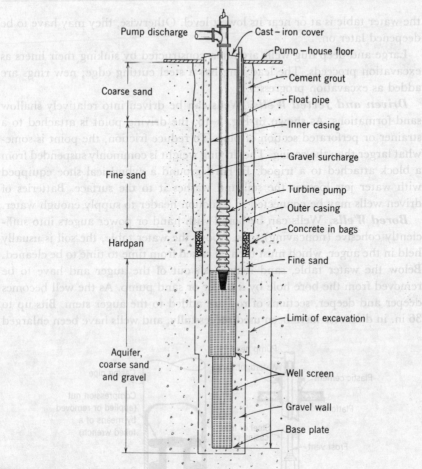

**Figure 5-12.**

**Gravel-packed well with deep-well turbine pump. (After Wisconsin State Board of Health.)**

operations as to the adequacy, suitability, and economics of proposed developments and the location of the works.

Well categories generally take their names from the methods by which wells are constructed. Shallow wells can be dug, driven, jetted, or bored.

**Dug Wells.** Small dug wells are generally excavated by hand. In loose overburden, they are cribbed with timber, lined with brick, rubble, or concrete, or cased with large-diameter vitrified tile or concrete pipe. In rock, they are commonly left unlined. Excavation is continued until water flows in more rapidly than it can be bailed out. Dug wells should be completed when

156/Groundwater Development

the water table is at or near its lowest level. Otherwise, they may have to be deepened later on.

Large and deep dug wells are often constructed by sinking their liners as excavation proceeds. The lead ring has a steel cutting edge; new rings are added as excavation progresses.

**Driven and Jetted Wells.** Wells can be driven into relatively shallow sand formations. As shown in Fig. 5-13, the driving point is attached to a strainer or perforated section of pipe. To reduce friction, the point is somewhat larger than the casing. The driving weight is commonly suspended from a block attached to a tripod. In hard ground a cylindrical shoe equipped with water jets loosens the soil and washes it to the surface. Batteries of driven wells may be connected to a suction header to supply enough water.

**Bored Wells.** Wells can be bored with hand or power augers into sufficiently cohesive (noncaving) soils. Above the water table, the soil is usually held in the auger, which must then be raised from time to time to be cleaned. Below the water table, sand may wash out of the auger and have to be removed from the bore hole by a bailer or sand pump. As the well becomes deeper and deeper, sections of rod are added to the auger stem. Bits up to 36 in. in diameter have been used successfully, and wells have been enlarged

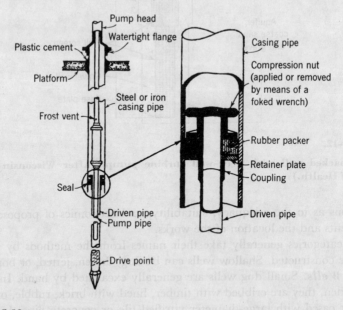

**Figure 5-13.**

**Driven well and its sanitary protection. (After Iowa State Department of Health.)**

in diameter up to 48 in. by reaming. A concrete, tile, or metal casing is inserted in the hole and cemented in place before the strainer is installed.

**Drilled Wells.** High-capacity, deep wells are constructed by drilling. As the water-bearing materials vary so widely there can be no one method of drilling that can be adopted under all conditions. The method of drilling is selected to suit the particular conditions of a site. The systems of drilling used in water-well construction are based on either the percussion or the rotary principle.

**Collector Wells.** A *collector well* consists of a central shaft of concrete caisson some 15 ft in internal diameter and finished off below the water table with a thick concrete plug. From this shaft, perforated radial pipes 6 or 8 in. in diameter and 100 to 250 ft long are jacked horizontally into water-bearing formation through ports near the bottom of a caisson. The collector pipes may be installed and developed in the same manner as for ordinary wells.

**Pumps.** Many types of well pumps are on the market to suit the wide variety of capacity requirements, depths to water, and sources of power. In the United States almost all well pumps are driven by electric motors.

Domestic systems commonly employ one of the following pumps: (1) for lifts under 25 ft, a small reciprocating or piston pump; (2) for lifts up to 125 ft, a centrifugal pump to which water is lifted by recirculating part of the discharge to a jet or ejector; and (3) for lifts that cannot be managed by jet pumps, a cylinder pump installed in the well and driven by pump rods through a *jack* mounted at the well head. Systems of choice normally incorporate pressure tanks for smooth pressure-switch operation. The well itself may provide enough storage to care for differences between demand rates in the house and flow rates from the aquifer. This is why domestic wells are seldom made less than 100 ft deep even though the water table may lie only a few feet below the ground surface. Deep wells and pump settings maintain the supply when groundwater levels sink during severe droughts or when nearby wells are drawn down steeply.

Large-capacity systems are normally equipped with centrifugal or turbine pumps driven by electric motors. A sufficient number of pump bowls are mounted one above the other to provide the pressure necessary to overcome static and dynamic heads at the lowest water levels. For moderate quantities and lifts, *submersible* motors and pumps, assembled into a single unit, are lowered into the well. The water being pumped cools the compact motors normally employed. Large-capacity wells should be equipped with suitable measuring devices. Continuous records of water levels and rates of withdrawal permit the operator to check the condition of the equipment and the behavior of the source of supply. This is essential information in the study and management of the groundwater resource.

**Development.** Steps taken to open up or enlarge flow passages in the formation in the vicinity of the well are called *development*. Thorough development of the completed well is essential regardless of the method of construction used to obtain higher specific capacities, to increase effective well radius, and to promote efficient operation over a longer period of time. This can be achieved in several ways. The method selected depends on the drilling method used and on the formation in which the well is located. The most common method employed is overpumping, that is, pumping the well at a higher capacity than the design yield. Temporary equipment can provide the required pumping rates. Pumping is continued until no sand enters the well. Other methods used include flushing, surging, high-velocity jetting, and back-washing. Various chemical treatments and explosives are used in special circumstances.

**Testing.** After a well is completed, it should be tested to determine its characteristics and productivity. Constant-rate and step-drawdown pumping tests are used for this purpose. The test should be of sufficient duration; the specific capacity of a well based on a one-hour test may be substantially higher than that based on a 1-day test. Longer duration is also required to detect the effect of hydraulic boundaries, if any. The extent to which the specific yield would decrease depends on the nature and the effectiveness of the boundaries.

**Sanitary Protection of Wells.** The design and construction of a well to supply drinking water should incorporate features to safeguard against contamination from surface and subsurface sources. The protective measures vary with the geologic formations penetrated and the site conditions. The well should be located at such a distance from the possible sources of pollution (for example, wells used for the disposal of liquid wastes or artificial recharge; seepage pits; and septic tanks) that there is no likelihood of contaminated water reaching the well. The casing should be sufficiently long and watertight to seal off formations that have undesirable characteristics. Failure to seal off the annular space between the casing and well hole has been responsible for bacterial contamination in many instances. The casing should be sealed in place by filling the open space around the casing with cement grout or other impermeable material down to an adequate depth. This prevents seepage of water vertically along the outside of the pipe. A properly cemented well is shown in Fig. 5-14. The well casing should extend above the ground. The top of the well should contain a watertight seal; the surface drainage should be away from it in all directions.

An essential final step in well completion is the thorough disinfection of the well, the pump, and the piping system. Although the water in the aquifer itself may be of good sanitary quality, contamination can be introduced into the well system during drilling operations and the installation of other ele-

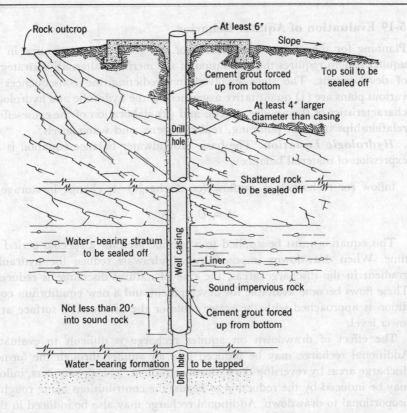

**Figure 5-14.**
**Drilled well and its sanitary protection. (After Iowa State Department of Health.)**

ments of the system. Periodic disinfection of the well during the drilling is a good practice and should be encouraged. In the case of an artificially gravel-packed well, all gravel-pack material should be sterilized before being placed in the well. Solution strengths of 50–200 ppm chlorine are commonly used for sterilizing wells. The effectiveness of disinfection should be checked after the completion of the work. Disinfection of the system is also necessary after repairs of any part of the system.

*Maintenance.* Good maintenance extends the life of a well. The maintenance of the yield of a well depends on (a) the well construction, (b) the quality of water pumped (water may be corroding or encrusting), and (c) the interference from neighboring wells. If the performance of the well declines, renovation measures should be undertaken that may include mechanical cleaning, surging, and chemical treatment.

## 5-19 Evaluation of Aquifer Behavior

Planning for the optimum utilization of the groundwater resource in an aquifer system requires the evaluation of the merits of alternative strategies of development. The steps involved in predicting the consequences of various plans are (1) quantitative assessment of the hydraulic and hydrologic characteristics of the aquifer system, and (2) elaboration of the cause-effect relationships between pumping, replenishment, and water levels.

***Hydrologic Equation.*** The basic groundwater balance equation is an expression of material balance:

inflow (or recharge) = outflow (or discharge) ± change in storage

$$I = O \pm \Delta S$$

This equation must be applied to a specific area for a specific period of time. When drawdowns imposed by withdrawals reduce the hydraulic gradient in the discharge areas, the rate of natural discharge is reduced. These flows become available for development and a new equilibrium condition is approached with the water table or the piezometric surface at a lower level.

The effect of drawdown on aquifer recharge is difficult to evaluate. Additional recharge may be induced into an aquifer through the former discharge areas by reversing the hydraulic gradient. In leaky aquifers, inflow may be induced by the reduction of heads, the contributions being roughly proportional to drawdown. Additional recharge may also be induced in the recharge areas if drawdown causes a dewatering in areas where recharge was limited because the aquifer was full. This is referred to as the capture of *rejected recharge*.

***Safe Yield of an Aquifer.*** The yield of an aquifer depends on (1) the characteristics of the aquifer, (2) the dimensions of the aquifer and the hydraulic characteristics of its boundaries, (3) the vertical position of each aquifer and the hydraulic characteristics of the overlying and underlying beds, and (4) the effect of proposed withdrawals on recharge and discharge of the aquifer. Thus it is evident that the safe yield of an aquifer is not necessarily a fixed quantity, and it is not strictly a characteristic of the groundwater aquifer.[38] It is a variable quantity dependent on natural hydrogeologic conditions, and on recharge and discharge regimes. Safe yield has been defined in a variety of ways, each definition laying emphasis on a particular aspect of groundwater resource development. These include, within economic limits, (1) development to the extent that withdrawals

---

[38] Am. Soc. of Civil Engrs., Ground-Water Basin Management, *Manual of Engineering Practice No. 40* (1961), p. 52.

balance recharge, and (2) development to the extent that change in the quality of groundwater allows.

## 5-20 Groundwater Quality Management

In a majority of cases when polluted water has been drawn from wells, the contamination was introduced at the well site, indicating faulty construction. There are, however, numerous examples of contamination of groundwater caused by disposal of wastes. Once groundwater is contaminated, the impairment of the groundwater resource is long-lasting, and recovery is extremely slow.

To predict where the contaminating fluids will go requires a three-dimensional geologic, hydrodynamic, and geochemical analysis. The principles governing the subsurface dissemination of polluted fluids in natural systems are not well established. The rate and extent of the spread of pollution are controlled by (a) the characteristics of the source of pollution, (b) the nature of rock formations in the unsaturated and saturated zones, and (c) the physical and chemical properties of the contaminant. The phenomena governing the disposition of the contaminant are capillary attraction, decay, adsorption, dispersion, and diffusion.

There have been numerous examples of contamination of groundwater by wastes allowed to seep into ground, wastes discharged into pits and ponds, and leaks from holding tanks and sewers. The safe distance from a polluting source of this type is determined to a large degree by the velocity of percolation through the unsaturated zone and by the lateral movement once the contamination reaches groundwater. Water-table aquifers, being near the surface and having a direct hydrologic connection to it, are more subject to contamination than are deeper-lying artesian aquifers.

The discharge of wastes into streams has had both direct and indirect effect on the quality of groundwater. The polluted rivers that cross recharge areas of artesian aquifers tapped by wells have affected the quality of their discharge. The aquifers that are replenished by infiltration from polluted streams will eventually be contaminated by soluble chemical wastes carried in the stream. Induced contamination of an aquifer can result when the cone of depression of a discharging well intersects a polluted river. This is frequently the case in coastal areas in wells located near streams containing brackish water. Artificial recharge with river water of poorer quality than that found in the aquifer will ultimately result in the deterioration of the quality of groundwater.

***Biological Contamination.*** Because of increasing numbers of septic tanks and growing use of effluents from wastewater treatment plants for artificial recharge of aquifers, possible contamination of groundwater by bacteria and viruses needs consideration. Filtration through granular material improves the biological quality of water. A 10-ft downward percolation in fine sand is capable of removing all bacteria from water. The length of time bacteria and viruses may survive and the distance they may travel through specific rock materials in different subsurface environmental conditions are uncertain. They seem to behave in a manner similar to the degradable and adsorbable contaminants. Romero[39] finds that under favorable conditions some bacteria and viruses may survive up to at least 5 years in the underground environment. However, the distances traveled in both the saturated and unsaturated media are surprisingly short when reasonable precautions are taken in disposal. The principal determinant of the distance traveled seems to be the size of the media. Romero gives diagrams that may be used to evaluate the feasibility of disposing of biologically contaminated wastes in saturated and unsaturated granular media. The danger of bacterial pollution is greater in fractured rocks, cavernous limestones, and gravel deposits where the granular materials have no filtering capacity. The distances traveled will be higher in areas of influence of discharging or recharging wells because higher velocities are present. The higher rates of artificial recharge and greater permeability of artificial recharge basins enable bacteria to be carried to a greater depth.

**Subsurface Disposal of Liquid Wastes.**[40] Subsurface space may be used to an increasing degree for the disposal of wastes. At present, the oil industry pumps nearly 20 million barrels of salt water per day into subsurface formations from which oil has been extracted. Some highly toxic chemical wastes are disposed of underground. The use of an aquifer as a receptacle of toxic waste materials is justified only if it has little or no value as a present, or potential, source of water supply. Further, there should not be any significant risk of contaminating other aquifers or of inducing fractures in the confining formations. Recharging of groundwater by injection or spreading of reclaimed municipal wastewaters is an accepted practice that will undoubtedly be further developed in the future.

---

[39] J. C. Romero, The Movement of Bacteria and Viruses through Porous Media, *Ground Water*, **8**, 2 (1970).

[40] P. T. Flawn, *Environmental Geology*, Harper and Row (1970).

six

# water transmission

## 6-1 Transmission Systems

Supply conduits, or aqueducts,[1] transport water from the source of supply to the community and so form the connecting link between collection works and distribution systems. Source location determines whether conduits are short or long, and whether transport is by gravity or pumping. Depending on topography and available materials, conduits are designed for open-channel or pressure flow. They may follow the hydraulic grade line as canals dug through the ground, flumes elevated above the ground, grade aqueducts laid in balanced cut and cover at the ground surface, and grade tunnels penetrating hills; or they may depart from the hydraulic grade line as pressure aqueducts laid in balanced cut and cover at the ground surface, pressure tunnels dipping beneath valleys or hills, and pipelines of fabricated materials following the ground surface, if necessary over hill and through dale, sometimes even rising above the hydrau-

---

[1] The word *aqueduct* comes from the Latin *aqua*, water, and *ducere*, to lead or conduct. It describes all artificial channels that transport water. Engineers often apply the word more specifically to covered masonry conduits built in place. Because they lacked pressure-resisting materials, the Romans constructed aqueducts tapping high-lying clean sources of water and conveyed it along the hydraulic grade line to the city, where it was distributed by gravity.

lic grade line.[2] The profile and typical cross-sections of a supply conduit are shown in Fig. 6-1. Static heads and hydraulic grade lines are indicated for pressure conduits.

## 6-2 Fluid Transport

The hydraulic design of supply conduits is concerned chiefly with (1) resistance to flow in relation to available and needed heads or pressures and (2) required and allowable velocities of flow relative to cost, scour, and sediment transport. In long supply lines, *frictional* or *surface resistance* offered by the pipe interior is the dominant element. *Form resistance* responsible for losses in transitions and appurtenances is often negligible. In short transport systems, on the other hand, form resistance may be of controlling importance.

*A Rational Equation for Surface Resistance.* The most nearly rational relationship between velocity of flow and head loss in a conduit is also one of the earliest. Generally referred to as the Darcy-Weisbach formula,[3] it is actually written in the form suggested by Weisbach, rather than Darcy, namely:

$$h_f = f(l/d)(v^2/2g) \qquad (6\text{-}1)$$

where $h_f$ is the head loss[4] in a pipe of length $l$ and diameter $d$ through which a fluid is transported at a mean velocity $v$; $g$ is the acceleration of gravity; and $f$ is a dimensionless friction factor. In the more than 100 years of its existence, use, and study, this formulation has been foremost in the minds of engineers concerned with the transmission of water as well as other fluids. That this has often been so in a conceptual rather than a practical sense does not detract from its importance.

*An Exponential Equation for Surface Resistance.* Because of practical shortcomings of the Weisbach formula, engineers have resorted to

---

[2] The Colorado River Aqueduct of the Metropolitan Water District of Southern California is 242 miles long and includes 92 miles of grade tunnel, 63 miles of canal, 54 miles of grade aqueduct, 29 miles of inverted siphons, and 4 miles of force main. The Delaware Aqueduct of New York City comprises 85 miles of pressure tunnel in three sections. Pressure tunnels 25 miles long supply the metropolitan districts of Boston and San Francisco. The supply conduits of Springfield, Mass., are made of steel pipe and reinforced-concrete pipe; those of Albany, N.Y., of cast-iron pipe.

[3] H. P. G. Darcy, *Recherches experimentales relatives au mouvement de l'eau dans les tuyaux* (Experimental Investigations on the Flow of Water in Pipes), Paris, 1857. Julius Weisbach, *Lehrbuch der Ingenieur- und Machinen-Mechanik* (Manual of Engineering and Machine Mechanics), Burnswick, Germany, 1845. Darcy's name has appeared before in footnote 16, Chap. 5. Weisbach (1806–1871) taught engineering mechanics at the School of Mines in the Erzgebirge city of Freiberg near Dresden, Germany.

[4] What engineers call head loss or lost head is more specifically the energy lost by a unit weight of water because of surface resistance within the conduit, mechanical energy being converted into nonrecoverable heat energy.

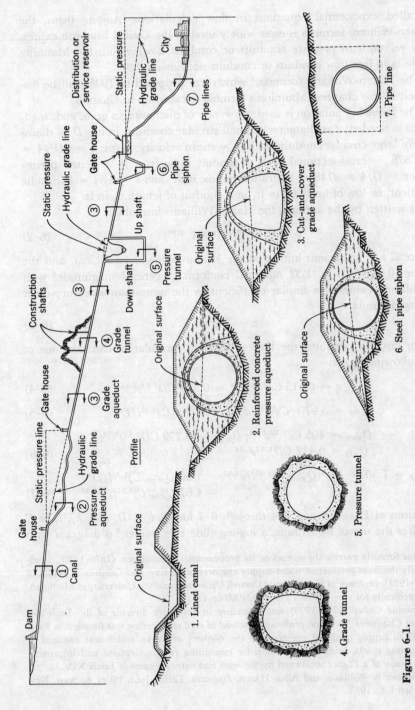

**Figure 6-1. Profile and typical cross-sections of a water-supply conduit.**

so-called exponential equations in flow calculations. Among them, the Hazen-Willians formula is most widely used in the United States to express flow relations in pressure conduits or conduits flowing full, the Manning formula in free-flow conduits or conduits not flowing full.

The Hazen-Williams formula,[5] which was proposed in 1905, will be discussed in this chapter; Manning's formula is taken up in Chap. 8.

The following notation is used: $Q$ = rate of discharge, in gpm, gpd, mgd, or cfs as needed; $d$ = diameter of small circular conduits, in in.; $D$ = diameter of large circular conduits, in ft; $v$ = mean velocity, in fps; $a = \pi D^2/4 = \pi d^2/576$ = cross-sectional area of conduit in sq ft; $r = a/$wetted perimmeter $= D/4 = d/48$ = hydraulic radius, in ft; and $s = h_f/l$ = hydraulic gradient, or loss of head $h_f$, in ft, in a conduit of length, $l$, in ft.

As written by the authors, the Hazen-Williams formula is

$$v = Cr^{0.63}s^{0.54}(0.001^{-0.04}) \qquad (6\text{-}2)$$

where $C$ is a coefficient known as the Hazen-Williams coefficient, and the factor $(0.001^{-0.04}) = 1.32$ makes $C$ conform in general magnitude[6] with established values of a similar coefficient in the more-than-a-century-older Chezy[7] formula

$$v = C\sqrt{rs} \qquad (6\text{-}3)$$

For circular conduits, the Hazen-Williams formulation can take one of the following forms:

$$v = 0.115\ Cd^{0.63}s^{0.54} = 0.550\ CD^{0.63}s^{0.54} \qquad (6\text{-}4)$$

$$h_f = 5.47(v/C)^{1.85}l/d^{1.17} = 3.02(v/C)^{1.85}l/D^{1.17} \qquad (6\text{-}5)$$

$$Q_{gpd} = 405\ Cd^{2.63}s^{0.54};\ Q_{mgd} = 0.279\ CD^{2.63}s^{0.54};$$
$$Q_{cfs} = 0.432\ CD^{2.63}s^{0.54} \qquad (6\text{-}6)$$

$$h_f = 1.50 \times 10^{-5}(Q_{gpd}/C)^{1.85}l/d^{4.87} = 10.6(Q_{mgd}/C)^{1.85}l/D^{4.87}$$
$$= 4.67(Q_{cfs}/C)^{1.85}l/D^{4.87} \qquad (6\text{-}7)$$

Solutions of Eqs. 6-2 and 6-4 through 6-7 for $Q$, $v$, $r$, $D$, $d$, $s$, $h_f$, $l$, or $C$ requires the use of logarithms, a log-log slide rule, tables,[8] a diagram with

---

[5] This formula carries the names of its proponents, Allen Hazen (1870–1930), intellectually the most productive water-supply engineer of his day, and Gardner S. Williams (1866–1931), professor of hydraulics at Cornell University and the University of Michigan.

[6] Specifically for $r = 1$ ft and $s = 1$ ft/1000 or 1‰.

[7] Antoine Chézy (1718–1798) was a teacher in and later director of the École des Ponts et Chaussées. This first professional school of civil engineering was founded in Paris in 1747 to supply qualified engineers to the ministry of works, which was engaged in constructing roads, bridges, and canals for expediting communications and improving the economy of a France weakened by the wars and extravagances of Louis XIV.

[8] Gardner S. Williams and Allen Hazen, *Hydraulic Tables*, John Wiley & Sons, New York, 3rd Ed., 1933.

logarithmic scales (such as Fig. 2 in the Appendix), or an alignment chart.

The weakest element in the Hazen-Williams formula is the estimate of $C$ in the absence of measurements of loss of head and discharge or velocity.

Values of $C$ vary for different conduit materials and their relative deterioration in service. They vary somewhat also with size and shape. The values listed in Table 6-1 reflect more or less general experience.

### Table 6-1

*Values of the Hazen-Williams Coefficient $C$ for Different Conduit Materials and Age of Conduit*

| | Age | |
|---|---|---|
| Conduit Material | New | Uncertain |
| Cast-iron pipe, coated (inside and outside) | 130 | 100 |
| Cast-iron pipe, lined with cement or bituminous enamel | 130[a] | 130[a] |
| Steel, riveted joints, coated | 110 | 90 |
| Steel, welded joints, coated | 140 | 100 |
| Steel, welded joints, lined with cement or bituminous enamel | 140[a] | 130[a] |
| Concrete | 140 | 130 |
| Wood stave | 130 | 130 |
| Cement-asbestos and plastic pipe | 140 | 130 |

[a] For use with the nominal diameter, i.e., diameter of unlined pipe.

For purposes of comparison, the size of a noncircular conduit can be stated in terms of the diameter of a circular conduit of equal carrying capacity. For identical values of $C$ and $s$, multiplication of Eq. 6-2 by the conduit area $a$ in square feet and equating the resulting expression to Eq. 6-5, the diameter of the equivalent conduit becomes

$$D = 1.53 \, a^{0.38} r^{0.24} \qquad (6\text{-}8)$$

Variation in the hydraulic elements of circular and noncircular conduits with depth of flow is discussed in Chap. 8.

### Example 6-1.

The pipe-flow diagram at the end of this volume establishes the numerical relationships between $Q$, $v$, $d$, and $s$ for a value of $C = 100$. Conversion to other magnitudes of $C$ is simple because both $v$ and $Q$ vary directly as $C$. Show the mathematical and graphical basis of this diagram.

1. Written in logarithmic form, Eq. 6–6 is (a) $\log Q = 4.61 + 2.63 \log d + 0.54 \log s$, or (b) $\log s = -8.54 - 4.87 \log d + 1.85 \log Q$.

A family of straight lines of equal slope is obtained, therefore, when $s$ is plotted against $Q$ on log-log paper for specified diameters $d$. Two points define each line. Pairs of coordinates for a 12-in. pipe, for example, are (a) $Q = 100,000$ gpd, $s = 0.028‰$; and (b) $Q = 1,000,000$ gpd, $s = 2.05‰$.

2. Written in logarithmic form, Eq. 6–4 is $\log v = 1.0607 + 0.63 \log d + 0.54 \log s$. If the diameter $d$ is eliminated from the logarithmic transforms of Eqs. 6–6 and 6–4, (a) $\log Q = 0.180 + 4.17 \log v - 1.71 \log s$ and (b) $\log s = 0.105 + 2.43 \log v - 0.585 \log Q$.

A family of straight lines of equal slope is obtained when $s$ is plotted against $Q$ on log-log paper for specified velocities $v$. Two points define each line. Pairs of coordinates for a valocity of 1 fps, for example are (a) $Q = 100,000$ gpd, $s = 1.5‰$ and (b) $Q = 10,000,000$ gpd, $s = 0.10‰$.

*Example 6–2.*

A tunnel of horseshoe shape (Fig. 6–3, for example) has a cross-sectional area of 27.9 sq ft and a hydraulic radius of 1.36 ft. Find the diameter, hydraulic radius, and area of the hydraulically equivalent circular conduit.

By Eq. 6–8, the diameter $D = 1.53 \times (27.9)^{0.38} \times (1.36)^{0.24} = 5.85$ ft; the hydraulic radius $r = D/4 = 1.46$ ft; and the area $a = (\pi D^2)/4 = 26.7$ sq ft.

It should be noted that neither the cross-sectional area nor the hydraulic radius of this equivalent circular conduit is the same as that of the horseshoe section proper.

*Form Resistance.* Pipeline transitions and appurtenances add *form* resistance to *surface* resistance. Head losses are stepped up by changes in cross-sectional geometry and changing directions of flow. Expansion and contraction exemplify geometric change; elbows and branches, directional change. Valves and meters as well as other appurtenances may create both geometrical and directional change. With rare exceptions, head losses are expressed either in terms of velocity heads, such as $kv^2/2g$, or as equivalent lengths of straight pipe, $l_e = kv^2/2gs = kD/f$. The outstanding exception is the loss on sudden expansion or enlargement called the Borda[9] loss $(v_1 - v_2)^2/2g$, where $v_1$ is the velocity in the original conduit and $v_2$ the velocity in the expanded conduit; even it, however, is sometimes converted, for convenience, into $kv^2/2g$. Because continuity as $a_1v_1 = a_2v_2$ equates $k_1v_1^2/2g$ with $(v_1^2/2g)(1 - a_1/a_2)^2$, the loss at the point of discharge of a pipeline into a reservoir (making $a_2$ very large in comparison with $a_1$) equals ap-

---

[9] Named after Jean Charles Borda (1733–1799), French military engineer.

proximately $v_1^2/2g$; consequently, there is no recovery of energy. In all but special cases like this, $k$ must be determined experimentally. When there is no experimental information, the following values of $k$ give useful first approximations on likely losses:

|  | Value of $k$ |  | Value of $k$ |
|---|---|---|---|
| Sudden contraction[a] | 0.3–0.5 | Valve (open), gate | 0.2 |
| Entrance,[b] sharp | 0.5 | With reducer and | 0.5 |
| Well-rounded | 0.1 | increaser |  |
| Elbow,[c] 90° | 0.5–1.0 | Globe | 10 |
| 45° | 0.4–0.75 | Angle | 5 |
| 22.5° | 0.25–0.5 | Swing check | 2.5 |
| Tee, 90° take-off | 1.5 | Meter, venturi | 0.3 |
| Straight run | 0.3 | Orifice | 1.0 |
| Coupling | 0.3 |  |  |

[a] Varying with area ratios.

[b] An additional decrease of head of $v^2/2g$ ($k = 1$) is required to establish motion. This energy per unit weight may or may not be reconverted to pressure energy pending on downstream geometry.

[c] Varying with radius ratios.

**Hydraulic Transients.** Transmission lines are subjected to transient pressures when valves are opened or closed or when pumps are started or stopped (Sec. 6–9). Water hammer and surge are among such transient phenomena.[10]

Water hammer is the pressure rise accompanying a sudden change in velocity. When velocity is decreased in this way, energy of motion must be stored by elastic deformation of the system. The sequence of phenomena that follows sudden closure of a gate, for example, is quite like what would ensue if a long, rigid spring, traveling at uniform speed, were suddenly stopped and held stationary at its forward end. A pressure wave would travel back along the spring as it compressed against the point of stoppage. Kinetic energy would change to elastic energy. Then the spring would vibrate back and forth. In a pipe, compression of the water and distention of the pipe wall replace the compression of the spring. The behavior of the pressure wave and the motion of the spring and the water are identically described by the differential equations for one-dimensional waves. Both systems would vibrate indefinitely, were it not for the dissipation of energy by internal friction.

Water hammer is held within bounds in small pipelines by operating them at moderate velocities, because the pressure rise in pounds per square inch

[10] G. R. Rich, *Hydraulic Transients*, McGraw-Hill Book Co., New York, 1951.

cannot exceed about fifty times the velocity expressed in feet per second. In larger lines the pressure is held down by arresting flows at a sufficiently slow rate to allow the relief wave to return to the point of control before pressures become excessive. If this is not practicable, pressure-relief or surge valves are introduced.

Very large lines, 6 ft or more in equivalent diameter, operate economically at relatively high velocities. However, the cost of making them strong enough to withstand water hammer would ordinarily be prohibitive if the energy could not be dissipated slowly in surge tanks. In its simplest form, a surge tank is a standpipe at the end of the line next to the point of velocity control. If this control is a gate, the tank accepts water and builds up back pressure when velocities are regulated downward. When demand on the line increases, the surge tank supplies immediately needed water and generates the excess hydraulic gradient for accelerating the flow through the conduit. Following a change in the discharge rate, the water level in a surge tank oscillates slowly up and down until excess energy is dissipated by hydraulic friction in the system.

## 6-3 Capacity and Size of Conduits

With rates of water consumption and fire demand known, the capacity of individual supply conduits depends on their position in the waterworks system and the choice of the designer for (1) a structure of full size or (2) duplicate lines staggered in time of construction.

Minimum workable size, as already stated, is one controlling factor in the design of tunnels. Otherwise, size is determined by hydraulic and economic considerations. For a gravity system, that is, where pumping is not required, controlling hydraulic factors are available heads and allowable velocities. Head requirements include proper allowances for drawdown of reservoirs and maintenance of pressure in the various parts of the community, under conditions of normal as well as peak demand. Reservoir heads greater than necessary to transport water at normal velocities may be turned into power when it is economical to do so.

Allowable velocities are governed by the characteristics of the water carried and the magnitude of the hydraulic transients. For silt-bearing waters, there are both lower and upper limits of velocity; for clear water, only an upper limit. The minimum velocity should prevent deposition of silt; it lies in the vicinity of 2 to 2.5 fps. The maximum velocity should not cause erosion or scour, nor should it endanger the conduit by excessive water hammer when gates are closed quickly. Velocities of 4 to 6 fps are common, but the upper limit lies between 10 and 20 fps for most materials of which supply conduits are built and for most types of water carried. Unlined canals impose greater restrictions. Silting and scouring are discussed in connection with self-cleansing velocities in sewers (Sec. 8-3) and with the design of grit chambers (Sec. 13-15). Some of the formulations advanced there are also applicable here.

The size of force mains and of gravity mains that include power generation is fixed by the relative cost or value of the conduit and the cost of pumping or power.

When aqueducts include more than one kind of conduit, the most economical distribution of the available head among the component classes is effected when the change in cost $\Delta c$ for a given change in head $\Delta h$ is the same for each kind. The proof for this statement is provided by Lagrange's method[11] of undetermined multipliers. As shown in Fig. 6-2 for three components of a conduit with an allowable, or constrained, head loss $H$, the Lagrangian requirement of $\Delta c_1/\Delta h_1 = \Delta c_2/\Delta h_2 = \Delta c_3/\Delta h_3$ is met when parallel tangents to the three $c:h$ curves identify, by trial, three heads $h_1$, $h_2$, and $h_3$ that satisfy the constraint $h_1 + h_2 + h_3 = H$.

## 6-4 Multiple Lines

Although masonry aqueducts and tunnels of all kinds are best designed to the full projected capacity of the system, this is not necessarily so for pipe lines. Parallel lines built a number of years apart may prove to be more economical. Cost, furthermore, is not the only consideration. It may be expedient to lay

[11] Developed by the French mathematician Joseph-Louis Lagrange (1736–1813).

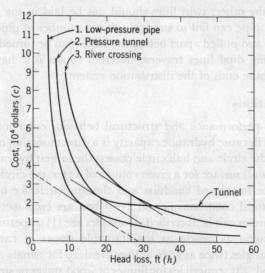

**Figure 6-2.**
**Lagrangian optimization of conduit sections by parallel tangents.**

*Example 6-3.*

Given the costs and losses of head shown in Fig. 6-2 for three sections of a conduit, find the most economical distribution of the available head $H = 60$ ft between the three sections:

By trial,
$h_1 = 13.5$     $c_1 = 2.0 \times 10^4$
$h_2 = 19.0$     $c_2 = 2.1 \times 10^4$
$h_3 = 27.5$     $c_3 = 2.2 \times 10^4$

$H = 60.0$     $C = 6.3 \times 10^4$

more than one line (1) when the maximum pipe size of manufacture is exceeded—36 in. in the case of centrifugal cast-iron pipe, for example; (2) when possible failure would put the line out of commission for a long time; and (3) when pipe location presents special hazards—floods, ice, and ships' anchors endangering river crossings or submarine pipes and cave-ins rupturing pipe lines in mining areas, for example.

Twin lines generally cost 30 to 50% more than a single line of equal capacity. If they are close enough to be interconnected at frequent intervals, gates should be installed in the bridging pipes to keep most of the system in operation during repairs to affected parts. However, if failure of one line will endanger the other, twin lines should not be laid in the same trench. Thus, cast-iron pipe can fail so suddenly that a number of pipe lengths will be undermined and pulled apart before the water can be turned off. Another reason for having dual lines traverse different routes is to have them feed water into opposite ends of the distribution system.

## 6-5 Cross-Sections

Both hydraulic performance and structural behavior enter into the choice of cross-section. Because hydraulic capacity is a direct function of the hydraulic radius, and the circle and half circle possess the largest hydraulic radius or smallest (frictional) surface for a given volume of water, the circle is the cross-section of choice for closed conduits and the semicircle for open conduits whenever structural conditions permit. Next best are cross-sections in which circles or semicircles can be inscribed. Examples are (1) trapezoids approaching half a hexagon as nearly as maintainable slopes of canals in earth permit; (2) rectangles twice as wide as they are deep for canals and flumes of masonry or wood; (3) semicricles for flumes of wood staves or steel; (4) circles for pressure aqueducts, pressure tunnels, and pipelines; and (5) horseshoe sections for grade aqueducts and grade tunnels.

Internal pressures are best resisted by cylindrical tubes and materials strong in tension; external earth and rock pressures (not counterbalanced by internal pressures) by horseshoe sections and materials strong in compression. By design, the hydraulic properties of horseshoe sections are only slightly poorer than are those of circles. Moreover, their relatively flat *invert* makes for easy transport of excavation and construction materials in and out of the aqueduct. As shown in Fig. 6-3, four circular arcs are struck to form the section: a circular arc rising from the *springing line* of the arch at half depth, two lateral arcs struck by radii equaling the height of the *crown* above the invert, and a circular arc of like radius establishing the bottom.

## 6-6 Structural Requirements

Structurally, closed conduits must resist a number of different forces singly or in combination: (1) internal pressure equal to the full head of water to which the conduit can be subjected; (2) unbalanced pressures at bends, contractions, and closures; (3) water hammer or increased internal pressure caused by sudden reduction in the velocity of the water—by the rapid closing of a gate or shutdown of a pump, for example; (4) external loads in the form of backfill and traffic; (5) their own weight between external supports (piers or hangers); and (6) temperature-induced expansion and contraction.

Internal pressure, including water hammer, creates transverse stress or *hoop tension*. Bends and closures at dead ends or gates produce unbalanced pressures and *longitudinal stress*. When conduits are not permitted to change length, variations in temperature likewise create longitudinal stress. External loads and foundation reactions (manner of support), including the

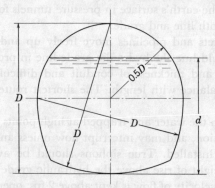

**Figure 6-3.**
**Common proportions of horseshoe sections.**

weight of the full conduit, and atmospheric pressure (when the conduit is under a vacuum), produce *flexural stress*.

In jointed pipes, such as bell-and-spigot cast-iron pipes, the longitudinal stresses must either be resisted by the joints or relieved by motion. Mechanical joints offer such resistance. The resistance of lead and lead substitute joints in bell-and-spigot cast-iron pipe to being pulled apart can be estimated from Prior's observational equation:[12]

$$p = \frac{3800}{d+6} - 40; \qquad P = \left(\frac{3000}{d+6} - 31\right)d^2 \qquad (6\text{-}9)$$

where $p$ is the intensity of pressure in psig and $P$ is the total force in lb.

Tables of standard dimensions and laying lengths are found in professional manuals, specifications of the American Water Works Association, and publications of manufacturers and trade associations.

## 6-7 Location

Supply conduits are located in much the same way as railroads and highways.

*Line and Grade.* The invert of a grade aqueduct or grade tunnel is placed on the same slope as the hydraulic grade line. Cut and fill, as well as cut and cover, are balanced to maintain a uniform gradient and reduce haul. Valleys and rivers that would be bridged by railroads and highways may be bridged also by aqueducts. Such indeed was the practice of ancient Rome; but modern aqueducts no longer rise above valley, stream, and hamlet except where a bridge is needed primarily to carry road or railway traffic. Pressure conduits have taken their place. Sometimes they are laid in trenches as sag pipes to traverse valleys and pass beneath streams; sometimes they strike deep below the earth's surface in pressure tunnels for which geological exploration fixes both line and grade.

Pressure aqueducts and pipelines move freely up and down slopes. For economy they should hug the hydraulic grade line in profile and a straight line in plan. Size and thickness of conduit and difficulty of construction must be kept in balance with length. The shortest route is not necessarily the cheapest.

Air released from the water and trapped at high points reduces the waterway, increases friction, and may interrupt flow unless an air relief valve or vacuum pump is installed. True siphons should be avoided if possible. However, if the height of rise above the hydraulic grade is confined to less than 20 ft and the velocity of flow is kept above 2 fps, operating troubles will

[12] J. C. Prior, Investigation of Bell-and-Spigot Joints in Cast-Iron Water Pipes, *Ohio State Univ. Eng. Exp. St. Bull.* **87** (1935).

be few. For best results, the line should leave the summit at a slope less than that of the hydraulic gradient.

In practice, possible locations of supply conduits are examined on available maps of the region; the topographic and geologic sheets of the U.S. Geological Survey are useful examples. Route surveys are then carried into the field. Topography and geology are confirmed and developed in needed detail, possibly by aerial surveys, borings, and seismic exploration. Rights of way, accessibility of proposed routes, and the nature of obstructions are also identified. The use of joint rights-of-way with other utilities may generate economies.

**Vertical and Horizontal Curves.** In long supply lines, changes in direction and grade are effected gradually in order to conserve head and avoid unbalanced pressures. Masonry conduits built in place can be brought to any desired degree of curvature by proper form work. Cast-iron and other sectional pipelines are limited in curvature by the maximum angular deflection of standard lengths of pipe at which joints will remain tight. The desired curve is built up by the necessary number of offsets from the tangent. Sharper curves can be formed by shorter or shortened pipes. The smaller the pipe, the sharper can be the deflection. Welded pipelines less than 15 in. in diameter are sufficiently flexible to be bent in the field. The ends of larger steel pipe must be cut at an angle that depends on the type of transverse joint, the thickness of the steel plate, and the size of the pipe.

For sharp curves, transitions, and branches, special fittings are often built up or manufactured of the same materials as the main conduit.

**Depth of Cover.** Conduits that follow the surface of the ground are generally laid below the frost line, although the thermal capacity and latent heat of water are so great that there is little danger of freezing so long as the water remains in motion. To reduce the external load on large conduits, only the lower half may be laid below frost. Along the forty-second parallel of latitude, which describes the southern boundaries of Massachusetts, upper New York, and Michigan in the United States, frost seldom penetrates more than 5 ft beneath the surface; along the forty-fifth parallel the depth increases to 7 ft. The following equation approximates Shannon's[13] observations of frost depth:

$$d = 1.65 \, F^{0.468} \qquad (6\text{--}10)$$

where $d$ is the depth of frozen soil in inches and $F$, the freezing index, is the algebraic difference between the maximum positive and negative cumulative departures, $\Sigma(T_d - 32)$, of the daily mean temperatures ($T_d$) from 32°F. Accumulation, as shown in Fig. 6-4, begins with the first day on which a

[13] W. L. Shannon, Prediction of Frost Penetration, *J. New Eng. Water Works Assoc.*, **59**, 356 (1945).

176/Water Transmission

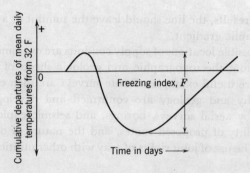

**Figure 6-4.**
**Determination of the freezing index of soils as the cumulative departure of the mean daily temperature from 32°F. (After Shannon.)**

freezing temperature is recorded. In concept, the freezing index is analogous to the *degree day*, which describes the heat requirements of buildings during the heating season. In the absence of daily readings, the value of $F$ may be approximated, in North America, from the mean monthly temperatures as follows:

$$F = (32n - \Sigma T_m)30.2 \qquad (6\text{-}11)$$

Here $n$ is the number of months during which the temperature is less than 32°F. $T_m$ is the sum of the mean temperatures during each of these months, and 30.2 is the mean number of days in December, January, February, and March.

Pipes laid at depths of 2 to 3 ft are safe from extremes of heat and ordinary mechanical damage, but it is wise to go to 5 ft in streets or roads open to heavy vehicles. Otherwise, structural characteristics of conduits determine the allowable depth of cover or weight of backfill. Some conduits may have to be laid in open cut to keep the depth of backfill below the maximum allowable value.

## 6-8 Materials of Construction

Selection of pipeline materials is based on carrying capacity, strength, life or durability, ease of transportation, handling and laying, safety, availability, cost in place, and cost of maintenance. Various types of iron, steel, reinforced concrete, and asbestos cement are most used for water transmission pipes, but plastic and reinforced fiberglass pipes are now being made in the smaller sizes. Other materials may come into use in the future.

*Carrying Capacity.* The initial value of the Hazen-Williams coefficient $C$ hovers around 140 for all types of well-laid pipelines but tends to be somewhat higher for reinforced-concrete and asbestos-cement lines and to drop to a normal value of about 130 for unlined cast-iron pipe. Cast-iron and steel pipes lined with cement or with bituminous enamel possess coefficients of 130 and over on the basis of their nominal diameter; improved smoothness offsets the reduction in cross-section.

Loss of capacity with age or, more strictly, with service depends on (1) the properties of the water carried and (2) the characteristics of the pipe. Modern methods for controlling aggressive water promise that the corrosion of metallic pipes and the disintegration of cement linings and of reinforced-concrete and asbestos-cement pipe will be held in check very largely, if not fully, in the future. In the present state of the art, however, it is not yet possible to estimate how $C$ will change with length of service.

Cement and bituminous-enamel linings and reinforced-concrete and asbestos-cement pipes do not, as a rule, deteriorate significantly with service.

***Strength.*** Steel pipes can resist high internal pressures, but large lines cannot withstand heavy external loads or partial vacuums unless special measures are taken to resist these forces. Cast-iron and asbestos-cement pipes are good for moderately high water pressures and appreciable external loads, provided that they are properly bedded. Prestressed reinforced-concrete pipe is satisfactory for high water pressures. All types of concrete pipe can be designed to support high external loads.

***Durability.*** Experience with all but coated, cast-iron pipe has been too short and changes in water treatment have been too many to give us reliable values on the length of life of different pipe materials. The corrosiveness of the water, the quality of the material, and the type and thickness of protective coating all influence the useful life of the various types of water pipes. External corrosion (soil corrosion) is important, along with internal corrosion. Pipes laid in acid soils, sea water, and cinder fills may need special protection.

***Transportation.*** When pipelines must be built in rugged and inaccessible locations, their size and weight become important. Cast-iron pipe is heavy in the larger sizes; steel pipe relatively much lighter. The normal laying-length of cast-iron pipe is 12 ft;[14] that of steel pipe is 20 to 30 ft. Both prestressed and cast reinforced-concrete pipe are generally fabricated in the vicinity of the pipeline. The sections are 12 and 16 ft long and very heavy in the larger sizes. A diameter smaller than 24 in. is unusual. Asbestos-cement pipe comes in laying-lengths of 18 ft. Its weight is about one quarter that of cast-iron pipe of like diameter.

***Safety.*** Breaks in cast-iron pipes can occur suddenly and are often quite destructive. By contrast, steel and reinforced-concrete pipes fail slowly, chiefly

[14] Lengths of 16 ft, 5 m (16.4 ft), 18 ft, and 20 ft are also available in different types of bell-and-spigot pipe.

by corrosion. However, steel pipelines may collapse under vacuum while they are being drained. With proper operating procedures, this is a rare occurrence. Asbestos-cement pipe fails suddenly, much like cast-iron pipe.

**Maintenance.** Pipelines of all sizes and kinds must be watched for leakage or loss of pressure—outward signs of failure. There is little choice between materials in this respect. Repairs to precast concrete pipe are perhaps the most difficult, but they are rarely required. Cast-iron and small welded-steel pipes can be cleaned by scraping machines and lined in place with cement to restore their capacity. New lines and repaired lines should be disinfected before they are put into service.

**Leakage.** All pipelines should be tested for tightness as they are constructed. Observed leakage is often expressed in gallons per day per inch diameter (nominal) and mile of pipe; but gallons per day per foot of joint is a more rational concept. The test pressure must naturally be stated. To make a leakage test, the line is isolated by closing gates and placing a temporary header or plug at the end of the section to be tested. The pipe is then filled with water and placed under pressure, the water needed to maintain the pressure being measured by an ordinary household meter. Where there is no water, air may be substituted. Losses are assumed to vary with the square root of the pressure, as in orifices.

The allowable leakage of bell-and-spigot cast iron, and cement-asbestos pipe that has been carefully laid and well tested during construction is often set at

$$Q = nd\sqrt{p}/1850 \qquad (6\text{--}12)$$

where $Q$ is the leakage in gallons per hour, $n$ is the number of joints in the length of line tested, $d$ is the nominal pipe diameter in inches, and $p$ is the average pressure during test in pounds per square inch, gage. A mile of 24-in. cast-iron pipe laid in 12-ft lengths and tested under a pressure of 64 psig, for example, can be expected to show a leakage of $Q = (5280/12) \times 24 \times \sqrt{64}/1850 = 46$ gal/hr. Considering that the pipe has a carrying capacity of 250,000 gal/hr at a velocity of 3 fps, the expected leakage from joints is relatively small.

When steel pipe is laid under water with mechanical couplings, small leaks are hard to detect, and allowances as high as 6 gpd per foot of transverse joint may have to be made.

## 6.9 Appurtenances

To isolate and drain pipeline sections for test, inspection, cleaning, and repairs, a number of appurtenances, or auxiliaries, are generally installed in the line (Figs. 6–5 and 6–6; also see Fig. 7–5).

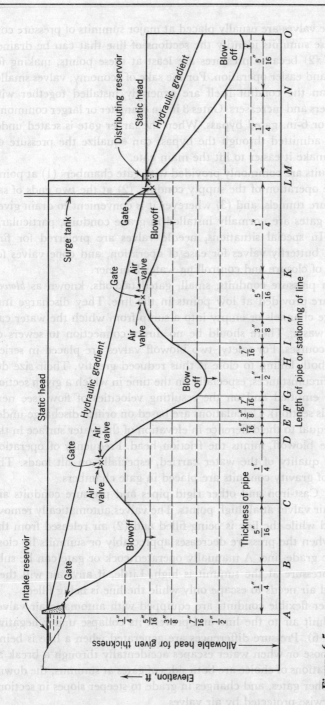

**Figure 6-5.** Profile of pipeline showing pipe thickness and location of gates, blowoffs, and air valves. (Not to scale, vertical scale magnified.)

***Valves.*** Gate valves are usually placed at major summits of pressure conduits (1) because summits identify the sections of line that can be drained by gravity and (2) because pressures are least at these points, making for cheaper valves and easier operation. For the sake of economy, valves smaller in diameter than the conduit itself are generally installed together with necessary reducers and increasers. Gates 8 in. in diameter or larger commonly include a 4-in. or 6-in. gated bypass. When the larger gate is seated under pressure, water admitted through the bypass can equalize the pressure on both sides and make it easier to lift the main gate.

Gravity conduits are commonly provided with gate chambers (1) at points strategic for the operation of the supply conduit, (2) at the two ends of sag pipes and pressure tunnels, and (3) wherever it is convenient to drain given sections. Sluice gates are normally installed in grade conduits, particularly in large ones. In special situations, needle valves are preferred for fine control of flow, butterfly valves for ease of operation, and cone valves for regulating time of closure and controlling water hammer.

***Blowoffs.*** In pressure conduits, small, gated takeoffs, known as *blowoff* or *scour valves*, are provided at low points in the line. They discharge into natural drainage channels or empty into a sump from which the water can be pumped to waste. There should be no direct connection to sewers or polluted water courses. For safety, two blowoff valves are placed in series. The chance of both failing to close is thus reduced greatly. Their size depends on local circumstances, especially on the time in which a given section of line is to be emptied and on the resulting velocities of flow (see next paragraph of this section). Calculations are based on orifice discharge under a falling head, equal to the difference in elevation of the water surface in the conduit and the blowoff, minus the friction head. Frequency of operation depends on the quality of the water carried, especially on silt loads. The drainage gates of gravity conduits are placed in gate chambers.

***Air Valves.*** Cast-iron and other rigid pipes and pressure conduits are equipped with air valves at all high points. The valves automatically remove (1) air displaced while the line is being filled and (2) air released from the flowing water when the pressure decreases appreciably or summits lie close to the hydraulic grade line. A manually operated cock or gate can be substituted if the pressure at the summit is high. Little, if any, air will then accumulate, and air needs to escape only while the line is being filled.

Steel and other flexible conduits are equipped with automatic air valves that will also admit air to the line and prevent its collapse under negative pressure (Fig. 6-6). Pressure differences are generated when a line is being drained on purpose or when water escapes accidentally through a break at a low point. Locations of choice are both sides of gates at summits, the downstream side of other gates, and changes in grade to steeper slopes in sections of line not otherwise protected by air valves.

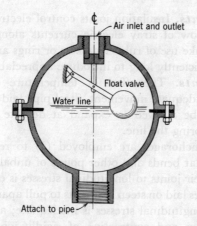

**Figure 6-6.**
**Air inlet and release valve.**

The required valve size is related to the size of the conduit, and to the velocities at which the line is emptied. The following ratios of air valve to conduit diameter provide common but rough estimates of needed sizes.

| | |
|---|---|
| For release of air only | 1:12 or 1 in. per ft |
| For admission as well as release of air | 1:8 or 1½ in. per ft |

An approximate calculation will show that under a vacuum of 48 in. of water, an automatic air valve, acting as an injection orifice with a coefficient of discharge of 0.5 under a head of $4/(1.3 \times 10^{-3}) = 3080$ ft of air of specific gravity $1.3 \times 10^{-3}$, is expected to admit about $0.5\sqrt{2g \times 3080} = 220$ cfs of air per square foot of valve. If the diameter ratio is 1:8, the displacement velocity in the conduit can be as high as $220/64 = 3.5$ fps without exceeding a vacuum of 48 in. of water. A similar calculation will show the rate of release of air. The amounts of air that can be dissolved by water at atmospheric pressure are about 2.9% by volume at 32°F and 1.9% at 77°F, changing in direct proportion to the pressure. Accordingly, they are doubled at two atmospheres or 14.7 psig.

An analysis of air-inlet valves for steel pipelines by Parmakian[15] takes the compressibility of air into account and combines equations for safe differential pressures of cylindrical steel pipe, pipe flow, and air flow.

**Manholes.** Access manholes are spaced 1000 to 2000 ft apart on large conduits. They are helpful during construction and serve later for inspection and repairs. They are less common on cast-iron and asbestos-cement lines than on steel and concrete lines.

---

[15] John Parmakian, Air-Inlet Valves for Steel Pipe Lines, *Trans. Am. Soc. Civil Engrs.*, **115**, 438 (1950). More exact equations are also derived in this paper.

***Insulation Joints.*** Insulation joints control electrolysis by introducing resistance to the flow of stray electric currents along pipelines. Modern insulation joints make use of rubber gaskets or rings and of rubber-covered sections of pipe sufficiently long to introduce appreciable resistance.

***Expansion Joints.*** The effect of temperature changes is small if pipe joints permit adequate movement. Steel pipe laid with rigid transverse joints must either be allowed to expand at definite points or be rigidly restrained by anchoring the line.

***Anchorages.*** Anchorages are employed (1) to resist the tendency of pipes to pull apart at bends and other points of unbalanced pressure when the resistance of their joints to longitudinal stresses is exceeded; (2) to resist the tendency of pipes laid on steep gradients to pull apart when the resistance of their joints to longitudinal stresses is inadequate; and (3) to restrain or direct the expansion and contraction of rigidly joined pipes under the influence of temperature changes.

Anchorages take many forms as follows: (1) for bends—both horizontal and vertical—concrete buttresses or *kick blocks* resisting the unbalanced pressure by their weight, much as a gravity dam resists the pressure of the water behind it, taking into consideration the resistance offered by the pipe joints themselves, by the friction of the pipe exterior, and by the bearing value of the soil in which the block is buried; (2) steel straps attached to heavy boulders or to bedrock; (3) lugs cast on cast-iron pipes and fittings to hold tie rods that prevent movement of the pipeline; (4) anchorages of mass concrete on steel pipe to keep it from moving, or to force motion to take place at expansion joints inserted for that purpose—the pipe being well bonded to the anchors, for example, by angle irons welded onto the pipe; and (5) gate chambers so designed of steel and concrete that they hold the two ends of steel lines rigidly in place.

In the absence of expansion joints, steel pipe must be anchored at each side of gates and meters in order to prevent their destruction by pipe movement. In the absence of anchors, flanged gates are sometimes bolted on one side to the pipe—usually the upstream side—and on the other side to a cast-iron nipple connected to the pipe by means of a sleeve or expansion joint.

***Other Appurtenances.*** These may include (1) air-relief towers at the first summit of the line to remove air mechanically entrained as water flows into the pipe entrance; (2) surge tanks at the end of the line to reduce water hammer created by operation of a valve at the end of the line; (3) pressure-relief valves or overflow towers on one or more summits to keep the pressure in the line below a given value by letting water discharge to waste when the pressure builds up beyond the design value; (4) check valves on force mains to prevent backflow when pumps shut down; (5) self-acting shutoff valves triggered to close when the pipe velocity exceeds a predetermined value as a result of an accident; (6) altitude-control valves that shut off the inlet to service reservoirs, elevated tanks, and standpipes before overflow levels are reached (Fig. 7-15); (7) pressure-reducing valves to keep pressures at safe levels in low lying areas; and (8) venturi or other meters and recorders to measure the flows.

## 6-10 Pumps and Pumping Stations

Pumps and pumping machinery serve the following purposes in water systems: (1) lifting water from its source (surface or ground), either immediately to the community through high-lift installations, or by low lift to purification works; (2) boosting water from low-service to high-service areas, to separate fire supplies, and to the upper floors of many-storied buildings; and (3) transporting water through treatment works, backwashing filters, draining component settling tanks, and other treatment units, withdrawing deposited solids and supplying water (especially pressure water) to operating equipment.

Today most water and wastewater pumping is done by either centrifugal pumps or propeller pumps. These are usually driven by electric motors, less often by steam turbines, internal combustion engines, or hydraulic turbines. How the water is directed through the impeller determines the type of pump. There is (1) *radial flow* in open- or closed-*impeller pumps*, with volute or turbine casings, and single or double suction through the eye of the impeller, (2) *axial flow* in *propeller pumps*, and (3) *diagonal flow* in *mixed-flow, open-impeller pumps*. Propeller pumps are not centrifugal pumps. Both can be referred to as *rotodynamic* pumps.

Open-impeller pumps are less efficient than closed-impeller pumps, but they can pass relatively large debris without being clogged. Accordingly, they are useful in pumping wastewaters and sludges. *Single-stage pumps* have but *one impeller*, and *multistage pumps* have *two or more*, each feeding into the next higher stage. Multistage turbine well pumps may have their motors submerged, or they may be driven by a shaft from the prime mover situated on the floor of the pumping station.

In addition to centrifugal and propeller pumps, water and wastewater systems may include (1) displacement pumps, ranging in size from hand-operated pitcher pumps to the huge pumping engines of the last century built as steam-driven units; (2) rotary pumps equipped with two or more rotors (varying in shape from meshing lobes to gears and often used as small fire pumps); (3) hydraulic rams utilizing the impulse of large masses of low-pressure water to drive much smaller masses of water (one half to one sixth of the driving water) through the delivery pipe to higher elevations, in synchronism with the pressure waves and sequences induced by water hammer;[16] (4) jet pumps or jet ejectors, used in wells and dewatering operations, introducing a high-speed jet of air or water through a nozzle into a constricted section of pipe; (5) air lifts in which air bubbles, released from upward-directed air pipe, lift water from a well or sump through an eductor pipe; and (6) displacement ejectors housed in a pressure vessel in which water (especially wastewater) accumulates and from which it is displaced through an eductor pipe when a float-operated valve is tripped by the rising water and admits compressed air to the vessel.

## 6-11 Pump Characteristics

Pumping units are chosen in accordance with *system heads* and *pump characteristics*. As shown in Fig. 6-7, the system head is the sum of the static and dynamic heads against the pump. As such, it varies with required flows and with changes in storage and suction levels. When a distribution system lies between pump and distribution reservoir, the system head responds also to fluctuations in demand. Pump characteristics depend on pump size, speed, and design. For a given speed $N$ in revolutions per minute, they are determined by the relationships between the rate of discharge, $Q$, usually in gallons per minute, and the head $H$ in feet, the efficiency $E$ in percent, and the power input $P$ in horsepower. For purposes of comparison, pumps of given geometrical design are characterized also by their specific speed $N_s$, the hypothetical speed of a homologous (geometrically similar) pump with an impeller diameter $D$ such that it will discharge 1 gpm against a 1-ft head. Because discharge varies as the product of area and velocity, and velocity varies as $H^{1/2}$, $Q$ varies as $D^2 H^{1/2}$. But velocity varies also as $\pi DN/60$. Hence $H^{1/2}$ varies as $DN$, or $N$ varies as $H^{3/4}/Q^{1/2}$, and the specific speed becomes

$$N_s = NQ^{1/2}/H^{3/4} \qquad (6-13)$$

Generally speaking, pump efficiencies increase with pump size and capacity. Below specific speeds of 1000 units, efficiencies drop off rapidly. Radial-flow pumps perform well between specific speeds of 1000 and 3500 units; mixed-flow pumps in the range of 3500 to 7500 units; and axial-flow pumps after that up to 12,000 units. As shown in Eq. 6-13, for a given $N$, high-capacity, low-head pumps have the highest specific speeds. For double-suction pumps, the specific speed is computed for half the capacity. For multistage pumps, the head is distributed between the stages. In ac-

---

[16] The driving water may or may not be derived from the same source as the driven water. Where it is not, a dangerous cross-connection may be established.

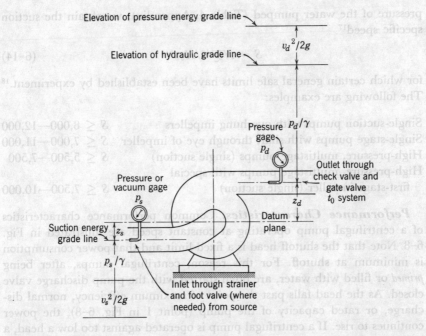

**Figure 6-7**

Head relationships in pumping systems. System head $H = (z_d - z_s) + (p_d/\gamma - p_s/\gamma) + (v_d^2/2g - v_s^2/2g)$. For suction lift, $p_s/\gamma$ and $z_s$ are negative, and $v_s^2/2g$ is positive. For suction pressure, $p_s/\gamma$ and $v_s^2/2g$ are positive, and $z_s$ is negative.

cordance with Eq. 6–13, this keeps the specific speed high and with it, also, the efficiency.

*Cavitation.* Specific speed is an important criterion, too, of safety against cavitation, a phenomenon accompanied by vibration, noise, and rapid destruction of pump impellers. Cavitation occurs when enough potential energy is converted to kinetic energy to reduce the absolute pressure at the impeller surface below the vapor pressure of water at the ambient temperature. Water then vaporizes and forms pockets of vapor that collapse suddenly as they are swept into regions of high pressure. Cavitation occurs when inlet pressures are too low or pump capacity or speed of rotation is increased without a compensating rise in inlet pressure. Lowering a pump in relation to its water source, therefore, reduces cavitation. If we replace the head $H$ in Eq. 6–13 by $H_{sv}$, the net positive inlet or suction head, namely, the difference between the total inlet head (the absolute head plus the velocity head in the inlet pipe), and the head corresponding to the vapor pressure of the water pumped (Table 4, Appendix), we obtain the suction specific speed[17]

$$S = NQ^{1/2}/H_{sv}^{3/4} \qquad (6\text{-}14)$$

for which certain general safe limits have been established by experiment.[18] The following are examples:

| | |
|---|---|
| Single-suction pumps with overhung impellers | $S \leq 8{,}000\text{--}12{,}000$ |
| Single-stage pumps with shaft through eye of impeller | $S \leq 7{,}000\text{--}11{,}000$ |
| High-pressure, multistage pumps (single suction) | $S \leq 5{,}500\text{--}7{,}500$ |
| High-pressure, multistage pumps with special first-stage impeller (single suction) | $S \leq 7{,}500\text{--}10{,}000$ |

**Performance Characteristics.** Common performance characteristics of a centrifugal pump operating at constant speed are illustrated in Fig. 6-8. Note that the shutoff head is a fixed limit and that power consumption is minimum at shutoff. For this reason, centrifugal pumps, after being *primed* or filled with water, are often started with the pump discharge valve closed. As the head falls past the point of maximum efficiency, normal discharge, or rated capacity of the pump (point 1 in Fig. 6-8), the power continues to rise. If a centrifugal pump is operated against too low a head, a motor selected to operate the pump in the head range around maximum efficiency may be overloaded. Pump delivery can be regulated (1) by a valve on the discharge line, (2) by varying the pump speed mechanically or electrically, or (3) by throwing two or more pumps in and out of service to best advantage.

What happens when more than one pumping unit is placed in service is shown in Fig. 6-8 with the help of a curve for the system head. Obviously, pumping units can operate only at the point of intersection of their own head curves with the system head curve. In practice, the system head varies over a considerable range at a given discharge (Fig. 6-9). For example, where a distributing reservoir is part of a system and both the reservoir and the source of water fluctuate in elevation, (1) a *lower* curve identifies head requirements when the reservoir is empty and the water surface of the source

---

[17] $H_{sv} = (p_s/\gamma) + v_s^2/2g - (p_w/\gamma)$, where $p_s/\gamma$ is the absolute pressure, $v_s$ the velocity of the water in the inlet pipe, and $p_w$ the vapor pressure of the water pumped, $\gamma$ being the specific weight of water and $g$ the gravity constant. The energy grade line is at a distance $h_s = (p_a/\gamma) - (p_s/\gamma) + v_s^2/2g$ from the eye of the impeller to the head delivered by the pump where $p_a$ is the atmospheric pressure. The ratio $H_{sv}/h_s$, where $h = (p_a/\gamma) - h_s$, is called the cavitation parameter.

[18] G. F. Wislicensus, R. M. Watson, and I. J. Karassik, Cavitation Characteristics of Centrifugal Pumps Described by Similarity Considerations, *Trans. Am. Soc. Mech. Engrs.*, **61**, 170 (1939); also Hydraulic Institute, *Standards for Centrifugal Pumps*, 10th ed., 1955.

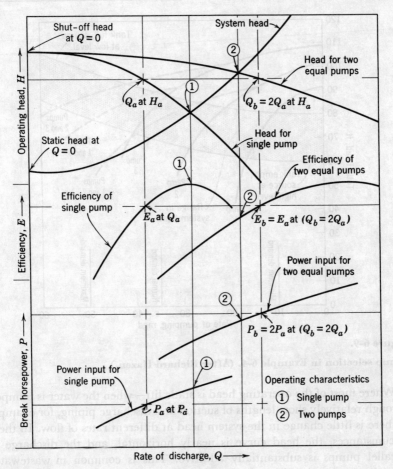

**Figure 6-8.**

**Performance characteristics of single and twin centrifugal pumps operating at constant speed.**

is high, and (2) an *upper* curve establishes the system head for a full reservoir and a low water level at the source. The location and magnitude of drafts also influence system heads. Nighttime pressure distributions may be very different from those during the day. How the characteristic curves for twin-unit operation are developed is indicated in Fig. 6-8. It should be noted that the two identical pumping units have not been selected with an eye to highest efficiency of operation in parallel. Characteristic curves for other multiple units are developed in the same way from the known curves of individual units.

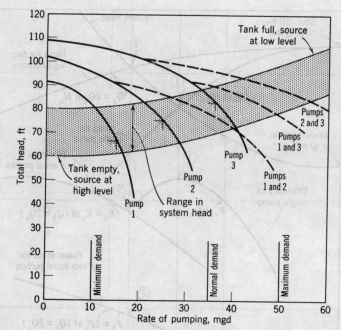

**Figure 6-9.**

**Pump selection in Example 6-4. (After Richard Hazen.)**

Where most of the operating head is static lift—when the water is pumped through relatively short lengths of suction and discharge piping, for example—there is little change in the system head at different rates of flow. In these circumstances, the head curve is nearly horizontal, and the discharge of parallel pumps is substantially additive. This is common in wastewater pumping stations in which the flow is lifted from a lower to an immediately adjacent higher level. Examples are pumping stations along intercepters or at outfalls.

By contrast, friction may control the head on pumps discharging through long force mains, and it may not be feasible to subdivide flows between pumping units with reasonable efficiency. Multispeed motors or different combinations of pumps and motors may then be required.

Because flows from a number of pumps may have to be fed through a different piping system than flows from any single unit, it may be necessary to develop "modified" characteristic curves that account for losses in different combinations of piping.

## Example 6–4.

A mill supply drawing relatively large quantities of water from a river is to deliver them at a fairly low head.[19] The minimum demand is 10 mgd, the normal 35 mgd, and the maximum 50 mgd. The river fluctuates in level by 5 ft, and the working range of a balancing tank is 15 ft. The vertical distance between the bottom of the tank and the surface of the river at high stage is 60 ft. The friction head in the pumping station and a 54-in. force main rises from a minimum of 1 ft at the 10-mgd rate to a maximum of nearly 20 ft at the 50-mgd rate. Make a study of suitable pumping units.

Hazen's solution of this problem is shown in Fig. 6–9. Three pumps are provided: No. 1 with a capacity of 15 mgd at 66-ft head; No. 2 with 25 mgd at 78-ft head; and No. 3 with 37 mgd at 84-ft head. Each pump has an efficiency of 89% at the design point.

The efficiencies at the top and bottom of the working range are listed in Table 6–2.

## Table 6-2
*Pumping Characteristics of System in Example 6–4*

| Pumps in service, No. | 1 | 2 | 3 | 1 & 2 | 1 & 3 | 2 & 3 |
|---|---|---|---|---|---|---|
| Rate of pumping, mgd | 10 | 21 | 33.5 | 27 | 36 | 42 |
| Head, ft | 81 | 83 | 88 | 85 | 90 | 93 |
| Efficiency, % | 80 | 88 | 88 | 71, 86 | 35, 87 | 68, 84 |
| Rate of pumping, mgd | 15 | 25 | 37 | 34 | 43.5 | 49.5 |
| Head, ft | 66 | 78 | 84 | 80 | 85 | 89 |
| Efficiency, % | 89 | 89 | 89 | 82, 88 | 71, 89 | 79, 87 |
| Rate of pumping, mgd | 16.5 | 28.5 | 40.5 | 40 | 49.5 | 56.5 |
| Head, ft | 62 | 66 | 73 | 73 | 79 | 84 |
| Efficiency, % | 88 | 84 | 86 | 88, 88 | 83, 88 | 79, 89 |

Centrifugal pumps are normally operated with discharge velocities of 5 to 15 fps. The resulting average outlet diameter (in inches) of the pump, called the pump size, is $0.2\sqrt{Q}$ where $Q$ is the capacity of the pump in gallons per minute.

[19] Richard Hazen, Pumps and Pumping Stations, *J. New England Water Works Assoc.*, **67**, 121 (1953).

seven

# water distribution

## 7-1 Systems of Distribution

Apart from a few scattered taps and takeoffs along their feeder conduits, distribution systems for public water supplies are networks of pipes within networks of streets. Street plan, topography, and location of supply works, together with service storage, determine the type of distribution system and the type of flow through it. Although service reservoirs are often placed along lines of supply, where they may usefully reduce conduit pressures (Fig. 6–5), their principal purpose is to satisfy network requirements. Accordingly they are, in fact, components of the distribution system, not of the transmission system.

***One- and Two-Directional Flow.*** The type of flow creates the four systems, sketched in Fig. 7–1. Hydraulic grade lines and residual pressures within the areas served, together with the volume of distribution storage, govern pipe sizes within the network. It is plain that flows from opposite directions increase system capacity. There is two-directional flow in the main arteries when a pumped or gravity supply, or a service reservoir, feeds into opposite ends of the distribution system; or through the system to elevated storage in a reservoir, tank, or standpipe situated at the far end of the area of greatest water demand. Volume and location of service storage depend on topography and water needs (Sec. 7–8).

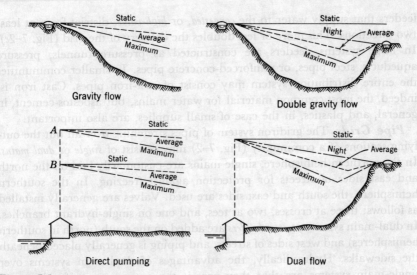

**Figure 7-1.**
**One- and two-directional flow in distribution systems.**

*Distribution Pattern.* Two distribution patterns emerge from the street plan: (1) a branching pattern on the outskirts of the community, in which ribbon development follows the primary arteries of roads (Fig. 7–2a), and (2) a gridiron pattern within the built-up portions of the community where streets crisscross and water mains are interconnected (Figs. 7–2b and 7–2c). Hydraulically, the gridiron system has the advantage of delivering water to any spot from more than one direction and of avoiding dead ends. The system is strengthened by substituting for a central feeder a loop or belt of

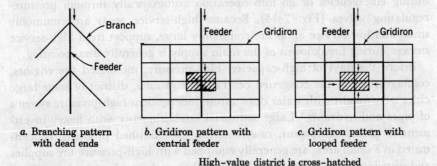

High-value district is cross-hatched

**Figure 7-2.**
**Patterns of water-distribution systems.**

feeders that supply water to the *congested*, or *high-value*, district from at least two directions. This more or less doubles the delivery of the grid (Fig. 7–2c). In large systems, feeders are constructed as pressure tunnels, pressure aqueducts, steel pipes, or reinforced-concrete pipes. In smaller communities the entire distribution system may consist of cast-iron pipes. Cast iron is, indeed, the most common material for water mains, but asbestos-cement, in general, and plastics, in the case of small supplies, are also important.

**Pipe Grids.** The gridiron system of pipes stretching over all but the outlying sections of a community (Fig. 7–2) may consist of *single* or *dual mains*. In the northern hemisphere, single mains are customarily laid on the north and east sides of streets for protection against freezing. In the southern hemisphere, the south and east sides are used. Valves are generally installed as follows: three at crosses, two at tees, and one on single-hydrant branches. In dual-main systems, *service headers* are added on the south (north in southern hemispheres) and west sides of streets, and piping is generally placed beneath the sidewalks. Hydraulically, the advantages of dual-main systems over single-main systems are that they permit the arrangement of valves and hydrants in such ways that breaks in mains do not impair the usefulness of hydrants and do not *dead-end* mains.

Dual-main systems must not be confused with dual water supplies: a high-grade supply for some purposes and a low-grade supply for others.

**High and Low Services.** Sections of the community too high to be supplied directly from the principal, or *low-service*, works are generally incorporated into separate distribution systems with independent piping and service storage. The resulting *high services* are normally fed by pumps that take water from the main supply and boost its pressure as required. Areas varying widely in elevation may be formed into intermediate districts or zones. Gated connections between the different systems are opened by hand during emergencies or go into operation automatically through pressure-regulating valves (Fig. 7–14). Because high-service areas are commonly small and low-service areas are commonly large, support from high-service storage during breakdowns of the main supply is generally disappointing.

Before the days of high-capacity, high-pressure, motorized fire engines, conflagrations in the congested central, or *high-value*, district of some large cities were fought with water drawn from independent high-pressure systems of pipes and hydrants.[1] Large industrial establishments, with heavy investment in plant, equipment, raw materials, and finished products, concentrated in a small area, are generally equipped with high-pressure fire supplies and distribution networks of their own. When such supplies are drawn from sources of questionable quality, some regulatory agencies enforce rigid

[1] Boston, Mass., still maintains a separate *fire supply*.

separation of private fire supplies and public systems. Others prescribe *protected cross-connections* incorporating backflow preventers that are regularly inspected for tightness (Sec. 7-9).

**Service to Premises.** Water reaches individual premises from the street main through one or more service pipes tapping the distribution system. The building supply between the public main and the takeoffs to the various plumbing fixtures or other points of water use is illustrated in Fig. 7-3, and the remainder of the system in Fig. 7-4. Small services are made of cement-lined iron or steel, brass of varying copper content, admiralty metal, copper, and plastics such as polyethylene (PE) or polyvinyl chloride (PVC). Because lead and lead-lined pipes may corrode and release lead to the water, they are no longer installed afresh. For large services, coated or lined cast-iron pipe is often employed. For dwellings and similar buildings, the minimum desirable size of service is ¾ in. Pipe-tapping machines connect services to the main without shutting off the water. They also make large connections within water-distribution systems.

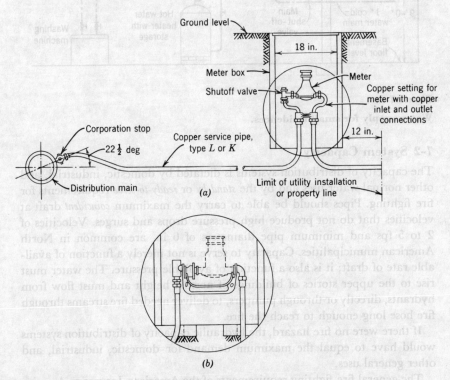

**Figure 7-3.**
(a) **Typical house service;** (b) **alternate method of mounting meter.**

194/*Water Distribution*

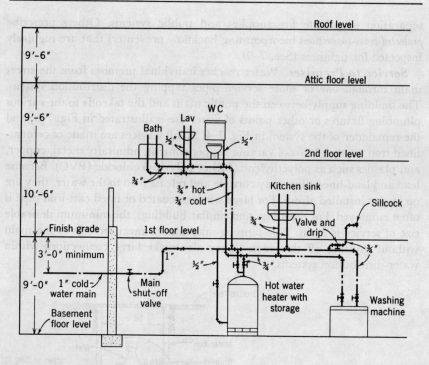

**Figure 7-4.**
**Water supply for small residences.**

## 7-2 System Capacity

The capacity of distribution systems is dictated by domestic, industrial, and other normal water uses and by the *stand-by* or *ready-to-serve* requirements for fire fighting. Pipes should be able to carry the maximum *coincident* draft at velocities that do not produce high pressure drops and surges. Velocities of 2 to 5 fps and minimum pipe diameters of 6 in. are common in North American municipalities. Capacity to serve is not merely a function of available rate of draft; it is also a function of available pressure. The water must rise to the upper stories of buildings of normal height and must flow from hydrants, directly or through pumpers, to deliver needed fire streams through fire hose long enough to reach the fire.

If there were no fire hazard, the hydraulic capacity of distribution systems would have to equal the maximum demand for domestic, industrial, and other general uses.

The general fire-fighting requirements of the American Insurance Association[2] are summarized in the following schedule:

---

[2] *Standard Schedule for Grading Cities and Towns.*

1. Within the central, congested, or high-value, district of North American communities: (a) for communities of 200,000 people or less, $Q = 1020\sqrt{P}(1-0.01\sqrt{P})$, $Q$ being the fire draft in gpm and $P$ the population in thousands; and (b) for populations in excess of 200,000, $Q = 12,000$ gpm with 2000 to 8000 gpm in addition for a second fire.

2. For residential districts with: (a) small, low buildings—$\frac{1}{3}$ of lots in block built upon, $Q = 500$ gpm; (b) larger or higher buildings, $Q = 1000$ gpm; (c) high-value residences, apartments, tenements, dormitories, and similar structures, $Q = 1500$ to $3000$ gpm; and (d) three-story buildings in densely built-up sections, $Q = $ up to $6000$ gpm.

3. Proportion or amount of estimated flow to be concentrated, if necessary, on one block or one very large building: (a) in the high-value district, $\frac{2}{3}$; (b) in compact residential areas, $\frac{1}{4}$ to $\frac{1}{2}$; and (c) for detached buildings, 500 to 750 gpm.

Table 7-1 shows the relatively large standby capacity prescribed.

To these requirements for fire fighting must be added a coincident demand of 40 to 50 gpcd in excess of the average consumption rate for the area under consideration. In small communities or limited parts of large-distribution systems, pipe sizes are controlled by fire demand plus coincident draft. In the case of main feeder lines and other central works in large communities or large sections of metropolitan systems, peak hourly demands may determine the design.

## 7-3 System Pressure

For normal drafts, water pressure at the street line must be at least 20 psig (46 ft) to let water rise three stories and overcome the frictional resistance of the house-distribution system, but 40 psig is more desirable. Business blocks are supplied more satisfactorily at pressures of 60 to 75 psig. To supply their upper stories, tall buildings must boost water to tanks on their roofs or in their towers, and often, too, on intermediate floors.

Fire demand is commonly gaged by the *standard fire stream*: 250 gpm issuing from a $1\frac{1}{8}$-in. nozzle at a pressure of 45 psig at the base of the tip. When this amount of water flows through $2\frac{1}{2}$-in. rubber-lined hose, the frictional resistance is about 15 psi per 100 ft of hose. Adding the hydrant resistance and required nozzle pressure of 45 psig then gives the pressure needs at the hydrant shown in Table 7-2.[3] A standard fire stream is effective to a height of 70 ft and has a horizontal carry of 63 ft.

[3] J. R. Freeman, Experiments Relating to Hydraulics of Fire Streams, *Trans. Am. Soc. Civil Engrs.*, **21**, 303 (1889). As a young man, Mr. Freeman (1855-1932) studied the performance of fire nozzles in the hydraulic laboratory of Hiram F. Mills, at Lawrence, Mass., brought order into the operations of fire departments and fire insurance agencies, and became a leading hydraulic engineer. He promoted the study of hydraulic models in America.

## Table 7-1
Required Fire Flow, Fire Reserve, and Hydrant Spacing Recommended by the American Insurance Association

| Population | Fire Flow gpm | Fire Flow mgd | Duration, hr | Fire Reserve, mg | Area per Hydrant, sq ft Engine Streams | Area per Hydrant, sq ft Hydrant Streams |
|---|---|---|---|---|---|---|
| 1,000 | 1,000 | 1.4 | 4 | 0.2 | 120,000 | 100,000 |
| 2,000 | 1,500 | 2.2 | 6 | 0.5 | — | 90,000 |
| 4,000 | 2,000 | 2.9 | 8 | 1.0 | 110,000 | 85,000 |
| 6,000 | 2,500 | 3.6 | 10 | 1.5 | — | 78,000 |
| 10,000 | 3,000 | 4.3 | 10 | 1.8 | 100,000 | 70,000 |
| 13,000 | 3,500 | 5.0 | 10 | 2.1 | — | — |
| 17,000 | 4,000 | 5.8 | 10 | 2.4 | 90,000 | 55,000 |
| 22,000 | 4,500 | 6.5 | 10 | 2.7 | — | — |
| 28,000 | 5,000 | 7.2 | 10 | 3.0 | 85,000 | 40,000[b] |
| 40,000 | 6,000 | 8.6 | 10 | 3.6 | 80,000 | — |
| 60,000 | 7,000 | 10.1 | 10 | 4.2 | 70,000 | — |
| 80,000 | 8,000 | 11.5 | 10 | 4.8 | 60,000 | — |
| 100,000 | 9,000 | 13.0 | 10 | 5.4 | 55,000 | — |
| 125,000 | 10,000 | 14.4 | 10 | 6.0 | 48,000 | — |
| 150,000 | 11,000 | 15.8 | 10 | 6.6 | 43,000 | — |
| 200,000[a] | 12,000 | 17.3 | 10 | 7.2 | 40,000 | — |

[a] For populations over 200,000 and local concentration of streams, see outline of AIA requirements.

[b] For fire flows of 5000 gpm and over.

## Table 7-2
Hydrant Pressures for Different Lengths of Fire Hose

| Length of hose, ft | 100 | 200 | 300 | 400 | 500 | 600 |
|---|---|---|---|---|---|---|
| Required pressure, psig | 63 | 77 | 92 | 106 | 121 | 135 |

Because hydrants are normally planned to control areas within a radius of 200 ft, Table 7-2 shows that direct attachment of fire hose to hydrants (hydrant streams) calls for a residual pressure at the hydrant of about 75 psig. To maintain this pressure at times of fire, system pressures must approach

100 psig. This has its disadvantages, among them danger of breaks and leakage or waste of water approximately in proportion to the square root of the pressure. Minimum hydrant pressures of 50 psig cannot maintain standard fire streams after passing through as little as 50 ft of hose.

Motor pumpers commonly deliver up to 1500 gpm at adequate pressures. Capacities of 20,000 gpm are in sight, with single streams discharging as much as 1000 gpm from 2-in. nozzles. To furnish domestic and industrial draft and keep pollution from entering water mains by seepage or failure under a vacuum, fire engines should not lower pressures in the mains to less than 20 psig. For large hydrant outlets, the safe limit is sometimes set at 10 psig. In a real way, modern fire-fighting equipment has eliminated the necessity for pressures much in excess of 60 psig, except in small towns that cannot afford a full-time, well-equipped fire department. Pumpers have increased system capacity in the ratio of $\sqrt{p - 20}/\sqrt{p - 75}$, where $p$ is the normal dynamic pressure of the network.

## 7-4 System Components

Pipes, gates, and hydrants are the basic elements of reticulation systems (Fig. 7-5). Their dimensioning and spacing rest on experience normally

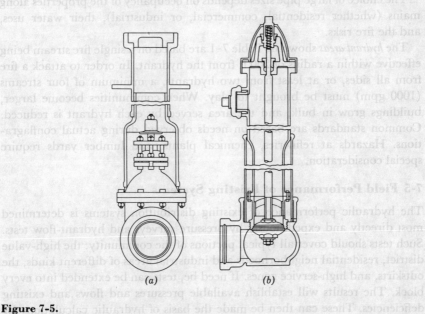

**Figure 7-5.**
(a) **Gate valve and extendable valve box;** (b) **post fire hydrant with compression valve.**

precise enough in its minimum standards to permit roughing in all but the main arteries and feeders. Common standards include the following:

### Pipes

| | |
|---|---|
| Smallest pipes in gridiron | 6-in. |
| Smallest branching pipes (dead ends) | 8-in. |
| Largest spacing of 6-in. grid (8-in. pipe used beyond this value) | 600 ft |
| Smallest pipes in high-value district | 8-in. |
| Smallest pipes on principal streets in central district | 12-in. |
| Largest spacing of supply mains or feeders | 2,000 ft |

### Gates

| | |
|---|---|
| Largest spacing on long branches | 800 ft |
| Largest spacing in high-value district | 500 ft |

### Hydrants

| | |
|---|---|
| Areas protected by hydrants | (See Table 7–1) |
| Largest spacing when fire flow exceeds 5000 gpm | 200 ft |
| Largest spacing when fire flow is as low as 1000 gpm | 300 ft |

The choice of large pipe sizes depends on occupancy of the properties along mains (whether residential, commercial, or industrial), their water uses, and the fire risks.

The *hydrant areas* shown in Table 7–1 are based on a single fire stream being effective within a radius of 200 ft from the hydrant. In order to attack a fire from all sides, or at least from two hydrants, a minimum of four streams (1000 gpm) must be brought to play. When communities become larger, buildings grow in bulk, and the area served by each hydrant is reduced. Common standards are based on needs observed during actual conflagrations. Hazards at refineries, chemical plants, and lumber yards require special consideration.

## 7-5 Field Performance of Existing Systems

The hydraulic performance of existing distribution systems is determined most directly and expeditiously by pressure surveys and hydrant-flow tests. Such tests should cover all typical portions of the community: the high-value district, residential neighborhoods and industrial areas of different kinds, the outskirts, and high-service zones. If need be, tests can be extended into every block. The results will establish available pressures and flows and existing deficiencies. These can then be made the basis of hydraulic calculations for extensions, reinforcements, and new gridiron layouts. Follow-up tests can show how nearly the desired changes have been accomplished.

**Pressure Surveys and Hydrant-Flow Tests.** Pressure surveys yield the most rudimentary information about networks. If they are conducted both at night (minimum flow) and during the day (normal demand), they will indicate the hydraulic efficiency of the system in meeting common requirements; but they will not establish the probable behavior of the system under stress—during a serious conflagration, for example.

Hydrant-flow tests commonly include (1) observation of the pressure at a centrally situated hydrant during the conduct of the test; and (2) measurement of the combined flow from a group of neighboring hydrants. Velocity heads in the jets issuing from the hydrants are usually measured by hydrant pitot tubes. If the tests are to be significant, (1) the hydrants tested should form a group such as might be called into play in fighting a serious fire in the district under study; (2) water should be drawn at a rate that will drop the pressure enough to keep it from being measurably affected by normal fluctuations in draft within the system; and (3) the time of test should coincide with drafts (domestic, industrial, and the like) in the remainder of the system, reasonably close to *coincident* values.

The requirements of the American Insurance Association (Sec. 7-2) are valuable aids in planning hydrant-flow tests. A layout of pipes and hydrants in a typical flow test is shown in Fig. 7-6, and observed values are summarized in Table 7-3.

## Table 7-3
*Record of a Typical Hydrant-Flow Test*

| Conditions of Test | Observed Pressure at Hydrant 1 psig | Discharge Velocity Head psig | Calculated Flow ($Q$), gpm | Remarks |
|---|---|---|---|---|
| All hydrants closed | 74 | ... | ... | All hydrant outlets are 2½ in. in diameter. |
| Hydrant 2 opened, 1 outlet | — | 13.2 | 610 | Total $Q$ = 2980 gpm |
| Hydrant 3 opened, 2 outlets | ... | 9.6 | 2 × 520 | Calculated engine streams = 4200 gpm |
| Hydrant 4 opened, 1 outlet | ... | 16.8 | 690 | |
| Hydrant 5 opened, 1 outlet | 46 | 14.5 | 640 | |
| All hydrants closed | 74 | ... | ... | |

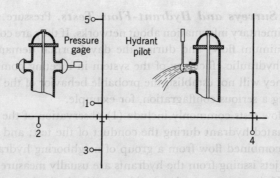

**Figure 7-6.**
**Location of pipes and hydrants in flow test and use of hydrant pitot and pressure gage. See Table 7-3 and Fig. 7-7.**

This table is more or less self-explanatory. The initial and residual pressure was read from a Bourdon gage at hydrant 1. Hydrants 2, 3, 4, and 5 were opened in quick succession, and their rates of discharge were measured simultaneously by means of hydrant pitots. A test such as this does not consume more than 5 min, if it is conducted by a well-trained crew.

Necessary hydrant-flow calculations may be outlined as in Example 7–1 for the flow test recorded in Table 7–3.

---

*Example 7–1.*

For outlets of diameter $d$ in., the discharge $Q$ in gpm is $Q = 30cd^2\sqrt{p}$, where $p$ is the pitot reading in psig and $c$ is the coefficient of hydrant discharge.[4] For smooth well-rounded $2\tfrac{1}{2}$-in. outlets, $c = 0.9$ and $Q = 170\sqrt{p}$.

---

Pressure-discharge relations established in this test are illustrated in Fig. 7–7. If the true static pressure is known, a more exact calculation is possible, although the additional labor involved is seldom justified. In accordance with the common hydraulic analysis of Borda's mouthpiece, a pressure gage inserted in a hydrant in juxtaposition to the hydrant outlet to be opened will also record the discharge pressure otherwise measured by hydrant pitots.

Hydrant tests are sometimes made to ascertain the capacity of individual hydrants and advertise it to firemen (particularly to engine companies

---

[4] Because $Q = cav$, where $c$ is the hydrant discharge coefficient, $a$ is the hydrant outlet area, and $v$ is the discharge velocity [$2.3p = v^2/(2g)$]. Here $Q$ is measured in cfs, $a$ in sq. ft, and $v$ in fps. The value of $c$ varies from 0.9 for well-rounded, smooth outlets to 0.7 for sharp outlets projecting into the barrel.

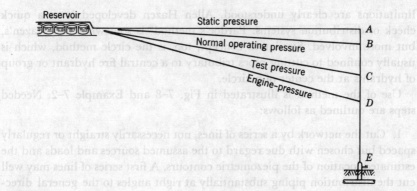

**Figure 7-7.**
Pressure and discharge relations established by hydrant-flow test. See Fig. 7-6 and Table 7-3.
  *A*: Static water table.
  *B*: No hydrant discharge. Pressure = 74 psig; pressure drop $p_0$ due to coincident draft $Q$.
  *C*: Hydrant discharge. Pressure = 46 psig; pressure drop $p_1 = (74 - 46) = 28$ psi accompanies discharge of $Q_1 = 2980$ gpm.
  *D*: Engine streams. Pressure = 20 psig; pressure drop $p_2 = (74 - 20) = 54$ psi accompanies discharge $Q_2 = 4200$ gpm.
  *E*: Hydrant 1, recording residual pressure of hydrant groups shown in Fig. 7-6.

summoned from neighboring towns) by painting the bonnet a suitable color. The weakness of this practice is its restriction of flow measurements to single hydrants. In fire fighting, groups of hydrants are normally brought into action. Tests of individual hydrants may be quite misleading.

## 7-6 Office Studies of Pipe Networks

No matter how energetically distribution systems are field-tested, needed extensions and reinforcements of old networks and the design of new ones can be adequately identified only by office studies. Necessary analysis presupposes familiarity with processes of hydraulic computation, including analog and high-speed digital computers. Even without large computers, however, the best processes can be so systematized as to make their application a matter of simple arithmetic and pipe-flow tables, diagrams, or slide rules.

Useful methods of analysis are (1) sectioning, (2) relaxation, (3) pipe equivalence, (4) computer programming, and (5) electrical analogy.

*Sectioning.* Sectioning is an approximate and, in a sense, exploratory method, simple in concept and application, and widely useful provided its limitations are clearly understood. Allen Hazen developed it as a quick check of distribution systems. Pardoe's method[5] is somewhat like Hazen's, but more involved. Similar in concept, too, is the circle method, which is usually confined to cutting pipes tributary to a central fire hydrant or group of hydrants at the center of a circle.

Use of the method is illustrated in Fig. 7-8 and Example 7-2. Needed steps are outlined as follows:

1. Cut the network by a series of lines, not necessarily straight or regularly spaced but chosen with due regard to the assumed sources and loads and the estimated location of the piezometric contours. A first series of lines may well cut the distribution piping substantially at right angles to the general direction of flow, that is, perpendicular to a line drawn from the supply conduit to the high-value district (Fig. 7-8). Further series may be oriented in some other critical direction, for example, horizontally and vertically in Fig. 7-8. For more than one supply conduit, the sections may be curved to intercept the flow from each conduit.

2. Estimate how much water must be supplied to areas *beyond* each section. Base estimates on a knowledge of the population density and the general characteristics of the zone—residential, commercial, and industrial. The water requirements comprise (*a*) the normal, coincident draft, here called the domestic draft, and (*b*) the fire demand (Table 7-1). Domestic use decreases progressively from section to section, as population or industry is left behind;

[5] W. S. Pardoe, *Eng. News-Rec.*, **93**, 516 (1924).

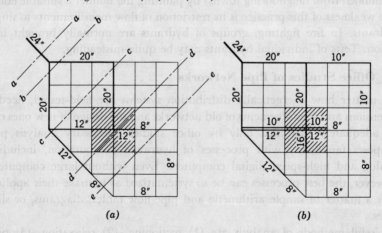

**Figure 7-8.**

Plan of network analyzed by method of sections. Example 7-2. (*a*) Existing system; (*b*) recommended system. Unless otherwise indicated, pipe diameters are 6 in. The high-value district is cross-hatched.

fire demand remains the same until the high-value district has been passed, after which it drops to a figure applicable to the type of outskirt area.

3. Estimate the distribution-system capacity at each section across the piping. To do this (a) tabulate the number of pipes of each size cut. Count only pipes that deliver water in the general direction of flow; and (b) determine the average available hydraulic gradient or frictional resistance, which depends on the pressure to be maintained in the system and the allowable pipe velocity.[6] Ordinarily, hydraulic gradients lie between 1 and 3‰, and velocities range from 2 to 5 fps.

4. For the available, or desirable, hydraulic gradient, determine the capacities of existing pipes and sum them for total capacity.

5. Calculate the deficiency or difference between required and existing capacity.

6. For the available, or desirable, hydraulic gradient, select the sizes and routes of pipes that will offset the deficiency. General familiarity with the community and studies of the network plan will aid judgment. Some existing small pipes may have to be removed to make way for larger mains.

7. Determine the velocities of flow. Excessive velocities may make for dangerous water hammer. They should be avoided, if necessary, by lowering the hydraulic gradients actually called into play.

8. Check important pressure requirements against the plan of the reinforced network.

The method of sections is particularly useful (1) in preliminary studies of large and complicated distribution systems, (2) as a check on other methods of analysis, and (3) as a basis for further investigations and more exact calculations.

*Example 7–2.*

Analyze the network of Fig. 7–8 by sectioning. The hydraulic gradient available within the network proper is estimated to lie close to 2‰. The value of $C$ in the Hazen-Williams formula is assumed to be 100, and the domestic (coincident) draft, in this

---

[6] To illustrate, for a level region, a distance of 25,000 ft from the junction of the supply conduit with the network to the high-value district, a pressure of 70 psig at the junction, an available pressure drop to 20 psig for engine streams, a requisite system capacity of 17 mgd, and a variation in pipe sizes from 6 in. to 24 in.: the available hydraulic gradient is $(70-20) \times 2.308/25 = 4.6$‰, the carrying capacity is that of a 30-in. pipe, and the velocities lie between 1.9 fps for 6-in. pipes and 4.7 fps for 24-in. pipes. If the velocity in 24-in. pipes is to be reduced to 3 fps, the hydraulic gradient must be lowered to 2‰ and the network strengthened by the addition of pipes. The reinforced network must possess the carrying capacity of a 36-in. pipe.

case, only 150 gpcd. The fire demand is taken from Table 7–1. Calculations are shown only for the first three sections.

Section *a-a*. Population 16,000. Demands (mgd): domestic, 2.2; fire, 5.6; total, 7.8. Existing pipes: one 24-in.; capacity, 6.0 mgd. Deficiency: $7.8 - 6.0 = 1.8$ mgd. If no pipes are added, the 14-in. pipe must carry 7.8 mgd. This it will do with a loss of head of 3.2‰ at a velocity of 3.8 fps (Hazen-Williams diagram).

Section *b-b*. Population and flow as in *a-a*. Total demand, 7.8 mgd. Existing pipes: 2, 20-in. @ 3.7 mgd = 7.4 mgd. Deficiency, $7.8 - 7.4 = 0.4$ mgd. If no pipes are added, existing pipes will carry 7.8 mgd with a loss of head of 2.2‰ at a velocity of 2.8 fps.

Section *c-c*. Population 14,000. Demands (mgd): domestic, 2.0; fire, 5.6; total, 7.6. Existing pipes: one 20-in. @3.7 mgd; two 12-in. @ 1.0 mgd = 2.0 mgd; five 6-in. @ 0.16 = 0.8 mgd; total, 6.5 mgd. Deficiency: $7.6 - 6.5 = 1.1$ mgd. Pipes added: two 10-in. @ 0.6 = 1.2 mgd. Pipes removed: one 6-in. @ 0.2 mgd. Net added capacity: $1.2 - 0.2 = 1.0$ mgd. Reinforced capacity: $6.5 + 1.0 = 7.5$ mgd.

The reinforced system (equivalent pipe,[7] 26.0 in.) will carry 7.6 mgd with a loss of head of 2.1‰.

**Relaxation.** A method of relaxation, or controlled trial and error, was introduced by Hardy Cross,[8] whose procedures are followed here with only a few modifications. In applying a method of this kind, calculations become speedier if pipe-flow relationships are expressed by an exponential formula with unvarying capacity coefficient, and notation becomes simpler if the exponential formula is written $H = kQ^n$, where, for a given pipe, $k$ is a numerical constant depending on $C$, $d$, and $l$, and $Q$ is the flow, $n$ being a constant exponent for all pipes.[9] Two procedures may be involved, depending on whether (1) the quantities of water entering and leaving the network or (2) the piezometric levels, pressures, or water-table elevations at inlets and outlets are known.

In balancing heads by correcting assumed flows, necessary formulations are made algebraically consistent by arbitrarily assigning *positive signs* to *clockwise flows* and associated head losses, and *negative signs* to *counterclockwise flows* and associated head losses. For the simple network shown in Fig. 7–9a, inflow $Q_i$ and outflow $Q_0$ are equal and known, inflow being split between

---

[7] The equivalent pipe will carry 7.5 mgd on a hydraulic gradient of 2‰.

[8] Hardy Cross, Analysis of Flow in Networks of Conduits or Conductors, *Univ. Illinois Bull.* 286 (1936). Hardy Cross (1885–1959) was Professor of Civil Engineering first at the University of Illinois and later at Yale University. The method discussed here was translated by him from structural to hydraulic use.

[9] In Eq. 6-6, for example, $Q = 405\,Cd^{2.63}s^{0.54}$, or $sl = H = kQ^{1.85}$ for pipes with given values of $C$, $d$, and $l$.

*Office Studies of Pipe Networks*/205

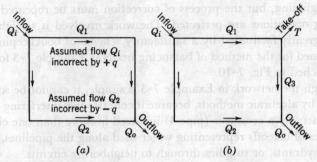

**Figure 7-9.**
**Simple network illustrating (a) the derivation of the Hardy-Cross method and (b) the effect of changing flows.**

two branches in such manner that the sum of the balanced head losses $H_1$ (clockwise) and $-H_2$ (counterclockwise) or $\Sigma H = H_1 - H_2 = 0$. If the assumed split flows $Q_1$ and $-Q_2$ are each in error by the same small amount $q$, $\Sigma H = \Sigma k(Q + q)^n = 0$. Expanding this binomial and neglecting all but its first two terms, because higher powers of $q$ are presumably very small, $\Sigma H = \Sigma k(Q + q)^n = \Sigma k Q^n + \Sigma n k q Q^{n-1} = 0$, whence

$$q = -\frac{\Sigma k Q^n}{n \Sigma k Q^{n-1}} = -\frac{\Sigma H}{n \Sigma (H/Q)} \qquad (7\text{-}1)$$

If a takeoff is added to the system as in Fig. 7-9b, both head losses and flows are affected.

In balancing flows by correcting assumed heads, necessary formulations become algebraically consistent when *positive signs* are arbitrarily assigned to *flows towards junctions* other than inlet and outlet junctions (for which water-table elevations are known) and *negative signs* to *flows away from* these *intermediate junctions*, the sum of the balanced flows at the junctions being zero. If the assumed water-table elevation at a junction, such as the takeoff junction in Fig. 7-9b, is in error by a height $h$, different small errors $q$ are created in the individual flows $Q$ leading to and leaving from the junction. For any one pipe, therefore, $H + h = k(Q + q)^n = kQ^n + h$, where $H$ is the loss of head associated with the flow $Q$. Moreover, as before, $h = nkqQ^{n-1} = nq(H/Q)$ and $q = (h/n)(Q/H)$. Because $\Sigma(Q + q) = 0$ at each junction, $\Sigma Q = -\Sigma q$ and $\Sigma q = (h/n)\Sigma(Q/H)$, or $\Sigma Q = -(h/n)\Sigma(Q/H)$. Therefore,

$$h = -\frac{n \Sigma Q}{\Sigma(Q/H)} \qquad (7\text{-}2)$$

The corrections $q$ and $h$ are only approximate. After they have been applied once to the assumed flows, the network is more nearly in balance than it was at the beginning, but the process of correction must be repeated until the balancing operations are perfected. The work involved is straightforward, but it is greatly facilitated by a satisfactory scheme of bookkeeping such as that outlined for the method of balancing heads in Example 7–3 for the network sketched in Fig. 7–10.

Although the network in Example 7–3 is simple, it cannot be solved conveniently by algebraic methods, because it contains two interfering hydraulic constituents: (1) a crossover (pipe 4) involved in more than one circuit and (2) a series of takeoffs representing water used along the pipelines, fire flows through hydrants, or supplies through to neighboring circuits.

*Example 7–3.*

Balance the network of Fig. 7–10 by the method of balancing heads. The schedule of calculations (Table 7–4) includes the following:

Columns 1–4 identify the position of the pipes in the network and record their length and diameter. There are two circuits and seven pipes. Pipe 4 is shared by both circuits. One star indicates this in connection with Circuit I; a double dagger does so with Circuit II. This dual pipe function must not be overlooked.

Columns 5–9 deal with the assumed flows and the derived flow correction. For purposes of identification the hydraulic elements $Q$, $s$, $H$, and $q$ are given a subscript zero.

Column 5 lists the assumed flows $Q_0$ in mgd. They are preceded by positive signs if they are clockwise and by negative signs if they are counterclockwise. The distribution of flows has been purposely misjudged in order to highlight the balancing operation. At each junction the total flow remaining in the system must be accounted for.

Column 6 gives the hydraulic gradients $s_0$ in ft per 1000 ft (‰) when the pipe is carrying the quantities $Q_0$ shown in Col. 5. The values of $s_0$ can be read directly from tables or diagrams of the Hazen-Williams formula.

Column 7 is obtained by multiplying the hydraulic gradients ($s_0$) by the length of the pipe in 1000 ft; that is, Col. 7 = Col. 6 × (Col. 3/1000). The head losses $H_0$ obtained are preceded by a positive sign if the flow is clockwise and by a negative sign if counterclockwise. The values in Col. 7 are totaled for each circuit, with due regard to signs, to obtain $\Sigma H$.

Column 8 is found by dividing Col. 7 by Col. 5. Division makes all signs of $H_0/Q_0$ positive. This column is totaled for each circuit to obtain $\Sigma(H_0/Q_0)$ in the flow-correction formula.

Column 9 contains the calculated flow correction $q_0 = -\Sigma H_0/(1.85 \times \Sigma H_0/Q_0)$. For example, in Circuit I, $\Sigma H_0 = -16.5$, $\Sigma(H_0/Q_0) = 43.1$; and $(-16.5)/(1.85 \times 43.1) = -0.21$; or $q_0 = +0.21$. Because pipe 4 operates in both circuits, it draws a correction from each circuit. However, the second correction is of opposite sign. As a part of Circuit I, for example, pipe 4 receives a correction of $q = -0.07$ from Circuit II in addition to its basic correction of $q = +0.21$ from Circuit I.

Columns 10–14 cover the once-corrected flows. Therefore, the hydraulic elements $(Q, s, H, \text{ and } q)$ are given the subscript one. Column 10 is obtained by adding, with due regard to sign, Cols. 5 and 9; Cols. 11, 12, 13, and 14 are then found in the same manner as Cols. 6, 7, 8, and 9.

Columns 15–19 record the twice-corrected flows, and the hydraulic elements $(Q, s, H, \text{ and } q)$ carry the subscript two. These columns are otherwise like Cols. 10 to 14.

Columns 20–23 present the final result, Cols. 20 to 22 corresponding to Cols. 15 to 18 or 10 to 12. No further flow corrections are developed because the second flow corrections are of the order of 10,000 gpd for a minimum flow of 200,000 gpd, or at most 5%. To test the balance obtained, the losses of head between points $A$ and $D$ in Fig. 7–10 via the three possible routes are given in Col. 23. The losses vary from 25.0 to 25.5 ft. The average loss is 25.3 ft and the variation about 1%.

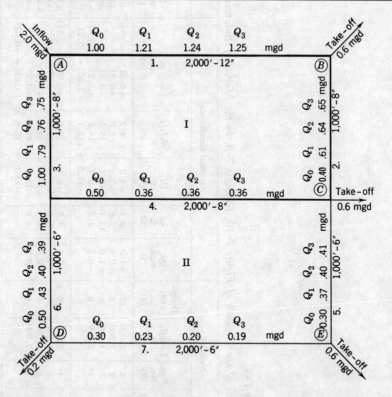

**Figure 7-10.**

**Plan of network analyzed by the method of balancing heads. Example 7-3.**

208 / Water Distribution

**Table 7-4**
Analysis of the Network of Figure 7–10, Example 7–3, by the Method of Balancing Heads

| Circuit No. (1) | Pipe No. (2) | Network Length, ft (3) | Diameter, in. (4) | $Q_0$, mgd (5) | $s_0$, ‰ (6) | Assumed Conditions $H_0$, ft (7) | $H_0/Q_0$ (8) | $q_0$, mgd (9) | $Q_1$, mgd (10) | $s_1$, ‰ (11) | First Correction $H_1$, ft (12) | $H_1/Q_1$ (13) | $q_1$, mgd (14) |
|---|---|---|---|---|---|---|---|---|---|---|---|---|---|
| I | 1 | 2,000 | 12 | +1.0 | 2.1 | +4.2 | 4.2 | +0.21 | +1.21 | 3.0 | +6.0 | 5.0 | +0.03 |
|  | 2 | 1,000 | 8 | +0.4 | 2.8 | +2.8 | 7.0 | +0.21 | +0.61 | 6.1 | +6.1 | 10.0 | +0.03 |
|  | 3 | 1,000 | 8 | −1.0 | 15.1 | −15.1 | 15.1 | +0.21 | −0.79 | 9.8 | −9.8 | 12.4 | +0.03 |
|  | 4ᵃ | 2,000 | 8 | −0.5 | 4.2 | −8.4 | 16.8 | +0.21 | −0.36 | 2.3 | −4.6 | 12.8 | +0.03 |
|  |  |  |  |  |  | −16.5 ÷ (43.1 × 1.85) = −0.21 |  |  |  |  | −2.3 ÷ (40.2 × 1.85) = −0.03 |  |  |
| II | 4ᵃ | 2,000 | 8 | +0.5 | 4.2 | +8.4 | 16.8 | −0.07 / −0.21ᵇ | +0.36 | 2.3 | +4.6 | 12.8 | −0.03ᵇ |
|  | 5 | 1,000 | 6 | +0.3 | 6.6 | +6.6 | 22.0 | +0.07 | +0.37 | 9.8 | +9.8 | 26.5 | +0.03 |
|  | 6 | 1,000 | 6 | −0.5 | 16.9 | −16.9 | 33.8 | +0.07 | −0.43 | 12.9 | −12.9 | 30.0 | +0.03 |
|  | 7 | 2,000 | 6 | −0.3 | 6.6 | −13.2 | 44.0 | +0.07 | −0.23 | 4.1 | −8.2 | 35.6 | +0.03 |
|  |  |  |  |  |  | −15.1 ÷ (116.6 × 1.85) = −0.07 |  |  |  |  | −6.7 ÷ (104.9 × 1.85) = −0.03 |  |  |

| Network | | | | Second Correction | | | | | Result | |
|---|---|---|---|---|---|---|---|---|---|---|
| Circuit No. (1) | Pipe No. (2) | Length, ft (3) | Diameter, in. (4) | $Q_2$, mgd (15) | $s_2$, % (16) | $H_2$, ft (17) | $H_2/Q_2$ (18) | $q_2$, mgd (19) | $Q_3$, mgd (20) | $s_3$, % (21) | $H_3$, ft (22) | Loss of Head $A$–$E$ (23) |
| I | 1 | 2,000 | 12 | +1.24 | 3.1 | +6.2 | 5.0 | +0.01 | +1.25 | 3.2 | +6.4 | 1. Via pipes 1, 2, 5, 25.0 ft |
|   | 2 | 1,000 | 8  | +0.64 | 6.6 | +6.6 | 10.3 | +0.01 | +0.65 | 6.8 | +6.8 | 2. Via pipes 3, 4, 5, 25.3 ft |
|   | 3 | 1,000 | 8  | −0.76 | 9.1 | −9.1 | 12.0 | +0.01 | −0.75 | 8.9 | −8.9 | 3. Via pipes 3, 6, 7, 25.5 ft |
|   | 4[a] | 2,000 | 8  | −0.36 | 2.3 | −4.6 | 12.8 | +0.01 | −0.36 | 2.3 | −4.6 | |
|   |   |   |   |   |   | −0.9 | (40.1 × | 1.85) = −0.01 |   |   | −0.3 | |
|   |   |   |   |   |   | ÷ |   |   |   |   |   | |
| II | 4[c] | 2,000 | 8  | +0.36 | 2.3 | +4.6 | 12.8 | +0.01[b] | +0.36 | 2.6 | +4.6 | |
|   | 5 | 1,000 | 6  | +0.40 | 11.3 | +11.3 | 28.2 | +0.01 | +0.41 | 11.8 | +11.8 | |
|   | 6 | 1,000 | 6  | −0.40 | 11.3 | −11.3 | 28.2 | +0.01 | −0.39 | 10.8 | −10.8 | |
|   | 7 | 2,000 | 6  | −0.20 | 3.1 | −6.2 | 31.0 | +0.01 | −0.19 | 2.9 | −5.8 | |
|   |   |   |   |   |   | −1.6 ÷ (100.2 × | 1.85) = −0.01 |   |   |   | −0.2 | |

[a] Pipe serves more than one circuit; first consideration of this pipe.
[b] Corrections in this column are those calculated for the same pipe in the companion circuit; they are of opposite sign.
[c] Second consideration of this pipe.

$Q$ = flow in mgd; $s$ = slope of hydraulic gradient or friction loss in ft per 1000 (‰) by the Hazen-Williams formula for $C$ = 100. $H$ = head lost in pipe (ft).

$q$ = flow correction in mgd; $q = -\dfrac{\Sigma H}{1.85 \Sigma (H/Q)}$; $Q_1 = Q_0 + q_0$; $Q_2 = Q_1 + q_1$; $Q_3 = Q_2 + q_2$.

## Example 7–4.

Balance the network of Fig. 7–11 by the method of balancing flows. Necessary calculations are given in Table 7–5.

The schedule of calculations includes the following:

Columns 1 to 5 identify the pipes at the three *free* junctions.

Columns 6 and 7 give the assumed head loss and the derived hydraulic gradient that determines the rate of flow shown in Col. 8 and the flow-head ratio recorded in Col. 9 = (Col. 8/Col. 6).

Column 10 contains the head correction $h_o$ as the negative value of 1.85 times the sum of Col. 8 divided by the sum of Col. 9, for each junction in accordance with Eq. 7–2. A subsidiary head correction is made for *shared* pipes as in Example 7–2.

Column 11 gives the corrected head $H_1 = H_0 + h_0$ and provides the basis for the second flow correction by determining $s_1$, $Q_1$, and $Q_1/H_1$ in that order.

**Pipe Equivalence.** In this method, a complex system of pipes is replaced by a single hydraulically equivalent line. The method cannot be applied directly to pipe systems containing crossovers or takeoffs. However, it is frequently possible, by judicious skeletonizing of the network, to obtain significant information on the quantity and pressure of water available at

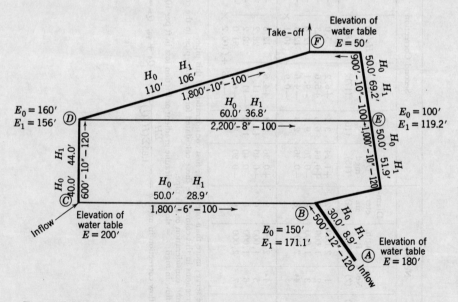

**Figure 7-11.**

**Plan of network analyzed by the method of balancing flows, Example 7-4.**

## Table 7-5
Analysis of the Network of Figure 7–11 by the Method of Balancing Flows[a,b]

| Junction Letter | Pipe | Length, ft | Diameter, in. | C | $H_0$, ft | $s_0$, ‰ | $Q_0$, mgd | $Q_0/H_0$ | $h_0$, ft | $H_1$, ft |
|---|---|---|---|---|---|---|---|---|---|---|
| (1) | (2) | (3) | (4) | (5) | (6) | (7) | (8) | (9) | (10) | (11) |
| B | AB | 500 | 12 | 120 | +30 | 60.0 | +7.33 | 0.244 | −21.1 | +8.9 |
|   | BE | 1000[c] | 10 | 120 | −50 | 50.0 | −4.12 | 0.082 | −21.1 + 19.2 | −51.9 |
|   | CB | 1800 | 6 | 100 | +50 | 27.8 | +0.66 | 0.013 | −21.1 | +28.9 |
|   |    |      |    |     |     |      | 1.85 × (+3.87) ÷ | 0.339 = | +21.1 |  |
| D | CD | 600 | 10 | 120 | +40 | 66.7 | +4.8 | 0.120 | +4.01 | +44.0 |
|   | DE | 2200[d] | 8 | 100 | −60 | 27.3 | −1.37 | 0.023 | +4.01 + 19.2 | −36.8 |
|   | DF | 1800 | 10 | 100 | −110 | 61.1 | −3.82 | 0.037 | +4.01 | −106.0 |
|   |    |      |    |     |     |      | 1.85 × (−0.39) ÷ | 0.180 = | −4.01 |  |
| E | BE | 1000[e] | 10 | 120 | +50 | 50.0 | +4.12 | 0.082 | −19.2 + 21.1 | +51.9 |
|   | DE | 2200[f] | 8 | 100 | +60 | 27.3 | +1.37 | 0.023 | −19.2 − 4.01 | +36.8 |
|   | EF | 900 | 10 | 100 | −50 | 55.6 | −3.64 | 0.073 | −19.2 | −69.2 |
|   |    |      |    |     |     |      | 1.85 × (+1.85) ÷ | 0.178 = | +19.2 |  |

[a] Only the first head correction is calculated for purposes of illustration.
[b] The basic data for this illustrative example are those used by C. E. Carter and Scott Keith, *J. New Eng. Water Works Assoc.*, **59**, 273 (1945).
[c] First consideration of pipe BE.  [d] First consideration of pipe DE.
[e] Second consideration of pipe BE.  [f] Second consideration of pipe DE.

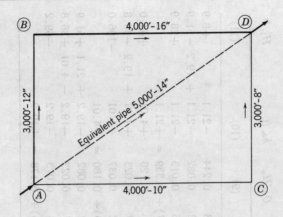

**Figure 7-12.**
**Plan of network analyzed by the method of equivalent pipes. Example 7-5.**

important points, or to reduce the number of circuits to be considered. In paring the system down to a workable frame, the analyst can be guided by the fact that pipes contribute little to flow (1) when they are small, 6 in. and under in most systems and as large as 8 or 10 in. in large systems and (2) when they are at right angles to the general direction of flow and there is no appreciable pressure differential between their junctions in the system.

Pipe equivalence makes use of the two hydraulic axioms: (1) that head losses through pipes in series, such as $AB$ and $BD$ in Fig. 7-12, are additive; and (2) that flows through pipes in parallel, such as $ABD$ and $ACD$ in Fig. 7-12, must be so distributed that the head losses are identical.

---

*Example 7-5.*

Find an equivalent pipe for the network of Fig. 7-12. Express $Q$ in mgd; $s$ in ‰; $H$ in ft; and assume a Hazen-Williams coefficient $C$ of 100.

1. *Line ABD.*  Assume $Q = 1$ mgd
   (a) Pipe $AB$, 3000 ft, 12 in.; $s = 2.1$; $H = 2.1 \times 3 = 6.3$ ft
   (b) Pipe $BD$, 4000 ft, 16 in.; $s = 0.52$; $H = 0.52 \times 4 = 2.1$ ft
   (c) $\qquad\qquad\qquad\qquad\qquad\qquad$ Total $H = 8.4$ ft
   (d) Equivalent length of 12-in. pipe: $1000 \times 8.4/2.1 = 4000$ ft
2. *Line ACD.*  Assume $Q = 0.5$ mgd
   (a) Pipe $AC$, 4000 ft, 10 in.; $s = 1.42$; $H = 1.42 \times 4 = 5.7$ ft
   (b) Pipe $CD$, 3000 ft,  8 in.; $s = 4.2$; $H = 4.2 \times 3 = 12.6$ ft
   (c) $\qquad\qquad\qquad\qquad\qquad\qquad$ Total $H = 18.3$ ft
   (d) Equivalent length of 8-in. pipe: $1000 \times 18.3/4.2 = 4360$ ft

3. *Equivalent line AD.*  Assume $H = 8.4$ ft
   (a) Line $ABD$, 4000 ft, 12 in.; $s = 8.4/4.00 = 2.1$;  $Q = 1.00$ mgd
   (b) Line $ACD$, 4360 ft, 8 in.; $s = 8.4/4.36 = 1.92$; $Q = 0.33$ mgd
   (c)                            Total $Q = 1.33$ mgd
   (d) Equivalent length of 14-in. pipe: $Q = 1.33$, $s = 1.68$, $1000 \times 8.4/1.68 = 5000$ ft.
   (e) Result:   5000 ft of 14-in. pipe.

Necessary calculations are as follows:

1. Because line $ABD$ consists of two pipes in series, the losses of head created by a given flow of water are additive. Find, therefore, from the Hazen-Williams diagram the frictional resistance $s$ for some reasonable flow (1 mgd), (a) in pipe $AB$ and (b) in pipe $BD$. Multiply these resistances by the length of pipe to obtain the loss of head $H$. Add the two losses to find the total loss $H = 8.4$ ft. Line $ABD$, therefore, must carry 1 mgd with a total loss of head of 8.4 ft. Any pipe that will do this is an equivalent pipe. Because a 12-in. pipe has a resistance $s = 2.1‰$ when it carries 1 mgd of water, a 12-in. pipe, to be an equivalent pipe, must be $1000 \times 8.4/2.1 = 4000$ ft long.

2. Proceed in the same general way with line $ACD$ to find a length of 4360 ft for the equivalent 8-in. pipe.

3. Because $ABD$ and $ACD$ together constitute two lines in parallel, the flows through them at a given loss of head are additive. If some convenient loss is assumed, such as the loss already calculated for one of the lines, the missing companion flow can be found from the Hazen-Williams diagram. Assuming a loss of 8.4 ft, which is associated with a flow through $ABD$ of 1 mgd,[10] it is only necessary to find from the diagram that the quantity of water that will flow through the equivalent pipe $ACD$, when the loss of head is 8.4 ft (or $s = 8.4/4.36 = 1.92‰$) amounts to 0.33 mgd. Add this quantity to the flow through line $ABD$ (1.0 mgd) to obtain 1.33 mgd. Line $AD$, therefore, must carry 1.33 mgd with a loss of head of 8.4 ft. If the equivalent pipe is assumed to be 14 in. in diameter, it will discharge 1.33 mgd with a frictional resistance $s = 1.68‰$, and its length must be $1000 \times 8.4/1.68 = 5000$ ft. Thence, the network can be replaced by a single 14-in. pipe 5000 ft long.

No matter what the original assumptions for quantity, diameter, and loss of head, the calculated equivalent pipe will perform hydraulically in the same way as the network it replaces.

Different in principle is the operational replacement of every pipe in a given network by equivalent pipes with identical diameters and capacity coefficients, but variable length. The purpose, in this instance, is to simplify subsequent calculations. For the Hazen-Williams relationship, Eq. 6-7, for

---

[10] It was, therefore, really unnecessary to specify the length and diameter of the equivalent pipe $ABD$.

example, $l_e = (100/C)^{1.85}(d_e/d)^{4.87} l$, where $l_e$ is the length of a pipe of diameter $d_e$ and discharge coefficient $C_e = 100$ and $l$, $d$, and $C$ are the corresponding properties of the existing pipe. Wanted values of $l_e$ can be found readily from a logarithmic plot of $l_e/l$ against $d_e/d$ at given values of $C$, Fig. 3, Appendix.

## 7-7 Computer Programming

High-speed digital computers can be programmed to solve network problems in a number of different ways. Convergence formulas need not be introduced as such. Instead, the computer can be assigned the task of adjusting the water table or pressure at each junction not controlled by a service reservoir until the *circuit laws* discussed in connection with the method of relaxation are satisfied throughout the system.[11] These laws can be summarized as follows:

1. At each junction $\Sigma Q_{inflow} = \Sigma Q_{outflow}$.
2. In each circuit $\Sigma H = 0$.
3. In each pipe $H = kQ^n$ or $Q = (H/k)^{1/n}$.

To program the operation, number each pipe and junction and identify pipe ends by junction numbers; then tabulate pipe resistances, junction pressures (including assumed values where pressures are unknown), and net inflows at each junction (zero at all but entrance and exit points of the system), and feed the tabulated information into the computer. The computer instructions are then as follows: Calculate by *circuit law 3* the total flow into the first junction for which the water-table elevation is unknown; adjust the assumed value until the total inflow and outflow are balanced in accordance with *circuit law 1*; proceed in sequence to the remaining junctions; and readjust the first water-table elevation. Repeat the cycle of operations until *all circuit laws* are satisfied. Contemplated network changes can be printed on separate tapes to avoid re-input of the entire program.

Camp and Hazen[12] built the first electric analyzer designed specifically for the hydraulic analysis of water distribution systems. Electric analyzers use nonlinear resistors, called fluistors in the McIlroy[13] analyzer, to simulate pipe resistances. For each branch of the system, the pipe equation, $H = kQ^{1.85}$, for example, is replaced by an electrical equation, $V = K_e I^{1.85}$, where $V$ is the

[11] J. H. Dillingham, Computer Analysis of Water Distribution Systems, *Water and Sewage Works* series, Jan. through May 1967.
[12] T. R. Camp and H. L. Hazen, Hydraulic Analysis of Water Distribution Systems by Means of an Electric Network Analyzer, *J. New Eng. Water Works Assoc.*, **48**, 383 (1934).
[13] M. S. McIlroy, Direct-Reading Electric Analyzer for Pipeline Networks, *J. Am. Water Works Assoc.*, **42**, 347 (1950); also Water-Distribution Systems Studied by a Complete Electrical Analogy, *J. New Eng. Water Works Assoc.*, **45**, 299 (1953).

voltage drop in the branch, $I$ is the current, and $K_e$ is the nonlinear-resistor coefficient suited to the pipe coefficient $k$ for the selected voltage drop (head loss) and amperage (water flow) scale ratios. If the current inputs and take-offs are made proportional to the water flowing into and out of the system, the head losses will be proportional to the measured voltage drops. Some large, rapidly developing communities have found it economical to acquire electric analyzers suited to their own systems.

## 7-8 Service Storage

The three major components of service storage are (1) equalizing, or operating, storage, (2) fire reserve, and (3) emergency reserve.

*Equalizing, or Operating, Storage.* Required equalizing, or operating, storage can be read from a demand rate curve or, more satisfactorily, from a mass diagram similar in concept to the Rippl diagram (Fig. 4–3). As shown in Fig. 7–13 for the simple conditions of steady inflow, during 12 and 24 hr respectively, the amount of equalizing, or operating, storage is the sum of the maximum ordinates between the demand and supply lines. To construct such a mass diagram, proceed as follows: (1) from past measurements of flow, determine the draft during each hour of the day and night for typical days (maximum, average, and minimum); (2) calculate the amounts of water drawn up to certain times, that is, the cumulative draft; (3) plot the cumulative draft against time; (4) for steady supply during 24 hr, draw a

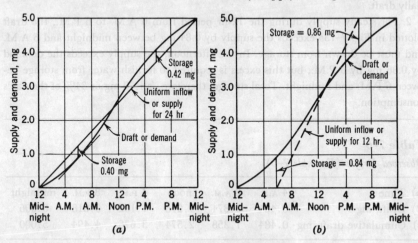

**Figure 7-13.**
Determination of equalizing, or operating, storage by mass diagram. See Example 7-6. Uniform inflow, or supply, (a) extending over 24 hr; (b) confined to 12 hr. (a) Total storage = 0.40 + 0.42 = 0.82 mg; (b) total storage = 0.84 + 0.86 = 1.70 mg.

straight line diagonally across the diagram, as in Fig. 7–13a. Read the storage required as the sum of the two maximum ordinates between the draft and the supply line; and (5) for steady supply during 12 hr—by pumping, for example—draw a straight line diagonally from the beginning of the pumping period to its end—for example, from 6 A.M. to 6 P.M., as in Fig. 7–13b. Again read the storage required as the sum of the two maximum ordinates.

Steady supply at the rate of maximum daily use will ordinarily require an equalizing storage between 15 and 20% of the day's consumption. Limitation of supply to 12 hr may raise the operating storage to an amount between 30 and 50% of the day's consumption.

*Example 7–6.*

Determine the equalizing, or operating, storage for the drafts of water shown in Table 7–6 (1) when inflow is uniform during 24 hr and (2) when flow is confined to the 12 hr from 6 A.M. to 6 P.M.

1. For steady supply during 24 hr, the draft plotted in Fig. 7–13a exceeds the demand by 0.40 mg by 6 A.M. If this excess is stored, it is used up by 11 A.M. In the afternoon, the demand exceeds the supply by 0.42 mg by 6 P.M. and must be drawn from storage that is replenished by midnight. Hence the required storage is the sum of the morning excess and afternoon deficiency, or 0.82 mg. This equals 16.4% of the daily draft.

2. For steady supply during the 12-hr period from 6 A.M. to 6 P.M., the draft plotted in Fig. 7–13b exceeds the supply by 0.84 mg between midnight and 6 A.M. and must be drawn from storage. In the afternoon, the supply exceeds the demand by 0.86 mg by 6 P.M., but this excess is required to furnish water from storage between 6 P.M. and midnight. Total storage, therefore, is 1.70 mg or 34% of the day's consumption.

*Table 7-6*
*Observed Drafts (Example 7–6)*

| (a) Time | 4 A.M. | 8 A.M. | noon | 4 P.M. | 8 P.M. | midnight |
|---|---|---|---|---|---|---|
| (b) Draft, mg | 0.484 | 0.874 | 1.216 | 1.102 | 0.818 | 0.506 |
| (c) Cumulative draft, mg | 0.484 | 1.358 | 2.574 | 3.676 | 4.494 | 5.000 |

***Fire Reserve.*** Basing its recommendations on observed durations of serious conflagrations, the American Insurance Association recommends that distributing reservoirs be made large enough to supply water for fighting a serious conflagration for 10 hr in communities of more than 6000 people and for 8, 6, and 4 hr in places with 4000, 2000, and 1000 people respectively. The resulting fire reserve, shown in Table 7-1, may not always be economically attainable, and design values may have to be adjusted downward to meet local financial abilities. Changing community patterns, moreover, may make for changing requirements in the future.

**Emergency Reserve.** The magnitude of this storage component depends on (1) the danger of interruption of reservoir inflow by failure of supply works and (2) the time needed to make repairs. If shutdown of the supply is confined to the time necessary for routine inspections during the hours of minimum draft, the emergency reserve is sometimes made no more than 25% of the total storage capacity, that is, the reservoir is assumed to be drawn down by one fourth its average depth. If supply lines or equipment are expected to be out of operation for longer times, higher allowances must be made. The American Insurance Association bases its rating system on an emergency storage of 5 days at maximum flow.

**Total Storage.** The total amount of storage is desirably equal to the sum of the component requirements. In each instance, economic considerations dictate the final choice. In pumped supplies, cost of storage must be balanced against cost of pumping, and attention must be paid to economies effected by operating pumps more uniformly and restricting pumping to a portion of the day only. In all supplies, cost of storage must be balanced against cost of supply lines, increased fire protection, and more uniform pressures in the distribution system.

*Example 7-7.*

For a steady gravity supply equal to the maximum daily demand, a 10-hr fire supply, and no particular hazard to the supply works, find the storage to be provided for a city of 50,000 people using an average of 7.5 mgd of water.

The equalizing storage is 15% of 7.5 mg, or 1.13 mg; and the fire reserve (Table 7-1) is 3.90 mg. The resulting subtotal is $1.13 + 3.90$ or 5.03 mg. Because the emergency reserve is one fourth of the total storage, the subtotal is three fourths of the total storage, and the total storage is 5.03/0.75, or 6.70 mg.

**Location of Storage.** As shown in Sec. 7-1 and Fig. 7-1, location as well as capacity of service storage is an important factor in the control of distribution systems. A million gallons of elevated fire reserve, suitably sited in reference to the area to be protected, is equivalent, for example, to the addition of a 12-in. supply main. The underlying reasoning is that drawing this volume of water in a 10-hr fire, flow is provided at a rate of $(24/10) \times 1 = 2.4$

mgd. This is the amount of water a 12-in. pipe can carry at a velocity less than 5 fps. Why this must be neighborhood storage is explained by the high frictional resistance of more than $10\%_0$ accompanying such use.

**Elevation of Storage.** Storage reservoirs and tanks operate as integral parts of the system of pumps, pipes, and connected loads. In operation all the parts respond to pressure changes as the system follows the diurnal and seasonal demands. Ideally the storage elevation should be such that the reservoir "floats" on the system, neither emptying nor standing continuously full. In systems with inadequate pipes or pumps, or having a storage reservoir that is too high, the hydraulic gradient may at times of peak demand fall below the bottom of the reservoir. When this occurs, the full load falls on the pumps and system pressures deteriorate suddenly.

**Types of Distributing Reservoirs.** Where topography and geology permit, service reservoirs are formed by impoundage, balanced excavation and embankment, or masonry construction (Fig. 7-14). To protect the water against chance contamination and against deterioration by algal growths stimulated by sunlight, distributing reservoirs should be covered. Roofs need not be watertight if the reservoir is fenced. Open reservoirs should always be fenced. Where surface runoff might drain into them, they should have a marginal intercepting conduit.

Earthen reservoirs, their bottom sealed by a blanket of clay or rubble masonry and their sides by core walls, were widely employed at one time. Today, lining with concrete slabs is more common. Gunite, a sand-cement-water mixture, discharged from a nozzle or gun through and onto a mat of reinforcing steel, has also been employed to line or reline them. Plastic sheets protected by a layer of earth have also been used to build inexpensive but watertight storage basins. Roofs are made of wood or concrete. Beam and girder, flat-slab, arch, and groined-arch construction have been used. Where concrete roofs can be covered with earth, both roof and water will be protected against extremes of temperature.

Inlets, outlets, and overflows are generally placed in a gate house or two. Circulation to ensure more or less continuous displacement of the water and to provide proper detention of water after chlorination may be controlled by baffles or subdivisions between inlet and outlet. Overflow capacity should equal maximum rate of inflow. Altitude-control valves on reservoir inlets (Fig. 7-15) will automatically shut off inflow when the maximum water level is reached. An arrangement that does not interfere with draft from the reservoir includes a bypass with a swing check valve seating against the inflow.

Where natural elevation is not high enough, water is stored in concrete or steel standpipes and elevated tanks. In cold climates, steel is most suitable. Unless the steel in reinforced-concrete tanks is prestressed, vertical cracks, leakage, and freezing will cause rapid deterioration of the structure. Ground-

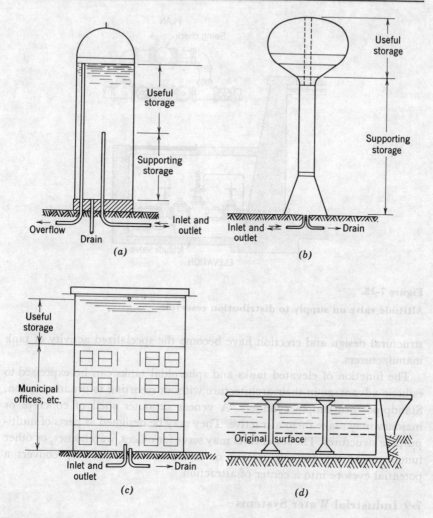

**Figure 7-14.**
**Types of service reservoirs.** (*a*) Standpipe; (*b*) and (*c*) elevated tanks; (*d*) ground-level service reservoir.

level storage in reinforced concrete or steel tanks in advance of automatic pumping stations is an alternative.

The useful capacity of standpipes and elevated tanks is confined to the volume of water stored above the level of wanted distribution pressure. In elevated tanks, this level generally coincides with the tank bottom; in standpipes, it may lie much higher. Steel tanks are welded or riveted. Their

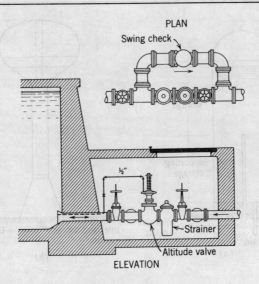

**Figure 7-15.**
**Altitude valve on supply to distribution reservoir.**

structural design and erection have become the specialized activity of tank manufacturers.

The function of elevated tanks and spheroidal tanks can be expressed to esthetic advantage in their architecture without resorting to ornamentation. Standpipes are simple cylinders. A veneer or outer shell of concrete or masonry may make them attractive. They may be designed as parts of multi-purpose structures. The lower level may serve for offices, warehouses, or other functions. At the top, sightseeing or restaurant facilities may convert a potential eyesore into a center of attraction.

## 7-9 Industrial Water Systems

Large industrial establishments, with a heavy investment in plant, equipment, raw materials, and finished products, concentrated in a small area, are generally equipped with high-pressure fire supplies and distribution networks of their own. Because such supplies may be drawn from sources of questionable quality, some regulatory agencies require rigid separation of all private fire supplies and public distribution systems. Others permit the use of *protected cross-connections* and require their regular inspection for tightness. How the two sources of supply can be divorced without denying the protective benefit and general convenience of a dual supply to industry is illustrated in Fig. 7-16. Ground-level storage and pumping are less advantageous.

*Maintenance of Distribution Systems*/221

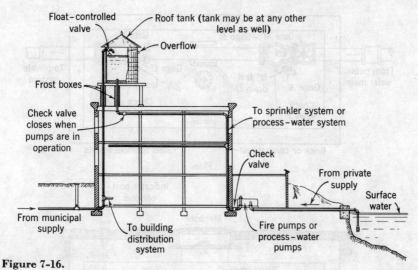

**Figure 7-16.**
**Use of industrial water supply without cross-connection. (After Minnesota State Board of Health.)**

A widely approved arrangement of double check valves in vaults accessible for inspection and test by the provision of valves, gages, and bleeders is shown in Fig. 7–17. No outbreak of waterborne disease has been traced to approved and properly supervised cross-connections of this kind. Automatic chlorination of the auxiliary supply can introduce a further safeguard.

## 7-10 Management, Operation, and Maintenance of Distribution Systems

For intelligent management of distribution storage, reservoir levels must be known at all times of day and night. Where levels cannot be observed directly by gages or floats, electrically operated sensors and recorders can transmit wanted information to operating headquarters.

Well-kept records and maps of pipes and appurtenances are essential to the efficient operation and maintenance of distribution systems. To avoid the occasional discharge of roiled water, piping should be flushed systematically, usually through hydrants. Dead ends need particular attention; a bleeder on the dead end will counteract the effects of sluggish water movements. Disinfecting newly laid pipe, or pipe newly repaired, is important (see Chap. 17). Control of pipe corrosion and the cleaning and relining of water mains are discussed in Sec. 16–13.

There is little flow through service pipes at night, and they may freeze in very cold weather. If water mains themselves are placed at a reasonable

222/Water Distribution

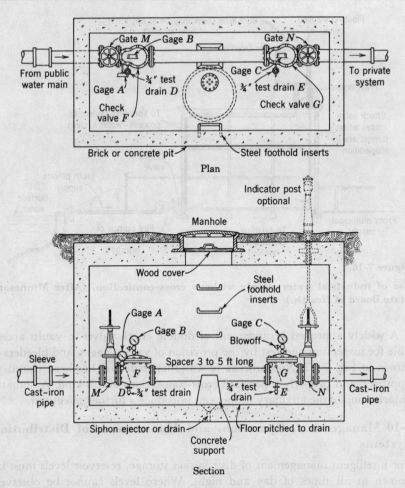

**Figure 7-17.**
**Cross-connection between municipal water supply and private (industrial) water supply protected by double check-valve installation. To test installation: (1) close gates $M$ and $N$; (2) open test drain $D$, and observe gages $A$ and $B$; (3) open test drain $E$, and observe gage $C$. If check valves $F$ and $G$ are tight, gage $A$ will drop to zero; gages $B$ and $C$ will drop slightly owing to compression of rubber gaskets on check valves $F$ and $G$.**

depth and enough flow is maintained in the system, they should not freeze. Pipes deprived of adequate cover by the regrading of streets or subjected to protracted and exceptionally cold spells can be protected by drawing water from them through services. Pipes exposed on bridges or similar crossings should be insulated. Large and important lines may be heated where exposure is severe. In very cold climates, water and sewer pipes are often laid in a heated boxlike conduit, known as a *utilidor*.

Frozen pipes are usually thawed by electricity. A transformer connected to an electric power circuit, or a gasoline-driven generator of the electric-welding type, supplies the current: 100 to 200 amperes at 3 to 10 volts for small pipes up to several thousand amperes at 55 or 110 volts for large mains. The current applied is varied with the electrical resistance and the melting point of the pipe metals. Nonmetallic jointing and caulking compounds and asbestos-cement or plastic pipes obstruct current flow. Electric grounds on interior water piping, or the piping itself, must be disconnected during thawing operations. Grounds are needed but are an annoyance when they carry high voltages into the pipes and shock workmen. Pipes and hydrants can also be thawed with steam generated in portable boilers and introduced through flexible block-tin tubing.

Loss of water by leakage from distribution systems and connected consumer premises should be kept under control by leakage surveys.

# eight

# flow in sewers and their appurtenances

## 8-1 Nature of Flow

Hydraulically, wastewater collection differs from water distribution in the following three essentials: (1) sewers, although most of them are circular pipes, normally flow only partially filled and hence as open channels, (2) tributary flows are almost always unsteady and often nonuniform, and (3) sewers are generally required to transport substantial loads of floating, suspended, and soluble substances with little or no deposition, on the one hand, and without erosion of channel surfaces on the other hand. To meet the third requirement, sewer velocities must be self-cleansing yet nondestructive.

As shown in Sec. 2-2, the design period for main collectors, intercepters, and outfalls may have to be as much as 50 years because of the inconvenience and cost of enlarging or replacing hydraulic structures of this nature in busy city streets. The sizing of needed conduits becomes complicated if they are to be self-cleansing at the beginning as well as the end of the design period. Although water-distribution systems, too, must meet changing capacity requirements, their hydraulic balance is less delicate; the water must transport only itself, so to speak. It follows that velocities of flow in water-distribution systems are important economically rather than functionally and can be allowed to vary over a wide range of magnitudes without markedly affecting system performance. In contrast, performance of wastewater systems is tied, more or less rigidly, to inflexible hydraulic gradients and so becomes functionally as well as economically important.

## 8-2 Flow in Filled Sewers

In the absence of precise and conveniently applicable information on how channel roughness can be measured and introduced into theoretical formulations of flow in open channels, engineers continue to base the hydraulic design of sewers, as they do the design of water conduits (Sec. 6–2), on empirical formulations. Equations common in North American practice are the Kutter-Ganguillet[1] formula of 1869 and the Manning[2] formula of 1890. In principle, these formulations evaluate the velocity or discharge coefficient $c$ in the Chézy formula of 1775 in terms of invert slope $s$ (Kutter-Ganguillet only), hydraulic radius $r$, and a coefficient of roughness $n$. The resulting expressions for $c$ are:

$$c_{\text{Kutter-Ganguillet}} = \frac{(41.65 + 2.81 \times 10^{-3}/s) + 1.811/n}{(41.65 + 2.81 \times 10^{-3}/s)(n/r^{1/2}) + 1} \quad (8\text{-}1)$$

$$c_{\text{Manning}} = 1.486\, r^{1/6}/n \quad (8\text{-}2)$$

Of the two, Manning's equation is given preference in these pages, because it satisfies experimental findings fully as well as the mathematically clumsier Kutter-Ganguillet formula. Moreover, it lends itself more satisfactorily to algebraic manipulation, slide-rule computation, and graphical representation.

Introducing Manning's $c$ into the Chézy formula, the complete Manning equation[3] reads:

$$v = (1.49/n)\, r^{2/3} s^{1/2} \quad (8\text{-}3)$$

and is seen to resemble the Hazen-Williams formula of Sec. 6–2. Indeed, the Hazen-Williams equation could be used instead. Values of $V_0 = 1.49 R^{2/3}$, $Q_0 = 1.49\, AR^{2/3}$, and $1/V_0$, the reciprocal of $1.49 R^{2/3}$, are listed in Table 9 of the Appendix to speed calculations.[4] The table is based on a generalization of Manning's formula in terms of the ratio $S^{1/2}/N$, where $S/N^2$ is, in a sense, the relative slope for varying conduit sizes and roughness coefficients. The

---

[1] E. Ganguillet and W. R. Kutter, *Flow of Water in Rivers and Other Channels*, translated by Rudolf Hering and J. G. Trautwine, John Wiley & Sons, New York, 1888.

[2] Named after Robert Manning (1816–1897), an Irish engineer who discussed the limitations of the formula $v = kr^{2/3} s^{1/2}$.

[3] Because the value of $n$ is normally good to no more than two significant figures, 1.49 and, indeed, 1.5 should replace the commonly quoted numerical constant 1.486.

[4] Capital letters are chosen here to denote the hydraulic elements of conduits flowing full, and lower-case letters for partially filled sections.

ratio $S^{1/2}/N$ appears, too, in formulations for the flow of stormwaters over land and into street inlets.

Choice of a suitable roughness coefficient is of utmost importance. The following ranges in values are recommended by the Joint Committee of the American Society of Civil Engineers and the Water Pollution Control Federation:[5] (1) for vitrified-clay, concrete, asbestos-cement, and corrugated steel pipe with smooth asphaltic lining, a coefficient ranging from 0.010 to 0.015 for strong sewage, 0.013 being a common design value for sanitary sewers and (2) for corrugated-steel pipes often used in culverts, a coefficient of 0.018 to 0.022 when asphalt coatings and paving cover 25% of the invert section, or 0.022 to 0.026 for uncoated pipe with ½-in. corrugations. All but corrugated pipes show little difference between values suitable for the Manning and Kutter-Ganguillet equations.

*Example 8-1.*

1. Given a 12-in. sewer, $N = 0.013$, laid on a grade of 4.05‰ (ft per 1000 ft), find its velocity of flow and rate of discharge from Table 9, Appendix.

Because $S^{1/2}/N = (4.05 \times 10^{-3})^{1/2}/1.3 \times 10^{-2} = 4.90$, $V = 4.90 \times 0.590 = 2.89$ fps, and $Q = 4.90 \times 0.463 = 2.27$ cfs.

2. Given a velocity of 3 fps for this sewer, find its (minimum) gradient for flow at full depth.

Because $NV = 3.9 \times 10^{-2}$, $S = (3.9 \times 10^{-2} \times 1.696)^2 = 4.37$‰. This is shown also in Table 10, Appendix.

Minimum grades $S$ and capacities $Q$ of sewers ($N = 0.013$) up to 24 in. in diameter flowing full at velocities of 2.0, 2.5, 3.0, and 5.0 fps are listed for convenience of reference in Table 10 of the Appendix.

## 8-3 Limiting Velocities of Flow

Wastes from bathrooms, toilets, laundries, and kitchens are flushed into sanitary sewers through *house* or *building* sewers. Sand, gravel, and debris of many kinds enter storm drains through curb and yard inlets. Combined sewers carry mixtures of the two. Heavy solids are swept down sewer inverts like the *bed load* of streams. Light materials float on the water surface. When velocities fall, heavy solids are left behind as bottom deposits, while light materials strand at the water's edge. When velocities rise again, gritty substances and the flotsam of the sewer are picked up once more and carried

---

[5] Design and Construction of Sanitary and Storm Sewers, *Am. Soc. Civil Engrs., Manuals of Engineering Practice*, No. 37, 84 (1969); *Water Poll. Control Fed., Manual of Practice* No. 9.

along in heavy concentration. There may be erosion. Within reason, all of these happenings should be avoided, insofar as this can be done. Each is a function of the tractive force of the carrying water that should be better known than it commonly is.

Conceptually, the drag exerted by flowing water on a channel is analogous to the friction exerted by a body sliding down an inclined plane. Because the volume of water per unit surface of channel equals the hydraulic radius of the channel,

$$\tau = \gamma r s \qquad (8\text{-}4)$$

where $\tau$ is the intensity of the tractive force, $\gamma$ the specific weight of water at the prevailing temperature, $r$ the hydraulic radius of the filled section, and $s$ the slope of the invert or loss of head in a unit length of channel when flow is steady and uniform and the water surface parallels the invert. Substituting $rs = (v/c)^2$ in accordance with the Chézy equation (Eq. 6-3), for example,

$$\tau = \gamma (v/c)^2 \qquad (8\text{-}5)$$

and the tractive force intensity is seen to vary as the square of the velocity of flow $v$ and inversely as the square of the Chézy coefficient, $c$.

**Transporting Velocities.** The velocity required to transport waterborne solids is derived from Eq. 8-4, with the help of Fig. 8-1.

For a layer of sediment of unit width and length, thickness $t$, and porosity ratio $f'$, the drag force $\tau$ exerted by the water at the surface of the sediment and just causing it to slide down the inclined plane equals the frictional resistance $R = W \sin \alpha$, where $W = (\gamma_s - \gamma)t(1 - f')$ is the weight of the sediment in water and $\alpha$ is the friction angle. Accordingly, $\tau = (\gamma_s - \gamma) \times t(1 - f') \sin \alpha$, and Eq. 8-4 becomes

$$\tau = (\gamma_s - \gamma)t(1 - f') \sin \alpha = k(\gamma_s - \gamma)t \qquad (8\text{-}6)$$

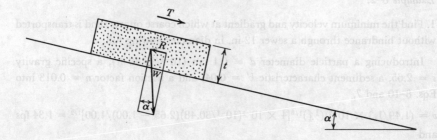

**Figure 8-1.**

**Forces acting on sediment of unit width and length and thickness $t$. $T$ is the tractive force per unit surface area. $R$, the resisting force, is a function of the weight of the sediment ($W$) and the friction angle ($\alpha$).**

Here $k = (1 - f') \sin \alpha$ is an important characteristic of the sediment. For single grains, the volume per unit area $t$ becomes a function of the diameter of the grains $d$ as an inverse measure of the surface area of the individual grains exposed to drag or friction. Thus $k'$ replaces $k$ when $d$ replaces $t$.

It follows from Eqs. 8–4 and 6 that the invert slope at which sewers will be self-cleansing is

$$s = (k'/r)[(\gamma_s - \gamma)/\gamma] d \qquad (8-7)$$

and that, in accordance with the Chézy equation,

$$v = c[k'd(\gamma_s - \gamma)/\gamma]^{1/2} \qquad (8-8)$$

where the value of $c$ is chosen with full recognition of the presence of deposited or depositing solids and expressed, if so desired, in accordance with any other pertinent capacity or friction factor. Examples are $(a)$:

$$v = [(8k'/f)gd(\gamma_s - \gamma)/\gamma]^{1/2} \qquad (8-9)$$

derived by Camp from studies by Shields[6] and obtainable from Eq. 8–8 by introducing the Weisbach-Darcy friction factor $f = 8g/c^2$; and $(b)$:

$$v = (1.49/n) \, r^{1/6} [k'd(\gamma_s - \gamma)/\gamma]^{1/2} \qquad (8-10)$$

in terms of a Manning evaluation of $c = 1.49(r^{1/6}/n)$, where $n$ is the Manning friction factor. When convenient, the ratio $(\gamma_s - \gamma)/\gamma$ can be replaced by the closely equal term $(s_s - 1)$, where $s_s$ is the specific gravity of the particles (solids) composing the deposit.

Applicable magnitudes of $k$ range from 0.04 for initiating scour of relatively clean grit to 0.8 or more for full removal of sticky grit. Their actual magnitude can be found only by experiment.

---

*Example 8–2.*

1. Find the minimum velocity and gradient at which coarse quartz sand is transported without hindrance through a sewer 12 in. in diameter flowing full.

Introducing a particle diameter $d = 0.1$ cm ($10^{-1}/30.48$ ft), a specific gravity $s = 2.65$, a sediment characteristic $k' = 0.04$, and a friction factor $n = 0.013$ into Eqs. 8–10 and 7,

$v = (1.49/1.3 \times 10^{-2})(\frac{1}{4})^{1/6}[4 \times 10^{-2}(10^{-1}/30.48)(2.65 - 1.00)/1.00]^{1/2} = 1.34$ fps

and

[6] A. Shields, Anwendung der Aenlichkeitsmechanik und der Turbulenzforschung an die Geschiebebegung (Application of Similitude Mechanics and Turbulence Research to Bed-Load Movement), *Mitt. der Preuss. Versuchsanstalt fur Wasserbau und Schiffbau*, No. 26, Berlin (1936). Shields showed that particles at the sediment surface were left undisturbed so long as $v^2\alpha < gd(\gamma_s - \gamma)/\gamma$, where $\alpha = 0.10$ approximately.

$$s = [4 \times 10^{-2}/(\tfrac{1}{4})](10^{-1}/30.48)[(2.65 - 1.00)/1.00] = 0.87 \times 10^{-3}$$

For a sediment characteristic $k' = 0.8 = 20 \times 0.04$, the required velocity is $\sqrt{20} = 4.5$ times as much and the required slope 20 times as steep.

2. If flows in storm and combined sewers are given velocities of 3.0 and 5.0 fps respectively, find the diameter of sand or gravel moved.

In accordance with Eq. 8-10, $d$ varies as $v^2$. Hence for $v = 3.0$ fps, $d = 10^{-1}(3.0/1.34)^2 = 0.50$ cm (small-sized gravel), and for $v = 5.0$ fps, $d = 10^{-1}(5.0/1.34)^2 = 1.4$ cm (large gravel).

**Damaging Velocities.** Of the ceramic materials used in sewers, vitrified tile and glazed brick are very resistant to wear; building brick, asbestos-cement, and concrete are less so. Abrasion is greatest at the bottom of conduits, because grit, sand, and gravel are heavy and travel along the invert. The bottom arch or invert of large concrete or brick sewers is often protected by vitrified tile liners, glazed or paving brick, or granite blocks.

Clear water can flow through hard-surfaced channels, such as good concrete conduits, at velocities higher than 40 fps without harm. Stormwater runoff, on the other hand, has to be held down to about 10 fps in concrete sewers and drains, because it usually contains abrading substances in sufficient quantity to wear away even well constructed, hard concrete surfaces. The magnitude of the associated tractive force is given by Eq. 8-5.

## 8-4 Flow in Partially Filled Sewers

In the upper reaches of sanitary sewerage systems, sewers receive relatively little wastewater. Depths of flow are reduced, because minimum pipe sizes (8 in. in North America) are dictated not by flow requirements alone but also by cleaning potentials. The lower portions of the system, too, do not flow full. Even when the end of their design period is reached, they are filled only spasmodically during times of maximum flow. Discharge ratios may vary, indeed, from as little as 4:1 at the end of the design period to as much as 20:1 at the beginning.

Hydraulic performance of the upper reaches of sanitary sewers is improved by steeper than normal grades, even though velocities of, for example, 3.0 rather than 2.5 or 2.0 fps produce still lower depths of flow. Why this is so is shown later. Requisite capacities of storm and combined sewers are even more variable. The hydraulic performance of partially filled as well as filled sections must, therefore, be well understood, especially in reference to the maintenance of self-cleansing velocities at expected flows.

The variables encompassed by a flow formula such as Manning's, namely, $q$ or $v$, $r$ or $a/p$ ($p$ = wetted perimeter), $s$ or $h/l$, and $n$, constitute the *hydraulic elements* of conduits. For a given shape and a fixed coefficient of roughness and invert slope, the elements change in absolute magnitude with the depth $d$ of the filled section. In the case of Manning's formulation, generalization in terms of the ratio of each element of the filled section (indicated here by a lower-case letter) to the corresponding elements of the full[7] section (indicated here by a capital letter) confines all ratios, including velocity and capacity ratios, to ultimate dependency on depth alone. Thus

$$v/V = (N/n)(r/R)^{2/3} \qquad (8-11)$$

and

$$q/Q = (N/n)(a/A)(r/R)^{2/3} \qquad (8-12)$$

Of the elements normally included in diagrams or tables, area and hydraulic radius are static, or elements of shape; roughness, velocity, and discharge are dynamic, or elements of flow. Except for roughness, the basis of their computation is explained in Table 11 of the Appendix. Variation of roughness with depth was observed by Willcox[8] on 8-in. sewer pipe, by Yarnell and Woodward[9] on clay and concrete drain tile 4 to 12 in. in diameter, and by Johnson[10] in large (Louisville, Ky.) sewers flowing at low depths. Figure 4 in the Appendix is a conventional diagram of the basic hydraulic elements of circular sewers. The two sets of curves included for $v/V$ and $q/Q$ mark the influence of a variable ratio of $N/n$ on these dynamic hydraulic elements.[11] It is important to note that velocities in partially filled, circular sections equal or exceed those in full sections whenever sewers flow more than half full and roughness is considered to vary with depth; moreover, where changes in roughness are taken into account, velocities equal to or greater than those in full sections are confined to the upper 20% of depth only. Nevertheless, sewers flowing between 0.5 and 0.8 full need not be placed on steeper grades to be as self-cleansing as sewers flowing full. The reason is that velocity and discharge are functions of tractive-force intensity, which depends on the friction coefficient as well as the flow velocity. Needed ratios of $v_s/V$, $q_s/Q$, and $s_s/S$, where the subscript $s$ denotes cleansing equaling that obtained in the full section, can be computed with the help of

[7] For circular sewers, flow at half depth has the same hydraulic radius as the full sewer.

[8] E. R. Willcox, A Comparative Test of the Flow of Water in 8-Inch Concrete and Vitrified Clay Sewer Pipe, *Univ. Wash. Eng. Exp. Sta. Ser. Bull.* 27 (1924).

[9] D. L. Yarnell and S. M. Woodward, The Flow of Water in Drain Tile, *U.S. Dept. Agr. Bull.* 854 (1920).

[10] C. F. Johnson, Determination of Kutter's $n$ for Sewers Partly Filled, *Trans. Am. Soc. Civil Engrs.*, **109**, 240 (1944).

[11] The values of $N/n$ in Table 11 and Fig. 4 of the Appendix are taken from Design and Construction of Sanitary and Storm Sewers, *Am. Soc. Civil Engrs., Manuals of Engineering Practice* No. 37, 87 (1969); and *Water Pollution Control Fed., Manual of Practice*, No. 9.

Eq. 8-4 on the assumption that equality of tractive-force intensity implies equality of cleansing, or $\tau = T = \gamma rs = \gamma RS$; whence

$$s_s = (R/r)S \qquad (8\text{-}13)$$

and

$$v_s/V = (N/n)(r/R)^{2/3}(s_s/S)^{1/2} = (N/n)(r/R)^{1/6} \qquad (8\text{-}14)$$

or

$$q_s/Q = (N/n)(a/A)(r/R)^{1/6} \qquad (8\text{-}15)$$

What these equations imply is illustrated in Fig. 5 of the Appendix and in Example 8-3.

## Example 8-3.

An 8-in. sewer is to flow at 0.3 depth on a grade ensuring a degree of self-cleaning equivalent to that obtained at full depth at a velocity of 2.5 fps. Find the required grades and associated velocities and rates of discharge at full depth and 0.3 depth. Assume that $N = 0.013$ at full depth.

1. From Table 10, Appendix, find for full depth of flow and $V = 2.5$ fps, $Q = 0.873$ cfs, and $S = 5.20\%$.
2. From Fig. 4 or Table 11, Appendix, find for 0.3 depth, $a/A = 0.252, r/R = 0.684$ (or $R/r = 1.46$), $v/V = 0.776$, $q/Q = 0.196$, and $N/n = 0.78$; and from Table 11, Appendix, find $(r/R)^{1/6} = 0.939$.

Hence, at 0.3 depth and a grade of 5.20‰, $v = 0.776 \times 2.5 = 1.94$ fps for $n = N$, or 1.51 fps for $N/n = 0.78$, and $q = 0.196 \times 0.873 = 0.171$ cfs for $n = N$, or 0.133 cfs for $N/n = 0.78$.

For self-cleaning flow, however, $s_s = 1.46 \times 5.20 = 7.6\%$, by Eq. 8-13; $v_s = 0.939 \times 2.5 = 2.35$ for $n = N$, or $0.78 \times 2.35 = 1.83$ fps for $N/n = 0.78$, by Eq. 8-14; and $q_s = 0.252 \times 0.939 \times 0.873 = 0.207$ cfs for $n = N$, or $0.78 \times 0.207 = 0.161$ cfs for $N/n = 0.78$, by Eq. 8-15.

Figure 5 of the Appendix confirms that minimum grades are enough so long as circular sewers flow more than half full. However, for granular particles, when flows drop to 0.2 depth, grades must be doubled for equal self-cleansing; at 0.1 depth they must be quadrupled. Expressed in terms of the Weisbach-Darcy friction factor $f$, Eqs. 8-14 and 15 become, respectively,

$$v_s/V = (F/f)^{1/2} \qquad (8\text{-}16)$$

and

$$q_s/Q = (a/A)(F/f)^{1/2} \qquad (8\text{-}17)$$

*Example 8-4.*

An 8-in. sewer is to discharge 0.161 cfs at a velocity as self-cleaning as a sewer flowing full at 2.5 fps. Find the depth and velocity of flow and the required slope.

1. From Example 8-3, $Q = 0.873$ cfs and $S = 5.20‰$. Hence, $q_s/Q = 0.161/0.873 = 0.185$.
2. From Fig. 5, Appendix, for $N = n$ and $q_s/Q = 0.185$, $d_s/D = 0.25$, $v_s/V = 0.91$, and $s/S = 1.70$. Hence $v_s = 0.91 \times 2.5 = 2.28$ fps, and $s_s = 1.70 \times 5.20 = 8.8‰$.
3. From Fig. 5, Appendix, for $N/n$ variable and $q_s/Q = 0.185$, $d_s/D = 0.30$, $v_s/V = 0.732$ and $s/S = 1.46$. Hence, $v_s = 0.732 \times 2.5 = 1.83$ fps, and $s_s = 1.46 \times 5.20 = 7.6‰$.

---

Egg-shaped sewers and cunettes (Fig. 9-6) were introduced, principally in Europe, to provide enough velocity for dry-weather flows in combined sewers. The hydraulic elements of these sewers and of horseshoe-shaped and box sewers can be charted in the same way as circular sewers are in Figs. 4 and 5 of the Appendix.

## 8-5 Flow in Sewer Transitions

Although flow in sewers is both *unsteady* (changing in rate of discharge) and *nonuniform* (changing in velocity and depth), these factors are normally taken into account only at sewer transitions. This is so because it is not practicable to identify with needed accuracy the variation in flow with time in all reaches of the sewerage system and because the system is designed for maximum expected flow in any case.

Sewer transitions include (1) changes in size, grade, and volume of flow; (2) free and submerged discharge at the end of sewer lines; (3) passage through measuring and diversion devices; and (4) sewer junctions. Of these, sewer transitions at changes in size or grade are most common. How they affect the profile of the water surface and energy gradient is shown, greatly foreshortened, in Fig. 8-2. Here $h_e$ is the loss in energy or head, $h_s$ the drop in water surface, and $h_i$ the required invert drop. For convenience of formulation, these changes are assumed to be concentrated at the center of the transition. The energy loss $h_e$ is usually small. In the absence of exact information, it may be considered proportional to the difference, or change, in velocity heads, that is, $h_e = k(h_{v_2} - h_{v_1}) = k\Delta h_v$. According to Hinds,[12] the proportionality factor $k$ may be as low as 0.1 for rising velocities and 0.2 for falling velocities, provided flow is in the *upper alternate stage* (Sec. 8-6). For

---

[12] Julian Hinds, The Hydraulic Design of Flume and Siphon Transitions, *Trans. Am. Soc. Civil Engrs.*, **92**, 1423 (1928).

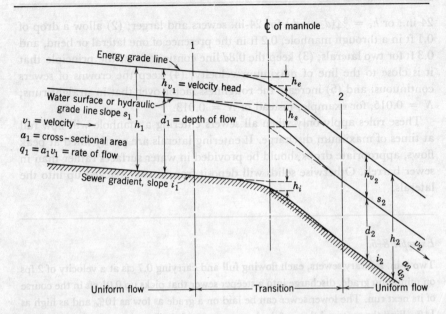

**Figure 8-2.**
**Changes in hydraulic and energy grade lines at a transition in size or grade of sewer.**

the *lower alternate stage* (Sec. 8-6), $k$ may be expected to increase approximately as the square of the velocity ratios. Camp[13] has suggested a minimum allowance of 0.02 ft for the loss of head in a transition of this kind. However, if there is a horizontal curve in the transition, more head will be lost.

The required invert drop $h_i$ follows from the relationships demonstrated in Fig. 8-3. There $h_2 + h_e = h_1 + h_i$, or

$$h_i = (h_2 - h_1) + h_e = \Delta(d + h_v) + kh_v \qquad (8\text{-}18)$$

The calculated change $h_i$ will be positive for increasing gradients and negative for sharply decreasing gradients. A positive value calls for a drop in the invert, a negative value for a rise. However, a rise would obstruct flow, and the invert is actually made continuous. The elevation of the water surface in the downstream sewer is thereby lowered, and the waters in the sewers entering the transition are drawn down towards it.

Rules of thumb sometimes followed by engineers in place of computations reflect average conditions encountered in practice. They may not always be justified by circumstances. Common rules for drops in manholes at changes in size are (1) make the invert drop $h_i = \frac{1}{2}(d_2 - d_1)$ for sewers smaller than

[13] T. R. Camp, Design of Sewers to Facilitate Flow, *Sewage Works J.*, **18**, 3 (1946).

24 in., or $h_i = \frac{3}{4}(d_2 - d_1)$ for 24-in. sewers and larger; (2) allow a drop of 0.1 ft in a through manhole, 0.2 ft in the presence of one lateral or bend, and 0.3 ft for two laterals; (3) keep the $0.8d$ line continuous on the principle that it is close to the line of maximum velocity; (4) keep the crowns of sewers continuous; and (5) increase the roughness factor over that in straight runs; $N = 0.015$, for example, instead of $N = 0.013$.

These rules apply only when all sewers entering a manhole will flow full at times of maximum discharge. If entering laterals are partly filled at peak flows, appropriate drops should be provided in water surfaces rather than in sewer barrels. Otherwise solids will deposit from sewage backed up into the laterals.

*Example 8–5.*

Two 8-in. sanitary sewers, each flowing full and carrying 0.7 cfs at a velocity of 2 fps on minimum grade, discharge into a steeper sewer that picks up 0.01 cfs in the course of its next run. The lower sewer can be laid on a grade as low as 10‰ and as high as 14‰. Find the required slope of the lower sewer and the invert drop in the transition.

1. From Table 9, Appendix, an 8-in. sewer flowing full will carry 1.41 cfs on a slope of 13.6‰ with a velocity of 4.04 fps if $N = 0.013$. Pertinent information is, therefore, as follows: $d_1 = 0.67$ ft, $v_1 = 2.00$ fps, $h_{v_1} = 0.062$ ft, $d_1 + h_{v_1} = 0.73$ ft; $d_2 = 0.67$ ft, $v_2 = 4.04$ fps, $h_{v_2} = 0.254$ ft, $d_2 + h_{v_2} = 0.92$ ft; and $\Delta h_v = 0.19$ ft, $\Delta(d + h_v) = 0.19$ ft. Assuming a loss of head $h_e = 0.2 \Delta h_v = 0.038$ ft, Eq. 8–18 gives the required drop in invert, $h_i = 0.19 + 0.04 = 0.23$ ft.

2. A 10-in. sewer laid on a grade of 10‰ has a capacity of 2.19 cfs and velocity of 4.02 fps when flowing full. From Fig. 4 of the Appendix, for $N/n \times q/Q = 0.644$, $d/D = 0.65$, and $N/n \times v/V = 0.92$, or $d = 6.5$ in., and $v = 3.69$ fps. Hence, the upper sewers remaining unchanged, $d_2 = 0.542$ ft, $v_2 = 3.69$ fps, $h_{v_2} = 0.21$ ft, $d_2 + h_{v_2} = 0.75$ ft, and $\Delta h_v = 0.15$ ft, $\Delta(d + h_v) = 0.02$ ft. Assuming a loss of head $h_e = 0.2 \Delta h_v = 0.036$ ft, Eq. 8–18 states that $h_i = 0.02 + 0.04 = 0.06$ ft.

## 8–6 Alternate Stages and Critical Depths

In the analysis of transitions, the designer must often find the alternate stage or depths of open-channel flow and their mergence into flow at critical depth. Referred to the sewer invert, the energy grade line shown in Fig. 8–2 is situated at a height

$$h = d + h_v = d + v^2/2g = d + q^2/(2ga^2) \qquad (8\text{–}19)$$

above this datum,[14] where the cross-sectional area $a$ of the conduit is a function of its depth, $d$. Accordingly, Eq. 8–19 is a cubic equation in terms of $d$. Two of its roots are positive and, except at critical depth, identify respectively the *upper alternate stage* and the *lower alternate stage* at which a given discharge rate $q$ can be associated with a given energy head $h$. The two stages fuse into a single *critical stage* for conditions of maximum discharge at a given total energy head, or minimum total energy head at a given discharge (Fig. 8–3 and Eq. 8–19). For uniform flow, the water surface parallels the invert ($s = i$). Open channel flow at near-critical stage is unstable and depth of flow is uncertain and fluctuating.

Equation 8–19 can be generalized by expressing its three components as dimensionless ratios. Bringing $q$ to one side and multiplying both sides by $(1/D)(a/A)^2$ give the following straight-line relationship:

$$(q/A\sqrt{gD})^2 = 2(a/A)^2(h/D - d/D) \tag{8-20}$$

Again, capital letters denote the hydraulic elements of the full section, and lower-case letters those of the partially filled section. As stated above, given the energy head, the maximum rate of discharge obtains at critical depth $d_c$. The associated specific head, $h_c/D$, is determined analytically by differentiating Eq. 8–20 with respect to $d$ and equating the result to zero.

For a trapezoidal channel of bottom width $b$ and with side slopes $z$ (horizontal to vertical),

$$h_c/D = d_c/D + \tfrac{1}{2}(d_c/D)(b + zd_c)/(b + 2zd_c). \tag{8-21}$$

[14] If $v$ is the mean velocity of flow, the kinetic energy head is actually greater than $v^2/2g$ by 10 to 20%, depending on the shape and roughness of the channel. But this fact is not ordinarily taken into account in hydraulic computations. (See footnote 13.)

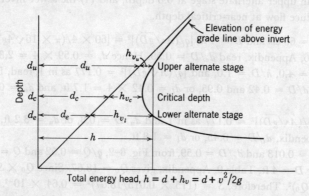

**Figure 8-3.**

**Alternate stages of flow and total energy head at constant rate of discharge in open-channel flow.**

For a rectangular channel ($z = 0$), therefore,

$$h_c/D = 3/2 d_c/D \quad \text{or} \quad h_c = 3/2 d_c \quad (8\text{--}22)$$

whence

$$v_c^2/2g = h = d_c \quad \text{or} \quad v_c = \sqrt{g d_c} \quad (8\text{--}23)$$

For a circular cross-section, finally:

$$h_c/D = \tfrac{1}{8}\left\{[10(d_c/D) - 1] + \frac{\tfrac{1}{4}\pi + \tfrac{1}{2}\sin^{-1}[2(d_c/D) - 1]}{\sqrt{(d_c/D)[1 - (d_c/D)]}}\right\} \quad (8\text{--}24)$$

Substituting values of $d_c/D$ varying by tenths from 0.1 to 0.9 yields the numerical results for $h/D$, $v_c/\sqrt{gD}$, and $[q/(A\sqrt{gD})]^2$ shown in Table 12 of the Appendix. A plot of Eq. 8–20 for circular conduits is shown in Fig. 6 of the Appendix.

The critical depth line in a closed conduit is seen to be asymptotic to the line $d/D = 1.0$, that is, there is neither a critical nor an alternate stage for an enclosed conduit flowing full.

---

*Example 8–6.*

The use of Fig. 6, Appendix, construction of which is simple and straightforward, can be exemplified as follows:

Given a discharge of 60 cfs in a 4-ft circular sewer, find (1) the critical depth, (2) the alternate stages for an energy head of 4 ft, (3) the lower alternate stage associated with an upper alternate stage at 0.8 depth, and (4) the sewer invert slope that would produce flow at near-critical depth.

1. For $q = 60$ cfs and $D = 4$ ft, $[q/(A\sqrt{gD})]^2 = [60 \times 4/(\pi \times 16\sqrt{4g})]^2 = 0.177$. From Fig. 6, Appendix, read $d_c/D = 0.59$. Hence $d_c = 0.59 \times 4 = 2.36$ ft.

2. For $h = 4.0$, $h/D = 1.0$, and $[q/(A\sqrt{gD})]^2 = 0.177$ as in 1, read, from Fig. 6, Appendix, $d/D = 0.42$ and 0.95, or $d_l = 0.42 \times 4 = 1.7$ ft, and $d_u = 0.95 \times 4 = 3.8$ ft.

3. For $[q/(A\sqrt{gD})]^2 = 0.177$ as in 1 and $d_u/D = 0.8$, or $d_u = 3.2$ ft, read, from Fig. 6, Appendix, $d_l/D = 0.45$, or $d_l = 1.8$ ft.

4. For $N = 0.013$ and $d_c/D = 0.59$, from Fig. 8–2, $q/Q = 0.52$ and $Q = 60/0.52 = 115$ cfs. For $D = 4$ ft, Table 9, Appendix, gives $Q_0 = 18.67$. Since $Q_0 \times S^{1/2}/N = Q$, $S = (QN/Q_0)^2$. Therefore, $S = (115 \times 0.013/18.67)^2 = 0.64 \times 10^{-2}$. This is less than a 1% grade. Since many streets have slopes in excess of this, large sewers not infrequently operate at the lower alternate depth, or high-velocity stage.

## 8-7 Length of Transitions

Transition from one to the other alternate stage carries the flow through the critical depth. Passage from the upper alternate stage (*a*) to the critical depth or (*b*) through it, to the lower alternate stage or to free fall produces nonuniform (accelerating) flow and a *drawdown curve* in the water surface. Passage from the lower to the upper alternate stage creates the hydraulic jump. Reduction in velocity of flow (*a*) by discharge into relatively quiet water or (*b*) by weirs and other flow obstructions, dams up the water and induces nonuniform (decelerating) flow and a *backwater curve* in the water surface. For economy of design, the size of conduit must fit conditions of flow within the range of transient depths and nonuniform flow. If initial and terminal depths of flow are known, the energy and hydraulic grade lines can be traced either by stepwise calculation or by integration (graphical[15] or analytical[16]). Both stem from the fact that the change in slope of the energy grade line must equal the sum of the changes in the slopes of (1) the invert, (2) the depth of flow, and (3) the velocity head.

For stepwise calculation of the length of conduit between the cross-sections of given depth (Fig. 8-4),

$$s\Delta l = i\Delta l + \Delta(d + h_v)$$

or

$$\Delta l = \Delta(d + h_v)/(s - i) \qquad (8\text{-}25)$$

Flow being steady, the rate of discharge is constant, and the velocity of flow at given depths is known. For a given invert slope $i$, therefore, only $s$ needs to be calculated. This is generally done by introducing the average hydraulic elements of each conduit reach into a convenient flow formula. Averages of choice are ordinarily arithmetic means, but geometric or harmonic means will also give defensible results. Necessary calculations are shown in Example 8-7 for a backwater curve and in Example 8-8 for a drawdown curve.

---

*Example 8-7.*

A 10-ft circular sewer laid on a gradient of 0.5‰ discharges 106 cfs into a pump well. The water level in this well rises, at times, 10 ft above the invert elevation of the in-

[15] H. A. Thomas, *Hydraulics of Flood Movement*, Carnegie Institute of Technology, Pittsburg, Pa., 1934, and V. T. Chow, *Open-Channel Hydraulics*, McGraw-Hill, N.Y., 1959.

[16] M. E. von Seggern, Integrating the Equation of Non-Uniform Flow, *Trans. Am. Soc. Civil Engrs.*, **115**, 71 (1950); see also G. J. Keifer and H. H. Chu, Backwater Functions by Numerical Integration, *Trans. Am. Soc. Civil Engrs.*, **120**, 429 (1955).

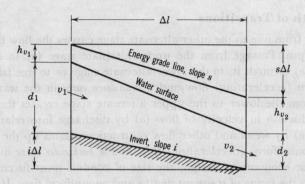

**Figure 8-4.**
**Flow conditions changing with increasing velocity.**

coming sewer. Trace the profile of the water surface in the sewer. Assume a coefficient of roughness of 0.012 for the full sewer.

A 10-ft sewer on a grade of $5 \times 10^{-4}$ has a capacity of 400 cfs by Manning's formula. The value of $q/Q$, therefore, is $106/400 = 0.265$, and $d/D$ from Fig. 4, Appendix, is 0.04 for variable $N/n$. Hence the initial depth of flow is $0.40 \times 10 = 4.0$ ft, and the terminal depth 10 ft. The reach in which depths change by chosen amounts is given by Eq. 8-25. Calculations are systematized in Table 8-1. The depth is seen to change from 4.0 to 10.0 ft in 14,440 ft, or slightly under 3 miles.

## Example 8-8.

A 10-ft circular sewer laid on a gradient of 0.5‰ discharges freely into a water course. Trace the profile of the water surface in the sewer when it is flowing at maximum capacity without surcharge.

As shown in Example 8-7, the full capacity of this sewer is 400 cfs for $N = 0.012$. To discharge in free fall, the flow must pass through the critical depth. Because $[Q/(A\sqrt{gD})]^2 = [400/(78.5\sqrt{10g})]^2 = 8.07 \times 10^{-2}$, $d_c = 0.47 \times 10 = 4.7$ ft, from Fig. 6 or Table 12, Appendix.

The reach in which the depth changes from 10 ft to 4.7 ft is calculated in Table 8-2 in accordance with Eq. 8-25 and as in Example 8-7. The drawdown is seen to extend over a length of 23,600 ft, or over 4 miles, between the full depth of the sewer and a critical depth of 4.7 ft. There is a further short stretch of flow between the point of critical depth and the end of the sewer. This additional distance is relatively small, namely, $4d_c = 18.8$ ft.

## Table 8-1

Calculation of Backwater Curve (Example 8–7)

| $d$ (1) | $d/D$ (2) | $a/A$ (3) | $r/R$ (4) | $N/n$ (5) | $a$ (6) | $r$ (7) | $v$ (8) | $h_v \times 10^2$ (9) | $d + h_v$ (10) |
|---|---|---|---|---|---|---|---|---|---|
| 10.0 | 1.00 | 1.000 | 1.000 | 1.00 | 78.5 | 2.50 | 1.35 | 2.83 | 10.028 |
| 8.0 | 0.80 | 0.858 | 1.217 | 0.89 | 67.5 | 3.04 | 1.57 | 3.83 | 8.038 |
| 6.0 | 0.60 | 0.626 | 1.110 | 0.82 | 49.1 | 2.78 | 2.16 | 7.23 | 6.072 |
| 4.0 | 0.40 | 0.373 | 0.857 | 0.79 | 29.3 | 2.14 | 3.62 | 20.3 | 4.203 |

| $n \times 10^2$ (11) | $nv \times 10^2$ (12) | Average $r$ (13) | Average $r^{2/3}$ (14) | Average $nv \times 10^2$ (15) | $s \times 10^5$ (16) | $(s-l) \times 10^5$ (17) | $\Delta(d+h_v)$ (18) | $\Delta l$ (19) | $\Sigma \Delta l$ (20) |
|---|---|---|---|---|---|---|---|---|---|
| 1.20 | 1.62 | | | | | | | | 0 |
| | | 2.77 | 1.97 | 1.87 | 4.07 | −45.9 | −1.990 | 4,330 | |
| 1.35 | 2.12 | | | | | | | | 4,330 |
| | | 2.91 | 2.04 | 2.63 | 7.53 | −42.5 | −1.966 | 4,620 | |
| 1.46 | 3.15 | | | | | | | | 8,950 |
| | | 2.46 | 1.82 | 4.32 | 25.5 | −24.5 | −1.869 | 7,630 | |
| 1.52 | 5.50 | | | | | | | | 16,580 |

Column 1: Depths between initial depth of 4 ft and terminal depth of 10 ft are assumed to increase by 2 ft.
Column 2: Col. 1 ÷ 10 (the diameter of the sewer).
Columns 3, 4, and 5: $a/A$, $r/R$, and $N/n$ read from Fig. 4 in Appendix
Column 6: Col. 3 × 78.5 (the area of the sewer).
Column 7: Col. 4 × 2.50 (the hydraulic radius of the sewer).
Column 8: 106 (the rate of flow)/Col. 6.
Column 9: $v^2/2g$ for Col. 8.
Column 10: Col. 9 + Col. 1.
Column 11: 0.012 (Manning's $N$ for sewer)/(Col. 5).
Column 12: Col. 11 × Col. 8.
Column 13: Arithmetic mean of successive pairs of values in Col. 7.
Column 14: (Col. 13)$^{2/3}$.
Column 15: Arithmetic mean of successive pairs of values in Col. 12.

Column 16: (Col. 15/1.489 × Col. 14)², i.e., $s = (nv/1.49r^{2/3})^2$.
Column 17: Col. 16 − 50.
Column 18: Difference between successive pairs of values in Col. 10.
Column 19: Col. 18/Col. 17 × 10⁻⁵, i.e., $\Delta l = \Delta(d + h_v)/(s - i)$.
Column 20: Cumulative values of Col. 19.

## Table 8-2

*Calculation of Drawdown Curve (Example 8–8)*

| $d$ | $d/D$ | $a/A$ | $r/R$ | $N/n$ | $a$ | $r$ | $v$ | $h_v$ | $d + h_v$ |
|---|---|---|---|---|---|---|---|---|---|
| (1) | (2) | (3) | (4) | (5) | (6) | (7) | (8) | (9) | (10) |
| 4.7 | 0.47 | 0.463 | 0.960 | 0.79 | 36.4 | 2.40 | 11.0 | 1.88 | 6.58 |
| 6.0 | 0.60 | 0.626 | 1.110 | 0.82 | 49.1 | 2.78 | 8.15 | 1.02 | 7.02 |
| 8.0 | 0.80 | 0.858 | 1.217 | 0.89 | 67.5 | 3.04 | 5.93 | 0.54 | 8.54 |
| 10.0 | 1.00 | 1.000 | 1.000 | 1.00 | 78.5 | 2.50 | 5.10 | 0.40 | 10.40 |

| $n \times 10^2$ | $nv \times 10^2$ | Average | | | $s \times 10^4$ | $(s - i) \times 10^4$ | $\Delta(d + h_v)$ | $\Delta l$ | $\Sigma \Delta l$ |
|---|---|---|---|---|---|---|---|---|---|
| | | $r$ | $r^{2/3}$ | $nv \times 10^2$ | | | | | |
| (11) | (12) | (13) | (14) | (15) | (16) | (17) | (18) | (19) | (20) |
| 1.52 | 16.8 | | | | | | | | 0 |
| | | 2.59 | 1.88 | 14.3 | 26.2 | 21.2 | 0.44 | 210 | |
| 1.46 | 11.9 | | | | | | | | 210 |
| | | 2.91 | 2.04 | 9.9 | 10.65 | 5.65 | 1.52 | 2,690 | |
| 1.35 | 7.99 | | | | | | | | 2,900 |
| | | 2.77 | 1.97 | 7.1 | 5.90 | 0.90 | 1.86 | 20,700 | |
| 1.20 | 6.12 | | | | | | | | 23,600 |

Beyond the critical depth, the hydraulic drop terminating in free fall is a function of velocity distribution. Flow is supercritical and depth decreases. At the free outfall, pressure on the lower as well as the upper nappe is atmospheric when the nappe is ventilated. Within the conduit, calculated ratios of the terminal depth to the critical depth[17] normally range between ⅔ and ¾. The critical depth itself lies upstream at a distance of about $4d_c$.

[17] For a closer definition of conditions, see B. A. Bakhmeteff, Hydraulic Drop as a Function of Velocity Distribution, *Civil Engr.*, **24**, 64 (Dec. 1954).

## 8-8 Hydraulic Jumps and Discontinuous Surge Fronts

When a conduit steep enough to discharge at supercritical velocities and depths is followed by a relatively flat channel in which entering velocities and depths cannot be maintained, a more or less abrupt change in velocity and depth takes the form of a hydraulic jump. Whereas alternate depths are characterized by equal specific energies ($d + h_v$), sequent depths are characterized by equal pressure plus momentum. In accordance with the momentum principle, illustrated in Fig. 8-5, the force producing momentum changes, when equated to the momentum change per unit volume, establishes sequent depths $d_1$ (lower) and $d_2$ (upper) such that the velocity $v$, associated with the depth $d$, is determined by the equation[18]

$$(v_1/\sqrt{gd_1})^2 = \tfrac{1}{2}(d_2/d_1)[1 + (d_2/d_1)] = \mathbf{F}^2 \qquad (8\text{–}26)$$

where **F** is the Froude number and $\sqrt{gd_1}$ is the celerity of an elementary gravity wave, or

$$d_2/d_1 = \tfrac{1}{2}[(1 + 8\mathbf{F}^2)^{1/2} - 1] \qquad (8\text{–}27)$$

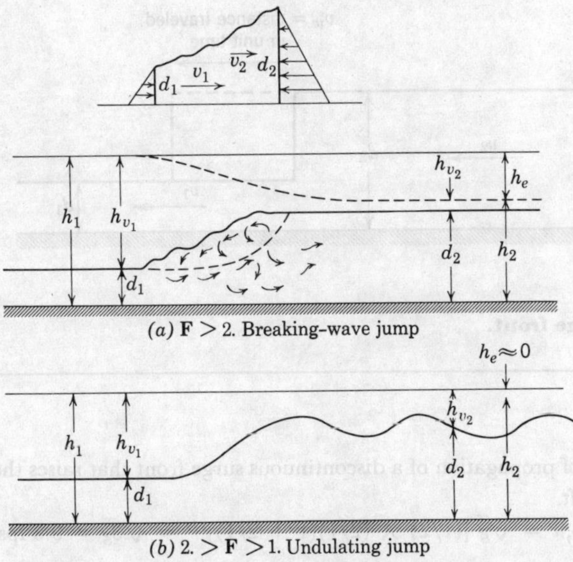

$d_1$
$v_1$
$v_2$
$d_2$
$h_1$
$h_{v_1}$
$h_{v_2}$
$h_e$
$d_2$
$h_2$
$d_1$
(a) F > 2. Breaking-wave jump
$h_e \approx 0$
$h_1$
$h_{v_1}$
$h_{v_2}$
$d_2$
$h_2$
$d_1$
(b) 2. > F > 1. Undulating jump


**Figure 8-5.**

**Profiles of hydraulic jumps.**

[18] Hunter Rouse, Ed., *Engineering Hydraulics*, John Wiley & Sons, New York, 1950, p. 72. The force per unit width producing the momentum changes is $\tfrac{1}{2}(\rho g d_1^2 - \rho g d_2^2)$ where $\rho$ is the mass density of the water and $g$ the gravity constant. The momentum change per unit volume is $q\rho(v_2 - v_1)$, where $q$ is the rate of flow. Equating the two and eliminating $q$ and $v_2$ by the continuity equation $q = v_1 d_1 = v_2 d_2$ leads directly to Eq. 8–26.

As shown by Rouse,[18] depths change (1) with substantially no loss of head in a series of undulations when $2 > \mathbf{F} > 1$ and (2) with appreciable head loss and a breaking wave when $\mathbf{F} > 2$. For cross-sections other than rectangles of unit width, all terms in Eq. 8-27 have numerical coefficients that must be determined experimentally.

As shown in Fig. 8-6, the momentum principle can be adduced to identify also the propagation of discontinuous waves in open channel flow. Waves of this kind may rush through conduits when a sudden discharge of water from a localized thunderstorm or the quick release of a large volume of industrial wastewater, for example, enters a drainage system. In cases such as these, the volume of water undergoing a change in momentum in unit time and unit channel width is $(v_w - v_1)d_1$. The celerity of propagation, which is the wave velocity or speed or propagation of the surge front, relative to the fluid velocity[19] being $c = v_w - v_1$, it follows that

$$(c/\sqrt{gd_1})^2 = \tfrac{1}{2}(d_2/d_1)(1 + d_2/d_1) \tag{8-28}$$

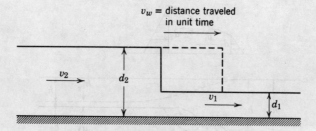

**Figure 8-6.**
**Profile of surge front.**

*Example 8-9.*

Find the rate of propagation of a discontinuous surge front that raises the flow depth from 1 ft to 2 ft.

By Eq. 8-28, $c = \sqrt{g}\,[(1/2) \times (2/1)(1 + 2/1)]^{1/2} = \sqrt{3g} = 9.8$ fps.

## 8-9 Bends, Junctions, and Overfalls

In common with many other uncertainties of fluid behavior in hydraulic systems, head losses in bends are often approximated as functions of the

---

[19] Equating force to momentum change $\tfrac{1}{2}gd_2^2 - \tfrac{1}{2}gd_1^2 = (v_2 - v_1)(v_w - v_1)d_1$ in a channel of unit width. Continuity of flow, moreover, requires that $v_2 d_2 = v_w(d_2 - d_1) + v_1 d_1$.

velocity head $v^2/2g$. For small sewers, the radius of an optimal circular curve of the center line of the sewer is reported to be three to six times the sewer diameter.[20]

The conjoining of the flow patterns of two or more sewers normally involves curvature as well as impact effects. Generalization of resulting head losses is difficult. Where predictions of surface profiles are important, in the case of large sewers, for example, model studies are advisable. Otherwise, the usual procedure of working upstream from a known point, especially a control point, should identify the water-surface profile reasonably well.

Storm flows in excess of interceptor capacity are often diverted into natural drainage channels through overfalls or side weirs. Needed weir lengths depend on the general dimensions and hydraulic characteristics of the sewer and the nature and orientation of the weir itself. Understandably, side weirs paralleling the direction of flow must be longer than weirs at right angles to it.

For the conditions of flow outlined in Fig. 8-7, Bernoulli's theorem gives the following relationship when head loss is based on Manning's formula:

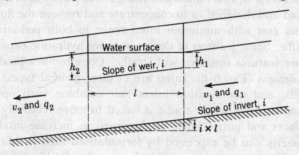

**Figure 8-7.**
**Flow over a side weir.**

$$(v_1^2/2g) + il + h_1 = (v_2^2/2g) + h_2 + l\left(\frac{nv}{1.49r^{2/3}}\right)^2$$

Hence

$$h_2 - h_1 = \frac{q_1^2 - q_2^2}{2ga^2} + il - n^2 l\left[\frac{q_1 + q_2}{2 \times 1.49ar^{2/3}}\right]^2 \quad (8\text{-}29)$$

The parameters $a$ and $r$ are based on average dimensions of the filled channel; those obtaining at the center of the weir, for example. Approximating the flow over the weir, $Q$, by $clh^{3/2}$,

$$Q = cl[\tfrac{1}{2}(h_1 + h_2)]^{3/2} \quad \text{and} \quad h_2 + h_1 = 2[(q_1 - q_2)/(cl)]^{2/3} \quad (8\text{-}30)$$

[20] A. G. Anderson and L. G. Straub, Hydraulics of Conduit Bends, St. Anthony Falls Hydraulic Laboratory, University of Minnesota, *Bulletin* No. 1 (1948).

Given $q_1, q_2, a, r, i, n$, and $h_2$, values of $l$ and $h_1$ are then determined by trial, as shown in Example 8–10. A formulation of this kind was first suggested by Forchheimer.[21]

*Example 8–10.*

Given $q_1 = 30$ cfs; $q_2 = 16$ cfs; $a = 32$ sq ft; $r = 1.6$ ft; $i = 10^{-4}$; $n = 1.25 \times 10^{-2}$; and $h_2 = 0.50$ ft.

Find $l$ and $h$; assume $c = 3.33$.

By Eq. 8–29: $h_1 = 0.49022 - 8.04 \times 10^{-5} l$. By Eq. 8–30: $h_1 = 5.20 l^{-2/3} - 0.5$.
Hence $(0.99022 - 8.04 \times 10^{-5} l) l^{2/3} = 5.20$ or $(12,320 - l) l^{2/3} = 64,700$.

$$l = 12 \text{ ft. and } h_1 = 0.49 \text{ ft.}$$

## 8-10 Street Inlets

As stated in Sec. 9–3, street inlets admitting storm waters to drainage systems are so placed and designed as to concentrate and remove the flow in gutters at minimum cost with minimum interference to both pedestrian and vehicular traffic. Some features of design improve hydraulic capacity but are costly; other features interfere with traffic. Compromises produce a wide variety of designs (Fig. 8–8). Inlets are of three general types: curb inlets, gutter inlets, and combination inlets that combine curb openings with gutter openings. Only where traffic is forced to move relatively slowly may gutter surfaces and gutter inlets be depressed to increase intake capacity. Gutter capacity can be expressed by formulations such as Manning's, the coefficient of roughness being suitably increased to 0.015 or more.

The intake capacity of inlets, particularly curb inlets, increases with decreasing street grade and increasing crown slope. However, curb inlets with diagonal deflectors in the gutter along the opening become more efficient as grades become steeper. Gutter inlets are more efficient than curb inlets in capturing gutter flow, but clogging by debris is a problem. Combination inlets are better still, especially if gratings are placed downstream from curb openings. Debris accumulating on the gratings will then deflect water into the curb inlets. Gratings for gutter inlets are most efficient when their bars parallel the curb. If crossbars are added for structural reasons, they should be kept near the bottom of the longitudinal bars. Depression of inlets, especially curb inlets, enhances their capacity. Long shallow depressions are as effective as short deep ones. If a small flow is allowed to outrun the inlet, the relative intake of water is greatly magnified. Significant

---

[21] Philip Forchheimer, *Hydraulik*, B. Teubner, Leipzig, p. 406, 1930.

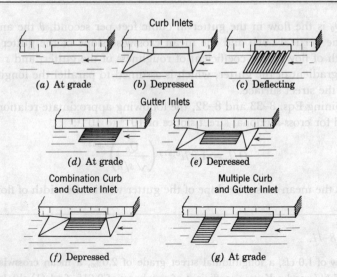

**Figure 8-8.**
Types of street inlets.

economies are effected, therefore, by small carry-over flows and their acceptance by downgrade inlets.

Model studies and street tests have produced empirical formulas for flow into gutter inlets and curb inlets with and without depressions.[22] A relationship for curb openings without depressions is

$$Q/l = 4.82 \times 10^{-3} d\sqrt{gd} \qquad (8\text{-}31)$$

or

$$d = (35.1/g^{1/3})(Q/l)^{2/3} \qquad (8\text{-}32)$$

Here $Q$ is the discharge into the inlet in cubic feet per second, $l$ the length of the opening in feet, $g$ the gravitational acceleration in feet per second squared, and $d$ the depth of gutter flow at the curb in inches. If $d$ can be calculated by Manning's formula, the equation for a gutter of wedge-shaped cross-section is

$$d = 0.1105\,\frac{(1 + \sec\theta)^{1/4}}{\tan^{5/8}\theta}\left(\frac{Q_0}{\sqrt{s}/n}\right)^{3/8} \qquad (8\text{-}33)$$

[22] W. H. Li, J. C. Geyer, G. S. Benton, and K. K. Sorteberg, Hydraulic Behavior of Storm-Water Inlets—Parts I and II, *Sewage and Ind. Wastes*, 23, 34, 722 (1951); also see *The Design of Storm-Water Inlets*, Storm Drainage Research Committee, Johns Hopkins University, Baltimore, Md., 1956, and J. J. Cassidy, Generalized Hydraulics of Grade Inlets, *Highway Research Board, Highway Research Record*, 123, 36 (1966).

where $Q_0$ is the flow in the gutter in cubic feet per second, $\theta$ the angle between the vertical curb and the mean crosswise slope of the gutter within the width of flow, $n$ the coefficient of roughness of the gutter, and $s$ the hydraulic gradient of the gutter, which is assumed to parallel the longitudinal slope of the street surface.

Combining Eqs. 8–33 and 8–32, the following approximate relationship is obtained for cross-sectional street slopes of $10^{-3}$ to $10^{-1}$:

$$Q/l = 1.87 i^{0.579} \left( \frac{Q_0}{\sqrt{s/n}} \right)^{0.563} \tag{8-34}$$

Here $i$ is the mean crosswise slope of the gutter within the width of flow.

## Example 8–11.

For a flow of 1.0 cfs, a longitudinal street grade of 2.0‰, a mean crosswise street grade of 5.6%, and a Kutter coefficient of roughness of 0.015, find (1) the length of an undepressed curb inlet required to capture 90% of the flow, and (2) the maximum depth of flow in the gutter.

1. By Eq. 8–34: $Q/l = 1.87(5.6 \times 10^{-2})^{0.579}[1.0/\sqrt{2 \times 10^{-2}}/1.5 \times 10^{-2})]^{0.563} = 0.10$, or $l = 10Q = 10 \times 0.9 \times Q_0 = 10 \times 0.9 \times 1.0 = 9$ ft.
2. By Eq. 8–32: $d = 3.48 g^{1/3} \times 0.10^{2/3} = 2.4$ in.

# nine

# wastewater removal

### 9-1 Drainage of Buildings

The water distributed to basins, sinks, tubs, bowls, and other fixtures in dwellings and other buildings, and to tanks and other equipment in industrial establishments, is collected as *spent water* by the drainage system of the building and run to waste (Fig. 9–1). Plumbing fixtures are arranged singly or in batteries. They empty into substantially horizontal *branches* or *drains* that must not flow full or under pressure if tributary fixtures are to discharge freely and their protecting *traps* are not to become unsealed. The horizontal drains empty into substantially vertical *stacks*. These, too, must not flow full if wastewaters are not to back up into fixtures on the lower floors. The drainage stacks discharge into the *building drain* which, 5 ft outside of the building, becomes the *building sewer* (or *house sewer*) and empties into the street sewer (Figs. 9–2 and 9–3).

Traps are either part of the drainage piping or built into fixtures such as water closets. The traps hold a water seal that obstructs, and essentially prevents, foul odors and noxious gases, as well as insects and other vermin, from passing through the drainage pipes and sewers into the building. Discharging fixtures send water rushing into the drains and tumbling down the stacks; air is dragged along, and air pressures above or below atmospheric would be created within the system and might unseal the traps were it not for the pro-

# 248/Wastewater Removal

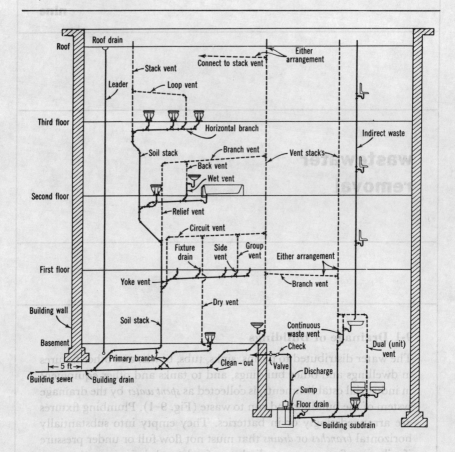

**Figure 9-1.**
**Building drainage system.**

vision of *vents*. These lead from the traps to the atmosphere and thereby equalize the air pressures in the drainage pipes.

The wastewater from fixtures and floor drains below the level of the public sewer must be lifted by ejectors or pumps (Fig. 9–1). Sumps or receiving tanks facilitate automatic operation. Sand and other heavy solids from cellars or yards are kept out of the drainage system by *sand interceptors*, grease by *grease interceptors*, and oil by *oil interceptors*. To act as *separators* or *traps*, these generally take the form of small settling, skimming, or holding tanks.

Cast iron, galvanized steel or wrought iron, lead, brass, and copper piping are employed for drains and vents above ground; cast iron alone is used for drains laid below ground. Building sewers are constructed of vitrified-clay, concrete, or cast-iron pipe. Stormwater from roofs and paved areas taken

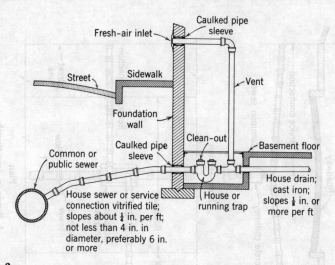

**Figure 9-2.**
**Connecting building drainage system to sewer. House, or running, trap may be installed or omitted.**

into a *property drain* is discharged into the street gutter or directly into the storm sewer. In combined systems, roofwater may be led into the house drain, and water from yard areas into the house sewer. Otherwise, storm runoff travels over the ground, reaches the street gutter, flows along it, enters a stormwater inlet or a catch basin, and is piped to a manhole, whence it empties into the drainage system.

In separate systems, connections to the wrong sewer, in violation of common regulations, carry some stormwater into sanitary sewers and some domestic wastewater into storm drains. The dry-weather flow of combined sewers is primarily wastewater and groundwater; the wet-weather flow is predominantly storm runoff. The first flush of stormwater will scour away deposited solids, including much putrescent organic matter.

## 9-2 Collection of Spent Waters

A system of sanitary sewers is shown in Fig. 1–3a. Because about 70% of the water brought into a community must be removed as *spent water*, the average flow in sanitary sewers is about 100 gpcd in North America. Variations in water use step up the maximum hourly rate about threefold. Illicit stormwater and groundwater magnify the required capacity still further, and a design value of 500 gpcd is not uncommon.

Sanitary sewers are fouled by the deposition of waste matters unless they impart self-cleaning velocities of 2 to 2.5 fps. Except in unusually flat country,

250/Wastewater Removal

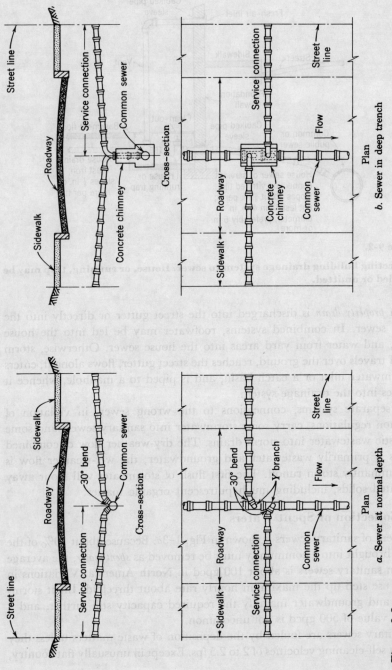

Figure 9-3.
Service connections to public sewer.

sewer grades are made steep enough to generate these velocities when the sewers are running reasonably full.[1] Nevertheless, there will be deposition of solids. To find and remove them, sewers must be accessible for inspection and cleaning. Except in large sewers, manholes are built at all junctions with other sewers, and at all changes in direction or grade. Straight runs between manholes can then be rodded out effectively if intervening distances are not too great. Maxima of 300 or 400 ft for pipes less than 24 in. in diameter are generally specified, but effective cleaning is the essential criterion. For larger sewers, distances between manholes may be upped to as much as 600 ft. Sewers so large that workmen can enter them for inspection, cleaning, and repair are freed from these restrictions, and access manholes are placed quite far apart either above the center line or tangential to one side. Introduction of flexible cleaning devices has encouraged the construction of curved sewers of all sizes, especially in residential areas.

A plan and profile of a sanitary sewer and its laterals are shown in Fig. 9–4, together with enlarged sections of sewer trenches and manholes. On short runs (<150 ft) and temporary stubs of sewer lines, terminal cleanouts are sometimes substituted for manholes. They slope to the street surface in a straight run from a Y in the sewer or in a gentle curve that can be rodded out. To keep depths reasonably small and pumping reasonably infrequent in very flat country and in other unusual circumstances, sewers are laid on flat grades, in spite of greater operating difficulties.

The smallest public sewers in North America are normally 8 in. in diameter. Smaller pipes clog more frequently and are harder to clean. Vitrified clay is the material of choice for small sewers, prefabricated concrete for large sewers. Vitrified-clay sewer pipe, 4 to 36 in. in diameter, is ordinarily 3 to 6 ft long. Unreinforced-concrete sewer pipe, 4 to 24 in. in diameter, is generally 2 to 4 ft long. Preformed joints made of resilient plastic materials increase the tightness of the system. To reduce the infiltration of groundwater sewers laid without factory-made joints in wet ground must be underdrained, or made of cast iron, asbestos cement, or other suitable materials. Cast-iron and asbestos-cement pipes are long and their joints are tight. Underdrainage is by porous pipes or clay pipes laid with open joints in a bed of gravel or broken stone beneath the sewer. Underdrains may serve during construction only or become permanent adjuncts to the system and discharge freely into natural water courses. Sewage seeping into permanent underdrains may foul receiving waters. As stated before, grit or other abrading materials will wear the invert of concrete sewers unless velocities are held below 8 to 10 fps. Very large sewers are built in place, some by tunneling. Hydraulically and structurally, they share the properties of grade aqueducts (Fig. 6–1).

[1] Half full or more in circular sections, because the hydraulic radius of a semicircle equals that of a circle.

252 / Wastewater Removal

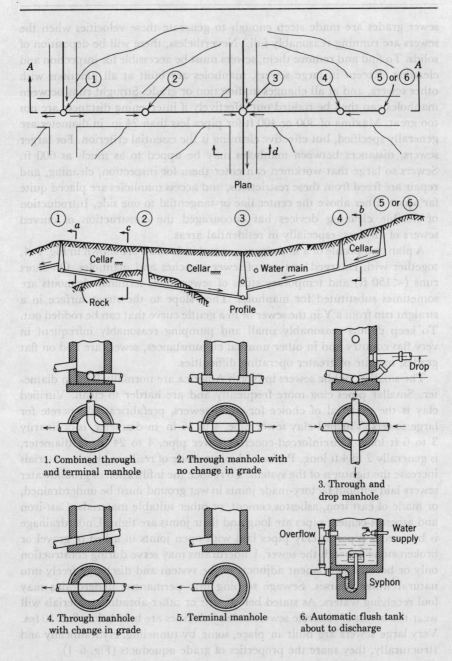

Figure 9-4a.
Plan, profile, and constructional details of sanitary sewers.

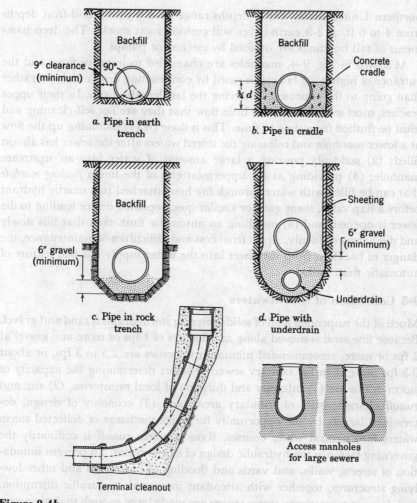

**Figure 9-4b.**
**Sewer trenches, access manholes, and terminal cleanouts.**

Sewers are laid deep enough (1) to protect them against breakage by traffic shock, (2) to keep them from freezing, and (3) to permit them to drain the lowest fixture in the premises served. Common laying depths are 3 ft below the basement floor and 11 ft below the top of building foundations (12 ft or more for basements in commercial districts), together with an allowance of ¼ in. per ft (2%) for the slope of the building sewer.[2] In the

---

[2] At this slope a 6-in. sewer flowing full will discharge about 300 gpm or 40 cfm at a velocity of 3.5 fps.

northern United States, cellar depths range from 6 to 8 ft and frost depths from 4 to 6 ft. A 2-ft earth cover will cushion most shocks. The deep basements of tall buildings are drained by ejectors or pumps.

As shown in Fig. 9–4, manholes are channeled to improve flow, and the entrance of high-lying laterals is eased by constructing drop manholes rather than going to the expense of lowering the last length of run. In their upper reaches, most sewers receive so little flow that they are not self-cleaning and must be flushed from time to time. This is done by (1) damming up the flow at a lower manhole and releasing the stored waters after the sewer has almost filled; (2) suddenly pouring a large amount of water into an upstream manhole; (3) providing at the uppermost end of the line a *flushing manhole* that can be filled with water through fire hose attached to a nearby hydrant before a flap valve, shear gate, or similar quick-opening device leading to the sewer is opened; and (4) installing an automatic flush tank that fills slowly and discharges suddenly. Apart from cost and difficulties of maintenance, the danger of backflow from the sewer into the water supply is a bad feature of automatic flush tanks.

## 9-3 Collection of Stormwaters

Much of the suspended load of solids entering storm drains is sand and gravel. Because fine sand is moved along at velocities of 1 fps or more and gravel at 2 fps or more, recommended minimum velocities are 2.5 to 3 fps, or about 0.5 fps more than for sanitary sewers. Factors determining the capacity of storm drains are (1) intensity and duration of local rainstorms, (2) size and runoff characteristics of tributary areas, and (3) economy of design, determined largely by the opportunity for quick discharge of collected storm waters into natural water courses. Rate of storm runoff is ordinarily the governing factor in the hydraulic design of storm drains. To prevent inundation of streets, walks, and yards and flooding of basements and other low-lying structures, together with attendant inconvenience, traffic disruption, and damage to property, storm sewers are made large enough to drain away, rapidly and without becoming surcharged, the runoff from storms shown by experience to be of such intensity and frequency as to be objectionable. The heavier the storm, the greater but less frequent is the potential inconvenience or damage; the higher the property values, the more sizable is the possible damage. In a well-balanced system of storm drains, these factors will have received proper recognition for the kind of areas served: residential, mercantile, industrial, and mixed. For example, in high-value mercantile districts with basement stores and stockrooms, storm drains may be made large enough to carry away surface runoff from all but unusual storms, estimated to occur only once in 5, 10, 20, 50, or even 100 years, whereas the drains in suburban residential districts are allowed to be surcharged by all storms larger than the 1- or 2-year storm.

Until there are storm drains in a given area and the area itself is developed to its ultimate use, runoff measurements are neither possible nor meaningful. Accordingly, the design of storm sewers is normally based not on analysis of recorded runoff but on (1) analysis of storm rainfalls—their intensity or rate of precipitation, duration, and frequency of occurrence—and (2) estimation of runoff resulting from these rainfalls in the planned development.

Storm sewers are occasionally surcharged and subjected to pressures, but usually no more than their depth below street level. Nevertheless, they are designed for open-channel flow and equipped with manholes in much the same way as sanitary sewers. In North American practice, the minimum size of storm sewers is 12 in., to prevent clogging by trash of one kind or another. Their minimum depth is set by structural requirements rather than basement elevations. Surface runoff enters from street gutters through *street inlets* or *catch basins* (Fig. 9-5) and *property drains*. Size, number, and placement of street inlets govern the degree of freedom from flooding of traffic ways and pedestrian crossings. To permit inspection and cleaning, it is preferable to discharge street inlets directly into manholes. Catch basins are, in a sense, enlarged and trapped street inlets in which debris and heavy solids are held back or settle out. Historically, they antedate street inlets and were devised to protect combined sewerage systems at a time when much sand and gravel were washed from unpaved streets. Historically, too, the air in sewers, called *sewer gas*, was once deemed dangerous to health; this is why catch basins were given water-sealed traps. Catch basins need much maintenance; they should be cleaned after every major storm and may have to be oiled to prevent production of large crops of mosquitoes. On the whole, there is little reason for continuing their use in modern sewerage systems.

## 9-4 Combined Collection of Wastewaters and Stormwaters

In combined sewerage systems, a single set of sewers collects both domestic and industrial wastewater and surface runoff from rainfall (Fig. 1-3b). Because stormwaters often exceed sewage flows by 50 to 100 times, the accuracy with which rates of surface runoff can be estimated is generally less than the difference between rates of stormwater and combined-sewage flows. Accordingly, most combined sewers are designed to serve principally as storm drains. However, they must be placed as deep as sanitary sewers. Backing-up and overflow of combined sewage into basements and streets is obviously more objectionable than the surcharge of drains that carry nothing but stormwater. Combined sewers are given velocities up to 5.0 fps to keep them clean.

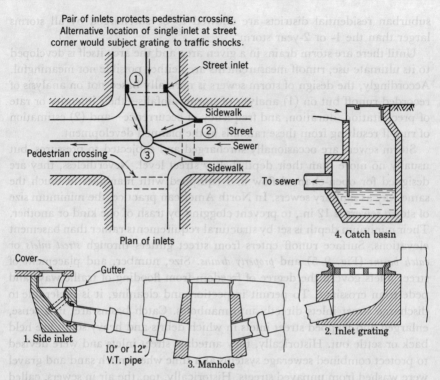

**Figure 9-5.
Street inlets and their connection to a manhole.**

The wide range of flows in combined sewers requires the solution of certain special problems, among them choice of a cross-section that will ensure self-cleaning velocities for both storm and dry-weather flows; design of self-cleaning *inverted siphons*—also called *sag pipes* and *depressed sewers*—dipping beneath the hydraulic grade line as they carry wastewater across a depression or under an obstruction; and provision of stormwater overflows in intercepting systems.

*Cross-sections.* Departures from circular cross-sections are prompted by hydraulic as well as structural and economic considerations. Examples are the *egg-shaped* sections and *cunettes* illustrated in Fig. 9-6. Two circular sewers, an underlying sanitary sewer, and an overlying storm drain are fused into a single egg-shaped section. The resulting hydraulic radius is nearly constant at all depths. Cunettes[3] form troughs dimensioned to the dry-

---

[3] The main drains of Paris, France, made famous by Victor Hugo's reference to them in *Les Miserables*, were constructed from 1833 onward. They were made sufficiently large (6 ft high and at least 2 ft 6 in. wide) to permit laborers to work in comfort. Their conversion from 1880 onward into combined sewers necessitated the addition of *cunettes*.

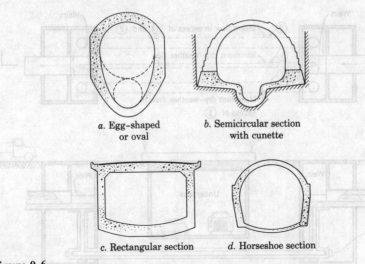

**Figure 9-6.**
**Sections of storm and combined sewers.**

weather flow. *Rectangular sections* are easy to construct and make for economical trenching with low head-room requirements. *Horseshoe sections* are structurally very satisfactory; egg-shaped sections are not. Large outfall sewers have been built as pressure tunnels.

*Inverted Siphons.* Siphons flow full and under pressure, and the velocities in them are relatively much more variable than in open channels, where depth and cross-section change simultaneously with flow. To keep velocities up and clogging by sediments down, two or more parallel pipes are, therefore, thrown in and out of operation as flows rise and fall. The pipes dispatch characteristic flows at self-cleaning velocities of 3 fps for pipes carrying sanitary sewage and 5 fps for pipes conveying storm or combined sewage. The smallest pipe diameter is 6 in., and the choice of pipe material is adjusted to the hydrostatic head under which it must operate.

Figure 9-7 shows a simple example: low dry-weather wastewater flows are passed through the central siphon; high dry-weather flows and storm flows spill over weirs into lateral siphons to right and left. The three siphons combine to equal the capacity of the approach sewer. Weir heights are fixed at depths reached by characteristic flows in the approach sewer and inlet structure. Flows are reunited in a chamber in advance of the outlet sewer.

*Interceptors.* Intercepting sewers (Fig. 1-3b) are generally designed to carry away some multiple of the dry-weather flow in order to bleed off as much stormwater and included wastewater as can be justified by hygienic, esthetic, and economic considerations. Where rainfalls are intense and sharp, as in most of North America, it is not possible to lead away much stormwater

258/*Wastewater Removal*

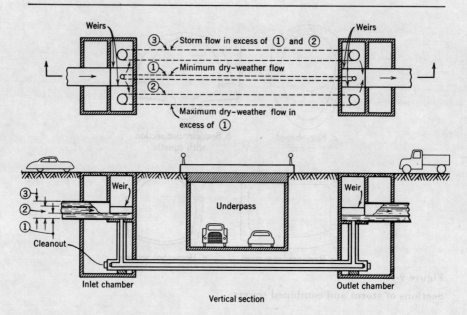

**Figure 9-7.**
**Inverted siphon or suppressed sewer for combined sewage.**

through reasonably proportioned interceptors. Consequently, they are designed to transport not much more than the maximum dry-weather flow, or 250 to 600 gpcd. A more informative measure of interceptor capacity in excess of average dry-weather flow is the rate of rainfall or runoff they can accept without overflowing. Studies of rainfalls in the hydrological surroundings of communities in the United States usually lead to the conclusion that most precipitation in excess of 0.1 in. is spilled; that spills can occur as frequently as half a dozen times a month; and that interception is not improved greatly by going even to ten times the dry-weather flow. However, where rains are gentle and long, as in the United Kingdom, six times the dry-weather flow comprises much of the runoff from rainfall and becomes a useful design factor.

The total yearly pollution reaching an interceptor-protected body of water is a significant fraction (3%) of the total annual volume of sanitary sewage. During the periods when spilling occurs, a very high percentage of the sanitary wastewater can be carried through combined sewer overflows. If solids have accumulated in the sewer during the interval between rains, these may be washed out also. Thus overflows from combined sewers can be heavily charged with solids. They present a serious pollution problem that will be difficult to correct. Detention, settling, and chlorination are useful, and under some circumstances it may be desirable to work toward full separation of sanitary wastewater from surface runoff.

**Retarding Basins.** Interception can be improved by introducing into combined systems retarding devices—for example, up-system detention basins or equalizing tanks. Constructed in advance of junctions between submains and interceptors, they store flows in excess of interceptor capacity until they are filled. After that, they continue to retard and equalize flows in lesser degree, but they do function as settling basins for the removal of gross and unsightly settleable matter (Fig. 9-8). Depending on local conditions, detention periods as short as 15 minutes can be quite effective as settling basins and more so as chlorine-contact basins. Operating ranges extend from the dry-weather flowline of the interceptor to the crown of the conjoined combined sewer. After storms subside, the tank contents are flushed or lifted into the interceptor, and the accumulated solids eventually reach the treatment works. Where much stormwater is carried as far as the treatment plant—in the British Isles, for example—stormwater standby tanks, which serve as primary sedimentation tanks, become useful adjuncts to the works.

**Overflows.** The amounts of water entering interceptors at junctions with submains must be controlled. Only as much should be admitted as individual interceptor reaches can carry without being surcharged. Higher flows must be diverted into stormwater overflows. As shown in Fig. 9-9, admission and diversion can be regulated hydraulically or mechanically.

Hydraulic separation of excess flows from dry-weather flows is accomplished by devices such as the following: (1) diverting weirs in the form of side spillways leading to overflows, with crest levels and lengths so chosen as to spill excess flows that, figuratively speaking, override the dry-weather flows, which follow their accustomed path of the interceptor (Fig. 9-9a); (2) leaping weirs, essentially *gaps* in the floor of the channel over which excess flows jump under their own momentum, while dry-weather flows tumble through the gap into the interceptor (Fig. 9-9b); (3) siphon spillways that carry flows in

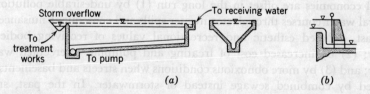

(a) (b)

**Figure 9-8.**

**Stormwater detention tank and outlet weir. After Imhoff and Fair. (a) Longitudinal and transverse section of tank. (b) Two-level outlet weir: lower level for dryweather flow to treatment works, upper level for stormwater overflow to receiving water.**

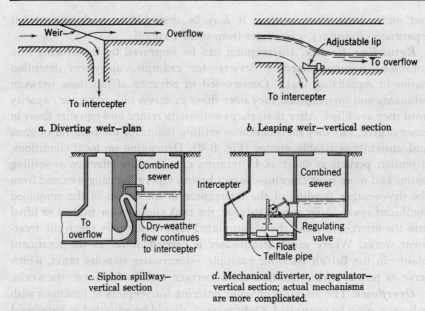

**Figure 9-9.**
**Regulation of stormwater overflow.**

excess of intercepter capacity into the overflow channel (Fig. 9–9c); and (4) mechanical devices, diversion of stormwater flows generally being regulated by float-operated control valves activated by flow levels in the interceptor (Fig. 9–9d).

## 9-5 Choice of Collecting System

Apart from questions of economy, the combined system of sewerage is at best a compromise between two wholly different objectives: water carriage of wastes and removal of flooding runoff. In the life of growing communities, initial economies are offset in the long run (1) by undesirable pollution of natural water courses through stormwater spills and consequent nuisance or, at least, debased esthetic and recreational values of receiving bodies of water; (2) by increased cost of treating and pumping intercepted wastewater; and (3) by more obnoxious conditions when streets and basements are flooded by combined sewage instead of stormwater. In the past, small streams, around which parks and other recreational areas could have grown, have been forced into combined sewerage systems because pressing them into service as receiving waters had degraded them into open sewers. By contrast, a separate system of sewerage can exploit natural water courses hydraulically by discharging stormwater into them through short runs of storm drains while preserving their esthetic and recreational assets. However, they may have to be channelized if they are to perform well. In the United States, large sums of money are being expended in sewer separation and related constructions in order to protect water courses from combined overflows. The construction of combined sewers was forbidden by most health departments at the turn of the century.

## 9-6 Design Information

The amount and detail of local information required for the design of sewers and drains are large. Special surveys are generally made to produce needed maps and tables as follows: (1) detailed plans and profiles of streets to be sewered; (2) plans and contour lines of properties to be drained; (3) sill or cellar elevations of buildings to be connected; (4) location and elevation of existing or projected building drains; (5) location of existing or planned surface and subsurface utilities; (6) kind and location of soils and rock through which sewers and drains must be laid; (7) depth of groundwater table; (8) location of drainage-area divides; (9) nature of street paving; (10) projected changes in street grades; (11) location and availability of sites for pumping stations, treatment works, and outfalls; and (12) nature of receiving bodies of water and other disposal facilities.

Much of the topographic information needed is assembled for illustrative purposes in Fig. 9-10 for a single sanitary sewer in a street also containing a storm drain. Aerial maps are useful.

Variations in flow to be handled by *sanitary sewers* are determined by (1) anticipated population growth and water use during a chosen design period and (2) fluctuations in flow springing from normal water use (Sec. 2-9). Choice of the design period itself will depend on anticipated population increases and interest rates (Sec. 2-2). By contrast, the design period for *storm drains* and *combined sewers* is important principally in connection with expected effects of drainage-area development on runoff coefficients and magnitudes of flood damage. Required storm-drain capacity is primarily a matter of probable runoff patterns. Because storms occur at random, adopted values may be reached or exceeded as soon as storm sewers and drains have been laid.

**Sanitary Sewers.** Although anticipated wastewater volumes and their hourly, daily, and seasonal variations (Sec. 2-10) determine design capacity, the system must function properly from the start. Comparative flows are shown in the following schedule.

BEGINNING OF DESIGN PERIOD. (*a*) Extreme minimum flow = $\frac{1}{2}$ minimum daily flow. Critical for velocities of flow and cleanliness of sewers. (*b*) Minimum daily flow = $\frac{2}{3}$ of average daily flow. Critical for subdivision of units in treatment works.

# 262 / Wastewater Removal

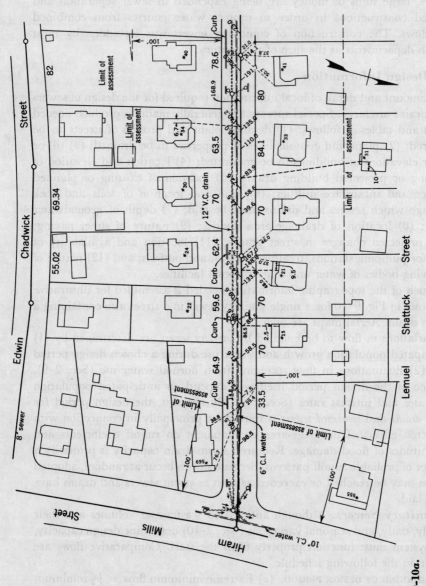

Figure 9-10a. Plan of sanitary sewer shown in profile in Figure 9-10b.

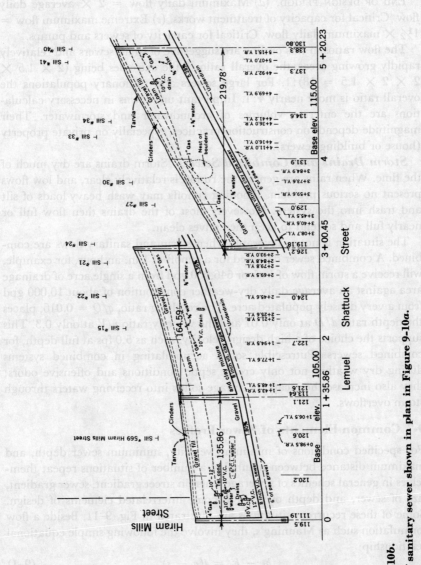

**Figure 9-10b.** Profile of sanitary sewer shown in plan in Figure 9-10a.

BEGINNING AND END OF DESIGN PERIOD. (c) Average daily flow at beginning of design period = ½ average daily flow at end of period. Critical for velocities of flow in force mains.

END OF DESIGN PERIOD. (d) Maximum daily flow = 2 × average daily flow. Critical for capacity of treatment works. (e) Extreme maximum flow = 1½ × maximum daily flow. Critical for capacity of sewers and pumps.

The flow ratios in this outline are suggestive of small sewers and relatively rapidly growing areas, the overall ratio of the extremes being (2 × 1.5 × 2 × 2 × 1.5 = 18):1. For large sewers and stationary populations the overall ratio is more nearly 4:1. Important unknowns in necessary calculations are the entering volumes of groundwater and stormwater. Their magnitude depends on construction practices, especially on private property (house or building sewers).[4]

*Storm Drains and Combined Sewers.* Storm drains are dry much of the time. When rains are gentle, the runoff is relatively clear, and low flows present no serious problem. Flooding runoffs may wash heavy loads of silt and trash into the system. However, most of the drains then flow full or nearly full and so tend to keep themselves clean.

The situation is not so favorable when storm and sanitary flows are combined. A combined sewer designed for a runoff of 1 in. an hour, for example, will receive a storm flow of 1 cfs or 646,000 gpd from a single acre of drainage area against an average daily dry-weather contribution of about 10,000 gpd from a very densely populated acre. The resulting ratio, $q/Q = 0.016$, places the depth ratio $d/D$ at only 0.07 and the velocity ratio $v/V$ at only 0.3. This supports the choice of a high design velocity, such as 5.0 fps at full depth, for combined sewers. Putrescible solids accumulating in combined systems during dry weather not only create septic conditions and offensive odors; they also increase the escape of sewage solids into receiving waters through storm overflows.

## 9-7 Common Elements of Sewer Profiles

For specified conditions of minimum velocity, minimum sewer depth, and maximum distance between manholes, a number of situations repeat themselves in general schemes of sewerage wherein street gradient, sewer gradient, size of sewer, and depth of sewer become interrelated elements of design. Some of these recurrent situations are illustrated in Fig. 9-11. Beside a flow formulation such as Manning's, they involve the following simple equational relationship:

$$h_1 - h_2 = l(g - s) \qquad (9\text{-}1)$$

[4] J. C. Geyer and J. J. Lentz, *An Evaluation of the Problems of Sanitary Sewer System Design*, Final Report of the Residential Sewerage Research Project of the Federal Housing Administration, Technical Studies Program, 1964.

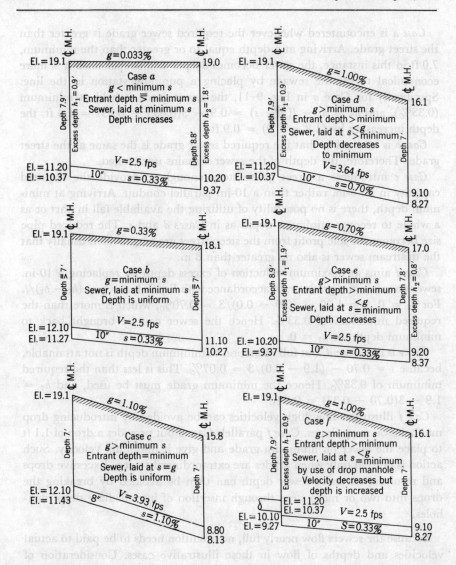

**Figure 9-11.**

**Common elements of sewer design. Required $q = 1.2$ cfs ($Q = 1.2$ cfs); $V = 2.5$ fps; $N = 0.012$; $l = 300$ ft. Minimum depth to crown = 7.0 ft; 8-in. sewer $Q = 1.2$ cfs; $S = 0.84\%$; $V = 3.4$ fps; 10-in. sewer $V = 2.5$ fps; $S = 0.33\%$; $Q = 1.36$ cfs.**

where $h_1$ and $h_2$ are sewer depths *in excess of minimum requirements*, $l$ is the distance between manholes, and $g$ and $s$ are respectively the street and sewer grades. Conditions of flow are stated in the legend accompanying Fig. 9-11.

Case *a* is encountered whenever the required sewer grade is greater than the street grade. Arriving at a depth equal to or greater than the minimum, 7.0 ft in this instance, the sewer becomes deeper and deeper until it is more economical to lift the sewage by placing a pumping station in the line. Specifically for Case *a* in Fig. 9–11, the sewer grade is held at minimum (0.33%), and $h_2 = h_1 - l(g - s) = 0.9 - 3(0.033 - 0.33) = 1.8$ ft, the depth increasing by $(1.8 - 0.9) = 0.9$ ft.

Case *b* is unusual in that the required sewer grade is the same as the street grade. Therefore, the depth of the sewer remains unchanged.

Case *c* introduces a street grade steep enough to provide the required capacity in an 8-in. rather than a 10-in. parallel conduit. Arriving at minimum depth, there is no possibility of utilizing the available fall in part or as a whole to recover minimum depth as in Cases *d* and *e*. The reduced pipe size becomes the sole profit from the steep street grade, provided usually that the upstream sewer is also no greater than 8 in.

Case *d* aims at maximum reduction of excess depth by replacing a 10-in. sewer on minimum grade or, in accordance with Eq. 9–1, $s = g - (h_1 - h_2)/l$. For $h_2 = 0$, $s = 1.00 - (0.9 - 0.0)/3 = 0.70\%$, which is more than the required minimum of 0.33%. Hence the sewer can be brought back to minimum depth, or $h_2 = 0$.

Case *e* is like Case *d*, but full reduction to minimum depth is not attainable, because $s = 0.70 - (1.9 - 0.0)/3 = 0.07\%$. This is less than the required minimum of 0.33%. Hence the minimum grade must be used, and $h_2 = 1.9 - 3(0.70 - 0.33) = 0.8$ ft.

Case *f* illustrates how high velocities can be avoided by introducing drop manholes on steep slopes. Case *f* parallels Case *d* but provides a drop of 1.1 ft to place the sewer on minimum grade and give it minimum velocity. Such action is normal only when grades are extraordinarily steep. Excessive drops and resulting excessive sewer depth can then be avoided by breaking the drops into two or more steps through insertion of intermediate drop manholes.

Because the sewers flow nearly full, no attention needs to be paid to actual velocities and depths of flow in these illustrative cases. Consideration of actual depths and velocities of flow is generally restricted to the upper reaches of sewers flowing less than half full. There, the designer may have to forgo self-cleaning grades for lateral sewers. Otherwise, they might reach the main sewer at an elevation below that of the main itself, and the main would have to be lowered to intercept them. Normally this would be expensive.

Generally speaking, the designer should try for the fullest possible exploitation of the capacity of minimum-sized sewers before joining them to larger sewers. The implications are demonstrated in Fig. 9–12. There

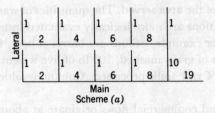
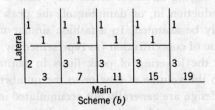

**Figure 9-12.**
**Relative utilization of the capacity of lateral sewers.**

Scheme *a* keeps lateral flows from joining the main conductor until as many as 10 units of flow have accumulated, whereas no lateral carries more than 3 units in Scheme *b*. Moreover, Scheme *b* exceeds 10 units in two sections for which the required capacities in Scheme *a* are still only 6 and 8 units respectively.

## 9-8 Capacity Design in Sanitary Sewerage

Systems of sanitary sewers receive the water-borne wastes from households, mercantile and industrial establishments, and public buildings and institutions. In addition, groundwater enters by infiltration from the soil and, too often, illicit property drain connections and leaking manhole covers increase the flow. Accordingly, their requisite capacity is determined by the tributary domestic and institutional population, commercial water use, industrial activity, height of groundwater table, tightness of construction, and enforcement of rainwater separation.

It is generally convenient to arrive at unit values of domestic flows on the basis of population density and area served; but it would also be possible to develop figures for the number of people per front foot in districts of varying occupancy and make *sewer length* rather than *area served* the criterion of capacity design. Length (sometimes coupled with diameter) of sewer, indeed, offers a perhaps more rational basis for the estimation of groundwater infiltration. Unit values for flows from commercial districts are generally expressed in terms of the area served. The quantities of wastewaters produced by industrial operations are more logically evaluated in terms of the units of daily production, for example, gallons per barrel of beer, 100 cases of canned goods, 1000 bushels of grain mashed, 100 lb of live weight of animals slaughtered, or 1000 lb of raw milk processed. Common values are suggested in Sec. 2–8.

Peak domestic and commercial flows originate at about the same hour of the day but travel varying distances before they reach a given point in the system. Hence a reduction in, or damping of, the peak of the cumulative flows must generally be assumed. In a fashion similar to the reduction in flood flows with time of concentration (as represented by the size and shape of drainage area), the lowering of peak flows in sanitary sewers is conveniently related to the volume of flow or to the number of people served, and unit values of design are generally not accumulated in direct proportion to the rate of discharge or to the tributary population.

## 9-9 Layout and Hydraulic Design in Sanitary Sewerage

Before entering upon the design of individual sewer runs, a preliminary layout is made of the entire system. Sanitary sewers are placed in streets or alleys in proper reference to the buildings served, and terminal manholes of lateral sewers generally lie within the service frontage of the last lot sewered.

Sewers should slope with the ground surface and follow as direct a route to the point of discharge as topography and street layout permit. To this purpose, flow in a well-designed system will normally take the path of surface runoff. Stormwater infiltration through manhole covers is kept down by placing sanitary sewers under the crown of the street.

In communities with alleys, the choice of location will depend on the relative advantages of alleys and streets. Alley location is often preferable in business as well as residential districts.

Instructions for preliminary layouts are as follows: (1) show all sewers as single lines; (2) insert arrows to indicate flow direction; (3) show manholes as circles at all changes in directions or grade, at all sewer junctions, and at intermediate points that will keep manhole spacing below the allowable maximum; and (4) number manholes for identification. Alternate layouts will determine the final design. The avoidance of pumping is not as important as it formerly was, because of the availability of "off-the-shelf" pumping stations equipped with pumps that do not clog readily.

The hydraulic design of a system of sanitary sewers is straightforward and is readily carried to completion in a series of systematic computations, as in Example 9-1.

*Example 9-1.*

Determine the required capacity and find the slope, size, and hydraulic characteristics of the system of sanitary sewers shown in the accompanying tabulation (Table 9-1) of their location, areas and population served, and expected sewage flows.[5]

1. Capacity requirements are based on the following assumptions:
(a) Water consumption: Average day, 95 gpcd; maximum day, 175% of average; maximum hour, 140% of maximum day.
(b) Domestic wastewater: 70% of water consumption; maximum is 285 gpcd for 5 acres decreasing to 245 gpcd for 100 acres or more.
(c) Groundwater: 30,000 to 50,000 gpd per mile of sewer for low land and 20,000 to 35,000 gpd per mile of sewer for high land; 0.14 to 0.15 cfs per 100 acres for 8-in. to 15-in. sewers in low land and 0.09 to 0.11 cfs per 100 acres in high land. These figures would be lowered by using preformed joints.
(d) Commercial wastewater: 25,000 gpd per acre = 3.88 cfs per 100 acres.
(e) Industrial wastewater: Flow in accordance with industry.
2. Hydraulic requirements are as follows:
(a) Minimum velocity in sewers: 2.5 fps (actual).
(b) Kutter's coefficient of roughness $N = 0.015$ includes allowances for change in direction and related losses in manholes except for (c) below.
(c) Crown of sewers is made continuous to prevent surcharge of upstream sewer.
3. Design procedures are as follows:
Columns 1-4 identify the location of the sewer run. The sections are continuous.
Columns 5-8 list the acreage immediately adjacent to the sewer.
Column 9 gives the density of the population per domestic acre.
Column 10 = Col. 9 × Col. 8.
Columns 11-13 list the accumulated acreage drained by the sewer. For example, in Sec. *b*, Col. 13 is the sum of Col. 8 in Secs. *a* and *b*, or $(40 + 27) = 67$.
Column 14 gives the average population density for the total tributary area. For example, in Sec. *b*, Col. 14 = $(40 \times 27 + 27 \times 19)/(40 + 27) = 23.8$.
Column 15 = Col. 14 × Col. 13.

[5] The numerical values shown in this example are taken from computations for the sewerage system of Cranston, R. I., by the firm of Fay, Spofford, and Thorndike as reported in *Eng. News-Rec.*, **123**, 419 (1939). Some of the values there given do not agree in detail with values suggested in this book.

## Table 9-1
### Illustrative Computations for a System of Sanitary Sewers[a] (Example 9–1)

| Section (1) | Location of Sewer | | | | Adjacent Area | | | | | | Total Tributary Area | | | | |
|---|---|---|---|---|---|---|---|---|---|---|---|---|---|---|---|
| | Street (2) | Stations or Limits | | Total acres (5) | Industrial acres (6) | Commercial acres (7) | Domestic acres (8) | Population | | Industrial acres (11) | Commercial acres (12) | Domestic acres (13) | Population | |
| | | From (3) | To (4) | | | | | Per acre (9) | Total (10) | | | | Per acre (14) | Total (15) |
| a | A Ave. | B Ave. | C St. | 49 | 5 | 4 | 40 | 27 | 1,080 | 5 | 4 | 40 | 27.0 | 1,080 |
| b | D Ave. | C St. | E St. | 37 | 3 | 7 | 27 | 19 | 513 | 8 | 11 | 67 | 23.8 | 1,593 |
| c | F St. | G St. | H St. | 29 | 8 | 1 | 20 | 25 | 500 | 16 | 12 | 87 | 24.1 | 2,093 |
| d | I St. | J St. | K St. | 63 | — | 10 | 53 | 21 | 1,113 | 16 | 22 | 140 | 22.9 | 3,206 |

| Section (1) | Maximum Volume of Sewage, cfs | | | | | Design Profile | | | | | | | | | |
|---|---|---|---|---|---|---|---|---|---|---|---|---|---|---|---|
| | Industrial (16) | Commercial (17) | Domestic (18) | Infiltration (19) | Total (20) | Size, in. (21) | Slopes, ‰ (22) | Capacity, cfs (23) | Velocity, fps | | Depth of flow, in. (26) | Length, ft (27) | Invert Elevation | | Cut | |
| | | | | | | | | | Full (24) | Actual (25) | | | Upper End (28) | Lower End (29) | Upper End (30) | Lower End (31) | Average (32) |
| a | .156 | .155 | .440 | .044 | 0.795 | 8 | 0.8 | 0.82 | 2.35 | 2.72 | 6.37 | 850 | 120.00 | 113.20 | 7.50 | 11.50 | 9.50 |
| b | .248 | .429 | .650 | .086 | 1.413 | 10 | 0.7 | 1.42 | 2.22 | 3.02 | 8.1 | 1,260 | 113.03 | 104.21 | 11.67 | 8.50 | 10.08 |
| c | .496 | .468 | .852 | .115 | 1.931 | 12 | 0.45 | 2.23 | 2.45 | 2.84 | 9.7 | 1,880 | 104.04 | 95.58 | 8.67 | 12.00 | 10.33 |
| d | .496 | .858 | 1.300 | .178 | 2.832 | 15 | 0.3 | 3.35 | 2.35 | 2.72 | 12.0 | 1,760 | 95.33 | 90.05 | 12.25 | 11.00 | 11.63 |

[a] *Note.* Sections of sewers rather than individual runs between manholes are shown in this example in order to include major changes in required capacity and consequent size.

Column 16 lists values obtained in a survey of industries in the areas served.

Column 17 = Col. 12 × 0.0388.

Column 18 = Col. 15 × (245 to 285) × (1.547 × 10$^{-6}$). For example, in Sec. $a$, 1080 × 264 × (1.547 × 10$^{-6}$) = 0.440 cfs.

Column 19 = Sum of Col. 5 times rate of infiltration. For example, in Sec. $a$, 49 × 0.09/100 = 0.044 cfs.

Column 20 = Sum of Cols. 16–19.

Columns 21–29 record the size of sewer for required capacity and available, or required, grade together with depth and velocity of flow. For example, in Sec. $a$, an 8-in. sewer laid on a grade of 6.8/850 = 0.008 or 0.8% will discharge $Q$ = 0.82 cfs at a velocity of 2.35 cfs when it flows full. Hence for $q/Q$ = 0.795/0.82 = 0.971, $d/D$ = 0.796, $v/V$ = 1.16, or $d$ = 8 × 0.796 = 6.37 in. and $v$ = 2.35 × 1.16 = 2.72 fps.

Columns 28–31 are taken from profiles of streets and sewers.

Column 28, Sec. $b$, shows a drop in the manhole of (113.20 − 113.03) = 0.17 ft compared with Col. 28, Sec. $a$. This allows for a full drop of 0.17 × 12 = 2 in. to offset the increase in the diameter of the sewer from 8 in. to 10 in.

Column 32 = arithmetic mean of Cols. 30 and 31.

## 9-10 Capacity Design in Storm Drainage

As shown in Sec. 3–13, the rational method of estimating runoff from rainfall provides a common hydrological basis for the capacity design of storm-drainage systems. The axiom of design is $Q$ = $cia$ (Eq. 3–18). Accordingly, the designer must arrive at the best possible estimates of $c$, the runoff-rainfall ratio, and $i$, the rainfall intensity, the area $a$ being determined by measuring tributary surfaces. As stated in Sec. 3–13, both $c$ and $i$ are variable in time. Because of this, storm flows reaching a given point in the drainage system are compounded of waters falling within the time of concentration.

**Time of Concentration.** The time of concentration is composed of two parts: (1) the inlet time, or time required for runoff to gain entrance to a sewer, and (2) the time of flow in the sewerage system.

The inlet time is a function of (1) surface roughness offering resistance to flow; (2) depression storage delaying runoff and often reducing its total; (3) steepness of areal slope, governing speed of overland flow; (4) size of block or distance from the areal divide to the sewer inlet determining time of travel; (5) degree of direct roof and surface drainage reducing losses and shortening inlet times; and (6) spacing of street inlets affecting elapsed times of flow. In large communities, in which roofs shed water directly to sewers, and runoff from paved yards and streets enters the drainage system through closely spaced street inlets, the time of overland flow is usually less than

5 min. In districts with relatively flat slopes and greater inlet spacing, the time may lengthen to 10 to 15 min.

Except for rains of considerable duration, relatively little of the water entering inlets at the time of peak flow originates from unpaved areas. Therefore, the time between the peak rainfall rate and the peak flow at an inlet is often less than one minute. Applying short inlet times and associated high intensities to paved areas only, will sometimes produce the best estimates of inlet flows.

Time of concentration for overland flow has been formulated by Kerby:[6]

$$t = [\tfrac{2}{3}l(n/\sqrt{s})]^{0.467} \qquad (9\text{--}2)$$

where $t$ is the inlet time in minutes, $l \leqslant 1200$ ft is the distance to the farthest tributary point, $s$ is the slope, and $n$ is a retardance coefficient analogous to the coefficient of roughness. Suggested values of $n$ are:

| Type of Surface | $n$ |
|---|---|
| Impervious surfaces | 0.02 |
| Bare packed soil, smooth | 0.10 |
| Bare surfaces, moderately rough | 0.20 |
| Poor grass and cultivated row crops | 0.20 |
| Pasture or average grass | 0.40 |
| Timberland, deciduous trees | 0.60 |
| Timberland, deciduous trees, deep litter | 0.80 |
| Timberland, conifers | 0.80 |
| Dense grass | 0.80 |

For $l = 500$ ft, $s = 1.0\%$, and $n = 0.1$, for example, $t = 1000^{0.467} = 29$ min.

As a matter of arithmetic, the time of flow in the system equals the sum of the quotients of the length of constituent sewers and their velocity when flowing full. Ordinarily, neither time increase, as sewers are filled, nor time decrease, as flood waves are generated by rapid discharge of lateral sewers, is taken into account.

**Runoff Coefficients.** Runoff from storm rainfall is reduced by evaporation, depression storage, surface wetting, and percolation. Losses decrease with rainfall duration. Runoff-rainfall ratios, or shedding characteristics, rise proportionately. As stated in Sec. 3–13, the coefficient $c$ may exceed unity because it is the ratio of a peak runoff rate to an average rainfall rate. Ordinarily, however, $c$ is less than 1.0 and approaches unity only when drainage areas are impervious and high-intensity storms last long enough.

---

[6] W. S. Kerby, Civil Engineering, **29**, 174 (1959). Equation 9–2 can be approximated by the dimensionally consistent expression: $t = 4.5[(l/g)(n/\sqrt{s})]^{1/2}$. Also see Eq. 9–3.

The choice of meaningful runoff coefficients is difficult. It may be made a complex decision. The runoff coefficient for a particular time of concentration should logically be an average weighted in accordance with the geometric configuration of the area drained, but fundamental evaluations of $c$ and $i$ are generally not sufficiently exact to warrant this refinement.

The choice of a suitable runoff coefficient is complicated not only by existing conditions, but also by the uncertainties of change in evolving urban complexes. Difficult to account for are the variations in runoff-rainfall relations to be expected in given drainage areas along with variations in rainfall intensities in the course of major storms. Fundamental runoff efficiency is least at storm onset and improves as storms progress. Graphic and equational relationships proposed by different authorities are shown in Fig. 9–13. However, they are not actually very helpful. Weighted average coefficients are calculated for drainage areas composed of districts with different runoff efficiencies.

Least arduous is acceptance of the fact that the degree of imperviousness of a given area is a rough measure of its shedding efficiency. Streets, alleys, side and yard walks, together with house and shed roofs, as the principal impervious components, produce high coefficients; lawns and gardens, as the principal pervious components, produce low coefficients. To arrive at

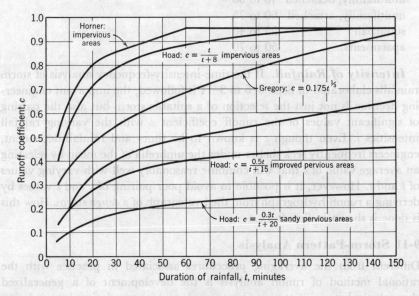

**Figure 9-13.**

**Variation in runoff coefficients with duration of rainfall and nature of area drained.**

a composite runoff-rainfall ratio, a weighted average is often computed from information such as the following:[7]

| Description of Area Component | Runoff Coefficient, % | Description of Area Component | Runoff Coefficient, % |
|---|---|---|---|
| Pavement, asphalt and concrete | 70 to 95 | Lawns, sandy, flat (2%) | 5 to 10 |
| | | steep (7%) | 15 to 20 |
| Pavement, brick | 70 to 85 | heavy, flat (2%) | 13 to 17 |
| Roofs | 75 to 95 | steep (7%) | 25 to 35 |

Resulting overall values for North American communities range between limits not far from the following:

| Areas | Overall Runoff Coefficient, % | Areas | Overall Runoff Coefficient, % |
|---|---|---|---|
| Business: | | Industrial, light | 50 to 80 |
| downtown | 70 to 95 | heavy | 60 to 90 |
| neighborhood | 50 to 70 | Parks, cemeteries | 10 to 25 |
| | | Playgrounds | 20 to 35 |
| Residential: | | Unimproved land | 10 to 30 |
| single-family | 30 to 50 | | |
| multifamily, detached | 40 to 60 | | |
| multifamily, attached | 60 to 75 | | |
| suburban | 25 to 40 | | |
| apartments | 50 to 70 | | |

**Intensity of Rainfall.** If the time-intensity-frequency analysis of storm rainfalls elaborated in Secs. 3-6 to 3-11 is followed, the important engineering decision is not just the selection of a suitable storm but also the pairing of significant values of the runoff coefficient $c$ with the varying rainfall intensities $i$. Even though $c$ is known to be time- and rainfall-dependent, engineers frequently seek shelter under the umbrella of the mean by selecting an average value of $c$ that will combine reasonably well with varying values of $t$ and $i$. However, it is possible to avoid poor pairing of $c$ and $i$ values by deriving a runoff hydrograph from the hyetograph of a *design* storm. How this is done is shown in Sec. 9-11.

## 9-11 Storm-Pattern Analysis

Different from the averaging procedures associated in practice with the rational method of runoff analysis is the development of a generalized chronological storm pattern or hyetograph and its translation into a design

---

[7] *Am. Soc. Civil Engrs.* Design and Construction of Sanitary and Storm Sewers, *Manuals of Engineering Practice*, No. 37 p. 51 (1969).

runoff pattern by subtracting rates of (1) surface infiltration, (2) depression storage, and (3) surface detention during overland flow. The runoff hydrograph obtained is routed through overland flow, gutter flow, and flow in building drains, catch basins, and component sewers.[8] Generalization of rainfall information by converting an intensity-duration-frequency curve into a hyetograph is illustrated in Fig. 9-14. An advanced peak in rainfall intensity is assumed in this case at ⅜ the time distance or storm duration from the beginning of appreciable precipitation, selection of a suitable fraction being based on specific rainfall experiences.[9]

Figure 9-14 illustrates results obtained in the application of conversion and routing procedures developed by Tholin and Keifer[8] for the city of Chicago. Necessary calculations are based on (1) infiltration-capacity curves shown in Fig. 9-14; (2) depth of depression storage assumed at ¼ and ½ inch, for example, and normally distributed about this mean depth, 50% of the area covered by depressions lying within 20% of the mean depth ($\sigma = \pm 14\%$); and (3) surface detention computed by Izzard's equation.[10]

$$D = 0.342[(7 \times 10^{-4} i + c_r)/s^{1/3}](lQ)^{1/3} \qquad (9\text{-}3)$$

where $D$ is the surface detention in inches of depth, $s$ is the slope of the ground, $l$ is the distance of overland flow in feet, $Q$ is the overland supply in inches per hour (cfs per acre), $i$ is the intensity of rainfall in inches per hour, and $c_r$ is a coefficient of roughness varying downward from $6.0 \times 10^{-2}$ for pervious areas of turf, through $3.2 \times 10^{-2}$ for bare, packed pervious areas, and $1.2 \times 10^{-2}$ for pavements, to $1.7 \times 10^{-2}$ for flat, gravel roofs.

## 9-12 Empirical Formulations

Variations of $c$ and $i$ with time (time being a function of watershed area $a$ and of other factors such as surface slope $s$) can be incorporated into overall runoff formulations developed empirically for given localities (Sec. 3-14). An example is McMath's formula $Q = cia^{4/5}s^{1/5}$ for St. Louis, Mo. If rainfall intensity $i$ and resulting runoff $ci$ are properly determined, overall formulas should yield results comparable with the rational method.

As storm drains are actually constructed in a given community, it becomes possible to design stormwater systems for adjacent unsewered areas on the basis of (1) actual runoff measurements conducted in times of heavy rainfall

---

[8] A. L. Tholin and C. J. Keifer, The Hydrology of Urban Runoff, *Trans. Am. Soc. Civil Engrs.*, **125**, 1308 (1960).

[9] In Example 3-2, the maximum rainfall was clocked between the 30th and 35th minute from the beginning of a 120-minute storm: $32.5/120 = 2.2/8$.

[10] C. F. Izzard, Hydraulics of Runoff from Developed Surfaces, *Proc. Highway Research Bd.*, **26**, 129 (1946). See also Y. S. Yu and J. S. McNown, Runoff from Impervious Surfaces, *Jour. Hydraulic Research*, **2**, 14 (1964).

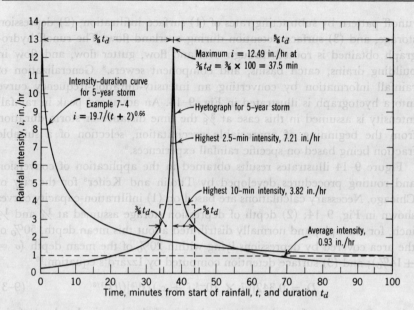

**Figure 9-14.**
**Hyetograph derived from rainfall intensity-duration frequency curve.**

or (2) surcharge experience with recorded storm intensities. In necessary calculations, it is important to identify possible downstream effects on surcharge.

### 9-13 Layout and Hydraulic Design in Storm Drainage

The layout of storm drains and sanitary sewers follows much the same procedure. Street inlets must be served as well as roof and other property drains connected directly to the storm sewers. How inlets are placed at street intersections to keep pedestrian crossings passable is indicated in Fig. 9-5. To prevent the flooding of gutters or to keep flows within inlet capacities, street inlets may also be constructed between the corners of long blocks. Required inlet capacity (Sec. 8-10) is a function of tributary area and its pertinent runoff coefficient and rainfall intensity.

Separate storm drains should proceed by the most direct route to outlets emptying into natural drainage channels. Easements or rights of way across private property may shorten their path. Manholes are included in much the same way and for much the same reasons as for sanitary sewers.

Surface topography determines the area tributary to each inlet. However, it is often assumed that lots drain to adjacent street gutters and thence to the sewers themselves. Direct drainage of roofs and areaways reduces the inlet time and places a greater load intensity on the drainage system. Necessary computations are illustrated in Table 9–2, which accompanies Example 9–2.

*Example 9–2.*

Determine the required capacity and find the slope, size, and hydraulic characteristics of the system of storm drains shown in the accompanying tabulation of location, tributary area, and expected storm runoff.

Capacity requirements are based on the rainfall curves included in Fig. 9–14. The area is assumed to be an improved pervious one, and the inlet time is assumed to be 20 min. Hydraulic requirements include a value of $N = 0.012$ in Manning's formula and drops in manholes equal to $\Delta(d + h_v) + 0.2\Delta h_v$ (Eq. 8–18) for the sewers flowing full.

Columns 1–4 identify the location of the drains. The runs are continuous.

Column 5 records the area tributary to the street inlets discharging into the manhole at the upper end of the line.

Column 6 gives cumulative area tributary to a line. For example, in Line 2, Col. 6 is the sum of Col. 6, Line 1, and Col. 5, Line 2, or $(2.19 + 1.97) = 4.16$.

Columns 7 and 8 record the times of flow to the upper end of the drain and in the drain. For example, the inlet time to Manhole 1 is estimated to be 20 min, and the time of flow in Line 1 is calculated to be $340/(60 \times 3.94) = 1.5$ min from (Col. 15)/(60 × Col. 14). Hence the time to flow to the upper end of Line 2 is $(20 + 1.5) = 21.5$ min.

Column 9 is the mean intensity of rainfall during the inlet time (time of flow to the upper end) read from Fig. 9–14.

Column 10 is the weighted mean runoff coefficient for the tributary area.

Column 11 is (Col. 9) × (Col. 10). For example, for Line 1, $2.56 \times 0.508 = 1.30$.

Column 12 = (Col. 11) × (Col. 6). For example, the runoff entering Line 1 is $1.30 \times 2.19 = 2.85$ cfs.

Columns 13–16 record the chosen size and resulting capacity and flow velocity of the drains for the tributary runoff and available or required grade. For example, in Line 1, a grade of 6.42‰ and a flow of 2.85 cfs call for a 12-in. drain. This drain will have a capacity of 3.09 cfs and flow at a velocity of 3.94 fps.

Columns 17–21 identify the profile of the drain. Col. 17 is taken from the plan or profile of the street; Col. 18 = (Col. 17) × (Col. 14); Col. 19 is obtained from Eq. 9–18, the required drop in Manhole 2 being $\Delta(d + h_v) + 0.2\Delta h_v = [(1.5 + 0.17) - (1.0 + 0.24)] + 0.2(0.17 - 0.24) = 0.42$ ft; and Col. 21 = (Col. 20) − (Col. 19), Col. 20 furthermore being (Col. 21) − (Col. 19) for the entrant line. For example, for Line 2, $(84.28 - 0.42) = 83.86$ and subsequently $(83.86 - 0.92) = 82.94$.

## Table 9-2
### Illustrative Computations for a Storm Drainage System (Example 9-2)

| Line Number (1) | Location of Drain | | | Tributary Area, acres, a | | Time of Flow, min | | Mean Rainfall Intensity, i in./hr (9) | Weighted Mean Runoff Coefficient, c (10) |
|---|---|---|---|---|---|---|---|---|---|
| | Street (2) | Manhole Number | | Increment (5) | Total (6) | To Upper End (7) | In Drain (8) | | |
| | | From (3) | To (4) | | | | | | |
| 1 | A | 1 | 2 | 2.19 | 2.19 | 20.0 | 1.5 | 2.56 | 0.508 |
| 2 | A | 2 | 3 | 1.97 | 4.16 | 21.5 | 1.9 | 2.45 | 0.518 |
| 3 | B | 3 | 4 | 3.05 | 7.21 | 23.4 | 1.3 | 2.35 | 0.532 |

| Line Number (1) | Runoff, cfs, Q | | Design | | | | | Profile | | | |
|---|---|---|---|---|---|---|---|---|---|---|---|
| | Per acre ci (11) | Total (12) | Diameter, in. (13) | Slope, ‰ (14) | Capacity, cfs (15) | Velocity, fps (16) | Length, ft (17) | Fall, ft (18) | Drop in M.H., ft (19) | Invert Elevation | |
| | | | | | | | | | | Upper End (20) | Lower End (21) |
| 1 | 1.30 | 2.85 | 12 | 6.42 | 3.09 | 3.94 | 340 | 2.18 | 0.00 | 86.46 | 84.28 |
| 2 | 1.27 | 5.28 | 18 | 2.71 | 5.93 | 3.35 | 340 | 0.92 | 0.42 | 83.86 | 82.94 |
| 3 | 1.25 | 9.02 | 24 | 1.50 | 9.48 | 3.02 | 440 | 0.66 | 0.46 | 82.48 | 81.82 |

## 9-14 Hydraulic Design of Combined Sewers

The capacity design of combined sewers allows for maximum rate of wastewater flow in addition to stormwater runoff. If entering rainwater is confined to roof water, the wastewater flows are considerable items in required sewer capacities. The resulting system is sometimes called a roof-water system rather than a combined system. If the full runoff from storms of unusual intensity is carried away by the system, wastewater flows become relatively insignificant items in required combined sewer capacity.

## 9-15 Operation and Maintenance of Drainage Systems

The principal problem in the operation and maintenance of sewers is the prevention and relief of stoppages. Tree roots and debris accumulation are the main causes. Important, too, in areas of cohesionless soil is the entrance of sand and gravel through leaky joints and pipe breaks. Cement, mortar, and lime-mortar joints will not keep out roots as effectively as bitumastic hot-poured or factory-installed rubber joints. The plastic and other newer jointing materials are promising, but their long-time performance is still to be evaluated. Understandably, debris is more likely to accumulate in the upper reaches of sewers where flows are low and unsteady. Sharp changes in grade and junctions at grade are danger points. Grease from eating places, oil from service stations, and mud from construction sites, often discharged intermittently and in high concentration, are leading offenders. Well-scheduled sewer flushing is an obvious answer when system design cannot be altered.

In arctic (permafrost) regions, sewers as well as water pipes may have to be placed in *utilidors*. The transfer of all utilities to such structures is being considered, too, in the rebuilding of old cities and the building of new ones. Effective inclusion of sanitary sewers may call for the introduction of pressurized systems. Storm sewers can probably not be accommodated economically.

# ten

# water-quality management

## 10-1 Objectives of Water-Quality Management

Water-quality management has received only passing notice in the first chapters of this book. Henceforward it becomes the central theme around which revolve prescriptions for the exploitation, preservation, and reclamation of those properties of water—physical, chemical, and biological—that are responsible for its extraordinary importance in urban and industrial societies. Within the *use cycle* of water in dwellings and manufactories, water-quality management places upon organized communities the obligation to seek out and purify for distribution natural waters of suitable fitness. Within the *waste cycle* it obliges cities, towns, and industries to send back to the common water resource spent waters or wastewater effluents of acceptable quality. Assigned to water-quality management thereby is a dual yet essentially unitary responsibility for both water supply and wastewater disposal that can prosper only when quality management establishes and honors reasonable and common objectives and necessary standards.

At one end of the quality spectrum of water lie objectives and standards for safe and palatable drinking waters; at the other end are quality requirements for spent waters or wastewater effluents to be introduced into receiving bodies of water or to be disposed of in other ways. Between the two fall quality criteria for bathing, fishing, shellfish harvesting, and irrigation, and for industrial waters of many kinds. They, too, are of concern to this chapter.

Water-quality management, as part of water-resource management, shares a need for public and technological support that is normally available only in a well-disciplined and industrially mature society. In furtherance of quality control, moreover, there must be adequate information not only on the nature and capacity of natural sources of water but also on their physical, chemical, and biological quality. Within the wider meaning of *water-quality management*, finally, there must be an understanding by engineers of the common properties of the many kinds of water on the earth: of brooks and rivers, of lakes and oceans, and of waters welling from the ground and falling from the sky (Fig. 10-1). "A river," in the words of Justice Oliver Wendell Holmes[1]—to which can be added, in good conscience, a lake, a spring, or a well—"is more than an amenity, it is a treasure."

## 10-2 Natural Waters

The source of water determines its inherent quality (Fig. 10-1 and Sec. 1-3). Rainwater absorbs the gases and vapors normally present in the atmosphere (oxygen, nitrogen, carbon dioxide, and rare gases [Sec. 12-1]) and sweeps particulates out of the air when droplets form about them. Salt nuclei (principally chlorides) reach the atmosphere from ocean spray and freshwater cataracts. Once the rain wets the earth's surface, however, it starts to acquire the properties of surface runoff. In *normal* times the composition of *surfacewaters* varies with the topography and vegetation of the catchment area and with

[1] Supreme Court of the United States, No. 16 Original—October Term, 1930. Delaware River Case. Opinion delivered May 4, 1931.

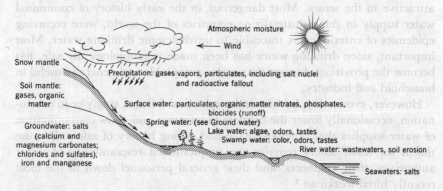

**Figure 10-1.**
**Characteristic properties of natural waters in the hydrological cycle. (See Figure 3-1 for comparisons.)**

land use and management. Both mineral and organic particulates are picked up by erosion, together with soil bacteria and other organisms, while salts and soluble substances are taken into solution. Natural and synthetic fertilizers enter the water along with biocide residues, even though the binding power of soils is remarkably strong. In times of *drought* much of the water flowing in surface channels is derived from underground sources traceable to rainfall that has seeped into the soil. In times of *flooding rainstorms* and *snowmelt*, lands not ordinarily eroded by runoff, and flood plains not usually occupied by surface streams, may contribute large amounts of silt to stream flows. Characteristic additions to water in lakes and ponds are algal and other growths that may produce odors and tastes. Swamp waters contain decaying vegetation that intensifies their color, odor, and taste. Cities and industries add wastes of many kinds.

*Groundwaters* absorb gases of decomposition and degradable organic matter within the pores of the soil through which they percolate. In the living earth rich in organic matter oxygen is removed from groundwaters and carbon dioxide is added. The pH is lowered and some of the soil minerals are dissolved. Calcium and magnesium carbonates, sulfates, and chlorides enter the water and increase its hardness. Iron and manganese, too, may be rendered soluble. Among gases of decomposition in the pores of rich soils are hydrogen sulfide and methane as well as carbon dioxide. Natural filtration of groundwaters removes organic matter and microbic life. Salts remain in solution.

## 10-3 Drinking Water

To slake man's thirst, drinking water must be wholesome and palatable. Accordingly it must not only be free from disease-producing organisms and poisonous or otherwise physiologically undesirable substances, but also attractive to the senses. Most dangerous, in the early history of communal water supply in the industrializing countries of the world, were recurring epidemics of enteric fevers traceable to unwholesome drinking water. Most important, since drinking water has been made microbiologically safe, has become the provision of water that is also acceptable and generally useful in household and industry.

However, even today, human and mechanical failures, singly or in combination, occasionally lower the barriers to infection and allow contamination of water supplies that otherwise have had a long history of safety. Because this is so, water safety remains the unquestioned responsibility of water authorities, their engineers, and their general personnel down to the most recently hired workman.[2]

[2] An example of inadequate water discipline is the outbreak of typhoid fever at Croydon (London), England in 1937. There a workman who happened to be a typhoid carrier failed to obey sanitary rules at work in a well shaft and contaminated the supply causing

Five categories of parasitic organisms infective to man are found in water: bacteria, protozoa, worms, viruses, and fungi. Some of these complete their life cycle by passage through an intermediary aquatic host; others are merely transported by water from man to man.

**Bacterial Infections.** The principal bacterial waterborne diseases of the middle latitudes, typhoid fever and cholera,[3] are two highly specific infections that exacted their awful toll of sickness and death in the cities emerging from the industrial revolution. Paratyphoid (salmonellosis[4]) and bacillary dysentry (shigellosis[5]) as well as hemorrhagic jaundice (leptospirosis), are waterborne diseases in a less direct sense.

**Typhoid Fever and Cholera.** At the turn of the present century the annual death rate from *typhoid fever* still averaged 30 per 100,000 in United States communities. By that time the principal epidemic focus of cholera had more or less retreated to Bengal (India and East Pakistan), where cholera could retain a foothold because of favorable climatic and social conditions. Not until water disinfection by compounds of chlorine (1908 in the United States) and by chlorine itself (1911 in the United States) was added to the armamentarium of sanitary engineers was the incidence of waterborne typhoid fever driven substantially to the vanishing point at less than 1 per million in organized communities of the United States and Europe.[6,7]

**Paratyphoid Fever and Bacillary Dysentery.** Records of waterborne paratyphoid, namely, *typhoidlike* fevers or salmonelloses, are few.[8]

**Leptospirosis.** Leptospirosis, which is also known as hemorrhagic jaundice or Weil's disease, is traceable to swimming or wading in polluted canals, streams, and lakes. Rats and dogs are among the carriers of the spirochetes causing the fairly large group of associated diseases.

---

341 cases of typhoid fever with 43 deaths. Another example is the outbreak of typhoid fever at Rochester, N.Y., in 1940 in which about 30,000 cases of mild enteritis and 5 cases of typhoid fever followed the inadvertent opening of a valve serving an emergency connection between the polluted industrial water supply of the city and its drinking-water supply.

[3] From the Greek *typhos*, meaning stupor arising from fever; and *chole*, bile.

[4] Named for the American bacteriologist Salmon.

[5] Named for the Japanese bacteriologist Shiga.

[6] From 1946 to 1960 there were, in the United States, 39 outbreaks of waterborne typhoid fever, 11 outbreaks with 563 cases of shigellosis, and 4 outbreaks with 24 cases of salmonellosis.

[7] The epidemic of typhoid fever in Zermatt, Switzerland, is of interest because the speed of modern travel permitted victims of the disease to scatter far and wide during its incubation period.

[8] In Madera, Cal., in 1965, both *shigellae* and *salmonellae* were implicated in an outbreak of 2500 cases of infection traced to the contamination of a well by irrigating an adjacent pasture with unchlorinated wastewater effluent. The attack rate was as high as 53.6% for water drawn from a 370-ft well.

***Protozoal Infections.*** Although it is estimated that between 1 and 10% of the United States population are carriers of amebic cysts, the reported incidence of waterborne amebic dysentery (amebiasis) has been low.[9] Among the probable reasons are (1) the relatively small number of cysts excreted by carriers and (2) the relatively large size and weight of the cysts, which account for their natural removal from water by sedimentation as well as by filtration. Relatively massive incursion of pollution into water distribution systems by backflow from house drainage-systems and by cross-connections with unsafe water supplies has generally been associated with waterborne outbreaks.

***Worm Infections.*** The eggs and larvae of intestinal worms may reach water courses from human and animal carriers either directly or in washings from the soil. The eggs or larvae involved are relatively small in number, and the organisms themselves are relatively large in size. Hence worm infections are sporadic and occur only under grossly insanitary conditions or through gross mismanagement of wastewater disposal systems.[10] Irrigation of crops that are consumed raw may transmit any one of the common intestinal worms; irrigation of grasslands may infect cattle and, through them, man.

In the spread of schistosomiasis, which results from the improper disposal of human feces, infection of man does not take place in the water itself, but when larvae released by their snail hosts are forced into the skin from shrinking water droplets as bathers or waders emerge from infected waters. Because the United States and Europe are free from the specific snail hosts of pathogenic schistosomes, they are also free of schistosomiasis. However, the snail hosts and larvae of schistosomes that cause *swimmer's itch*, a skin disease (cercarial dermatitis), do occur in some parts of the United States. They are transported from one body of water to another by infected water fowl.

In some parts of the world the minute crustacean *Cyclops*[11] ingests the larvae of the guinea worm, a nematode or roundworm that infects man through drinking water and is released by him to water again when skin ulcers filled with larvae break while he wades or swims in freshwater.

***Viral Infections.*** Aside from the virus of infectious hepatitis, which is still to be isolated, described, and classified as a specific living entity, mem-

---

[9] Ameba, or amoeba, from the Greek *amoibe*, change, because the cell moves about by changing its shape. Only three major waterborne epidemics of amebic dysentery are on record in the United States. The first was confined to the patrons of two hotels; the second, to firemen and spectators at a stockyard fire; and the third, to workers in an industrial establishment. In the period from 1946 to 1960 there were 2 outbreaks with 36 cases.

[10] Cases in point are the massive infection of the population of Darmstadt, Germany, in 1947 with the roundworm *Ascaris lumbricoides* through spray-irrigation of vegetable crops, and the infection of beef cattle and man with tapeworm.

[11] Named after the one-eyed giant of the Odyssey because of its single eyelike spot.

bers of half a dozen virus groups, constituting over 100 different strains, are known to be excreted in the feces of infected persons. Isolation of these enteric organisms is especially common during the summer months, when streams are normally low. Yet the number of waterborne outbreaks of virus diseases has been small. Moreover, only two of the six groups—the Echo and Reo groups but neither the Polio, Coxsackie A and B, and Adeno groups, nor the hepatitis virus—produce enteric symptoms in man. Why water has not played a more significant role in the spread of these infections awaits explanation. Their small size is favorable to water transport, and the relatively small numbers in which they are probably excreted—apparently of the order of $10^6$ daily per infected person rather than the $10^{11}$ for typhoid organisms—may militate against their completing the cycle from man to man via the water route.

Of the few reasonably well-documented[12,13] waterborne outbreaks of virus infections, two possibly waterborne outbreaks of poliomyelitis in the United States and Canada, and the apparently waterborne outbreak of infectious hepatitis at New Delhi, India, in 1955–56, are commonly listed in the epidemiological roster.

Other gastrointestinal upsets, apparently waterborne, possibly of viral origin, and associated with heavy pollution of water supplies, more especially during periods of severe and prolonged drought, are on record in the United States as well as abroad.[14] Although parasitic fungi occur in water, they do not appear to infect man through water.

## 10-4 Infections from Water-Related Sources

If infections conveyed by excrement as well as by wastewaters and their sludges are taken into account, the ways of spreading enteric disease increase in number and the list of possible infections becomes longer.

Common modes of transmission other than through drinking water are (1) through watercress, or shellfish harvested from or stored in sewage-polluted water (typhoid, paratyphoid, bacillary dysentery, and infectious hepatitis); (2) through vegetables and fruits contaminated by feces, sewage, or sewage sludge (typhoid, paratyphoid, the dysenteries, parasitic worms,

---

[12] Gerald Berg, Ed., *Transmission of Viruses by the Water Route*, John Wiley and Sons, New York, 1966. From 1946 to 1960 the annual number of presumptively waterborne outbreaks of hepatitis in the United States was 1.5, and the number of cases 62.

[13] Engineering Evaluation of Virus Hazard in Water, Comm. on Env. Quality Management, *J. San. Eng. Div. ASCE*, **96**, SA 1, 111, 1970.

[14] Epidemic nonbacterial gastroenteritis affected 5 to 10% of the population (principally children) in the Ruhr Valley during the drought of 1959–60. Arc. Hyg. Bakteriol., **145**, 302 (1961). The United States averaged 8.4 outbreaks and about 900 cases of waterborne gastroenteritis annually between 1946 and 1960. An additional 16 outbreaks with 5160 cases were classified as diarrhea during this 15-year period.

and infectious hepatitis); (3) through exposure to soil contaminated by human dung (hookworm); (4) through all manner of food contaminated by flies and other vermin that feed also on human fecal matter (typhoid, paratyphoid, the dysenteries, and infectious hepatitis); (5) through milk and milk products contaminated by utensils that have been washed in polluted water (typhoid, paratyphoid, and bacillary dysentery); (6) through fish and crayfish from polluted waters eaten raw too soon after salting (flukes and tapeworms); and (7) through bathing or other exposure to polluted waters (leptospirosis and schistosomiasis).

Other chains of infection link (1) tuberculosis to the milk of cows infected by drinking from polluted streams running through pastures below tuberculosis sanitaria; (2) eye, ear, nose, and throat infections to heavily patronized bathing pools even though their waters are well treated and contain appreciable amounts of free chlorine; and (3) enteric diseases, in general, to lack of hand-washing facilities and to primitive methods of excreta disposal.

## 10-5 Reduction of Infections by Water-Quality Management

The accomplishments of water-quality control are best exemplified for water supply, for which cause and effect are usually quite clear. For wastewater this is not so, because there can be no sewerage without water supply and the results of sewerage are masked by the impact of water supply. For discernible effects, one must look, instead, at the record of enteric disease in rural and urban areas. To this purpose Leach and Maxcy[15] recorded the rates of typhoid fever per 100,000 population in communities of different sizes or types, shown in Table 10-1. These statistics they interpreted as showing relatively good sanitary protection (1) in rural areas because of lack of contact and (2) in larger communities because of good community sanitation, including both water supply and wastewater disposal.

An example of what can be accomplished by the introduction and intelligent management of public water supplies is offered by the Commonwealth of Massachusetts. In this state an engineering division was organized in 1886 "to protect the purity of inland waters." Among its accomplishments was an increase in public water supplies and with it the striking reduction in typhoid fever portrayed in Fig. 10-2.

An observation that should be made in the interest of developing countries is that the introduction of water supplies imposes a peculiarly heavy responsibility on water authorities for strict and effective supervision of water quality. Otherwise, drinking water may become the disseminator of enteric infections in large-scale epidemics such as occurred in North America during the second

[15] C. N. Leach and K. F. Maxcy, The Relative Incidence of Typhoid Fever in Cities, Towns, and Country Districts of a Southern State, *Public Health Rept.*, **41**, 705 (1926).

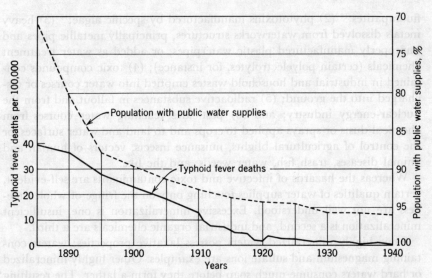

**Figure 10-2.**
Reduction of typhoid fever accompanying the proliferation of public water supplies in the Commonwealth of Massachusetts. (After Whipple and Horwood.)

*Table 10-1*
*Typhoid Fever and Size or Type of Community*[a]

| Size or type | rural | 500–1000 | 1000–2500 | 2500–5000 |
|---|---|---|---|---|
| Morbidity | 52 | 443 | 307 | 180 |
| Size | 5000–10,000 | | 10,000–25,000 | >25,000 |
| Morbidity | 165 | | 118 | 63 |

[a] Morbidity = cases per 100,000 population.

half of the nineteenth century. Important, too, is the realization that epidemics of enteric infections like typhoid fever leave large numbers of chronic carriers in their wake. What should be deduced from American experience is that a rising tide of waterborne disease must have preceded the high-water mark reached by typhoid fever before careful management of public water supplies during the first half of the twentieth century could "dam the flooding tide."

## 10-6 Waterborne Poisons and Other Health-Connected Properties of Water

A variety of poisons may conceivably find their way into water supplies, among them (1) toxic substances leached from mineral formations such as fluorapatites;[16] (2) phytotoxins manufactured by specific algae;[17] (3) heavy metals dissolved from waterworks structures, principally metallic pipes and improperly manufactured plastic waterpipes, or added as water treatment chemicals (certain polyelectrolytes, for instance); (4) toxic compounds contained in industrial and household wastes emptied into water courses or discharged into the ground; (5) radioactive substances in fallout and from the nuclear-energy industry; and (6) pesticides reaching water courses from chemical dusts or sprays applied to crops and to land and water surfaces for the control of agricultural blights, nuisance insects, vectors of human and animal diseases, trash fish, water weeds, and the like.

Whereas the hazards of infective and toxic contaminants are self-evident, certain qualities of water supplies touching only on the fringe of wholesomeness are not well understood. Excessive mineralization is one, insufficient mineralization is a second, and industrial organic chemicals are a third.

Some highly mineralized waters possess laxative properties; waters containing magnesium and sulfate ions are examples. Other highly mineralized or hard waters consume much soap before they form a lather. The resulting causticity irritates the skin of sensitive persons, and *winter chapping* and *dishpan hands* may become chronic complaints.

Iodide and fluoride offer striking examples of insufficient mineralization. Glaciated or otherwise heavily leached soils may not contain enough iodine to satisfy physiological requirements. Goiter is then endemic. However, the remedy for iodine deficiency does not seem to be the introduction of iodides into drinking water, even though this was done at one time with apparent success. Distribution of iodide tablets to school children and, later, the iodization of table salt have proved to be fully as effective and more economical. By contrast, fluoridation of water for the control of dental caries will continue to be justified until a more economical and manageable method of furnishing physiologically needed amounts is brought forward (Table 10-4). Water of high sodium content, including some softened waters, may be troublesome to persons on low-salt diets. Why soft waters appear to be correlated with more cardiovascular disease than moderately hard and fluoridated waters needs further study.

Literally hundreds of new chemical compounds are being introduced into our environment daily. Few of these are assessed for their potential impact

---

[16] The mottling of tooth enamel observed, for instance, in the mid- and southwest United States, is caused by excessive concentrations of fluorides in water. By contrast, small amounts of fluoride are ingested safely and lower the prevalence of dental caries.

[17] The sudden death of cattle after drinking water that supported luxuriant growths of blue-green algae such as *Anabaena flos aquae* and *Microcystis aeruginosa* appears to be caused by a cyclic polypeptide. Ordinarily water of this kind would be so repulsive to man that he would not drink it.

on the health of man, particularly on the synergistic effect they may have when acting together or in concert with other kinds of environmental insults. The impact of the long-term ingestion of low levels of pollutants of water is very difficult to ascertain. Studies made in Holland showed that cancer death rates in municipalities with a water system tended to be lower than in those without a water system and that municipalities receiving their drinking water from polluted rivers had a higher cancer death rate than those taking their water from purer underground sources.[18]

## 10-7 Palatability

To be palatable, water must be significantly free from color, turbidity, taste, and odor, of moderate temperature in summer and winter, and well aerated. At least four human perceptions respond to these qualities: the senses of sight (color and turbidity), taste, smell (odor), and touch (temperature). If the pleasant sound of running water is considered one of its qualities, the sensory appeal of water becomes complete.

**Color and Turbidity.** Color is usually of vegetable provenance, like the *meadow-tea* Thoreau saw in the running brooks of New England. However, water may also become discolored by industrial wastes, natural iron and manganese, and the products of corrosion. To appeal to visitor as well as native, the color of a given water should be low. Yet the accustomed color of a supply may have been quite high in the past, without eliciting comment. Much the same can be said of turbidity. A muddy water, roiled by suspended clay, is more obnoxious to those who do not live where, according to Mark Twain, "a tumblerful of river water contains an acre of land." Turbidity comes from eroding clay banks but also from industrial wastes, products of corrosion, and growths of algae and other plankton organisms.

**Tastes and Odors.** The words taste and odor are often used loosely and interchangeably. Actually there are but four tastes—sour, salt, sweet, and bitter—strictly confined in their perception to the taste buds of the tongue. Odors appear to be without limit in number and are known to change in quality as the concentration of the odorous compounds, or the intensity of their smell, is varied. However, careful screening of odors suggests that there may be certain fundamental odors from which all odors could be compounded. The smallest number in any classification is four: sweet or fragrant, sour or acid, burnt or empyreumatic, and goaty or caprylic.[19]

Generally speaking, tastes and odors should not be sufficiently intense to impress themselves on the user without his knowingly searching for them.

[18] D. A. Okun, Alternatives in Water Supply, *Jour. American Water Works Assoc.*, **61**, 5, 215 (May 1969).

[19] E. C. Crocker, *Flavor*, McGraw-Hill, New York, 1945.

## 10-8 Drinking-Water Quality Standards

In the United States the standards promulgated and revised from time to time by the U.S. Public Health Service are widely accepted by governmental agencies and public utilities, even though conformance is obligatory only for water-supply systems serving common carriers (railways, ships, buses, and aircraft engaged in interstate commerce) and other enterprises subject to Federal quarantine regulations.

The Public Health Service Drinking Water Standards of 1962 include (1) general rules relating to water sources and their protection and (2) specific rules defining required bacteriological quality and acceptable limits of significant physical and chemical characteristics of water delivered to the consumer.

*Bacteriological Characteristics.* For detailed information on the Public Health Service Drinking Water Standards for bacteriological quality, the reader is referred to the official document itself. The 1962 standards for bacteriological quality imply, in general, that to be acceptable, drinking water must not contain more than one *coliform* organism in 100 milliliters (ml). The *coliform* group of organisms is chosen as an indication of dangerous pollution, because the group contains members that are excreted in large numbers from the human intestinal tract. Moreover, they are normally nonpathogenic, unable to multiply outside the human body, and readily identified and enumerated by simple laboratory techniques. The number of bacteriological samples collected from representative points throughout the water-distribution system must be in keeping with the size of the population at risk. Typical minimum monthly numbers are prescribed as follows:

| Population served (thousands) | 1–2 | 10 | 50 | 100 | 900 | 2000 | 4500 |
|---|---|---|---|---|---|---|---|
| Number of samples | 2 | 12 | 50 | 95 | 300 | 400 | 500 |

The U.S. Public Health Service Standards are not necessarily meaningful in other parts of the world. They would be so only if the prevailing endemicity of waterborne infections, the water-use habits of the people, and their economic conditions were substantially the same as they were in the United States in 1962.[20]

*Physical Characteristics.* The physical characteristics of water supplied to a community should be examined at least once a week, samples again being drawn from representative points throughout the system. Turbidity, color, odor, and taste should not be so high as to offend the senses of sight, taste, or smell. Maximum acceptable values for surfacewaters used without

[20] H. A. Thomas, Jr., A Mathematical Model for the Discussion of Social Standards for Control of the Environment, "The Animal Farm," *Jour. Am. Water Works Assoc.*, 56, 9, 1087 (Sept. 1964).

treatment other than disinfection are 5 units of turbidity, 15 units of color, and a threshold odor number of 3. For filtered surfacewaters and for groundwaters, these values are excessively high.

**Chemical Characteristics.** Important chemical characteristics should be determined at least twice a year. More frequent analyses are required when there are reasonable doubts about the constancy of recorded information or when the supply is fluoridated.

Drinking water should not contain impurities in hazardous concentrations, be excessively corrosive, or retain treatment chemicals in excessive concentrations. Substances that are possibly deleterious physiologically should not be permitted to reach the consumer.

When, in the judgment of the reporting agencies and certifying authorities, other more suitable supplies are or can be made available, chemical impurities should not be present in quantities above the concentrations shown in Table 10-2. The presence of toxic substances in excess of the concentrations listed constitutes grounds for rejection of the supply.

**Fluoride.** Fluoride naturally present in drinking water should not average more than the upper limits in Table 10-3; concentrations greater than twice the optimum values constitute grounds for rejection of the supply.

Where fluorides are added to drinking water, their average concentration should be kept within the upper and lower control limits shown in Table 10-3.

**Radioactivity.** Water supplies must not contain more than 3 $\mu\mu$c per l (micromicrocuries or picocuries per liter) of radium-226, nor more than 10 $\mu\mu$c per l of strontium-90, a bone seeker. If they do, surveillance of total intake of radioactivity from all sources should meet the requirements of the Federal Radiation Council. When $\alpha$-emitters and strontium-90 are essentially absent, the gross $\beta$ concentrations must not exceed 1000 $\mu\mu$c per l unless more complete information shows that exposures lie within the limits set by the Federal Radiation Council.

## 10-9 Industrial Water-Quality Standards

Industries may set stricter or more lenient quality standards than are commonly subscribed to by municipalities. Thus the common criteria for cooling waters are normally broader or less specific than for process waters and boiler feed waters. Generally prescribed for *cooling waters* is that they shall neither generate scales or sludges in process or ancillary plant equipment nor support slimes, insect larvae, mussels, or other aquatic organisms in conduits, tanks, and related portions of the cooling system. Cooling waters can be conserved by recycling. Salts and insoluble debris are then kept from building up to unwanted amounts by resorting to continuous or periodic blowdown. Salt

## Table 10-2
*Maximum Concentration of Chemical Substances in Drinking Water*

| Substance | Property | Maximum Concentration, mg/l Allowable | Grounds for Rejection |
|---|---|---|---|
| Alkylbenzenesulfonate (ABS) | taste-producing | 0.5 | — |
| Arsenic (As) | toxic | 0.01 | 0.05 |
| Barium (Ba) | toxic | — | 1.0 |
| Cadmium (Cd) | toxic | — | 0.01 |
| Chloride (Cl) | taste-producing | 250.0 | — |
| Chromium (hexavalent $Cr^{6+}$) | toxic | — | 0.05 |
| Copper (Cu) | taste-producing | 1.0 | — |
| Cyanide (CN) | toxic | 0.01 | 0.2 |
| Fluoride (F)[a] | toxic | — | 1.4–2.4 |
| Iron (Fe) | taste- and color-producing | 0.3 | — |
| Lead (Pb) | toxic | — | 0.05 |
| Manganese (Mn) | taste- and color-producing | 0.05 | — |
| Nitrate ($NO_3$)[b] | toxic | 45 | — |
| Phenols | taste-producing | 0.001 | — |
| Selenium (Se) | toxic | — | 0.01 |
| Silver (Ag) | toxic | — | 0.05 |
| Sulfate ($SO_4$) | taste-producing | 250.0 | — |
| Zinc (Zn) | taste-producing | 5.0 | — |
| Carbon chloroform extract (CCE)[c] | taste-producing | 0.2 | — |
| Total solids | laxative | 500.0 | — |

[a] Varying with air temperatures and equal to twice the optimum values shown in Table 10-3.

[b] When nitrates are known to exceed this amount, the public should be warned that the water may be dangerous for infant feeding.

[c] The carbon chloroform extract may possibly contain toxic substances.

and brackish waters are used as well as fresh water. In fact, about 20% of the water supplied to U.S. industry is brackish. Most of it comes from the sea, but some is pumped from the ground. Saline waters are most readily accept-

## Table 10-3
*Allowable and Recommended Concentrations of Fluoride in Drinking Water*

| Annual Average of Maximum Daily Air Temperatures, F* | Recommended Control Limits in Fluoride Concentrations, mg/l | | |
|---|---|---|---|
| | Lower | Optimum | Upper |
| 50.0–53.7 | 0.9 | 1.2 | 1.7 |
| 53.8–58.3 | 0.8 | 1.1 | 1.5 |
| 58.4–63.8 | 0.8 | 1.0 | 1.3 |
| 63.9–70.6 | 0.7 | 0.9 | 1.2 |
| 70.7–79.2 | 0.7 | 0.8 | 1.0 |
| 79.3–90.5 | 0.6 | 0.7 | 0.8 |

* Based on temperature data obtained for a minimum of 5 years.

able if they are to be used in *once-through* cooling. However, the inland disposal of brackish waters is difficult—so difficult, indeed, as to prohibit drawing heavily on this resource. Deep-well injection may offer the only satisfactory means of disposal.

Most *process waters* must be cleaner than cooling waters. Municipal supplies are generally good enough for process use but not necessarily for boiler feed. By contrast, about 60% of the water developed by industry itself has to be purified in order to meet process needs. In tanneries the proportion is as little as 16%; in the manufacture of photographic supplies it is 100%. Process waters, too, may be either recycled or reused in other processes for which their quality is or can be made suitable. Removal of secondary contaminants may then offer opportunities for salvaging heat and useful product or process components.

To illustrate the scope of quality objectives, the production of canned goods, milk, meats, beverages, and ice normally imposes sanitary requirements on process waters that surpass drinking-water standards. In general this is so because drinking water can be protected by chlorination, and because foods must usually be free from color, tastes, and odors other than their own. Excessive hardness and trace concentrations of iron, manganese, and other metals may also be objectionable. Many paper-making processes, for instance, cannot tolerate even small amounts of iron, manganese, or hardness. By contrast, breweries, distilleries, and bakeries prefer hard waters. In electroplating, trace substances may destroy brightness, corrosion resistance, and other primary properties of deposited metals. Brackish waters that are satisfactory for cooling may be too corrosive for other uses; the chloride tolerance of steel rolling mills, for example, is only 150 mg per l.

294/Water-Quality Management

**Table 10-4**
Quality Tolerances for Industrial Process Waters*

| Industry | Turbidity | Color | Hardness as mg/l of CaCO$_3$ | Alkalinity | Fe + Mn, mg/l | Total Solids, mg/l | Other |
|---|---|---|---|---|---|---|---|
| Food products | | | | | | | |
| Baked goods | 10 | 10 | † | | 0.2 | | a |
| Beer | 10 | | | 75-150 | 0.1 | 500-1000 | a, b |
| Canned goods | 10 | | 25-75 | | 0.2 | | a |
| Confectionery | | | | | 0.2 | 100 | a |
| Ice | 5 | 5 | | | 0.2 | 300 | a, c |
| Laundering | | | 50 | 30-50 | 0.2 | | |
| Manufactured products | | | | | | | |
| Leather | 20 | 10-100 | 50-135 | 135 | 0.4 | | |
| Paper | 5 | 5 | 50 | | 0.1 | 200 | d |
| Paper pulp | 15-50 | 10-20 | 100-180 | | 0.1-1.0 | 200-300 | e |
| Plastics, clear | 2 | 2 | | | 0.02 | 200 | |
| Textiles, dyeing | 5 | 5-20 | 20 | | 0.25 | | f |
| Textiles, general | 5 | 20 | 20 | | 0.5 | | |

* Stated values are general averages only. There is much local variance. Information is derived from *Manual on Industrial Water*, American Society for Testing Materials.
† Some hardness is desirable.
a Must conform to standards for potable water.
b NaCl no more than 275 mg per l.
c SiO$_2$ no more than 10 mg per l; Ca and Mg bicarbonates are troublesome; sulfates and chlorides of Na, Ca, and Mg each no more than 300 mg per l.
d No slime formation.
e Noncorrosive.
f Constant composition; residual alumina no more than 0.5 mg per l.

Specific quality standards for industrial process waters are identified in Table 10–4.

## 10-10 Bathing Waters

The examination of swimming-pool and other bathing waters must have somewhat different objectives from the examination of drinking water. There are distinctions, too, between safeguarding water in swimming pools and at bathing beaches, more particularly saltwater beaches. Swimming-pool water is readily amenable to purification, including maintenance of disinfecting concentrations of chlorine or other halogens; the water at bathing beaches and similar places is not. The use of bathing beaches and outdoor pools, on the other hand, is normally confined to the warm season of the year. Bright sunlight and low incidence of respiratory infections may then combine with less crowding of bathers at most bathing beaches to decrease health hazards.

In the absence of satisfactory tests for nose, mouth, and throat organisms, as well as skin organisms, the 24-hr, 37°C agar plate count serves a useful purpose in the examination of *swimming-pool waters*. It is generally supplemented by tests for coliform bacteria as measures of the effectiveness of bather supervision, all bathers being required to take a cleansing shower before entering the pool or returning to it after using the toilet.

Standards recommended by the Joint Committee on Bathing Places of the Conference of State Sanitary Engineers and the American Public Health Association[21] are examples of responsible thinking.

Reasoning statistically, the bacteriological standards for bathing waters should be more stringent, the longer the bathing season and the more likely the ingestion of bathing water (by small children, for instance). Accordingly, saltwater bathing should be relatively safer and has so been declared in England, for example.[22] Judgment of needed water quality should be based also on the prevalence of enteric disease in the drainage area, information obtained in pollution surveys, and bacteriological examinations. Most studies of disease incidence among bathers are inconclusive.

*Natural bathing waters* should be free from *schistosome cercariae* that can infect man or cause skin irritation (swimmer's itch), and from *leptospirae* of all types. Except for eye infections, possible relationships between virus diseases and bathing waters remain obscure.

---

[21] Joint Committee on Swimming Pools in cooperation with the U.S. Public Health Service, *Suggested Ordinances and Regulations Covering Public Swimming Pools*, American Public Health Association, 1964.

[22] Medical Research Council, *Sewage Contamination of Bathing Beaches in England and Wales*, H.M. Stationery Office, London, 1959; *Bull. Hyg.*, **35**, 635 (1960).

## 10-11 Fishing and Shellfish Waters

Ellis and others have listed five categories or groups of impurities as hazardous to fish life:[23] (1) matter that settles, such as sawdust, and deprives fish of natural foods by depositing a pollutional carpet on the bottom of streams and lakes; (2) substances that exert sufficient oxygen demand to lower the dissolved oxygen (DO) content below the level needed to support fish in their normal spawning, foraging, migrating, and other activities at all stages of development. To flourish at normal water temperatures, most food fish require at least 4, and trout at least 5, mg of DO per ml. As water temperatures rise, the biochemical oxygen demand (BOD) increases, the DO saturation value declines, and the rate of respiration of the fish and their threshold of asphyxiation go up. For a temperature rise of 10°C, for example, the oxygen intake of goldfish increases more than threefold and the point of asphyxiation of trout almost twofold.[24] In the absence of adequate DO, moreover, fish are more susceptible to metallic poisons and other hazards; (3) compounds that lift the pH above 8.4 or drop it below 6.8, more or less, may be directly lethal, and pH changes may throw out of balance the tolerances of fish to high temperatures and low DO concentrations. Acid wastes are especially detrimental; (4) among wastes that increase salinity and, with it, osmotic pressure, are oil-well brines. A specific conductance per cm of 150 to 500 mho $\times$ $10^{-6}$ at 25°C, with a maximum of 1000 to 2000 mho $\times$ $10^{-6}$, is considered permissible; and (5) wastes that contain specifically toxic substances include insecticides such as the chlorinated hydrocarbons and organic phosphorus compounds that have become of special interest because of their wide use in pest control. The death of minnows offers a direct and meaningful screening test for tolerable limits and threshold limits of toxic substances (Sec. 10–21).

Sometimes overlooked in the management of public waters is the fact that some wastewaters contribute to the fertility of aquatic meadows in which fish browse for food.

*Shellfish Waters.* Shellfish—more specifically oysters, clams, and mussels—grown in polluted waters have been responsible for important outbreaks of disease, including typhoid fever and infectious hepatitis. The reasons for this are understandable. The bivalves pass large volumes of water through their gills and other organs and strain out food particles that include living organisms of many kinds. Moreover, some shellfish are eaten raw.

[23] M. M. Ellis, Detention and Measurement of Stream Pollution, *U.S. Bur. Fisheries Bull.*, **22**, 365 (1937); also M. M. Ellis, B. A. Westfall, and M. D. Ellis, Determination of Water Quality, *Dept. Interior Res. Rep.*, **9** (1946).

[24] B. A. Southgate, *Treatment and Disposal of Industrial Wastes*, H.M. Stationery Office, London, 1948.

Accordingly, the saline waters of tidal estuaries in which they are cultivated must be unpolluted.

## 10-12 Irrigation Waters

The quality of irrigation waters is of interest in relation to (1) resource developments in which available waters are exploited in parallel for agricultural and municipal purposes, (2) schemes in which the waters made available for urban use are derived wholly or in part from the underflow of irrigated fields, and (3) wastewater disposal by irrigating agricultural areas either by direct discharge from the drainage system or by diversion of sewage-polluted receiving waters. Narrower in its implications, yet of broad concern, is the quality of municipal waters applied to the parks, lawns, and gardens of the community. McKee and Wolf[25] have summarized the properties of irrigation waters excellently in a fashion suited to the watering of most plants under most conditions.

Because the introduction of wastewater systems accompanies an advancing economy, it is generally possible to limit or dispense with utilization of the fertilizing constituents of wastewaters in order to protect the public health. Accordingly most health authorities prohibit the irrigation of vegetables, garden truck, berries, or low-growing fruits with partially treated or undisinfected municipal wastewaters. The watering of vineyards or orchards where windfalls or fruit lie on the ground is also forbidden. Only nursery stock; vegetables raised exclusively for seed purposes; cotton; and field crops such as hay, grain, rice, alfalfa, fodder corn, cow beets, and fodder carrots are allowed to be watered with municipal wastewaters. However, milk cows and goats are not permitted to be pastured on irrigated land moist with such wastewaters and must be kept away from irrigation ditches that carry them. Even when produce from irrigated areas is to be cooked before consumption, irrigation with wastewaters must be stopped at least a month prior to harvest. Commercial canning of irrigated crops is sometimes permitted under proper control by health authorities. Where water is scarce, full reclamation of wastewaters for watering rather than fertilizing may be justified. The use of wastewater sludges as fertilizers or soil builders is discussed in Sec. 20-16.

## 10-13 Objectives of Water and Wastewater Examination

Basic to successful water-quality management is an understanding of the manageable properties of water under the wide variety of conditions in

[25] J. E. McKee and H. W. Wolf, *Water Quality Criteria*, 2nd ed. (Cal.), State Water Quality Control Board, Publication No. 3A, 1963, p. 106. For further information on agricultural water requirements, the reader is referred to *Handbook 60*, U.S. Department of Agriculture, February 1954, and to T. R. Camp, *Water and Its Impurities*, Reinhold, New York, 1963, p. 130.

which it is found on the earth and put to use by man. Accordingly, water is examined to identify its salient properties and, if need be, their amenity to change. This is not a routine undertaking, even though it may appear to be so. The properties of a given source of water vary with (1) its hydrology, in time and season, as well as distance of passage over and through the soil, and (2) its use, in flow through collection, transmission, purification, and distribution works. In a similar fashion, but in stepped-up intensity, the properties of a given wastewater vary with (1) its use, in flow through collection, treatment, and disposal works, and (2) the hydrology of the bodies of water into which the wastewaters are discharged, in time and season, as well as distance of passage over and through the soil.

Most pronounced are the variations in properties of water and wastewater during purification and treatment and, with them, the requirements for analytical supervision or monitoring of water quality and response. Depending on the objectives of quality management, monitoring information may be provided by (1) automatic sensing or sampling and measuring devices, (2) samples collected in the field and carried to the laboratory for analysis, (3) functional testing of existing or contemplated operational procedures, and (4) research procedures through which new insight is gained into the behavior of water under promising conditions of quality management or control.

*Surveys, Sampling, and Analysis.* A good way to anticipate the probable composition of water samples in preparation for their analysis and to explain the analytical results obtained is to become familiar with the conditions under which these waters occur in nature. The scope of field surveys is dictated by circumstances. A single survey, no matter how carefully made, shows only what conditions were at the times and places of sampling, no more.

*Surveys.* Field surveys are called *sanitary surveys* when they identify watershed conditions affecting or endangering the sanitary quality of water for water-supply purposes; *pollutional surveys* when they determine the effects of wastewaters upon receiving bodies of water; and *industrial-waste surveys* when they establish the volumes and characteristics of effluents from industrial establishments.

*Field surveys* normally include, in addition, observation of gross qualities of the water source, such as growths of water weeds and algal scums, unsightly floating substances, sludge banks and bottom sediments, massive occurrence of fungi and other pollutional populations, and conditions offensive to the senses of smell and sight. In addition, they afford an opportunity to measure properties, such as temperature,[26] and fix chemical constituents, such as

[26] Temperature is informationally so important that it should be recorded whenever water is under observation or test.

carbon dioxide ($CO_2$) and dissolved oxygen (DO), that may change during transportation and storage of samples or otherwise affect laboratory determinations and results.

*Sampling.* A well-conducted field survey may compensate partially, but not wholly, for a paucity of analytical information, by suggesting the possibilities of future happenings. It does not show, as do repeated sampling and analysis, the actual fluctuations in quality taking place from day to day or month to month. Moreover, frequent sampling permits the assessment of mean values and their variance as well as the degree of fluctuation in water quality. Meaningful and reliable sampling assures the validity of analytical findings. To this purpose, samples must fairly represent the body of water or wastewater from which they are taken, and there must be no significant change in the samples between the times of collection and analysis. Special sampling equipment may be required, and samples may have to be composited in time, location, and volume. *Grab samples* may give grossly misleading information. The interpretation of analytical findings is helped or reinforced by parallel sampling not only of component waters and wastewaters but also of environmental constituents such as surface scums, pipe flushings and growths, and bank and bottom flora and fauna, together with sediments or deposits. Meteorological and hydrological observations, including temperature, records of antecedent rainfall and runoff, and droughts and floods, add important information. Both air and water temperatures are informative.

*Analysis.* *Laboratory examinations* are concerned with the analysis of samples collected in the field or in treatment works and from sampling points in water-distribution systems or wastewater-collecting systems. Familiarity with the environment of specific waters and wastewaters and with sampling conditions is helpful in the choice of laboratory tests and in the interpretation of test results. The analysis of a single sample, no matter how complete, also traces but a single cross-sectional pattern of water or wastewater qualities at the time of sampling. Nevertheless certain groups of tests performed on a single sample, such as the tests for the various forms of nitrogen (Sec. 10-17), do provide inferential information on the pollutional history of the water examined. Ordinarily, only multiple sampling and analysis will establish quality profiles in time and space.

## 10-14 Methods of Examination

If the results obtained by different laboratories are to be comparable and have legal validity, suitable methods of collection and analysis must be agreed on by the profession and accepted by government. In the United States, *Standard Methods for the Examination of Water and Wastewater*[27] have been

---

[27] 12th Ed., New York, 1965; herafter referred to as *Standard Methods.*

prepared, approved, and published jointly by the American Public Health Association, American Water Works Association, and Water Pollution Control Federation, and these methods are used in support of water quality standards at all levels of responsible government.

Water-quality standards normally identify the concentration of component properties shown by experience or scientific judgment to be safe, desirable, and acceptable, and to be attainable from available water sources. Effluent-quality standards serve as criteria for the maintenance of acceptable conditions in receiving bodies of water or on land areas. Together these standards must rest on a broad understanding of the regional water and land economies and on the cost of available treatment methods.

## 10-15 Standard Tests

Although many of the tests employed in the examination of water and wastewater samples are identical for both kinds of water, the information sought through their analysis may have very different objectives. The essential purpose of an analysis of municipal water is to detemine raw-water quality, probable need for and response to purification, possible change during distribution, and usefulness in households and industry. The essential purpose of a wastewater analysis is to find the composition, concentration, and condition of the raw wastewater, possible effects on the collecting system, probable response to treatment, and possible influences on receiving bodies of water or land.

Domestic waters are characterized as safe or unsafe, pure or impure, palatable or unpalatable, hard or soft, corrosive or stable, and sweet or saline, as the case may be. Industrial waters are categorized in terms of the purposes they are intended to serve. Wastewaters or effluents are said to be putrescible or nonputrescible, strong or weak, and fresh or septic, to mention but one characteristic in each of the three general categories: composition, concentration, and condition.

Some of the tests included in water and wastewater analyses are more useful and more generally applicable than others. Some are of long standing and forge important bonds with the past; others are more recent in concept. Some give direct information on specific constituents; others are inferential in character.

## 10-16 Examination of Water

Generally speaking, the tests included in the analysis of natural and treated waters in the absence of gross pollution fall into one of the following more or less overlapping categories:

1. Tests measuring or reflecting the *safety and wholesomeness* of water: (a) tests for contamination as measured by the presence of members of the

*coliform group of organisms*, sometimes supplemented by other tests such as *plate counts* in the analysis of swimming-pool waters; (b) tests for toxic forms and quantities of arsenic, boron, bromide, cyanide, chromium, lead, and selenium, and for nitrate (methemoglobinemia) and fluoride (mottled enamel of the teeth); (c) tests for physiologically beneficial amounts of iodide (endemic goiter) and fluoride (dental caries), and possibly dangerous concentrations of organics in carbon-chloroform extracts (CCE), such as benzpyrene and chlorinated hydrocarbons; (d) tests for pollution indicated, for example, by the relative amounts of organic, albuminoid, nitrite, and nitrate nitrogen, and by chloride, sulfate, phosphate, and surfactants; (e) tests for laxative properties in the form of magnesium and sulfate; (f) tests for radioactivity and radioactive substances, especially strontium; and (g) tests for residual chlorine.

2. Tests measuring or reflecting the *palatability* or *esthetic acceptability* of water: (a) temperature, turbidity, color, taste, and odor—sometimes supplemented by (b) microscopic examination for plankton and tests for residues and metals, chloride, active chlorine, hydroxide, tannin and lignin, CCE, sulfite, and hydrogen sulfide, which may explain the origin of observed turbidity, color, odor, and taste.

3. Tests measuring the *economic usefulness* of water—most of the tests listed in *Standard Methods* but depending for their selection on the use to which the water is to be put. For ordinary *municipal purposes* the following tests are important: (a) hardness in relation to soap consumption, hot-water heating, and steam making; (b) dissolved oxygen (DO) as well as hydrogen ion concentration (pH) and carbon dioxide, together with other substances affecting the corrosion of metals; (c) iron, manganese, and other metals as such; and (d) hydrogen sulfide as well as carbon dioxide in connection with the destruction of cement and concrete.

For *industrial purposes* tests may place emphasis on any one or more of the tests normally applied to municipal supplies and in addition, one or more tests identified by specific requirements of the industry (Table 10–4).

4. Tests related in particular to *water treatment processes:* examples, *in addition to pertinent tests for the substances removed or destroyed, or to be removed or destroyed*, are (a) tests for alkalinity, pH, carbon dioxide, and aluminum, iron, and polyelectrolytes in connection with coagulation; (b) tests for odor or taste and copper after the destruction of algae by copper compounds; (c) tests for chlorine demand, pH, and active chlorine before and after chlorination; (d) tests for stability, phosphate, silica, pH, hydroxide, and tannin and lignin following treatment for corrosiveness; (e) tests for hydroxide, sodium, potassium, and silica, in connection with water softening; (f) tests for pH, carbon dioxide, and dissolved oxygen in relation to deferrization and demanganization; (g) tests for pH, hydrogen sulfide, methane, carbon dioxide,

and dissolved oxygen before and after aeration; and (h) tests for salinity in reference to desalination.

5. *Functional tests* performed for the purpose of identifying wanted accomplishments, including (a) some tests conveniently listed also in the preceding paragraph, namely, tests for chlorine demand supported by tests for pH and chlorine residuals, and tests for stability; (b) tests for adsorption of odors, tastes, and other substances on activated carbon; (c) tests (called jar tests) for optimal coagulation with alum, iron, and polyelectrolytes; (d) tests for lime-soda softening of hard waters or for ion-exchange response; (e) tests for desalination by electroosmosis and reverse osmosis; and (f) tests for resistance to straining and filtration. In a sense, the threshold-odor test is also functional, in that it measures both a degree of perception and a response.

## 10-17 Examination of Wastewater

Generally speaking, the tests included in the analysis of raw sewage and industrial wastewaters, treatment plant effluents, and waters polluted by one or more of these, fall into one of the following more or less overlapping categories:

1. Tests measuring or reflecting the *concentration* or strength of sewage and industrial wastewaters: (a) tests for solid matter in its various stages and hence for the potential offensiveness of sewage to the sense of sight—total, suspended, dissolved, and settleable solids, grease, and, in the case of treatment plant effluents, turbidity; and (b) tests for organic matter and, in view of the putrescibility of organic matter, for the potential offensiveness of wastewaters to the sense of smell—volatile components of the total, suspended, dissolved, and settleable solids, biochemical oxygen demand (BOD), chemical oxygen demand (COD), total organic carbon (TOC), sulfide, organic nitrogen, odor, surfactants, and grease. Together these tests measure or reflect the strength of municipal and industrial wastewaters in terms of solids and organics.

2. Tests measuring the *composition* of wastewaters in terms of specific substances or types of substances in addition to those included in the preceding paragraph: (a) tests for the various forms of nitrogen—ammonia, organic (Kjeldahl), nitrite, and nitrate nitrogen; (b) tests for phosphate and other fertilizing substances; (c) tests for dissolved oxygen, chloride, sulfide, acidity, and alkalinity; (d) tests for radioactivity and radioactive substances; and (e) bioassays for acutely toxic wastes.

3. Tests measuring the *condition* of wastewaters and explaining the *progress of decomposition* of organic substances in wastewaters, effluents, and receiving waters: (a) physical, chemical, and biochemical tests—DO, BOD (including relative stability), COD, sulfide, odor, nitrogen in its various forms, pH

value, and temperature; and (b) biological tests—growths of microscopic and macroscopic indicators of pollution, and bacteria (including coliform organisms).

4. Tests related in particular to *treatment processes:* (a) most commonly, tests for the removal of suspended and settleable solids, and for BOD and COD, and other tests such as those for nitrogen in its various forms; (b) in connection with disinfection, most significantly, tests for chlorine demand, pH value, and active chlorine as well as bacteriological tests; and (c) for effluents, most usefully, tests for suspended solids, DO, BOD, and COD, and, possibly, instead of BOD, the relative stability.

5. *Functional tests*, among them, for example, (a) tests for chlorine demand; (b) tests for BOD, in itself a functional test predicting the oxygen requirements of effluents and the degradability of specific substances such as synthetic detergents, more especially when samples of receiving water are employed in making necessary dilutions; (c) tests for the rate of oxygen demand of raw or clarified wastewaters, that is, following primary treatment; (d) tests for optimal coagulation; and (e) tests for adsorption or desalination by ion exchange in water renovation (sometimes called tertiary treatment, that is, treatment following biological or other secondary treatment).

## 10-18 Examination of Wastewater Sludges and Benthal Sediments

Examination of sludges and sediments (1) deposited from wastewaters in collecting systems, (2) generated in biological treatment processes, and (3) commonly removed by physical and chemical treatment processes or accumulated in rivers, lakes, and estuaries, has much the same objectives as the examination of wastewaters from which the sediments and sludges settle or in which they are formed. Accordingly, the tests fall into the following categories:

1. Tests recording the *concentration* of the sludge or sediment: (a) tests for solid matter or water content, and for potential offensiveness to the sense of sight—total residue, moisture, and specific gravity; and (b) tests for organic matter and potential offensiveness to the sense of smell—volatile residue and grease content.

2. Tests identifying the *composition* of the sludge or sediment: (a) tests for organic content—volatile residue, grease, including hydrocarbon and fatty-matter content, BOD, and total and organic nitrogen; (b) tests for pH value, alkalinity, and acidity of the sludge liquor, including water-soluble, water-insoluble, and volatile fatty acids; (c) tests for ammonia; and (d) tests for radioactivity.

3. Tests measuring the *condition* of the sludge or sediment: (a) tests for appearance—descriptive color and odor; (b) chemical tests—BOD and

nitrogen; and (c) biological tests—bottom flora and fauna as a function of the pollutional status of the waters examined.

4. Tests relating to *treatment processes*, more especially to the activated sludge process in the treatment of municipal wastewaters: (a) tests for suspended matter and its settleability and (b) tests for relative sludge volume and density—sludge volume index and sludge density index.

5. *Functional tests*, for instance, (a) tests of the settleability of sediments and sludges; (b) tests of their aerobic degradability by BOD or Warburg determinations; (c) tests of their anaerobic degradability by determination of their rate of gas production and its composition, pH changes, and production of organic acids; (d) tests of their chlorine demand; (e) tests of the heat content of dried sediments and sludges; (f) tests for the filtrability of sludge in anticipation of vacuum filtration or drying; and (g) tests for the viability of crustacea (amphipods) and midge larvae on sediments and sludge; and (h) bioassay of sediments and sludges employing algae and water fleas.

## 10-19 Expression of Analytical Results

The results of chemical analysis are commonly expressed in *milligrams per liter* (mg per l) or as *parts per million* (ppm). For industrial wastes of high specific gravity, however, it is customary to use milligrams per liter only and to add the specific gravity. For concentrations below 1 mg per l, *micrograms per liter* ($\mu$g per l) or *parts per billion* (ppb) may be substituted; for high concentratives (above $10^5$ mg per l), *percentage values* are useful. Other useful expressions are *quantity units*, such as *pounds per day*, equal to lb per 24 hr = 5.39 times mg per l times cfs = 8.3 times mg per l times mgd; and the *population equivalent* (more especially for the BOD) equal to pounds per 24 hr and divided by the pertinent per capita release (such as 0.17 lb per 24 hr per capita for BOD). Mineral analyses may be reported as concentrations of ions, expressing the results in milliequivalents per liter (me per l), equal to the sum of the atomic weights of constituent atoms divided by the number of charges normally associated with the different ions.

Hydrogen ion concentration is stated in terms of the *pH value*, or negative logarithm of the hydrogen ion concentration. Specific conductivity measurements for total dissolved solids, for example, are expressed in mhos or $\mu$mhos per cm, mhos being the reciprocal of resistance, or ohms$^{-1}$. Odor intensity is recorded as the reciprocal of the dilution ratio with odor-free water at which the odor remains just faintly discernible. Color and turbidity are reported in *units of color* and *units of turbidity* corresponding to standard color and turbidity simulants based on 1 mg per l of platinum in potassium chloroplatinate for color and 1 mg per l of silica in *diatomaceous* or *fuller's earth* for turbidity. Bacteriological results are reported, respectively, as *plate counts per milliliter*; *most probable numbers* (MPN) of coliform organisms *per*

100 ml when the serial-dilution techinque is employed; and as coliform colonies per 100 ml when the membrane filter (MF) technique is used. Algae and other plankton organisms are counted as individual cells, or their size, area, or bulk is measured in standard units 20 $\mu$ in length, 400 $\mu^2$ in area, or 8000 $\mu^3$ in volume.[28]

The *specific gravity* of a sludge or sediment is calculated as the ratio of a given volume of sample to the weight of an equal volume of distilled water. The *sludge volume index* (SVI) of activated sludge is the volume (in milliliters) occupied by the sludge containing a unit weight (in grams) of suspended solids after settling the aerated liquor for 30 min; the *sludge density index* (SDI) is the percentage weight of suspended solids in the sludge also after settling the aerated liquor for 30 min. Accordingly, the sludge density index is 100 times the reciprocal of the sludge volume index.[29]

## 10-20 Interpretation of Analyses

The scope of tests employed in the examination of water is purposely broad. The uses of water are many, and the nature of the waste matter discharged into water is varied. Many of the tests introduced in the course of time are interrelated, some throwing light on others. Considered in suitable combinations, they supply the body of knowledge needed to answer the questions proposed in the first two sections of this chapter. However, there are no fixed rules for the interpretation of water analyses. The essential competency that distinguishes sanitary engineers from other civil or hydraulic engineers is one of broad understanding of (1) the composition of water in nature and the significance and behavior of its components and (2) the conditions imposed on water and its many constituents by (a) precipitation, runoff, and percolation, (b) pollution and natural purification, (c) collection, purification, distribution, and use as water, and (d) collection, treatment, and dispersal as wastewater.

[28] The standard unit was proposed as an areal unit by G. C. Whipple, Experience with the Sedgwick-Rafter Method, *Technol. Quart.*, **IX** (Dec. 1896). Whipple (1866–1924) became Professor of Sanitary Engineering at Harvard University in 1911, after a career in practice. He was an expert on waterborne typhoid fever and one of the first American engineers to take fruitful interest in what he called the microscopy of drinking water.

[29] The sludge volume index was proposed by F. W. Mohlman in an editorial of *Sewage Works J.*, **6**, 119 (1934); the sludge density index by Wellington Donaldson in Some Notes on the Operation of Sewage Treatment Works, *Sewage Works J.*, **4**, 48 (1932). Mohlman was Director of Laboratories for the Chicago Sanitary District during many years of intensive investigation of treatment processes in advance of and following construction of large wastewater treatment works. Donaldson was associated with George W. Fuller as a consultant before he became chief of operations for the evolving treatment works of New York City.

## 10-21 Bioassay of Toxic Wastes

Although toxic wasteproducts may originate in household and agriculture as well as in industry, the bioassay of toxicants is performed most frequently on industrial wastewaters because of the variety and quantity in which they are discharged by manufacturing plants. That the concentration of possibly toxic substances is often small makes their detection, separation, and measurement difficult, and sometimes impossible of practical accomplishment. The bioassay then becomes the procedure of choice, and fingerling fish become the preferred test organism. However, it may be desirable to select other aquatic test organisms in special situations.

A governing principle of bioassays is to adopt testing procedures that fit local conditions, employ local varieties of fish or other aquatic animals, and draw experimental water from the receiving water itself. Necessary dissolved-oxygen concentrations are maintained (1) by aerating the water in the test vessels, either in advance of or during the assay, or (2) by renewing the test solution at suitable intervals of time. Volatility, instability, and possible detoxification of the waste under test determine which method of assuring adequate oxygen concentrations to adopt. Other conditions of exposure to be taken into account are temperature, pH changes at different waste concentrations, and synergisms between potential toxicants present in the experimental water and in the added wastewaters.

Many species of fish perform satisfactorily—among them sunfish, bass, trout, salmon, true minnows, and suckers in fresh waters; sticklebacks and killifish in estuarine waters; and members of the genus *Fundulus* in salt waters. The test population is generally 10 in number. Its members are somewhat less than 3 in. in length, and the longest specimen is desirably no more than 1.5 times as long as the smallest. Enough experimental water is provided to assure 1 l of fluid for no more than 2-g weights of fish. Test periods are usually 24 and 48 hr but often also 96 hr long. Signs of distress are noted if that is of interest, and the number of organisms succumbing during exposure is determined at the end of each specified test period.

In assays of acute toxicity the *median tolerance limit*, $TL_m$, is determined as equal to the concentration of the waste in a suitable diluent (experimental water) at which 50% of the test organisms have survived for a specified period of exposure. The $TL_m$ can be read from a semilogarithmic plot of waste concentration (logarithmic) relative to percentage survival (arithmetic), or it is calculated as

$$\log TL_m = \log c_2 + \frac{p_2 - 50}{p_2 - p_1} [\log c_1 - \log c_2] \qquad (10\text{-}1)$$

where $c$ is the concentration of the waste as a percentage by volume, $p$ is the percentage survival of test organisms for a given test period, and the numerical subscripts are related to the observations lying closest to the median (50%) survival. Concentrations of nonaqueous wastes are expressed as milligrams per liter or parts per million by weight added to the experimental water.

Acceptable assays of responses during long-term exposures at low degrees of toxicity remain to be worked out (Sec. 10-6).

## 10-22 Composition of Municipal Wastewater

Wastes from households, hotels, hospitals, restaurants, offices, and mercantile buildings are composited within the sewerage system and produce relatively constant per capita amounts of suspended solids, organic matter, and other substances of special concern in the disposal of sewage. If the nature and capacity of existing industries are known, it is possible to add estimates of the per capita distribution and population equivalent of the industrial wastes. The population equivalent is defined as the ratio of the amount of suspended solids, putrescible matter in terms of BOD, or other significant substances or properties issuing from a given industry, to the per capita amounts of the respective substances normally found in domestic wastewaters. If significant substances are already present in the water supply of the community, in groundwater entering the drainage system, or in storm runoff collected incidentally or purposefully (by combined systems, for example), additional per capita allowances or population equivalents must be brought into play. Common averages for domestic wastewaters are included in Table 10-5. Appreciable departures from these figures must be expected. All of them are affected by the wealth and habits of the population. Dissolved and total solids are a function, too, of the hardness and general mineral content of the water supply and the infiltering groundwater. The BOD population equivalent of stormwater carried into *combined* systems varies widely. A background figure for the average per capita BOD of combined municipal wastewaters lies in the vicinity of 1.4 times the BOD of domestic wastewaters.

Human body wastes (feces and urine) are excreted in quantities varying with age, sex, and nutrition. Fecal matter contains food residues, the remains of bile and intestinal secretions, cellular substances from the alimentary tract, and bacterial cell masses in large amounts (about a quarter of the weight of feces). Bulky foods usually contain much cellulose and indigestible materials derived from roughage. The feces of reasonably well-fed people average about 90 g per capita daily, or 20.5 g on a dry-weight basis.

Of the organic matter in average domestic sewage, about 40% is made up of nitrogenous substances, 50% of carbohydrates, and 10% of fats. The daily contribution of ether-soluble matter is set at 10 to 15 g per capita. The bacterial content of wastewater, its seasonal variation, and its significance as an indicator of pollution are discussed in Secs. 21-3 and 21-4.

### Table 10-5
*Average per Capita Solids, BOD, and COD of Domestic Wastewaters*[a,b]

| State of Solids | Solids | | | 5-Day, 20 C BOD | COD |
|---|---|---|---|---|---|
| | Mineral | Organic | Total | | |
| (1) | (2) | (3) | (4) | (5) | (6) |
| Suspended | 25 | 65 | 90 | 42 | 41 |
| Settleable | 15 | 39 | 54 | 19 | 16 |
| Nonsettleable | 10 | 26 | 36 | 23 | 25 |
| Dissolved | 80 | 80 | 160 | 12 | 16 |
| Total | 105 | 145 | 250 | 54[c] | 57 |

[a] Values are grams per capita daily (1 g per capita = 2.2 lb per 1000 population).
[b] The use of home garbage grinders increases these averages.
[c] 76, i.e., 1.4 × 54 for combined municipal wastes.

The strength or concentration of municipal wastewaters depends on domestic and industrial water use, the general nature and tightness of the system, and the degree of purposeful or clandestine admission of roof and other stormwaters. In the United States the volume of domestic wastewater averages 100 gpcd. Because 1 g of waste substance per capita daily equals $265/Q$ mg per l, where $Q$ is the sewage flow in gallons per capita per day, 1 g per capita daily = 2.65 mg per l, and the average strength of domestic sewage becomes of the order shown in Table 10-6. For flows larger or smaller than 100 gpcd, the concentrations of the various substances are changed more or less proportionately, unless the municipal water or infiltering groundwater contains unusually large amounts of solid matter. The pollutional load remains substantially unchanged, but dilution does reduce the settleability of the suspended solids.

The condition of wastewater at its treatment works or at the outfall of the sewerage system is a function of the time of travel and temperature of the wastewater, the time depending on the length and grade of the collecting system. Long lines, low grades (sluggish flow), and high temperatures destroy the freshness of wastewater. Fresh domestic wastewaters have little odor, are gray in color, and contain dissolved oxygen; their solids retain much of their original bulk. As putrescible constituents decompose, the carrying waters become stale and eventually septic (DO exhausted). The wastewaters turn black and have a foul odor (hydrogen sulfide); floating and suspended solids are disintegrated.

### Table 10-6
*Average Composition of Domestic Wastewater*[a]

| State of Solids | Solids | | | 5-Day, 20 C BOD | COD |
|---|---|---|---|---|---|
| | Mineral | Organic | Total | | |
| (1) | (2) | (3) | (4) | (5) | (6) |
| Suspended | 65 | 170 | 235 | 110 | 108 |
| Settleable | 40 | 100 | 140 | 50 | 42 |
| Nonsettleable | 25 | 70 | 95 | 60 | 66 |
| Dissolved | 210 | 210 | 420 | 30 | 42 |
| Total | 275 | 380 | 655 | 140[b] | 150 |

[a] Values are in mg per l.
[b] 200, i.e., 1.4 × 140 for combined municipal wastewaters.

In the course of the day wastewater fluctuates in strength as well as flow (Fig. 2-4). The volume of flow reaches a maximum about noon, yet its strength is also greatest at this hour. Collection of representative samples is not a simple matter. Daily samples must be composited in proportion to flows if they are to be of average strength. The depth at which samples are taken from conduits is often important, because solids settle and move along the bottom, not unlike the bed load of streams.

eleven

# unit operations and treatment kinetics

### 11-1 Purpose and Prospect of Water and Wastewater Treatment

The variety of operations that bring about the purification of water and wastewater in nature and are called on to do so at an accelerated pace in man-made works is large, and the phenomena with which water-quality management must occupy itself are many. This may explain the high degree of empiricism that characterizes much of the information on which the design of treatment works and their operation must still be based. Yet it has become possible to identify, by carefully controlled laboratory experiments, the fundamental nature of the purification responses of water and to interpret them mathematically. In an important and far-reaching sense, therefore, the concepts and models that have been created for this purpose offer the prospect of codifying patterns of purification behavior shared in common and laying a foundation of purification theory that will strengthen the art of water treatment by giving it the support of a more perfect science.

It is the purpose of the present chapter to introduce the student of water-quality management *first* to the fundamental ways and means, or unit operation, by which water purification is effected and *second* to the formulation of the rates of change and conditions of equilibrium that govern these actions or operations. By proceeding in this way, it will become evident that the study of water treatment is marked by "a quest for unity behind the appearance of the many."[1]

## 11-2 Unit Operations

The concept of unit operations first found expression in the analysis of common procedures in chemical engineering. As suggested by A. D. Little,[2] "Any chemical process, on whatever scale conducted, may be resolved into a coordinated series of what may be called 'unit actions'. . . . The number of these basic unit operations is not very large, and relatively few of them are involved in any particular process." The kind of thinking that emerged from this early pronouncement of principle has contributed greatly to the development of chemical engineering and, in the course of time, also to the advancement of water and wastewater treatment. Among its principal contributions are (1) a better understanding of inherent processes and capabilities in water and wastewater treatment, (2) the development of mathematical and simple physical models or analogs of treatment mechanisms and their use in identifying the basic components of treatment plant design, and (3) the coordination of effective treatment procedures to attain wanted plant performance and effluent quality.

The output of water-treatment works is a *quality water*, that of waste treatment plants an *acceptable effluent*. Available unit operations serving these purposes are classified and discussed in succession from *gas* to *ion, solid*, and *nutrient transfer*, each in its own realm.

Most water- and wastewater-treatment processes bring about changes in the concentration of a specific substance by moving the substance either into or out of the water or wastewater itself. This is called phase transfer. The principal phases are gas, liquid, and solid, but it is possible to identify other phases as well—the vapor phase, for instance—and to recognize specific states—the dissolved and colloidal states, for example.

The analysis of unit operations is generally approached in one of the following ways: (1) through development of a mathematical model and (2) through construction or conceptualization of a simple physical model producing the wanted reaction (Sec. 18–13).

---

[1] One of the four attitudes of mind that Sir John Wolfenden has suggested marked the Greek thinker, the others being a search for clarity, a sense of wonder, and a demand for the rational explanation of things.

[2] A. D. Little, Report to the Corporation of MIT, 1915, quoted in *Silver Anniversary Volume, Am. Inst. Chem. Engrs.*, 7 (1933). Mr. Little (1863–1935) was a pioneer in chemical engineering and industrial management. See A. S. Foust, L. A. Wenzel, C. W. Clump, L. Maus, and L. B. Anderson, *Principles of Unit Operations*, John Wiley & Sons, New York, 1964, p. 4.

## 11-3 Gas Transfer

In gas transfer gases are released or desorbed from water or absorbed or dissolved by water through its exposure to the air or to other atmospheres under normal, increased, or reduced pressures.

Examples are (1) the addition of oxygen by spray or bubble aeration for deferrization and demanganization (water) and for the creation or maintenance of aerobic conditions (wastewater); (2) the removal of carbon dioxide, hydrogen sulfide, and volatile, odorous substances by spray or bubble aeration for odor and corrosion control; (3) the addition of ozone, chlorine, or chlorine dioxide in ozone towers or gas chlorinators for disinfection or odor destruction; (4) the removal of oxygen by evacuation in degasifiers for corrosion control; and (5) the recarbonation of lime-softened water. The release of methane and carbon dioxide from wastewater sludges undergoing decomposition is also a form of gas transfer. Conversely, the oxygenation of water for deferrization and demanganization is also a form of chemical precipitation.

## 11-4 Ion Transfer

Ion transfer can be effected by chemical coagulation, chemical precipitation, ion exchange, and adsorption.

*Chemical Coagulation.* To bring about *chemical coagulation*, floc-forming chemicals are normally added to water and wastewater for the purpose of enmeshing or combining with settleable or filterable but, more particularly, with otherwise nonsettleable or nonfilterable suspended and colloidal solids to form rapidly settling and readily filterable aggregates, or flocs. The coagulants themselves are soluble, but they are precipitated by transfer of their ions to substances in or added to the water or wastewater. In water purification the floc formed is subsequently removed by sedimentation or filtration; in wastewater treatment, floc removal by sedimentation is usually a terminal operation, except when filtration is included as a form of advanced treatment. The most common coagulants are aluminum and iron salts, which, upon solution, form trivalent aluminum and ferric ions. The precipitating ions are provided by naturally present alkalinity or, more rarely, by alkaline chemical additives such as soda ash.

Examples are (1) the addition of aluminum sulfate to water and of ferric chloride to water or wastewater to coagulate colloids and (2) the addition of polyelectrolytes to balance aggregation.[3] Dosing, mixing, and flocculating

---
[3] The addition of coagulating chemicals to wastewater is erroneously, but generally, referred to as chemical precipitation. The term can be traced to the early practice of adding lime to wastewaters rich in iron wastes to produce settling flocs.

or stirring are useful adjunct operations. Byproducts are chemical sludges and their included impurities.

**Chemical Precipitation.** In *chemical precipitation*, dissolved substances are thrown out of solution. The added chemicals are soluble and the ions released react with ions in the water or wastewater to form precipitates.

Examples are (1) flocculation of iron by the addition of lime to iron-containing water or wastewater, the reaction being carried to completion by dissolved oxygen; (2) precipitation of iron and manganese from water by aeration, the reaction being one of oxidation by dissolved oxygen; (3) softening water by the addition of lime to precipitate carbonate hardness and of soda ash to precipitate noncarbonate hardness; and (4) removal of fluoride ions from water by the addition of tricalcium phosphate, or by their precipitation along with magnesium ions in water softening. Here also, dosing, mixing, and flocculating or stirring are needed adjunct operations, and the byproducts are chemical sludges and their included impurities.

**Ion Exchange.** In ion-exchange operations, specific ions in water are exchanged for complementary ions that are part of the complex of a solid exchange medium.

Examples are (1) the exchange of calcium and magnesium ions for sodium ions by passage of water through a bed of sodium zeolite, which is regenerated by brine (base or cation exchange); (2) the exchange of sodium and potassium ions as well as calcium and magnesium ions by synthetic organic cation exchangers and adsorption of the acids produced on other synthetic organic anion exchangers, the cation exchanger being regenerated with acid and the anion exchanger with sodium carbonate. The precipitation of iron and manganese on manganese zeolite and the regeneration of the zeolite with potassium permanganate is, in a sense, an example of surface or contact precipitation rather than ion exchange. Byproducts are spent washes.

**Adsorption.** In adsorption, interfacial forces remove ions and molecules (*adsorbates*) from solution and concentrate them at the interface of *adsorbents*.

Examples are the adsorption of odor- and taste-producing ions and molecules on beds of granular carbon or on powdered activated carbon suspended in water and removed by sedimentation or filtration. Granular carbon beds are regenerated by leaching. Spent powdered activated carbon is normally wasted. Byproducts are leaching fluids or powdered carbon with contained impurities.

## 11-5 Solute Stabilization

Water is stabilized by a variety of operations in which objectionable solutes are converted into unobjectionable forms without removal.

Examples are (1) the chlorination of water for the oxidation of hydrogen sulfide into sulfate; (2) the liming of water or passage of water through chips of marble, limestone, or dolomite for the conversion of carbon dioxide in excess of equilibrium requirements into soluble bicarbonate;[4] (3) the recarbonation of water softened by excess-lime treatment to convert excess lime into bicarbonate; (4) the superchlorination of water or addition of chlorine dioxide for the oxidation of odor-producing substances; (5) the removal of excess chlorine by reducing agents such as sulfur dioxide; (6) the addition to water of complex phosphates to keep iron in solution; and (7) the addition to water of lime, complex phosphates, or sodium silicate to protect metallic surfaces by deposit coatings or otherwise to reduce the corrosive action of water.

## 11-6 Solids Transfer

In order of decreasing size, solids are removed from water by straining, sedimentation, flotation, and filtration.

*Straining.* Screens and racks strain out floating and suspended solids larger in size than their openings. The rakings and screenings are removed for treatment and disposal. Shredding devices combined with racks and screens convert coarse rakings and screenings into fine solids, which are normally returned to the water to be removed later conjointly with other suspended solids by sedimentation.

Examples are (1) the removal from water of leaves, sticks, and other debris by racks and screens, and the straining out of algae by microscreens; and (2) the removal from wastewater of coarse suspended and floating solids by racks, and of finer suspended solids by screens. Byproducts are the rakings and screenings removed for disposal by burial, incineration, or digestion. In-place comminution of solids by cutting screens is an indirect method of straining. The solids become part of the wastewater sludge and are treated and disposed of with it. Filtration is, in part, a screening or straining operation.

*Sedimentation.* To permit the removal of solids from water by sedimentation, the carrying and scouring powers of flowing water, which are functions of its velocity, are reduced until suspended particles settle by gravity to the bottom of holding tanks or basins and are not resuspended by scour.

Examples are (1) the removal of sand and heavy silt from water in settling basins; (2) the collection of heavy mineral solids from wastewaters by differential sedimentation and scour (grit chambers); (3) the removal of

---

[4] *Marble*, a crystalline limestone, and *limestone* itself consist primarily of calcium carbonate; *dolomite* contains both calcium and magnesium carbonate. Lime is calcium oxide (*quick*, *burnt*, or *caustic* lime) or calcium hydroxide (*slaked* or *hydrated* lime).

settleable, suspended wastewater solids in settling tanks; and (4) the removal from water and wastewater of nonsettleable substances rendered settleable by coagulation or precipitation. The byproducts, known as sediment, grit, or sludge, must be removed from the sedimentation devices for treatment and disposal.

*Flotation.* In flotation operations, the transporting power of flowing water is reduced by quiescence or the suspending power of water is overcome by quiescence and sometimes by the addition of *flotation agents*. Substances naturally lighter than water or rendered lighter than water by flotation agents rise to the water surface and are skimmed off. Flotation agents include fine air bubbles and chemical compounds that, singly or in combination, are often hydrophobic wetting and foaming agents.

Examples are (1) the removal of grease and oil from wastewaters with or without the benefit of aeration in skimming tanks or tanks serving the primary purpose of sedimentation; (2) the release of fine bubbles of air into wastewaters either by diffusion of compressed air or by desorbing air dissolved in the wastewater through reduction of the pressure of the overlying atmosphere, the fine air bubbles attaching themselves to suspended particles, imparting buoyancy to them, and lifting them to the surface; and (3) the addition to water and wastewater of flotation agents that attach themselves to suspended solids or attach suspended solids to bubbles of air and lift the particles to the surface. Byproduct skimmings or foam must be removed from the flotation devices and disposed of. Examples of flotation agents are the anionic, neutral, or cationic detergents, and oils, greases, resins, and glues.

*Filtration.* Conceptually, filtration combines straining, sedimentation, and interfacial contact to transfer suspended solids or flocs onto grains of sand, coal, or other granular materials from which the solids or flocs must generally be removed later.

Examples are (1) the filtration of water through naturally permeable formations in groundwater recharge and of wastewater through natural soils in surface or subsurface irrigation; (2) the slow filtration of water and wastewater through beds of sand (usually in natural deposits) that are cleaned by scraping or are allowed to rest and reaerate between dosings; and (3) the rapid filtration of water through beds of sand, coal, or other granules singly or in combinations of stratified layers, the accumulated impurities being scoured from the filter by water alone, air followed by water, or water concurrently with mechanical rakes. The wash water is a byproduct that may be reclaimed or may have to be treated before discharge into a drainage system, into a receiving body of water, or onto land. A fourth example is the filtration of water through relatively thin layers of diatomaceous earth that is normally discarded after each filter run.

## 11-7 Nutrient or Molecular Transfer

In the natural purification of water and wastewater, saprobic organisms convert complex, principally organic, substances into living cell material and simpler or more stable matter, including gases of decomposition; and photosynthetic organisms convert simple, principally inorganic, substances into cell material with the aid of sunlight, with oxygen being a byproduct of their activity.

Examples are (1) the essentially aerobic destruction or stabilization of suspended and dissolved organic matter by saprobic organisms[5] multiplying in polluted receiving bodies of water, (2) the essentially anaerobic destruction or stabilization (also by saprobic organisms) of organic substances deposited on the bottom of polluted receiving bodies of water in appreciable thickness, and (3) the production of algae and large aquatic vegetation in the presence of simple plant nutrients and sunlight. These three actions may occur concurrently in polluted waters, and they do so quite generally in stabilization ponds for wastewaters. Bottom deposits and algal or other blooms are byproducts.

*Interfacial Contact.* Interfacial contact is provided by biologically active flocs or slimes of living organisms that have been generated under aerobic conditions. Putrescible, principally finely divided and dissolved, nutrients are transferred to the flocs, film, slimes, or cell interfaces. Some of the nutrients promote the growth of living cells, some provide energy to the living system. Soluble and stable end products of biological activity are returned to the water.

Examples are (1) the biological treatment of sewage on trickling filters, (2) the aeration of wastewater in activated-sludge units, (3) the filtration of water through slow sand filters (only in part), and (4) the treatment of wastewater on intermittent sand filters and on irrigation fields (likewise in part only). Byproducts are the biological films unloaded intermittently from trickling filters as settleable solids called trickling-filter humus, and more or less continuously from activated sludge units as excess activated sludge. Secondary settling is a common adjunct operation.

## 11-8 Miscellaneous Operations

Miscellaneous operations include the following:

*Disinfection* of water and wastewater by heat, light, or chemicals that kill living, potentially infectious organisms.

*Copper sulfating* of water to control algae.

[5] Saprobic organisms feed on dead organic matter; plantlike members are called *saprophytes;* animallike members are called *saprozoa.*

*Fluoridation* of water for the reduction of dental caries.

*Desalination* of water by *thermal* processes, such as evaporation and freezing, and by *diffusion* or *dialytic* processes, such as electro-osmosis or reverse osmosis.

## 11-9 Solids Concentration and Stabilization

These are operations in which byproduct sludges from water or wastewater works are prepared for disposal. Water-works sludges, however, are seldom so putrescible as to require treatment before discharge onto drying beds or into lagoons, drainage systems, or receiving bodies of water. Component operations include the following:

*Thickening* concentrates sludges by stirring them long enough to form larger, more rapidly settling aggregates with smaller water content.

An example is the thickening of activated sludge to increase its solids concentration 3- to 6-fold in 8 to 12 hr of stirring, with the addition of chlorine, if necessary, to impede decomposition. Displaced sludge liquor is the byproduct.

*Centrifuging* concentrates sludges run into a centrifuge intermittently or continually to separate solids from the suspending sludge liquor, which becomes a byproduct.

*Chemical conditioning* coagulates sludges and improves their dewatering characteristics.

Examples are the addition of ferric chloride and polyelectrolytes to wastewater sludge to be dewatered on vacuum filters.

*Elutriation* washes out of sludges substances that interfere physically or economically with chemical conditioning and vacuum filtration.

An example is the reduction in the alkalinity of digested wastewater sludge and, with it, in the amounts of chemicals that need to be added in advance of filtration. The elutriating water is a byproduct.

*Biological flotation* lifts sludges to the surface by gases of decomposition. This concentrates the sludge.

An example is the flotation of primary sludge in 5 days at 35°C and the withdrawal of the subnatant, which is a byproduct.

*Vacuum filtration* withdraws moisture from a layer of sludge by suction, the sludge to be *dewatered* being supported on a porous medium, such as coiled springs, or cloth on screening.

An example is the dewatering of chemically conditioned, activated sludge on a continuous, rotary, vacuum drum filter. A sludge paste or cake is produced. The sludge liquor removed is a byproduct.

*Air drying* removes moisture from sludges run onto beds of sand or other granular materials. Included moisture evaporates into the air and drains to the drying bed.

An example is the air drying of well-digested sewage sludge on sand beds, a spadable, friable sludge cake being produced. The byproduct is the liquor reaching the underdrains.

*Heat drying* drives off sludge moisture.

An example is the drying of vacuum-filtered, activated sludge in a continuous flash drier. If sludge is to be marketed, its residual moisture content must generally be reduced to less than 10%.

*Sludge digestion* is the anaerobic decomposition of sludges. Digestion is accompanied by gasification, liquefaction, stabilization, destruction of colloidal structure, and concentration, consolidation, or release of moisture. The gases produced generally include, besides carbon dioxide, combustible methane and, more rarely, hydrogen.

Examples are (1) the digestion of settled solids in septic tanks (single- or double-storied) and (2) the digestion in heated, separate tanks of wastewater solids removed from primary or secondary settling tanks, or both. The byproduct combustible gas components are often employed for digestion-tank stirring or for heating, air compression, and other plant purposes. Sludge liquor is the other byproduct.

*Dry combustion or incineration* leads to the ignition and incineration of heat-dried sludges at high temperatures, alone or with added fuel.

Examples are (1) the incineration of heat-dried sludges and (2) the burning of heat-dried sludges on the lower hearths of a multiple-hearth furnace, on the upper hearths of which the sludge to be incinerated is being dried. The end product of incineration is a mineral ash. The stack gases and quenching waters are byproducts.

*Wet combustion* oxidizes wet sludges at temperatures of about 540°F and air pressures of 1200 to 1800 psig. The effluent suspension and exhaust gases are byproducts.

Other unit operations of sludge treatment include conditioning by heating, freezing, or physical flotation and dewatering by pressure filtration (filter pressing).

## 11-10 Coordination of Unit Operations

The unit operations of water purification and wastewater treatment are introduced into treatment works in many different combinations and sequences to meet (1) in water purification works, existing conditions of raw-water quality and requirements of pure-water quality, and (2) in wastewater treatment works, prevailing situations of influent concentration, composition, and condition and specifications of effluent quality. The selection and elaboration of the unit operations to be employed constitutes the *process* or *systems design* of the treatment works.

## Table 11-1
Common Attributes of Water Affected by Conventional Unit Operations and Water Treatment Processes

| Attribute (a) | Aeration (b) | Coagulation and Sedimentation (c) | Lime-Soda Softening and Sedimentation (d) | Slow Sand Filtration without (c) (e) | Rapid Sand Filtration Preceded by (c) (f) | Disinfection (Chlorination) (g) |
|---|---|---|---|---|---|---|
| Bacteria | 0 | ++ | (+++)[1,2] | ++++ | ++++ | ++++ |
| Color | 0 | +++ | 0 | ++ | ++++ | 0 |
| Turbidity | 0 | +++ | (++)[2] | ++++[3] | ++++ | 0 |
| Odor and taste | ++[4] | (+) | (++)[2] | ++ | (++) | ++++[5] −−[6] |
| Hardness | + | (−−)[7] | ++++[11] | 0 | (−−)[7] | 0 |
| Corrosiveness | +++[8] −−−[9] | (−−)[10] |  | 0 | (−−)[10] | 0 |
| Iron and manganese | +++ | +[12] | (++) | ++++[12] | ++++[12] | 0 |

(1) When very high pH values are produced by excess lime treatment; (2) by inclusion in precipitates; (3) but filters clog too rapidly at high turbidities; (4) not including chlorophenol tastes; (5) when break-point chlorination is employed or superchlorination is followed by dechlorination; (6) when (5) is not employed in the presence of intense odors and tastes; (7) some coagulants convert carbonates into sulfates; (8) by removal of carbon dioxide; (9) by addition of oxygen when it is low; (10) some coagulants release carbon dioxide; (11) variable, some metals are attacked at high pH values; (12) after aeration.

**Water Purification.** In order to direct attention to feasible combinations of water-treatment operations, the attributes of water affected by some conventional unit operations and processes of treatment are identified in Table 11-1. There the relative degree of effectiveness of each unit operation is indicated by the number of plus signs (+) up to a limit of four; adverse effects are shown by minus signs (−) also to degree; and indirect effects are shown by parentheses placed around the signs. Limitations and other factors are explained in footnotes.

Combined operations in water treatment are illustrated in Fig. 11-1.

**Wastewater Treatment.** Why the unit operations of wastewater treatment must be applied in different combinations to meet particular conditions

320 / Unit Operations and Treatment Kinetics

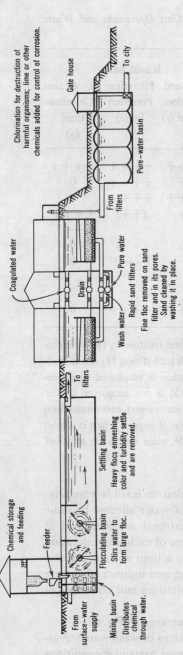 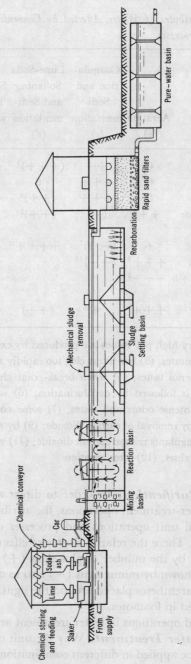

*a.* Filtration plant including coagulation, settling, filtration, chlorination, corrosion control, and pure-water storage.

*b.* Softening plant including addition of softening chemicals, settling, recarbonation, filtration, and pure-water storage.

**Figure 11-1.**
**Typical water-treatment plants.**

of wastewater disposal is evident from Table 11-2. The degree of treatment that can be accomplished by component operations and processes may be gaged from Table 11-2.

### Table 11-2
*Efficiencies of Sewage Treatment Operations and Processes Expressed as Percent Removals*

| Treatment Operation or Process (a) | 5-day, 20 C BOD (b) | Suspended Solids (c) | Bacteria (d) | COD (e) |
|---|---|---|---|---|
| Fine screening | 5–10 | 2–20 | 10–20 | 5–10 |
| Chlorination of raw or settled sewage | 15–30 | ... | 90–95 | ... |
| Plain sedimentation | 25–40 | 40–70 | 25–75 | 20–35 |
| Chemical precipitation | 50–85 | 70–90 | 40–80 | 40–70 |
| Trickling filtration preceded and followed by plain sedimentation | 50–95 | 50–92 | 90–95 | 50–80 |
| Activated-sludge treatment preceded and followed by plain sedimentation | 55–95 | 55–95 | 90–98 | 50–80 |
| Stabilization ponds | 90–95 | 85–95 | 95–98 | 70–80 |
| Chlorination of biologically treated sewage | — | ... | 98–99 | ... |

Discrepancies between generalized values and recorded results are to be expected. Efficiencies cannot be high, for example, when treatment plants are overloaded and part of the wastewater is bypassed, or when sludge cannot be fully treated and some of it must be wasted into the effluent channel. Attention must be paid also to byproduct *sludge liquor* from separate sludge digestion tanks, mechanical sludge-dewatering equipment, sludge elutriation and concentration, or inefficient drying beds. The liquor often has a high BOD and spoils the performance of the plant if it is introduced into the effluent without treatment.

Although any desired degree of purification can be reached by combinations of sequent unit operations, economic considerations ultimately govern the process design. As a general rule, treatment works that include secondary treatment by trickling filters or activated-sludge units may be expected to turn out an effluent having a BOD of 10 to 40 mg per l and a suspended solids content of less than 40 mg per l. Such effluents ordinarily are stable for 10 days or more, in part because of their incipient nitrification. Primary and partial treatment can be secured by modification as well as by selection of different treatment processes.

Combinations of unit operations often used in wastewater treatment practice are illustrated in Fig. 11-2.

**Sludge Treatment.** Common combinations of unit operations in the category of solids concentration and stabilization are (1) digestion of plain-sedimentation sludge followed by air drying; (2) concentration and chemical conditioning of activated sludge in advance of vacuum filtration; and (3) incineration of a mixture of trickling-filter humus and plain-sedimentation sludge after digestion, elutriation, chemical conditioning, and vacuum filtration (see Sec. 20-1 and Fig. 20-1). The waste liquor from wastewater sludges is often putrescible and high in solids. Coagulation and concentration of the removed floc may be necessary in preparing the effluent for discharge.

The common sequences of unit operations employed in wastewater treatment has suggested that some operations (screening, sedimentation, and chemical flocculation or precipitation, for example) constitute *primary treatment;* that other, normally subsequent, operations (notably those associated with trickling filtration and activated-sludge treatment) constitute *secondary treatment;* and that still other operations (not yet clearly so categorized, but possibly including filtration and adsorption, for instance) constitute *tertiary treatment* or *advanced waste treatment.* Whether this classification will prove satisfactory in the long run is open to question.

## 11-11 Water Renovation

There is growing evidence that the need for water by the municipalities, industries, and agriculture of some areas is outstripping the supply of natural waters, that is, waters that have passed through the evaporative and precipitative phases of the hydrological cycle. For this reason the profession is being encouraged to discover ways and means for a fuller purification of all waters that have completed their use cycle in the service of man. On the one hand, pursuance of nature's way of purification by evaporation of all needed waters appears to remain outside the economic reach of present-day society. On the other hand, neither conventional wastewater treatment nor conventional water-purification processes promise to accomplish the equivalent of water evaporation and condensation. The removal of 90 to 95% of the suspended solids, BOD, and COD of wastewaters before the discharge of effluents into receiving waters is not considered enough, nor is the subsequent natural purification of receiving waters for reuse by man accepted as sufficiently rigorous to assure the safety and palatability as well as the general usefulness of such waters. There are new pollutants, among them the wastes from the synthesis of organic chemicals and residues from the dissemination of biocidal chemicals, in the environment. New, also, is the lengthening life

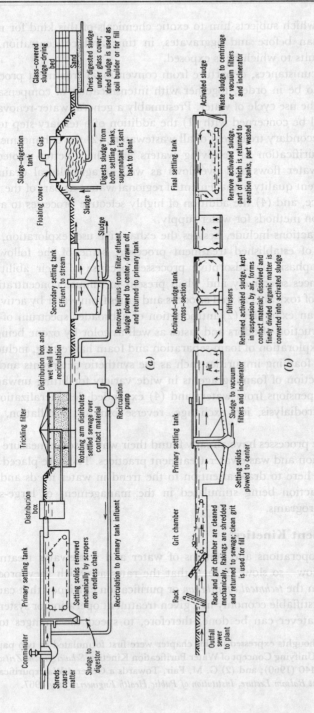

**Figure 11-2.**
Typical sewage-treatment plants. (a) Trickling filter including comminution, final settling, and digestion and drying of sludge. (b) Activated-sludge plant including coarse screening, grit removal, plain sedimentation, contact treatment, and final settling. Sludge is partly dewatered by centrifugation or on vacuum filters and then incinerated.

span of man, which subjects him to exotic chemicals of this kind for many more years than before and aggravates, in time and concentration, the pollutional insults to which he is exposed.

In these circumstances, departure from conventional treatment processes is considered to be in order, together with intensification of compensatory action within the use cycle of water. Presumably a general water-renovation technology will be concerned with (1) the addition of a tertiary step to the primary and secondary treatment of all wastewaters, (2) the enhancement of the natural purification of receiving waters and their better protection against stormwater flows and overflows as well as agricultural drainage, (3) more stringent quality management of regional waters as part of the total natural resource, and (4) the addition of highly selective processes to available purification methods for water supply.

Component actions include, besides the exhaustive use, exploration, and intensification of established treatment processes, some of the following: (1) greater emphasis on adsorptive processes because of their ability to remove substances selectively and when present in minute concentrations, the adsorption of toxicants as well as odors and tastes from water by activated carbons being an example; (2) introduction of a broader spectrum of oxidants, the destruction of odors and tastes as well as color by ozone being an example; (3) exploration of foam separation and foam harvesting, including the removal of foaming impurities such as the synthetic detergents and the purposeful injection of foaming agents in wide variety for lifting unwanted solutes and suspensions from water; and (4) expanded demineralization of water by electrodialysis, ion exchange, reverse osmosis, distillation, and freezing.

Most of these processes have already found their way in some measure into water-purification and wastewater-treatment practices. They are placed in a single category here to draw attention to the trend in water needs and the contemplated action being stimulated in the management of large-scale water-quality programs.

## 11-12 Treatment Kinetics[6]

The principal operations and systems of water and wastewater treatment are generally slow—so slow, in fact, that the rates at which they proceed generally govern the *technical equilibrium* or purification efficiency that can be attained with justifiable economy in a given treatment operation or system of operations. Whatever can be done, therefore, to speed the changes to be

[6] Many of the thoughts expressed in this chapter were first formulated in two papers: (1) G. M. Fair, A Unifying Concept of Water Purification Kinetics, *Schweizerische Zeitschrift für Hydrologie*, **22**, 440 (1960); and (2) G. M. Fair, Towards a General Waterpurification Equation, *The First Balsam Lecture, Institution of Public Health Engineers*, July 1967.

wrought bears directly on the space requirements of necessary treatment systems. Spatial needs, in turn, determine the kind and costs of required structures.

Studies of the time dependence of purification processes and of observed variations under different conditions of exposure point to the following kinetic concepts: (1) *reaction kinetics*, governing not only simple, normally rapid *chemical reactions* but also complex, often slow, and encumbered *biochemical reactions* and changes in the population of associated living organisms; and (2) *transfer kinetics*, governing the transport of substances to and across *phase boundaries* or *interfaces* between adjacent phases, and further distinguishable as the kinetics of *diffusion, sorption, ion exchange*, and *conjunction*, as expressions of transfer that takes place in water and wastewater treatment. Only the kinetics of diffusion, conjunction, and sorption are discussed in the present chapter.

Within limits, the effectiveness of reaction kinetics can generally be enhanced by raising the *temperature* of the mix, and the effectiveness of transfer kinetics in ways such as the following: (1) enlargement of the phase boundary either *statically* through expansion of the interfacial area relative to the volume of water undergoing treatment or *dynamically* through rapid renewal or clearance of the interface by useful *power dissipation* and (2) steepening the *transfer gradient* and with it the *driving force* across the phase boundary either directly by changing the relative concentrations or concentration gradients within the system or indirectly through the introduction of countercurrent flow, flow recirculation, or similar methods.

Ordinarily both reaction kinetics and transfer kinetics need to be supported by interlocking formulations that measure responses to a wide range of physical conditions. Some of these responses can be expressed directly in terms of ambient temperatures or indirectly in terms of their influence on the properties of water substance, for instance; others must account, in suitable terms, for complex longitudinal, that is, time-dependent and process-dependent, environmental changes.

Examples of progressive *natural purification* and of purposeful treatment that can be formulated in terms of *reaction kinetics* are the biochemical oxygen demand (BOD) exerted in the natural purification of streams; the natural *die-away* of pathogens and pollutional indicator organisms in clean and polluted waters; induced and accelerated die-away of pathogens and other living things by heat, chemicals, and destructive radiations; and both aerobic and anaerobic stabilization of wastewaters and wastewater sludges.

Categories of progressive *natural purification* and of purposeful treatment that can be formulated in terms of *transfer kinetics* are exemplified by gas transfer to and from water, wastewater, and wastewater sludges; nutrient transfer to populations of living organisms; adsorption of odor- and taste-producing substances on activated carbon; removal of suspended matter by sedimentation, upflow, and filtration; flocculation for coagulation and precipitation; leaching or washing of flocs or sludges; and ion exchange in the removal of salinity and hardness.

## 11-13 Time of Exposure

Stated in mathematical terms, the mean time of exposure of water to purifying forces is $C/Q$ or $l/v$, where $C$ is the volumetric capacity of the treatment unit, commonly a tank or basin which may or may not contain a contact medium, $Q$ is the rate of flow, $l$ is the length or depth of the treatment unit or reactor or the capacity per unit cross-sectional area, and $v$ is the velocity through the free cross-sectional area $a$ of the unit at right angles to the direction of flow, or the rate of discharge per unit cross-sectional area.[7] Because the path of flow traversed by water during treatment is not the same for each of its molecules, statistical averages of the time of exposure must be adduced to identify the general experience. In this sense median, modal, and arithmetic mean flowing-through periods are each in its own way significant measures of length of exposure. Magnitudes of displacement times range from seconds—in spray aeration, for instance—to minutes in trickling filters and rapid sand filters, hours in activated-sludge units, days in the natural purification of flowing waters, weeks in the anaerobic digestion of wastewater sludges, and months and possibly years in the stabilization of benthal deposits in streams, lakes, and seas.

As charted by dyes, electrolytes, or isotopes, the time of passage of a single dose of suitable tracers through a basin or bed is descibed by frequency curves such as those shown in Fig. 11-3.[8] Under normal conditions of flow, there is both longitudinal displacement and mixing. On the assumption of instantaneous and perfect mixing in the unit as a whole, the change in concentration $dc$ of the tracer in a time $dt$ is proportional to the concentration $c$ remaining after time $t$, or

$$-dc/dt = cQ/C \qquad (11\text{-}1)$$

where $Q$ is the rate of flow, $C$ is the volumetric capacity of the unit, and $C/Q$ is the nominal detention time $t_d$ in the unit. It follows that

$$-\int_{c_0}^{c} (dc/c) = (1/t_d)\int_{0}^{t} (dt) \qquad (11\text{-}2)$$

---

[7] Engineers usually find it convenient to think in terms of the face or approach velocity, which is normally smaller than the interstitial velocity of a filter, for example.

[8] H. A. Thomas, Jr. and J. E. McKee, Longitudinal Mixing in Aeration Tanks, *Sewage Works J.*, **16**, 42 (1944).

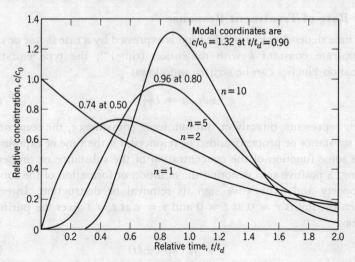

**Figure 11-3.**

**Time of passage and dispersal of a labeled volume of water through a basin or bed as a measure of longitudinal mixing.**

Integrating between the limits $c_0$ at $t = 0$ and $c$ at $t = t$, $\ln c_0 - \ln c = -t/t_d$, or

$$c/c_0 = \exp(-t/t_d) \tag{11-3}$$

Assuming that the degree of mixing taking place in the unit, whether inherent or induced, can be represented by sequent flows with instantaneous and perfect mixing through subdivisions of the treatment unit, Eq.11-3 takes the following form:

$$c/c_0 = [n/(n-1)!]\, n^{n-1}(t/t_d)^{n-1} \exp(-nt/t_d) \tag{11-4}$$

Maximum and minimum analysis of Eq. 11-4 then identifies the relative maximum concentration as

$$c/c_0 = [n/(n-1)!](n-1)^{n-1} \exp[-(n-1)]$$

or, by Stirling's formula[9] for approximating factorials, as

$$c/c_0 = n/\sqrt{2\pi(n-1)} \tag{11-5}$$

and the time required to reach maximum concentration as the relative modal time as

$$t/t_d = (n-1)/n \tag{11-6}$$

[9] Named for James Stirling, seventeenth-century Scottish mathematician, but discovered by Abraham De Moivre (1667–1754), French-English mathematician.

## 11-14 Rate of Treatment Response

If the time dependency of purification is expressed by a rate factor or specific reaction-rate constant $k$ with dimension $(\text{time})^{-1}$, the type equation of purification kinetics can be written as follows:

$$\pm dy/dt = k\phi(y) \tag{11-7}$$

where $y$ represents, directly or through some equivalence, the concentration of the substance or property added or removed, $t$ is the time of exposure, and $\phi(y)$ is some function of the concentration of the substance or property remaining, a positive sign denoting the addition or formation of the substance or property and a negative sign its removal or destruction. Integration between the limits $y = 0$ at $t = 0$ and $y = y$ at $t = t$ gives the purification function

$$y = y_0 \exp(\pm kt) \tag{11-8}$$

when $\phi(y) = y$ and

$$y = y_0[1 - \exp(\pm kt)] \tag{11-9}$$

when $\phi(y) = y_0 - y$.

Equation 11-9 was borrowed in first-order form in 1925 from chemical kinetics by Streeter and Phelps[10] to describe the course of the biochemical oxygen demand of polluted waters.

Much of the water-treatment literature has been concerned with the identification of useful formulations of $\phi(y)$. Important examples are Chick's law for the destruction of bacteria by disinfectants;[11] Adeney and Becker's determination of the rate of solution of nitrogen and oxygen in water;[12] Whitman's and Lewis and Whitman's principles of gas absorption;[13] Fair and Moore's formulation of sludge digestion;[14] Iwasaki's approximation of

---

[10] H. W. Streeter and E. B. Phelps, A Study of the Pollution of the Ohio River, Part III, Factors Concerned in the Phenomena of Oxidation and Reaeration, U.S. Public Health Service, *Publ. Health Bull.*, **146** (1925); but see also E. B. Phelps, Disinfection of Sewage and Sewage Effluents, *U.S. Geol. Surv., Water Supply Paper* 229 (1909).

[11] Harriet Chick, Investigation of the Law of Disinfection, *J. Hyg.*, **8**, 92 (1908). Equation 11-8.

[12] W. E. Adeney and H. G. Becker, The Determination of the Rate of Solution of Atmospheric Nitrogen and Oxygen in Water, *Phil. Mag.*, **38**, 317 (1919). Equation 11-9.

[13] W. G. Whitman, The Two-Film Theory of Gas Absorption, *Chem. Metal. Eng.*, **29**, 146 (1923); W. K. Lewis and W. G. Whitman, Principles of Gas Absorption, *Ind. Eng. Chem.*, **16**, 1215 (1924).

[14] G. M. Fair and E. W. Moore, Heat and Energy Relations in the Digestion of Sewage Solids, Part III, Mathematical Formulation of the Course of Digestion, *Sewage Works J.*, **4**, 433 (1932). Autocatalytic first-order equation.

filtration effects;[15] Fair, Moore, and Thomas' evaluation of the rates of benthal decomposition;[16] and Ives' characterization of filtration performance.[17] These and other processes are discussed individually in succeeding chapters of this book.

It may be postulated that the time-rate factor $k$ will stay constant in those specific processes in which operation is continuous and all conditions of exposure remain unaltered. This is seldom true in water and wastewater treatment. The trend in essentially all operations is eventually toward a dropoff in $k$, rather than an increase or even constancy, as a technical equilibrium of purity, degree of treatment, population change, or saturation concentration is approached. Among changing factors are longitudinal variations in the response of included impurities to treatment, disparities in the nature and extent of the active surfaces with which the flowing water successively comes into contact, differences in temperature, and fluctuations in the hydraulics of the treatment system, including quantity-rate, velocity, and turbulent, transitional, or viscous states of flow.

Time of exposure is often a derivative rather than a fundamental factor. This is illustrated, for example, by the filtration of water through beds of granular materials. In these it is $l/v$, namely, the depth, $l$, or thickness of the layers traversed by specified amounts of water, $v = Q/a$, that is the determinant, and the rate of change is described in an equal, but more pertinent, sense by $dy/d(l/v)$ rather than $dy/dt$. The mass-transfer performance of sedimentation basins under identical loads is similarly a matter of general basin hydraulics rather than detention or displacement time alone.

When treatment response drops off during the course of exposure or while the most responsive impurities are removed or destroyed, a *response factor* attaches to the reaction velocity $k$ in relation either to time $t$ or to the relative amount of treatment accomplished, $y/y_0$. An example of modifying $k$ in relation to time is

$$k = k_0/(1 + rt) = k_0/[1 + r(l/v)] \qquad (11\text{-}10)$$

where $k_0$ is the initial, or starting, rate of reaction and $r$ is a coefficient of retardation.[16] An example of modifying $\phi(y)$ in relation to $y_0$ is

$$k = k_0(1 - y/y_0)^n \qquad (11\text{-}11)$$

---

[15] V. T. Iwasaki, Some Notes on Filtration, *J. Am. Water Works Assoc.*, **29**, 1596 (1937). Equation 11-9.

[16] G. M. Fair, E. W. Moore, and H. A. Thomas, Jr., The Natural Purification of River Muds and Pollution Sediments, *Sewage Works J.*, **13**, 1227 (1941). Equation 11-9 modified.

[17] K. J. Ives, Discussion of paper by D. M. Fox and J. L. Cleasby, Experimental Evaluation of Filtration Theory, *Proc. Am. Soc. Civil Engrs.*, **93**, SA 3, 138 (1967).

where $n$ is the response coefficient of the substances to be removed or destroyed. What Eqs. 11-10 and 11 say is that, as purification progresses, the relative rate of purification $k/k_0$ is a function of the relative amount of material remaining to be removed, changed, or destroyed.

Introducing Eqs. 11-10 and 11 and integrating,

$$y = y_0[1 - (1 + rt)^{-k_0/r}]$$

or

$$y = y_0[1 - (1 + nk_0 t)^{-1/n}] \qquad (11\text{-}12)$$

A generalized plot of Eq. 11-12 is shown in Fig. 11-4. Illustrated are (1) the time rate of purification in terms of the dimensionless products $k_0 l$ and $k_0 t$ and the ratios of impurities remaining or removed relative to their initial concentration and (2) the relative response of the impurities in the course of time as measured by the dimensionless factor $n$ in relation to the initial rate coefficient $k_0$. A value $n = 0$ represents uniform response throughout the treatment process, that is, no change in $k_0$ with time. The larger the value of $n$, the poorer is the relative response to treatment as time progresses. Thus 50% removal is accomplished at time factors of $k_0 t = 0.5, 1, 1.5,$ and 3.75, and removals of 56, 68, 75, and 95% are accomplished at the time factor $k_0 t = 3.0$ when the response factor $n$ is 3, 1.5, 1, and 0, respectively.

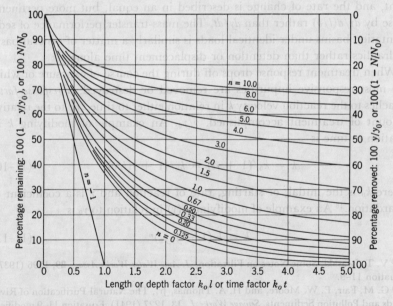

**Figure 11-4.**
**Generalized time rates of purification.**

In terms of Eq. 11–12, the relative amount of a given substance removed or added in a given time can be expressed as the dimensionless ratio

$$(y_0 - y)/y_0 = (1 + nk_0 t)^{-1/n} \qquad (11-13)$$

and the half-life of the reaction $t_{1/2}$, or time required for 50% completion of the reaction becomes

$$t_{1/2} = (2^n - 1)/(nk_0) \qquad (11-14)$$

For integral values of the response coefficient $n$ from $n = -1$ to $n = 4$, Eqs. 11–13 and 14 then yield the following diverse but fundamentally homogeneous results.

| Response Coefficient | Purification Ratio | Half Life of the Reaction | Nature of the Reaction[18] |
|---|---|---|---|
| $n$ | $(y_0 - y)/y_0$ | $t_{1/2}$ | |
| $-1$ | $1 - k_0 t$ | $0.500/k_0$ | Zero order |
| $0$ | $\exp(-k_0 t)$ | $0.693/k_0$ | First order |
| $1$ | $1/(1 + k_0 t)$ | $1.000/k_0$ | Decreasing |
| $2$ | $1/\sqrt{1 + (2k_0 t)}$ | $1.500/k_0$ | rates of |
| $4$ | $1/\sqrt[4]{1 + (4k_0 t)}$ | $3.750/k_0$ | response |

Where living organisms—bacteria and algae, for example—remove impurities by using them as nutrients, this operation becomes an important parameter of the purification process. Substrate utilization is evidenced by the growth of the populations of responsible organisms and the accompanying increase in weight of the biomasses generated by them. For a population increase $dy$ and a substrate utilization $ds$, substitution of the proportionate population growth $dy/ds = y'$ in Eq. 11–9 and integration between the limits $y_0$ at $t = 0$ and $y$ at $t = t$, identifies the substrate utilization as

$$s = (L/y')[1 - (1 + nk_0 t)^{-1/n}] \qquad (11-15)$$

where $L$ is the saturation population or limiting value of $y$. For logistic growth of the population (Sec. 2–4), moreover

$$s_{\text{logistic}} = (L/y')[(1 + m)^{-1} - (1 + m \exp kt)^{-1}] \qquad (11-16)$$

where $m$ is a coefficient of retardation in much the same sense as in the logistic growth of human populations.

[18] These are the well-known reaction orders of chemical kinetics. The percentage removal or transfer of impurities is $100y/y_0$, the percentage remaining being $100(1 - y/y_0)$. The time $t_p$ in which a proportion $p = y/y_0$ of the reaction is completed is $t_p = [(1-p)^{-n} - 1]/(nk_0)$.

## 11-15 Temperature Effects

The response of chemically and biologically activated rate processes to varying operating temperatures of treatment systems is generally identified, within viable or otherwise justifiable temperature ranges, by the van't Hoff-Arrhenius equation, namely,

$$d(\ln k)/dT = E/(RT^2) \qquad (11\text{-}17)$$

in which $k$, as before, is the reaction rate constant, $T$ the temperature in degrees Kelvin, $E$ the activation energy, and $R$ the gas constant. Integration of Eq. 11-17 between the limits $T$ and $T_0$ gives

$$k/k_0 = \exp\left[\frac{E}{R}\frac{(T - T_0)}{TT_0}\right] \qquad (11\text{-}18)$$

where the subscript zero denotes a reference temperature. In a sense this equation is a measure of chemical reaction opportunity and is complementary to Eqs. 11-22 and 23, which connote physical reaction opportunity (Sec. 11-17). As shown by Eq. 11-22, equations of this type may also be temperature-dependent, because viscosity, in this instance, and other important properties of water (density, surface tension, and vapor pressure) change with temperature.[19]

For a temperature difference of 10°C, the ratio of the respective reaction rate constants is often referred to as the quotient $Q_{10}$.

## 11-16 Kinetics of Diffusion

Substances introduced into water tend to penetrate, that is, to diffuse into the space between the water molecules. In *liquid water*, the mean pore space between molecules is about a third of the total and the molecules occur singly or in groups as molecules of $H_2O$ and as hydrogen ($H^+$) and hydroxyl ($OH^-$) ions. In *ice*, the molecules occupy, uniquely, a larger volume, and this reduces the density of ice to 0.93. In the *vapor state* the molecules are widely separated. The rate of diffusion of a given substance increases with its temperature and concentration. *Gases* expand and diffuse easily into air and other gases. They diffuse also into liquid water and through ice. *Liquids* diffuse into liquid water when *miscible* and through ice. For the absorption and precipitation of gases in water see Sec. 12-4.

**Molecular Diffusion of Dissolved Substances.** Even without mechanical mixing, the concentration of substances in true solution in water,

---

[19] As illustrated, for example, by Stokes' law for settling particles, sedimentation and, with it, filtration of water are also temperature-dependent in terms of both water density and viscosity. See Allen Hazen, On Sedimentation, *Trans. Am. Soc. Civil Engrs.*, **53**, 63 (1904).

both molecules and ions, will eventually become uniform. However, this equalization process or diffusion is extremely slow. Fick's law[20] of diffusion, which is analogous to the law of heat conduction (Sec. 20-15), states that the rate of diffusion $\partial W/\partial t$ across an areal boundary $dydz$ is proportional to the concentration gradient $\partial c/\partial x$ of the substance from a point of higher concentration to one of lower concentration, or

$$\partial W/\partial t = -k_d(\partial c/\partial x)\, dy\, dz \qquad (11\text{-}19)$$

Here $W$ is the weight of dissolved substance, $t$ the time, $c$ the concentration, $x$ the distance in the $x$ direction, $dy\, dz$ the area through which the molecules must pass, and $k_d$ a proportionality factor or *coefficient of molecular diffusion*. The magnitude of $k_d$ decreases as the molecular weight increases and changes with temperature in accordance with the van't Hoff-Arrhenius relationship (Sec. 11-15). For gases, furthermore, $k_d$ varies as the square root of their density. Because the concentration gradient decreases as diffusion takes place, Fick's law is written as a partial differential equation. Solution of the equation is by a Fourier series such as the Black and Phelps[21] series:

$$c_t = c_s - 0.811(c_s - c_0)[\exp(-K_d) + {}^1\!/_9 \exp(-9K_d) \\ + {}^1\!/_{25} \exp(-25K_d) + \cdots] \qquad (11\text{-}20)$$

Here $c_s$ is the saturation concentration of the dissolved substance, $c_0$ and $c_t$ are its concentrations at time zero and time $t$, respectively, and

$$K_d = \pi^2 k_d t/4x^2 \qquad (11\text{-}21)$$

The coefficient of diffusion is usually expressed as grams of solute diffusing through 1 cm$^2$ in 1 hr when the concentration gradient is 1 g per ml per linear cm. The dimensions of $k_d$ are then $[l^2 t^{-1}]$, and $K_d$ is dimensionless. A few coefficients of molecular diffusion are listed in Table 13 of the Appendix, in which $k_d$ is given in square centimeters per hour and square centimeters per day.

For oxygen dissolved in water, the diffusion coefficient is $9.4 \times 10^{-2}$ cm$^2$ per hr at 25°C, and changes approximately as follows within normal water temperatures:

$$k_d = (8.7 \times 10^{-2}) \times 1.016^{T_c - 20} = 8.7 \times 10^{-2} \exp[0.159(T_c - 20)]\ \text{cm}^2/\text{hr} \qquad (11\text{-}22)$$

The diffusion coefficient of carbon dioxide in water is $6.2 \times 10^{-2}$ cm$^2$ per hr at 18°C.

[20] A law proposed by Adolph Fick (1829–1901), German physiologist.

[21] W. M. Black and E. B. Phelps, Report to the Board of Estimate and Apportionment, New York, 1911. Black was an officer in the Corps of Engineers; Phelps was a member of the U.S. Public Health Service and later professor of public-health engineering at Columbia University and at the University of Florida.

## Example 11-1.

The initial dissolved oxygen concentration of a quiescent body of water 1 ft (30.48 cm) deep is 3.0 mg/l. Find the concentration after $9\frac{1}{3}$ days if the temperature of the water is 17.5°C, on the assumption that the rate of absorption of oxygen from the atmosphere is sufficiently fast to saturate the surface layer of the water in a relatively short time.

For a coefficient of diffusion $k_d = 8.7 \times 10^{-2} \times 1.016^{17.5-20} = (8.4 \times 10^{-2})$ cm²/sec, Eq. 11-21 states that

$$K_d = [\pi^2(8.4 \times 10^{-2}) \times 9.3 \times 24]/[4(30.48)^2] = 5 \times 10^{-2}$$

At 17.5°C the dissolved oxygen saturation value of the water exposed to the atmosphere is 9.5 mg/l (Table 17, Appendix) and Eq. 11-20 gives

$$c_t = 9.5 - 0.811(9.5 - 3.0)[\exp(-0.05) + \tfrac{1}{9}\exp(-0.45) \\ + \tfrac{1}{25}\exp(-1.25) + \ldots]$$

or

$$c_t = 9.5 - 5.3 \times 1.036 = 5.5 \text{ mg/l}$$

That is, it takes $9\frac{1}{3}$ days to increase the oxygen concentration at the 1-ft level by $5.5 - 3.0 = 2.5$ mg/l when the increase is by molecular diffusion alone.

## 11-17 Conjunction Kinetics

Conjunctions can take place at size levels of ions, molecules, colloids, and suspensions. For colloids—in a more general sense, therefore, particles suspended in a liquid—von Smoluchowski[22] has shown that the conjunction of individual particles is the result of *perikinetic* and *orthokinetic* motion.[23] Because the regulation of hydraulic flow patterns contributes little to perikinetic conjunction but is the primary and manageable factor in orthokinetic conjunction, as well as being directly significant in the chemical coagulation of water,[24] von Smoluchowski's mathematical model of the orthokinetic

---

[22] Miron Smoluchowski, Drei Vorträge über Diffusion, Brownsche Molekularbewegung and Koagulation von Kolloidteilchen (Three Lectures on Diffusion, Brownian Motion, and Coagulation of Colloidal Particles), *Phys. Z.*, **17**, 557 (1916); Versuch einer mathematischen Theorie der Koagulationskinetik Kolloider Lösungen (Trial of a Mathematical Theory of the Coagulation Kinetics of Colloidal Solutions), *Z. Physik. Chem.*, **92**, 129, 155 (1917).

[23] From the Greek *peri*, near; *ortho*, proper or correct; and *kinetic*, of or resulting from motion, and referring, respectively, to Brownian motion, the zigzag random motions of small particles suspended in a fluid, and their more systematic hydraulic transport.

[24] As shown later, the principles of orthokinetic conjunction are essential parts of other operations, too.

coagulation of colloidal solutions is of much importance in water and wastewater treatment.[25]

For $n_i$ particles of diameter $d_i$ and $n_j$ particles of diameter $d_j$ suspended in a liquid in which the velocity gradient in a given direction $z$ is $dv/dz$ and constant, von Smoluchowski arrives at the following number of conjunctions $J_{ij}$ of $i$ with $j$ particles:

$$J_{ij} = \tfrac{1}{6} n_i n_j (d_i + d_j)^3 \, dv/dz \tag{11-23}$$

If the velocity gradient is not constant throughout the system, the point velocity differential $dv/dz$ must be replaced by the temporal mean velocity gradient $\overline{dv/dz}$, which is given the symbol $G$ as a shear gradient. For average conditions of shear, therefore,

$$J_{ij} = \tfrac{1}{6} n_i n_j G (d_i + d_j)^3 \tag{11-24}$$

As a velocity or shear gradient, $G$ has the dimensions $[G] = [lt^{-1}/l]$ or $[t^{-1}]$ and is generally expressed in sec$^{-1}$. Its magnitude is a function of the useful power input, $P$, relative to the volume, $C$, of the fluid and a proportionality factor, $\mu$, which has the same dimensions as the absolute viscosity and is equal to it in laminar flow. Dimensionally, therefore, $G = \phi(P, C, \mu)$, or designating dimensional relations by square brackets $[G] = [P^x, C^y, \mu^z]$ and introducing the fundamental units of mass, $m$, length $l$, and time $t$ of the various parameters into this equation, $[lt^{-1}l^{-1}] = [t^{-1}] = [m^{x+z}l^{2x+3y-z}t^{-3x-z}]$, and solving for $x$, $y$, and $z$,

$$G = \sqrt{P/(C\mu)} \tag{11-25}$$

as shown by Camp and Stein.[25] Equation 11-24 then becomes

$$J_{ij} = \tfrac{1}{6} n_1 n_2 \sqrt{P/(\mu C)} (d_i + d_j)^3 \tag{11-26}$$

and states that the number of contacts per unit time and volume increases with the number and size of particles, the power input per unit volume, and the temperature of the fluid ($\mu^{-1}$). As particles conjoin, however, their number decreases rapidly. The effective or relative binding energy also declines, and the shearing action of the fluid may eventually overbalance any further tendency for particle aggregation and even break up agglomerates that have been well formed.

The *power dissipation function* $P/C = \mu G^2$ is an important unifying concept in modern water and wastewater treatment operations or processes. If $G$, the mean temporal shear or velocity gradient, is combined with the mean or dis-

[25] See also T. R. Camp and P. C. Stein, Velocity Gradients and Internal Work in Fluid Motion, *J. Boston Soc. Civil Engrs.*, **30**, 219 (1943).

placement time, $t_d = C/Q$, where $Q$ is the rate of flow, the following dimensionless measure of conjunction opportunity is obtained:

$$Gt_d = (C/Q)\sqrt{P/(\mu C)} = \sqrt{(PC)/\mu}/Q \qquad (11\text{-}27)$$

It follows that $Gt_d$ is, in a sense, a ratio of power-induced rate of flow to displacement-induced rate of flow. Other implications can be read into modifications of these relationships. Some of them are developed in connection with specific treatment units.[26] For a given value of $Gt_d$, for instance, the loading of flocculation units is

$$Q/C = 1/t_d = \sqrt{P/(\mu C)}/Gt_d \qquad (11\text{-}28)$$

thus implying that the hydraulic loading of flocculation units is not merely a function of their relative capacity or detention time but, more significantly, a function also of useful power input and of viscosity. Power input thereby becomes a manageable variable in numerous treatment operations.

Generally speaking, power is dissipated usefully in treatment processes (1) for the purpose of increasing or decreasing the size or concentration of substances within a given phase or (2) for transferring substances from one phase to another: from the solid to the liquid phase or the reverse, and from the liquid to the gas phase or the reverse. Because quantitative change is effected by conjunction contact and because transfer or phase change is promoted by interfacial contact, the creation, renewal, or energizing of needed conjunctions or interfaces has become a governing component of modern treatment procedures.

## 11-18 Adsorption Kinetics and Equilibria

Ordinary, or physical, adsorption is usually rapid. It is reversible, and a condition of equilibrium between adsorbed and dissolved *adsorbate* is reached soon after contact with the *adsorbent*. However, when the adsorbent is a porous solid, like granular rather than powdered activated carbon, or when the concentration of the adsorbate is small, full contact may be limited by diffusional or other transport processes. The attainment of equilibrium is then delayed correspondingly.

*Adsorption Kinetics.* The adsorption of solutes by porous adsorbents proceeds consecutively as follows: (1) the adsorbate moves from the bulk of the solution to the outer shell of the adsorbent; and (2) adsorbents with relatively large exterior surface area—powdered activated carbon, for example— remove much of the sorbate, whereas adsorbents with relatively small

---

[26] Experience has shown that optimum values of $G$ for flocculation of common coagulants, for example, appear to lie between 20 and 25 sec$^{-1}$ with useful detention periods of 10 to 100 min, or useful magnitudes of the dimensionless product $Gt_d$ between $10^4$ and $10^5$ (Sec. 15-2).

exterior surface area and relatively large interior surface area—granular activated carbon, for example—remove little of the sorbate at their exterior surface. The bulk of the sorbate is carried into and through the pores of the adsorbent by *intraparticle* transport or diffusion.[27] Only when power input is low does transport of the adsorbate to the adsorbent become a controlling factor.

A study of the observational data included in Fig. 11-5 suggests that the rate of adsorptive removal of responsive impurities is closely approximated by Eq. 11-12 in the following form:

$$y/y_0 = 1 - (1 + nk_0 t)^{-1/n} \qquad (11\text{-}29)$$

where $y/y_0$ is the proportion of adsorbate removed in $t$ days.

**Conditions of Equilibrium.** Positions of equilibrium in adsorption are given by *adsorption isotherms*, which relate the quantity adsorbed per unit of adsorbent to the concentration of adsorbate (Fig. 11-6). Two mathematical formulations of these isotherms are in common use for adsorption from aqueous solution. One, the Freundlich[28] equation, is

$$y/m = Kc^{1/n} \qquad (11\text{-}30)$$

in which $y/m$ is the quantity adsorbed by a unit weight of adsorbent, $c$ is the equilibrium concentration of adsorbate in solution, and $K$ and $n$ are empirical constants. Values of $n$ are normally greater than unity, suggesting that adsorption is relatively more efficient at low concentrations. The logarithmic form of the Freundlich equation,

$$\log (y/m) = \log K + 1/n \log c \qquad (11\text{-}31)$$

indicates a linear variation of log $(y/m)$ with log $c$. Accordingly, logarithmic plots of adsorption data can be used for correlation and interpolation and for evaluation of $K$ and $n$.

The second equation, the Langmuir[29] equation, is based on assumptions of

---

[27] With powdered carbon <300-mesh in size of separation, solution transport seems to be the rate-determining step at most reasonable stirring speeds. By contrast, intraparticle transport is normally the slowest and, therefore, rate-determining step in the uptake of organic substances, even in rapidly stirred systems. W. J. Weber, Jr., and J. C. Morris, Kinetics of Adsorption on Carbon from Solution, *J. Sanit. Eng. Div., Am. Soc. Civil Engrs.,* **89**, 31 (1963); Adsorption in Heterogeneous Aqueous Systems, *J. Am. Water Works Assoc.,* **56**, 447 (1964).

[28] H. Freundlich, *Kapillarchemie*, Akademische Verlagsgesellschaft, 1922, p. 232, revised from van Bemmelen, *Mitt. Landw. Versuchsst.*, **35** (1888).

[29] Developed by Irving Langmuir (1881–1957), noted scientist who received the Nobel Prize for his work on surface films (Sec. 4-12).

*338/Unit Operations and Treatment Kinetics*

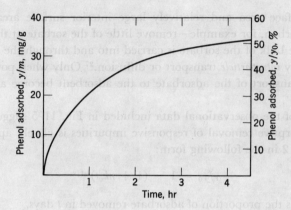

**Figure 11-5.**
Rate of adsorption of phenol by granular activated carbon. Dosage 35 mg per l of 100/140-mesh (0.13-mm) granular carbon. Initial phenol concentration 2.3 mg per l; 4-hr concentration 1.15 mg per l.

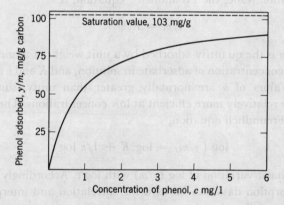

**Figure 11-6.**
Equilibrium adsorption of phenol by granular activated carbon. Room temperature, 50/60-mesh (0.273-mm) granular carbon.

a monomolecular layer of adsorbent, uniformity of adsorbent surface, and no interaction between adsorbate molecules. It has the form

$$y/m = (ab \cdot c)/(1 + a \cdot c) \qquad (11\text{-}32)$$

The constants $a$ and $b$, although they have theoretical significance within the limits of the assumptions, are usually determined empirically. The equation predicts a linear variation of $y/m$ with $c$ at low concentrations when $a \cdot c \ll 1$

and a limiting adsorptive capacity, $y/m = b$, when $a \cdot c \gg 1$ at high concentrations.

The Langmuir equation can be converted into linear form in a number of ways. A convenient representation for evaluating the limiting adsorptive capacity, $b$, is

$$\left(\frac{y}{m}\right)^{-1} = \frac{1}{b} + \frac{1}{ab}\left(\frac{1}{c}\right) = \frac{1}{b}\left[1 + \frac{1}{a}\left(\frac{1}{c}\right)\right] \qquad (11-33)$$

The extent of adsorption of a particular adsorbate on an adsorbent is also dependent on the temperature and on the presence of other adsorbates in the solution. Adsorption usually decreases with increasing temperature, but its extent is not readily predictable for adsorption from aqueous solutions and has not been found to be significant for the range of temperatures ordinarily encountered in water treatment. If other adsorbates are present, they tend to decrease adsorption of a particular adsorbate by competing for space on the surface.

As a unit operation, adsorption is generally not confined to the removal of single substances. However, the interactions at play may be competitive. Rapid adsorption of one substance conceivably reduces the number of sites remaining open to other substances. An example of mutual inhibition in the sorption of competing adsorbates is the less effective transport of phenolic odors from water that contains much organic color. The overall rate of removal of organic matter may drop off in such circumstances.

Generally speaking, power is dissipated usefully in treatment processes (1) for the purpose of increasing or decreasing the size or concentration of substances within a given phase or (2) for transferring substances from one phase to another: from the solid to the liquid phase or vice versa, and from the liquid to the gas phase or vice versa. Because quantitative change is effected by conjunction contact and because transfer or phase change is promoted by interfacial contact, the creation, renewal, or energizing of needed conjunctions or interfaces has become a governing component of modern treatment procedures.

## 11-19 Countercurrent Operation

Figure 11-7 illustrates the advantages to be gained by countercurrent operation of suitable transfer processes. To be transferred, in this example, is heat from water flowing through the annulus of a large outer pipe to cold water moving through a coaxial inner pipe. A heat exchanger on a heated sludge-digestion tank is a close example.

In *cocurrent* operation a high initial temperature difference $[(T_{H_0}) - (T_{C_0})]$ between the influent hot and cold waters is eventually reduced to a lower

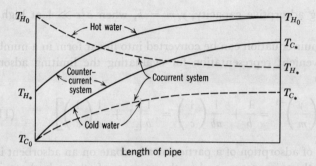

**Figure 11-7.**
Cocurrent and countercurrent heat exchange (not to scale). Subscript $H$ denotes heating medium; subscript $C$, water being heated; subscript 0, initial temperatures; subscript asterisk, final temperatures.

temperature difference $[(T_{H_*}) - (T_{C_*})]$ at a final temperature $(T_{C_*})$ somewhere between the temperatures at the outset.

In countercurrent operation, by contrast, the same temperature difference between the influent hot and cold waters $[(T_{H_0}) - (T_{C_0})]$ is replaced by a gain in the temperature $[(T_{C_*}) - (T_{H_*})]$ of the effluent cold water over the effluent hot water.

The flow of heat at any point along the coaxial pipes is given by the equation

$$dq = k\Delta T \, dA \tag{11-34}$$

where $q$ is the rate of heat flow (British thermal units per hour, for instance), $k$ is the overall heat transfer coefficient (such as British thermal units per [hour] [square foot] [degree F]), $\Delta T$ is the temperature drop or difference (degrees F), and $A$ is the area of the transfer surface (square feet). Accordingly, the transfer of heat in the way it is done in the heat exchangers of sludge digesters is understandably much more effective in countercurrent operation than in cocurrent operation.

Other simple examples of countercurrent operations are (1) ozonizing towers, in which water droplets move downward against upward-flowing ozonized air, (2) upward passage of water through a bed or column of hydraulically stratified sand (Sec. 14-24), and (3) the adsorption of organics onto a bed or column of unigranular activated carbon that has been in service for some time (Sec. 15-5). The elutriation of digested wastewater sludge prior to chemical coagulation is a somewhat less complete (stepwise) countercurrent process (Sec. 20-8).

## 11-20 Recirculation

In a sense, recirculation is a form of countercurrent operation that adds uniformity and flexibility especially to wastewater treatment operations. Besides distributing the load of applied impurities more effectively, it affords an opportunity for smoothing out the rate of applied flow through adjustment of the rate of recirculation (Fig. 19–5). In the best circumstances the rate of flow can, indeed, be held substantially constant. At the same time the quality of the effluent is altered appreciably by recirculation. This can be exemplified by contrasting the operation of a treatment unit that receives a unit quantity of applied water on a once-through basis with a unit that treats the same quantity of applied water on a recirculation basis when a unit quantity of effluent is added to the incoming flow. In the first instance the effluent has been produced by exposure of all the influent for the full treatment time. In the second instance the effluent is a composite of ½ the influent exposed for ½ the treatment time, ¼ the influent exposed for the full treatment time, ⅛ the influent exposed for 3⁄2 the treatment time, 1⁄16 the influent exposed for twice the treatment time, and the like. It is seen that the two effluents are by no means identical, although the average exposure of the influent to treatment is the same with and without recirculation of effluent.

Both flow equalization and load equalization are important. Units may be flooded, and units may be laid still. Treatment efficiencies drop in either event. Moreover, excessive flows may have both direct and indirect consequences in that conditions of exposure as well as times of exposure are altered unfavorably. Units may, indeed, be overwhelmed when pollutional loadings rise suddenly and intensely. Shocks of toxic wastes are cases in point.

If the rate of inflow is $I$ and the rate of recirculation is $R$, the recirculation ratio is $R/I$, and the average number of passages, $N$, of the inflow through a treatment unit is

$$N = (I + R)/I = 1 + (R/I) \qquad (11\text{-}35)$$

In accordance with the reasoning pursued in the preceding paragraph, furthermore:

$$N = 1\frac{I}{I+R} + 2\frac{I}{I+R}\left(\frac{R}{I+R}\right) + 3\frac{I}{I+R}\left(\frac{R}{I+R}\right)^2 \cdots$$

or

$$N = \frac{I}{I+R}\left[1 + 2\frac{R}{I+R} + 3\left(\frac{R}{I+R}\right)^2 \cdots\right] = 1 + \frac{R}{I} \qquad (11\text{-}36)$$

If the removability of impurities is assumed to decrease as the number of passages increases, a weighting factor, $f_w < 1$, must be introduced into Eq. 11-36 to obtain a satisfactory expression for the average number of *effective* passes, $N_e$, of the putrescible matter through the treatment unit (see also Sec. 11-14). For a constant value of $f_w$, which is not necessarily so, Eq. 11-36 then becomes

$$N_e = \frac{I}{I+R}\left[1 + 2f_w\left(\frac{R}{I+R}\right) + 3f_w^2\left(\frac{R}{I+R}\right)^2 \cdots \right]$$

or

$$N_e = \frac{1 + (R/I)}{[1 + (1-f_w)(R/I)]^2} \qquad (11\text{-}37)$$

If $N_e$ and $R/I$ are the dependent variables, and $f_w$ is constant, $N_e$ reaches a maximum value at $dN_e/df_w = 0$, $d^2N_e/(df_w)^2$ being negative, or

$$R/I = (2f_w - 1)/(1 - f_w) \qquad (11\text{-}38)$$

In practice, recirculation has established itself as an important alternative in the operation of trickling filters at high rates of wastewater dosage. A modicum of recirculation is built into the conventional activated-sludge process in its own right, and higher rates of recirculation are bound to find special uses.

# twelve

# aeration and gas transfer

## 12-1 Sources of Gases in Water

Within the hydrological cycle, freshwater is exposed to the earth's atmosphere in falling rain and snow, and in runoff from rainfall and snowmelt gathered into brooks and rivers, ponds, lakes, and reservoirs. In reduced volume, freshwaters are exposed also to ground air within the voids of soils through which seepage waters flow. From the free atmosphere, surface waters absorb mainly oxygen and nitrogen; in smaller amounts, carbon dioxide, hydrogen sulfide, and other gases released to the atmosphere (1) by household and industrial operations (mainly the combustion of fuels) and (2) by the respiration of living things ranging from man and the higher animals to the saprophytes responsible for the degradation of organic matter. From the ground air, groundwaters may absorb methane, hydrogen sulfide, and large amounts of carbon dioxide, all of them gases of decomposition that accumulate in the ground when plants die, the stubble of crops is left to rot, leaves fall, and organic waste substances are destroyed by bacteria, molds, and other microorganisms of the teeming soil. Concurrently, groundwaters may surrender their dissolved oxygen to the saprophytes. If all of the available oxygen disappears, decomposition becomes anaerobic. Similar changes take place, also, in the stagnant depths of ponds, lakes, and reservoirs and in tidal estuaries in which organic detritus is laid down in benthal deposits.

Some apparently clean ponds and lakes and the backwaters of streams may, at times, support luxuriant growths of algae and related organisms; some of them contain volatile oils responsible for bad odors and tastes. During daylight the algae and other plants or plantlike organisms absorb carbon dioxide from the environment and release oxygen to it in the course of *photosynthesis*. At night, *respiration* reverses the absorption and desorption of these gases.

From what has been said, it is clear that the discharge of putrescible or decomposable organic matter into natural waters by households and industry and its entrance into these waters as decaying vegetation or as fertilizing elements through runoff from agricultural lands increases the aquatic food supply and with it the generation of gases of decomposition, while drawing heavily on available oxygen resources.

## 12-2 Objectives of Gas Transfer

Gas-transfer operations serve a multitude of purposes in both water and wastewater treatment, occupy a unique place in water-quality management, and are important factors in the pollution and self-purification of natural waters (Sec. 21–5). Although gas transfer is a physical phenomenon in which gas molecules are exchanged between a liquid and a gas at a gas-liquid interface, the physical operation is accompanied, more often than not, by chemical, biochemical, and biological as well as biophysical changes. These consequences may, indeed, be the primary purpose of the operation. The objectives of gas transfer are correspondingly varied in concept and manifold in the ways of their attainment. In most instances the shared engineering objective of aeration is either the removal of gases and other volatile substances from water or their addition to water, or both at the same time. In some instances, however, air may also be injected into water solely for purposes of agitation. Gas exchange then becomes incidental. Examples are aerated grit chambers (Sec. 13–15) and flocculating chambers (Sec. 15–3).

Aeration for gas exchange in its simplest and most direct form has the following aims: (1) addition of oxygen to oxidize dissolved iron and manganese in waters drawn from the ground and, in wide measure, to maintain wanted oxygen tension in wastewater treatment and disposal, including both natural and induced aeration of polluted waters; (2) removal of carbon dioxide to reduce corrosion and interference with lime-soda softening; (3) removal of hydrogen sulfide to eliminate odors and taste, decrease the corrosion of metals and disintegration of cement and concrete, and lessen interference with chlorination; (4) removal of methane to prevent fires and explosions; and (5) removal of volatile oils and similar odor- and taste-producing substances released by algae and other microorganisms.

Aeration is but one form of gas exchange. There are others in which water is exposed not to natural air but to (1) a specific pure gas, (2) air surcharged with a specific gas, or (3) air or gas, including air-gas mixtures, at pressures above or below atmospheric and possibly at high or low temperatures. Examples are (1) addition of carbon dioxide from flue gas or carbon dioxide generators to *recarbonate* lime-softened water that has purposely been overtreated (Sec. 16-19) or to promote or complete the removal of hydrogen sulfide (Sec. 12-10); (2) addition of ozone from ozone generators or chlorine gas from chlorine dispensers, for either the disinfection of waters (Secs. 17-8 and 17-14) or the destruction of odors and tastes in waters; and (3) removal of corrosion-promoting oxygen as well as other gases (degasification) by spraying water into a vacuum chamber at ordinary temperatures or at elevated temperatures (Sec. 16-13).

## 12-3 Absorption and Desorption of Gases

The absorption and desorption, precipitation, or release of gases by water finds expression in the general gas law:

$$pV = NRT \qquad (12\text{-}1)$$

where $p$ is the absolute pressure of the gas, $V$ is its volume, $N$ is the number of moles, $R$ is the universal gas constant ($8.3136 \times 10^7$ dyne-cm per gram-mole and deg C absolute, or 1546 lb-ft per lb-mole and deg F absolute), and $T$ is the absolute temperature ($273.1 +$ deg C) [Kelvin], or ($459.7 +$ deg F) [Rankin]. Component and amplifying concepts and formulations comprise Avogadro's hypothesis,[1] laws named after Dalton,[2] Henry,[3] Boyle,[4] Charles,[5] Gay-Lussac,[6] and Graham,[7] and also Boltzmann's molecular constant.[8]

If attention is directed specifically to water in contact with an atmosphere of air and other gases or gas mixtures, the most important relationships can be summarized as follows:

---

[1] Amadeo Avogadro, Conte di Quaregna (1776–1856), Italian physicist, whose *Essay on a Way of Determining the Relative Masses of Elementary Molecules of Bodies and Proportions in Which They Enter into These Combinations* was published in 1811. Avogadro's law states that equal volumes of different gases at the same pressure and temperature contain the same number of molecules, namely $6.02 \times 10^{23}$ molecules per mole, a mole being a mass numerically equal to the molecular weight.

[2] John Dalton (1766–1844), British chemist and physicist. $pV = V(p_1 + p_2 + \ldots)$.

[3] Joseph Henry (1797–1788), American physicist, first secretary and director of the Smithsonian Institution. See Eq. 12-2.

[4] Robert Boyle (1627–91), British physicist. $p_1 V_1 = p_2 V_2$ at constant $T$.

[5] J. A. C. Charles (1746–1823), French physicist. $V_1/V_2 = T_1/T_2$ at constant $p$.

[6] J. L. Gay-Lussac (1778–1850), French chemist and physicist. See footnote 12.

[7] Thomas Graham (1805–69), British chemist. $k_{d1}/k_{d2} = \sqrt{\rho_2/\rho_1}$.

[8] Ludwig Boltzmann (1844–1906), Austrian physicist. $N$ in Eq. 12-1.

The *solubility* of a gas depends on (1) its partial pressure in the atmosphere in contact with the water, (2) the water temperature, and (3) the concentration of impurities in the water.

The rate of solution and precipitation of a gas is controlled by (1) the degree of undersaturation or supersaturation of the water, (2) the water temperature, and (3) the interfacial area of gas contact and water exposure, including the prevention, by movement of the atmosphere and the water, of the buildup of stationary gas and water films at the gas-water interface.

The rate of gas dispersion in water depends upon the rate of (1) molecular diffusion, (2) eddy diffusion by convection, and (3) eddy diffusion by agitation.

Implicit in these statements is that the gas does not react chemically with the water. Of the gases in the earth's atmosphere, oxygen, nitrogen, carbon dioxide, and the rare gases are of this kind, although carbon dioxide does react to the extent of about 1% to form carbonic acid ($H_2CO_3$). Among other gases of significance in natural and treated water and in wastewaters and products of wastewaters such as digesting sludge, methane and hydrogen are inert; hydrogen sulfide is less so; and chlorine is strongly reactive.

In accordance with Dalton's law of partial pressures, the molecules of each gas in a gas mixture exert the pressure they would if they were present alone and the sum of these partial pressures equals the total pressure; that is, $pV = V\Sigma p$. In accordance with Henry's law, moreover, the saturation concentration of a gas in a liquid, such as water, is directly proportional to the concentration, or partial pressure, of the gas in the atmosphere in contact with the liquid; that is,

$$c_s = k_s p \qquad (12\text{-}2)$$

where $c_s$ is the saturation concentration of the gas in the water, $p$ the partial pressure of the gas in the gas phase, and $k_s$ the proportionality constant, or *coefficient of absorption*. The units of $c_s$ are conveniently milliliters per liter, for $p$ as a proportionality pressure or volume, and for $k_s$ in milliliters per liter. Atmospheric pressure, namely, 1 atm or 760 mm (29.2 in.) Hg, is commonly specified for $k_s$, and $p$ is then the pressure in atmospheres. The solubilities of a number of gases important in water and wastewater engineering are shown in Table 17 of the Appendix. Volumes are converted to weights on the basis of Avogadro's hypothesis[1] that equal volumes of ideal gases contain, at the same temperature and pressure, equal numbers of molecules. At standard temperature and pressure (0°C and 760 mm Hg, or 32°F and 29.92 in. Hg), the molal volume of any gas is 22,412 ml per g-mole or 359 cu ft per lb-mole. The gas volume at a given temperature and pressure is reduced to standard conditions by means of Eq. 12-3, as

$$V_0 = [(p - p_w)/p_0](T_0/T)\,V \qquad (12\text{-}3)$$

Here $p$, $V$, and $T$ are as in Eq. 12-1; $p_w$ is the vapor pressure of water; and the subscript zero denotes standard conditions.

*Example 12-1.*

What is the equilibrium concentration of oxygen in pure water at 0°C exposed to air under a barometric pressure of 760 mm?

From Table 17 (Appendix), $k_s = 49.3$ ml/l. Because dry air normally includes 20.95% of oxygen by volume and air in contact with water is generally saturated with water vapor, the partial pressure of the oxygen is $0.2095 \times (760 - 4.58) = 157$ mm Hg, 4.58 being the vapor pressure of water at 0°C. The volume concentration of oxygen, therefore, is $49.3 \times 157/760 = 10.2$ ml/l, and because 1 ml of oxygen weighs $2 \times 16 \times 10^3/22{,}412 = 1.43$ mg, the weight concentration of oxygen is $1.43 \times 10.2 = 14.6$ mg/l.

Water is saturated with a gas when the proportionality implied in Henry's law is fully effective. Rising temperatures decrease the saturation value, as do the salts of hard and brackish waters. A higher altitude or a falling barometer reduces solubility in the ratio of observed to standard pressure, the approximate change in pressure with altitude being 1% for every 270 ft of elevation.

Saturation values for oxygen in fresh and brackish waters at different temperatures are given in Table 20 in the Appendix. North Atlantic Ocean water of substantially full strength has a chloride content of nearly 18,000 mg per l, and its dissolved-oxygen saturation is about 82% that of fresh water. The DO saturation of domestic wastewaters[9] is about 95% that of clean water.

*Example 12-2.*

Find the approximate DO saturation value at 10°C and 760 mm of (1) distilled water and (2) salt water containing 18,000 mg/l of chloride; also (3) the percentage saturation of these waters when their DO content is 5 mg/l.

From Table 4 (Appendix), the vapor pressure of water at 10°C is 9.21 mm, and (1) $c_s = [(0.680 - 0.006)(760 - 9.21)]/(10 + 35) = 11.2$ mg/l; (2) $c_s = 11.2(1 - 9 \times 10^{-6} \times 1.8 \times 10^4) = 9.4$ mg/l; and (3) for $c_s = 11.2$ mg/l, the percentage saturation is $100 \times 5/11.2 = 44.7\%$; for $c_s = 9.4$ mg/l, it is $100 \times 5/9.4 = 53.2\%$.

---

[9] W. A. Moore, The Solubility of Atmospheric Oxygen in Sewage, *Sewage Works J.*, **10**, 241 (1938).

## 12-4 Rates of Gas Absorption and Desorption

If it is postulated that the *rate of gas absorption* is proportional to its degree of undersaturation (or saturation deficit) in the absorbing liquid,

$$dc/dt = K_g(c_s - c_t) \tag{12-4}$$

where $dc/dt$ is the change in concentration, or rate of absorption transport, or transfer, at time $t$; $c_s$ the saturation concentration at a given temperature; $c_t$ the concentration at time $t$; and $K_g$ a proportionality factor for existing conditions of exposure. Integration between the limits $c_0$ at $t = 0$ and $c_t$ at $t = t$ then yields the basic equation

$$c_t - c_0 = (c_s - c_0)[1 - \exp(-K_g t)] \tag{12-5}$$

where $K_g$ increases with temperature and the degree of mixing of the gas and liquid, that is, the rate of renewal of the gas-liquid interface and the degree of eddy diffusion. The temperature effect follows the van't Hoff-Arrhenius relationship (Sec. 11-15), but mixing effects are difficult to define unless useful power expenditure can be identified (Sec. 11-17). Because the molecules of gas must pass through the gas-liquid interface, $K_g$ is also a function of $A/C$, the area of interface per unit volume of liquid. Accordingly $K_g = k_g A/C$, where $k_g$ is the *gas-transfer coefficient*. For absorption, $c_0 < c_t < c_s$, and both $c_t - c_0$ and $c_s - c_0$ are positive. For desorption, $c_s < c_t < c_0$, and both $c_t - c_0$ and $c_s - c_0$ are negative. An implicit assumption is that the rate of gas transfer across the gas-liquid interface, rather than the rate of diffusion of the dissolved gas within the liquid, is the controlling factor. Becker[10] reports the following values of $k_g \, \theta^{T\text{C}-20}$ in centimeters per hour for the absorption of oxygen, nitrogen, and air from bubbles in the temperature range 3.5 to 35°C: $O_2$, 32.3 × 1.018$^{T\text{C}-20}$; $N_2$, 34.0 × 1.019$^{T\text{C}-20}$; air, 32.1 × 1.019$^{T\text{C}-20}$. Understandably, these coefficients apply only for conditions of exposure obtaining in Becker's experiments. Values can be both higher and lower in different circumstances.

According to the two-film theory of Lewis and Whitman,[11] boundary films form at the interface within both the liquid and the gas, and rate of passage through them is governed by the thickness of the films (Fig. 12-1). Film thickness itself is fundamentally a function of kinematic viscosity but it can be decreased by stirring or agitating the main body of gas or liquid. In accordance with Sec. 11-16, the rate of diffusion through the films depends on

---

[10] H. G. Becker, Mechanism of Absorption of Moderately Soluble Gases in Water, *Ind. Eng. Chem.*, **16**, 1220 (1924).

[11] W. K. Lewis and W. C. Whitman, Principles of Gas Absorption, *Ind. Eng. Chem.*, **16**, 1215 (1924). Other theories are the penetration theory and the boundary layer theory.

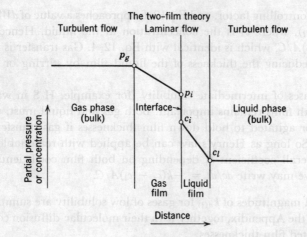

**Figure 12-1.**
**Pressure and concentration gradients in gas and liquid films at a gas-liquid interface.**

the area of the interface and the concentration gradient within the component films, or

$$dW/(A\,dt) = -k_{d(g)}(p_g - p_i) = -k_{d(l)}(c_i - c_l) \quad (12\text{--}6)$$

where $dW/(A\,dt)$ is the weight of gas passing through a unit area in a unit time; $p_g$ and $p_i$ are respectively the partial pressures of the gas in the main body of the gas and at the interface; $c_i$ and $c_l$ are respectively the concentration of the gas at the interface and in the main body of the liquid; and $k_{d(g)}$ and $k_{d(l)}$ are respectively the film-diffusion or transfer coefficients in the gaseous and liquid phases. These coefficients possess the dimension of velocity ($lt^{-1}$). Equilibrium obtains at the interface. Hence $p_i$ is a function of $c_i$ and equals $c_i/k_s$ when Henry's law applies. Three general situations are encountered:

1. The gas is highly soluble in the liquid (for example, $NH_3$ in water). In these circumstances $c_i$ is large even when $p_i$ is small. Passage of the gas molecules across the gas film then becomes the controlling factor, and Eq. 12–6 approaches a value of $dW/(A\,dt) = k_{d(g)}p_g$ because $p_i$ is negligible. Hence $dc/dt = k_{d(g)}p_g A/C$, and it follows that gas transfer can be promoted by reducing the thickness of the gas film by moving or stirring the gas.

2. The solubility of the gas in the liquid is low (for example, $O_2$, $N_2$, and $CO_2$ in water). For these conditions $c_i$ practically equals $c_s$, the concentration at which the dissolved gas is in equilibrium with the gas in the atmosphere (saturation concentration). Passage of gas molecules through the liquid film

is then the controlling factor, and Eq. 12-6 approaches a value of $dW/(A\,dt) = k_{d(l)}(c_s - c_l)$, where $c_l$ is the concentration in the liquid. Hence $dc/dt = k_{d(l)}(c_s - c_l)A/C$, which is identical with Eq. 12-4. Gas transfer is then promoted by reducing the thickness of the liquid film by stirring or agitating the liquid.

3. For gases of intermediate solubility (for example, $H_2S$ in water), the effect of both films remains important. Both gas and liquid must, therefore, be stirred or agitated to hold down film thicknesses if gas transfer is to be promoted. So long as Henry's law can be applied with reasonable satisfaction, an overall coefficient $k$, depending on both film coefficients, can be used, and we may write $dc/dt = -k(c_s - c_l)A/C$.

Observed magnitudes of $k_{d(l)}$ for gases of low solubility are summarized in Table 18 in the Appendix, together with their molecular diffusion coefficients and calculated film thicknesses.

If the concentration gradient across the interface of a gas bubble is constant, the diameter of the bubble will decrease linearly with time. The rate of decrease is proportional to the gas-transfer coefficient and concentration gradient and inversely proportional to the molecular weight of the gas. Turbulence, temperature, and concentration of dissolved substances all play a part. The higher the concentration of dissolved substances, the lower is the transfer coefficient.

The rate at which gas is adsorbed by a falling drop of water or a rising bubble of gas is greatest at the moment of formation. It decreases rapidly thereafter, because the film of water at the interface increases in thickness in the absence of internal liquid motion.

In contrast to absorption, the *rate of desorption, precipitation, release, or dissolution of a gas* from a liquid becomes proportional to its degree of oversaturation in the liquid or the saturation surplus. It follows that the equations for rates of absorption also apply to rates of dissolution. As stated before, the fact that the saturation concentration, $c_s$, will be less than the observed concentrations, $c_0$ and $c_t$, makes for negative differences.

---

## Example 12-3.

In an experiment on the removal of carbon dioxide from water sprayed into the air in spherical droplets 0.55 cm in diameter, the initial supersaturation of the water with carbon dioxide was 27.5 mg/l. After 1 sec of exposure this was reduced to 11.5 mg/l. Find the coefficient of gas transfer.

Because $c_s - c_0 = -27.5$ mg/l and $c_t - c_s = 11.5$ mg/l, $c_t - c_0 = -16$ mg/l. In accordance with Eq. 12-5, therefore, $(-16) = -27.5[1 - \exp(-K_g \times 1)]$, and

$K_g = 0.872$ sec$^{-1}$. The droplet volume per unit surface area being $0.55/6 = 9.17 \times 10^{-2}$ cm, $k_{d(g)} = 0.872 \times 9.17 \times 10^{-2} = 8.00 \times 10^{-2}$ cm/sec. The transfer coefficient then equals $8.00 \times 10^{-2} \times 3600 = 288$ cm/hr.

## 12-5 Types of Aerators

Four types of aerators are in common use: (1) gravity aerators, (2) spray aerators, (3) diffusers, and (4) mechanical aerators. Their aim is to create extensive, new, and self-renewing interfaces between air and water, to keep interfacial films from building up in thickness, to optimize the time of gas transfer, and to accomplish these aims with a minimal expenditure of energy.

*Gravity Aerators* (Fig. 12-2). Characteristic designs include *cascades*, in which the available fall is subdivided into a series of steps; *inclined planes*,

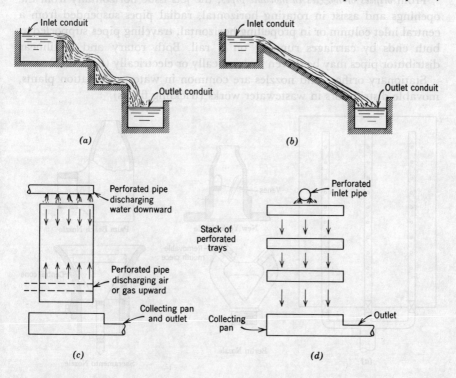

**Figure 12-2.**

**Gravity aerators. (a) Cascade; (b) inclined apron possibly studded with riffle plates; (c) tower; with countercurrent flow of air (gas) and water; (d) stack of perforated pans possibly containing contact media.**

usually studded with *riffle* plates set into the planes in herringbone fashion and breaking up the sheet of water that would otherwise form; *vertical stacks*, through which droplets fall and updrafts of air ascend in countercurrent flow; and *stacks of perforated pans or troughs*, often filled with contact media such as coke or stone, the water dropping freely from pan to pan or trough to trough and trickling over the surfaces of the contact media present.

**Spray Aerators.** Spray or pressure aerators spray droplets of water into the air from stationary or moving orifices or nozzles. A whirling motion imparted to the droplets makes for turbulence at the air-water interface.

From *orifices or nozzles in stationary pipes* (Figs. 12-3 and 4), the water rises either vertically or at an angle and falls onto a collecting apron, a contact bed, or a collecting basin serving some other useful purpose. Longer exposure in vertical jets is offset in some measure by freer access of air to the trajectory of inclined jets.

From *orifices or nozzles in movable pipes*, the jets issue horizontally from the openings and assist in rotating horizontal, radial pipes suspended from a central inlet column or in propelling horizontal, traveling pipes supported at both ends by carriages running on a rail. Both rotary and rectilinear distributor pipes may be driven hydraulically or electrically (Sec. 19-6).

Stationary orifices and nozzles are common in water-purification plants, movable distributors in wastewater works (trickling filters).

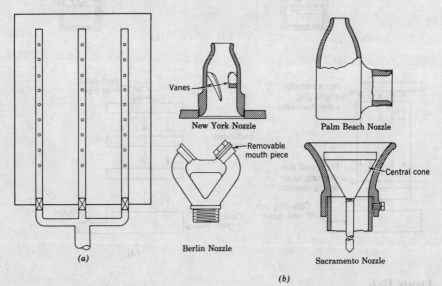

**Figure 12-3.**
Spray aerator and nozzles. (a) Nozzled aerator; (b) aerator nozzles. The coefficients of discharge of these nozzles vary from 0.85 to 0.92.

***Air Diffusers*** (Fig. 19-7). Most air diffusers or injection aerators bubble compressed air into water through orifices or nozzles in air piping, diffuser plates or tubes, or *spargers*. Ascending bubbles acquire smaller terminal velocities than would drops falling freely in air through the same distance. This increases the exposure time of air bubbles but reduces turbulence at the bubble interface. Spiral or cross-current flow can lengthen the path of travel of both air and water. It is conveniently induced by injecting air along one side of the tank (Fig. 19-4).

Air diffusion is employed in water as well as wastewater treatment. Its best known application is in the activated-sludge process of wastewater treatment. Less well known is the operation, especially during hot summer months, of floating compressors capable of (1) raising the oxygen content of receiving waters that are overloaded with waste matters and might become septic and (2) destroying their stratification (Fig. 21-10). During the winter season the equipment may be kept at work in inland harbors, where air blown into the waters will prevent the formation of sheet ice and keep the harbor open for shipping.

***Mechanical Aerators.*** Of the many different kinds of mechanical aerators, the following are of special interest to the designer of simple systems: *submerged paddles* that circulate the water in aeration chambers and renew its air-water interface; *surface paddles or brushes* that dip lightly into aeration chambers but far enough to circulate their waters, release air bubbles, and throw a spray of droplets onto their water surface; *propeller blades* that whirl at the bottom of a central downdraft tube in an aeration chamber and aspirate air into the water; and *turbine blades* that cap a central updraft tube in an aeration chamber and spray droplets over its water surface. See Fig. 19-4.

Mechanical aerators of these kinds are employed principally in the treatment of wastewaters by the activated-sludge process. Here a function of mechanical as well as injection aeration as fully as important as that of gas transfer is to keep the activated floc in mobile and useful suspension.

Unless they are housed, gravity and pressure aerators for water works may have to be bypassed in winter to keep the water from freezing. Wastewater, however, is normally warm enough not to congeal during the brief interval of its exposure in advance of trickling filtration. During high winds, rising jets of water may have to be throttled down or shut off entirely to keep their spray within the boundaries of the aerator. Aerator spaces, especially enclosed spaces, should be well ventilated not only to create effective differentials in gas concentrations between the two phases, but also to prevent (1) asphyxiation of operating or repair crews and visitors by carbon dioxide, (2) their poisoning by hydrogen sulfide, and (3) formation of explosive mixtures of methane with air.

## 12-6 Factors Governing Gas Transfer

What has just been said about aerators and their aims is supported most immediately by the common gas-transfer equations. In accordance with Eq. 12-5, for example, $c_t = c_0 + (c_s - c_0)\{1 - \exp[-k_g(A/C)\,t]\}$, and transfer can be optimized, no matter what its direction, (1) by generating the largest practicable area, $A$, of interface between a given water volume, $C$, and air, a pure gas, an air-gas mixture, or a mixture of gases; (2) by preventing the buildup of thick interfacial films, or by breaking them down to keep the transfer coefficient, $k_g$, high; (3) by inducing as long a time of exposure, $t$, as possible; and (4) by ventilating the aerator and its components well enough to maintain the highest possible driving force or concentration difference, $(c_s - c_t)$ for absorption and $(c_t - c_s)$ for desorption. These, then, are the manageable variables.

Values of the proportionality constant $K_g = (A/C)k_g$ must normally be determined experimentally and verified in plant-scale tests.[12] A plot of $y = \log(c_t - c_s)$ against $x = t$ should yield a straight line with an intercept $y_0 = \log(c_0 - c_s)$ at $t = 0$ and a slope $(\log y_2 - \log y_1)/(t_2 - t_1) = K_g$.

## 12-7 Design of Gravity Aerators

The controlling element in gravity aerators (Fig. 12-2) is the available head. It can be put to use in a single or in multiple descent. At a given instant the rate of free fall $dh/dt = v = gt$ and $\int_0^h dh = g\int_0^h t\,dt$, or $h = \frac{1}{2}gt^2$, where $h$ is the height of fall in feet, $t$ the time in seconds, $v$ the velocity in feet per second, and $g$ the acceleration of gravity in feet per (second)$^2$. It follows that, in a single descent through a height $h$, the elapsed time is $t = \sqrt{2h/g}$ and that in $n$ descents through the same vertical distance

$$t = n\sqrt{2h/(ng)} = \sqrt{2nh/g} \qquad (12\text{-}7)$$

In other words, $t$ is proportional to $\sqrt{n}$.

---

*Example 12-4.*

Find the time of exposure of water falling through a distance of 9 ft (1) in single descent, (2) in 4 descents.

1. For $n = 1$ and $h = 9$, Eq. 12-7 states that $t = \sqrt{2 \times 9/32.2} = 0.75$ sec.
2. For $n = 4$, $\sqrt{n} = 2$, and $t = 2 \times 0.75 = 1.5$ sec.

---

[12] P. D. Haney, Theoretical Principles of Aeration, *J. Am. Water Works Assoc.*, **46**, 353 (1954).

However, it should be said that the quality of exposure is usually poorer in multiple descents, because droplets do not necessarily break away from jets of falling water as soon as they strike the air.

## 12-8 Design of Fixed-Spray Aerators

The hydraulic performance of *fixed-spray pressure aerators* involves three principal parts: orifice or nozzle behavior, including applicable ballistic principles; windage or wind effects, shared with spray cooling; and pipe friction associated with multiple takeoffs along the line of flow, shared with multiple tank inlets and filter underdrains (Secs. 13–15 and 14–13).

Wind effects are variable. However, even relatively gentle winds are more influential than air resistance, which does not reduce the height or time of rise by as much as 10% for heights under 15 ft and exposures under 2 sec. That is why the resistance offered by calm air can usually be neglected in aerator calculations. The time of exposure also governs the distance droplets are carried by the wind because

$$l = 2c_D v_w t_r \tag{12-8}$$

where $l$ is the distance, $c_D$ the coefficient of drag (about 0.6), and $v_w$ the wind velocity.

As shown in Fig. 12–4, flow through a perforated or nozzled pipe decreases stepwise at each opening or takeoff. If the diameter remains unchanged, the resistance to flow within the perforated section equals approximately the resistance that would be offered to the full entrant flow by one third the length of perforated pipe. Thus for the idealized case of a slotted pipe and $s = kQ^n$ (Sec. 7–6), the flow at a distance $(L - l)$ from the end of the pipe is $Q = Q_e(L - l)/L$ and $h_f = \int_0^l s \, dl = k(Q_e/L)^n \int_0^l (L - l)^n \, dl = (s_e/L^n) \int_0^l (L - l)^n \, dl$, or

$$h_f = [s_e/(n + 1)][L - (L - l)^{n+1}/L^n] \tag{12-9}$$

Here the subscript $e$ denotes entrant flow and resistance to it. For the Chezy formula ($n = 2$) and $l = L$, $h_f = \tfrac{1}{3} s_e L$, as stated at the outset and implied also geometrically by the parabolic nature of the curve for $s$ versus $l$.[13] Actual losses may be reduced by the recovery of velocity head.

For good ventilation, sprays should not overlap more than necessary to ensure spatial economy. In water-purification works, nozzles are commonly placed 2 to 12 ft on centers, and nozzled aerators occupy 50 to 150 sq ft of area per mgd capacity.

---

[13] The reader will remember that the area inside a parabola equals two thirds the area of the circumscribed rectangle. This make the area outside the parabola one third the area of the pertinent rectangle.

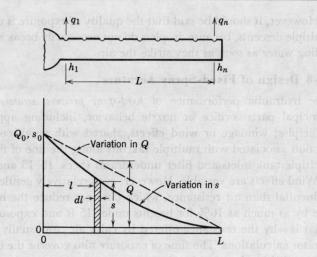

**Figure 12-4.**
**Frictional resistance to uniformly decreasing flow (idealized).**

## 12-9 Design of Movable-Spray Aerators

The hydraulic performance of *movable-spray aerators* is governed by the power required to revolve rotary distributors or keep straight-line (traveling) distributors in motion, and the sizing of nozzles and fanning out of sprays for uniform distribution of the applied waters, both of them under varying hydraulic loadings. Neither ballistic nor wind effects are important, because the sprays issuing horizontally lie only 9 to 15 in. above the surface of the trickling filters they commonly serve.

Distributors become self-propelling under hydrostatic heads of 18 to 30 in. When fully effective, the impulse transmitted by the jets to the distributor is $(\gamma Q/g)v = \gamma Q c_v \sqrt{2h/g} = 28$ to 36 ft-lb per sec at a flow rate of 1 mgd, that is, about 0.05 to 0.07 hp. Losses in supply piping within the sprinkling unit itself together with mechanical friction in moving parts are estimated at 25%. When, in the absence of recirculation, flows (night flows, for instance) drop and become so weak that they cannot keep the distributor in motion, a dosing tank must be interpolated to accumulate and discharge minimal operative flows (Sec. 19-6). Otherwise the distributor must be driven by a motor.

## 12-10 Design of Injection Aerators

The air supply of injection aerators is made available from the ambient atmosphere through (1) air filters that clean incoming air to protect compressors against abrasion and air-distribution systems against fouling and clogging; (2) air compressors that place the air under wanted pressure; (3) measuring devices (orifice or venturi meters) that measure and often record rates of air flow; (4) air piping that conveys the compressed air to points of use; and (5) diffusers, spargers, nozzles, or orifices through which the air is injected into the waters under treatment. Economical designs normally hold friction losses in air-supply systems to about 25% of the depth of submergence of the terminal air-distribution units. Accordingly, the air pressure at the blowers (psig) approximates 0.54 × distribution depth (ft). Because the terminal velocity of rising bubbles is close to 1 fps, the time of exposure in seconds is numerically much the same as the depth in ft.

Aeration tanks are commonly 10 to 15 ft deep and can maintain good transverse (often spiral) circulation at ratios of tank width to tank depth up to 2:1, if the cross-section is shaped to avoid stagnant corners. The nominal detention period of aeration units is least in water purification operations (usually 10 to 30 min), most in biological wastewater treatment (seldom as low as 1 hr, usually up to 6 hr, and sometimes up to 24 hr on an average). The area (square feet) occupied by tanks aerating 1 mgd of water is normally $9300/(dt)$, where $d$ is the depth in feet and $t$ the time in minutes. Applied air ranges from 0.01 to 0.15 cu ft of free air per gallon of water being purified, but may exceed 1.0 cu ft per gallon in the treatment of wastewaters. A value of 0.134 identifies volume for volume use. Required power and power consumption change correspondingly (Sec. 19-13).

## 12-11 Mechanical Aerators

Mechanical surface aerators are widely used for many applications in modern wastewater treatment.[14] They transfer atmospheric oxygen to the liquid by surface renewal and exchange and provide the mixing necessary for distribution of the aerated liquid throughout the tank and for maintaining suspended solids in suspension where this is necessary. Mechanical aerators have found favor because they dispense with the compressors, air-cleaning devices, piping, and diffusers, and their maintenance, which are necessary with injected-air systems, and because they fit into a wide variety of tank shapes, not being bound to the depth and width strictures common with diffused-air systems.

Various types of vertical-shaft mechanical aerators have been developed in addition to the horizontal shaft Kessener brush aerator and cage rotors developed in Europe (see Fig. 19-4). Many configurations operate at the surface of the tank, spraying the liquid into the air by means of blades, or entraining air by creating vortices. Such units are limited to applications in

[14] Water Pollution Control Federation, Aeration in Wastewater Treatment, *Manual of Practice No. 5, J. Water Pollution Control Federation,* **41**, 1863 (1969); **41**, 2026 (1969); and **42**, 51 (1970).

relatively shallow tanks. Incorporating a draft tube on a spray aerator allows the unit great latitude in the depth of tank into which it is fitted. Down-draft tube aerators operate by entraining air at the surface and discharging the air-liquid mixture at the bottom of the tank. Turbine mixers immersed in the liquid have been combined with air spargers set underneath the turbine, the turbine serving to shear the large air bubbles emitted from the sparger.

The oxygen-transfer efficiency of mechanical aerators, about 3 lbs of oxygen per horsepower hr, is of the same order of magnitude as for diffused air systems, except that the efficiency of vertical shaft aerators tends to drop with larger installations. Full-scale testing of mechanical aerators is necessary for untried units, with performance specifications carefully spelled out. Such tests, however, because they use water rather than wastewaters, may not always provide data directly applicable in a treatment plant. Model tests may be helpful in assessing the significance of liquid characteristics on aeration efficiency.[14]

## 12-12 Removal of Specific Gases

Methane, carbon dioxide, and hydrogen sulfide are found in water as well as wastewater. The order of their respective solubilities is relatively low, moderately high, and very high.

**Methane.** Methane, $CH_4$ (Table 17, Appendix), also called marsh gas and methyl hydride, is the simplest member of the *paraffin* hydrocarbons. In nature, methane is a common constituent of natural gas from oil and gas wells and of *fire damp* in coal mines. Accordingly, it may be present in deep well and mine waters. In wastewater treatment, it is the most common component of sludge gas.

A gas like methane, which is only slightly soluble in water, does not react with it, has a low boiling point, is normally absent from the ambient atmosphere, and is readily swept out of water by aeration. Good ventilation will promote its desorption and prevent the accumulation of explosive air-methane mixtures. The explosive range extends from 5.6 to 13.5% of methane by volume.

**Carbon Dioxide.** The removal of carbon dioxide from water is more complex than that of methane. The solubility of carbon dioxide, $CO_2$, is high (Table 17, Appendix), and somewhat less than 1% of it reacts, at ordinary pressures and concentrations, with water to form carbonic acid, $H_2CO_3$, and its ionization products $HCO_3^-$ and $CO_3^=$. However, the boiling point of $CO_2$ is low, and its concentration in normal atmospheres is small. In some waters $H_2CO_3$ may be the only acid present; in all waters it has a strong influence on the hydrogen-ion concentration, which is lowered as the gas is removed.

Neither $H_2CO_3$ nor its ionization products can be desorbed as such by aeration.

The partial pressure of $CO_2$ in the atmosphere is normally about 0.033%, making for a saturation value of 0.56 mg per l in water at 20°C (78°F) and atmospheric pressure. Good ventilation will promote desorption of the gas and prevent its accumulation in asphyxiating concentrations. Exposure of water droplets to air for 2 sec ordinarily lowers the $CO_2$ concentration in the water by 70 to 80%, and efficiencies as high as 90% are not unusual.

As a means of $CO_2$ removal, aeration enters into competition with the neutralization of $CO_2$ by lime (CaO), especially in conjunction with the lime-soda process of water softening (Sec. 16-9). Because the rate of $CO_2$ desorption varies directly with its supersaturation in water, aeration is particularly efficient at high concentrations. By contrast, the efficiency of chemical neutralization is substantially independent of concentration. For this reason treatment may be optimized by aerating at high concentrations and liming at low concentrations. The dividing line generally lies at about 10 mg per l. Where water must be pumped for aeration, comparisons should be based on power costs versus chemical costs. Soda ash ($Na_2CO_3$) can take the place of lime, but it is more expensive in equivalently effective amounts.

For recarbonation, namely, the addition of $CO_2$ to water or wastewater for the purpose of taking lime into solution, see Sec. 16-8.

**Hydrogen Sulfide.** The desorption of hydrogen sulfide, $H_2S$, is even more complex than the removal of $CO_2$. The solubility of $H_2S$ is more than twice as high (Table 17, Appendix), and the dissolved gas ionizes to form the sulfides $HS^-$ and $S^=$, neither of which can be removed directly by aeration. Moreover, aeration desorbs $CO_2$ faster than it does $H_2S$, and the consequent rapid rise in pH affects the ionization equilibria of $H_2S$ and steps up the $HS^-$ and $S^=$ concentration. A further complication is that the oxygen introduced by aeration reacts with $H_2S$ to throw down elemental sulfur, $2 H_2S + O_2 \rightarrow 2 H_2O + 2 S\downarrow + 2 \times 3300$ cal. To be sure, there is a decrease in $H_2S$, but the sulfur remaining in the water exerts a chlorine demand.[15] Usually this is undesirable. Moreover, the sulfur may be oxidized by aeration to sulfate, $2 S + 3 O_2 \rightarrow 2 SO_3 + 2 \times 4540$ cal, and it may be reduced back to $H_2S$ when the water becomes anaerobic—for example, in the dead ends of water distribution systems and in parts of wastewater systems, such as force mains. Elemental sulfur, which imparts a milky-blue cast to water, can be removed by chemical coagulation and filtration.

---

[15] H. P. Black and J. P. Goodson, Jr., The Oxidation of Sulfides by Chlorine in Dilute Aqueous Solutions, *J. Am. Water Works Assoc.*, **44**, 309 (1952).

A low pH favors H₂S removal. To this purpose $CO_2$ can be bubbled through the water in advance of aeration. When flue gas is made the source, $CO_2$ concentrations of about 10% by volume are aimed for; the $CO_2$ is taken into solution in proportionate amounts; the hydrogen ion concentration is increased; and the concentration of $H_2S$ relative to other dissolved sulfides rises. As a result sulfides are flushed out as $H_2S$. After that has been accomplished, unwanted $CO_2$ must be removed, possibly in a second aerator. The response of sulfide water is so variable that aerator design should be based on laboratory experiments, followed by plant-scale tests. It may be desirable to experiment with oxidizing chemicals, such as potassium permanganate ($KMnO_4$).

Hydrogen sulfide is an explosive and extremely toxic gas. Brief exposure (30 min or less) to concentrations as low as 0.1% by volume in air may terminate fatally. Marked symptoms of nausea, headache, and dizziness may follow exposure to 5% of the lethal concentration. Of much interest, too, is the growth of sulfur bacteria on the walls of aerators and in piping systems.

The useful removal and addition of gases other than methane, carbon dioxide, and hydrogen sulfide, including the all-important addition of oxygen, are considered in other chapters of this book.

## 12-13 Removal of Odors and Tastes

The tastes and odors ascribed to algal growths are thought to be caused by essential oils included in their cells. To be desorbed, these oils must first be released to the water. A three-phase system is probably involved, because the volatile principles must be translated in succession from oil to water to air. An oil-air interface is possible but hardly probable in significant magnitude. The boiling points of most essential oils are much higher than those of most gases, and the difficulty of volatilizing most algal odors and tastes is correspondingly greater. Odor removal of but 50% has been reported, for instance, when *Synura* was the odor- and taste-producing organism.[16] On the other hand, appreciable reductions in odors from industrial wastes have been achieved by spray aeration at high pressures.[17]

The taste and odor principles of phenolic substances and many other industrial chemicals are seldom volatile enough to be removed by aeration. The boiling point of phenol itself, $C_6H_5OH$, for example, is 182°C, whereas substances with boiling points above 0°C are probably not desorbed from water at its common temperatures unless the solubility of the given substance is quite low. Even chlorine, which has a boiling point of −34°C, is so soluble that it does not respond well to aeration. Moreover, it hydrolyzes, ionizes, and

[16] Committee Report, Aeration of Water, *J. Am. Water Works Assoc.*, **47**, 873 (1955).
[17] *Water Treatment Plant Design*, American Water Works Assoc., New York, 1969.

reacts with other substances that may be present. Marginal chlorination of phenolic waters may intensify odor and taste troubles without improving the volatility of the chlorophenols formed. Other treatment methods, including breakpoint chlorination, must be resorted to for their control (Sec. 17–12).

A few odors and tastes are reduced in intensity after water has been aerated and allowed to stand for a day or two. Slow oxidation of organic substances may be responsible for observations of this kind. On the other hand, odors due to chlorine reaction products have been known to intensify on standing.

**thirteen**

# screening, sedimentation, and flotation

## 13-1 Treatment Objectives

The unit operations discussed in this chapter are designed to remove successively smaller suspended matters from water and wastewaters. Leaves, the flotsam of lakes and streams, the jetsam of drainage systems, and other sizeable clogging substances respond to removal from water and wastewater by *screening*. Relatively heavy suspended matter visible to the naked eye can normally be removed from water and wastewaters by *sedimentation*. Light and fine municipal and industrial wastewater solids respond selectively to removal by *flotation*.

## 13-2 Screening

Screening intercepts particles that are larger than the smallest openings of wire mesh, parallel bars (racks), or other screen patterns developed for use (1) in surfacewater intakes, (2) ahead of water or wastewater pumps, and (3) in conjunction with wastewater treatment units and outfalls for untreated wastewaters.

**Racks and Screens on Water Intakes.** Trash *racks* are often included in dams and other intake structures. To facilitate cleaning, they are placed on slopes of 3 to 6 vertical to 1 horizontal below a working platform from which long-tined rakes can engage the bars to pull up the rakings. Three intake structures equipped with more or less self-cleaning *screens* are illustrated in Fig. 13–1: (1) a vertical screen constructed either of wire mesh or solely of vertical wires or bars in order to deprive leaves and other debris of purchase against horizontal elements, (2) an upward-flow screen that can be tilted into a vertical position for cleaning, and (3) a surface-wash screen.[1] A common arrangement of a different kind is to slide a pair of removable screens into vertical grooves in the walls and bottom of an inlet channel and to leave one in place while the other is being lifted out to flush off the collected debris with high-velocity water jets. Intake screens are normally constructed to 2- to 4-mesh screening, more rarely of screening with 6 meshes to the inch. An exception is fine-mesh, rotating drum screens, called microstrainers.[2] In one such screen, pairs of warp and weft stainless steel wires create 160,000 openings per square inch, approximately 23 $\mu$ in size. In tests at Denver, Colorado, lake water flowing radially outward through the screen deposited in excess of 90% of its filter-clogging microscopic organisms and 25% of its turbidity on the screen. About 3% of the water strained had to be recycled as washwater to keep the strainer in service.[3]

**Wastewater Racks and Screens.** A number of wastewater racks and screens are shown in Fig. 13-2. Coarse racks of steel bars are given clear openings $1\frac{1}{2}$ to $2\frac{1}{2}$ in. wide or wider. Fine-rack openings may be as small as $\frac{1}{2}$ in. Screens are usually expected to collect waste matters down to $1/16$ in. in size, but some screens have openings as small as $1/32$ in. in their smallest, that is, controlling, dimension. They may be many inches long. Racks are cleared by hand with long-handled rakes, or by mechanical scrapers (Fig. 13-2a and b). To expand the rack area, the bars are placed on a slope of 1 vertical to 1, 2, or 3 horizontal. Cage racks are arranged in pairs (in series) and lifted from the wastewater channel for cleaning. Screens are rotated through the water as endless bands, disks, or drums and are cleaned by brushes, jets of water, or blasts of air (Fig. 13-1d and e). Hydraulic requirements are (1) that the approach velocity in the raking or screening channel not fall below a self-cleaning value (1.25 fps) nor rise so high as to dislodge rakings or screenings (2.5 fps); and (2) that the loss of head through the

[1] J. W. Cunningham, Waterworks Intakes and the Screening of Water, *J. Am. Water Works Assoc.*, **23**, 258 (1931). Other screening arrangements are described in the following papers: V. C. Lischer and H. O. Hartung, Intakes on Variable Streams, *J. Amer. Water Works Assoc.*, **44**, 873 (1952); and W. E. Whitlock and R. D. Mitchell, Hydraulically Backwashed Stationary Screens for Surface Water, *J. Amer. Water Works Assoc.*, **50**, 1337 (1958).

[2] P. L. Boucher, Microstraining, *J. Inst. Water Engineers*, **5**, 561 (1951); G. R. Evans, Review of Experiences with Microstrainer Installations, *J. Am. Water Works Assn.*, **49**, 541 (1957); and P. L. Boucher, Microstraining and Ozonation of Water and Wastewater, *Proc., 22nd Ind. Waste Conf.*, Purdue Univ., Publ. No. 129, 771 (1967).

[3] G. W. Turre, Use of Microstrainer Unit at Denver, *J. Am. Water Works Assoc.*, **51**, 354 (1959).

364/Screening, Sedimentation, and Flotation

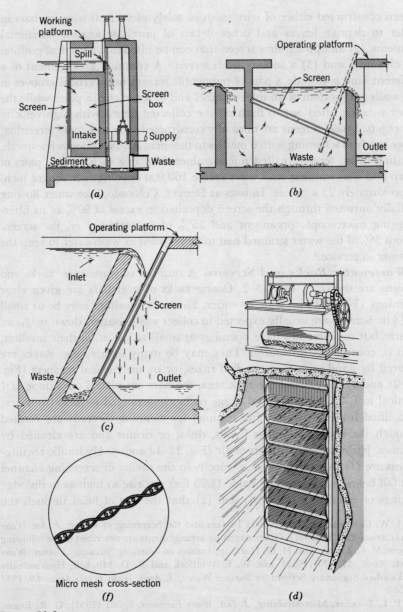

**Figure 13-1.**
Racks and screens on water intakes. (a) Vertical rack cleaned by hand-operated long-tined rake; (b) inclined screen with upward flow (screen can be tilted for cleaning); (c) self-flushing inclined screen; (d) mechanically operated fine screen; (e) microstrainer; (f) cross-section of twinned mesh fabric for microstrainer. [(a) to (c) after J. W. Cunningham; (e) and (f) after P. L. Boucher.]

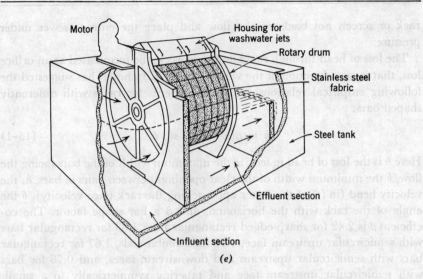

**Figure 13-1—continued**

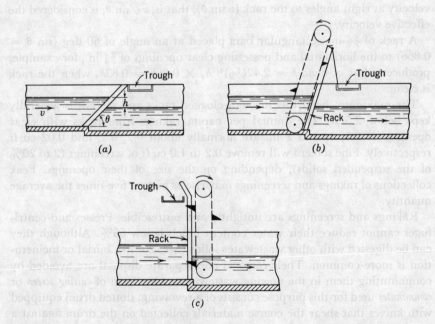

**Figure 13-2.**
**Wastewater racks and screens.** (*a*) Hand-cleaned inclined rack; (*b*) and (*c*) mechanically cleaned racks.

rack or screen not back up the flow and place the entrant sewer under pressure.

The loss of head through racks and screens can be formulated as an orifice loss, that is, as a function of the velocity head. Kirschmer[4] has suggested the following empirical relationship and coefficients for racks with differently shaped bars:

$$h = \beta(w/b)^{4/3} h_v \sin \theta \qquad (13\text{-}1)$$

Here $h$ is the loss of head in feet, $w$ the maximum width of the bars facing the flow, $b$ the minimum width of the clear openings between pairs of bars, $h_v$ the velocity head (in ft) of the water approaching the rack (face velocity), $\theta$ the angle of the rack with the horizontal, and $\beta$ a bar shape factor. The coefficient $\beta$ is 2.42 for sharp-edged rectangular bars, 1.83 for rectangular bars with semicircular upstream face, 1.79 for circular rods, 1.67 for rectangular bars with semicircular upstream and downstream faces, and 0.76 for bars with semicircular upstream face and tapering symmetrically to a small, semicircular, downstream face (teardrop). The geometric mean of the horizontal, longitudinal, approach velocity $v$ and the component of the velocity at right angles to the rack ($v \sin \theta$), that is, $v\sqrt{\sin \theta}$, is considered the effective velocity.

A rack of ⅜-in. rectangular bars placed at an angle of 60 deg ($\sin \theta = 0.866$) to the horizontal and possessing clear openings of ¾ in., for example, produces a loss of head $h = 2.42(½)^{4/3} h_v \times 0.866 = 0.83 h_v$ when the rack is clean.

The maximum head loss through clogged racks and screens is generally kept below 2.5 ft. The annual per capita rakings from racks with clear openings of 0.5, 1, and 2 in. are normally about 0.2, 0.1, and 0.02 cu ft respectively. Fine screens will remove 0.2 to 1.0 cu ft of screenings (2 to 20% of the suspended solids), depending on the size of their openings. Peak collections of rakings and screenings may rise as high as five times the average quantity.

Rakings and screenings are unsightly and putrescible. Presses and centrifuges cannot reduce their water content much below 65%. Although they can be digested with other wastewater solids (Sec. 20–13), burial or incineration is more common. Their removal and separate disposal are avoided by comminuting them in the flowing wastewaters. One form of *cutting screen* or *comminuter* used for this purpose consists of a revolving, slotted drum equipped with knives that shear the coarse materials collected on the drum against a comb (Fig. 13–3a). The solids are chopped down until the wastewater carries

[4] O. Kirschmer, Untersuchungen uber den Gefallsverlust an Rechen (Investigations of Head Loss in Racks), *Trans. Hydraulic Inst.*, Munich, R. Oldenbourg, 21 (1926).

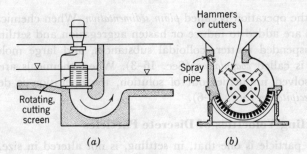

**Figure 13-3.**
(a) **Cutting screen or comminuter. (Chicago Pump Co.)** (b) **Shredder or disintegrator. (Jeffrey Co.)**

them through the 3/16-in. to 3/8-in. slots of the drum and into the effluent channel. The returned solids are too small to clog pumps or to float at the water surface. Grinding rakings and screenings after they have been removed from the wastewater and returning them to the flow are also common practices (Fig. 13-3b). To permit the discharge of domestic wastewaters into a pressurized system of small-diameter conduits, a storage-grinder-pumping unit has been developed under the auspices of the American Society of Civil Engineers.

## 13-3 Sedimentation

The erosion of the land by runoff from rainstorms carries vast amounts of soil and debris into streams and other water courses. To the resulting suspensions of mineral soil and organic debris are added community and industrial wastes that are transported through their collecting systems, spilled from storm drains and the overflows of combined systems of sewerage, or left within the treated effluents from separate and combined sewers. Some of the eroded particles and the jetsam of human life and enterprise are heavy enough to settle when flood waters subside, often to be picked up again at rising river stages, to be redeposited farther downstream in successive waves, and eventually to reach the ocean. In this way mud and sludge banks are formed, destroyed, and shifted downstream. When river waters come to rest in ponds, lakes, and impoundages, gravitational forces also succeed in pulling fine or otherwise light particles to the bottom. There they accumulate and build up deltas in inland bodies of water as well as in the sea.

In water and wastewater treatment, *sedimentation* or the removal, by gravitational settling, of suspended particles heavier than water is perhaps the most widely useful operation. When the impurities are separated from the suspending fluid by gravitation and natural aggregation of the settling particles, the operation is called *plain sedimentation*. When chemical or other substances are added to induce or hasten aggregation and settling of finely divided suspended matter, colloidal substances, and large molecules, the operation is called *coagulation* (Sec. 16-3). When chemicals are added to throw dissolved impurities out of solution, the operation is described as *chemical precipitation* (Sec. 16-6).

## 13-4 Settling Velocities of Discrete Particles

A discrete particle is one that, in settling, is not altered in size, shape, or weight. In falling freely through a quiescent fluid, such a particle accelerates until the frictional resistance, or drag, of the fluid equals the weight of the particle in the suspending fluid. Thereafter, the particle settles at a uniform (terminal) velocity, which is an important hydraulic attribute or characteristic of the particle. Accordingly,

$$F_I = (\rho_s - \rho) gV \qquad (13\text{-}2)$$

where $F_I$ is the impelling force, $g$ the gravity constant, $V$ the volume of the particle, and $\rho_s$ and $\rho$ are respectively the mass density of the particle and the fluid.

The drag force $F_D$ of the fluid, on the other hand, is a function of the dynamic viscosity $\mu$ and mass density $\rho$ of the fluid, and of the velocity $v_s$ and a characteristic diameter $d$ of the particle. To be fully representative, this diameter must reflect (1) the orientation of the particle relative to its direction of motion, represented, for example, by its cross-sectional area or projected area at right angles to motion, and (2) the relative frictional surface of the particle in contact with the fluid, represented, for example, by its surface area in relation to its volume. Dimensionally, therefore, $F_D = \phi(v_s, d, \rho, \mu)$ or, designating dimensional relations by square brackets, $[F_D] = [v_s^x d^y \rho^p \mu^q]$. Introducing the fundamental units of mass $m$, length $l$, and time $t$ of the various parameters into this equation, $[mlt^{-2}] = [m^{p+q} l^{x+y-3p-q} t^{-x-q}]$, and solving for $x, y,$ and $p$ in terms of $q$,

$$F_D = v_s^2 d^2 \rho \phi(v_s d\rho/\mu) = v_s^2 d^2 \rho \phi(\mathbf{R}) \qquad (13\text{-}3)$$

where $\mathbf{R}$ is the Reynolds number and the values of $\rho$ and $\mu$ are given in Table 3 of the Appendix.

This dimensionally derived relationship for the frictional drag has been verified experimentally.

By substituting the cross-sectional, or projected, area $A_c$ at right angles to the direction of settling for $d^2$, the dynamic pressure $\rho v_s^2/2$ for $\rho v_s^2$, and Newton's *drag coefficient* $C_D$ for $\phi(\mathbf{R})$,

$$F_D = C_D A_c \rho v_s^2/2 \qquad (13\text{-}4)$$

The magnitude of $C_D$ is not constant, but varies with **R** as shown in Fig. 13-4. For spheres, the observational relationships between $C_D$ and **R** are approximated by the following equation (upper limit of $\mathbf{R} = 10^4$):

$$C_D = \frac{24}{\mathbf{R}} + \frac{3}{(\mathbf{R})^{1/2}} + 0.34 \tag{13-5}$$

Equations 13-2 and 4 can be combined to establish a general relationship for the settling or rising of free and discrete particles, as follows:

$$v_s = \{(2g/C_D)[(\rho_s - \rho)/\rho](V/A_c)\}^{1/2} \tag{13-6}$$

or, for spherical particles, $V = (\pi/6)d^3$ and $A_c = (\pi/4)d^2$,

$$v_s = \{(^4/_3)(g/C_D)[(\rho_s - \rho)/\rho]d\}^{1/2}$$

or, approximately,

$$v_s \simeq [(^4/_3)(g/C_D)(s_s - 1)d]^{1/2} \tag{13-7}$$

Here $s_s$ is the specific gravity of the particle and $d = {}^3/_2 V/A_c = 6 V/A$, where $A$ is the surface area of the particle.

For *eddying resistance* at high Reynolds numbers ($\mathbf{R} = 10^3$ to $10^4$), $C_D$ assumes a value of about 0.4, and

$$v_s \simeq [3.2\, g(s_s - 1)d]^{1/2} \tag{13-8}$$

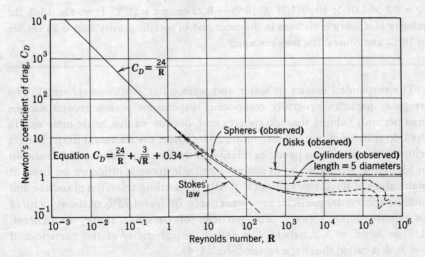

**Figure 13-4.**

**Newton's coefficient of drag for varying magnitudes of Reynolds number. [Observed curves after T. R. Camp, *Trans. Am. Soc. Civil Engrs.*, 103, 897 (1946).]**

For viscous resistance at low Reynolds numbers ($\mathbf{R} < 0.5$), $C_D = 24/\mathbf{R}$, and Eq. 13–6 reads as follows:

$$v_s = (g/18)[(\rho_s - \rho)/\mu]\, d^2$$

and, approximately,

$$v_s \simeq (g/18)[(s_s - 1)/\nu]\, d^2 \tag{13-9}$$

Equation 13–9 is Stokes' law,[5] $\nu$, being the kinematic viscosity with values shown in Table 3 of the Appendix.

To span the region between the Stokes range and the turbulent range, the curves shown in Fig. 13–5 are useful.

*Example 13-1.*

Find the settling velocity in water at 20°C of spherical silica particles of specific gravity 2.65: (1) $5 \times 10^{-3}$ cm in diameter and (2) $10^{-1}$ cm in diameter.

1. From Table 3, Appendix, $\nu = 1.010 \times 10^{-2}$ cm²/sec at 20°C, and $1.310 \times 10^{-2}$ at 10°C. From Eq. 13–9, $v_s = 981(2.65 - 1.00)(5 \times 10^{-3})^2/(18 \times 1.01 \times 10^{-2}) = 0.222$ cm/sec. Hence $\mathbf{R} = (2.22 \times 10^{-1})(5 \times 10^{-3})/(1.01 \times 10^{-2}) = 1.1 \times 10^{-1}$, and Stokes' law applies.

2. From Fig. 13–5, $v_s = 0.2$ at 10°C, whence, in proportion to $\nu$ at 10 and 20°C, $v_s = 0.2 \times 1.01 \times 10^{-2}/(1.31 \times 10^{-2}) = 0.23$ cm/sec at 20°C. From Fig. 13–5, the velocity of a particle $10^{-1}$ cm in diameter and of specific gravity 2.65 is 21 cm/sec at 10°C, and Stokes' law does not apply.

The suspended matter in water and wastewater is seldom spherical. The irregular particles generally composing suspensions possess greater surface area per unit volume than do spheres and, because of this, settle more slowly than do spheres of equivalent volume. Moreover, the frictional drag changes with the orientation of particles relative to the direction of motion. As shown in Fig. 13–4, irregularities in shape exert their greatest influence on drag at high values of $\mathbf{R}$. At low values ($\mathbf{R} \leq 10$), the settling velocities of rodlike and disklike spheroidal particles are respectively 78% and 73% of the velocity of an equal-volume sphere.[6] For particles of irregular shape, in general, $A/V = 6/\psi d = S/d$, where $\psi$ is called the sphericity of the particle and $S = 6/\psi$ is called the shape factor (Sec. 14–4).

[5] Stokes derived his law from theoretical considerations of the motion of a spherical pendulum in a fluid (*Trans. Cambridge Phil. Soc.*, **8**, 287 [1845]).

[6] John S. McNown and Jamil Malaika, Effects of Particle Shape on Settling Velocity at Low Reynolds Numbers, *Trans. Am. Geophys. Union*, **31**, 74 (1950).

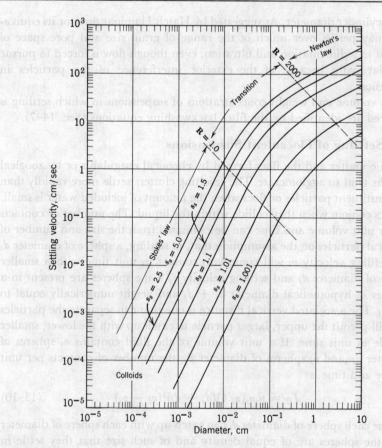

**Figure 13-5.**
**Settling and rising velocities of discrete spherical particles in quiescent water at 10°C. For other temperatures, multiply the Stokes values by $\nu/(1.31 \times 10^{-2})$, where $\nu$ is the kinematic viscosity at the stated temperature.**

## 13-5 Hindered Settling of Discrete Particles

When a discrete particle settles through a liquid in *free fall*, the liquid displaced by the particle moves upward through an area large enough not to interpose friction. In *hindered* settling, by contrast, particles are spaced so closely that the friction rises as the velocity fields around the individual particles interfere. In laboratory observations of attained velocities, the walls of narrow containers in which even a single particle is settling may have a similar hindering influence. However, this *wall effect* becomes negligible at high Reynolds numbers or when the particle diameter is less than about 1%

of the cylinder diameter. As suggested by Hatch,[7] laminar flow, or its equivalent, may persist over much of the range of grain size and pore space of interest in sedimentation and filtration, even though flow is forced to pursue irregular paths because of the exterior interference of the particles in suspension.

The volume and weight concentrations of suspensions in which settling is hindered are identified by the filter backwashing equations (Sec. 14-7).

## 13-6 Settling of Flocculent Suspensions

Organic matter and the flocs formed by chemical coagulants or by zoogleal growths tend to agglomerate. The resulting clusters settle more rapidly than the constituent particles or flocs when the amount of included water is small.

Flocs conjoin when they collide within the liquid. The number of contacts $J_{ij}$ per unit volume and time can be estimated from the size and number of spherical particles on the assumption that, in settling, a sphere of diameter $d_i$ and settling velocity $v_i$ will come into contact in unit time with a smaller sphere of diameter $d_j$ and settling velocity $v_j$ if the spheres are present in a cylinder of hypothetical diameter $d_i + d_j$ and height numerically equal to $v_i - v_j$. The associated vertical distance is one that can separate the particles and still permit the upper, larger particle to catch up with the lower, smaller particle in unit time. If a unit volume of the fluid contains $n_i$ spheres of diameter $d_i$, and $n_j$ spheres of diameter $d_j$, the number of contacts per unit volume and time is

$$J_{ij} = n_i n_j (\pi/4)(d_i + d_j)^2 (v_i - v_j) \qquad (13\text{-}10)$$

because each sphere of diameter $d_i$ can catch up with each sphere of diameter $d_j$. If the spheres are of equal density and of such size that they settle in accordance with Stokes law, the difference in settling velocities is $v_i - v_j = (g/18)[(s_s - 1)/\nu](d_i^2 - d_j^2)$ and the relationship

$$J_{ij} = n_i n_j (\pi/72) g[(s_s - 1)/\nu](d_i + d_j)^3 (d_i - d_j) \qquad (13\text{-}11)$$

implies that contact, and, with it, possible aggregation, is greatest for a large concentration of particles of large size, large relative weight, and large size difference in a liquid of small viscosity (for example, water of high temperature). More precisely, $d_i$ and $d_j$ are the diameters of the sphere of influence of van der Waals attractions (Sec. 16-2).

Most floc aggregations formed in water and wastewater are relatively fragile. As they grow in size, velocity gradients across them grow larger and may break them up at some *limiting* size. As a rule, flocculent suspensions

---

[7] L. P. Hatch, Flow of Fluids through Granular Material, *Trans. Am. Geophys. Union*, **24**, 536 (1943).

entering settling tanks in water and wastewater treatment works have not yet reached this limit, and sedimentation is improved materially by further floc growth (Sec. 15-2). Floc formation, depending on the nature of the suspended solids and the settling process, may be self-induced or induced by chemical coagulation (Sec. 16-3). In either case controlled stirring should promote floc growth (Sec. 15-3).

## 13-7 Efficiency of an Ideal Settling Basin

For convenience of discussion, a continuous-flow basin can be divided into four zones: (1) an inlet zone, in which influent flow and suspended matter disperse over the cross-section at right angles to flow; (2) a settling zone, in which the suspended particles settle within the flowing water; (3) a bottom zone, in which the removed solids accumulate and from which they are withdrawn as underflow; and (4) an outlet zone, in which the flow and remaining suspended particles assemble and are carried to the effluent conduit (Fig. 13-6).

The paths taken by *discrete* particles settling in a horizontal-flow, rectangular or circular basin are shown in Fig. 13-6. They are determined by the vector sums of the settling velocity $v_s$ of the particle and the displacement velocity $v_d$ of the basin. All particles with a settling velocity $v_s \geq v_0$ are removed, $v_0$ being the velocity of the particle falling through the full depth $h_0$ of the settling zone in the detention time $t_0$. Also, $v_0 = h_0/t_0$, $t_0 = C/Q$, and $C/h_0 = A$, where $Q$ is the rate of flow, $C$ the volumetric capacity of the

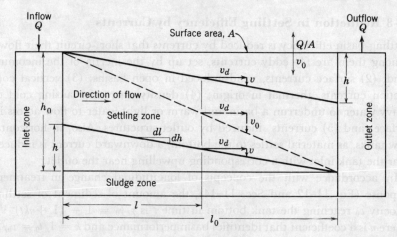

**Figure 13-6.**
**Settling paths of discrete particles in a horizontal-flow tank (idealized). The capacity of the settling zone is $C$, and its surface area is $A$.**

settling zone, and $A$ its surface area. Therefore, $v_0 = Q/A$ is the surface loading or overflow velocity of the basin. In vertical-flow basins, particles with velocity $v_s < v_0$ do not settle out. By contrast, such particles can be removed in horizontal-flow basins if they are within vertical striking distance $h - vt_0$ from the sludge zone. If $y_0$ particles possessing a settling velocity $v_s \leq v_0$ compose each size within the suspension, the proportion $y_0/y$ of particles removed in a horizontal-flow tank becomes

$$y/y_0 = h/h_0 = (v_s t_0)/(v_0 t_0) = v_s/v_0 = v_s/Q/A \qquad (13\text{-}12)$$

These relationships follow also from the geometry of Fig. 13-6.

Derived by Hazen[8] in somewhat different fashion, Eq. 13-12 states that, for *discrete* particles and *unhindered* settling, basin efficiency is solely a function of the settling velocity of the particles and of the surface area of the basin relative to the rate of flow, which, in combination, constitute the surface loading or overflow velocity. The efficiency is otherwise independent of basin depth and displacement time or detention period. It follows that particles with settling velocity $v_s \geq v_0$ are removed and that particles with velocity $v_s < v_0$ can be fully captured in horizontal-flow basins if false bottoms, trays, or stacks of hexagonal tubes or similar structures are inserted at vertical intervals $h = v_s t_0$. The larger the number of inserts the smaller can be the settling velocity of particles. Conceptually, therefore, filters are approached by settling basins with a very large number of inserts.

## 13-8 Reduction in Settling Efficiency by Currents

Settling-basin efficiency is reduced by currents that short-circuit their flows. Among them are (1) eddy currents, set up by the inertia of the incoming fluid; (2) surface currents, wind-induced in open basins; (3) vertical convection currents, thermal in origin; (4) density currents, causing cold or heavy water to underrun a basin and warm or light water to flow across its surface; and (5) currents induced by outlet structures. Also, in horizontal flow tanks, as material settles to the bottom, a downward current is induced near the tank inlet with a corresponding upwelling near the outlet.

In accordance with the concepts of longitudinal change in treatment response (Eq. 11-12 and Sec. 11-14), the amount of sediment of settling velocity $v_0$ reaching the tank bottom in time $t$ is $y/y_0 = 1 - (1 + nkt)^{-1/n}$, where $n$ is a coefficient that identifies basin performance and $k = 1/t_0 = v_0/h_0$ is a coefficient that characterizes the settleability of the sediment in terms of the time $t_0$ required for a particle with settling velocity $v_0$ to settle through the

[8] Allen Hazen, On Sedimentation, *Trans. Am. Soc. Civil Engrs.*, **53**, 63 (1904).

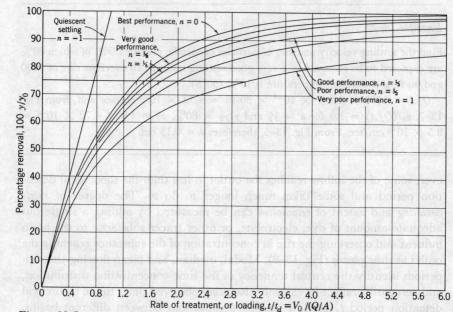

**Figure 13-7.**
**Performance curves for settling basins of varying effectiveness (after Hazen footnote 8).**

filled depth $h_0$ of the basin. The basin coefficient $n$ has a lower limit of zero and an upper limit of 1. Because $t/t_0 = v_0/(Q/A)$, moreover,

$$y/y_0 = 1 - [1 + nv_0/(Q/A)]^{-1/n} \quad (13\text{-}13)$$

The validity of this equation is supported by Hazen's theory of sedimentation.[8] Settling curves for various values of $n$ are shown in Fig. 13-7.

From a mathematical analysis of longitudinal mixing in settling tanks, Thomas and Archibald[9] have concluded that the value of $n$ can be approximated by the ratio of the difference between the mean and modal flowing-through periods to the mean flowing-through period (Fig. 13-8). Common values of $n$ are 0, ⅛, ¼, ½, and 1 respectively for ideal, very good, good, poor, and very poor performance.

## 13-9 Short-Circuiting and Basin Stability

In an ideal basin displacement is steady and uniform, and each unit volume of fluid is detained for a time $t_d = C/Q$. Even in well-designed basins, how-

[9] H. A. Thomas and R. S. Archibald, Longitudinal Mixing Measured by Radioactive Tracers, *Trans. Am. Soc. Civil Engrs.*, **117**, 839 (1952).

## Example 13-2.

Find the settling velocity and size of particles of specific gravity 1.001, of which 80% are expected to be removed in a very good settling basin at an overflow rate of 1000 gpd/sq ft, if the water temperature is 10°C (50°F).

$Q/A = 1000 \times 1.547 \times 10^{-6} \times 30.48 = 4.72 \times 10^{-2}$ cm/sec and, from Fig. 13-7, $v_0/(Q/A) = 1.8$ for $n = \frac{1}{8}$ and $y/y_0 = 80\%$, $v_0 = 1.8 \times 4.72 \times 10^{-2} = 8.5 \times 10^{-2}$ cm/sec. From Fig. 13-5, therefore, $d = 0.15$ cm.

---

ever, some of the inflow reaches the outlet in less than the theoretical detention period and some takes much longer to do so. The degree of *short-circuiting* and extent of *retardation* can be measured by adding a single but adequate amount of dye, electrolyte, or other tracer substance to the basin influent and observing the rise in concentration of the substance reaching the outlet as time passes (Fig. 13-8). Modal, median, and mean flowing-through periods identify the central tendency of the time-concentration distribution, and percentiles reflect its variance. Relating observed times to the theoretical detention period $t_d$ permits making comparisons between different basins. For the complementary concept of *longitudinal mixing*, see Sec. 11-13.

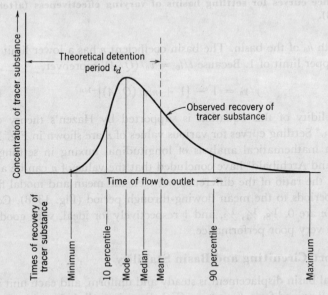

**Figure 13-8.**
**Dead spaces and short-circuiting in a settling basin are reflected in the concentration and time of recovery of tracer substances.**

Spaces in which the flow rotates upon itself receive no suspended solids, do no work, and reduce the effective capacity of the basin while shortening the flowing-through times relative to the theoretical detention period, $t_d$. In the absence of such currents, the ratio of the mean time to $t_d$ must equal unity. In the absence of short-circuiting, the mean, median, and mode must coincide. Short-circuiting is characterized, therefore, by the ratio of the mode or median to the mean being less than unity or by the ratio of the difference between the mean and mode, or the mean and median, to the mean being large.

If there is some interchange of flow between ineffective spaces and active portions of the tank, the time-concentration curve becomes unduly long, because small amounts of tracer material are released but slowly for capture in the effluent. The proportion of tracer substance arriving at the outlet in a given time equals the ratio of the area under the frequency curve as far as the given time to the total area or total dose of tracer substance. Percentile ratios then identify the degree of variability of exposure to sedimentation.[10] If the time-concentration curve of a basin does not reproduce itself reasonably well in repeated tests, flow through the tank is not stable, and tank performance may be erratic.

## 13-10 Scour of Bottom Deposits

As suggested by Ingersoll, McKee, and Brooks,[11] fine, light, and flocculent solids settling from coagulated waters, biologically treated wastewaters, and the like may be lifted from the sludge zone when $v_s = (\tau/\rho)^{1/2}$. Here $\tau$ is the shear stress at the liquid-sludge interface and $\rho$ is the density of the supernatant water. Because $\tau/\rho = grs$ in accordance with Eq. 8-4, Sec. 8-3, and $rs = (f/8g) v_d^2$ in the Weisbach-Darcy equation, $(\tau/\rho)^{1/2} = v_s = v_d(f/8)^{1/2}$, where $v_d$ is the displacement velocity, and

$$v_d = [(8/f)(\tau/\rho)]^{1/2} = (8/f)^{1/2} v_s \qquad (13\text{-}14)$$

It follows that $v_d$ should be kept well below 18 $v_s$ for $f = 2.5 \times 10^{-2}$. Accordingly, the ratio of length to depth in rectangular basins or of surface area $A$ to cross-sectional area $a$ must be kept below

$$A/a = l_0/h_0 = (v_d t_d)/(v_0 t_0) = (8/f)^{1/2}(t_d/t_0)(v_s/v_0) \qquad (13\text{-}15)$$

for $(t_d/t_0) = 1$. Then, where $v_s = v_0$, $l_0/h_0 \leq 18$.

[10] D. Thirumurthi, A Breakthrough in the Tracer Studies of Sedimentation Tanks, *J. Water Poll. Control Fed.*, **41**, R405 (1969).

[11] A. C. Ingersoll, J. E. McKee, and N. H. Brooks, Fundamental Concepts of Rectangular Settling Tanks, *Trans. Am. Soc. Civil Engrs.*, **121**, 1179 (1956).

The channel velocity initiating scour of particles deposited in sewers and hence also in grit chambers is shown in Eq. 8–9, Sec. 8–3. To avoid scouring velocities, the ratio of length to depth, or surface area to cross-sectional area, must be kept below a value of

$$l_0/h_0 = A/a = v_d t_d/v_0 t_0 = (t_d/t_0)[(6k/f)C_D]^{1/2} \qquad (13\text{–}16)$$

where $t_d/t_0$ equals unity for an ideal channel.

*Example 13–3.*

Find for alum floc ($s_s = 1.1$), $10^{-1}$ cm in diameter, the displacement velocity at which the floc can be removed without danger of resuspension and the length-to-depth ratio of the settling unit in which the removal can be effected. Assume a Weisbach-Darcy friction factor $f = 3.0 \times 10^{-2}$ and a temperature of 10°C.

By Eq. 13–14, $v_d/v_s = [8/(3 \times 10^{-2})]^{1/2} = 16.3$; and from Fig. 13–5, $v_s = 3.0$ cm/sec. Accordingly, $v_d = 3.0 \times 16.3 = 48.9$ cm/sec $= 1.60$ fps; and by Eq. 13–15, $l_0/h_0 = 16.3(t_d/t_0) = 16.3$ for an ideal basin ($t_d/t_0 = 1.0$).

That the removal efficiency of some suspensions in settling tanks is stepped up by gentle stirring is a not-uncommon observation. The potential of such useful power dissipation can be evaluated by the ratio of expected conjunctions in terms of von Smoluchowski's orthokinetic flocculation (Eq. 11–23), on the one hand, and quiescent settling (Eq. 13–11), on the other hand. This is also a useful consideration in sludge thickening (Sec. 20–5).

## 13-11 Elements of Tank Design

Each of the four functional zones of sedimentation basins and flotation tanks (Fig. 13–6) presents special problems of hydraulic and process design that depend on the behavior of the suspended matter within the tank, during removal, and after deposition as sludge or scum.

Size, density, and flocculating properties of the suspended solids, together with their tendency to entrain water, determine the geometry of the settling or rising zone. Their concentration by volume and the contemplated length of storage establish the dimensions of the bottom zone and the scum zone. In wastewater treatment, both the settling zone and the bottom and scum zones must take the possible putrescence of liquid and solids into account if the liquid is not to become septic and if gas-lifted solids are to be kept out of the effluent and away from the tank surface. Putrescence and excessive accumulations are avoided by removing sludge more or less continuously.

Removal devices affect tank design as well as tank operation. Thermal convection currents and wind-induced currents are held in check by housing or covering the tanks. The proper number of units is a matter of wanted flexibility of operation and economy of design.

The use of flocculating agents may add as many as three ancillary functions to settling units by requiring (1) rapid distribution of the agent through the water to be treated, (2) adequate reaction time or time for floc growth, and (3) return of floc to the influent for the sake of promoting flocculation.

## 13-12 Upflow Clarification

Unless vertical flow is introduced for the specific purpose of separating large or heavy, fast-settling particles from small or light, slow-settling particles by differential sedimentation, vertical-flow basins normally combine flocculation or contact filtration (also called sludge-blanket filtration) with sedimentation. If differential sedimentation or solids classification is the primary purpose, the small or light, discrete particles are purposely carried over the effluent weir of the tank while the large or heavy particles settle to the tank bottom. This is done in some grit chambers, for example. If floc buildup and removal is the primary function, entering particles are not intended to remain discrete, but to conjoin into aggregates that are eventually withdrawn from the upflowing water.

For floc buildup, vertical flow tanks (Fig. 13-9) are differentiated zonally. Inlet and sludge zone are in close contact, and the flocculation zone is occupied in part or as a whole by a cloud or blanket of flocs, not unlike a fluidized filter bed in its general nature (Sec. 14-7). Rising flocs or particles

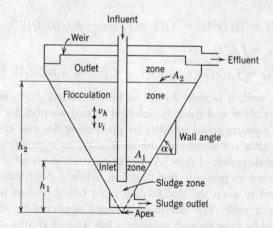

**Figure 13-9.**
**Vertical section through conical or pyramidal upflow tank.**

come into contact with settling flocs or particles and with a stationary cloud of flocs or particles in equilibrium with their hydraulic environment. An outlet zone at the top of the tank allows for some upward and downward displacement of the flocculation zone. In this respect it forms a buffer between tank and outlet.

In sludge blanket filtration, hydraulic operations are aimed at the (1) control of floc growth, (2) positioning of the flocculation-zone or floc-blanket surface, and (3) regulation of the intensity of floc shear. Hydraulic control is exerted by proper dissipation of hydraulic power and adjustment of residence time in the flocculation or contact zone.

The power dissipated, $P$, is $\rho g h_f Q$, where $\rho$ is the mass density of the fluid, $g$ the gravity constant, $h_f$ the head loss in passage through a zone of depth $h_2 - h_1$ (Fig. 13-9), and $Q$ the rate of flow. The useful loss of head (Eq. 14-14) equals the weight in water of the suspended floc, or $h_f = [(\rho_s - \rho)/\rho](1 - f_e)(h_2 - h_1) = (s_s - 1)(1 - f_e)(h_2 - h_1)$, where $\rho_s$ and $s_s$ are respectively the mass density and specific gravity of the flocs, and $f_e$ is the relative pore space of the flocculation zone. For a zonal volume or capacity $C$, cross-sectional area $A$, and wall angle $\alpha$, $C = \int A dh = 4 \cot^2\alpha \int h^2 dh = \frac{4}{3} \cot^2\alpha (h_2^3 - h_1^3)$ for a square pyramidal tank and $\pi \cot^2\alpha \int h^2 dh = (\pi/3) \cot^2\alpha (h_2^3 - h_1^3)$ for a conical tank. The detention time $t_d$ is $f_e C/Q$, and $f_e = (v/v_s)^{1/5} = (v_h/v_s)^{1/4}$, $v_s$, $v_h$, and $v$ being the settling, interstitial (hindered settling), and face velocities of the particles and fluid, respectively (Eq. 14-20). If these values are introduced into the relevant expressions for wanted and limiting velocity gradients, $G = \sqrt{P/\mu C}$ (Eq. 11-25) and the product of gradients and exposure time $t_d$ is formed to identify contact opportunity,

$$G = [(g/\nu)(s_s - 1)(1 - f_e)(h_2 - h_1)/(C/Q)]^{1/2} \quad (13\text{-}17)$$

and, because $t_d = f_e C/Q$,

$$Gt_d = f_e[(g/\nu)(s_s - 1)(1 - f_e)(h_2 - h_1)(C/Q)]^{1/2} \quad (13\text{-}18)$$

Here $\mu/\rho = \nu$, and, if $G$ and $Gt_d$ are to be controlled, $f_e$, $h$, and $C$ are the manageable variables and must be selected to suit wanted floc growth and clarification. To assure zonal stability by preventing floc rise and escape, the cross-sectional area (or width) of the tank may, as a practical matter, be enlarged in the direction of flow. At the same time, a useful initial $G$ value should be imposed to promote floc building, while a destructive terminal $G$ value is avoided to keep the floc formed from being broken up and swept over the effluent weir. For adequate contact opportunity, the period of residence should be long enough to accomplish wanted results.

Cross-sectional area is increased in the direction of flow by providing a wall angle of 45° to 65° with the horizontal (2 cot $\alpha$ = 2.00 to 0.93) to create a diameter of circular tanks or width of square tank as large as 2.00 to 0.93 times the distance from the apex. At a wall angle of 63° 26', incidentally, diameter $D$ and width $B$ equal the apical distance; that is, $D = B = h$.

## Example 13–4.

Given an upward flow of 0.5 cfs (323,000 gpd) in a square 45° pyramidal tank, with an effective flocculating zone between 2 ft above the apex (1 ft above the tank bottom) and 9 ft above the apex, find (1) the average $G$ and $Gt$ values ($\overline{G}$ and $\overline{Gt}$) and the limiting $G$ values ($G_2$ and $G_1$) at the top and bottom of the flocculating zone, and (2) the average settling velocities and diameters of the flocs when the average floc concentration is 40%, the termperature 50°F, and the average specific gravity of the floc 1.10. Assume that the settling is hindered and approximated by Stokes' law (Sec. 13–4).

1. From Table 3, Appendix, $\nu = 1.41 \times 10^{-5}$ (ft)$^2$/sec at 50°F or 1.31 stokes at 10°C; moreover, $g = 32.2$ ft/(sec)$^2$, $(s_s - 1) = 0.10$; $(1 - fe) = 0.40$; and $Q = 0.5$ cfs. For the recurrent product $[(g/\nu)(s_s - 1)(1 - fe)(h_2 - h_1)]^{1/2} = \{[32.2/(1.41 \times 10^{-5})]10^{-1}(4 \times 10^{-1})(9 - 2)\}^{1/2} = 8.00 \times 10^2$ and $(C/Q)^{1/2} = [4/3 \ (9^3 - 2^3)/0.5]^{1/2} = 43.9$, or $t_d = 1160$ sec. Consequently $\overline{G} = 8.00 \times 10^2/43.9 = 18.2$ sec$^{-1}$ by Eq. 13–16; and $Gt_d = 6 \times 10^{-1} \times 8.00 \times 10^2 \times 43.9 = 2.1 \times 10^4$ by Eq. 13–17.

For $G$ in upper 1 ft, $h_2 - h_1 = 1$, and $h_2{}^3 - h_1{}^3 = 729 - 512 = 217$, whence, in proportion to $\overline{G}$, $G = [(1/7)/(217/721)]^{1/2} \times 18.2 = 12.5$ sec$^{-1}$. For $G$ in lower 1 ft, $h_2 - h_1 = 1$, and $h_2{}^3 - h_1{}^3 = 3^3 - 2^3 = 19$, whence again, in proportion to $\overline{G}$, $G = [(1/7)/(19/721)]^{1/2} \times 18.2 = 42.5$ sec$^{-1}$.

2. For the upper 1 ft, the average value of $h$ at the center of gravity is 8.63 and the average area $A = 4 \times (8.63)^2 = 298$ sq ft. Hence the face velocity $v = (0.5/298)$ 30.5 = $5.12 \times 10^{-2}$ cm/sec. For $f_e = 0.6$, $fe^5 = 7.78 \times 10^{-2} = v/v_s$ and $v_s = 5.12 \times 10^{-2}/(7.78 \times 10^{-2}) = 0.657$ cm/sec, or $d = 4.0 \times 10^{-2}$ cm by Stokes' law as an approximation to hindered settling. For the lower 1 ft, the average value of $h$ is 1.63 ft and the average area $A$ is $4(1.63)^2 = 10.6$ sq ft. Hence $v = 1.44$ cm/sec and $v_s = 18.5$ cm/sec, or $d = 2.1 \times 10^{-1}$ cm by Stokes' law as an approximation to hindered settling.

If the upflow tank serves as a flocculation as well as a settling unit, the magnitudes of $G$ and $Gt_d$ suggest (1) that the flocs should grow well at average values of $\overline{G} = 18.2$ sec$^{-1}$ and $Gt_d = 2.1 \times 10^4$; (2) that the floc formed is not likely to be destroyed by shear in the upper foot of the tank, where $G = 42.5$ sec$^{-1}$; and (3) that, to settle into the lower foot, the floc must grow until it has a settling velocity of 18.5 cm/sec or its diameter is $2.1 \times 10^{-1}$ cm approximately. The surface loading of the upflow tank is $323,000/(2 \times 9)^2 = 690$ gpd/sq ft—therefore, not far from the normal loading of settling units.

## 13-13 General Dimensions of Settling Tanks

Horizontal-flow tanks and upward-flow tanks have been constructed in great variety. Thumbnail sketches of representative designs are shown in Figs. 13-10 and 11. Circular, square, or rectangular in plan, they vary in depth from 7 to 15 ft, 10 being a preferred value. Circular tanks are as much as 200 ft in diameter, but they are generally held to a 100-ft maximum to reduce wind effects. Square tanks are generally smaller. A side length of 70 ft is common. Rectangular tanks have reached lengths of almost 300 ft, but a 100-ft limit is generally imposed on them, too. The width of mechanically cleaned, rectangular tanks is dictated in part by the available length of scrapers. This is 16 ft for wooden scrapers, but scrapers can be operated in parallel in the same tank. A width of 30 ft is not unusual. The diameter of

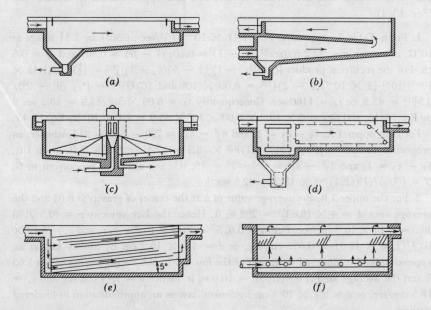

**Figure 13-10.**

Representative designs of horizontal and vertical displacement settling tanks. (a) Rectangular tank with longitudinal flow. Tank is thrown out of operation for cleaning. Solids are flushed to sump for removal from the dewatered tank. (b) Tank a, equipped with a single tray. (c) Circular tank with radial flow. Solids are plowed to central sump and withdrawn during operation. The rotary mechanism carries plows. (d) Tank a, mechanized. (e) Tank a filled with tubular settlers at small angle (5°) to the horizontal. (f) Tank provided with pipe grid for distribution of entering flow and with large-angle (45° to 60°) tubes near surface of tank. Solids are either scraped or flushed to influent or effluent end, thence to sump to be withdrawn.

General Dimensions of Settling Tanks/383

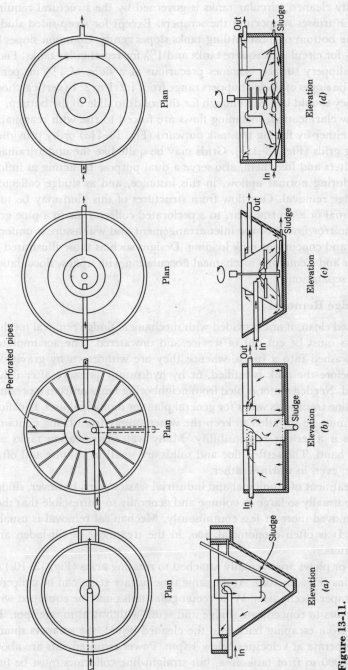

Figure 13-11.
Representative designs of upflow settling tanks. (a) Flaring circular or square tank. Sludge is removed during operation; in this case hydrostatically. (b) Circular tank with central mixing and flocculating chamber. (c) Circular, flaring tank with central mixing and flocculating chamber, intermittently collects sludge. (Spaulding Precipitator.) (d) Circular tank with central mixing and flocculating chamber and built-in sludge recirculation. (Accelator.)

mechanically cleaned, circular tanks is governed by the structural requirements of the trusses supporting the scrapers. Except for steep-sided sludge hoppers, the bottom of most settling tanks slopes gently. Common slopes lie close to 8% for circular or square tanks and 1% for rectangular tanks. Foothold on a slippery surface becomes precarious at a slope of 1½ in. per ft (12.5%). The slopes of sludge hoppers range from 1.2:1 to 2:1 (vertical:horizontal). They should be steep enough for the solids to slide to the bottom.

In upflow clarification, incoming flows are forced to rise with reasonable uniformity either by flaring the tank outward (Fig. 13-11$a$) or by providing distributing grids (Fig. 13-11$b$). Grids may be quite like the underdrainage piping of filters and like them, also serve a dual purpose by acting as inflow structures during normal upflow, in this instance, and as sludge collectors during sludge removal. Overflow from structures of this kind may be to a circumferential or a central weir, to a perforated collector, or to a pipe grid that is the mirror image of the inlet arrangement and will assure a uniform rate of rise and concurrent tank loading. Designs such as those illustrated in Figs. 13-11$c$ and $d$ combine mechanical flocculation with upflow flocculation and sludge separation.

### 13-14 Sludge Removal

To be flushed clean, if not provided with mechanical sludge removal mechanisms, tanks must be cut out of service and unwatered. The accumulated solids are washed into a sump, whence they are withdrawn by gravity or pumping before the tank is refilled, or by hydrostatic pressure after it has been refilled. Needed water is bled from neighboring tanks or from a pressure line. If the line transports water for general plant or municipal uses, backflow preventers must be installed to keep the water supply from being contaminated. This is a serious responsibility. Many water-purification tanks are cleaned by hand. The settled floc and solids are small in volume and often quite stable, even in warm weather.

In the treatment of municipal and industrial wastewaters, however, sludge deposits are usually so large in volume and generally so putrescible that they must be removed more or less continuously. Mechanical removal is usually warranted. It is often economical, too, in the treatment of silt-laden and softened waters.

Scrapers or plows are normally attached to rotating arms (Fig. 13-10$c$) or endless chains (Fig. 13-10$d$). At the same time surface scum can be collected with the scraper mechanism. Wide rectangular tanks may be equipped with cross-conveyors to concentrate sludge and scum withdrawal in one spot. To keep solids from escaping back into the cleaned liquid, the scrapers should preferably operate at velocities below 1 fpm. Power requirements are about 1 hp for 10,000 sq ft of tank area, but straight-line collectors must be furnished with motors about 10 times that strong in order to master the starting load. Sludge pipes operating much like suction cleaners can take the place of plows when the sludge is feathery and light (Fig. 13-10c).

Sludge is withdrawn from vertical-flow tanks before the sludge blanket rises high enough to be carried over the effluent weir. Hydraulic and biochemical stability are important considerations. Sludge recirculation by pumping or by a system of baffles may promote flocculation (Fig. 13-11).

By contrast, two-story tanks are constructed to separate accumulating solids from the flowing wastewaters and keep them fresh (Fig. 13-12). The solids are commonly held long enough to be digested (Sec. 20-13). By design, rising gas bubbles and sludge particles cannot escape from the sludge hopper into the settling compartment.

## 13-15 Inlet Hydraulics

For high efficiency, inlets must distribute flow and suspended matter more or less equally into batteries of tanks and within individual tanks. For hydraulic equality, the dividing flow must encounter equal frictional resistances or be subjected to a controlling loss of head, that is, a head large in comparison with the frictional resistances between inlets or inlet openings. The water levels of parallel tanks are held at the same elevation by regulating the outflow (Sec. 13-16). If suspended matter travels along the bottom of the influent conduit, hydraulic equality does not necessarily ensure loading equality. Necessary adjustments can then be made only by trial.

In reference to Fig. 13-13, the principles of flow regulation can be identified as follows. The flow originating at $I$ in Fig. 13-13a traverses identical paths before it is discharged at $A$, $B$, $C$, and $D$. Consequently flow is evenly distributed between the tanks and within them except when the water-surface elevation of the two tanks is not the same. Equality is then confined to points in the individual tanks.

The flow originating at $I$ in Fig. 13-13b is subdivided in such manner that the discharge $q_n$ through any inlet orifice is held to $mq_1$, where $m < 1$ and $q_1$

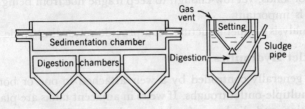

**Figure 13-12.**
**Two-storied Imhoff or Emscher tank.**

is the discharge through the first orifice. The head on the $n$th orifice must then be

$$h_n = kq_n^2 = k(mq_1)^2 = m^2 h_1 \qquad (13\text{-}19)$$

or, if $h_f$ is the head lost between points of discharge (1 and $n$), $h_n = h_1 - h_f = m^2 h_1$, and

$$h_1 = h_f/(1 - m^2) \qquad (13\text{-}20)$$

The magnitude of $h_f$ can be estimated from friction losses and velocity changes. Wanted flow distribution by the piping shown in Fig. 12-3 and in filter underdrains (Sec. 14-13) can also be obtained in this way.

*Example 13-5.*

The inlet farthest from the point of supply of a settling tank is to discharge 99% of the flow delivered by the nearest inlet. Find, in terms of the friction head $h_f$, (1) the required head loss through the nearest inlet and (2) the associated head loss through the farthest inlet.

1. From Eq. 13-20, $h_1 = h_f/[1 - (0.99)^2] = 50.3\ h_f$.
2. From Eq. 13-19, $h_n = (0.99)^2 h_1 = 0.980\ h_1 = 49.3\ h_f$.

Baffle boards in front of inlet openings will destroy much of the kinetic energy of the incoming water and assist in distributing the flow laterally and vertically over the basin. Training or dispersion walls perforated by holes or slots (Fig. 13-13) operate on the principle demonstrated in Eq. 13-19 by introducing a controlling head loss. Frictional resistance in advance of the openings is a function of the velocity head of the eddy currents. Baffles of this kind promote flow stability. However, by creating and destroying velocity, baffles can quickly build up head losses that are not offset by increased basin efficiency. Conduit and orifice velocities should be high enough to prevent deposition of solids, yet low enough to keep fragile floc from being destroyed when that is important.

Model analysis of inlet structures can provide rewarding guide lines.

## 13-16 Outlet Hydraulics

Outflow is generally controlled by a weir attached to one or both sides of single- or multiple-outlet troughs. If weirs in adjacent tanks are placed at the same elevation and discharge freely over the same length, the loading of equal basins should stay within the limits of inflow variation. If effluent weirs are submerged, the degree of submergence will vary along the trough. Draw-

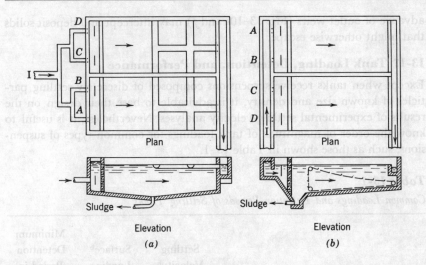

**Figure 13-13.**
**Inflow and outflow structures of settling tanks. (a) Uniformity of inflow is secured by equality of resistance. (b) Uniformity of inflow is secured by control of resistance.**

off then becomes unequal and induces short-circuiting, unless a training wall similar to the inlet training wall again introduces a controlling loss of head.

Outlet troughs are lateral spillways. Required dimensions are given by the drawdown curve of the water surface in the trough in much the same way as for the washwater troughs of rapid filters (Sec. 14–17).

Weir length relative to surface area determines the strength of outlet currents. By experience, this should lie below 30,000 gpd per ft to avoid surges. To prevent nonuniform discharge caused at low rates of flow by wind, obstruction to flow by stranded solids, and slight variations in weir level, the weir may be subdivided into triangular notches. This will also ventilate the nappe of the jets issuing from the notches and avoid changes in rates of flow caused by departures from free flow. Adjustable weir plates can be leveled from the water level in the tank.

In wastewater treatment a scum board, or shallow baffle plate, paralleling outlet weirs, will keep floating solids, grease, and oil from stranding on the weirs or reaching the effluent. In ordinary circumstances, variations in flow are of little concern in the design and operation of settling tanks. Provided that the maximum design flow is not exceeded, the mass of water they contain is large enough to smooth out normal flow variations. In grit chambers, on the other hand, allowable variations fall within relatively narrow limits and call for the use of flow-control devices (Sec. 13–18). Tubular inserts in advance of outlet weirs (Fig. 13-10e and f) may intercept and deposit solids that might otherwise escape.

## 13-17 Tank Loading, Detention, and Performance

Except when tanks receive suspensions composed of discretely settling particles of known size and density, it is advisable to base their design on the results of experimental settling-velocity analyses. Nevertheless, it is useful to know the order of magnitude of tank loadings for common types of suspensions, such as those shown in Table 13-1.

*Table 13-1*

Common Loadings and Detention Periods of Settling Tanks

| Composition of Suspension | Specific Gravity | Size (cm) | Settling Velocity[b] at 10°C (cm/sec) | Surface[b] Loading (gpd/sq ft) | Minimum Detention Period in 10-ft Basin (hr) |
|---|---|---|---|---|---|
| Alum and iron floc containing water | 1.002 | $\geq 10^{-1}$ | $8.3 \times 10^{-2}$ | 1800 | 1.0 |
| Activated sludge[c] | 1.005 | $10^{-1}$ | $2 \times 10^{-1}$ | 1200 | 1.5 |
| Calcium precipitate[d] | 1.2 | $10^{-2}$ | — | 900 | 2.0 |
| Wastewater organics | $\geq 1.001$ | $10^{-1}$ | $4.2 \times 10^{-2}$ | 900 | 2.0 |
| Silt, clay | 2.65 | $\geq 10^{-3}$ | $6.9 \times 10^{-3}$ | 150 | 12 |

[a] 1 cm per sec = 21,200 gpd per sq ft, and surface loading varies directly with velocity; also, 1 cm per sec = $8.47 \times 10^{-3}$ hr of detention per ft of basin depth, and the detention period varies directly with depth and inversely with velocity.

[b] Surface loading is that commonly used in practice.

[c] Settling is hindered by about 10% at solids concentration of 2500 mg/l. In practice, loading is reduced from a maximum of 4000 to 1200 gpd/sq ft to allow for poorly settling sludge.

[d] Specific gravity on dry basis < 1.2.

Departure from the ideal either decreases permissible loadings and increases requisite detention periods or lowers tank efficiencies. Hazen's theory of sedimentation will give a clue to attainable or required values. In general, coagulated and lime-softened waters, as well as municipal and industrial wastewaters subjected to plain sedimentation or coagulation in primary tanks, are provided detention periods of about 2 hr or surface loadings of 900 gpd per sq ft in tanks 10 ft deep above the sludge zone.

In wastewater treatment, capacities smaller than normal are employed for the removal of coarse solids only—for example, when effluents are discharged into receiving waters of large capacity, or when wastewaters are settled in advance of further treatment such as chemical precipitation or activated-sludge treatment. Capacities larger than normal are provided (1) when BOD removal as well as suspended-solids removal (Fig. 13-14) is to be high, (2) when settling tanks serve as buffers against surges of municipal and industrial wastewaters (especially toxic wastewaters), and (3) when abnormally large amounts of combined sewage are to be chlorinated or receive some other form of treatment in settling units.

**Tank Performance.** Most settling basins incorporated in water-purification works include or follow coagulation or chemical precipitation. Tank performance depends in considerable measure on the effectiveness of these preparatory processes and is not a unique measure of settling efficiency. Only when turbid river waters are subjected to plain sedimentation is it common practice to report the percentage removal of suspended solids. Otherwise, interest centers on the removal of such items as color, turbidity, hardness, and iron.

The weight of suspended matter in turbid river waters may run into thousands of milligrams per liter. Recorded efficiencies of removal by plain sedimentation vary widely for different rivers and different stretches of the same river. As little as 30% and as much as 75% of the suspended matter may settle out in 1 hr, and as little as 50% and as much as 90% in 2 hr. Very fine silt may not settle out even during many months of storage.

By contrast, the settling tanks of wastewater treatment works generally offer plain sedimentation only. Performance is usually expressed as percentage removal of suspended solids, BOD and COD. Typical settling curves

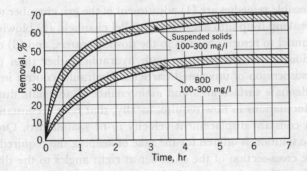

**Figure 13-14.**
**Removal of suspended solids and biochemical oxygen demand from sewage by plain sedimentation in primary tanks.**

for primary tanks are shown in Fig. 13-14. The Great Lakes-Upper Mississippi River Board of Public Health Engineers[12] has developed the following normally allowable surface loadings for settling tanks required to remove given percentages of BOD:

| BOD removal, % | 20 | 24 | 27 | 30 | 32 | 34 | 36 | 37 |
|---|---|---|---|---|---|---|---|---|
| Loading, gpd/sq ft | 2200 | 1800 | 1500 | 1200 | 1000 | 800 | 520 | 400 |

Further load reduction is of little consequence.

## 13-18 Grit Chambers

Gritty substances originate in stormwater runoff, industrial wastes, pumpage from excavations, and groundwater seepage. Grit is removed in advance of pumps and treatment units to prevent wear of machinery and unwanted accumulation of heavy inert matter in inverted siphons, settling tanks, and sludge-digestion units.

Grit chambers are intended to remove inert substances preferentially, because exclusion of decomposable solids simplifies grit disposal. Purposeful inclusion of organic matter converts grit chambers into *detritus tanks*. The organic matter may subsequently be separated from the grit and returned to the flow.

Most grit chambers are constructed as fairly shallow, elongated channels that capture particles with a specific gravity of 2.65 and a diameter of $2 \times 10^{-2}$ cm. Depth of flow is normally governed by the size of the outfall sewer. Except for the space assigned to grit storage, the invert of the chamber is made continuous with that of the outfall sewer.

Selective deposition of heavy inert particles is complicated by fluctuations in the rate of flow, especially by the large fluctuations accompanying storm rainfalls. Possible remedies are (1) adjustment of the grit chamber to differing flows by breaking it up into a series of parallel channels, (2) blowing air into the grit channel to return organics to the flowing waters, and (3) combining sedimentation of wanted particles with hydraulic rather than pneumatic scour or resuspension of unwanted particles. Hydraulic control implies provision of adequate surface area and maintenance of adequate displacement velocity. Fluctuations in flow require, ideally, that both a constant value of $Q/A$ and a constant displacement velocity $v_d$ be maintained. Ordinarily a compromise solution is offered for the sake of keeping the required structure simple. The cross-section of the chamber at right angles to the direction of flow is made uniform throughout its length, and the displacement velocity is held substantially constant at all depths of flow by placing a flow-control

---

[12] *Recommended Standards for Sewage Works*, adopted by Great Lakes-Upper Mississippi River Board of State Sanitary Engineers, Health Education Service, Albany, N.Y., 1968.

device, such as a *proportional-flow weir*, a *vertical throat*, or a *standing-wave flume*, at the end of the chamber. Deposition of wanted large and heavy particles is ensured by making the area of the water surface at maximum flow large enough. Selective movement and resuspension of smaller and lighter particles that settle as flow is reduced is engendered by keeping up the scouring action of the flowing water. In theory, a channel with flaring catenary sides should keep channel velocities the same at all depths of flow. However, this makes for difficult and uneconomical construction and operation.

Two outlet control devices are shown in Fig. 13-15: a proportional-flow weir and an adjustable throat. A standing-wave flume—a Parshall flume, for instance—could be used instead. Its advantage is that it can measure flow without requiring special calibration. The discharge $Q$ through devices such as these is a simple function of channel depth or head, $Q = kh^n$, where $k$ and $n$ are numerical constants. To have these control devices keep the displacement velocity in the chamber constant, therefore, the width $w$ of the chamber must be so chosen that $Q = kh^n = v_d \int_0^h w\, dh$. As shown by Camp[13] this condition is satisfied when

$$w = nkh^{n-1}/v_d \qquad (13\text{-}21)$$

[13] T. R. Camp, Grit Chamber Design, *Sewage Works J.*, **14**, 368 (1942).

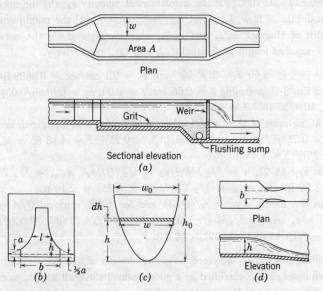

**Figure 13-15.**

Grit chamber and outlet control devices. (*a*) Twin-compartment grit chamber with weir control. (*b*) Proportional flow weir plate (Sutro weir). Narrowing opening at base is replaced by rectangular notch. (*c*) Hypothetical cross-section of channel controlled by throat. (*d*) Outlet throat. $Q = kbh^{3/2}$.

If flow is controlled by a Sutro weir,[14] for example,

$$Q = cb\sqrt{2ga}\,(h - \tfrac{1}{3}a) = k(h - \tfrac{1}{3}a) = kh' \qquad (13\text{-}22)$$

Here $c$ is about 0.62 for unsymmetrical weirs, that is, sharp-edged weirs with one vertical side, and 0.61 for symmetrical sharp-edged weirs; $k$ is a weir characteristic; and $l/b = 1 - (2/\pi)\tan^{-1}[(h/a) - 1]^{1/2}$, $h$, $l$, $a$, and $b$ being the weir dimensions shown in the symmetrical weir of Fig. 13-15$b$, or corresponding to them in an unsymmetrical weir. The grit-channel cross-section can be rectangular because $w = k/v_d$ is constant, the exponent $n$ equaling unity.

If flow is controlled by a throat, $Q = kbh^{3/2}$, or $n = {}^3/_2$, and $w = {}^3/_2(kbh^{1/2}/v_d) = {}^3/_2(Q/hv_d)$. This is the equation of a parabola, and the channel must be parabolic in cross-section or approach a parabola closely.

## Example 13-6.

Two grit channels, controlled by outlet throats 3 in. wide, are to remove particles of specific gravity $s_s = 2.65$ and diameter $d = 2 \times 10^{-2}$ cm from combined sewage. The maximum rate of flow, $Q_{\max}$, is 15 cfs; the minimum, $Q_{\min}$, is 3 cfs. Find (1) the settling velocity $v_s$ at 10°C; (2) the displacement velocity $v_d$; (3) the maximum and minimum depths of flow, $h_{\max}$ and $h_{\min}$, respectively; (4) the maximum and minimum width of channel, $w_{\max}$ and $w_{\min}$, respectively; and (5) the maximum and minimum required length of channel, $l_{\max}$ and $l_{\min}$, respectively.

1. From Fig. 13-5 for $d = 2 \times 10^{-2}$ cm, $v_s = 2.1$ cm/sec = 0.0689 fps.

2. From Eq. 8-9, assuming $k = 0.06$ and $f = 0.03$, $v_d = [8(0.06/0.03) \times 32.2 \times 1.65(2 \times 10^{-2}/30.48)]^{1/2} = 0.75$ fps.

3. For a discharge of $Q = kbh^{3/2}$, where $b = 0.25$ ft and $k$ approximates 3.5, $h = [Q/(3.5 \times 0.25)]^{2/3}$ and $h_{\max} = [7.5/(3.5 \times 0.25)]^{2/3} = 4.18$ ft, or $h_{\min} = [1.5/(3.5 \times 0.25)]^{2/3} = 1.43$ ft.

4. From Eq. 13-21, $w = {}^3/_2\, kbh^{1/2}/v_d = {}^3/_2\,(Q/hv_d)$, $w_{\max} = {}^3/_2\, 7.5/(4.18 \times 0.75) = 3.59$ ft, and $w_{\min} = {}^3/_2\, 1.5/(1.43 \times 0.75) = 2.10$ ft.

5. From Eq. 13-12, $y/y_0 = v_s/(Q/A)$, and for 100% removal $Q/A = v_s$, or ${}^2/_3\, hwv_d/(lw) = v_s$ and $l = {}^2/_3\, h(v_d/v_s)$. Hence $l_{\max} = {}^2/_3\, 4.18\,(0.75/0.0069) = 30.3$ ft, and $l_{\min} = {}^4/_3\, 1.43\,(0.75/0.069) = 10.3$ ft.

The settling zone must, therefore, be given a length of 30.3 ft.

If the channel can be classified as a good basin, its overall length, to ensure 75% removal of wanted particles, should be 1.7 times as long, or 52 ft.

---

[14] E. Soucek, H. E. Howe, and F. T. Mavis, Sutro Weir Investigations Furnish Discharge Coefficients, *Eng. News Rec.*, **117**, 679, 904 (1936).

The amount of grit collected by grit chambers varies from 1 to 12 (average 4) cu ft per million gallons treated and depends on the topography, surface cover, type of roadway and sidewalk, size of intercepter, and intensity of storm rainfall on the drainage area. Daily maxima of 10 to 30 cu ft per million gallons and as much as 80 cu ft per million gallons have been reported. Grit is generally stored by dropping the bottom of the chamber 6 to 18 in. below the calculated invert of the inlet and outlet channels. In small plants the accumulated solids are removed from the unwatered channels by hand or by flushing the grit onto a disposal area. In large plants some type of mechanical grit conveyor generally removes grit without emptying the compartment.

## 13-19 Flotation

In current practice flotation is essentially a unit operation of wastewater treatment rather than water treatment. In *natural flotation*, oil, grease, or other substances lighter than water are allowed to rise naturally to the water surface of quiescent tanks, where they are skimmed off in ways analogous to the removal of sludge from settling tanks (Sec. 13–14). In *air flotation*, particles heavier than water are lifted to the surface with the help of air and flotation reagents not unlike coagulation reagents in sedimentation. Flotation operations of this kind are borrowed largely from the metallurgical industry and are employed principally in the treatment of industrial wastes.

**Natural Flotation.** The natural gravitational rising of discrete particles is the obverse of natural gravitational settling. Stokes' law applies without change in formulation (Eq. 13–9). As shown in Example 13–7, the Stokes velocity acquires a negative sign.

---

*Example 13–7.*

Find the rising speed in water at 20°C of a spherical particle with a specific gravity of 0.80 and a diameter of $5 \times 10^{-3}$ cm.

By Eq. 13–9, $v = 981(0.80 - 1.00)(5 \times 10^{-3})^2/(18 \times 1.01 \times 10^{-2}) = -2.69 \times 10^{-2}$ cm/sec. Moreover, $\mathbf{R} = 2.69 \times 10^{-2} \times 5 \times 10^{-3}/(1.01 \times 10^{-2}) = 1.3 \times 10^{-2}$, and Stokes' law applies.

---

**Air Flotation.** Air is introduced into flotation tanks either by diffusion or mechanical dispersal of extraneous air, or by the precipitation of air dissolved in the fluid. Dispersed bubbles are normally about 1 mm in diameter, whereas bubbles precipitated from solution may be less than one tenth that size. Necessary air volumes may be generated by supersaturating the fluid under treatment with air. Exposure to air under pressures of 5 to 25 psig for

60 to 30 sec, respectively, will do this. If the suspending fluid has been saturated with air at atmospheric pressure, it must normally be placed under a vacuum of 9 to 10 in. Hg. After pressure aeration, flotation can proceed at atmospheric pressure.

Equation 13-23 shows the time of rise $t$ of an expanding air bubble that obeys Stokes' law.

$$t = -\,^3/_5 \frac{(h^{5/3} - h_0^{5/3})}{kd_0^2(h_0 - 34)^{2/3}} = \frac{10.8(h_0^{5/3} - h^{5/3})}{g(s_s - 1)d_0^2(h_0 - 34)^{2/3}} \quad (13\text{-}23)$$

Here $h_0$ is the depth of water in which a bubble of minimum diameter $d_0$ is released. As the bubble rises, its volume $V_0$ and diameter $d_0$ increase to $V$ and $d$ respectively, where $V_0 p_0 = Vp$, as the pressure decreases isothermally.

The density of moist air in grams per liter is

$$\rho_a = 1.2929(273.1/T_C)[(p_a - 0.3783 p_w)/760] \quad (13\text{-}24)$$

where $T_C$ is the absolute temperature in degrees centigrade, $p_a$ is the barometric pressure in millimeters Hg, and $p_w$ is the vapor pressure of the moisture in the air in millimeters Hg.

Buoyancy can be imparted by air bubbles (1) when they are entrapped physically in flocculent or flocculated aggregates of particles and (2) when they are attracted to particles by interfacial forces and adhere to them. A three-phase (gas-liquid-solid) system is created in which the following three interfacial tensions exert governing effects: (1) the gas-solid tension, (2) the solid-liquid tension, and (3) the gas-liquid tension. Small bubbles have high capillary pressure, which may increase with gas absorption faster than the drop in pressure that accompanies the rise.

## 13-20 Flotation Reagents[15]

Surface-active additives may promote air flotation by altering the properties of the system. *Frothing* or *foaming reagents* reduce the surface tension and promote bubble formation at the water-air interface where a large and stable surface foam can support the particles lifted into it. Among useful reagents are the higher alcohols and pine oil. Depressing the surface tension of the suspending fluid too much may keep particles from adhering to the surface. *Collecting reagents* increase the interfacial tension between air and particle and decrease the interfacial tension between the particle and the liquid. Among the useful reagents are fatty acids and soaps. The usefulness of collectors is enhanced by *activating reagents* that react with molecules at the surface of particles to produce a surface on which the collecting reagent can be ad-

---

[15] R. S. Burdon, *Surface Tension and the Spread of Liquids*, Cambridge University Press, Cambridge, England, 1949.

sorbed. *Depressing reagents* accomplish the opposite task. In the proper combination, activators and depressants will promote the flotation of desired phases.

## 13-21 Flotation and Skimming Tanks

The operation of *air-flotation tanks* is supported by (1) air injectors and pressure tanks, pumps, and regulators in the case of compressed-air flotation, and (2) air injectors and vacuum pumps in the case of vacuum flotation. In both instances air bubbles lift buoyant particles to the surface while particles too heavy to be lifted sink to the bottom. Mechanical skimmers and scrapers remove the accumulating solids. A flow diagram is sketched in Fig. 13-16.

Important concepts in air-flotation performance are air-to-solids ratios in relation to (1) effluent solids and (2) float solids concentrations. Pilot plant studies are essential undertakings prior to design.

**Skimming Tanks.** Skimming tanks are commonly long, trough-shaped structures. Surface area, in accordance with the principles of sedimentation and flotation, is a governing factor. Detention periods seldom exceed 3 min. In the structure shown in Fig. 13-17, air is blown into the sewage from diffusers situated in the tank bottom.[16] This keeps heavy solids from settling, lifts light solids, and captures them in the surface froth. Vertical baffle walls separate the tanks into a central aerated channel and two lateral stilling chambers in which oil and grease gather at the surface. The baffles are slotted near the flowline to give entrance to the stilling compartments. Settleable matter slides back into the central channel. There it is moved forward and eventually delivered to the inclined outlet from the tank. The outlet velocity is high

[16] Karl Imhoff, *Taschenbuch der Stadtentwässerung (Pocket Book of Urban Frainage)*, 21st ed., R. Oodenbourg, Munich, 1965.

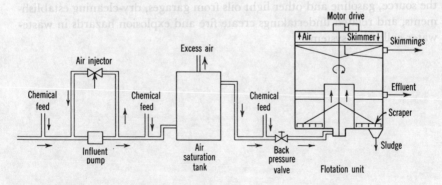

**Figure 13-16.**

**Flow diagram of compressed-air flotation unit.**

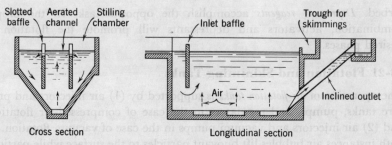

**Figure 13-17.**
**Aerated skimming tank. (After Imhoff.)**

enough to resuspend the solids. Oil and grease are drawn off from time to time into a channel leading to a separator. Air requirements are about 0.03 cu ft per gallon of sewage. Where there is no advantage in aerating the wastewater and the scum is not unsightly, skimming can be combined with sedimentation in tanks equipped with scum-removal mechanisms.

The grease from domestic or municipal wastewaters is generally too polluted to be of commercial value and is best added to digestion tanks. Skimmings containing much mineral oil are best buried or burned, together with rakings and screenings. The volume of skimmings from municipal wastewaters approximates 0.1 to 6.0 cu ft per million gallons or 0.003 to 0.2 cu ft per capita annually. Some industrial wastes—wool-scouring wastes, for example—are rich in grease that can be recovered for sale. Flotation agents are normally added only to industrial wastewaters.

Grease traps and oil separators are employed in some industries and in many large kitchens. They are designed as small skimming tanks or skimming pots with a submerged inlet and bottom outlet. Unless they are separated at the source, gasoline and other light oils from garages, dry-cleaning establishments, and related undertakings create fire and explosion hazards in wastewater-collection systems.

**fourteen**

# filtration

## 14-1 Natural and Managed Filtration

The seepage of rainfall and runoff into porous soils and rocks and the storage and movement of groundwaters in open-textured geological formations are important elements in the resource and quality management of water and wastewater. Although fine-textured granular materials remove pollutants, the water drawn from them is acceptable only when *natural filtration* together with *time lapse* between pollution and use bar the rapid transport of pollutants to springs, wells, and infiltration galleries.

Natural and managed recharge of aquifers with surface or spent waters may serve one or more of the following purposes: (1) water conservation by underground rather than surface storage; (2) water quality improvement by storage and filtration; (3) replenishment of overdrawn aquifers; and (4) freshwater protection by the erection of groundwater barriers against saltwater intrusion.

That some groundwaters dissolve gases and solids in their passage through the soil and that acquired impurities may not be removed by natural filtration are shown in Secs. 12-1 and 12.

Early water filters were constructed as small sand beds in 1804 by John Gibb in Paisley, Scotland, and as large sand beds in 1828 by James Simpson, for the clarification of Thames River water by the Chelsea Water Company of London, England. Filters of this kind were not introduced into the United States until 1872, when

James P. Kirkwood was commissioned by Poughkeepsie, N.Y., to construct an *English* filter on the banks of the Hudson River.

Early wastewater filters evolved as natural sand beds from New England *sewage farms*. These as well as more sophisticated granular filters may be pressed into service with increasing frequency as higher-quality standards for wastewater effluents become more general in the protection of the water resource.

## 14-2 Granular Water Filters

Water filters constructed for cities and towns during most of the nineteenth century were *slow sand filters* that consisted of beds of *run-of-bank* sands provided with tile underdrains and with inlet and outlet wells. The beds were placed in operation by being filled from below to displace air from their pores, thereafter being kept flooded when in operation, even while the beds were being cleaned. Depending on the substances to be removed and their concentration, rates of filtration were seldom less than 1 mgad (million gallons per acre daily) or more than 10 mgad. In warm climates filter beds were usually left open to the sun and air. In cold climates they were generally covered to keep them from freezing.

The essential components of early slow sand filter beds are illustrated in Figs. 11-1$b$ and 14-1$a$. Their general features of construction and operation are summarized in Table 14-1. Also summarized in this table are the features of *rapid sand filters* introduced toward the end of the nineteenth century (Figs. 11-1$a$ and 14-1$b$). These could be operated about 30 times faster than the older filters. Although rapid filters soon began to be constructed in large numbers for the treatment of turbid river waters by combining alum or iron coagulation and sedimentation with rapid filtration, a number of places kept their slow sand filters going,[1] and they remain of interest where raw-water quality and economics of construction and operation favor their use. Aside from their modest hydraulic loading, the distinguishing features of slow sand filters are their nonvarying sand mixtures at all bed depths, the small effective size and large coefficient of nonuniformity of their grains, and the associated selective removal and accumulation of raw-water impurities at the surface of the bed and within its top inch or two. Because of this, slow sand filters can be kept going for long periods of time, (1) if the surface of the partially dewatered bed is raked after about 2 weeks and again a week or so later to break through the surface accumulations and (2) if the top inch or two of the bed and the surface accumulations are scraped off at the end of a month and transported out of the filter for winter storage either before or after washing, or for immediate washing and return to the bed.

[1] London, England, Springfield, Mass., and Hartford, Conn., among them.

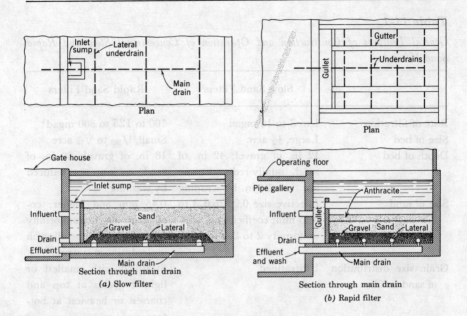

**Figure 14-1.**
**Diagrammatic sections through and simplified plans of (a) a slow and (b) a rapid sand filter.**

The incorporation of rapid filter units into municipal water works was encouraged by the successful studies of coagulation and rapid filtration by George W. Fuller at Louisville, Ky., in 1895. The first sizable, municipal, rapid filter-plant was designed by him for Little Falls, N.J., in 1909.

Because rapid sand filters operate at many times the rate of slow sand filters, they need to be cleaned many times more often. As shown in Table 14-1, filter runs are normally no more than 1 day long. Distinguishing features of rapid filters, therefore, are their relatively small size and the ways in which they must be cleaned because they accumulate impurities rapidly and at nearly all depths. As suggested in Fig. 14-1b, the filter units are washed from below with previously filtered water retained in storage for that purpose. The underdrainage system then doubles as a washwater distribution system. The sand cleans itself when the bed has been *fluidized* by the rising water. Scour is intensified in one of the following ways: (1) by stepping up the velocity or rate of backwash per unit area sufficiently (*high-velocity wash*); (2) by directing jets of water into the fluidized bed (*surface scour*); (3) by blowing air upward through the bed before or during fluidization (*air scour*); and (4) by stirring the fluidized bed mechanically (*mechanical scour*).

## Table 14-1
*General Features of Construction and Operation of Conventional Slow and Rapid Sand Filters*[a]

|  | Slow Sand Filters | Rapid Sand Filters |
|---|---|---|
| Rate of filtration | 1 to 3 to 10 mgad | 100 to 125 to 300 mgad[b] |
| Size of bed | Large, ½ acre | Small, $1/100$ to $1/10$ acre |
| Depth of bed | 12 in. of gravel; 42 in. of sand, usually reduced to no less than 24 in. by scraping | 18 in. of gravel; 30 in. of sand, or less; not reduced by washing |
| Size of sand | Effective size 0.25 to 0.3 to 0.35 mm; coefficient of nonuniformity 2 to 2.5 to 3 | 0.45 mm and higher; coefficient of nonuniformity 1.5 and lower, depending on underdrainage system |
| Grain size distribution of sand in filter | Unstratified | Stratified with smallest or lightest grains at top and coarsest or heaviest at bottom |
| Underdrainage system | Split tile laterals laid in coarse stone and discharging into tile or concrete main drains | (1) Perforated pipe laterals discharging into pipe mains; (2) porous plates above inlet box; (3) porous blocks with included channels |
| Loss of head | 0.2 ft initial to 4 ft final | 1 ft initial to 8 or 9 ft final |
| Length of run between cleanings | 20 to 30 to 60 days | 12 to 24 to 72 hr |
| Penetration of suspended matter | Superficial | Deep |
| Method of cleaning | (1) Scraping off surface layer of sand and washing and storing cleaned sand for periodic resanding of bed; (2) washing surface sand in place by washer traveling over sand bed | Dislodging and removing suspended matter by upward flow or backwashing, which fluidizes the bed. Possible use of water or air jets, or mechanical rakes to improve scour |
| Amount of wash water used in cleaning sand | 0.2 to 0.6% of water filtered | 1 to 4 to 6% of water filtered |
| Preparatory treatment of water | Generally none | Coagulation, flocculation, and sedimentation |
| Supplementary treatment of water | Chlorination | Chlorination |

**Table 14-1**—*continued*

|  | Slow Sand Filters | Rapid Sand Filters |
|---|---|---|
| Cost of construction, U.S.A. | Relatively high | Relatively low |
| Cost of operation | Relatively low where sand is cleaned in place | Relatively high |
| Depreciation cost | Relatively low | Relatively high |

[a] The most common values are shown in boldface type.
[b] 125 mgad = 2 gpm per sq ft = 16 ft per hr = 125 m per day.

After the bed has been washed clean, the washwater is turned off and the sand settles back into place. The coarsest or heaviest grains reach the bottom first; the finest or lightest arrive last. This stratifies the bed and imposes a size limit on its finest or lightest particles.

Rapid filters have almost wholly displaced slow filters in North American practice because of their convenience of size, ability to remove turbidity in conjunction with coagulation and sedimentation, adaptability to changing raw-water quality, and overall economy of construction and operation under North American conditions.

The obtainable purification becomes very high when water is submitted, in succession, to rapid and slow filtration. However, this has been done principally where the treatment works began as slow filters and rapid units were added later either to compensate for deteriorating raw-water quality or to increase plant output.

Rapid filters enclosed in pressure tanks and operated under line pressures have been installed principally in industrial water systems and in the recirculation systems of swimming pools. Some of their operating parts are accessible only with difficulty, and their space requirements are often unfavorable.

In conventional down-flow sand filters, the entering water comes into contact with fine grains before coarse grains or with clogged portions of the bed before open ones. The design of granular water filters is being modified to permit the influent water to pass through coarse grains before it is finally polished by passing through fine grains. This is accomplished (1) by using several media of varying specific gravity, with large lighter particles on top or (2) by upward flow in a conventionally stratified filter. Process design, too, has been modified in a number of ways. Interesting is the optimization of the coagulation, flocculation, and filtration sequence, which may result in (1) a heavier than conventional loading of the filter for economic reasons or (2) other functional uses of the sequent components in ways somewhat different from the conventional.

## 14-3 Granular Wastewater Filters

Because they are putrescible and clog filters rapidly, *raw* municipal wastewaters are not amenable to treatment in conventional slow or rapid filtration plants. However, effluents from biological or chemical wastewater treatment works can be filtered as successfully as heavily polluted raw waters. Sand filtration then becomes a *tertiary* wastewater treatment operation (Sec. 11-11). As such it can provide improved protection to receiving bodies of water or prepare effluents for reuse in industry, agriculture, and recreation, and for recharge to the ground. Rapid filters serving these ends are patterned after rapid water filters but generally contain considerably coarser grains and operate at lower rates. Filter runs are often short, and beds may have to be heavily chlorinated or purged regularly with caustic or other chemicals to keep them open and sweet.

Under favorable geological and topographical conditions, natural sand deposits have been converted into *intermittent sand filters* to good purpose. Settled urban wastewaters are run onto them at rates averaging 40,000 to 120,000 gpd per acre, and biologically treated effluents at 400,000 to 800,000 gpd per acre. At such rates these filters produce excellent effluents.

An intermittent sand filter contains drainage pipes that are laid with open joints at depths of 3 to 4 ft and surrounded with layers of coarse stone and gravel graded from coarse to fine to keep sand out. In deep sand deposits the percolating waters may reach the groundwater table, and no effluent may come to view. Influent wastewater is piped to the beds for discharge onto a protective stone or concrete apron or into a concrete flume distributor. Flooding the bed to a depth of 1 to 4 in. delivers 25,000 to 100,000 gal to an acre of surface. One or more doses are applied each day for 7 to 20 min. Surface accumulations of solids are scraped off from time to time. Some sand is included in the scrapings, which are ordinarily buried or disposed of as fill. Filters are resanded when they become too shallow. In cold weather the beds are furrowed to keep them from freezing and opening up cracks through which the applied wastewater can escape with little treatment. To form protective sheets of ice spanning the furrows and keep the beds warm, furrowed beds are dosed deeply on cold nights.

## 14-4 Granular Filtering Materials

Rapid filters may be composed of natural silica sand, crushed anthracite (hard) coal, crushed magnetite (ore), and garnet sands.[2]

The natural granular sands and crushed minerals normally employed differ in size and size distribution, in shape and shape variation, and in

---

[2] Plastic spheres and granules manufactured to wanted specifications have been used experimentally.

density and chemical composition. Mean values and variances provide acceptable parameters (1) for describing the geometry of the materials, (2) for the rational prediction of their hydraulic behavior, and (3) for assessing the removal of impurities by filtration.

**Grain Size and Size Distribution.** The parameters of mean grain size and variance in grain size are commonly determined by sieving a representative sample of filter grains through a set of calibrated sieves.

Starting with the weight of sievings shaken downward through a vertical stack of successively finer sieves to the pan at the bottom, the portions of sand held between adjacent sieves are added in sequence, and the cumulative weights are recorded. After conversion into percentages by weight equal to or less than the rated size of the overlying coarser sieve, a summation curve can be plotted for purposes of generalization (Fig. 14-2).

The American (U.S.) standard sieve series (Table 13, Appendix) is referred to a sieve opening of 1 mm (generated by approximately 18 meshes

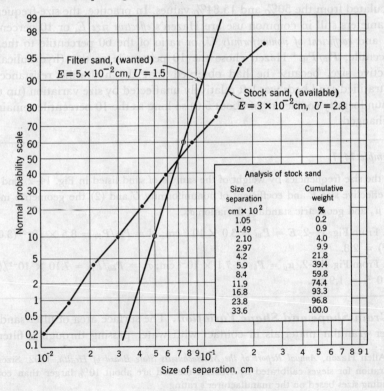

**Figure 14-2.**
**Grain size distribution of a stock sand and required sizing of a filter sand.**

to the inch). Thence, sieve openings stand successively in the ratio of $\sqrt[4]{2}$ to one another, the largest being 5.66 mm (produced by approximately $3\frac{1}{2}$ meshes to the inch) and the smallest 0.037 mm (generated by approximately 400 meshes to the inch). For less precise measurements, every second sieve in the series is removed; possibly even every second sieve in the reduced stack. Resulting ratios are $\sqrt[4]{2}:1$, $\sqrt{2}:1$, and $2:1$, but intermediate sieves within important size ranges may be left in place at the discretion of the analyst. U.S. standard sieves are normally calibrated directly by measuring the clear dimensions of a representative number of screen openings. This is known as the *manufacturer's rating*.[3]

For many natural filtering materials, the summation curve of weights below or equal to given sieve sizes approaches geometric normality. Analyses plotted on logarithmic probability paper then trace an almost straight line, in which interpolation is straightforward (Fig. 14–2), and from which geometric mean size $\mu_g$ and geometric standard deviation in size $\sigma_g$ can be read or calculated from the 50% and 15.84% values. In practice, the size-frequency parameters still in common use are Hazen's *effective size E*, or 10 percentile $P_{10}$, and *coefficient of nonuniformity U*, or ratio of the 60 percentile to the 10 percentile, $P_{60}/P_{10}$.[4] Hazen chose the 10 percentile as the (hydraulically) effective size, because he had observed that the hydraulic resistance of unstratified sand beds was left relatively unaffected by size variation (up to a nonuniformity coefficient of about 5.0) so long as the 10 percentile remained unchanged.

*Example 14–1.*

For the size frequencies by weight of the sample of sand listed in Fig. 14–2, find (1) the effective size $E$ and coefficient of nonuniformity $U$ and (2) the geometric mean size $\mu_g$ and geometric standard deviation $\sigma_g$.

1. From Fig. 14–2, $E = P_{10} = 3.0 \times 10^{-2}$ cm; $U = P_{60}/P_{10} = 8.5 \times 10^{-2}/(3.0 \times 10^{-2}) = 2.8$.
2. From Fig. 14–2, $\mu_g = P_{50} = 7.1 \times 10^{-2}$ cm; $\sigma_g = P_{50}/P_{16} = 7.10 \times 10^{-2}/(3.7 \times 10^{-2}) = 1.9$ cm.

**Grain Shape and Shape Variation.** The surface area of filter sand or other granular materials in contact with water passing through a filter is

---

[3] Allen Hazen, *Annual Report of the Massachusetts State Board of Health*, 1892. Sizes of separation for sieves calibrated by Hazen's method are about 10% larger than corresponding sizes based on the manufacturer's rating.

[4] Hazen called $U = P_{60}/P_{10}$ the uniformity coefficient, but this ratio is more logically a coefficient of nonuniformity, because it increases numerically with nonuniformity rather than uniformity.

both an operational and a hydraulic (frictional) determinant. Unfortunately there is as yet no satisfactory way to find this surface area by direct measurement. However, it is possible to establish the pore volume and complementary sand volume, $V$, of a filter without difficulty. Because the surface area $A$ of grains relative to the volume of water in the pore space is a function of $A/V$, it is this size characteristic that is normally employed in formulations of filter performance and referred to in identifying shape. Among useful measures of shape are the *sphericity*, $\psi$, and the *shape factor S*. Defined as the ratio of the surface area of the equivalent-volume sphere to the actual or true surface area, $\psi$ assumes values such as those listed below Fig. 14–3, in which

| Description | Sphericity, $\psi$ | Shape Factor, $S$ | Typical Porosity, $f$ |
|---|---|---|---|
| (a) Spherical | 1.00 | 6.0 | 0.38 |
| (b) Rounded | 0.98 | 6.1 | 0.38 |
| (c) Worn | 0.94 | 6.4 | 0.39 |
| (d) Sharp | 0.81 | 7.4 | 0.40 |
| (e) Angular | 0.78 | 7.7 | 0.43 |
| (f) Crushed | 0.70 | 8.5 | 0.48 |

particles of representative shapes are outlined. For a single truly spherical grain of diameter $d$, $\psi$ equals unity and, by definition, $A/V = \pi d^2/(\frac{1}{6}\pi d^3) = 6/d$. The shape symbols of true spheres are therefore a sphericity $\psi = 1.0$ and a shape factor $S = 6.0/\psi = 6.0$. These are respectively the largest possible value of $\psi$ and smallest possible value of $S$. The magnitudes of $\psi$ and $S$ in the table below Fig. 14–3 include measurements by Carman.[5]

The magnitude of $\psi$ for a given sample of sieved particulates can be estimated from Fig. 14–3 or derived from observed settling velocities of representative grains when the temperature of the test fluid and the specific weight of the particulates are known. If the settling velocity is $v_s$ and the grain diameter is $d_s$,

$$\psi = d_0/d_s = \sqrt{v_s/v_0} \qquad (14\text{--}1)$$

Here $d_0$ is the diameter of the equivalent volume sphere read from Fig. 14–2 or 3, and $v_0$ is its settling velocity.

The hydraulic properties of a bed or a layer of granular material do not depend on the ratio of surface area to volume of solid material, but on the ratio of surface area to volume of void space, $A/V_v$. This ratio is the reciprocal

---

[5] P. C. Carman, Fluid Flow through Granular Beds, *Trans. Inst. Chem. Engrs.* (*London*), **15**, 150 (1937). As a matter of incidental interest, Carman assigns the small value $\psi = 0.28$ to the sphericity of mica flakes. This makes for a large shape factor (21.4) and much hydraulic resistance.

406/Filtration

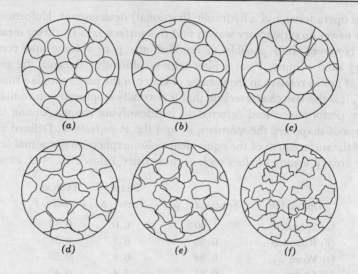

**Figure 14-3.**
**Sphericity and shape factors of granular (not flakelike) materials and typical porosities associated with them in stratified rapid sand-filter beds.**

of the familiar hydraulic radius (Sec. 14-8). For a bed or layer of perfect spheres of diameter $d$, the area-voids volume ratio, $A/V_v$, of equal spheres within a given cube of porosity $f$ can be calculated from the number of spheres, $n$. The following simple relationships hold true: $n =$ (total volume of spheres)/(volume of a single sphere); $A = n$ (surface area of a single sphere) $= n\pi d^2$; $V$ (total volume of spheres) $= (1 - f)$ (volume of given cube) $= \frac{1}{6}n\pi d^3$; volume of given cube $= V/(1-f)$; and $V_v =$ volume of voids $= f$ (volume of given cube) $= [f/(1-f)]V$, or

$$\frac{A}{V_v} = \frac{n\pi d^2}{\frac{1}{6}[f/(1-f)]n\pi d^3} = \frac{1-f}{f}\left(\frac{6}{d}\right) \qquad (14\text{-}2)$$

## 14-5 Preparation of Filter Sand

Natural *run-of-bank sand* may be too coarse, too fine, or too nonuniform for a projected filter. Within economical limits, specified sizing and uniformity can be obtained by *screening out* coarse grains and *washing out* fines.

If a filter sand is specified in terms of effective size and nonuniformity coefficient and a sieve analysis of the stock sand has been made (Fig. 14-2), the coarse and fine portions of stock sand to be removed are functions of $P_{10}$ and $P_{60}$, namely, the percentages of stock sand smaller than the desired effective size and the 60 percentile size, respectively. This can be explained as follows:

1. Because the sand lying between the $P_{60}$ and $P_{10}$ sizes constitutes half the specified sand, the percentage of usable stock sand is

$$P_{\text{usable}} = 2(P_{60} - P_{10}) \qquad (14\text{-}3)$$

2. Because the specified sand can contain but $1/10$ of the usable sand below the $P_{10}$ size, the percentage below which the stock sand is too fine for use is

$$P_{\text{too fine}} = P_{10} - 0.1\, P_{\text{usable}} = P_{10} - 0.2(P_{60} - P_{10}) \qquad (14\text{-}4)$$

provided that the grain size associated with the sand that is too fine is equal to or greater than the smallest size of sand to be included in the filter.

3. Because a percentage of stock sand equal to $P_{\text{usable}} + P_{\text{too fine}}$ has been accounted for, the percentage above which the stock sand is too coarse is

$$P_{\text{too coarse}} = P_{\text{usable}} + P_{\text{too fine}} = P_{10} + 1.8(P_{60} - P_{10}) \qquad (14\text{-}5)$$

Before sand is placed in a filter, fines can be washed from it in a sand or grit washer. Washers of the same kind will also cleanse sand removed from a filter—a slow sand filter, for example—and relieve grit of organic matter.

*Example 14-2.*

What must be done to the stock sand of Fig. 14–2 to convert it into a filter sand of effective size $5 \times 10^{-2}$ cm and nonuniformity coefficient 1.5? Assume a water temperature of $10°C$ and $S = 1.31 \times 10^{-2}$ poise.

1. From Fig. 14–2 the proportion of sand, $P_{10}$, less than the desired effective size of $5 \times 10^{-2}$ is 30% and the proportion less than the desired $P_{60}$ size of $5 \times 10^{-2} \times 1.5$ cm is 53.5%. Hence by Eq. 14–3 the proportion of usable sand equals $2(53.5 - 30.0) = 47.0\%$.

2. By Eq. 14–4 the percentage below which the stock sand is too fine is $30.0 - 4.7 = 25.3\%$, and the diameter of this sand from Fig. 14–2 is $4.4 \times 10^{-2}$ cm. From Fig. 13–5, the settling velocity of this sand, hence the overflow rate of a washer at low concentrations of sand, is 6.4 cm/sec or 94 gpm/sq ft at $10°C$ ($50°F$). However, as shown in Sec. 14–6, the hindered settling velocity is hypothetically 13.2 cm/sec by Stokes' law (Eq. 13–8). Hence the overflow rate at a norm of 50% void space is $(0.5)^5 \times 13.2 = 0.41$ cm/sec, or 60 gpm/sq ft (Eq. 14–20).

3. The percentage above which the stock sand is too coarse is $47.0 + 25.3 = 72.3$ by Eq. 14–5, and the diameter of this sand is $1.10 \times 10^{-1}$ cm from Fig. 14–2.

It follows that all stock sand finer than $4.4 \times 10^{-2}$ cm at one end of the size range and coarser than $1.10 \times 10^{-1}$ cm at the other end must be removed.

## 14-6 Hydraulics of Filtration

At the velocities commonly employed in granular water filters, flow is normally laminar and obeys Darcy's law, $v = Ks$, where $v$ is the face or approach velocity of the water above the sand bed, $s = h/l$ is the loss of head, $h$, in a depth of bed, $l$, and $K$ is Darcy's coefficient of permeability (Fig. 14-4). Identifiable components of Darcy's $K$ are the density, $\rho$, and viscosity, $\mu$, of the water, the porosity, $f$, of the bed, and the size and shape of the component sand grains that establish the surface area $A$ of the grains within the bed relative to their volume, $V$.

Specifically, the resistance of a clean bed of sand to the filtration of clean water is given by the Blake-Kozeny equation:[6]

$$h/l = (k/g)(\mu/\rho) v \frac{(1-f)^2}{f^3} (A/V)^2 \qquad (14\text{-}6)$$

Here, after all recognizable factors have been introduced into Darcy's coefficient of permeability, $k$ (Kozeny) becomes a residual dimensionless coefficient that assumes a magnitude close to 5.0 under most conditions of water filtration. The porosity factor, $(1-f)^2/f^3$, derives in part from the conversion of the approach velocity term, $v$, into an interstitial velocity term

[6] Credit for this formulation should go to Blake as well as Kozeny. C. F. Blake, The Resistance of Packing to Fluid Flow, *Trans. Am. Soc. Chem. Engrs.*, **14**, 415 (1922); and J. Kozeny, Ueber kapillare Leitung des Wassers im Boden (On Capillary Conduction of Water in the Soil), *Sitzungsber. Akad. Wiss.*, Vienna, Abt. IIIa, **136**, 276 (1927); also *Wasserkraft Wasserwirtschaft*, **22**, 67 (1927).

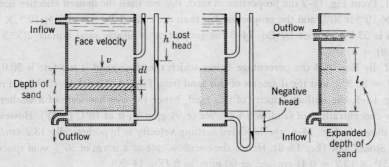

**Figure 14-4.**
**Head loss in filtration, and expansion in backwashing.**

$v/f$, and in part from identification of the hydraulic radius, $r$, of the interstitial channels as

$$r = \frac{\text{cross-sectional area of flow} \times \text{length of channel}}{\text{wetted perimeter of channel} \times \text{length of channel}}$$

$$= \frac{\text{volume of water in interstices}}{\text{surface area of sand}}$$

Because the volume of water in the interstices is $f$ (pore volume of the bed) and the bed volume is $V/(1-f)$,

$$r = \frac{f}{1-f}\frac{V}{A} = \frac{f}{1-f}\frac{\psi d}{6} \qquad (14\text{--}7)$$

In terms of the size and shape of its constituent sand grains, therefore, the rate of head loss becomes

$$h/l = (k/g)\,\nu\,v\,\frac{(1-f)^2}{f^3}\,(6/\psi d)^2 \qquad (14\text{--}8)$$

or

$$h/l = (k/g)\,\nu\,v\,\frac{(1-f)^2}{f^3}\,(S/d)^2 \qquad (14\text{--}9)$$

Here the kinematic viscosity $\nu$ has replaced the ratio $\mu/\rho$.

Equations 14–6, 8, and 9 can be derived from (1) the Weisbach-Darcy equation (Eq. 6–1),[7] (2) considerations entering into the settling of particles (Sec. 13–4), and (3) dimensional analysis.[8]

**Hydraulics of Stratified Beds.** In a clean filter, stratified by backwashing, the head loss is the sum of the losses in successive sand layers. Because the thickness of each of $n$ layers is closely proportional to the fractional weight, $p_i$, of sieved size $d_i$, Eq. 14–8, as proposed by Fair and Hatch,[9] becomes

$$h/l = (k/g)\,\nu\,v\,\frac{(1-f)^2}{f^3}\,(6/\psi)^2 \sum_{i=1}^{n} (p_i/d_i^2) \qquad (14\text{--}10)$$

---

[7] $h_f = f(l/d)(v^2/2g)$, $d = 4r$, $r = [f/(1-f)](V/A)$, $v = v/f$, and $f = 64/\mathbf{R}$ for laminar flow, $\mathbf{R}$ being $4(v/f)r\rho/\mu$.

[8] H. E. Rose, On the Resistance Coefficient—Reynolds Number Relationship for Fluid Flow through a Bed of Granular Material, *Proc. Inst. Mech. Engrs. London*, **153**, 154 (1945); **160**, 493 (1949).

[9] G. M. Fair and L. P. Hatch, Fundamental Factors Governing the Streamline Flow of Water through Sand, *J. Am. Water Works Assoc.*, **25**, 1551 (1933).

Here the diameter, $d_i$, is an average diameter such as the geometric mean diameter or square root of the product of the upper and lower sieve sizes representing a fraction $p_i$ of the analyzed sand. Normally the values of pertinent sizes and fractions are respectively those of the adjacent sieves employed and the proportion, $p_i$, of the sand trapped between them. However, it is possible to read usable values also from the plotted analysis (Fig. 14-2).

*Example 14-3.*

Assume that the straight line drawn in Fig. 14-2 describes the size distribution by manufacturer's rating of filter sand in a 24-in. bed operated at a rate of 3 gpm/sq

*Table 14-2*

Computation of $\sum_{i=1}^{n} p_i d_i^{-2}$ for Example 14-3

| Size of Sand, cm × $10^2$ | Sand Larger than Stated Size,[a] % | Sand Fraction Within Adjacent Sieve Sizes, $P_i$, % | Geometric Mean Diameter, $d_i$, cm × $10^2$ | $\dfrac{P_i}{d_i^2}$ |
|---|---|---|---|---|
| 3 | 0.0 | | | |
| | | 2 | 3.5 | 16 |
| 4 | 2.0 | | | |
| | | 8 | 4.5 | 40 |
| 5 | 10.0 | | | |
| | | 17 | 5.5 | 56 |
| 6 | 27.0 | | | |
| | | 23 | 6.5 | 52 |
| 7 | 50.0 | | | |
| | | 20 | 7.5 | 36 |
| 8 | 70.0 | | | |
| | | 20 | 8.9 | 25 |
| 10 | 90.0 | | | |
| | | 10 | 11.8 | 8 |
| 14 | 100.0 | — | | — |
| Sum | — | 100 | ... | 233 |

[a] It is assumed that the filter does not contain sand smaller than 0.03 cm or larger than 0.14 cm.

ft = 0.2 cm/sec; that the bed after backwashing settles back uniformly to a porosity[10] of 0.4; and that the sand has a sphericity of 0.8. Find (1) the head lost in the clean bed, (2) the head lost if the particles in the bed are all of the geometric mean size, and (3) the Darcy coefficient of bed permeability.

1. By Eq. 14–10, $h/l = (5/981)\ 1.01 \times 10^{-2} \times 0.2[(1 - 0.4)^2/0.4^3](6/0.8)^2 \times 233 = 0.76$, and $h = 2 \times 0.76 = 1.52$ ft.

2. For a geometric mean size of $7 \times 10^{-2}$ cm, find $1/d_i^2 = (7 \times 10^{-2})^{-2} = 204$, and $h/l = 0.76 \times 204/233 = 0.67$, making $h = 2 \times 0.67 = 1.33$ ft compared with 1.52 ft.

3. Calculate the Darcy coefficient of permeability as $K = 0.2/0.76 = 0.26$ cm/sec.

Values of $\sum_{i=1}^{n} p_i d_i^{-2}$ are computed in Table 14–2.

**Hydraulics of Unstratified Beds.** In an unstratified bed of homogeneously packed sand each component fraction, $p_i$, of size $d_i$ contributes its share to the total area, the individual area-volume ratios being $6/(\psi d_i)$. For uniform sphericity, therefore,[9]

$$A/V = 6/\psi \sum_{i=1}^{n} p_i/d_i \qquad (14\text{-}11)$$

and

$$h/l = (k/g)\ \nu\ v\ \frac{(1-f)^2}{f^3}\left[(6/\psi) \sum_{i=1}^{n} p_i/d_i\right]^2 \qquad (14\text{-}12)$$

*Example 14–4.*

If the natural sand of Fig. 14–2 is used in an unstratified slow sand filter 42 in. deep operated at 3 mgad $= 3.24 \times 10^{-3}$ cm/sec, the water temperature is 68°F (20°C) or $\nu = 1.01$ centistokes (Table 3, Appendix), the porosity is 0.35, and the sphericity is 0.8, find (1) the initial head loss following cleaning and replacement of all the sand; (2) the single diameter, $d_e$, that would produce the same hydraulic result as the mixed bed and where this size lies in the cumulative weight scale of Fig. 14–2; and (3) the Darcy coefficient of permeability.

1. Using the sizes and percentages shown in Fig. 14–2, find $\left(\sum_{i=1}^{n} \dfrac{p_i}{d_i}\right)^2 = (17.80)^2$.

---

[10] Porosity is calculated from measurements of depths, $l$, weight of sand per unit area of filter, $W$, and sand specific weight, $\gamma_s$, as $f = 1 - W/(l\gamma_s)$.

By Eq. 14-12, find $h/l = 5/981 \times 1.01 \times 10^{-2} \times 3.24 \times 10^{-3} [(1 - 0.35)^2/0.35^3]$ $(6/0.8)^2\, 17.8^2 = 2.9 \times 10^{-2}$ and $h = 2.9 \times 10^{-2} \times 42/12 = 0.10$ ft.

2. Determine the single equivalent size by equating $1/d_e$ to $\sum_{i=1}^{n} p_i/d_i$ or $d_e = 1/17.8 = 5.6 \times 10^{-2}$ cm. From Fig. 14-2, the amount of sand smaller than $5.6 \times 10^{-2}$ cm is 37%.

3. Find the Darcy coefficient of permeability as $K = 3.24 \times 10^{-3}/(29 \times 10^{-3}) = 0.11$ cm/sec.

### 14-7 Hydraulics of Fluidized Beds—Filter Backwashing

The operation of granular water filters is discontinuous. As described in Sec. 14-2, output is stopped before an excessive loss of head is built up in the clogging bed or the quality of the product water—its clarity, for instance—deteriorates. Output is not resumed until the bed has been backwashed thoroughly. The rate of backwash is made high enough (1) to fluidize the active portion of the bed and (2) to open the passageways between adjacent grains wide enough to allow floc and other residues deposited in the bed during filtration to escape with the overflowing washwater.

Conceptually, the hydraulics of fluidized beds differs from the hydraulics of filtration principally by the increase in pore volume of the expanded (lifted) sand. With the materials and degrees of expansion ordinarily used, however, the flow in a fluidized filter bed is laminar even though the suspended grains are in motion.

If a grain is to be kept in suspension, the frictional drag of the water rising past it must equal the pull of gravity on it; that is, the difference or head loss, $h$, between the bottom and top of any layer of thickness, $l_e$, must equal the weight in water of the suspended material.[9] For each unit area of filter,

$$h \rho g = l_e (\rho_s - \rho)\, g (1 - f_e) \qquad (14\text{-}13)$$

where $f_e$ is the porosity ratio of the expanded layer of thickness, $l_e$. Hence

$$(h/l_e) = [(\rho_s - \rho)/\rho](1 - f_e) \qquad (14\text{-}14)$$

In terms of Eq. 14-6, therefore,

$$h/l_e = \frac{\rho_s - \rho}{\rho}(1 - f_e) = \frac{k_e}{g} \nu v \frac{(1 - f_e)^2}{f_e^3} [(6/\psi)(1/d)]^2 \qquad (14\text{-}15)$$

Here $k_e$ is a function of $f_e$. By experiment it assumes a value of about 4 when the bed is just fluidized. If the $i$th layer consists of grains of size $d_i$, Eq. 14-15 reduces to

$$\frac{f_{ei}^3}{1-f_{ei}} = \frac{k}{g} \frac{\mu}{\rho_s - \rho} v (6/\psi d_i)^2 \qquad (14\text{-}16)$$

For such a layer, the ratio of the expanded depth, $l_{ei}$, to the unexpanded depth, $l_i$, is

$$l_{ei}/l_i = (1-f)/(1-f_{ei}) \qquad (14\text{-}17)$$

and the total expansion is

$$L_e = \Sigma l_{ei} = \Sigma \, l_i(1-f)/(1-f_{ei}) \qquad (14\text{-}18)$$

If the porosity ratio, $f$, of the unexpanded bed is constant for all layers, the unexpanded thickness of any layer, $l_i$, equals the full depth, $L$, times the percentage, $p_i$, of diameter $d_i$, and $L_e$ becomes[10]

$$L_e = L(1-f) \sum_{i=1}^{n} [p_i/(1-f_{ei})] \qquad (14\text{-}19)$$

The reciprocal of $(1-f_{ei})$ can be evaluated from the ratio $f_{ei}^3/(1-f_{ei})$ in Eq. 14-16 with the assistance of a table such as Table 14-3, or of a plot of $1/(1-f_{ei})$ against given magnitudes of $f_{ei}^3/(1-f_{ei})$.

Alternatively, it has been shown by experiment that the ratio of the measured settling velocity of a grain, $v_s$, to the face or hypothetical, viscous, settling velocity, $v$, equals about $(1/f_e)^5$, and hence that

$$v_s \simeq v/f_e^{5.0}, \qquad v \simeq v_s f_e^{5.0}, \qquad f_e \simeq (v/v_s)^{0.20}, \qquad \text{and} \qquad f_e^{5.0} \simeq v/v_s$$
$$(14\text{-}20)$$

**Table 14-3**
Values of $1/(1-f_e)$ Corresponding to Values of $f_e^3/(1-f_e)$ Ranging from 0.1 to 9.9

| $\dfrac{f_e^3}{1-f_e}$ | 0.0 | 0.1 | 0.2 | 0.3 | 0.4 | 0.5 | 0.6 | 0.7 | 0.8 | 0.9 |
|---|---|---|---|---|---|---|---|---|---|---|
| 0 | 0.00 | 1.62 | 1.89 | 2.10 | 2.28 | 2.44 | 2.59 | 2.74 | 2.88 | 3.01 |
| 1 | 3.14 | 3.27 | 3.40 | 3.52 | 3.65 | 3.78 | 3.89 | 4.01 | 4.13 | 4.24 |
| 2 | 4.35 | 4.47 | 4.58 | 4.70 | 4.81 | 4.93 | 5.05 | 5.16 | 5.27 | 5.38 |
| 3 | 5.49 | 5.60 | 5.71 | 5.82 | 5.92 | 6.03 | 6.14 | 6.24 | 6.35 | 6.46 |
| 4 | 6.57 | 6.68 | 6.78 | 6.88 | 6.99 | 7.10 | 7.20 | 7.31 | 7.41 | 7.52 |
| 5 | 7.62 | 7.73 | 7.83 | 7.94 | 8.04 | 8.15 | 8.25 | 8.35 | 8.46 | 8.56 |
| 6 | 8.67 | 8.77 | 8.88 | 8.98 | 9.08 | 9.18 | 9.29 | 9.39 | 9.49 | 9.60 |
| 7 | 9.70 | 9.81 | 9.91 | 10.01 | 10.11 | 10.21 | 10.32 | 10.42 | 10.52 | 10.62 |
| 8 | 10.72 | 10.83 | 10.93 | 11.03 | 11.14 | 11.24 | 11.35 | 11.45 | 11.56 | 11.66 |
| 9 | 11.76 | 11.86 | 11.96 | 12.06 | 12.16 | 12.27 | 12.37 | 12.47 | 12.58 | 12.68 |

The settling velocity, $v_s$, must be either measured by test or calculated by first obtaining a value $v$ from Eq. 13–7 and an appropriate sphericity from Fig. 14–3 for introduction into Eq. 13–6 before computation of $v_s$. Associated values of $f_e$ are given by Eq. 14–20.

*Example 14–5.*

Given the sand sizes and size frequencies shown in Cols. 3 and 5 of Table 14–4, calculate for a face velocity of 25 in./min, an unexpanded filter depth of 28 in., a porosity of 0.45, and a grain sphericity 0.78, the bed expansion at water temperature of (1) 25°C, and (2) 4°C, respectively, and (3) check the results obtained by applying the approximate method of calculation.

1. By Eq. 14–16,

$$\frac{f_{ei}^3}{1-f_{ei}} = \frac{4}{981}\left(\frac{0.895 \times 10^{-2}}{2.65 - 0.997}\right)\left(\frac{25 \times 2.54}{60}\right)\left(\frac{6}{0.78\,d_i}\right)^2 = \frac{1.38 \times 10^{-3}}{d_i^2}$$

Hence Col. 6, Table 14–4, is $1.38 \times 10^{-3} \times$ Col. 4. Column 7 is taken from Table 14–3. Column 8 is Col. 7 × Col. 5. By Eq. 14–19, therefore, $L_e = 28(1 - 0.45)(2.608) = 40$ in. and the percentage expansion is $100\,L_e/L = 100 \times 40/28 = 143\%$.

2. In proportion to $\mu$ in Eq. 14–16,

$$\frac{f_{ei}^3}{1-f_{ei}} = \left(\frac{1.38 \times 10^{-3}}{d_i^2}\right)\left(\frac{1.57 \times 10^{-2}}{0.895 \times 10^{-2}}\right) = \frac{2.43 \times 10^{-3}}{d_i^2}$$

Hence, as in the first instance, $L_e = 28(1 - 0.45)(3.269) = 50$ in. and the percentage expansion is $100\,L_e/L = 100 \times 50/28 = 180\%$.

3. For each value of $d_i$, Eq. 13–9 gives the hypothethical viscous (hindered) settling velocity, $v_0'$, as

$$v_{0i}' = \left(\frac{g}{18}\right)\frac{\rho_s - \rho}{\mu}\,d_i^2 = \frac{981}{18}\left(\frac{2.65 - 0.997}{0.895 \times 10^{-2}}\right)d_i^2 = 10^4 d_i^2$$

From Eq. 14–1, $v_{si} = v_{0i}'\,\psi^2$ and $v_{0i}' = 10^4\,\psi^2 d_i^2 = 6.1 \times 10^3 d_i^2$.
By Eq. 14–20,

$$f_{ei} = (v/v_{si})^{0.20} = [(25 \times 2.54/60)/(6.1 \times 10^3 d_{si}^2)]^{0.20}$$

Columns 12, 13, and 14 show the results of calculations. By Eq. 14–19, $L_e = 28(1 - 0.45)(2.56) = 39.4$ in. and the percentage expansion is $100\,L_e/L = 100 \times 39.4/28 = 141\%$, which is not significantly less than the value of 143% computed in the first instance.

Example 14–5 illustrates how strongly water temperature affects sand expansion during backwashing. Accordingly, it is necessary to vary the flow

### Table 14-4
Data and Computations for Example 14-5

| Sieve Mesh, per in. | Mfgr's Rating, cm × 10² | Sand Characteristics $d_i$ cm × 10² | $d_i^{-2} \times 10^{-2}$ | $p_i \times 10^2$ | Expansion at 25°C $\dfrac{f_{ei}^3}{1-f_{ei}}$ | $\dfrac{1}{1-f_{ei}}$ | $\dfrac{p_i}{1-f_{ei}}$ | $\dfrac{f_{ei}^3}{1-f_{ei}}$ | Expansion at 4°C $\dfrac{1}{1-f_{ei}}$ | $\dfrac{p_i}{1-f_{ei}}$ | Approximation at 25°C $\dfrac{p_i}{1-f_{ei}}$ | $\dfrac{1}{1-f_{ei}}$ | $\dfrac{p_i}{1-f_{ei}}$ |
|---|---|---|---|---|---|---|---|---|---|---|---|---|---|
| (1) | (2) | (3) | (4) | (5) | (6) | (7) | (8) | (9) | (10) | (11) | (12) | (13) | (14) |
| 80 | 1.75 | 1.91 | 26.4 | 0.1 | 3.64 | 6.18 | 0.006 | 6.40 | 9.08 | 0.009 | 0.864 | 7.35 | 0.007 |
| 65 | 2.08 | 2.48 | 16.2 | 2.9 | 2.23 | 4.62 | 0.134 | 3.90 | 7.46 | 0.187 | 0.781 | 4.57 | 0.133 |
| 48 | 2.95 | 3.22 | 9.65 | 8.5 | 1.38 | 3.50 | 0.298 | 2.34 | 4.74 | 0.402 | 0.707 | 3.41 | 0.290 |
| 42 | 3.51 | 3.82 | 6.85 | 11.0 | 0.93 | 3.05 | 0.335 | 1.66 | 3.96 | 0.435 | 0.660 | 2.94 | 0.326 |
| 35 | 4.17 | 4.54 | 4.85 | 21.5 | 0.67 | 2.69 | 0.580 | 1.18 | 3.37 | 0.725 | 0.613 | 2.58 | 0.555 |
| 32 | 4.95 | 5.40 | 3.42 | 31.0 | 0.47 | 2.39 | 0.742 | 0.83 | 2.92 | 0.907 | 0.574 | 2.35 | 0.729 |
| 28 | 5.89 | 6.42 | 2.42 | 13.5 | 0.33 | 2.15 | 0.290 | 0.59 | 2.57 | 0.347 | 0.536 | 2.16 | 0.294 |
| 24 | 7.01 | 7.64 | 1.71 | 9.0 | 0.236 | 1.97 | 0.177 | 0.415 | 2.30 | 0.207 | 0.497 | 1.99 | 0.179 |
| 20 | 8.83 | 9.09 | 1.21 | 1.8 | 0.167 | 1.82[a] | 0.033 | 0.294 | 2.09 | 0.037 | 0.487 | 1.95 | 0.035 |
| 16 | 9.91 | 10.76 | 0.86 | 0.7 | 0.120 | 1.82[a] | 0.013 | 0.209 | 1.91 | 0.013 | 0.487 | 1.95 | 0.014 |
| 14 | 11.68 | | | | | | 2.608 | | | 3.269 | | | 2.562 |

[a] $1/(1 - f_e)$ of unexpanded bed is $1/(1 - 0.45) = 1.82$. This is the smallest possible value. Sand larger than the 20-mesh sieve size is not lifted at the wash water rate and the temperature (25°C) obtaining in the example.

rate in line with the temperature or to measure and control expansion during washing. When the bed is fully fluidized, there is no further increase in head loss, and the rate of upflow can be kept constant thenceforward. A washwater rate controller can regulate flow automatically.

## 14-8 Removal of Impurities

The overall removal of solid impurities is brought about in various ways. Which one of them exerts the controlling influence depends on circumstances. At the beginning of a filter run fine-grained, nonstratified filters act primarily as strainers. Slow sand filters are examples. They normally accumulate most of the impurities at the sand surface and in the top 2 in. or so of sand. Large microbic populations, derived in the first instance from the applied water, multiply and flourish in the accumulating organic matter, nutrients, or *Schmutzdecke*.[11] To clean these simple structures in simple ways, the top sand and its *Schmutzdecke* are scraped off and removed or washed in place.

By contrast, well-designed and well-operated rapid filters, especially multilayered filters, are intended to remove impurities, principally flocculated impurities, down to the last layer of active grains that is cleared of accumulated floc when the bed is washed.

Most granular water filters remove particulates smaller than the passages between adjacent grains. This can be explained by the concept of a filter as possessing, in the aggregate, a relatively huge surface area or interface in contact with the water and its impurities that pass by. Surface forces are brought into play—among them van der Waals forces (Sec. 16-2)—that bind particles to the surfaces even though they may bear the same electrical charge as the filter grains.

The substances removed during filtration are distributed irregularly, that is, by chance, over the grain surfaces and generally are not dislodged by the passing water. Interstices, however, are narrowed down by accumulating deposits, and some of them are undoubtedly closed. Particulates entering pores still open are then transported deeper into the bed, until they reach grain sites still able to accept them. Only if they fail to find sites of this kind do they escape into the effluent. In fluidized beds the grain surfaces are scoured clear of deposits by repeated abrasive contact between moving grains. At the water velocities engendered during filtration even in partially clogged interstices there is no scour as such and no consistent downward displacement of filter deposits because of scour.[12]

[11] From the German *Schmutz*, dirt or impurity, and *Decke*, cover or layer.
[12] Because a backwashed filter normally settles and compacts at less than critical density, some downward displacement of grains and deposits may accompany rearrangement of grains under stress.

## 14-9 Kinetics of Filtration

As a rate process, filtration shares the common elements of purification kinetics. This was shown by Iwasaki[13] in 1937. Since then the basic purification equation $-dy/dt = k(y_0 - y)$ has been expanded by Ives[14] to include important factors in filter behavior. Among them is the identification of filtration kinetics as the kinetics of a transient operation responsible for progressive changes in the rate constant $k$ during the course of a filter run. Ives has expressed the changes as follows:

$$k = k_0 + c\sigma - \phi\sigma^2/(f_0 - \sigma) \qquad (14\text{-}21)$$

where $k_0$ is the initial rate constant, which, along with the coefficients $c$ and $\phi$, describes a specific exposure; $f_0$ is the porosity of the clean bed; and $\sigma$ is the specific deposit, or volume of deposited matter per unit filter volume. Equation 14–21 ks not entirely empirical. The second term depends on the increase in specific surface by initial deposits. The third term depends on the increase in interstitial velocity caused by the accumulation of deposits and the accompanying straightening of flow paths in pores and diminution of specific surface. As sketched in Fig. 14–5, $k$ rises to a maximum value during a breaking-in or maturation period. After that it falls consistently. During the rise the second term of Eq. 14–21 is dominant; after that the third term, which derives in essence from a materials balance in which input minus output must equal the volume of deposit stored; that is,

$$\partial\sigma/\partial t = -v(\partial C/\partial l) \qquad (14\text{-}22)$$

Here $t$ is the time from the start of filtration, $C$ is the concentration of suspended matter passing a given filter depth $l$, and $v$ is the rate of filtration per unit cross-sectional area of filter.

[13] T. Iwasaki, Some Notes on Filtration, *J. Am. Water Works Assoc.*, **29**, 1596 (1937).
[14] K. J. Ives, Rational Design of Filters, *Proc. Inst. Civil Engrs.*, London, **16**, 189 (1960); K. J. Ives and I. Sholji, Research on Variables Affecting Filtration, *Proc. Am. Soc. Civil Engrs.*, **91**, SA 4, 1 (1965).

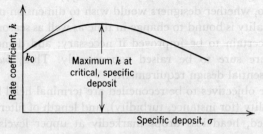

**Figure 14-5.**
**Rise and fall in rate coefficient $k$ with specific deposit in a filter run.**

Solutions of Eqs. 14–21 and 22 with a suitable form of Eq. 11–1 yields equations describing the distribution of suspended matter in the water and the specific deposit $\sigma$ through a depth $l$, with time $t$, as a function of $k_0$, $c$, $\phi$, $f_0$, and $v$ as determined in pilot tests that involve sampling at successive depths and times.

## 14-10 Filter Design

The dimensioning of filters and their appurtenances depends on entrant water quality; filter type; process and hydraulic loading; method and intensity of cleaning; nature, size, and depth of filtering materials; and prescribed quality of product water. Unless pilot-plant tests provide information on (1) the observed responses of a specific water to treatment in a plant being designed for this purpose and (2) the associated responses of the filter or filters examined, it is common to start with design values that have stood the test of experience and to modify these values in the light of new information and wanted objectives.

Some filters are expected to be no more than polishing units that separate residual floc from well-coagulated, well-settled water. Such filters can be relatively shallow, coarse-grained, high-rate units producing a clear effluent that is readily and reliably amenable to disinfection. Roughing filters—so called because they only prepare water for further filtration—may also be of this design. Placed in advance of slow sand filters, they have been operated with or without coagulation. By contrast, filters are normally designed as relatively deep, fine-grained, low-rate units if they are intended to (1) offer an effective barrier to waterborne pathogens, (2) treat water containing much floc, or (3) serve as reacting units for floc formation and removal after dosage of the applied water with a coagulant. Characteristic design values are shown in Table 14–1 for water filters and discussed in Sec. 14–3 for intermittent sand filters.

Even for waters of fairly constant composition, it is not yet possible to prescribe unique or optimizing combinations of filter depth, layered arrangements of filtering media (size, shape, and density), and rates of filtration. It is doubtful, too, whether designers would wish to dimension units narrowly. Raw-water quality is bound to change in time as well as season; preparatory processes are certain to be improved if necessary; and demands on plant productivity are sure to be raised concurrently. Thus, *built-in flexibility* becomes an essential design requirement.

Performance objectives to be reconciled are terminal head loss, standards of effluent quality (for instance, turbidity), and length of filter run. As grain size is increased, head loss drops markedly at upper levels, but effluent turbidity increases relatively rapidly. As rates of filtration are upped, head loss and turbidity rise relatively fast at all filter levels. In coarse-grained units, filter runs ended at a given head loss are relatively longer than in fine-grained units, but filter runs ended at a given turbidity are relatively shorter.

## 14-11 Bed Depth

The design depth of rapid filters is generally related to grain size, terminal head loss, and terminal turbidity or a suitable surrogate measure of terminal effluent quality. In an analysis of bed performance, Hudson[15] concluded that inadequately filtered water breaks through rapid filters when

$$Qd^3h/l = B \qquad (14\text{-}23)$$

Here $Q$ is the rate of filtration in gallons per minute per square foot of filter, $d$ is the sand size in centimeters, $h$ is the terminal loss of head in feet, $l$ is the depth of bed in inches, and $B$ is a breakthrough index that assumes the following magnitudes for different influent waters at 50°F:

| Response to Coagulation | Degree of Pretreatment | Value of $B$ |
| --- | --- | --- |
| Poor | Average | $4 \times 10^{-4}$ |
| Average | Average | $1 \times 10^{-3}$ |
| Average | High | $2 \times 10^{-3}$ |
| Average | Excellent | $6 \times 10^{-3}$ |

For temperatures other than 50°F, viscosity effects can be allowed for by multiplying the breakthrough index by $60/(T + 10)$.

*Example 14-6.*

Water receiving an average degree of pretreatment and responding in average fashion to coagulation is to be filtered at a rate of 3 gpm/sq ft through a layer of sand grains 0.1 cm in diameter. Find the requisite minimal depth of sand that will prevent breakthrough of turbidity at a terminal loss of head of 8 ft.

Because the expected value of the breakthrough index is $1 \times 10^{-3}$, Eq. 14-23 states that the depth of the layer must be

$$l = 1000 \times 8(1 \times 10^{-1})^3 3 = 24 \text{ in.}$$

Multilayered filters, often referred to as *mixed-media filters*, simulate countercurrent operation by being composed of successive layers of coarse but light filter grains on top of increasingly finer but heavier particles. Such

---

[15] H. E. Hudson, Jr., Declining-Rate Filtration, *J. Am. Water Works Assoc.*, 51, 1455 (1959).

filters must preserve their layered structure during backwashing and resettling. To this purpose, the light grains of largest size in an upper layer must rise higher and settle more slowly than the heavy grains of smallest size in the layer next below.

Equal expansion during backwashing is identified by Eq. 14-16, and equal rate of settling by calculations involving Eq. 14-20, because settling is hindered. In accordance with these equations,

$$d_u/d_l = (\psi_l/\psi_u)[(\rho_l - \rho)/(\rho_u - \rho)]^{1/2} \tag{14-24}$$

where the subscripts $u$ and $l$ respectively denote the largest grains within the upper layer and the smallest grains within the lower layer.

It follows that mixing during settling as well as during expansion determines the maximum allowable ratio of the grain sizes in the two layers. Because the conditions of flow and the specific shape of filter grains are normally uncertain, a value of $d_u/d_l$ smaller than that from Eq. 14-24 would be employed.

*Example 14-7.*

A layer of crushed anthracite with a sphericity of 0.70 and a density of 1.50 is to rest on a layer of sand with a sphericity of 0.80 and a density of 2.65. Find the maximum ratio of the diameter of the coarsest anthracite to the finest sand that will ensure both equal expansion and equal settling of the two materials at the common boundary. Assume that the density of the water is 1.00.

By Eq. 14-24, $d_u/d_l = (0.80/0.70)[(2.65 - 1.00)/(1.50 - 1.00)]^{1/2} = 2.1$ for equal expansion and settling. Therefore the grain-diameter ratio in the two layers must be less than 2.1.

## 14-12 Underdrainage Systems

Underdrainage systems of rapid filters perform two primary functions: (1) they collect the filtrate and send it on its way to a clear well or pure-water reservoir and (2) they distribute wash water to the bed during scouring operations. Because the rate of wash is many times that of filtration, the hydraulics of underdrains is governed by upflow requirements. If these are met, downflow distribution should be satisfactory. A secondary responsibility of underdrainage systems is to withdraw to waste chemical solutions added to filter beds from time to time, (1) to break up, loosen, and remove incrustants accumulated on filter grains and (2) to break up mudballs formed and built up near the cleavage plane between supporting grains and filtering and expanding grains.

Two types of underdrainage systems are in use in rapid filters: (1) pipe grids and (2) filter floors or false bottoms.

## 14-13 Pipe Grids

In their simplest form, pipe grids comprise a main, called a *manifold*, and perforated laterals (Fig. 14-1). Perforations are normally drilled into the laterals in a single row of orifices directed vertically downward or in two rows as pairs of orifices directed downward at angles of 45° on either side of the vertical (see Fig. 14-6). To protect them against corrosion, pipe grids are normally lined with cement or bitumastics and well coated on the outside. Their walls are generally so thick that water issuing from the orifices expands within the perforations, which then have a coefficient of discharge nearer 0.75 than 0.6.

Because the jets are broken up when they strike the filter bottom or the gravel surrounding the grid, the head loss equals the full driving head. This is important. Made large enough in reference to the system's head loss as a whole, the orifice loss can overshadow the remaining losses in magnitude. The hydraulic performance of the system then becomes much the same above each orifice and relatively uniform over the entire bed. In practice, the *controlling head loss* or driving head is set between 3 and 15 ft. This makes for a ratio of orifice area to bed area of $(3/60)(0.75\sqrt{2gh}) = 0.5\%$ to $0.2\%$ when wash rate is 36 in. rise per minute.

Other useful calculations for hydraulic equilibrium are (1) that the allowable friction loss, $h_f$, in a lateral with $n$ orifices must equal $(1 - m^2)$ times the driving head, $h_d$, on the first orifice reached by the washwater if the head on the last or $n$th orifice is to be no more than $m^2$ times the driving head (Eq. 13-19), and (2) that, in terms of lost head, and in the absence of side effects, the friction in a lateral of unvarying size is numerically equal to the friction exerted by the full incoming flow in passing through one third the length of the lateral (Eq. 12-9). For $m = 0.9$, for example, $h_f = (1 - 0.81) h_d = 0.57$ to 2.85 ft. Hydraulic studies of pipe grids must also cover entrance losses and losses in fittings. Reductions in flow by takeoff at orifices, moreover, produce hydraulic transients similar to the transients associated with sudden enlargement of flow channels. In accordance with Borda's loss function,[16] the reduction in velocity and recovery of velocity head produces a net head recovery at the $i$th orifice equal to

$$[(v_i^2/2g) - (v_{i+1})^2/2g)] - (v_i - v_{i+1})^2/2g = 2v_i v_{i+1}/2g \quad (14\text{-}25)$$

where $v_i$ and $v_{i+1}$ are the pipe velocities on approach to the $i$th and $(i + 1)$th orifices, respectively. It follows that there may be a net gain, not a net loss,

---

[16] Named for Jean Charles Borda (1733-99), French military engineer.

in head as the washwater travels down the lateral if $(1/g)\Sigma v_i v_{i+1} > \frac{1}{3} ls$, that is, the pipe friction.

As a rule, hydraulic systems are optimized when 25% of the overall energy loss is incurred in delivering required flows to the control points. In the case of pipe grids, the orifices are these points. Therefore the unused 75% of the total energy remains to be expended in driving the washwater in succession through the orifices, gravel, and fluidized bed.

Rules of thumb that make it possible, on first trial, to lay out a reasonably well-balanced underdrainage system for filters washed at rates of 6 to 36 in. per min follow:

1. Ratio of area of orifice to area of bed served: $[(1.5 \text{ to } 5) \times 10^{-3}]:1$.
2. Ratio of area of lateral to area of orifices served: $(2 \text{ to } 4):1$.
3. Ratio of area of main to area of laterals served: $(1.5 \text{ to } 3):1$.
4. Diameter of orifices: ¼ to ¾ in.
5. Spacing of orifices: 3 to 12 in. on centers.
6. Spacing of laterals: closely approximating spacing of orifices.

Twinned units halve washwater rate requirements of individual filters by washing the two component units in succession.

## 14-14 Filter Gravel

Perforated pipe grids are only one part of an underdrainage system. The other part comprises the stone and gravel that surround the grid and support the sand bed. This system of collecting and distributing waterways is seldom less than 10 or more than 24 in. thick. For particles sized by screening, the depth, $l$, in inches, of a component gravel layer of size $d$ in., where $d > \frac{3}{64}$ in., may be estimated from the following equation:

$$l = k(\log d + 1.40) \qquad (14\text{--}26)$$

Here $k$ varies numerically from 10 to 14.[17] Stones as large as 3 in., but generally no larger than 2 in., are placed near the pipes. To keep the screened gravel in place during backwashing, it should be carefully packed. Indeed, the larger gravel sizes should be packed into the filter by hand (Fig. 14–6).

*Example 14–8.*

A sand bed is to be supported on gravel $\frac{1}{10}$ to 2½ in. in size. Find the requisite depths of component gravel layers.

[17] J. R. Baylis, Filter Bed Troubles and Their Elimination, *J. New England Water Works Assoc.*, **51**, 17 (1937).

Equation 14–26 and a value of $k = 12$ establish the following sequences for sieve ratios of 2:1:

| Size, in. | $5/64$ | $5/32$ | $5/16$ | $5/8$ | $1\frac{1}{4}$ | $2\frac{1}{2}$ |
|---|---|---|---|---|---|---|
| Depth, in. | 3.5 | 7.1 | 10.7 | 14.4 | 17.0 | 21.6 |
| Increment, in. | 3.5 | 3.6 | 3.6 | 3.7 | 3.6 | 3.6 |
| Chosen depth, in. | $3\frac{1}{2}$ | $3\frac{1}{2}$ | $3\frac{1}{2}$ | $3\frac{1}{2}$ | $3\frac{1}{2}$ | $4 + 2 = 6$ |

Accordingly, the total depth is $23\frac{1}{2}$ in. (Compare with Fig. 14–6a.)

Because the behavior of fine gravel and coarse sand in a stratified bed is much the same, this arrangement of supporting gravel is by no means as stable as the designer may want it to be. For this reason, some works have introduced the arrangement illustrated in Fig. 14–6b. Here the progress is symmetrically from coarse to fine to coarse. At the sand-gravel interface, coarse sand slips into the interstices of the upper half, but there is no harm in this. The supporting gravel layer maintains itself no matter what the rate of washwater rise is, within reason. Effective depths of gravel are given by Eq. 14–26 with the following modifications: (1) all but the increments in

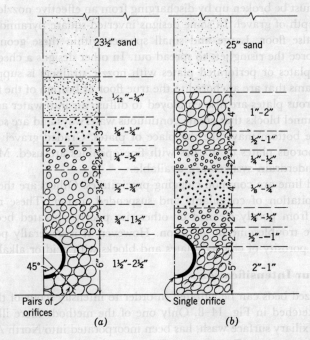

**Figure 14–6.**

(a) Asymmetrical and (b) symmetrical sequences of gravel below a sand bed.

depth of the largest gravel are halved; (2) the bottom halves together with the unhalved depth of the largest gravel increased by 1 in. are put in place; and (3) the other half of the gravel, together with the unhalved depth of the largest gravel, is superimposed on the bottom half (Fig. 14–6b).

If overall filter depths must be kept small, pipe perforations and gravel can be replaced by nozzles with slots or holes small enough to exclude filter grains but large enough in the aggregate to deliver and accept required flows. There is danger of galvanic corrosion when dissimilar metals are used for the nozzles and the piping.

## 14-15 Filter Floors

Filter floors, also called *false bottoms* or *false floors*, are intended to replace pipe grids and serve two functions: (1) to support the filter bed, possibly without stone and gravel in transitional layers below the filter bed proper, and (2) to create a single, boxlike waterway beneath the filter as the dispenser of washwater and the collector of filtered water (Fig. 14–7). The floor, depending on its thickness, is perforated by short tubes or orifices of such dimension as to introduce the *controlling loss of head* that will make for even distribution of washwater. The openings must be relatively small and closely spaced, and their jets must be broken up by discharging from an effective nozzle or into a suitable depth of gravel. In some designs inverted-square pyramids are cast into the false floor. Large and small spheres within these geometric depressions force the rising jets to spread out. In other designs a checkerboard of porous plates or perforated plates with nozzle strainers is supported on bolts or beams that are anchored to the true floor or bottom of the filter box. Similar porous plates are also employed to diffuse air into water and wastewater. Channel blocks that create continuous waterways and are set directly on the true bottom may take the place of plates. Graded gravel is not required if porous plates or nozzles with fine openings are used. Many proprietary underdrain systems are available.

Iron and lime are common clogging precipitates. Others are the result of after-precipitation of coagulants and suspended matter. These reach the false floor from poorly cleaned or otherwise poorly operated beds. Slime growths are troublesome on occasion. However, it is generally possible to restore the porosity of clogged plates and blocks with acid or alkali washes.

## 14-16 Scour Intensification

How fluidized beds can be agitated in order to intensify scour of their filter grains is sketched in Fig. 14–8. Only one of the methods there illustrated, namely, auxiliary surface wash, has been incorporated into North American plants for some time. By contrast, European engineers continue to favor auxiliary air scour as such and because it saves water. Mechanical scour has

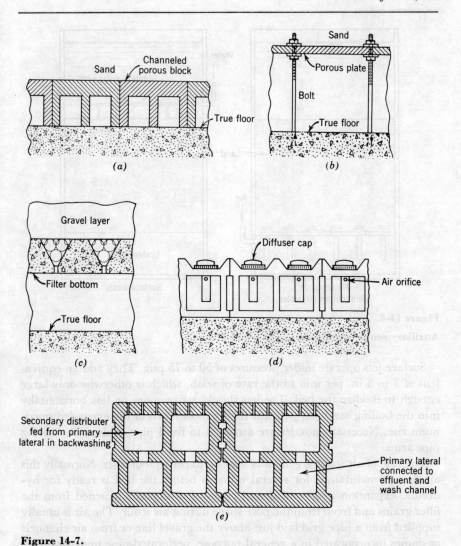

**Figure 14-7.**
**Filter floors. (a) Channeled porous block; (b) porous plate floor; (c) Wheeler bottom; (d) Camp filter underdrain (combination air/water backwash—Walker process); (e) Leopold duplex filter bottom.**

found little use since the early days of rapid sand filtration, then called *mechanical filtration*. However, new designs are made possible by individualized power sources. They may be worthy of restudy because, unlike air and water, mechanical rakes and related devices do not seek out paths of least resistance. Instead, they destroy points of resistance.

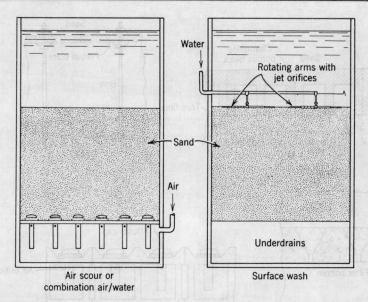

**Figure 14-8.**
**Auxiliary scour intensification in rapid filters.**

Surface jets operate under pressures of 50 to 75 psig. They add an equivalent of 3 to 5 in. per min to the rate of wash, which is otherwise only large enough to fluidize the bed. The jets should strike more or less horizontally into the boiling sand just beneath the topmost reach of the expanded maximum rise. Necessary nozzles are attached to fixed pipe grids or to moving pipe arms.

The rate of air scour is generally 3 to 5 cfm per sq ft of filter. Normally this air flow is maintained for several minutes before the bed is ready for hydraulic expansion and coincidental removal of deposits loosened from the filter grains and freed from the pore space during air scour. The air is usually supplied from a pipe grid laid just above the gravel line or from air channels or domes incorporated in a general-purpose, perforated-pipe underdrainage system or filter floor. Elements of air-grid design are much the same as for water grids. In both instances a controlling loss of head equal to about 75% of the total head supplied to the washing system will normally make for economy of design.

## 14-17 Washwater Troughs

Troughs spanning filters and receiving the spent washwater are intended to keep the upwelling water from establishing a significant slope to its surface and impressing a head differential on the washwater distribution system. The aim is an even washwater rate over every square inch of bed surface. Troughs are not common in European practice.

Hydraulically the troughs are not unlike those serving the effluent weirs of settling tanks and the side spillways of dams. A general relationship for the water-surface curve within washwater troughs and similar structures can be based on the momentum theorem[18] by making the following simplifying assumptions: (1) the kinetic energy of the water falling into the trough does not contribute to the longitudinal (displacement) velocity; (2) channel friction can be neglected in all but very long channels; (3) flow is essentially horizontal; and (4) the water-surface curve approximates a parabola. The forces changing the momentum then derive solely from the unbalanced static pressure forces $P_1$, $P_2$, and $P_3$ shown in Fig. 14–9. They have the following magnitudes: $P_1 = \frac{1}{2} wbh_0^2$; $P_2 = -\frac{1}{2} wbh_l^2$; and[19] $P_3 = wbli\,(\frac{2}{3} h_0 + \frac{1}{6} il + \frac{1}{3} h_l)$. Here $w$ is the unit weight of water; $b$, $l$, and $i$ are respectively the width, length and invert slope of the trough; and $h_0$ and $h_l$ are respectively the initial and terminal depths of the water. The change in momentum, $Qv_lw/g$, equals $wbh_c^3/h_l$, because $Q = bh_cv_c$, $v_l = h_cv_c/h_l$, and $v_c = \sqrt{gh_c}$. Here $Q$ is the rate of discharge, and $v_l$ and $v_c$ are the velocities of flow at the submerged depth $h_l$ and the critical depth $h_c$, respectively. Equating the sum of the forces to the change in momentum and solving for $h_0$,

$$h_0 = \sqrt{(2h_c^3/h_l) + (h_l - \tfrac{1}{3}il)^2} - \tfrac{2}{3}il \qquad (14\text{--}27)$$

For level inverts ($i = 0$) and for the critical depth $h_c^3 = Q^2/gb^2$, where $Q$ is the total rate of discharge and $b$ is the width of a rectangular trough,

$$h_0 = \sqrt{h^2 + \frac{2Q^2}{gb^2h_l}} = \sqrt{h_l^2 + 2\frac{h_0^3}{h_l}} \qquad (14\text{--}28)$$

When discharge is free, $h_l$ closely equals $h_c$, and

$$h_0 = h_c\sqrt{3} = 1.73\,h_c \qquad (14\text{--}29)$$

or

$$Q = 2.49\,bh_0^{3/2} \qquad (14\text{--}30)$$

Equations 14–28 and 29 hold also for troughs that are not rectangular in cross-section. In long troughs the frictional drawdown of the water surface can be approximated by assuming that turbulence increases the roughness factor about twofold.

In American practice lateral travel of the water overflowing into the troughs is commonly limited to between 2.5 and 3.5 ft; that is, the clear

---

[18] T. R. Camp, Lateral Spillway Channels, *Trans. Am. Soc. Civil Engrs.*, **105**, 606 (1940).
[19] Because the volume of water in the trough is $bl[h_0 + il - \tfrac{1}{2}il - \tfrac{1}{3}(h_0 + il - h_l)]$.

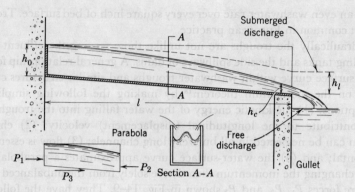

**Figure 14-9.**
**Water-surface curves in a washwater gutter.**

distance between gutters is held to between 5 and 7 ft in order to keep the head of water on the underdrains, and, with it, the rate of wash, nearly uniform. The height of troughs above the sand bed is determined by the degree of sand expansion. The trough must lie above the surface of the expanded sand if it is not to reduce the waterway open to the sand-water mixture, increase its upward velocity, and cause sand loss.

Troughs run the length or width of the filter. They discharge into a gullet, the hydraulics of which is like that of the tributary troughs except that flow increases stepwise instead of uniformly. Choice of trough cross-section depends somewhat on the material of construction. Rectangular, semicircular, semihexagonal, and semioctagonal shapes are common. Externally the semicircle interferes least with the upward streaming of washwater. Internally it possesses the best hydraulic properties as well.

---

*Example 14–9.*

Troughs 24 ft long, 18 in. wide, and 7 ft on centers are to serve a filter that is washed at a rate of 26 in./min. Find (1) the depth of the troughs if their invert is to be kept level and they are to discharge freely into the gullet, and (2) the height of the lip of the trough if the bed below it contains a 30-in. bed of sand and is expanded 50%.

1. For $Q = 24 \times 7 \times 26/(12 \times 60) = 6.1$ cfs, find, by Eq. 14–30, $h_0 = [6.1(2.49 \times 1.5)]^{2/3} = 1.32$ ft or, say, 16 in.

2. It follows that the troughs must be placed above the sand by a distance of $(0.5 \times 2.5 + 1.32) = 2.57$ ft, plus the depth of freeboard in the trough and its thickness.

## 14-18 Filter and Conduit Dimensions

The dimensioning of individual filter units in a plant of given size becomes a matter of economics once their number has been made large enough to routinize cleaning operations and allow for occasional repairs. Essential factors are the cost of the filter proper, its walls, and appurtenances. Lagrangian or maximum-minimum cost analyses (Sec. 6-3) will identify the number of beds to be used. However, final sizings must be confirmed by comparative designs. As stated before, twinned units can be introduced to keep the size of the washwater system within useful working limits.

Filters must be large enough to allow for time out of service for cleaning and repair. Rapid filters are thrown out of operation for about 10 min during each cycle—normally 1 day.[20] Further allowances must be made for the amount of filtered water consumed in washing the beds, on the order of 3 to 5% of the plant output, and for occasional repairs. Including a freeboard of 1 ft, the depth of both slow and rapid filter units is commonly 10 ft.

Pipes and other conduits, including valves and gates, are ordinarily designed to carry water at velocities close to the following:

|  | *fps* |
|---|---|
| Influent conduits carrying raw water | 3-6 |
| Influent conduits carrying flocculated water | 1-3 |
| Effluent conduits carrying filtered water | 3-6 |
| Drainage conduits carrying spent washwater | 4-8 |
| Washwater conduits carrying clean washwater | 8-12 |
| Filter-to-waste connections | 8-12 |

Where washwater tanks are used, they are made large enough to wash two filters in sequence. They must refill between washes. In large plants it is usually economical to pump washwater directly to the filters. In small plants washwater can be taken from distribution systems through pressure-reducing valves.

## 14-19 Filter Appurtenances

Filter appurtenances include manually, hydraulically, pneumatically, or electrically operated sluice gates and valves on the influent, effluent, drain, and washwater lines (where valves are inserted into pipe lines, butterfly valves are economical in cost and space); measuring devices such as venturi meters; rate controllers activated by a measuring device (Fig. 14-10); loss-of-head and rate-of-flow gages; sand-expansion indicators; washwater controllers and indicators; operating tables and water-sampling devices; water-quality-

---

[20] The time required to clean slow filters may be as much as 3 days in every 30-day period.

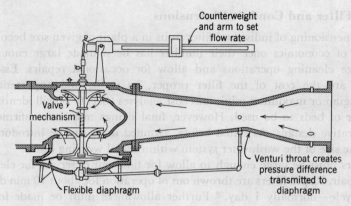

**Figure 14-10.**
**Rate of flow controller. (Simplex Valve and Meter Company.)**

monitoring devices; sand ejectors and washers; and washwater pumps and tanks. The larger the plant and the higher the rate of filtration, the more is the inclusion of mechanical and automatic operating aids justified. Flow regulators automatically open or close a valve mechanism to keep the discharge rate constant. Pressure differentials between the venturi throat and outlet are translated into valve movements by a balancing diaphragm or a piston.

Much of the equipment and piping of modern gravity filters is shown in Fig. 14–11. The *filter-to-waste* connection serves the purpose of wasting the water held in the filter after washing. It was called the *rewash connection* when filters were washed regularly with raw water. Today it is operated only when beds must be cleaned with deterging chemicals. The chemicals are usually drawn into the bed after distribution onto or over the surface, allowed to remain in contact with the grains until they are clean, and displaced from the bed through the filter-to-waste connection. The connection branches from the filter effluent in advance of the rate controller. Generally it is a valved pipe stub proportioned to discharge at about the maximum rate of filtration when the bed is clean. A waste connection of this kind should not endanger the normal filtrate by allowing polluted water to be drawn into the system.

## 14-20 Length of Filter Runs

The impurities transferred from the applied water to the filter together with their coagulating and precipitating agents clog its pores and increase the hydraulic loss of head. The time rate at which head loss rises depends on sand size, porosity, filtration rate, and amount and character of the suspended matter in the applied water. Relationships between these factors are best

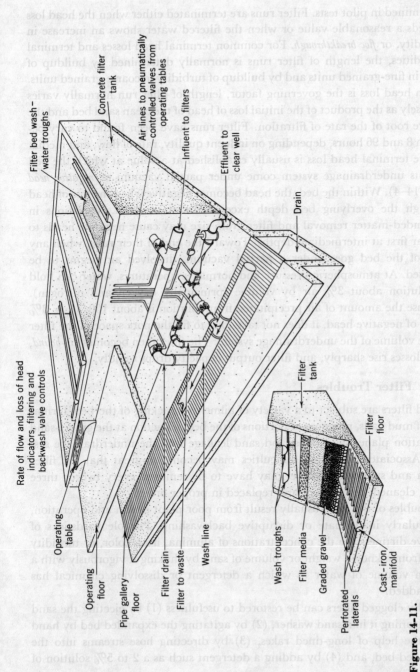

Figure 14-11.
Rapid filters and accessory equipment.

determined in pilot tests. Filter runs are terminated either when the head loss exceeds a reasonable value or when the filtered water shows an increase in turbidity, or *floc breakthrough*. For common terminal head losses and terminal turbidities, the length of filter runs is normally determined by buildup of head in fine-grained units and by buildup of turbidity in coarse-grained units. When head loss is the governing factor, length of filter run normally varies inversely as the product of the initial loss of head of the clean sand bed and the square root of the rate of filtration. Filter runs have been found to vary between 8 and 96 hours, depending on influent quality, rate of flow, and media.

The terminal head loss is usually established at a value at which the bed and its underdrainage system come under partial vacuum or *negative head* (Fig. 14-4). Within the bed, the head becomes negative when the loss of head through the overlying bed depth exceeds the static head. Variations in suspended-matter removal and filter grain size may cause negative heads to appear first at intermediate depths. Toward the end of their runs, when any part of the bed goes under a partial vacuum, dissolved air begins to be released. At atmospheric pressure and normal temperatures, water can hold in solution about 3% air by volume (principally oxygen and nitrogen). Because the amount of air precipitated from solution is about $100/34 = 3\%$ per ft of negative head, it does not take long to fill the pore space of the filter or the volume of the underdrainage system. The filter then becomes *air-bound*, head losses rise sharply, and filter output capacity drops rapidly.

## 14-21 Filter Troubles

Rapid filters are subject to a variety of ailments: cracking of the bed, formation of mud balls, plugging of portions of the bed, jet action at the gravel-sand separation plane, sand boils, and sand leakage into the underdrainage system. Associated operating difficulties may become so great that the filter media and supporting gravel may have to be removed every two or three years, cleaned, reclassified, and replaced in proper order.

Troubles of this kind usually result from poor plant design and operation, particularly inadequate or disruptive backwashing. Possible yardsticks of relative dirtiness are the concentrations of alumina, iron, color, or turbidity freed from a known weight or volume of sand by shaking it vigorously with a known volume of water to which a detergent or dissolving chemical has been added.

Badly clogged filters can be restored to usefulness (1) by ejecting the sand and cleaning it in a sand washer, (2) by agitating the expanded bed by hand with the help of long-tined rakes, (3) by directing hose streams into the expanded bed, and (4) by adding a detergent such as a 2 to 5% solution of caustic soda,[21] draining it off through the filter-to-waste connection, and washing the bed clean.

**Filter Cracks.** Because resistance to flow is least along the walls of a filter, less head is lost there, and resulting pressure differentials within the bed establish inward as well as downward flows; filter grains are pushed away from the walls; water short-circuits through resulting shrinkage cracks and fills them with caking suspended matter; and pressure differentials may open cracks also in the body of the bed. In coarse, well-compacted, clean sand, shrinkage is less than 1% at maximum head loss.

**Mud Balls.** Mud balls are conglomerations of coagulated turbidity, floc, sand, and other binders. Mud-ball concentration can be measured by washing a known volume of sand through a 10-mesh sieve, transferring the retained conglomerations to a graduated cylinder partly filled with water, and noting the volume of displaced water. In well-designed and well-operated filters, mud balls should not occupy more than about 0.1% of the volume of the top 6 in. of sand.

Newly formed mud balls are small and light. They collect on the surface of the bed after it has been backwashed and can be removed by suction. Otherwise they should be broken up by air scour or destroyed by one of the methods listed in connection with filter clogging. As they grow in size, they increase in density by accretion. Eventually they become heavy enough to migrate downward as far as the gravel. There they accumulate and clog portions of the bed. Backwashing then becomes uneven, and gravel is lifted and mixed into the sand along the margin of the clogged volume. Associated deterioration in both filter output and effluent quality can be rectified only by rebuilding the filter.

**Jetting and Sand Boils.** Even small differences in porosity and permeability of sand and gravel cause the first flush of backwash water to follow paths of least resistance and break through to the surface at scattered points. Within the zone of flow, the clogged and compressed sand is fluidized, back pressure is reduced, flow is increased, and water is jetted from the gravel into the sand. If jetting becomes severe, the sand boils up like quicksand, and gravel as well as sand is lifted to the surface. If, in defense against these happenings, backwash valves are opened slowly, the bed is given an opportunity to disintegrate from the surface downward. Surface wash or air scour in advance of final bed fluidization can help to loosen a clogged bed, yet leave the gravel where it belongs.

**Sand Leakage.** If both gravel layers and sand are properly sized and placed, sand cannot leak into the underdrainage system unless the layers of smallest gravel are displaced during backwashing.

---

[21] $\frac{1}{4}$ to 2 lb per sq ft of filter covered with 2 to 3 in. of water.

## 14-22 Plant Loading and Performance

In practice, plant performance as a whole rather than the performance of specific portions of filter plants is normally identified. This is understandable, because it is overall performance that establishes the quality of the product water.

*Bacterial Efficiency.* Modern, well-operated water-treatment plants, in which effluent residuals of free chlorine are maintained, deliver a product water consistently free of coliform organisms, no matter how polluted the raw water may have been. At one time this was not so, even in large plants. Today it may still not be so in small plants in which supervision of chlorination and other operations is intermittent and deficient. Accordingly, the experience of earlier times remains of interest. Based on information collected by its engineers, the U.S. Public Health Service[22] established the following empirical relationship between the effluent concentration, $E$, and the influent, or raw-water, concentration, $R$, of coliform organisms in water that had been subjected to purification:

$$E = cR^n, \quad \text{or} \quad \log E = \log c + n \log R \tag{14-31}$$

Here $c$ and $n$ are coefficients reflecting respectively the effluent concentration of coliform organisms for a given raw-water concentration, and the relative shift in effluent concentration with changing raw-water concentration. A low $c$ value represented high fundamental efficiency, and a low $n$ value great constancy of performance with varying raw-water quality.

*Removal of Color, Turbidity, and Iron.* On an average only 30% of the natural color in water is removed by slow sand filters themselves. However, a colorless water can be produced by both slow and rapid filters after suitable coagulation and settling and by rapid filters after the addition of polyelectrolytes immediately in advance of filtration. Turbidity responds well to slow sand filtration without the aid of coagulation. However, all but very small amounts of turbidity clog slow filters so quickly that waters with turbidities above about 40 units should not be applied to them.

Both slow and rapid sand filters remove oxidized or oxidizing iron and manganese. Manganese precipitates slowly and responds better to slow than to rapid filtration, unless it has been suitably coagulated (Sec. 16-9). If the sand is coated with manganese, precipitation is hastened by catalysis.

*Removal of Large Organisms.* The large microorganisms, including the algae and diatoms, are readily removed by filtration, but the odors and tastes associated with them may pass through unchanged in intensity unless treatment processes adapted to their removal or destruction are included in

---

[22] *U.S. Publ. Health Serv. Bull.*, **172** (1927); **193** (1929); also *U.S. Publ. Health Rep.*, **48**, 396 (1933). The equation is not unlike the Freundlich isotherm (Eq. 11-30).

the works. In the absence of turbidity, chlorophyllaceous organisms flourish on the surface of open filters. Mats of appreciable thickness may, indeed, build up on open slow filters. On sunny days, algal photosynthesis may release enough oxygen to lift sections of the mats and cause a rush of water through denuded spots. This makes for poor performance. The cells of diatoms interlock and form tenacious, often almost impenetrable, mats. This shortens filter runs.

The eggs and adults of the common intestinal parasitic worms, as well as the cysts of the pathogenic amebae, are relatively so large that they do not normally pass through beds of sand. By contrast, the cercariae of the blood flukes, although larger than amebic cysts, are sufficiently motile to wriggle through beds of sand of normal depth. Upward and downward currents created during scouring operations, moreover, may carry even large microorganisms deep into and through filter beds.

*Oxidation of Organic Matter.* Intermittent sand filters treating raw or settled wastewaters can remove bacteria and other particulates in acceptable degree. They can also oxidize organic matter quite completely. However, the rates of filtration at which good results are obtainable are woefully low. Only in rare circumstances is biological oxidation of organic matter a recognizable element in the purification of municipal water supplies (Sec. 14–3).

## 14-23 Diatomaceous-Earth Filters

Diatomaceous-earth filters have found use (1) as mobile units for water purification in the field and (2) as stationary units for swimming pools and general water supplies (Fig. 14–12). The filtering medium is a layer of diatomaceous earth built up on a porous *septum* when a slurry of diatomaceous earth is recirculated through the septum until a firm layer is formed on the septum. The resulting *precoat* is supported by the septum, which serves also as a drainage system. Water is strained through the precoat unless the applied water contains so much turbidity that the unit will maintain itself only if additional diatomaceous earth, called *body feed*, is introduced into the incoming water and preserves the open texture of the layer.[23]

Skeletons of diatoms 0.5 to 12 $\mu$ in size compose the diatomaceous earth mined from deposits laid down in ancient seas. Precoating requires 0.1 to 0.5 lb of diatomaceous earth per sq ft of septum. Body feed is added in a ratio close to 1.25:1 on a dry basis when waters contain inorganic silts. For organic slimes the ratio is stepped up to about 3:1. The filter operates at rates of 2.5 to 6 gps per sq ft. Backwashing rates of 7 to 10 gpm per sq ft remove spent filter cake.

[23] Task Group Report, Diatomite Filters for Municipal Use, *J. Am. Water Works Assoc.*, **57**, 157 (1965).

436/Filtration

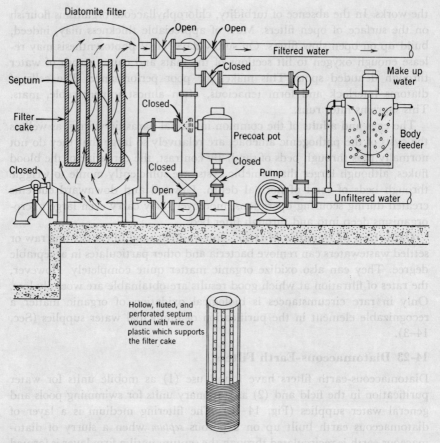

**Figure 14-12.
Diatomite filter system. (From Babbitt, Baumann, and Cleasby, *Water Supply Engineering*, 6th ed., p. 572.)**

**fifteen**

# flocculation, adsorption, desalination, and ion exchange

## 15-1 Treatment Objectives

The treatment processes described in this chapter are, in a sense, both physical and chemical operations, because they affect both the physical and the chemical composition of the waters subjected to them. There is *flocculation*, for example, when small, often colloidal, particles conjoin as a result of interparticle contact while the suspending fluid is being stirred or otherwise submitted to hydraulic shear; there is *adsorption* when particles of colloidal or molecular size are adsorbed on surfaces of activated carbon added to water as a powder or forming the granules of fixed or fluidized beds and exposing relatively immense interfacial areas to the carrying, suspending, or filtering water; there is *desalination* when *salt* is separated physically from water by induced passage through a selectively permeable membrane and when *water* is separated physically from salt by thermal conversion to the vapor or solid phase or by passage through a permeable membrane, and there is *demineralization* by ion exchange when ions are transported from water to solid ion-exchange media in fixed or fluidized beds.

## 15-2 Flocculation

A combination of mixing and stirring or agitation that produces aggregation is called *flocculation*, even though more specific terms such as coagulation and thickening may be preferred by physical chemists and chemical engineers to identify the origin of the particulates concerned. *Mixing* is the specific blending, mingling, or commingling of coagulating chemicals or materials with water or wastewater in order to create a more or less homogeneous single- or multiple-phase system, whereas *stirring* describes the disturbing of the flow pattern of a fluid in a mechanically orderly way for the purpose of effecting a dynamic redistribution of particles. Random rather than orderly turbulence is distinguished by the term *agitation*.

Generally speaking, mixing is a brief operation seeking a quick response, often in advance of stirring or agitation, whereas stirring and agitation are more protracted operations normally aiming at the conjunction of suspended particles or flocs, but sometimes intended to break up large flocs in order to maintain particle numbers. Aggregation of particles is referred to as *floc growth*; breakup of flocs, as *floc shear*.

## 15-3 Mixing and Stirring Devices

The sources of power for flocculating devices are gravitational, pneumatic, or mechanical. Generally speaking, mechanical and pneumatic devices are relatively flexible in power input; gravitational devices, relatively inflexible. Because of this, gravitational devices are seldom included in large plants, even though they may possess quite useful features.

**Baffled Channels.** Baffled channels are prime examples of gravitational mixing and stirring devices. Most other gravitational devices are relatively inefficient. Baffled channels differ from unobstructed open channels and, for that matter, from pipelines in that shear gradients or turbulence are not merely functions of frictional resistance to flow. Velocity gradients are purposely intensified by induced changes in the direction of flow (Fig. 15-1). For baffled channels of capacity $C$, in which a loss of head $h$ is incurred when the rate of flow is $Q$, the *useful* power input $P = Q\rho gh$, where $\rho g$ is the weight density of water. Accordingly the permissible channel loading at a given value of $Gt_d$ (Sec. 11-17) is

$$Q/C = \sqrt{Q\rho gh/\mu C}/(Gt_d) = \sqrt{Qgh/\nu C}/(Gt_d) \qquad (15\text{-}1)$$

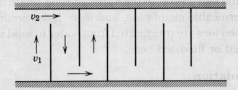

**Figure 15-1.**
**Baffled channel (schematic diagram). Plan of round-the-end baffles, or vertical section of over-and-under baffles.**

Here $g$ is the gravity constant and $\nu$ the kinematic viscosity of the fluid. Each foot of lost head is $62.4 \times 1.547/550 = 0.175$ hp per mgd or $62.4 \times 1.547/737.6 = 0.131$ kw per mgd. In practice, head losses commonly lie between 0.5 and 2 ft, velocities vary from 0.5 to 1.5 fps, and detention times run from 10 to 60 min. For $(n-1)$ equally spaced *over-and-under* or *around-the-end* baffles and for velocities $v_1$ and $v_2$ in the channels and baffle slots, respectively, the loss of head approaches $nv_1^2/2g + (n-1)v_2^2/2g$ in addition to normal channel friction. In arriving at this estimate the assumption is made that necessary velocities must be redeveloped at each change in direction of flow (Fig. 15-1).

*Example 15-1.*

Water zigzags through a baffled channel at a velocity of 0.5 fps and speeds up to 1.5 fps in the slots. There are 19 around-the-end baffles. Estimate (1) the loss of head, neglecting normal channel friction, (2) the power dissipated, and (3) the $G$ and $Gt_d$ values for a flow of 6.46 mgd (10.0 cfs), with a displacement time of 30 min. Assume a water temperature of 50°F, that is, $\mu = 2.74 \times 10^{-5}$ (lb force) (sec)/(sq ft), and calculate (4) the channel loading.

1. $h = 20 \times 0.25/2g + 19 \times 4/2g = 1.26$ ft.
2. $P = 10 \times 62.4 \times 1.26 = 790$ ft-lb/sec.
3. $G = \sqrt{790/(2.74 \times 10^{-5} \times 18{,}000)} = 40$ sec$^{-1}$; $C = 10 \times 30 \times 60 = 18{,}000$ cu ft; and $Gt_d = 40 \times 1800 = 7.2 \times 10^4$.
4. $Q/C = 6.46 \times 10^6/(1.8 \times 10^4) = 360$ gpd/cu ft.

*Note:* The tank consists of 20 channels each 2.5 ft wide, 8 ft deep on an average, and 45 ft long.

**Pneumatic Mixing and Stirring.** When air is injected or diffused into water after suitable compression, it normally expands isothermally. Accordingly, the work done by the air is $\int p\, dV$, where $p$ is the absolute pressure intensity and $V$ is the volume of air. Because $pV$ is constant ($p_a V_a$, for example), $V_a \int_{V_c}^{V_a} dV/V = p_a V_c \ln(V_a/V_c) = p_c V_a \ln(p_c/p_a)$, where the subscripts $a$ and $c$ denote free or atmospheric conditions and compressed conditions, respectively. If, for example, $Q_a$ cu ft per min of free air are injected into water from a diffuser situated $h$ ft below the water surface, the power dissipated usefully by the rising air bubbles is essentially $P = (14.7 \times 144 \times 2.303/60)\, Q_a \log[(h+34)/34]$ ft-lb per sec, or

$$P = 81.5\, Q_a \log[(h+34)/34] \tag{15-2}$$

The allowable loading $Q/C$ of an aerated flocculating tank or channel is found by substituting Eq. 15-2 in Eq. 11-28 and simplifying the resultant mathematical expression.

Compressed air is diffused into treatment units also in aeration for gas exchange (Sec. 12-10), in cleaning granular filters by air scour (Sec. 14-16), and in aerating and stirring activated-sludge units (Sec. 19-13).

**Mechanical Mixing and Stirring.** The impellers employed in mechanical mixing and stirring generate both mass flow and turbulence. Three types are in common use: (1) paddles, (2) turbines, and (3) propellers.

*Paddles* consist of blades attached directly to vertical or horizontal shafts (Fig. 15-2c,d). The moving blades (rotors) may be complemented by stationary blades (stators) that oppose rotational movement of the entire mass of water within the treatment unit and help to suppress vortex formation. However, stators are not often used in water or wastewater treatment practice. Paddles are rotated at slow to moderate speeds of 2 to 15 rpm. The currents generated by them are both radial and tangential.

*Turbines* comprise flat or curved blades attached by a connecting radius arm to a vertical or horizontal shaft. Operating in the middle range of speeds (10 to 150 rpm), they generate much the same kind of currents as do paddles (Fig. 15-2a).

*Propellers* are shaped like ships' screws. The blades are mounted on a vertical or inclined shaft and generate strong axial currents. Their speed is high—150 to 1500 rpm or more—and they may be placed off center in the treatment unit (Fig. 15-2b). Propellers are employed primarily in flash mixers.

Paddle flocculators, which are most widely used in modern plants, may be arranged so that the flow is parallel to horizontal shafts installed lengthwise in parallel or sequential basins, or at right angles to vertical shafts or horizontal shafts installed across the width of one or more tanks or compartments. Alternative installations are illustrated in *Water Treatment Plant Design*.[1]

*The useful power input for mixing and stirring* is lowered by the rotational movement of water masses as a whole and by vortex formation. Shear gradients become smaller because the velocity differential between the impeller and the water is narrowed, and power expended in changing water levels by vortex formation is not put to use for mixing purposes. Stators are useful adjuncts to all types of impellers. Combined mixing and settling devices are shown in Fig. 13-11c and d.

*The useful power input of an impeller* is a function of the impelling force $F_I$ and coefficient of drag $C_D$, that is, the drag force, $F_D = C_D F_I$ of the paddle, blade, or propeller, the stator, if any, and the tank wall; the relative velocity $v$ of

---

[1] *Water Treatment Plant Design*, American Water Works Assoc., New York, N.Y., 1969.

# Mixing and Stirring Devices/441

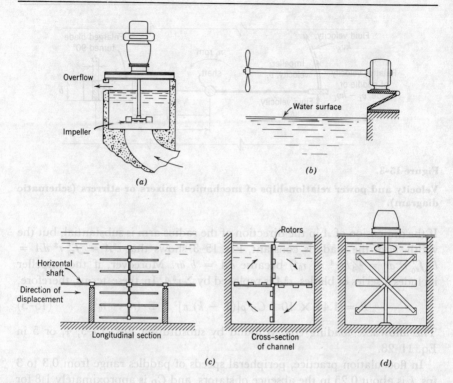

**Figure 15-2.**
**Mixing and stirring impellers.** (a) Flash mixer; (b) mixing propeller tilted into horizontal position; (c) paddle or blade mixer with horizontal shaft; (d) paddle mixer with vertical shaft.

the impeller and the fluid; and the area $A$ of the impeller blade (Fig. 15-3). The useful power input[2] $P = F_D v = C_D F_I v$, where $F_I = \rho A v^2 / 2$. Therefore

$$P = \tfrac{1}{2} C_D \, \rho A v^3 \tag{15-3}$$

If $k$ is the ratio of the fluid velocity to the impeller velocity $(v_i)$, the relative velocity of the blade $v = v_i - k v_i = (1 - k) v_i = 2\pi (1 - k) \, rn/60$. Here $r$ is the effective radius arm of the blade and $n$ is the number of revolutions per minute. It follows that the useful power expended by a single blade is

$$P = 5.74 \times 10^{-4} \, C_D \, \rho [(1 - k) \, n]^3 \, r^3 A \tag{15-4}$$

[2] Because power input identifies only geometric and kinematic similarity, chemical engineers prefer to express mixing and stirring relationships by a dimensionless power number $P = 2.16 \times 10^5 P/(32 \rho n^3 r^5) = 6.75 \times 10^3 P/(\rho n^3 r^5) = K \mathbf{R}^p \mathbf{F}^q$, where $K$, $p$, and $q$ are coefficients changing with the conditions of flow, $P$ is the power input, and $\mathbf{R}$ and $\mathbf{F}$ are the Reynolds and Froude numbers, respectively.

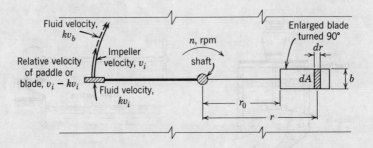

**Figure 15-3.**
**Velocity and power relationships of mechanical mixers or stirrers (schematic diagram).**

If the dimension of $A$ in the direction of the radius arm is substantial, but the width, $b$, of the blade is constant, Fig. 15-3 shows that $r^3 A = \int_{r_0}^{r} r^3 \, dA = b \int_{r_0}^{r} r^3 \, dr = \frac{1}{4} b(r^4 - r_0^4)$ because $dA = b \, dr$. Moreover, if the impeller includes a series of blades, $r^3 A$ is replaced by $\Sigma r^3 A$. In these terms, therefore,

$$P = 1.44 \times 10^{-4} \, C_D \, \rho [(1 - k) \, n]^3 \, b \, \Sigma \, (r^4 - r_0^4) \quad (15\text{-}5)$$

The allowable loading $Q/C$ is found by substituting Eqs. 15-3, 4, or 5 in Eq. 11-28.

In flocculation practice, peripheral speeds of paddles range from 0.3 to 3 fps, $k$ is about 0.25 in the absence of stators, and $C_D$ is approximately 1.8 for flat plates. At $n = 1$ to 5 rpm or more, paddles 8 ft in diameter have Reynolds numbers of $7.57 \times 10^4$ to $3.79 \times 10^5$, and the power number is directly proportional to $G^2$. Accordingly, $G$ becomes as meaningful as $P$. $G$ and $Gt$ values have been tapered from high to low for best results in specific cases.[3] Wherever possible the power dissipation function $P/C$ should be determined by measuring the actual torque and rotor speed.

*Example 15-2.*

A flocculator designed to treat 20 mgd is 100 ft long, 40 ft wide, and 15 ft deep. It is equipped with 12-in. paddles supported parallel to and moved by four 40-ft long horizontal shafts that rotate at a speed of 2.5 rpm. The center line of the paddles is 6.0 ft from the shaft, which is at middepth of the tank. Two paddles are mounted on each shaft, one opposite the other. If the mean velocity of the water is approximately one fourth the velocity of the paddles and their drag coefficient is 1.8, find (1) the velocity differential between the paddles and the water, (2) the useful power input

[3] T. R. Camp, Flocculation and Flocculation Basins, *Trans. Am. Soc. Civil Engrs.*, **120**, 1 (1955); Hydraulics of Mixing Tanks, *J. Boston Soc. Civil Engrs.*, **56**, 1(1969).

and the energy consumption, (3) the detention time, (4) the value of $G$ and the product $Gt_d$, and (5) the flocculator loading. Assume a water temperature of 50°F, $\mu = 2.74 \times 10^{-5}$ (lb force)(sec)/sq ft.

1. The paddle velocity is $2\pi rn = 2\pi \times 6 \times 2.5/60 = 1.57$ fps, and the velocity differential is $v = (1 - 0.25)1.57 = 1.18$ fps.

2. Because the area of the paddles is $A = 40 \times 2 \times 4 \times 1 = 320$ sq ft, and the coefficient of drag $C_D = 1.8$, the useful power input, by Eq. 15-3, is $P = 0.5 \times 1.8(62.4/32.2)320(1.18)^3 = 918$ ft-lb/sec, and $918/550 = 1.67$ hp, or 1.24 kw. The energy consumption per million gallons, therefore, is $1.67 \times 24/20 = 2.0$ hphr/mg, or 1.5 kwhr/mg treated. For electrical drive, there must be added the energy required to overcome mechanical friction and to provide for electrical losses in the lines and motor. (In practice, flocculators consume 2 to 6 kwhr/mg treated.)

3. Because the volume of the tank is $40 \times 100 \times 15 = 6 \times 10^4$ cu ft, the detention period $t_d = 6 \times 10^4 \times 7.48 \times 24 \times 60/(20 \times 10^6) = 32.5$ min.

4. By Eqs. 11-25 and 11-27, $G = [918/(2.74 \times 10^{-5} \times 6 \times 10^4)]^{1/2} = 23.7$ fps/ft and $Gt_d = 23.7 \times 32.5 \times 60 = 4.64 \times 10^4$.

5. $Q/C = 20 \times 10^6/(6 \times 10^4) = 333$ gpd/cu ft.

**Flocculator Performance.** The performance of mixing and stirring devices for flocculation is a function of (1) the substances to be removed and (2) the additives, if any, introduced to promote coagulation. Examples of common operational practices in successful chemical coagulation (Sec. 16-2) are, in the simplest cases, routine additions of standard amounts of chemicals; in more closely supervised cases, daily, hourly, or even more frequent adjustments of coagulant dosages to the results obtained in *jar tests* (Sec. 16-5).

However, if comparative performance is to be identified, analysis must turn in a different direction, at the hand of the fundamental equations (Sec. 11-14) of purification or process kinetics $dy/dt = k\phi(y)$, where $dy/dt$ is the rate of performance, $y$ the property, quality or manifestation under observation, $t$ the time, and $k$ the rate constant, which may depend on a number of factors governing the process kinetics in one way or another.[4]

## 15-4 Solid-Liquid Adsorption

Although solid-liquid adsorption plays an important part in a variety of water and wastewater treatment processes, it has found important *direct* application in water treatment and analysis in no more than two operations: (1) removing unwanted odors and tastes from drinking water and (2) concentrating minute amounts of organic compounds for the CCE test (Sec.

[4] H. E. Hudson, Jr., Physical Aspects of Flocculation, *J. Am. Water Works Assoc.*, **57**, 885 (1965).

10–6). In both of these, activated carbon is, so far, the *adsorbent* of choice. However, there is promise in applying the principles of adsorption in water renovation and to mixed and multiple purification functions in general.

**Selectivity of Adsorption.** In water and wastewater treatment, conventional adsorption technology normally lacks specificity. The accumulation of a substance as an adsorbate at an interface is a vectorial function of the forces of attraction and repulsion of the substance to and from the solution phase or from and to the solid phase, respectively. The affinity of adsorbents for adsorbates varies with the force fields at (or built up at) the liquid-solid interface. Affinity to the solution phase—often repulsion rather than attraction—cannot be modified. Substances repelled from solution are surface-active; that is, they decrease surface or interfacial tension, and tend to accumulate at an interface. This they do because of the dual constitution of their responding molecules or ions. Their hydrophilic part is attracted relatively more to the water phase than to the solid phase, and their hydrophobic part is repelled relatively more from the water phase than from the solid phase. Detergents are examples of surface-active substances, but most organic constituents of wastewaters exhibit some surface or interfacial activity. By contrast, simple ions tend to increase surface tension and are desorbed from interfaces. The failure of activated carbon to remove ions such as $Ca^{++}$, $Cl^-$, and $F^-$ from solution is an example. Understandably, the interfacial adsorption of organic molecules increases with their surface activity and their size. Understandably, too, interfacial adsorption depends on the structure of the adsorbent and not only on the relative magnitude of its interfacial area. Other surface properties being equal, adsorbents with large pore openings, for instance, should adsorb large molecules and colloidal particles preferentially. Macromolecular dyes are indeed removed from solution selectively by wide-pore adsorbents.

**Transfer Mechanisms.** The adsorption of solutes by porous adsorbents such as activated carbon proceeds consecutively as follows: (1) the adsorbate moves from the bulk of the solution to the outer shell of the adsorbent; (2a) powdered carbon and similar adsorbents have external surface areas large enough in the aggregate to remove much of the sorbate; (2b) in granular activated carbon and similar adsorbents, on the other hand, the bulk of the sorbate is carried into and through the pores of the adsorbent by intraparticle transport or diffusion; and (3) the sorbate is adsorbed specifically onto the sorption layer.

With powdered carbon < 300-mesh in size of separation, solution transport seems to be the rate-determining step at most reasonable stirring speeds. By contrast, intraparticle transport is normally the slowest and, therefore, the rate-determining step in the uptake of organic substances by granular carbon, even in rapidly stirred systems.[5] Only when power input is very low does transport of the adsorbate to the outer shell of the adsorbent become the controlling factor. For compounds of high molecular weight—ABS detergents in dilute solution, for instance—equilibrium may not be established even during weeks of vigorous stirring and continuous contact with activated carbon. *Overall adsorption rates* are observed to vary (1) as the square root of the initial solute concentration, (2) inversely as the square of particle size for a given weight of carbon, and (3) in keeping with a $Q_{10}$ value of about 1.3 or an energy of activation of 4270 cal per mole, which is indicative of a diffusion-controlled process.

Common to most studies of intraparticle diffusion is the finding that uptake varies almost proportionately to the square root of the time of exposure. Although the rate of uptake of organic solutes is low, *ultimate adsorptive capacities* can be high—of the order of 15 to 25% by weight, for instance. If the observed performance of low concentrations of solute ($\simeq$7 mg per l) is interpreted in terms of monolayer adsorption, required specific surface areas would have to be 200 to 300 m² per gram of carbon. Equilibrium adsorption from dilute solutions appears to be best described by the Langmuir adsorption isotherm (Eq. 11-32). Unlike the rate of adsorption, *adsorptive capacity* is not improved by high temperatures. Adsorptive capacity and rate of adsorption are illustrated in Figs. 11-6 and 5, respectively, for phenol as the adsorbate and granular activated carbon as the adsorbent.[6] A study of the observational data included in Fig. 11-5 suggests that the rate process is closely approximated by the general purification equation (Eq. 11-12) as follows:

$$y/y_0 = 1 - (1 + 3.6t)^{-0.25}$$

## 15-5 Activated Carbon

In the United States, G. L. Spalding showed in 1930 that powdered activated carbon can be applied successfully and economically to a public water supply. When it is applied to a storage reservoir and open settling basins it acts both as an adsorbent and as a means of reducing the passage of light and thus discouraging algal growths.

Activated carbon can be produced from a variety of carbonaceous raw materials. The most common sources are wood, peat, lignin, and pulp-mill char, the latter an industrial waste product. They are generally carbonized

---

[5] W. J. Weber, Jr., and J. C. Morris, Kinetics of Adsorption on Carbon from Solution, *J. Sanit., Eng. Div., Am. Soc. Civil Engrs.*, **89**, 31 (1963); Adsorption in Heterogeneous Aqueous Systems, *J. Am. Water Works Assoc.*, **56**, 447 (1964).

[6] These figures reproduce capacity and rate information respectively from *Advances in Water Pollution Research*, Vol. 2, Pergamon, London, 1962, Fig. 5, p. 240; and *Advanced Waste Treatment Research*, No. 9, U.S. Public Health Service, 1964, p. 24.

(charred) in the absence of air at a temperature of 600°C and activated by slow burning at 600 to 700°C, or by oxidation with suitable vapors or gases, such as steam or carbon dioxide, at 800 to 900°C. Granular activated carbon is generally of filter-sand size.

Powdered activated carbon is normally crushed to such size that 50% will pass a 300-mesh sieve and 95% a 200-mesh sieve—that is, 50 to 75 $\mu$, respectively. The adsorptive capacity is high because a pound of finely divided activated carbon contains some $10^{13}$ particles and a solid cubic foot of carbon particles presents a combined pore and external surface of about 10 sq miles to the water in which it is suspended.

At one time the adsorptive capacity of different carbons for pure phenol was made the basis for their general comparison. However, the *phenol value* or amount of carbon in milligrams per liter required to reduce 100 $\mu$g per l of phenol by 90% does not necessarily reflect the relative efficiency of a given carbon in removing substances other than phenol. Direct testing against the waters to be treated has therefore become the preferred procedure. Commercial carbons possess a phenol value of 15 to 30.

Although the Freundlich and Langmuir equations (Eqs. 11-30 and 32) can be used in comparative tests to identify the magnitudes of the pertinent adsorption coefficients of a given adsorbent, the equations are seldom fitted to observed information in practice. Instead, experimental carbon dosage is plotted against observed residual concentrations of the sorbate, and required dosages are read from curves fitted by eye. Either arithmetic or double logarithmic plots are used (Fig. 15-4 and Example 15-3).

**Adsorption of Odors and Tastes.** Within the range and intensity of odors and tastes generally encountered in water and wastewater, man's sensory responses to these stimuli obey the Weber-Fechner[7] law. According to this law, discrimination between odor and taste stimuli is confined to threshold ratios, not to threshold differences.[8] The common ratio of successive dilutions is approximately 1.4. It follows that man's discrimination between odor and taste intensities is relatively poor in comparison with his much more acute senses of sight and hearing.

The origin and nature of odors and tastes in drinking water are discussed in Sec. 10-7. The odors of municipal and other decomposable wastewaters are most pronounced when they, or the solids removed from them, become

---

[7] Ernst Heinrich Weber (1795-1878) and Gustav Theodor Fechner (1801-87), German natural philosophers, whose law states that a stimulus must increase geometrically if th sensation is to increase by an equal amount, that is arithmetically.

[8] This explains the *Standard Methods* series of odor tests in which threshold numbers of 1, 1.5, 2, 3, 4, 5, 7, 10, and 14, for instance, are assigned respectively to 200 ml of undiluted sample and samples of 130, 100, 67, 50, 40, 29, 20 and 14 ml diluted with odor-free water to 200 ml.

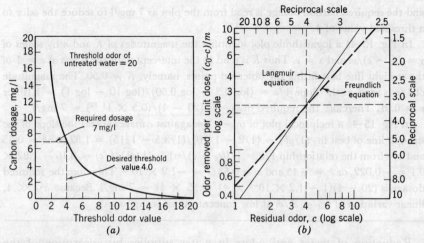

**Figure 15-4.**

**Determination of carbon dosage for odor control.** (a) Arithmetic plotting of experimental results; (b) double-logarithmic plotting for Freundlich equation and reciprocal plotting for Langmuir equation.

septic. However, there are industrial wastes with characteristic odors of their own that are not changed by septicity. Proper planning and management of water supplies and wastewater works will minimize odor and taste troubles.

To describe the adsorption of odors and tastes from water in terms of threshold values, the Freundlich and Langmuir equations (Eqs. 11-30 and 32) are normally translated into the following forms:

$$(c_0 - c)/m = Kc^{1/n} \tag{15-6}$$

$$\log[(c_0 - c)/m] = \log K + (1/n)\log c \tag{15-7}$$

$$(c_0 - c)/m = (ab \cdot c)/(1 + a \cdot c) \tag{15-8}$$

$$1/[(c_0 - c)/m] = 1/b + (1/ab)(1/c) \tag{15-9}$$

Here $c_0$ is the threshold odor or taste of the water to be treated, $c$ is the residual threshold value, $m$ is the concentration of the adsorbent, and $K$, $n$, $a$, and $b$ are coefficients, $b$ being the limiting adsorptive capacity of the adsorbent. Generalization of observed removals is important in research and in the identification of attainable adsorption.

*Example 15-3.*

Fig. 15-4a shows an arithmetic plot of carbon dosage $m$ in milligrams per liter against observed threshold odor values $c$. The threshold odor of the untreated water is 20, and the required carbon dosage is read from the plot as 7 mg/l to reduce the odor to a threshold value of 4.

In Fig. 15-4$b$ a logarithmic plot identifies the magnitudes of $K$ and $n$ by scales of $y = (c_0 - c)/m$ and $x = c$. Thus $K$ is read at the intercept with the $y$-axis at $c = 1$ of the straight line best fitting the observed points, namely, $K = 0.50$. The magnitude of $1/n$ is read as the slope $1/n = (\log 6.3 - \log 0.50)/(\log 10 - \log 1) = 1.10$, or $n = 0.91$. Therefore the required dosage is $(20 - 4)/(0.5 \times 1^{1.10}) = 7$ mg/l.

In Fig. 15-4$b$ a reciprocal plot of $(c_0 - c)/m$ against $c$ makes $(1/ab)$ the slope of the straight line of best fit, $(1/ab) = (1/2 - 1/10)/(1/3.5 - 1/15) = 1.82$ or $ab = 0.55$ and $1/b$ from the relationship $1/b = 1/[(c_0 - c)/m] - (1/ab)(1/c) = 0.16 - 1.82 \times 0.1 = -0.022$, or $b = -45$ and $a = 0.55/b = -1.2 \times 10^{-2}$. Therefore the required dosage is $(20 - 4)(1 - 1.2 \times 10^{-2} \times 4)/(0.55 \times 4) = 7$ mg/l. Because $a \cdot c \ll 1$, linear variation of $(c_0 - c)/m$ at low concentrations is supported.

---

Reduction of tastes and odors in water supplies by adsorption of the offending substances on activated carbon is probably the most important direct use of adsorption in water treatment. Columns or beds of granular activated carbon are employed (1) for concentrating organic pollutants from water for purposes of analysis or (2) for removal of the pollutants. Some of the removal of color-producing substances and other pollutants from water during coagulation may be the result of adsorption. Indeed, certain processes make use of this phenomenon to reduce silica, fluoride, or radioactive substances. Beds of granular activated carbon can be operated either as *fixed* or *packed* beds or as *fluidized* or expanded beds. Packed beds normally accumulate solids faster than do fluidized beds.[9]

**Process Technology.** It is quite likely that *fixed-bed* operation, which is in essence countercurrent operation, will turn out to be the most effective and efficient way of using carbon and other adsorbents. In the course of each operating cycle, entrant water then comes into contact with the adsorbent along a gradient of mounting residual activity until the most active carbon gives a final polish to the effluent water while the adsorption reaction is driven to ultimate saturation of the carbon.

Effluents from biological treatment processes may retain 20 to 40% of the applied COD in colloidal form. If these colloids are not to interfere with adsorption, they must be removed. Conventional coagulation will normally do this.

Partial regeneration of carbon by thermal volatilization or steam distillation of organic adsorbates is possible, but available regeneration procedures will have to be improved or new ones invented if adsorption is to become a widely useful operation in wastewater treatment. The use of multi-hearth

---

[9] Robert A. Taft Water Research Center, A Comparison of Expanded-Bed and Packed-Bed Adsorption Systems, *Report No. TWRC-2* (1968).

furnaces is a possibility.[10] Difficulties and costs of regeneration explain why powdered activated carbon continues to be used most widely.

Granular activated carbon can replace other filtering materials in structures not unlike present-day rapid filters. Beds of granular activated carbon can in fact be made to perform as both filters and adsorbents. However, activated-carbon filters must be somewhat deeper than sand filters, even though they may be operated at higher rates of flow per square foot of bed. For adsorption, the rate of flow per cubic foot rather than per square foot of bed is the important parameter in practice. Carbon grain sizes range from 0.8 to 3 mm or 14/40 mesh, and rates of 2.0 to 0.4 gpm per cu ft are obtained when beds 5 to 10 ft deep filter water at rates of 10 to 4 gpm per sq ft. Associated contact times are 2 to 8 min, and efficiencies of 50 to 90% collection of carbon-chloroform extractables (CCE) are obtained. Because of the many variables involved, pilot-plant scale tests are normally obligatory.

## 15-6 Desalting

The removal of dissolved solids from water is described as desalting, desalination, or salt-water conversion. The salinity of raw waters varies with their origin. The degree of salinity is normally expressed in milligrams per liter of dissolved solids, chloride ion, $Cl^-$, or common salt, NaCl. The following classification is based on the water source and its relative saltiness: (1) *mildly saline waters*: brackish mixtures of saline and sweet waters with salt concentrations of 1000 to 2000 mg per l of dissolved solids; (2) *moderately saline waters*: inland waters with salt concentrations of 2000 to 10,000 mg per l of dissolved solids; (3) *severely saline waters*: inland and coastal waters with salt concentrations of 10,000 to 30,000 mg per l of dissolved solids; and (4) *seawater*: offshore waters of the oceans and their seas with salt concentrations of 30,000 to 36,000 mg per l of dissolved solids, including 19,000 ppm Cl, 10,600 ppm Na, 1270 ppm Mg, 880 ppm S, 400 ppm Ca, 380 ppm K, 65 ppm Br, 28 ppm C, 13 ppm Sr, 4.6 ppm B, and others.

Inland seas that do not drain to river systems may be saltier than the ocean itself. The Great Salt Lake, Utah, and the Dead Sea are examples.

Desalting or demineralizing processes share as a common objective the removal of *salt* from saline waters. By contrast, the present section is concerned with separating *water* from saline waters. As shown in Table 15-1, which is a classification of saline-water conversion processes suggested by the National Academy of Sciences—National Research Council,[11] this is not a quibbling statement.

[10] D. G. Hager and M. E. Flentje, Removal of Organic Compounds by Granular Carbon Filtration, *J. Am. Water Works Assoc.*, **37**, 1440 (1965).

[11] *Desalination Research and the Water Problem*, NAS-NRC Publication 941, Washington, D.C., 1961, p. 24. For up-to-date information on desalination, see comprehensive *annual reports* of the Office of Saline Water, Washington, D.C.

**Table 15-1**
*Classification of Saline-Water Conversion Processes*

| Constituent Removed | Phase to Which Transported | Process Number and Kind |
|---|---|---|
| Salt | → Vapor | 1. (None known) |
| | → Liquid | 2. Electrodialysis<br>    Osmionic<br>    Thermal diffusion |
| | → Solid | 3. Ion exchange<br>    Adsorption on carbon electrode |
| Water | → Vapor | 1. Distillation |
| | → Liquid | 2. Solvent extraction<br>    Reverse osmosis |
| | → Solid | 3. Freezing<br>    Contact freezing<br>    Adsorption |

**Distillation.** Of the processes removing water from saline solutions, distillation is the oldest and, in terms of established plants, the most productive. It differs from other processes by its passage of water through the vapor phase. Of interest is the fact that the potential of this operation is readily assessed in terms of the minimum work to be accomplished by the heat energy applied. If, for example, needed energy is derived from steam, as it usually is, the minimum work requirement is a function of the temperature of the steam. Because saturated steam can provide about 205 Btu per lb of work energy when it is condensed at atmospheric pressure and rejected at a temperature of about 70°F, the expected volume of product water is 100 gal when 47 lb of steam are put to use efficiently. The commercial value of atmospheric-pressure steam thereby equals the price of about 2.5 kwhr of electrical energy. This figure is useful for first estimates of cost along with a realization that, in addition to final costs, the efficiency of energy production and utilization is the cost-controlling factor in distillation plants. Hence plant design is directed to (1) tapping the most economic sources of heat energy and (2) exploiting the most efficient processes of heat transfer. A cost target is offered, moreover, by the alternative possibility of supplying water from a natural freshwater source based, in each instance, on delivery of the water at the plant under the required systems head.

Three types of evaporators are illustrated in Fig. 15-5: a multiple-effect evaporator, a multistage flash evaporator, and a vapor-compression still. All of them recapture the latent heat of vaporization of water, namely, $(1094 - 0.56\,T)$ Btu per lb, in countercurrent circulation with steam.

*Multiple-effect evaporators* are relatively efficient, whereas single-effect evaporators are relatively inefficient. Each component unit of a multiple-effect evaporator is maintained in sequence at slightly lower pressure and temperature in order to permit the steam produced in one effect to become the source of heat in the next. Pound for pound, the amount of product water then approximates the number of effects (Fig. 15-5a).

*Multistage flash evaporation*, too, is accomplished at successively lower pressures and temperatures. The incoming water is warmed by the heat of condensation and only a small amount of heat energy is required to flash the preheated water in the reduced-pressure stage into steam (Fig. 15-5b).

*The vapor-compression process* relies on mechanical compression of the vapor to boost its temperature high enough, because $pv = RT$, to supply through its own condensation the heat necessary to evaporate the feed water (Fig. 15-5c). Once started, this process does not draw upon further heat energy, only upon mechanical energy. When electric motors are made the prime movers of compressors, only mechanical energy is supplied. However, internal combustion engines can assist in preheating the feed water in their cooling jackets and in exhaust-gas heat exchangers (Sec. 20-15).

There is much scale formation and corrosion as well as erosion at the high temperatures at which stills must be operated (Sec. 16-11).

**Solvent Extraction.** Organic solvents that are partially miscible with water can produce (1) a more concentrated brine and (2) an extract of solvent and freshwater from which freshwater can be withdrawn after the extract has been separated from the brine and heated to reduce the solubility of the water in the solvent.

**Reverse Osmosis.** Osmotic pressure drives water molecules through a permeable membrane from a dilute to a concentrated solution in search of equilibrium. This natural response can be reversed by placing the salt water under hydrostatic pressures higher than the osmotic pressure. Hence the term *reverse osmosis*. Because of its simplicity in concept and in execution, reverse osmosis appears to have considerable potential for wide application in water and wastewater treatment.

The critical element in the process is the cellulose acetate membrane which, though exceedingly thin, 100 to 150 microns, must be capable of withstanding the hydraulic pressures necessary for the process. These pressures range from 600 psi for brackish water to 1500 psi for seawater and

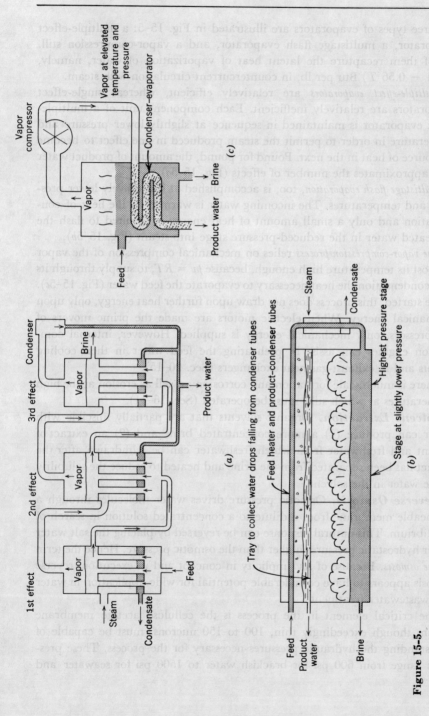

**Figure 15-5.** Schematic arrangement of evaporation. (a) Multiple-effect evaporator; (b) multistage flash evaporator; (c) vapor-compression still (Sec. 15-6, footnote 11).

require that the membranes be adequately supported. Four configurations have been developed.

The plate-and-frame design is similar to the conventional filter press, with the membranes being mounted on the two sides of solid plates into which product water channels have been cut. These plates, alternated with feed water frames, are set in a pressure vessel. This arrangement does not permit as large membrane surface areas per unit volume as do the other configurations.

Spiral-wound modules consist of a number of membrane envelopes, each having two layers of membrane separated by a porous, incompressible backing. The envelopes, with influent side spacer screens, are wound around a water-collection tube housed in carbon-steel pipes. The pressurized influent flows axially along the spacer screen, while product water passes through the membrane into the porous backing from which it is collected.

A tubular design has the membrane on the inside of multiple tubes, which are either porous or perforated.

The hollow fiber unit, shown in Figure 15-6, offers the greatest promise for economy by permitting the largest membrane surface area per unit volume of the unit. The fibers range in diameter from 50 to 200 microns and yet because they are tightly packed, they need no structural support.

*Freezing.* Water can be transported from brine to the solid phase as ice. That the latent heat of fusion, namely 144 Btu per lb, is small compared with the latent heat of vaporization would seem to be an advantage. However, even though the ice crystals formed are essentially pure water, the yield of product water is decreased (1) because some of it must be used to wash salt from the ice surfaces and (2) because heat is required to melt the ice crystals. In freezing, as in distillation, countercurrent operation conserves heat energy—in this instance (1) by cooling the feed water to the freezing point, after which a refrigerant is evaporated in direct contact with the feed, and (2) by countercurrent washing and melting of the ice crystals.

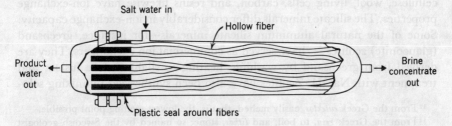

**Figure 15-6.**
**Illustration of hollow-fiber reverse-osmosis process. (Office of Saline Water.)**

*Contact freezing* makes use of two heat-transfer circuits of recycling hydrocarbons. The first circuit absorbs heat from the incoming salt water, transfers it in part to the fresh water, and loses it in part to the waste brine. The second circuit vaporizes the liquid hydrocarbon in contact with the salt water to freeze it; the vapor is then compressed, and the heat energy released is used to melt the ice. The vapor separating from the fresh water is repumped through the freeze chamber.

*Eutectic freezing* operates at the eutectic temperature[12] of the incoming water. Down to the eutectic point, only ice is formed. At the eutectic point, ice crystals nucleate and grow independently of salt crystals and other substances in the water. Further removal of heat does not continue to lower the temperature. Both ice and salt freeze. They can then be separated, because the ice floats and the salt sinks.

The potential recovery of bromine, chlorine, magnesium, and refined salt by salt-water conversion plants is of interest.

## 15-7 Ion Exchange and Ion Exchangers

Ion exchange is the reversible interchange of ions between a solid ion-exchange medium and a solution. Water softening by ion exchange is an important example. In industry both cation and anion exchangers are employed to prepare boiler feed-water, deionize or demineralize process waters, concentrate dilute solutions of electrolytes, and prepare chemical reagents.

To be effective, solid ion exchangers must (1) contain ions of their own, (2) be insoluble in water, and (3) provide enough space in their porous structure for ions to pass freely in and out of the solid. Cation exchangers have a negatively charged framework, but their pores contain cations that maintain electroneutrality. Anion exchangers carry just the opposite electrical charges.

Soils are important ion exchangers—among them clay soils, the humus produced by decaying vegetation, and the bottom sediments in rivers and lakes. Alumina, $SiO_2$, $MnO_2$, metal phosphates and sulfides, lignin, proteins, cellulose, wool, living cells, carbon, and resins likewise have ion-exchange properties. The silicate minerals differ considerably in ion-exchange capacity. Some of the natural aluminum silicate minerals—for instance, greensand (glauconite) zeolites[13]—have served as commercial ion exchangers. They are derived from greensand by washing, heating to slight surface fusion, and treatment with NaOH. Zeolites have also been synthesized by mixing solu-

---

[12] From the Greek *eutektos*, easily melted, that is, the lowest melting point possible.

[13] From the Greek *zein*, to boil, and *lithos*, stone; so named by the Swedish geologist Asel Cronstedt in 1756 because these natural minerals gave off their water of hydration or combination as steam when they were heated. Glauconite derives its name from th Greek *glaukos*, blue-green. Cronstedt is also known for his discovery of the element nickel

tions of sodium silicate and sodium aluminate, drying the resulting white gel, and crushing it to wanted size.

Modern ion-exchange technology began in 1935 with the discovery of synthetic ion-exchange resins. However, natural as well as synthetic zeolites have continued in commercial use. The synthetic ion exchangers are pervious, reasonably stable, and have high capacities (Table 15-2). Relatively large particles, 1 to 2 mm in diameter and synthesized with specific ionic functional groups, can be employed. The chemical structure of two synthetic ion-exchange resins is shown in Fig. 15-7. They act essentially as a single, large, many-charged ion in which thousands of atoms are linked into a three-dimensional network. The charge of the polymeric ion is neutralized by small ions of opposite charge within the network, where they are held in place by electrostatic forces without being bound to the ion-exchange matrix. Thus they are able to change places with ions in a solution. The charge in the matrix is allied to functional groups.

The cation exchanger of Fig. 15-7 contains $-SO_3^-$ groups. Different functional groups, such as $-COO^-$, $-PO_3H^-$, and $-C_6H_5O^-$, can be introduced instead. Exchangers with functional groups derived from a strong acid, $-SO_3^-$ from $H_2SO_4$, for instance, are called *strongly acidic*. Exchangers with carboxylic groups derived from a weak acid are referred to as *weakly acidic*.

Anion exchangers contain positively charged functional groups, among them quaternary ammonium ($-NR_3^+$), amino ($-NH_3^+$), imino ($-NRH_2^+$)

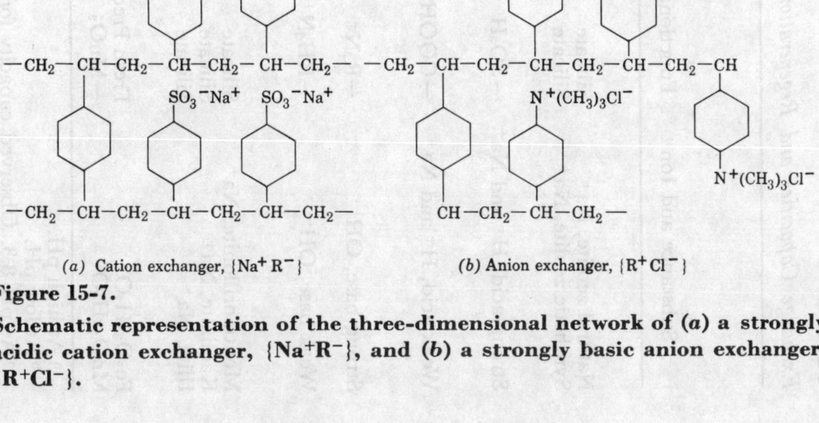

(a) Cation exchanger, {Na$^+$ R$^-$}  (b) Anion exchanger, {R$^+$ Cl$^-$}

**Figure 15-7.**
Schematic representation of the three-dimensional network of (a) a strongly acidic cation exchanger, {Na$^+$R$^-$}, and (b) a strongly basic anion exchanger, {R$^+$Cl$^-$}.

### Table 15-2
*Exchange Capacities and Regeneration Requirements of Natural and Synthetic Ion Exchangers*

| Substance and Ion | Functional Group | Exchange Capacity Wet, me/ml | Exchange Capacity Dry, me/g | Regeneration into $H^+$, $Na^+$, or $OH^-$ Form by | Trade Name |
|---|---|---|---|---|---|
| *A. Cation-exchange zeolites* | | | | | |
| Natural zeolite, $Na^+$ | Silicate | 0.15–0.3 | — | Excess NaCl | Zeolite |
| Synthetic zeolite, $Na^+$ | Silicate | 0.3–0.8 | ... | Excess NaCl | Zeolite |
| *B. Cation-exchange reins* | | | | | |
| Strong acid, $H^+$ and $Na^+$ | $-SO_3H$ | 4–5 | ... | Excess strong acid or NaCl | Zeo-Karb/225<br>Amberlite IR-120<br>Dowex 50 |
| Weak acid, $H^+$ and $Na^+$ | $-COOH$ | 4–5[a] | 8–10[a] | Acid or NaCl | Zeo-Karb/226<br>Amberlite IRC-50 |
| *C. Anion-exchange resins* | | | | | |
| Strong base, $OH-$ | $-R_3N^+$ | 1.2–1.4 | 3.5–4.5 | Excess strong base | Amebrlite IRA-400, 410<br>Dowex 1,2 |
| Weak base, $OH-$ | $-RH_2N^+$ | 2–2.5[b] | 4.5–5.5[b] | $Na_2CO_3$ | Amerlite IR<br>Dowex 3 |
| *D. Clays* | | | | | |
| Montmorillonite, $Na^+$ | Silicate | — | 0.8–1.0 | ... | ... |
| Kaolinite, $Na^+$ | Silicate | ... | 0.02–0.1 | ... | ... |
| Illite, $Na^+$ | Silicate | ... | 0.2–0.4 | ... | ... |
| *E. Hydrous oxides* | | | | | |
| $Fe_2O_3xH_2O$ | Fresh Precipitate | ... | 4[c] | ... | ... |
| $MnO_2xH_2O$ | $-MnO_2$ | ... | 15[c] | ... | ... |

[a] At high pH.
[b] At low pH.
[c] At pH 8.3. Observed capacity for divalent transition elements. In practice, exchange capacities are commonly expressed in kilograms per cubic foot, and regenerant requirements in ppounds per cubic foot.

phosphoninium ($—PR_3^+$), and sulfonium ($—SR_3^+$). A strongly basic exchanger is one that contains quaternary ammonium groups.

**The Ion-Exchange Process.** The ion-exchange process can be formulated as follows:

$$\{H^+R^-\} + Na^+ = \{Na^+R^-\} + H^+; \quad \text{and}$$
$$2\{Na^+R^-\} + Ca^{++} = \{Ca^{++}R_2^=\} + 2\,Na^+ \quad (15\text{-}10)$$

$$\{R^+Cl^-\} + OH^- = \{R^+OH^-\} + Cl^-; \quad \text{and}$$
$$2\{R^+OH^-\} + SO_4^= = \{R_2^{++}SO_4^=\} + 2\,OH^- \quad (15\text{-}11)$$

where $R^-$ and $R^+$ symbolize the negatively and positively charged network of the cation or anion exchanger, respectively.

Within the solution and the ion-exchange medium, a charge balance (electroneutrality) must be maintained; the number of charges, and not the number of ions, must stay constant within or on the exchanger granule. One $Ca^{++}$ ion displaces 2 $Na^+$ ions, for instance. Consequently the *exchange capacity* of an ion exchanger is expressed by the number of charges, that is, the equivalents of ions necessary to maintain electroneutrality within the solid phase for a given weight or volume of ion-exchange material. The capacity can be established analytically by determining the quantity of ions exchangeable when the exchange reaction is driven to completion. Because cation exchangers in the $\{H^+R^-\}$ form are essentially solid *polyprotonic acids*,[14] they can be titrated alkalimetrically (Fig. 15–8).

The design of ion-exchange processes is based on a knowledge of *ion-exchange equilibria*, which govern the distribution of ions between an aqueous phase and the ion-exchanger phase. Because ion-exchange reactions obey the mass law, the exchange reactions may be characterized as follows:

$$\frac{[Na^+R]\,[H^+]}{[H^+R]\,[Na^+]} = Q_{HR \to NaR} \quad (15\text{-}11)$$

and

$$\frac{[Ca^{++}R]\,[Na^+]^2}{[Na^+R]^2\,[Ca^{++}]} = Q_{NaR \to CaR} \quad (15\text{-}12)$$

The resulting equilibrium concentration quotients, $Q$, are not thermodynamic equilibrium constants and provide only a semiquantitative interpretation of the ion-exchange equilibrium. Because they characterize the relative selectivities of an ion exchanger for specific ions, the quotients are called selectivity coefficients. In the exchange of monovalent ions by other monovalent ions, the magnitudes of $Q$ are very close to unity. The affinity of

[14] Acids that can donate more than a single hydrogen ion (proton).

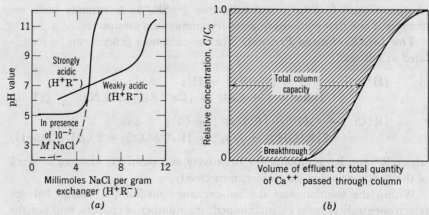

**Figure 15-8.**

**Determination of ion-exchange capacity.** (a) Alkalimetric titration of cation exchanger in the {H⁺R⁻} form provides information on the acidity of the exchanger and permits the determination of the analytical exchange capacity. (b) Determination of useful capacity of an ion-exchange column.

exchangers for bivalent ions is much larger than for monovalent ions; selectivity coefficients of 20 to 40 are typical. The mass law formulation for bivalent-monovalent exchange (Eq. 15-11) shows that a dilute $Ca^{++}$ ion solution in contact with a cation exchanger in the $Na^+$ form, $\{Na^+R\}$, drives the reaction to almost stoichiometric completion as written, that is, from left to right. However, the resulting exchanger in the $Ca^{++}$ form, $\{Ca^{++}R_2^=\}$, can be regenerated into the $Na^+$ form by treatment with a concentrated solution of $Na^+$.

It follows from Eq. 15-12 that the distribution of two ions of differing valences is strongly dependent on the concentration of the solution. In a hetero-ionic reaction, the selectivity of the exchanger for the bivalent ion over the monovalent ion rises with increasing dilution of the solution. An exchanger will selectively *sorb* $Ca^{++}$ ions with great preference from a mixed $Na^+$-$Ca^{++}$ solution when it is sufficiently dilute. In solutions of high concentration, however, the exchanger loses its selectivity. A cation exchanger can remove $Ca^{++}$ ions selectively from a dilute solution, that is, natural water, even when $[Ca^{++}] \ll [Na^+]$ in the solution. However, an exhausted exchanger in the $Ca^{++}$ form can be regenerated or reconverted into a $\{Na^+R^-\}$ exchanger by a concentrated brine solution or full-strength seawater. Exchange reactions are generally slower than reactions between electrolytes in solution. In reactions with synthetic resin exchangers, for example, the time of half-exchange is of the order of a few minutes when the reaction is carried out under the usual conditions of small-particle resins, small-diameter ions, approximately $10^{-3}$ $M$ solutions, and room temperature.

***Ion-Exchange Technology.*** Solutions are put into contact with ion exchangers either in batch or in column operations. In batch operations exchanger and solution are stirred in a vessel until equilibrium is reached. Because only a small portion of the exchange capacity is generally satisfied, batch processes are of limited usefulness. In column operations the solution flows through a vertical cylinder filled with the ion exchanger. The column can be considered a series of sequent batch operations, with progressively more ion-depleted solution coming into contact with less-exhausted ion exchanger, until exchange is ultimately complete. Multiple-equilibrium or near-equilibrium steps are established. Accordingly, even solute ions with unfavorable selectivity coefficients can be removed in quantity by this simple application of the mass-action principle. The reaction is driven to completion in much the same way as in adsorption chromatography or countercurrent operations.

The columnar uptake of $Ca^{++}$ ions by an exchanger in the $Na^+$ form ($Na^+R^-$) is shown schematically in Fig. 15–9. The concentration ratio

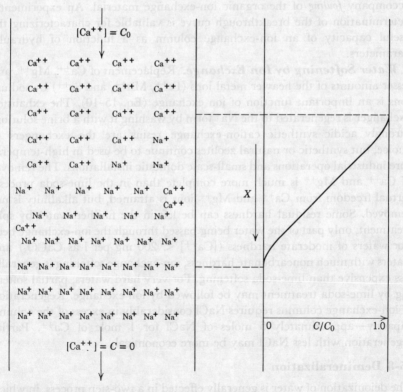

**Figure 15-9.**
**Columnar operation of an ion exchanger.**

($C/C_0$) or $[Ca^{++}]/[Ca^{++}_{initial}]$ along the length of the column is also indicated. The effluent from the column is free of $Ca^{++}$ ions, but it contains an equivalent concentration of $Na^+$ ions. Eventually the column will become saturated with $Ca^{++}$ ions, and $Ca^{++}$ ions will appear in the effluent at this *breakthrough point* for $Ca^{++}$. Fig. 15–8 shows, instead, the concentration ratio $C/C_0$ or $[Ca^{++}]/[Ca^{++}_{initial}]$ as a function of the volume or total quantity of $Ca^{++}$ passed through the column. The area enclosed by the breakthrough curve is a measure of the total capacity of the ion-exchange bed. However, the useful or *breakthrough capacity* is smaller in actual operations. Breakthrough capacity is influenced by exchanger particle size, column dimensions, flow rate, temperature, and solution composition. The breakthrough curve generally steepens with increase in column length, decrease in flow rate, and decrease in particle size. The curve is not necessarily symmetrical. There may be considerable *tailing* at the end and beginning of the curve when the bed contains air pockets or air channels, or deposits of colloidal substances (for example, iron oxide). Breakthrough capacity is then reduced. Similar effects accompany *fouling* of the organic ion-exchange material. An experimental determination of the breakthrough curve is valuable for characterizing the useful capacity of an ion-exchange column as a function of hydraulic parameters.

**Water Softening by Ion Exchange.** Replacement of $Ca^{++}$, $Mg^{++}$, and lesser amounts of the heavier metal ions ($Fe^{++}$, $Mn^{++}$, and $Sr^{++}$) by sodium ions is an important function of ion exchange (Eq. 15–10). The exhausted exchanger is regenerated to the $Na^+$ form by washing it with a brine solution. Strongly acidic synthetic cation-exchange resins are the exchangers of choice, but synthetic or natural zeolites continue to be used in high-temperature industrial operations and small-scale domestic installations. The removal of $Ca^{++}$ and $Mg^{++}$ is much more complete than in the lime-soda process; virtual freedom from $Ca^{++}$ and $Mg^{++}$ ions is attained, but alkalinity is not removed. Some residual hardness can be left in the finished water by split treatment, only part of the water being passed through the ion-exchange bed. For waters of moderate hardness ($[Ca^{++}] < 200$ mg per l as $CaCO_3$) and waters with much noncarbonate hardness, ion-exchange softening is generally less expensive than lime-soda softening. For very hard waters, partial softening by lime-soda treatment may be followed by ion exchange. Regeneration of ion-exchange columns requires NaCl considerably in excess of the column capacity—approximately 5 moles of NaCl for 1 mole of $Ca^{++}$. Partial regeneration with less NaCl may be more economical.

## 15-8 Demineralization

The deionization of water is generally effected in a two-step process, in which the water is passed successively through a cation exchanger in the $H^+$ form,

$\{H^+R^-\}$, and an anion exchanger in the $OH^-$ form, $\{R^+OH^-\}$. On entering the cation exchanger, all cations are exchanged for an equivalent quantity of $H^+$ ions. The effluent, actually a solution of the acids of the anions, enters the anion exchanger, where all anions are exchanged for hydroxide ions that neutralize the equivalent quantity of $H^+$ formed in the cation exchanger. An equivalent amount of water is produced. Mixed-bed exchangers are a more recent development in water demineralization. A single column contains a mixture of equivalent quantities of cation and anion exchangers. The effluent is generally superior in quality (lower in conductivity). To regenerate a mixed bed, the resins must be separated. This can be done by differential backwashing, because cation- and anion-exchange resins normally have different densities. Demineralizing may be as effective as distillation. However, nonelectrolytes (organic materials) are not quantitatively removed in the ion-exchange process, even though there may be some removal through adsorption.

The cation-exchange resins used in the demineralization process are regenerated with strong acids. $H_2SO_4$ is generally used, although it occasionally precipitates $CaSO_4$ in the exchange bed. If weak acids such as carbon dioxide and silicic acid, $Si(OH)_4$, are to be removed, strongly basic anion exchangers must be employed. These must be regenerated with sodium hydroxide. Some weakly basic anion exchangers can be regenerated with soda ash.

Demineralization produces high-quality water for industry, especially for service as makeup water in steam-power plants. Ion-exchange resins are commonly employed only for supplies with less than 500 mg per l of dissolved solids. The introduction of weak-electrolyte exchange resins of high capacity and high regeneration efficiency has raised the economy of treating waters up to concentrations of dissolved solids of 1000 to 2000 mg per l.

**Concentration of Ions.** Ion exchange fosters the concentration, isolation, and recovery of ionic materials from dilute solutions. There can be savings in treatment chemicals, and process waters can be recycled. Wastewaters are reduced in quantity and strength, and byproducts may possibly be worth reclaiming. For example, cation exchangers can recover and concentrate copper, zinc, and chromic acid from dilute washwaters for reuse in metallurgical processing; anion-exchange resins can concentrate cyanide and fatty acids from a number of different waste streams; and ion exchange can play an important role in the decontamination of radioactive wastes in general.

Ion exchange is highly selective. As yet, however, specific cations or anions—phosphate ions or heavy metal ions, for instance—cannot be held more strongly than others. However, it is possible to synthesize resins that incorporate *chelating*[15] reagents, such as EDTA, into the polymeric network. This should make exchangers more selective for specific metal ions such as nickel, copper, and cobalt.

Under well-chosen conditions, fluoride removal by hydroxylapatite, $Ca_{10}(PO_4)_6(OH)_2$, exemplifies a reasonably selective ion-exchange process. Hydroxylapatite is converted into fluoroapatite, $Ca_{10}(PO_4)_6F_2$, and $OH^-$ ions are released. Degreased, protein-free bone contains hydroxylapatite and will defluoridate water. The exchanger is regenerated with NaOH.

## 15-9 Ion-Exchange Membranes and Dialysis

Water can be desalinized electrochemically by *electrodialysis* through membranes selectively permeable to cations or anions. *Dialysis* is the fractionation of solutes made possible by differences in the rate of diffusion of specific solutes through porous membranes. *Semipermeable membranes* are thin barriers that offer easy passage to some constituents of a solution but are highly resistant to the passage of other constituents. Highly selective membranes have been prepared by casting ion-exchange resins as thin films. Membranes made from cation-exchange resins are cation-permeable; those made from anion-exchange resins are anion-permeable. Dialytic processes are common separation techniques in laboratory and industry. The recovery of caustic soda from industrial wastes, such as viscose *press liquor* from the rayon industry and mercerizing solutions, is an example of continuous-flow dialysis.

Diffusion of ions through membranes can be accelerated by applying a voltage across the membrane. The resulting *electrodialytic* separation with ion-exchange membranes permits the desalination of brackish waters. A shown schematically in Fig. 15-10, a series of chambers includes alternating anion-permeable and cation-permeable membranes, together with inert electrodes in the outermost compartments. Water introduced into alternate chambers is demineralized by passage of a direct current through the battery of compartments. The applied voltage drives anions toward the anode and out of cells with anion-permeable membranes on the anode side. However, the anions are trapped in the adjacent cell because it has a cation-permeable (anion-impermeable) membrane on the side facing the anode. In this way, compartments from which anions migrate toward the anode also lose cations toward the cathode, while alternate compartments retain both anions and cations. If saline water is fed to the ion-losing compartments and brine is bled from the ion-concentrating compartments, the water can be demineralized electrochemically in continuous flow. Power loss is minimized if the water is demineralized only partially to final concentrations of less than

[15] From the Greek *chele*, claw. A chelating agent such as ethylenediaminetetraacetate (EDTA) can attach itself to a central metallic atom so as to form a heterocyclic ring.

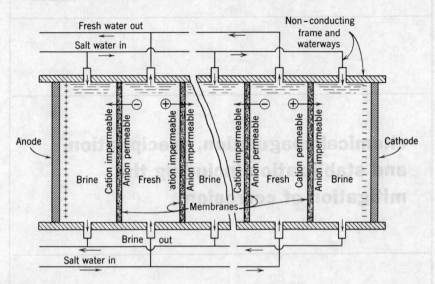

**Figure 15-10.**
**Schematic drawing of electrochemical desalting of water.**

500 mg per l in a multicompartment cell. Necessary power rises with increasing salinity. By contrast, the cost of distillation and freezing does so only slightly. For this reason only waters containing less than 5000 to 10,000 mg per l of dissolved solids are normally desalted electrochemically.

**sixteen**

# chemical coagulation, precipitation, and stabilization including the mitigation of corrosion

## 16-1 Nature of Inquiry

The present chapter is concerned with the *chemical* management of water and wastewater quality. It inquires into the nature and removal of colloids and solutes and deals with the responses of iron and manganese to the water environment in general, including the phenomenon of corrosion.

In an engineering sense, the chapter is addressed to (1) chemical coagulation by electrolytes; (2) chemical precipitation of hardness and carbonates (water softening); (3) chemical precipitation of iron and manganese, and of phosphates; (4) chemical stabilization of water; (5) control of corrosion; (6) storage and application of chemicals; and (7) disposal of chemical sludges and slurries.

## 16-2 The Colloidal State

Responsible for many of the characteristic properties of substances of colloidal size is their large specific surface. Thus 1 $cm^3$ of a colloid composed of $10^{18}$ cubical particles 10 m$\mu$ on a side has a surface area of 6500 sq ft. If the same volume is drawn into a square filament 10 m$\mu$ on a side in cross-section, the surface area is reduced to 4300 sq ft and if the same volume is flattened into a film 10 m$\mu$ in thickness, the surface area is still 2200 sq ft. It follows, therefore, that only one dimension of a colloid needs to be small to produce a large specific surface area.

Important surface properties of colloids are (1) their tendency to concentrate substances from the surrounding medium, that is, their *adsorptive* properties, and (2) their tendency to develop charge in relation to the surrounding medium, that is, their *electrokinetic* properties.

**Electrokinetic Properties of Colloids.** Colloidal particles are normally charged with respect to the surrounding medium, the sign and magnitude of the electrical charge being characteristic of the colloidal material and the composition of the medium. Thus, if electrodes from a direct-current (dc) source are placed in a colloidal dispersion, the particles migrate toward the pole of opposite charge at a rate proportional to the potential gradient set up in the solution. This is called *electrophoresis*.[1] Conversely, if the colloidal material is held fixed, the application of a dc potential causes the liquid to flow in a direction opposite to that in which the particles would normally move. This is called *electroosmosis*.[1]

In reference to natural waters near pH 7, colloidal silica, silicate minerals like the clays, the tealike organic color of natural waters, and most proteins are normally negatively charged, whereas the hydrous oxides of iron and aluminum are usually positively charged.

**Stability of Colloids.** When colloidal particles come into contact, they conjoin to form larger particles, flocculates, and precipitates, unless they are *stabilized* in some way. For *hydrophobic* colloids, like those of metals and most salts, the presence of charge causes the particles to repel one another and is the major stabilizing factor. *Hydrophilic* colloids, like starches and proteins, are stabilized primarily by bound layers of water. Hydrous ferric and aluminum oxides[2] and, to some degree, the clay minerals exhibit properties associated with both these classes of colloids. Charge and hydration are both important, making the behavior of these materials more complex than the behavior of typical colloids of either class.

In many cases the stability of colloids is found to depend on the magnitude of the *zeta potential*, $\zeta$, defined by the equation

$$\zeta = 4\pi\delta q/D \qquad (16-1)$$

in which $q$ is the charge on the particle (or the charge difference between the particle and the body of the solution), $\delta$ is the thickness of the layer around

---

[1] Greek *electron*, amber because it possesses the electrostatic properties of amber and, respectively, *phoresis*, being borne, or *osmos*, impulse.

[2] Produced when iron and alum are added to water for the coagulation of colloids and other finely divided substances.

the particle or the apparent distance of the charge separation, and $D$ is the dielectric constant of the medium. The zeta potential is thus a measure of both the charge on a colloidal particle and the distance into the solution to which the effect of the charge extends (Fig. 16-1). In many cases, hydrophobic colloids and, to some extent, the hydrous ferric or aluminum oxides are stable so long as the zeta potential exceeds a critical value. If it drops below this value, coagulation tends to occur slowly in the immediate neighborhood of the critical zeta potential and more rapidly the nearer it gets to zero.

Specific chemical equilibria of colloidal and dissolved substances and chemical interaction between impurities and coagulating metal ions as well as short-range chemical forces are sometimes more important than the electrostatic forces in affecting colloid stability. Adsorption of hydrolyzed metal ions onto dispersed colloids is an example. Physical measurements of the electrokinetic properties of colloids, such as electrophoretic mobility or zeta potential, cannot by themselves characterize coagulative behavior, even though they do elucidate some aspects of the mechanism. Occasionally coagulation is slow at zero zeta potential. Occasionally, too, effective coagulation takes place when the zeta potential is far away from zero.

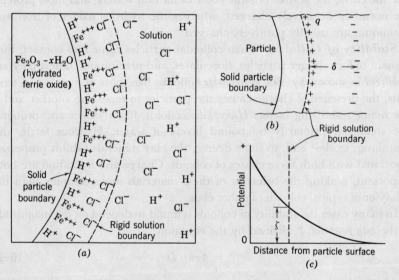

**Figure 16-1.**

**Electrical relations at surfaces of colloids. (a) Diagrammatic representation of charge distribution in neighborhood of surface of particle of ferric oxide solution formed by dispersing $FeCl_3$ in water. (b) Simple Helmholtz picture of particle surface, showing net charge $q$ and double-layer distance $\delta$. (c) Potential relations corresponding to charge distribution shown in (a). Note that the zeta-potential is the potential at the *rigid-solution* boundary with respect to the bulk solution. The ordinary range of $\zeta$ is 10 to 200 mv.**

**Coagulation of Colloids.** Entering into the coagulation of colloidal particles are two opposing forces: (1) an *attractive*, van der Waals *force* tending to draw the particles together, and (2) an *electrostatically repulsive force* tending to keep them apart. The van der Waals force produces the energy of attraction ($A$ in Fig. 16-2) which varies inversely with the square of the distance between two particles. The electrostatic force produces a repulsive energy ($R$ in Fig. 16-2), which decreases more or less exponentially with distance. Curves $A$ and $R$ are combined in curve $S$ as the resultant energy of interaction. A state of equilibrium is reached at the point of minimum potential energy, $P_m$, where the distance between the particles is approximately zero and the particles coalesce. To approach each other, the particles must surmount the *energy hill*, $E_b$. Only if the kinetic energy of the particles is large enough to do this will the colloids become unstable and coagulate. The position of the repulsion curve is affected by the electrolyte content of the solution. Added electrolytes increase the density of the opposing ionic atmosphere and steepen the repulsive energy curve. The effect of polyvalent ions on the repulsive energy curve is especially large. The curve can also be influenced in other ways that vary with the nature of the colloids and the adsorbability of coagulating ions onto the colloid.

Oppositely charged colloids tend to act toward one another like very highly charged ions. When they are mixed, there is mutual coagulation with neutralization of charge. Thus the effective removal of negatively charged

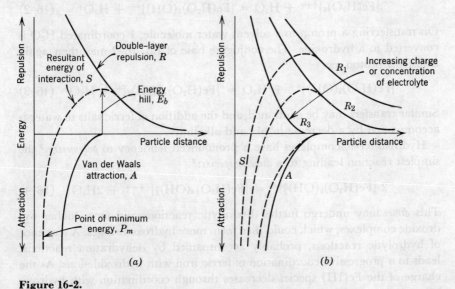

**Figure 16-2.**

**Energy of interaction between two colloidal particles in the form of attraction by van der Waals forces and repulsion by electrostatic forces.**

color colloids and clay turbidity by alum is sometimes due partly to interaction with positively charged colloidal aluminum oxide particles. However, chemical factors, such as the complexing abilities of ferric and aluminum ions, are also of importance in coagulation with alum or ferric salts. When the ions responsible for the charge on a colloid are hydrogen or hydroxyl ions (or other acid or basic ions), neutralization of charge can often be brought about by changes in pH. This explains the variations in ease of floc formation from alum or ferric chloride at different pH values.

## 16-3 Coagulation with Ferric Iron and Aluminum

The coagulating power of ions rises rapidly with their valence. Although charge has been cited as accounting for $Al^{+++}$ and $Fe^{+++}$ being many times as effective as $Ca^{++}$ in the coagulation of negative colloids, only relatively small concentrations of *free* $Fe^{+++}$ and $Al^{+++}$ with a charge of 3+ are actually present in solution during coagulation. Indeed, the effect of Fe(III) and Al(III) on coagulation is not brought about by the ions themselves, but by their hydrolysis products. By convention, Roman numerals are used to indicate the oxidation state of the element; for example, Fe(III) refers to all constituents that contain iron in oxidation states 3, whereas $Fe^{+++}$ refers specifically to ferric iron with a charge of 3.

*Hydrolysis of Fe(III) and Al(III).* The acid base equilibrium for *ferric iron* is

$$[Fe(H_2O)_6]^{+++} + H_2O = [Fe(H_2O)_5(OH)]^{++} + H_3O^+ \qquad (16\text{-}2)$$

On transferring a proton to a solvent water molecule, 1 coordinated $H_2O$ is converted to a hydroxide. The conjugate base of Eq. 16-2 may then again transfer a proton, or

$$[Fe(H_2O)_5(OH)]^{++} + H_2O = [Fe(H_2O)_4(OH)_2]^+ + H_3O^+ \qquad (16\text{-}3)$$

Similar transfers may be continued, and the addition of ferric salts to water is accompanied by a decrease in pH and alkalinity.

Hydroxo ferric complexes have a pronounced tendency to *polymerize*,[3] the simplest reaction leading to a *dimeric species*:[3]

$$2\,[Fe(H_2O)_5(OH)]^{++} = [Fe(H_2O)_8(OH)_2]^{++++} + 2H_2O \qquad (16\text{-}4)$$

This *dimer* may undergo further hydrolytic reactions and form higher hydroxide complexes, which could then form more hydroxo bridges. A sequence of hydrolytic reactions, probably accompanied by dehydration reactions, leads to a progressive coordination of ferric iron with hydroxide ions. As the charge of the Fe(III) species decreases through coordination with hydroxo

---

[3] From the Greek *polys*, many; *dis*, twice; and *meros*, part.

groups, there is less repulsion between the ions, and a greater tendency toward polymerization. Eventually, colloidal hydroxo polymers and, ultimately, insoluble hydrous ferric oxide precipitates are formed.

Aluminum(III) behaves in much the same way as iron(III), but its hydrolysis is apparently more complicated. Aluminum salts are easily hydrolyzed, and the acidity of the aquo-aluminum ion is below that of the ferric ion. In accordance with the concepts described for Fe(III), stepwise conversion of the positive aluminum ion into the negative aluminate ion must be assumed. The formation of polynuclear ionic hydroxo aluminum complexes is well substantiated. Hydrolysis increases progressively with the age of Al(III) and Fe(III) solutions, and coagulation effects of aged solutions and unaged solutions are quite different.

From what has been said about Fe(III) and Al(III), it is clear that the effect of these *multivalent* metal ions on coagulation is not generally brought about by the ions themselves but by their hydrolysis products. For a given metal-ion concentration, the rate and efficiency of coagulation depend on the pH of the medium.

**Coagulation with Hydrolyzed Fe(III) and Al(III).** The *dissolved* polynuclear hydroxo complexes that are intermediates in the hydrolytic transition of iron and aluminum ions to hydrous oxides are efficient coagulants. Under favorable conditions of pH, temperature, and time of aging, the metal ion hydrolysis products are more strongly sorbed on colloidal dispersions than are nonhydrolyzed metal ions. Figure 16–3a shows as a function of pH the critical coagulation concentration for a dispersion of amorphous $SiO_2$, that is, the minimal concentration of coagulant, Fe(III) in this instance, necessary to produce a floc. The following conclusions may be drawn: (1) at pH values below 3, where Fe(III) is not hydrolyzed, coagulation of $SiO_2$ is brought about by $Fe^{+++}$ ions; (2) at higher pH values, lower dosages of Fe(III) cause the coagulation of the $SiO_2$ sol, highly charged polynuclear hydroxo complexes being formed and strongly sorbed on the surfaces of the colloids where they are able to reverse the charge of the colloids when they are present at a somewhat higher concentration than the critical coagulation concentration; (3) it follows from the charge-reversal concentration that the hydrolyzed species, and not the free $Fe^{+++}$, causes charge reversal; (4) at large concentrations of Fe(III), ferric oxide is precipitated within the time period of observation; and (5) the stability and instability domains of an $SiO_2$ suspension in the presence of ferric iron are identified in Fig. 16–3 as Region I, in which $SiO_2$ is stable as a negatively charged sol; Region II, shaded, in which $SiO_2$ is coagulated by Fe(III); Region III, in which $SiO_2$ is restabilized as a positive sol by the adsorbed Fe(III) hydrolysis species; and Region IV, in which hydrous iron oxide precipitates and entraps some of the suspended $SiO_2$.

*470/Coagulation, Precipitation, and Stabilization*

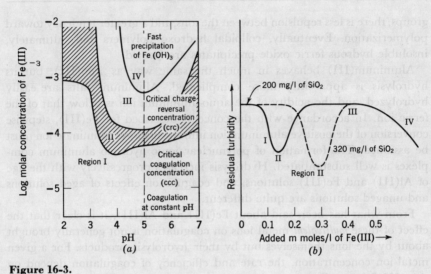

**Figure 16-3.**
**Coagulation of amorphous silica of diameter 15 $\mu$ with Fe(III). (a) Concentration-pH domain for sol stability (Region I); coagulation and charge reversal (Region II); stability of charge reversed sol (Region III) and of Fe(OH)$_3$ precipitation (Region IV). (b) Effect of dosage on residual turbidity at a constant pH of 5.**

Coagulation data of this kind can be obtained by running coagulation tests and plotting the information obtained. Suitable variables are residual turbidity and color in relation to coagulant dose. A coagulation curve at a constant and controlled pH of 5 is shown in Fig. 16–3b for the system decribed in Fig. 16–3a. Initial increments of Fe(III) added to the dispersion of negatively charged SiO$_2$ are seen to have little effect on the stability of SiO$_2$ in Phase I, which corresponds to Region I of Fig. 16–3a. In Phase II flocculation occurs as soon as the critical coagulation concentration is exceeded. In Phase III, however, stability of the sol is restored upon further addition of coagulant and as soon as the critical restabilization concentration is reached. Finally, in Phase IV, an excess of coagulant either precipitates hydrous ferric oxide or flocculates the positive sol because a critical concentration of anions has been established in the system. The charge-reversal concentration and, to a smaller extent, the critical-coagulation concentration depend on the concentration or surface area of the colloids to be coagulated. The coagulating dose may be proportional to the concentration of the colloidal impurities. In practice, however, coagulation is not always induced by dosages in the charge-reversal concentration range (Phase II). The rate of coagulation may be too slow, especially when dispersions are dilute. Instead of this, coagulation becomes successful only when concentrations of metal ions are high enough to precipitate metal hydroxides that *sweep* the colloidal dispersions together (Phase IV).

## 16-4 Coagulation with Polyelectrolytes

Polyelectrolytes are linear or branched chains of small identical subunits—sometimes of two or three different kinds of subunits. The subunits contain ionizable —COOH, —OH, —POP$_3$H$_2$, —NH$_2^+$, and R$_1$NR$_2^+$ groups. Polymers of this kind differ from nonelectrolyte polymers in the same way that low-molecular-weight electrolytes differ from nonelectrolytes. They are soluble in water, conduct electricity, and are affected by the electrostatic forces between their charges. Proteins, protamines, nucleic, pectic, and alginic acids, polysaccharides, and numerous poly acids are natural polyelectrolytes. Synthetic polyelectrolytes are formed by polymerization of simple substances, known as monomers. Long-chain molecules of this kind may be designed to contain from 2 to 3 to nearly $10^6$ subunits. Cationic, anionic, and ampholytic[4] polyelectrolytes can be prepared by choosing suitable monomers. Examples follow:

*Polyacrylate* exemplifies an anionic polyelectrolyte:

$$-CH_2-CH-CH_2-CH-CH_2-CH-$$
$$\quad\quad\;\; |\quad\quad\quad\;\; |\quad\quad\quad\;\; |$$
$$\quad\;\; COO^-\quad\; COO^-\quad\; COO^-$$

*Polyvinylpyridinium salts* exemplify cationic polyelectrolytes:

with pyridinium rings bearing $N^+\!-\!R$ groups on the $-CH_2-CH-CH_2-CH-CH_2-$ backbone.

*Polyamino acids*, such as polylysine-glutamic acid, exemplify polyampholites:

$$-NH-CH-CO-NH-CH-CO-NH-CH-CO-$$
$$\quad\quad\quad\;\; |\quad\quad\quad\quad\quad\;\; |\quad\quad\quad\quad\quad\;\; |$$
$$\quad\quad\; (CH_2)_4\quad\quad\quad (CH_2)_2\quad\quad\quad (CH_2)_4$$
$$\quad\quad\quad\;\; |\quad\quad\quad\quad\quad\;\; |\quad\quad\quad\quad\quad\;\; |$$
$$\quad\quad\quad NH_3^+\quad\quad\quad\; COO^-\quad\quad\quad NH_3^+$$

In principle these polymers are not unlike some polyelectrolytic colloids found in natural waters. Particle charge depends on the degree of ionization and, consequently, on the pH of the medium. Natural and synthetic macromolecules have a pronounced effect on the stability of colloidal dispersions.

Polyelectrolytes are successful flocculents in water and wastewater treatment and in sludge conditioning. The flocculation of bacteria with cationic polyelectrolytes is illustrated in Fig. 16-4.

---

[4] From the Greek, *amphi*, signifying both kinds, that is, containing both cationic and anionic groups; and *lytikos*, able to loose, loosing.

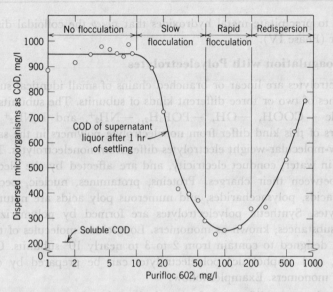

**Figure 16-4.**

**Flocculation of dispersed microorganisms by a cationic polyelectrolyte (Purifloc 602, Dow Chemical Company) at constant pH. With increasing amounts of added polyelectrolyte, a greater fraction of the microbial surface becomes covered until flocculation becomes optimal. Beyond this point there is mutual repulsion and redispersion of the cells. The curve has not been corrected for the COD contribution by the residual polyelectrolyte. However, 1 mg per l of Purifloc 602 has a COD equivalent of only 0.65 mg per l, and this would not significantly alter the shape of the curve. [After M. W. Tenney and Werner Stumm, Chemical Flocculation of Microorganisms in Biological Waste Treatment,** *J. Water Pollution Control Fed.*, **37, 1370 (1965).]**

Flocculation by polyelectrolytes proceeds by a bridging mechanism,[5] so that close-range chemical interaction, that is, formation of hydrogen and coordinative bonds, is much more important than electrostatic interaction. Accordingly flocculation may be efficient even when the polyelectrolytes carry a net charge of the same sign as the particles being flocculated. Negatively charged clays, for instance, may be flocculated expediently with negatively charged—that is, anionic—polyacrylamides. Coagulation of colloidal dispersions by neutral salts (for example, $CaCl_2$) is an entirely different matter. Dispersions agglomerate because the repulsive potential of the double layer surrounding the particles is reduced. The agglomerates produced by the two mechanisms possess different properties and, therefore, respond differently to settling and filtration.

[5] V. K. La Mer and T. Healy, Adsorption-Flocculation Reactions of Macromolecules at the Solid Liquid Interface, *Rev. Pure Appl. Chem.*, **13**, 113 (1963).

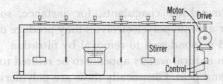

**Figure 16-5.**
Laboratory apparatus for determining optimum dosage of coagulating chemicals and needed stirring. A variable-speed motor is provided. Tyndall beams directed through the beakers can be of assistance. Measurements by T. R. Camp, Hydraulics of Mixing Tanks, 1969 Freeman Memorial Lecture, Boston Society of Civil Engineers, prove that the values of $G$ vary as $\nu^{1/2}n^{3/2}$ where $\nu$ is the kinematic viscosity of the water and $n$ is the number of revolutions per minute of the paddle. Stators introduced into 2-liter beakers increased the $G$ values about thirtyfold.

## 16-5 Coagulation Systems

Because waters and wastewaters vary and fluctuate widely in quality, needed coagulation and flocculation are best determined in practice by trial. Required coagulant dosage is usually found by *jar tests* performed in a laboratory stirring device (Fig. 16-5). Because coagulation depends on so many variables that are themselves interdependent, as many testing parameters as possible should be kept constant. The importance of (1) pH governing the nature of the coagulant or flocculent through the extent of hydrolysis and ionization and (2) determining the charge of colloidal impurities suggests that the pH be kept constant, too. The pH and alkalinity are changed implicitly when a coagulant is added, every millimole per liter of iron(III) or aluminum(III), for instance, decreasing the alkalinity by 3 me. The degree of stirring and the temperature of the water tested are important.

Coagulation may be improved by *coagulant aids*, that is, substances that increase the critical mass of colloids and speed up coagulation. Kinetically, for example, water with little turbidity may not coagulate as easily and as well as water of moderate turbidity. Coagulation may then be improved by adding colloids that carry a charge of the same sign as the natural turbidity of the water. Examples are bentonite, anionic polyelectrolytes, and activated silica.[6] Because the *critical* mass of colloids interacting with coagulants is increased by additives of this kind, coagulation is accelerated. Occasionally coagulant aids may reduce coagulant dosage by speeding the kinetics of the process. They may also improve the physical character of the flocs. In solu-

---

[6] Activated silica is a negatively charged polysilicate, that is, an anionic polyelectrolyte, which is prepared by partially neutralizing to a pH of 6 to 7 a concentrated sodium silicate solution. This is subsequently diluted. The polysilicate is thermodynamically unstable and must be freshly prepared.

tions containing metal-ion coagulants, for instance, some anions, polysilicates, and other anionic polyelectrolytes may produce dense agglomerates that settle fast and respond well to removal by filtration.

Organic color in natural waters appears to be related to humic substances that occur in soil and peat. Chemically these are polymeric compounds with carboxylic and phenolic functional groups. They may exist as large molecules or as colloids. Their true nature is yet to be fully explored. Chemical interaction accounts for color removal by Fe(III) or Al(III). At a given pH, required coagulant dosage is strictly proportional to the concentration of color constituents.

In the purification of municipal water supplies, coagulated impurities are normally removed by gravitational settling or upflow clarification in advance of filtration. Overall efficiency depends on optimal integration of component treatments. Both settling and filtration are governed in some degree by the compactness, size, density, sheer strength, and compressibility of the coagulates or flocs. How chemical and physical conditions such as these affect performance remains to be more fully established.

## 16-6 Chemical Precipitation

Natural waters and wastewaters acquire their chemical characteristics either within the hydrological cycle or during domestic and industrial use. Some of the acquired solids, liquids, and gases are taken into solution and maintain their integrity; others react with water or with one another in water to form new chemical species. Some solids may be left in suspension, some liquids may be immiscible and nonreactive, and some gases may be entrapped only mechanically.

Precipitation is essentially the opposite of dissolution. Understandably, therefore, precipitation, like solution, can often be based on the law of mass action and available equilibrium constants. However, conformance of this kind does not necessarily assure either stoichiometric completion of the precipitation reaction or the adequacy of added stoichiometric amounts. Amounts greater than those calculated may have to be called on to attain wanted performance. As a rule, precipitants should be chosen from among (1) chemicals normally occurring in natural waters—hydroxide and carbonate, for example; (2) relatively insoluble chemicals—iron(III) and aluminum(III), for instance; and, in a general sense, (3) safe chemicals that do not produce toxic residues—for instance, avoidance of $Ba^{++}$, because it is toxic and would be left in solution if sulfate were precipitated from drinking water as $BaSO_4$.

Most multivalent cations in water can be precipitated in predictable amounts as carbonates or hydroxides. This is fortunate, because their solubility governs important treatment processes, including chemical stabilization and the precipitation of $Ca^{++}$ and $Mg^{++}$, as well as other cations in water softening. However, knowing the solubility product is not enough. Thus the dissolving species, when $CaCO_3(s)$[7] is added to pure water, are $Ca^{++}$ and $CO_3^=$, and because $CO_3^=$ is a base its reaction products with water are predominantly $HCO_3^-$ and $OH^-$. In more general terms, $CaCO_3$ reacts with hydrogen ions, or an acid, as follows:

$$CaCO_3(s) + H^+ = Ca^{++} + HCO_3^- \qquad (16\text{-}6)$$

It is seen that lowering the pH by adding an acid increases the solubility of $CaCO_3$, whereas raising the pH by adding a base decreases it. Raising the pH, indeed, governs removal of $Ca^{++}$ by the lime-soda softening process (Sec. 16-8). Systemic and quantitative determinations of $CaCO_3$ solubility relations can be based on the equilibria of Table 16-1.

## 16-7 Chemical Stabilization

Within the frame of reference of this book, chemical stabilization is the adjustment of the pH, $[Ca^{++}]$, and alkalinity of water to its $CaCO_3$ saturation equilibrium. Because a stabilized water neither dissolves nor precipitates $CaCO_3$, it will neither remove coatings of $CaCO_3$ that may protect pipes against corrosion, nor lay down deposits of $CaCO_3$ that may clog pipes. Equilibrium conditions can be defined by the equations $CaCO_3(s) + H^+ = Ca^{++} + HCO_3^-$ and $[Ca^{++}][HCO_3^-]/[H^+] = K$. The $CaCO_3$ saturation point can be characterized also by $[H^+]_{eq} = [H^+]_s$ or $pH_{eq} = pH_s$, respectively the hydrogen ion concentration or pH at the hypothetical equilibrium or saturation with $CaCO_3$. Hence

$$[H^+]_{eq} = [Ca^{++}][HCO_3^-]/K \qquad (16\text{-}7)$$

Below $pH_{eq}$, or when $[H^+] > [H^+]_{eq}$, no $CaCO_3$ will be deposited. Some natural waters of low alkalinity and hardness and of high $CO_2$ content, as well as coagulated, ion-exchange-softened, or demineralized waters, fall into this category. Above $pH_{eq}$, or when $[H^+] < [H^+]_{eq}$, $CaCO_3$ will precipitate. Carbonate deposits may then accumulate in distribution mains, boilers, and other equipment, and on sand and gravel in water filters. To provide a measure of the stability of a given water, Langelier[8] proposed calling the difference between the measured pH and the calculated or determined equilibrium pH value ($pH_s = pH_{eq}$) the saturation index $I = pH - pH_s$. When $I = 0$, the water is in equilibrium; when $I$ is positive, the water is oversaturated; and when $I$ is negative, the water is undersaturated or aggressive.

---

[7] By convention, the symbol $(s)$ indicates that the substance is present as a solid. Incidentally, pure solid substances have an activity of 1 in equilibrium expressions.
[8] W. F. Langelier, Chemical Equilibria in Water Treatment, *J. Am. Water Works Assoc.*, 38, 169 (1946).

### Table 16-1
$CaCO_3$ Solubility and Carbonate Equilibria in Terms of Dissociation Constants, $pK = (-\log K)$

| | Reaction[a] | | Temperature, °C | | | | | |
|---|---|---|---|---|---|---|---|---|
| | | 5 | 10 | 15 | 20 | 25 | 40 | 60 |
| 1. | $CO_{2(s)} + H_2O = H_2CO_3^*$; $K$ | 1.20 | 1.27 | 1.34 | 1.41 | 1.47 | 1.64 | 1.8 |
| 2. | $H_2CO_3^* = HCO_3^- + H^+$; $K_1$ | 6.52 | 6.46 | 6.42 | 6.38 | 6.35 | 6.30 | 6.30 |
| 3. | $HCO_3^- = CO_3^{--} + H^+$; $K_2$ | 10.56 | 10.49 | 10.43 | 10.38 | 10.33 | 10.22 | 10.14 |
| 4a. | $CaCO_3(s) = Ca^{++} + CO_3^{--}$; $K_s$ | 8.09 | 8.15 | 8.22 | 8.28 | 8.34 | 8.51 | 8.74 |
| 4b. | $CaCO_3(s) + H^+ = Ca^{++} + HCO_3^-$; $(K_s/K_2)$ | −2.47 | −2.34 | −2.21 | −2.10 | −1.99 | −1.71 | −1.40 |
| 4c. | $CaCO_3(s) + H_2CO_3^* = Ca^{++} + 2 HCO_3^-$; $(K_sK_1/K_2)$ | 4.05 | 4.12 | 4.21 | 4.28 | 4.36 | 4.59 | 4.90 |
| 5. | $H_2O = H^+ + OH^-$; $K_w$ | 14.73 | 14.54 | 14.35 | 14.17 | 14.00 | 13.54 | 13.02 |

**Salinity Corrections for Equilibrium Constants[b]**

$$pK_1^I = pK_1 - (\mu)^{0.5}/[1 + 1.4(\mu)^{0.5}]$$
$$pK_2^I = pK_2 - 2(\mu)^{0.5}/[1 + 1.4(\mu)^{0.5}]$$
$$pK_s^I = pK_s - 4(\mu)^{0.5}/[1 + 3.9(\mu)^{0.5}] \qquad \log K_s^I/K_2^I = \log K_s/K_2 + 2.5(\mu)^{0.5}/[1 + 5.3(\mu)^{0.5}] + 5.5(\mu)^{0.5}]$$

**Estimation of $\mu$**

$$\mu \simeq 2.5 \times 10^{-5} S_d$$, where $S_d$ is the total dissolved solids content of the water, or
$$\mu \simeq 4H - T$$, where $H$ = total hardness in moles/l; and $T$ = alkalinity in eq/l ($10^3$ me/l)

[a] $H_2CO_3^*$ refers to the sum of dissolved $CO_2$ and $H_2CO_3$. Because $[H_2CO_3] \ll [CO_2]$, $[CO_2]$ is essentially equal to $[H_2CO_3^*]$.
[b] $K$ and $K^I$ are thermodynamic and operational constants, respectively. The operational constants can be used with mass law expressions containing concentration terms, with the exception of $H^+$, which is always expressed in activities. From Larson and Buswell, Calcium Carbonate Saturation Index and Alkalinity Interpretations, *J. Am. Water Works Assoc.*, **34**, 1664 (1942).

The equilibrium constant of Eq. 16-7 is given by the ratio of the mass-law expression for Reaction 4a to Reaction 3 in Table 16-1, namely,

$$[Ca^{++}][HCO_3^-]/[H^+] = K_s/K_2 \qquad (16-8)$$

and the pH for $CaCO_3$ saturation becomes

$$pH_s = \log K_s/K_2 - \log [Ca^{++}] - \log [HCO_3^-] \qquad (16-9)$$

The concentration of $HCO_3^-$ can be calculated from analytically more readily determinable parameters than $HCO_3^-$ itself. For example,

$$[HCO_3^-] = \{T - K_w/[H^+] + [H^+]\}/\{1 + 2K_2/[H^+]\} \qquad (16-10)$$

where $T$ is the total alkalinity (eq/l). Below pH 8.5, $[HCO_3^-] \approx T$.

*Example 16-1.*

Compute the saturation index, $I$, of a water having the following composition: pH = 7.55; calcium hardness = 150 mg/l (as $CaCO_3$); total hardness = 180 mg/l (as $CaCO_3$); and alkalinity = 160 mg/l (as $CaCO_3$). The temperature of the water is 10°C. In molar concentrations: $Ca^{++} = 1.5 \times 10^{-3}$ $M$ and the alkalinity $T = 3.2 \times 10^{-3}$ eq/l.

In accordance with Table 16-1, the ionic strength, $\mu$, of this water is approximately 0.006, $\log K_s/K_2 = 2.34$, and the salinity correction for $\mu = 0.006$ is $+0.134$. By Eq. 16-9, therefore, $pH_s = 2.34 + 0.134 + 2.824 + 2.495 = 7.79$, and $I = 7.55 - 7.79 = -0.24$ pH unit.

## 16-8 Precipitation of Hardness and Carbonates (Water Softening)

Calcium and magnesium ions are the principal hardness-forming constituents of water. In the lime-soda or Clark (1841)-Porter (1857) process of water softening, $Ca^{++}$ is precipitated as $CaCO_3$ and $Mg^{++}$ as $Mg(OH)_2$. However, it is important to recognize that other carbonic constituents are affected at the same time. Thus the solubility relations of $CaCO_3$ and $Mg(OH)_2$ shown in Figs. 16-6a and b suggest that removal of hardness and carbonate constituents is only achieved to best advantage (1) when the pH of the final solution is sufficiently high to depress the solubility of $CaCO_3$ and $Mg(OH)_2$, and (2) when $[Ca^{++}]_{final} = C_{T\,final}$ ($C_T$ = total dissolved carbon species: $[H_2CO_3] + [HCO_3^-] + [CO_3^=]$), a prerequisite for simultaneous precipitation of $Ca^{++}$ and carbonate with equal efficiency. The solubility of $[Mg^{++}]$ is governed by the solubility product of $Mg(OH)_2$ (Table 19, Appendix).

The pH is raised conveniently and economically with lime (CaO), or $Ca^{++} + 2\ OH^-$ after addition to water. To equalize the final stoichiometric concentrations of $Ca^{++}$ and $C_T$, a carbonate-bearing base such as $Na_2CO_3$ (soda ash) is added. The following condition must be met after lime and soda have been added:

$$[Ca^{++}]_{original} + [Ca^{++}]_{added\ lime} = C_{T\ original} + [C]_{added\ Na_2CO_3} \quad (16\text{-}11)$$

Requisite amounts of chemicals are dictated (1) by Eq. 16–11 and (2) by the desired degree of softening. Concentrations of residual $Ca^{++}$, $C_T$, and $Mg^{++}$ are given by the final pH of the solution, provided that solubility equilibria have been reached (Figs. 16–6a and b). Therefore the chemical dosage necessary to reach the pH value that produces the selected residual concentration levels of soluble $Ca^{++}$, $C_T$, and $Mg^{++}$ can either be calculated or determined experimentally. Experimentally, the conditions laid down in Eq. 16–11 are met by first adding lime or soda ash to adjust $[Ca^{++}]_{original}$ to $C_{T\ original}$, followed by simultaneous titration of a sample of the water to be softened from two burettes dispensing equimolar lime and soda solutions. Small and volumetrically equal increments of both solutions are added until the pH approximates the desired value.

---

*Example 16–2.*

A raw water has the following composition: total hardness = 215 mg/l (as $CaCO_3$) or $2.15 \times 10^{-3}\ M$; magnesium = 65 mg/l (as $CaCO_3$) or $0.65 \times 10^{-3}\ M$; alkalinity = 185 mg/l (as $CaCO_3$) or $3.7 \times 10^{-3}$ eq/l. Find the pH to which the water must be raised in the lime-soda softening, if the residual total hardness is not to exceed 25 mg/l (as $CaCO_3$).

Because $[Mg^{++}] + [Ca^{++}] \leqslant 2.5 \times 10^{-4}\ M$, the equation defining the solubility product of $Mg(OH)_2$ (Table 19, Appendix), states that

$$(K_{s(MgOH2)}/[OH^-]^2) + \left[ K_{s(CaCO3)} \left( 1 + \frac{[H^+]}{K_2} + \frac{[H^+]^2}{K_1 K_2} \right) \right] \leqslant 2.5 \times 10^{-4}\ M.$$

Using equilibrium constants for 25°C, the threshold pH is found to be $\simeq 10.5$. At equilibrium the finished water will therefore contain approximately $1 \times 10^{-4}\ M$ soluble $Ca^{++}$ and $1.5 \times 10^{-4}\ M$ soluble $Mg^{++}$, the total carbonic species being $C_T \simeq 1 \times 10^{-4}\ M$ and the alkalinity $T \simeq 5 \times 10^{-4}$ eq/l (Figs. 16–6a and b). These answers are valid only if equilibrium is attained. As a practical matter, allowable detention times are generally inadequate for this to happen, and the residual hardness exceeds the calculated value.

An approximate evaluation of requisite chemicals can be obtained from the stoichiometry of the following reactions:

$$H_2CO_3^* + Ca(OH)_2 = CaCO_3(s) + 2 H_2O \quad (16\text{--}12)$$

$$Ca^{++} + 2 HCO_3^- + Ca(OH)_2 = 2 CaCO_3(s) + 2 H_2O \quad (16\text{--}13)$$

$$Ca^{++} + Na_2CO_3 = CaCO_3(s) + 2 Na^+ \quad (16\text{--}14)$$

$$HCO_3^- + Ca(OH)_2 = CaCO_3(s) + OH^- + H_2O \quad (16\text{--}15)$$

$$Mg^{++} + 2 HCO_3^- + 2 Ca(OH)_2 = 2 CaCO_3(s) + Mg(OH)_2(s) + 2 H_2O \quad (16\text{--}16)$$

$$Mg^{++} + Ca(OH)_2 + Na_2CO_3 = CaCO_3(s) + Mg(OH)_2(s) + 2 Na^+ \quad (16\text{--}17)$$

The reactions are assumed to go to completion.

*Example 16–3.*

Calculate how much lime and soda ash are required to soften a raw water with a total hardness = 215 mg/l (as $CaCO_3$) or $2.15 \times 10^{-3}$ $M$; magnesium = 15.8 mg/l (as $Mg^{++}$) or $0.65 \times 10^{-3}$ $M$; $Na^+$ = 8 mg/l or $0.35 \times 10^{-3}$ $M$; $SO_4^-$ = 28.6 mg/l or $0.3 \times 10^{-3}$ $M$; $Cl^-$ = 10 mg/l or $0.285 \times 10^{-3}$ $M$; alkalinity = 185 mg/l (as $CaCO_3$) or $3.7 \times 10^{-3}$ eq/l; carbon dioxide = 25.8 mg/l (as $CO_2$) or $0.59 \times 10^{-3}$ $M$; pH = 6.7.

Strike an electroneutrality balance as in Fig. 16-6a, recognizing that, at pH 6.7, $[HCO_3^-] \gg [CO_3^-]$, a condition applying to virtually all hard and high-alkalinity waters. Subdivide the diagram into individual blocks after computing the required dosages as in Table 16-2.

Verify that Eq. 16-11 is fulfilled, that is, that $[Ca^{++}] + [\text{lime}] = C_T + [\text{soda}]$. By substitution, find that this is so: $1.5 \times 10^{-3} + 3.44 \times 10^{-3} = 4.29 \times 10^{-3} + 0.30 \times 10^{-3} = 4.59 \times 10^{-3}$.

*Example 16–4.*

Calculate the amounts of lime and soda ash required to soften a water of the following composition: $[Ca^{++}] = 5 \times 10^{-3}$ $M$; $Mg^{++} = 2 \times 10^{-3}$ $M$; $[HCO_3^-] = 4 \times 10^{-3}$ $M$; $H_2CO_3^* = 4 \times 10^{-3}$ $M$; $SO_4^- = 6 \times 10^{-3}$ $M$; $Cl^- = 0.8 \times 10^{-3}$ $M$; pH = 6.3.

To strike the electroneutrality balance shown in Fig. 16-6b, calculate the requisite amounts of chemicals shown in Table 16-3. To hasten and complete precipitation, add lime and soda in excess of calculated values.

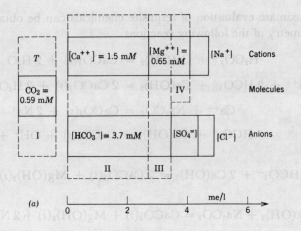

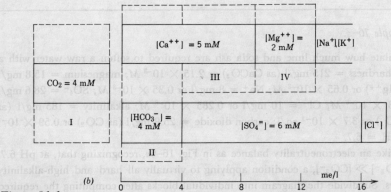

**Figure 16-6.**
**Cation-anion balance of waters to be softened by lime-soda process (Examples 16-3 and 16-4).**

**Table 16-2**
Computed Electroneutrality Balance (Example 16–3)

| Block No. | Applicable Equation | Requisite CaO, m$M$[a] | Requisite Na$_2$CO$_3$, m$M$ |
|---|---|---|---|
| I | 16–12 | 0.59 | — |
| II | 16–13 | 1.50 | ... |
| III | 16–16 | 0.70 | ... |
| IV | 16–17 | 0.30 | 0.30 |
| Total to be applied | | 3.09 | 0.30 |

[a] m$M$ = millimolar.

## Table 16-3
*Electroneutrality Balance (Example 16-4)*

| Block No. | Applicable Equation | Requisite CaO, m$M$ | Requisite Na$_2$CO$_3$, m$M$ |
|---|---|---|---|
| I | 16-12 | 4 | — |
| II | 16-13 | 2 | ... |
| III | 16-14 | ... | 3 |
| IV | 16-17 | 2 | 2 |
| Total | | 8 | 5 |
| Excess[a] | | 1 | 1 |
| To be applied | | 9 | 6 |

[a] In excess-lime treatment, carbonate equivalent to the excess of lime must be added. Otherwise the calcium ions in excess would defeat the purpose of softening. Carbon dioxide is often added in lieu of Na$_2$CO$_3$.

Examples 16-3 and 4, together with Eqs. 16-14 and 17, show that the amount of soda ash needed for efficient softening is a function of the so-called *noncarbonate hardness (NCH)*. The noncarbonate hardness is the total hardness not compensated by HCO$_3^-$ or CO$_3^=$ in a charge-balance equation; that is, $NCH$ (eq per l) = $2[Ca^{++}] + 2[Mg^{++}] - [HCO_3^-] - 2[CO_3^=]$. The $NCH$ has also been called permanent hardness, because it is not precipitated when water is heated to the boiling point. Because soda ash is relatively expensive in comparison with lime, some of the noncarbonate hardness is purposely not removed for the sake of economy. Instead it is left behind as the residual hardness of the finished water. When lime softening is combined with ion-exchange softening (Sec. 15-7), the lime serves the purpose of precipitating most of the carbonate hardness, and the ion exchange eliminates the remaining carbonate hardness as well as the noncarbonate hardness.

The residual hardness of lime-soda-softened waters is normally higher than its calculated value, because detention times can seldom be made long enough to ensure full precipitation. To speed precipitation, lime and soda ash are added to the raw water in excess of stoichiometric requirements. To shorten the settling time and remove precipitates more efficiently, coagulants—alum, activated silica, or polyelectrolytes, for instance—may be added. The solubilities of CaCO$_3$ and Mg(OH)$_2$ decrease, and their rates of precipitation increase, with rising temperatures. Therefore hot-process softening is commonly used in boiler-water treatment, and sometimes also to obtain waters of very low residual hardness (Fig. 16-7).

Waters softened by the lime-soda process are generally supersaturated with CaCO$_3$ and Mg(OH)$_2$. They can be *stabilized* by blowing CO$_2$ into them.

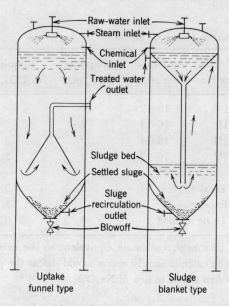

**Figure 16-7.**
**Equipment used in hot-process softening. (Courtesy Nalco Chemical Co.)**

Normally this is done before filtration. So-called secondary carbonation or recarbonation relieves supersaturation and reduces precipitation of $CaCO_3$ on sand grains and in pipelines. Small amounts of polyphosphates or metaphosphates are also helpful. Although phosphates form soluble complexes with $Ca^{++}$, too little phosphate is normally added to account for the observed retardation of precipitation. Moreover, polyphosphates do not prevent the precipitation of $Mg(OH)_2$.

Reduction in the concentrations of fluorides and silica[9] is an interesting side reaction of the excess lime treatment of high-magnesium waters. Coprecipitation of $F^-$ and silicate with $Mg(OH)_2$ forms nonstoichiometric fluoro- and silicato-hydroxo precipitates of $Mg^{++}$. Observed removals of $F^-$, in Ohio water supplies, for instance, approximate $7\sqrt{Mg^{++}}$ %.[10]

Numerous ways have been proposed for recovering lime from the sludge produced by softening. Precipitated $CaCO_3$ can be converted into $CaO$ by calcining, that is, by heating the dried solids to drive off $CO_2$. However, the calcined sludge must not contain much magnesium if it is to be recycled. The Hoover and the Lykken-Estabrook processes are examples of effective

---

[9] In natural waters soluble silica consists of $Si(OH)_4$, orthosilicic acid. Only at pH values above $9\pm$ are silicate anions formed.

[10] R. D. Scott, A. E. Kimberley, H. L. van Horn, L. F. Ey, and F. W. Waring, Jr., Fluoride in Ohio Water Supplies, *J. Am. Water Works Assoc.*, **29**, 9 (1937).

recovery.[11] The Hoover process does so by two-stage treatment. In the first stage only enough lime is added to precipitate the $Ca^{++}$ ions. In the second stage further amounts of lime precipitate the $Mg^{++}$ ions. The first-stage sludge is recovered; the second-stage sludge is wasted. The carbon dioxide produced during calcining can be used to recarbonate the softened water. In the Lykken-Estabrook process all the recovered precipitate is added to about 12% of the water being softened. At the resulting high pH both $Ca^{++}$ and $Mg^{++}$ and most of the $Mg(OH)_2$ in the added chemical are precipitated. Because of this the second-round sludge is wasted. Subsequently the overtreated (12%) portion of water is mixed with the mainstream and softens it.

## 16-9 Removal of Iron and Manganese

Iron and manganese are major components of the earth's crust. They occur naturally in groundwater and are common constituents of acid-mine drainage and industrial wastewaters. In water-distribution systems iron makes its appearance as a product of corrosion. In limnology and oceanography iron and manganese are essential elements of plant nutrition. Neither iron nor manganese is very soluble in water. Iron(III) oxide is the cause of *red water* in distribution systems; manganese oxides are the cause of *brown or black water*. Both make water unsuitable for laundering, and for dyeing, papermaking, and other manufacturing processes.

*Solubility of Fe and Mn.* Within the pH range of natural waters, soluble bivalent iron and manganese consist predominantly of $Fe^{++}$, $Mg^{++}$, $FeOH^+$, and $MnOH^+$. Within the common pH range (6 to 9) of carbonate-bearing waters, their solubility in the bivalent oxidation state is governed generally by the solubility products of their carbonates, not their hydroxides. Thus maximum soluble [Fe(II)] or [Mn(II)] depends on pH and bicarbonate content. Using the appropriate equilibrium constants of Table 16-1, and 19 (Appendix), a solubility diagram[12] for iron(II) and manganese(II) has been constructed in Fig. 16-8. Small quantities of sulfide ($H_2S$, $HS^-$, $S^=$) can limit Mn(II) and Fe(II) solubility even more than carbonate.

*Example 16-5.*

Derive an equation for the concentration of maximum soluble $Fe^{++}$ as a function of pH and alkalinity, $T$.

From Tables 19 (Appendix) and 16-1, $[Fe^{++}][CO_3^-] = K_s = 10^{-10.7}$ and $[H^+][CO_3^-]/[HCO_3^-] = K_2 = 10^{-10.3}$. By division, $[Fe^{++}][HCO_3^-]/[H^+] = K_s/K_2$, or

$$[Fe^{++}] = ([H^+]K_s)/(K_2[HCO_3^-]) \qquad (16\text{-}18)$$

[11] Charles Potter Hoover (1884–1950), for many years manager of the water softening and purification works at Columbus, Ohio.

[12] The effect of hydrolysis, that is, the formation of $FeOH^+$ and $MnOH^+$, has been neglected because it becomes significant only above pH 10.

Because the concentration of $HCO_3^-$ equals the alkalinity at pH values below about 8.5, and is also given in terms of $T$ by Eq. 16–10, Eq. 16–18 can be converted to read

$$[Fe^{++}] = (K_s/K_2)[H^+]\{1 + 2K_2/[H^+]\}/\{T - K_w/[H^+] + [H^+]\} \quad (16\text{–}19)$$

However, this equation is valid only within the pH range in which the solubility of $Fe^{++}$ is governed by the solubility product of $FeCO_3$. Above pH 10, more or less, solubility is controlled by the solubility product of $Fe(OH)_2$, and

$$[Fe^{++}] = K_{s[Fe(OH)_2]}/[OH^-]^2 \quad (16\text{–}20)$$

In water with an alkalinity of 10 mg/l (as $CaCO_3$), $T = 2 \times 10^{-4}$ eq/l and at a pH of 7.5, for example, the concentration of soluble $Fe^{++}$ cannot exceed $2.8 \times 10^{-4}$ $M$ (15.7 mg/l) in accordance with Eq. 16–20. The equation would say that $[Fe^{++}] = 7.9 \times 10^{-3}$ $M$, but this would not be true.

The solubility of Fe(III) in natural waters is generally governed by that of ferric hydroxide, $Fe(OH)_3$, or ferric oxide hydroxide, $FeOOH$. To evaluate it, the formation constants for soluble hydroxo-ferric complexes, $FeOH^{++}$, $FeOH_2^+$, $Fe_2(OH)_2^{++++}$, and $Fe(OH)_4^-$, must be considered together with the solubility product of hydrous ferric oxide. The resulting solubility relations are shown in Fig. 16–8. Within the common pH range, total soluble Fe(III) is seen not to exceed concentrations of approximately $10^{-3}$ mg per l. The solubility of $MnO_2$ is even lower than that of hydrous ferric oxide, and no soluble Mn(IV) can be detected within the pH range 3 to 10. In the absence of strong complex formers, Mn(III) does not occur as a dissolved species.

**Redox Reactions of Fe and Mn.** A comparison of the solubility relations of the bivalent metal ions with those of higher-valent oxidation states shows that oxidation of the ferrous and manganous ions to ferric oxide and higher-valent manganese oxides will render them insoluble. In the presence of dissolved oxygen, both Mn(II) and Fe(II) are thermodynamically unstable at all pH values of natural waters.[13]

$$2\ Fe^{++} + \tfrac{1}{2} O_2 + 5\ H_2O = 2\ Fe(OH)_3(s) + 4\ H^+$$
$$(\log K = 5.8;\ 25°C) \quad (16\text{–}21)$$

$$Mn^{++} + \tfrac{1}{2} O_2 + H_2O = MnO_2(s) + 2\ H^+$$
$$(\log K = 0;\ 25°C) \quad (16\text{–}22)$$

---

[13] These constants were computed from standard electrode potentials. Oxygen is entered into the equilibrium expressions as partial pressure, $pO_2$. The constants predict that the oxygenation reactions, at natural water pH, proceed virtually to stoichiometric completion.

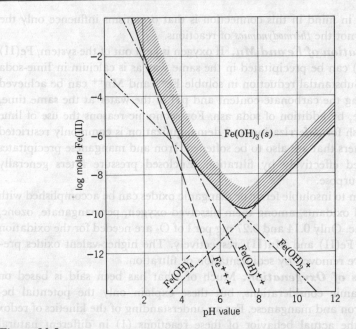

**Figure 16-8.**
**Solubility of Fe(III) at 25°C. Various Fe(III) components are in equilibrium with solid Fe(OH)$_3$. The line surrounding the shaded area identifies total soluble iron(III) as [Fe$^{+++}$] + [Fe(OH)$^{++}$] + [Fe(OH)$_2{}^+$] + [Fe(OH)$_4{}^-$]. The solubility contribution by Fe$_2$(OH)$_2{}^{+++}$ is negligible above pH 4 and is not included here.**

The oxygenation[14] reactions are accompanied by a reduction in pH. Redox-potential data suggest that all organic and some inorganic substances (for example, sulfide) are potential reductants for both Fe(III) and Mn(IV); equilibrium relations suggest that soluble iron and manganese occur in water only in the bivalent state and in the absence of oxygen. Thus iron and manganese are found in oxygen-free groundwater. Anaerobic, hypolimnetic layers of lakes and reservoirs may also hold substantial quantities in solution.[15] Biological activity may influence the chemical reactions indirectly, for example, by creating metabolites, capable of reducing Fe(OH)$_3$ or MnO$_2$. Some bacteria and algae are capable of depositing manganese and ferric oxides from ferrous and manganous solutions. Although their metabolism can be linked to such oxygenation processes, the autotrophic nature of many so-called iron and manganese bacteria has not been proved conclusively.

[14] Oxygenation implies oxidation by oxygen dissolved in water.
[15] J. J. Morgan and Werner Stumm, Multivalent Metal Oxides in Limnological Transformations as Exemplified by Iron and Manganese, *Proc. Second Intern. Conf. Water Pollution Res.*, Tokyo, 1964.

To be kept in mind in this connection is that organisms influence only the *kinetics* and not the *thermodynamics* of reactions.

**Precipitation of Fe and Mn.** If oxygen is kept out of the system, Fe(II) and Mn(II) can be precipitated in the same way as is calcium in lime-soda softening. Substantial reduction in soluble $Fe^{++}$ and $Mn^{++}$ can be achieved by increasing the carbonate content and pH of the water at the same time, for example, by addition of soda ash. For economic reasons the use of lime and soda ash for deferrization and demanginization is commonly restricted to hard waters that are also to be softened. Iron and manganese precipitates are removed effectively by filtration. Enclosed pressure filters generally serve this purpose.

Oxidation to insoluble ferric or manganic oxides can be accomplished with a variety of oxidants, among them dissolved oxygen, permanganate, ozone, and chlorine. Only 0.14 and 0.27 mg per l of $O_2$ are needed for the oxidation of 1 mg of Fe(II) and Mn(II), respectively. The higher-valent oxides precipitated are removed by sedimentation or filtration.

**Kinetics of Oxygenation.** Much of what has been said is based on thermodynamic considerations, but these explain only the potential behavior or iron and manganese. For an understanding of the kinetics of redox reactions, the actual behavior of these reactions (1) in different natural systems and (2) during deferrization and demanganization must also be examined. Representative kinetic experiments are brought together in Fig. 16-9. It is evident that the reaction rates are strongly pH-dependent. Oxygenation of Fe(II) is seen to take place only slowly, especially below pH 6.5. Oxidation of Mn(II) is still slower and becomes measurable only above pH 8.5. The rate of reaction of ferrous iron rises a hundredfold with a unit

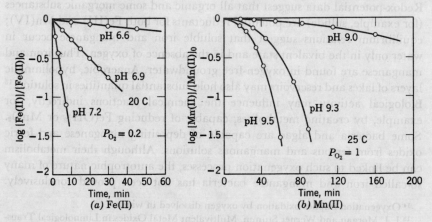

**Figure 16-9.**

**Oxygenation of iron (II) and manganese (II) in bicarbonate solutions.**

increase in pH. Catalysts (especially $Cu^{++}$ and $Co^{++}$), as well as anions that form complexes with Fe(III) (e.g., $HPO_4^-$), also speed up the reaction rate significantly.

Slowly formed $MnO_2$ becomes a sorbent for $Mn^{++}$ ions. Consequently the oxidation products of manganese oxygenation are generally nonstoichiometric compounds in various average degrees of oxidation ranging from $MnO_{1.3}$ to $MnO_{1.9}$ (30 to 90% oxidation to $MnO_2$). The relative proportions of Mn(II) and Mn(IV) in the solid phase depend strongly on pH and other variables. The suggested reaction pattern tells why a substantial fraction of the Mn(II) removed from solution is not oxidized, and explains the autocatalytic nature of the reaction. Filter sands coated with $MnO_2$ can be effective catalysts for $[Mn^{++}]$ oxygenation.

**Engineering Management of Oxidative Deferrization and Demanganization.** Oxidation of Fe(II) and Mn(II) is but one step in the removal process. Coagulation and sedimentation or filtration codetermine the overall removal rate, or the efficiency of removal of these constituents. It can be inferred from the kinetics of oxygenation that the oxidation reaction may be rate-controlling up to pH 7 for ferrous iron and pH 8.5 for manganous manganese.

Coagulation of iron(III) and manganese(IV) oxides is fast within neutral or slightly acid pH ranges, and removal of the oxides by sand filtration is best within the same range. In such circumstances catalysts may be of assistance—for example, $Cu^{++}$, which hastens the oxygenation of Fe(II); or oxidants, such as chlorine or permanganate, both of which oxidize Fe(II) and Mn(II) readily even in the acid pH range. Hydrous oxides of Fe(III) and Mn(IV) have high sorption capacities for bivalent metal ions. Both hydrous $MnO_2$ and $Fe(OH)_3$, for instance, tend to sorb $Mn^{++}$ and $Fe^{++}$ ions. Sorption capacities for $Mn^{++}$ at pH 8 are on the order of 1.0 and 0.3 mole of Mn(II) sorbed per mole of $MnO_2$ and $Fe(OH)_3$, respectively (Table 15–2). The tendencies for sorption of $Mg^{++}$ and $Ca^{++}$ ions are significantly smaller. Sorption of Fe(II) and Mn(II) onto both ferric and manganic oxide is an important feature of oxidative deferrization and demanganization. An example is the easier removal of Mn(II) from water containing substantial amounts of iron $\{[Fe(II)] > [Mn(II)]\}$ rather than Mn(II) alone. Removal then becomes predominantly a matter of sorption of $Mn^{++}$ on incipient precipitates of ferric oxide. In practice, removal of iron and manganese is generally hastened and made more efficient by letting water trickle downward over coke or crushed stone or rise upward through gravel or other relatively coarse, heavy materials. As the contact interfaces become coated with hydrous oxides of Fe(III) or Mn(IV), the removal of Mn(II) and Fe(II) by sorption becomes swifter and more complete. Sand and anthracite filters, too, must *mature*, that is, become coated with hydrous oxides, if they are to remove iron and manganese effectively. The alkalinity of limestones, dolomites, and magnesium oxides makes them not only useful contact media during the oxygenation of Mn(II) and Fe(II), but also powerful sorbents for unoxidized Mn(II) and Fe(II) after the surfaces have become coated with the products of oxidation. In a similar fashion, diatomaceous earth, clays, or zeolites that have been coated with $MnO_2$ by treating them with $Mn^{++}$ and $MnO_4^-$ provide excellent interfaces for iron and manganese removal. When the sorption capacity of surfaces of this kind has been exhausted, they can, in a sense, be *regenerated*. Permanganate will do this; that is, $3\{Mn(II) \cdot MnO_2\}(s) + 2\ MnO_4^- + 2\ H_2O = 8\ MnO_2(s) + 4\ H^+$. Sedimentation of oxide precipitates washed or escaping from the contact units can be speeded and improved by coagulation. Indeed, small amounts of iron and manganese will be removed from some waters by cationic polyelectrolytes, alum, and iron(III) alone.

## 16-10 Precipitation of Phosphate

Wastewater effluents, even though they may have been biologically treated, will still contain substantial amounts of phosphate, ammonia, and nitrate. All three are nutrients for the plankton and for littoral aquatic weeds. Heavy fertilization (eutrophication) may stimulate nuisance growths, especially in lakes, ponds, and similar waters of suitable depth (Sec. 21-11). When this is so, it may pay to remove phosphates from effluents by chemical precipitation. Multivalent metal ions will do this, Fe(III), Al(III), and Ca(II) being candidate additives. All of them are precipitants of phosphate rather than coagulants. Under favorable pH conditions the quantity of metal ions needed for precipitation obeys the stoichiometry of the reaction

$$Al^{+++} + H_nPO_4^{3-n} = AlPO_4(s) + n\ H^+ \qquad (16-23)$$

Hence 1 mole of Al(III) or Fe(III) is necessary to precipitate 1 mole of phosphorus. The pH dependence of the metal-ion-and-phosphate interaction is quantitatively in accord with the solubility relationships of the metal phosphate and the metal hydroxide, the hydrolysis of the metal ion, and the acid-base equilibria of the phosphate ions (Table 19-App.). Fig. 16-10 charts the solubility of $FePO_4(s)$ and $AlPO_4(s)$ from the equilibrium constants. The total concentration of soluble phosphorus in equilibrium with both solid $FePO_4$ and $AlPO_4$ is plotted as a function of pH.

Under neutral and alkaline conditions, $FePO_4$ and $AlPO_4$ are quite soluble; much hydroxide is coprecipitated with phosphate, and the stoichiometric efficiency of phosphate removal decreases with increasing pH. Because the acidity of $Al^{+++}$ is less than that of $Fe^{+++}$ (that is, $Al^{+++}$ hydrolyzes at a higher pH than does $Fe^{+++}$), the difference in pH values at mini-

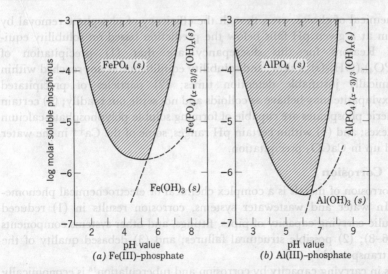

**Figure 16-10.**
Solubility of iron(III) and aluminum(III) phosphates. The solid lines trace the concentration of residual soluble phosphorus after precipitation by Fe(III) or Al(III), respectively, in concentrations equimolar to the original phosphorus concentration. Inside the shaded area pure metal phosphates are precipitated. Outside, toward higher pH values, mixed hydroxo-phosphato metal precipitates are formed.

mum $AlPO_4$ and $FePO_4$ solubility is about $+1$ unit. For the same reason, the difference in accompanying minimum solubilities of $AlPO_4$ and $FePO_4$ is about $-1$ unit, even though $AlPO_4$ has a slightly larger solubility product. Accordingly, proper pH adjustment can achieve a high degree of removal control. Polymeric phosphates and organic phosphates are also precipitated by Fe(III) and Al(III).

In the alkaline pH range, $Ca^{++}$ ions can precipitate phosphate. The solubility of phosphate is then controlled by the solubility equilibrium of hydroxylapatite, or

$$Ca_{10}(PO_4)_6(OH)_2(s) = 10\ Ca^{++} + 6\ PO_4^- + 2\ OH^-$$
$$(\log K \simeq -90;\ 25°C) \quad (16\text{-}24)$$

So-called calcium phosphate, $Ca_3(PO_4)_2(s)$, is not formed under such conditions. The magnitude of the equilibrium constant for Eq. 16-24 indicates that the addition of $Ca^{++}$ and proper pH adjustment could render the residual concentration of soluble phosphorus insignificant. Because $[Ca^{++}] \gg [P]$ in most waters, any base producing a pH greater than 10 will exceed the solubility product of hydroxylapatite. For economic reasons, lime is generally the chemical of choice, even though the efficiency of phosphate removal by calcium at a given pH falls below the prediction based on solubility equilibria. Reasons for this discrepancy are that (1) precipitation of $Ca_{10}(PO_4)_6(OH)_2(s)$ is slow, and solubility equilibrium is not reached within economically justifiable detention times; (2) particles of precipitated hydroxylapatite may behave as colloids and not settle out readily; (3) certain polymeric phosphates are capable of forming soluble polyphosphato-calcium complexes; and (4) within certain pH ranges, some of the $Ca^{++}$ in the water is used up in $CaCO_3$ precipitation.

## 16-11 Corrosion

The corrosion of metals is a complex chemical or electrochemical phenomenon. In water and wastewater systems, corrosion results in (1) reduced hydraulic carrying capacity of pipes, fittings, and other systems components (Sec. 6-8); (2) possible structural failures; and (3) debased quality of the water transported.

Loss in carrying capacity by corrosion and tuberculation[16] is economically as well as operationally important. A comprehensive survey of water systems before the days of successful corrective treatment of aggressive waters showed, for example, that tar-coated, cast-iron pipe carrying the relatively soft waters of New England lost about half its capacity in 30 years of service. Then, pumped supplies might see their power requirements greatly increased and gravity systems might have to meet failing capacities by adding new conduits. Capacity loss through other than direct corrosion is produced by scale formation, biological growths, and silt deposition. Other indirect effects of corrosion are the adsorption of organic substances on corrosion products and encouragement of microscopic growths, the creation of chlorine demands—often high enough to exhaust chlorine residuals, and the production of odors, tastes, and other objectionable changes in water quality. In systems that include iron piping, *red water* is produced. However, the appearance of red water is not necessarily a measure of the degree of corrosion. Depending on pH, electrolyte content, oxygen concentration, and velocity of flow, coatings and tubercles may incorporate different proportions of the corrosion products. In some circumstances corrosion, although extensive, may be *hidden*, because virtually all the corroded iron may be deposited on the interior of the pipe. In other circumstances corrosion, although low, may be *visible*, because the corroded iron remains in suspension within the flowing waters.

A dry-cell battery is a reasonable analog model of the corrosion of metals in water. The cell is composed of two electrodes (C and Zn in the common

[16] The formation of localized corrosion products scattered over the surface in the form of knoblike mounds.

flashlight battery) separated from each other by an electrolyte, such as a solution of $NH_4Cl$. In the cell there is chemical reduction at the cathode (the zinc electrode) and chemical oxidation at the anode, the carbon electrode. Electrical energy is generated by chemical reactions at each electrode. The more intense the reactions, the greater is the flow of current when the circuit is closed. Most metals contain minute amounts of impurities, and their surfaces are neither chemically nor physically homogeneous. Instead, they are composed, in essence, of arrays of microcells with anodic and cathodic microareas. Because of this, exposure of a metal to an aqueous solution permits chemical reductions and oxidations to take place. By convention, the electricity flowing through the metal, that is, the external circuit, is called a positive current even though the electrons moving through the metal carry a negative charge.

*The Corrosion Reaction.* The overall corrosion reaction can be formulated in single chemical equations. Table 16–4 lists some of them, together with pertinent values of electromotive force, emf. The larger the potential, the greater, in general, is the tendency of the reaction to proceed. For instance, iron is seen to be less corrodible than magnesium, but more so than copper. That the standard emf for gold is negative means that gold does not corrode in water.

By definition, the standard emf is the potential in volts of an ideal electrochemical cell in which the reaction under consideration takes place at unit activity of all reacting substances. Accordingly, the emf is the algebraic

*Table 16-4*

*Electromotive Force of Corrosion Reactions*[a]

| Reaction | $E^0$ (25°C), volts |
|---|---|
| $Mg(s) + 2H^+ = Mg^{++} + H_2(g)$ | +2.37 |
| $Zn(s) + 2H^+ = Zn^{++} + H_2(g)$ | +0.76 |
| $Zn(s) + H_2O = ZnO(s) + H_2(g)$ | +0.42 |
| $Zn(s) + \frac{1}{2}O_2(g) = ZnO(s)$ | +1.65 |
| $Fe(s) + 2H^+ = Fe^{++} + H_2(g)$ | +0.41 |
| $Fe(s) + \frac{1}{2}O_2(g) + 2H^+ = Fe^{++} + H_2O$ | +1.77 |
| $Fe(s) + H_2O + \frac{1}{2}O_2(g) = Fe(OH)_2(s)$ | +1.27 |
| $Cu(s) + 2H^+ = Cu^{++} + H_2(g)$ | −0.34 |
| $Cu(s) + \frac{1}{2}O_2(g) = CuO(s)$ | +0.66 |
| $Cu(s) + \frac{1}{2}O_2(g) + H_2O = Cu(OH)_2(s)$ | +0.62 |
| $Au(s) + \frac{3}{2}H_2O + \frac{3}{4}O_2(g) = Au(OH)_3(s)$ | −0.23 |

[a] Here (*s*) stands for solid and (*g*) for gas.

difference between the electrode potentials of any two half-reactions so combined as to cancel the electrons and yield the overall (cell) reaction.

When the emf is positive, the reaction tends to proceed in the direction shown in Table 16-4. However, reaction tendency is not reaction rate. Large positive values of the standard emf may not signify rapid corrosion, but negative values do imply that the reaction will *not* proceed under the conditions described. In a thermodynamic sense the metals commonly used in water and wastewater systems are bound to corrode in aqueous environments, that is, to return, in essence, to the condition in which they were won from the soil.

The composition of natural waters and the conditions of exposure of metals are so varied that present knowledge of the vagaries of the corrosion process cannot be fully explained. This is so especially in a quantitative sense. In these circumstances the following generalizations are meaningful only in a qualitative sense.

Depending on conditions of exposure, both oxidation and reduction reactions can be involved. Reactions associated with the existence of microcells are illustrated in Fig. 16-11. The *feedback* of electrons is seen to maintain the corrosion process. The corrosion reactions illustrated in Fig. 16-11 have been written to show passage of 2 electrons through the metal in each instance.

Contact between dissimilar metals, or the existence of areas of dissimilar oxidation potentials in the same metal, normally promotes corrosion. A galvanic cell is formed, and the rate of corrosion is stepped up. The anodic metal or area possessing the highest oxidation potential corrodes; the cathodic metal or area does not corrode. The relative behavior of different metals may be gaged from the electromotive force series of Table 16-4 or, better, from a

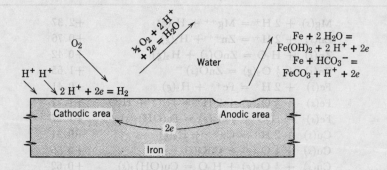

**Figure 16-11.**

**Corrosion cell on the surface of iron in water.**

## Table 16-5
*Galvanic Series of Metals and Alloys*

| | |
|---|---|
| Corroded (anodic or least noble) | Magnesium |
| | Zinc |
| | Aluminum (commercially pure) |
| | Steel or iron |
| | Cast iron |
| | Lead |
| | Tin |
| | Brasses |
| | Copper |
| | Bronzes |
| | Chromium-iron (passive) |
| | Silver |
| | Graphite |
| Protected (cathodic or most noble) | Gold |
| | Platinum |

galvanic series that takes into account the environmental conditions normally encountered in water. Table 16-5 is such a table.[17]

To give an example, zinc is anodic to both copper and iron, and iron, in turn, is anodic to copper. Among other things, this accounts for (1) the dezincification of yellow brass, an alloy of copper and zinc, (2) the reduced corrosion of iron in galvanized-iron pipes for as long as the zinc coating lasts, and (3) red or rusty water issuing from a bronze faucet attached to an iron pipe.

Corrosion of cast-iron pipes produces tubercles of rust and other precipitates above pits in the metal. The pipe surface becomes rough and flow is markedly impeded. Localized corrosion or pitting is common, too, in steel pipes. Moreover, they are so relatively thin that pits soon break through the pipe shell. Copper pipes, too, fail by pitting. Sand, dirt, mill scale, residual lubricants, suspended solids, and iron and manganese, as well as non-homogeneities at interfaces—scratches on the metal and differences in velocities of flow, for instance—establish concentration cells and promote pitting. Clean pipes and clean water are the answer. Copper corrodes fast in waters of low pH and high concentrations of sulfides, chlorides, and nitrates.

The rate of reduction of oxygen at the cathodic area, which depends on the rate at which oxygen has access to the area, governs the transfer of elec-

[17] F. N. Speller, *Corrosion*, McGraw-Hill, New York, 1951, p. 20.

trons from the metal to the solution. The area to which oxygen has easiest access tends to become the cathodic area; the area to which oxygen has access with difficulty becomes the anodic area. The resulting corrosion cell is called a differential-aeration cell or, in a more inclusive sense, a differential-concentration cell. In water supply and wastewater disposal such cells may take many forms. Examples of anodic areas, or areas sheltered against oxygen, are pits or depressions in the metal, areas underlying mill scale or products of corrosion, and areas below biological growths. In water pipes the rate of corrosion slows up in time, owing to the accumulation of rust or similar oxidation products. However, corrosion quickens at high velocities of flow, owing to the more rapid removal of corrosion-retarding substances and the replenishment of corrosion-promoting substances. On the other hand, high flow rates also carry more corrosion-retarding substances to surfaces. If these substances outweigh the corrosion-promoting substances that are present, rates of corrosion may indeed fall off while rates of flow are rising.

The presence of electrolytes other than hardness and alkalinity promotes corrosion because their peptizing action destroys passivating or otherwise protective oxide films. Chloride and sulfate are examples.

Corrosion is more rapid in acid than in neutral or alkaline solutions, but its ultimate extent is often larger in alkaline than in neutral solutions. In the range of pH values generally encountered in water treatment, faster corrosion in acid solutions is related more closely to the stability of protective films on the metal than to the actual rate of corrosion. Although $CO_2$ or $H_2CO_3$ does not participate in the electrochemical reaction, the concentration of free $CO_2$ is an important function of pH at given alkalinities.

Direct-current electricity corrodes the metal of the pole serving as the anode. Hence underground pipes are corroded by stray electrical currents at points where positive electricity leaves the pipe.

In water-distribution systems the deterioration of water quality by corrosion is most noticeable in dead ends. It is there that products of corrosion accumulate and there, too, that oxygen may be absent and sulfides present.

## 16-12 Control of Corrosion

Selection of corrosion-resistant materials and methods of corrosion control should be directed toward interrupting or otherwise modifying the cycle of corrosion. Commonly employed materials and methods include the following: (1) corrosion-resistant metals or alloys that either possess potentials in the emf range of noble metals or lay down protective coatings of dense oxides as they corrode, stainless steel, Monel Metal, tin, and copper being examples; (2) coatings and linings that bar both anodic and cathodic reactions by preventing escape of cations and denying access of water to the underlying metal,

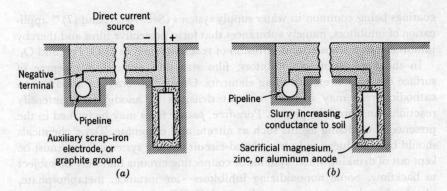

**Figure 16-12.**
Pipeline protected (a) by auxiliary anode and impressed voltage, (b) by sacrificial anode forming galvanic cell.

either metallic (for example, zinc, tin, and chromium) or nonmetallic coatings (for example, paints, plastics, and bituminous materials) being suitable; (3) deaeration for the removal of oxygen, accomplished by the direct application of a vacuum, by heating and degasification, or by passage of hot water over large surfaces of steel followed by removal of corrosion products by filtration (called *deactivation*); (4) cathodic protection, which, as shown in Fig. 16-12, can be provided by (a) using direct-current electricity to feed electrons into a metal and render it cathodic or (b) introducing into the system a metal higher in the electromotive series to become the anode and be corroded or sacrificed, electrical bleeding of distribution mains and other underground utilities and cathodic protection of steel water-storage tanks being examples of the first method, and the attaching of plates of zinc to the hulls or other underwater metal parts of ships, gates, and locks and the insertion of magnesium plugs into hot-water heaters being examples of the second;[18] (5) insulation by the creation of resistance to the flow of electrical currents as exemplified by (a) the insertion of insulating couplings or connectors between dissimilar metals to prevent generation of galvanic currents and (b) the use of insulating joints in water mains to oppose the flow of stray electrical currents;[19] (6) deposition of protective coatings, calcium carbonate

[18] Soil resistivity and changes in potential are formulated by H. H. Uhlig, *Corrosion and Corrosion Control*, John Wiley & Sons, New York, 1963, p. 352.

[19] Under some conditions the deposition of corrosion products will nullify the effect of insulating couplings. Where it is possible to avoid doing so, metallic pipes should not be laid in natural soils or fills through which current can skip around insulation couplings and joints.

coatings being common in water supply systems (Sec. 16-7); and (7)[20] application of inhibitors, namely substances that form protective films and thereby inhibit electron transfer and diffusion of reactants such as $H_2O$, $H^+$, and $O_2$.

In the application of inhibitors, film structure and relative degree of surface coverage are controlling elements. Greater coverage of anodic than cathodic areas may increase current densities at anodic sites, intensify reactions, and produce pitting. Pore-free, *passive* films may be formed in the presence of oxidizing agents such as nitrate and chromate. Toxic chemicals should be introduced only into closed-circuit cooling systems. They must be kept out of drinking-water systems or connecting circuits that may be subject to backflow. Some nonoxidizing inhibitors—for instance, metaphosphate, silicate, bicarbonate, organic color, and $CaCO_3$—permit dissolved oxygen to induce passivity.

Although calcium carbonate can be a good inhibitor, it performs well only if it is laid down properly in high concentrations of $Ca^{++}$ and $HCO_3^-$. Protection is poorer in the presence of $Cl^-$ and $SO_4^=$. It is better when the water is not stagnant. The Langelier saturation index (Sec. 16-7) does not predict how much $CaCO_3$ will be deposited. This is so because the pH of the solution in immediate contact with the metal is not the same as the pH at large, and because $CaCO_3$ deposition is affected by the concentration of $CaCO_3$, by electrochemical changes at the surface and rates of corrosion, and by the buffer capacity of the water and its rate of flow. Some inhibitors, among them polyphosphates and silicates, will reduce corrosion in some circumstances and promote it in others. Sequestering and reduction of detectable iron—by polyphosphates, for instance—only masks corrosion. In recirculated hot-water systems, moreover, polyphospahtes hydrolyze to nonprotective orthophosphates.

A summary of available methods of corrosion control leads to the conclusion that chemicals should be employed only on the basis of careful experimentation. Laboratory tests must be followed up in the field. Acknowledged, too, must be the difficulty of determining how much corrosion has taken place. Flow tests and water-quality determinations are the most instructive overall measures of the effectiveness of corrosion control in pipelines and distribution systems.

## 16-13 Handling, Storing, and Feeding Chemicals

The properties of chemicals commonly used in the treatment of water and wastewater are listed in the Appendix. Information such as this is basic to

---

[20] Deposition of dense, adherent, but slightly permeable silicate films is another possibility. Recommended is an initial dosage of 12 to 16 mg per l as $SiO_2$ for about a month, to be followed by the maintenance of a residual of 1 mg per l in remote parts of the distribution system.

Handling, Storing, and Feeding Chemicals/497

the design of handling, storage, and feeding facilities. Size and location of plant as well as available sources of supply and shipping facilities enter into decisions. In small plants, chemicals may be handled satisfactorily by simple hoisting equipment and two-wheeled trucks to be (1) stored on open floors in their shipping containers and (2) moved to feeding devices. Receiving and feeding weights may be determined by simple beam scales with platforms placed at floor level for convenience. Large plants generally require mechanical or pneumatic material-handling equipment to unload dry bulk chemicals from freight cars or automotive trucks and to transport them to storage bins, similar to grain-storage bins, whence they can flow by gravity to weighing and feeding machines (Fig. 16-13). In such plants liquids are pumped to storage and to feeding devices. Liquefied gases such as chlorine and ammonia are kept in factory-filled containers from which they flow under their container pressure.

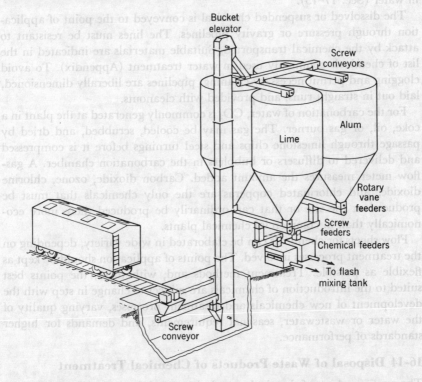

**Figure 16-13.**
**Unloading, storing, and feeding chemicals at treatment works. (Courtesy of Link-Belt Co.)**

498 / Coagulation, Precipitation, and Stabilization

With rare exceptions, the chemicals are dissolved or suspended in water before they are introduced into feed lines. Feeding devices regulate the amounts of chemicals to be added to the water or wastewater. Dry-feed machines control the dosage by the rate of volumetric or gravimetric displacement of dry chemicals. Volumetric and gravimetric machines generally plow, push, or shake the chemical off a receiving table, onto which the chemical flows from a hopper-shaped supply bin, as in a chicken-feed device (Fig. 16-14). Solution feed depends on the regulated displacement of liquid or dry chemicals that have been dissolved or suspended in water to produce solutions of known strength or slurries of known concentration. Measurement is by constant-head orifices, swinging pipes, or regulated pumps. Pumped flow may be proportioned automatically to the rate of flow of the water or wastewater to be treated. Gas feed is generally through pressure regulators and controlled orifices as a gas or through solution-feed devices after solution in water (Sec. 17–15).

The dissolved or suspended chemical is conveyed to the point of application through pressure or gravity pipelines. The lines must be resistant to attack by the chemical transported. Suitable materials are indicated in the list of chemicals commonly used in water treatment (Appendix). To avoid clogging and permit necessary cleaning, pipelines are liberally dimensioned, laid out in straight runs, and provided with cleanouts.

For the carbonation of water, $CO_2$ is commonly generated at the plant in a coke, oil, or gas burner. The gas may be cooled, scrubbed, and dried by passage through limestone chips and steel turnings before it is compressed and delivered to diffusers or bubblers in the carbonation chamber. A gas-flow meter measures the amount added. Carbon dioxide, ozone, chlorine dioxide, and chlorinated copperas are the only chemicals that must be produced at the site or that can ordinarily be produced there more economically than in commercial chemical plants.

Flow sheets for chemicals can be elaborated in wide variety, depending on the treatment processes involved. The points of application should be kept as flexible as possible. Treatment methods and, with them, the points best suited to the introduction of chemicals are subject to change in step with the development of new chemicals and treatment processes, varying quality of the water or wastewater, seasonal requirements, and demands for higher standards of performance.

## 16-14 Disposal of Waste Products of Chemical Treatment

The waste products of chemical coagulation, precipitation, and ion exchange include slurries, sludges, and liquids. Liquids are normally discharged into wastewater collection systems where available, but require treatment before discharge into receiving waters. In arid areas, the wastewaters are reclaimed for return to process after sedimentation. Treatment and disposal of waste solids are considered in Chapter 20.

## Disposal of Waste Products of Chemical Treatment/499

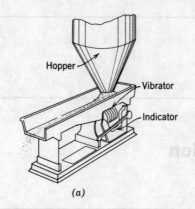

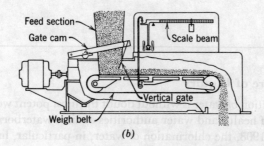

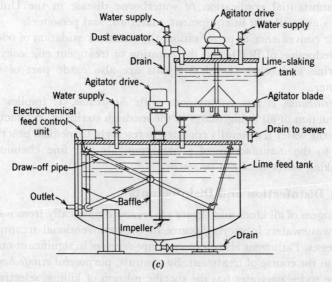

**Figure 16-14.**
**Chemical feeders. (a) Dry feeder with vibrator; (b) gravimetric-type dry feeder (Wallace and Tiernan, Inc.); (c) wet feeder with auxiliary lime-slaking tank (Courtesy of National Lime Association).**

**seventeen**

# disinfection

## 17-1 Nature of Inquiry

The disinfection of water is without doubt the most potent weapon in the hands of health and water authorities against waterborne infection. Since 1908, the chlorination of water, in particular, has led to the substantial eradication of waterborne disease in the United States at small cost in equipment, materials, and personnel.

For convenience of presentation, the chemical oxidation of odors, the reduction of BOD, and the boosting of treatment efficiency by chlorine and other chemical agents are also made part of this chapter.

*Sterilization* is not synonymous with *disinfection*. It implies the destruction of all living things in the medium sterilized. Production of sterile water is generally confined to research, to medical practice, and to the manufacture of pharmaceuticals and fine chemicals. Drinking water need not be sterile.

## 17-2 Disinfection and Disinfectants

Pathogens of all kinds and classes are removed physically from water and wastewaters in varying degree by most conventional treatment processes. Pathogens also *die away* or are *destroyed* in significant numbers in the course of treatment. By contrast, purposeful *disinfection* of water and wastewater has the specific mission of killing, selectively if necessary, those living organisms that can spread or transmit infection through or in water (Sec. 10–4). Within the context of this book, this primary concern is twofold: (1) to prevent direct transmission of disease to man through water and (2) to break the chain of disease and infection by destroying responsible infective agents before they reach the water environment. Disinfection of wastewaters is bound to receive increased attention as populations grow and recreation as well as urbanization is intensified. The safety of swimming-pool waters and of market shellfish and farm produce pose questions of their own (Secs. 10–10 to 12).

As shown in Sec. 10–4, three categories of human enteric pathogens are normally of consequence: bacteria, viruses, and amebic cysts. (In addition, helminths are of concern in wastewater and their residues.) Purposeful disinfection must be capable of destroying all three. Fortunately there are disinfectants that can do this.

To be of practical service, such water disinfectants must possess the following properties: (1) they must destroy the kinds and numbers of pathogens that may be introduced into municipal water or wastewater and do so within a practicable period of time, and over an expected range in water temperature, while meeting possible fluctuations in composition, concentration, and condition of the waters or wastewaters to be treated; (2) they must be neither toxic to man and his domestic animals nor unpalatable or otherwise objectionable in required concentrations; (3) they must be dispensable at reasonable cost, and safe and easy to store, transport, handle, and apply; (4) their strength or concentration in the treated water must be easily, quickly, and preferably automatically determinable; and (5) they must either persist within disinfected water in a sufficient concentration to provide reasonable *residual* protection against its possible recontamination before use; or, because this is not a normally attainable property, the disappearance of residuals must be a warning that recontamination may have taken place. Independent sentinels of recontamination have not yet been added to water.

As a rule the *concentration of disinfectants*, depending on their nature, is determined by physical measurements or chemical analyses, whereas their *disinfecting efficiency*—in all but experimental studies—is assessed by the reduction of indicator organisms (usually coliform organisms) to numbers implying statistically acceptable safety against possible infection. Correlative laboratory studies of the destruction of pathogens are performed either with laboratory cultures or with organisms harvested from carriers or cases of disease. To be equally significant with harvested organisms, cultured organisms must be known or shown to be equally resistant.

Water can be disinfected in a number of categorical ways. A list of the most practical ones includes heat and light, as well as chemicals.

## 17-3 Disinfection by Heat

Raising water to its boiling point will disinfect it. This is resorted to as an emergency measure in the form of *boil-water orders* by health and water authorities.

## 17-4 Disinfection by Light

Sunlight is a natural disinfectant, principally as a desiccant. Irradiation by ultraviolet light intensifies disinfection and makes it a manageable undertaking. The most common source of ultraviolet light is a mercury-vapor lamp constructed of quartz or special glass likewise transparent to the intense and destructive, invisible light of 2537 angstrom units ($10^{-8}$ cm) emitted by the mercury-vapor arc. To ensure disinfection, the water must be free from light-absorbing substances—phenolic and other aromatic compounds, including LAS, for instance—and from suspended matter that shades the organisms against the light; the time-intensity product of exposure must be adequate, and the water must be well mixed while it is being exposed in relatively thin films in order to counter its own absorptivity. Other forms of radiant and sonic energy are destructive to microorganisms, but they have yet to find engineering application in water disinfection. Neither light nor sound will produce or leave an identifiable residual disinfectant or monitoring substance.

Radiant energy occurs in discrete units or quanta, $E = hc\lambda$, where $E$ is the energy of a single quantum in ergs, $h$ is Planck's[1] constant ($6.62 \times 10^{-27}$ erg-sec), $c$ is the velocity of light ($3 \times 10^{10}$ cm per sec), and $\lambda$ is the wavelength of radiation in centimeters. By definition, a germicidal unit is an intensity of 100 mw per sq cm for radiations of wavelength 2537 A. The relative effectiveness of other radiations may be identified accordingly. The rate of attenuation of radiant energy is proportional to the radiation intensity, or $di/dl = -k_e i$, whence

$$p_e = (i_0 - i)/i_0 = 1 - \exp(-k_e l) \qquad (17\text{-}1)$$

Here $i$ is the radiation intensity, $l$ the length of the beam path, $i_0$ the initial intensity, $k_e$ the *coefficient of attenuation* or *extinction* (dimension $l^{-1}$), and $p_e$ the proportion of the energy absorbed.

The coefficient of attenuation of ultraviolet light of 2537 A varies from 0.03 cm$^{-1}$ for filtered waters to 0.2 cm$^{-1}$ for unfiltered waters supplied to municipalities in North America. Exposures of *Escherichia coli* to 3000, 1500, and 750 mw-sec per cm$^2$ are reported to produce 99.99, 99, and 90% kills.

---

[1] Max Karl Ernst Ludwig Planck (1858–1947), German physicist, whose name is attached to the fundamental law of quantum theory and so to the beginning of modern physics.

About 2% of the incident radiation of 2537 A from a source of ultraviolet light is reflected at the water surface. The reflectivity of aluminum is 90%.[2]

## 17-5 Disinfection by Chemicals

Exposing water long enough to adequate concentrations of chemicals of the following kinds will disinfect it.

*Oxidizing Chemicals.* These comprise (1) the halogens—chlorine, bromine, and iodine—released in suitable form from acceptable sources; (2) ozone; and (3) other oxidants such as potassium permanganate and hydrogen peroxide, but these are not as effective as the halogens and ozone.

Among the halogens, gaseous chlorine and a number of chlorine compounds are economically most useful. Bromine ($Br_2$) has been employed on a limited scale for the disinfection of swimming-pool waters; iodine has been used for the disinfection of swimming pools and small quantities of drinking water in the field. Tablets of tetraglycine hydroperiodide have been developed for field use. Ozone is a good but relatively expensive disinfectant that normally leaves no measurable monitoring residual. However, the ability of ozone to destroy a number of objectionable odors and to bleach color effectively is in its favor. Potassium permanganate, too, is a relatively expensive disinfectant.

It is an axiom of disinfection by chemicals that the oxidizing capacity of a compound is not necessarily a measure of its disinfecting efficiency. Thus hydrogen peroxide is a strong oxidant but a poor disinfectant.

*Metal Ions.* Silver ions are neither viricidal nor cysticidal in acceptable concentrations, but they are bactericidal. Disinfection at the low concentrations employed—as low as 15 µg per l (micrograms per liter, or closely, parts per billion)—is slow. This is a weakness of silver. Moreover, silver is costly at practicable concentrations. Copper ions are strongly algicidal but only weakly bactericidal.

*Alkalis and Acids.* Pathogenic bacteria do not last long in highly alkaline or highly acid waters, for example, at very high ($>11$) or very low ($<3$) pH values. The destruction of bacteria by caustic lime incidental to lime softening is an example (Sec. 16-8).

*Surface-Active Chemicals.* Among surfactants, the cationic detergents are strongly destructive, the anionic detergents only weakly so. The neutral detergents occupy an intermediate position. Detergents have been applied selectively as disinfectants only in the wash waters and rinse waters of eating establishments.

In summary, it can be said that for the routine disinfection of municipal and industrial waters, only chlorine is both efficient and reasonably cheap;

[2] Lewis R. Koller, *Ultraviolet Radiation*, John Wiley & Sons, New York, 1952.

ozone is efficient but relatively expensive and not persistent enough for monitoring purposes; and heat and ultraviolet light are relatively still more expensive and without monitoring properties. What makes heat especially useful is that it can usually be made available at a moment's notice in times of emergency. Chlorine and hypochlorite compounds are the disinfectants of choice in United States practice.

Although the disinfection of water and wastewater is aimed almost wholly at the destruction of single-celled organisms, the killing of the cercariae of the schistosomes and of the adult guinea worms infesting *Cyclops* are of much importance in some parts of the world (Sec. 10-4). Even if these large organisms are included, however, the protoplasmic mass involved is relatively so small that chemical disinfectants can perform successfully in minute concentrations in short periods of time (fractions of a milligram per liter—of free chlorine in otherwise clean water, for example—and no more than 15 min of contact). Important, too, is the fact that disinfecting chemicals such as chlorine are not toxic to man in the concentrations employed. Heavily chlorinated water can be ingested with impunity not only by man and the higher animals, but also by less highly organized living things, such as guppies and other aquatic organisms in balanced aquaria.

## 17-6 Theory of Chemical Disinfection

The disinfecting species of chlorine react with enzymes that are essential to the metabolic processes of bacterial cells.[3] Cells die when these key substances are inactivated. Enzyme destruction also remains the primary lethal mechanism of disinfectants when a radical process, such as heat, coagulates cell contents. Because enzymes are generated within the cell plasm, chemical disinfection of bacteria proceeds theoretically in two steps: (1) penetration of the cell wall and (2) reaction with the cell enzymes.

Factors governing the chemical disinfection technology of water fall essentially into the following three categories:

1. *The nature of the organisms to be destroyed and their concentration, distribution, and condition in the water to be disinfected.*

Although nonsporeforming bacteria are less resistant to disinfection than sporeforming bacteria, they are normally unimportant in water disinfection. Among the enteric bacteria, *Escherichia coli* appears to be somewhat more resistant than the pathogenic bacteria. This makes it a useful test organism. The cysts of *Entamoeba histolytica* are quite resistant. A number of enteric viruses are also measurably more resistant to chlorination than *Esch. coli.*

[3] W. E. Knox, P. K. Stumpf, D. E. Green, and V. H. Auerbach, The Inhibition of Sulfhydryl Enzymes as the Basis of the Bactericidal Action of Chlorine, *J. Bact.*, **55**, 451 1948.

Poliomyelitis virus Type 1 and Coxsackie A2 virus are examples. The virus of infectious hepatitis appears to be an especially hardy organism. However, this conclusion is based on inadequate evidence in reference to chemical disinfection. Adenovirus Type 3 has been found less resistant than *Esch. coli*. The concentration of organisms normally becomes important only when it is so high that oxidation of constituent cell matter competes for the disinfectant. When bacteria, such as the staphylococci, and viruses form clumps of cells, protected inner cells may remain untouched and viable. To be reached by disinfectants of average strength or intensity, the organisms must be distributed uniformly through the water and be shifted in their position. Stirring will do this.

2. *The nature, distribution, and concentration of the disinfecting substance and its reaction products in the water to be disinfected.*

Chlorine and hypochlorite compounds can form in water a number of chlorine species of quite different disinfecting efficiency. Chloride with zero disinfecting power is formed when chlorine reacts with reducing agents; chloramine is formed in reactions with nitrogen; and hypochlorite, free chlorine formed above pH 7, is only one eightieth as efficient as hypochlorous acid. To be of average strength or intensity, disinfectants must be uniformly distributed through the water. This, too, may require stirring the water.

3. *The nature and condition of the water to be disinfected.*

Suspended matter may shelter embedded organisms against chemical disinfectants as well as against destructive light rays. Organic matter uses up oxidizing chemicals. Other substances react with chemical disinfectants and change their structure. Some of the resulting compounds may be inefficient and some even innocuous. In water chlorination, pH is of controlling importance. The higher the temperature of the water to be disinfected, the more rapid is the kill; and the longer the time, the greater is the opportunity for destruction.

## 17-7 Kinetics of Chemical Disinfection

Under ideal conditions all cells of a single species of organism are discrete units equally susceptible to a single species of disinfectant; both cells and disinfectant are uniformly dispersed in the water; the disinfectant stays substantially unchanged in chemical composition and substantially constant in concentration throughout the period of contact; and the water contains no interfering substances. Under such conditions the rate of disinfection is a function of the time of contact, the concentration of the disinfectant, and the temperature of the water.[4]

[4] But see a later part of this section for possible effects of the number of organisms.

**Time of Contact.** When under ideal conditions an exposed cell contains a single active center vulnerable to a single unit of disinfectant, the time-rate of kill follows Chick's law of disinfection[5] (Sec. 11–12). This states that $y$, the number of organisms destroyed in unit time, is proportional to $N$, the number of organisms remaining, the initial number being $N_0$, or

$$dy/dt = k(N_0 - y) \qquad (17\text{-}2)$$

where $k$ is the coefficient of proportionality or the rate constant with dimension $[t^{-1}]$. By integration between the limits $y = 0$ at $t = 0$ and $y = y$ at $t = t$,

$$\ln[(N_0 - y)/N_0] = \ln(N/N_0) = -kt \quad \text{or} \quad N/N_0 = \exp(-kt) \qquad (17\text{-}3)$$

Therefore a plot of log $N/N_0$ against $t$ traces a straight line with a slope of $-k \log e = -k'$ and an intercept of 1 (or 100%) at $t = 0$. When $kt = 1$ or $k't = 0.4343$, the *surviving fraction* is 0.368.

Understandably, departures from Chick's law are not uncommon, even when test conditions are nearly ideal. Rate of kill, rather than being constant, may increase or decrease with time. Increase in rate of kill can be explained in at least two ways: (1) as a combination of slow diffusion of chemical disinfectants through the cell wall and a rate of kill accelerating with the accumulation of disinfectant within the cell, and (2) as the consequence of a time lag before the disinfectant can reach a lethal number of vital centers in the organism. Decrease in rate of kill is generally explained as a variation in cell resistance within the culture. However, declining concentrations of disinfectant, poor distribution of organisms and disinfectant, and other interfering factors may also account for it. The experience may be generalized by adding to $k$ or to $t$ coefficients that will linearize functional plots of $N/N_0$. A simple and often successful assumption is that $N/N_0$ varies logarithmically with time, that is, as $t^m$, where $m > 1$ when rates of kill rise in time and $m < 1$ when rates of kill fall in time (Fig. 17–1).

**Concentration of Disinfectant.** For changing concentrations of disinfectant, the observed disinfecting efficiency is generally approximated by the relationship

$$c^n t_p = \text{constant} \qquad (17\text{-}4)$$

Here $c$ is the concentration of the disinfectant, $t_p$ is the time required to effect a constant percentage kill of the organims, and $n$ is a coefficient of dilution or, according to van't Hoff, a measure of the order of the reaction. Examples of time-concentration relationships for chlorine as HOCl and 99% kill of *Esch. coli* and enteric viruses at 0 to 6°C are: $c^{0.86} t_p = 6.3$ for Coxsackie

---

[5] Harriet Chick, Investigation of the Laws of Disinfection, *J. Hyg.*, **8**, 92 (1908).

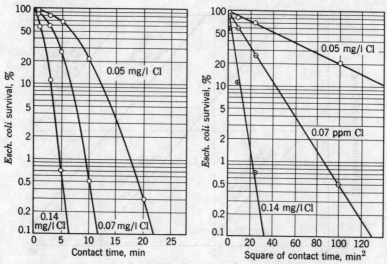

**Figure 17-1.**
**Length of survival of *Esch. coli* in pure water at pH 8.5 and 2 to 5°C.**

virus A2; $c^{0.86}t_p = 1.2$ for poliomyelitis virus 1; and $c^{0.86}t_p = 0.098$ for adenovirus 3—all in comparison with $c^{0.86}t_p = 0.24$ for *Esch. coli*.[6] When $n > 1$, the efficiency of the disinfectant decreases rapidly as it is diluted; when $n < 1$, time of contact is more important than dosage. When $n = 1$, concentration and time are of equal weight and a first-order reaction may be in progress.

Equation 17–4 is empirical. For a straight-line plot on double logarithmic paper of concentration versus contact time for fixed disinfection efficiency, the slope of the line is $(-1/n)$ (Fig. 17-2).

**Concentration of Organisms.** The concentration of organisms is seldom high enough to result in differential percentage kills. Where this is not so, at large differences in concentration of cell substance, for example,

$$c^q N_p = \text{constant} \qquad (17\text{--}5)$$

Here $c$ is the concentration of the disinfectant; $N_p$ the concentration of organisms that is reduced by a given percentage in a given time; and $q$ a coefficient of disinfectant strength. Like Eq. 17–4, this is an observational relationship only.

**Temperature of Disinfection.** If the rate of disinfection is determined by the rate of diffusion of the disinfectant through the cell wall or by the

---

[6] Gerald Berg, The Virus Hazard in Water Supplies, *J. New England Water Works Assoc.*, **78**, 79 (1964).

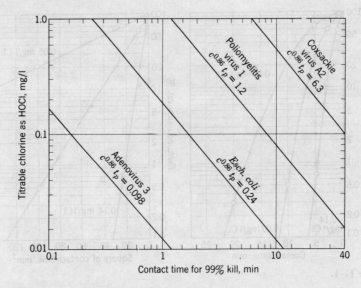

**Figure 17-2.**

**Time-concentration relationships in disinfection. Concentration of chlorine as HOCl required for 99% kill of *Esch. coli* and three enteric viruses at 0 to 6°C. (After Berg, Sec. 17-7, footnote 6.)**

rate of the reaction with an enzyme, temperature effects usually conform to the van't Hoff-Arrhenius relationship (Eq. 11-17). A convenient form is

$$\log \frac{t_1}{t_2} = \frac{E(T_2 - T_1)}{2.303 R T_1 T_2} = \frac{E(T_2 - T_1)}{4.56 T_1 T_2} \quad (17\text{-}6)$$

Here $T_2$ and $T_1$ are two absolute temperatures (normally degrees Kelvin) for which the rates are to be compared; $t_1$ and $t_2$ are the times required for equal percentages of kill at fixed concentrations of disinfectant; $E$ is the activation energy (normally in calories) and a constant characteristic of the reaction; and $R$ is the gas constant (1.99 cal per deg C, for example). For $T_2 - T_1 = 10$, the useful ratio $t_1/t_2$, called $Q_{10}$, is related to $E$ about as follows at normal water temperatures (Eq. 11-18):

$$\log Q_{10} = \log (t_1/t_2) = E/39{,}000 \quad (17\text{-}7)$$

The work of Butterfield and his associates[7] identifies the temperature dependence of disinfecting concentrations of aqueous chlorine and chloramines (Secs. 17-10 and 11) in destroying *Esch. coli* shown in Table 17-1.

[7] C. T. Butterfield, Elsie Wattie, Stephen Megregian, and C. W. Chambers, Influence of pH and Temperature on the Survival of Coliforms and Enteric Pathogens When Exposed to Free Chlorine, *U.S. Publ. Health Rept.*, **58**, 1837 (1943).

## Table 17-1
*Temperature Dependence of Disinfecting Concentrations of Aqueous Chlorine and Chloramines in the Destruction of* Esch. coli *in Clean Water*

| Type of Chlorine | pH | $E$, cal[a] | $Q_{10}$ |
|---|---|---|---|
| Aqueous chlorine | 7.0 | 8,200[b] | 1.65[b] |
|  | 8.5 | 6,400 | 1.42 |
|  | 9.8 | 12,000 | 2.13 |
|  | 10.7 | 15,000 | 2.50 |
| Chloramines | 7.0 | 12,000 | 2.08 |
|  | 8.5 | 14,000 | 2.28 |
|  | 9.5 | 20,000 | 3.35 |

[a] The higher the value of $E$, the slower is the reaction.

[b] The rate of reaction at pH 7.0 is relatively so fast that these values are probably unreliable. The magnitudes of $E$ throw some light on the nature of the disinfecting process for the chlorine species released in water (Sec. 11–15).

## 17-8 Disinfection by Ozone

Purposeful production of ozone—from the Greek word *ozein*, to smell—by the corona discharge of high-voltage electricity into dry air was introduced by the German electrical engineer Werner von Siemens (1816–92). Both in air and in water, ozone breaks down rapidly in the presence of oxidizable matter. It is corrosive and poisonous in strong concentrations in the atmosphere, and its photochemical genesis in conjunction with gasoline vapors from automobile exhausts is responsible for oxidant smogs which are eye, throat, and lung irritants. Contrary to popular belief and common usage of the word, ozone is a *toxic*, not a *tonic*, substance. The danger limit in treatment plant operation is commonly set at 0.2 mg of $O_3$ per m³ of air.

If ozone is to be employed effectively and efficiently as a deodorant, decolorant, and disinfectant of drinking water, its physical and chemical properties in water solution and their influence on pathogenic microorganisms need to be known over the full range of possible exposures. This is not yet the case. However, we can get a feeling for the usefulness of ozone and constraints on its use from information such as the following:[8]

The weight of ozone is about 2154 g per m³. At normal temperatures ozone residuals disappear rapidly from water. This is shown by the following

---

[8] Many of the underlying data are taken from D. C. Donovan, Treatment with Ozone, *J. Am. Water Works Assoc.*, 57, 1167 (1965).

observational relationship, which applies to residuals above 1 ppm as $Cl_2$:

$$p_t = 100 \exp(-0.275t) \qquad (17\text{-}8)$$

where $p_t$ is the percentage of residual ozone at time $t$. Incidentally, expressing ozone as chlorine is no longer recommended.

Only in the absence of organic matter does ozone follow the laws of ideal, that is, nonreacting, gases in water. The *distribution coefficient* of ozone between air and water, that is, the ratio of the equilibrium concentration of ozone in the liquid phase to that in the gas phase at like temperature and pressure, is then about 0.6 at 0°C and 0.2 at 20°C. Increasing either the total pressure of the system or the partial pressure of ozone in the air raises the concentration of ozone in water in direct proportion to these pressures. In the presence of oxidizable substances, their nature and concentration in water rather than the distribution coefficient govern the amount of entering ozone.

As a disinfectant ozone is said to possess an all-or-none property, implying that it produces essentially no disinfection below a critical concentration but substantially complete disinfection above that concentration. This property is illustrated in Fig. 17-3, in which the *decontamination factor* of ozone for *Esch. coli*, that is, the reciprocal of the surviving fraction, is seen to be substantially zero at a concentration of 0.42 ppm and $10^4$ at 0.53 ppm. For comparison, similar tests of the behavior of chlorine are seen to trace a more or less straight-line increase of the decontamination factor from 10 at a con-

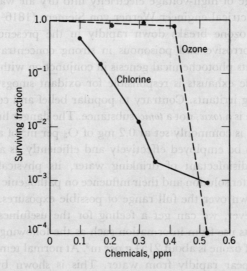

**Figure 17-3.**

**Comparative disinfection by ozone and chlorine.** [After R. S. Ingols and R. H. Fetner, *Proc. Soc. Water Treatment Exam.*, 6, 8 (1957).]

centration of 0.2 ppm of chlorine to $10^3$ at 0.48 ppm. It is also reported that coliforms and other bacteria are normally wiped out by the time an ozone residual can be detected. A first estimate of ozone requirements for the disinfection of clean water may be placed at 1 to 2 mg per l.

On the air side, modern ozonation plants (Fig. 17-4) comprise air cleaners, blowers, and refrigerative as well as adsorptive air driers. These condition and transport the air to be ozonized. On the electrical side, a transformer steps up the voltage to produce the corona in the air supply. On the water side, a contact chamber or ozone tower effects the transfer of ozone from the gas phase to the water phase. This is the key hydraulic and phase-transfer operation and, therefore, also the principal manageable component once the ozonizing unit has been selected. Because the absorption of ozone from the air into the water to be disinfected is a matter of contact opportunity, contact-chamber design aims at a maximization of (1) effective interface, (2) driving force or concentration differential, and (3) time of exposure with due consideration of advantages to be gained by countercurrent operation.

To be decided is whether droplets of water shall fall through a column of rising ozonized air, or bubbles of air shall be injected into a column of water flowing in the same or opposite direction; moreover, whether the air shall be injected into the entrant water or into the ozonizing tower itself; and, finally, whether mixing other than that inherent in the air-water streams shall be provided by mixing devices. Detention times are of the order of 10 min.

As a rule, capital and running costs of ozonation equipment cannot compete with those of comparable chlorination equipment for the treatment of a given water unless ozone is called upon and able to remove objectionable odors and tastes and reduce the color of the water more effectively than chlorine in combination with activated carbon and coagulants. Comparisons

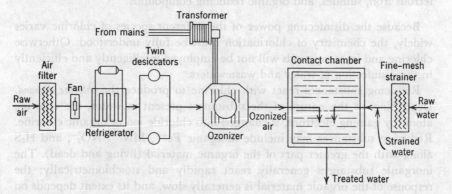

**Figure 17-4.**
**Flow diagram of the ozonizing of water. (After Donovan, Sec. 17-8, footnote 8.)**

of operating expenses derive from the cost of power versus the cost of chlorine and auxiliary chemicals in specific circumstances. The energy used in converting 0.5 to 1% of atmospheric oxygen into ozone is 0.025 to 0.030 kwhr per g of $O_3$. If water is to be deodorized and decolorized as well as disinfected, the required dosage may be nearer to 2 to 4 mg per l.

## 17-9 Disinfection by Chlorine

Chlorine was first used for day-in, day-out disinfection of a municipal water supply in America in 1908, when George A. Johnson and John L. Leal added chloride of lime to the water supply of Jersey City, N.J.[9]

The following substances are released when chlorine or its hypochlorite compounds are added to water:

1. Hypochlorous acid (HOCl), hypochlorite ion ($OCl^-$), and elemental chlorine ($Cl_2$). Distribution of the three species depends on pH. Elemental chlorine, from chlorine gas, lasts but a fleeting moment within the normal pH zone. The two prevailing species (HOCl and $OCl^-$) are referred to in practice as *free available chlorine*.

2. Monochloramine ($NH_2Cl$), dichloramine ($NHCl_2$), and nitrogen trichloride ($NCl_3$). Ammonia, or organic nitrogen, is essential to the production of these compounds. The distribution of the three species is again a function of pH. Nitrogen trichloride is not formed in significant amounts within the normal pH zone except when the *breakpoint* (Sec. 17–12) is approached. The two prevailing species, $NH_2Cl$ and $NHCl_2$, are referred to in practice as combined available chlorine.

3. Complex organic chloramines, especially in wastewaters.

4. Chloride, formed on reaction with reducing compounds, such as ferrous iron, sulfides, and organic reducing compounds.

Because the disinfecting power of the different species of chlorine varies widely, the chemistry of chlorination must be fully understood. Otherwise chlorine and its compounds will not be employed intelligently and efficiently in the disinfection of water and wastewaters.

Reducing substances react with chlorine to produce the *chlorine demand*. Depending on the nature of the substances present in water, the chlorine atom, by gaining electrons, is changed into chloride ion or organic chloride. Reducing substances may include inorganic $Fe^{++}$, $MN^{++}$, $NO_2^-$, and $H_2S$ along with the greater part of the organic material (living and dead). The inorganic substances generally react rapidly and stoichiometrically; the response of the organic material is generally slow, and its extent depends on

[9] J. L. Leal, The Sterilization Plant of the Jersey City Water Supply Company at Boonton, N.J., *Proc. Am. Water Works Assoc.*, p. 100 (1909).

how much available chlorine is present in excess of requirements. Because the organic material in drinking-water supplies is closely related to their natural color or stain, their probable organic chlorine demand may be estimated from the depth of color. In an analogous fashion, the organic chlorine demand of wastewaters bears some relation to their BOD or, more closely, to their COD.

These reactions are complicating factors in water chlorination. Enough chlorine must be added to take care of them so that sufficient free chlorine remains for the disinfecting reactions. To assure this, chlorine residuals remaining after a specific time of contact, rather than initial chlorine doses, are made standards of accomplishment or comparison. Ten minutes are specified in most testing. Because the chlorine demand is a function of temperature, concentration, and time, its determination must take all three factors into account.

## 17-10 Free Available Chlorine

The following equilibrium equations obtain when elemental chlorine is dissolved in pure water:

*Hydrolysis*:

$$Cl_2 + H_2O \rightleftharpoons HOCl + H^+ + Cl^- \qquad (17\text{-}9)$$

$$(HOCl)(H^+)(Cl^-)/(Cl_2) = K_h = 4.5 \times 10^{-4} (\text{mole}/l)^2 \text{ at } 25°C \qquad (17\text{-}10)$$

*Ionization*:

$$HOCl \rightleftharpoons H^+ + OCl^- \qquad (17\text{-}11)$$

$$(H^+)(OCl^-)/(HOCl) = K_i \quad \text{or} \quad (OCl^-)/(HOCl) = K_i/(H^+) \qquad (17\text{-}12)$$

Solutions of hypochlorites, such as chloride of lime and calcium hypochlorite, establish the same ionization equilibrium in water. Taking calcium hypochlorite as the example, the reactions leading up to equilibirum are

$$Ca(OCl)_2 \rightarrow Ca^{++} + 2OCl^- \qquad (17\text{-}13)$$

and

$$H^+ + OCl^- \rightleftharpoons HOCl$$

as in Eq. 17-11. The hydrolysis constant $K_h$, Eq. 17-10, is of such magnitude that no measurable concentration of $Cl_2$ remains in solution when the pH of the chlorinated water is more than about 3.0 and the total chloride concentration is less than about 1000 mg per l.

At ordinary water temperatures the hydrolysis of chlorine is essentially complete within a few seconds, and the ionization of hypochlorous acid produced is in essence an instantaneous, reversible reaction. The ionization constant $K_i$ varies in magnitude with temperature, as shown in Table 17-2. The percentage distribution of HOCl and OCl$^-$ at various pH values is shown in Fig. 7 of the Appendix. It is calculated from Eq. 17-12 and Table 17-2 as

$$\frac{(HOCl)}{(HOCl) + (OCl^-)} = \frac{1}{1 + (OCl^-)/(HOCl)} = \frac{1}{1 + K_i/(H^+)} \quad (17\text{-}14)$$

**Table 17-2**
*Values of the Ionization Constant of Hypochlorous Acid at Different Temperatures*

| Temperature, °C | 0 | 5 | 10 | 15 | 20 | 25 |
|---|---|---|---|---|---|---|
| $K_i \times 10^8$, moles/l | 1.5 | 1.7 | 2.0 | 2.2 | 2.5 | 2.7 |

At 20°C and pH 8, for instance, the percentage distribution of HOCl is $100 \times [1 + 2.5 \times 10^{-8}/10^{-8}]^{-1} = 100/3.5 = 29\%$. The chlorine species that constitute free available chlorine are identified in this way.

Observed colicidal efficiency (concentration of aqueous, or free available, chlorine required to kill 99% of *Esch. coli* in 30 min at 2 to 5°C) is illustrated in Fig. 17-5. That the two curves in Fig. 17-5 are mirror images suggests a higher killing efficiency for HOCl than for OCl$^-$ in the approximate ratio of

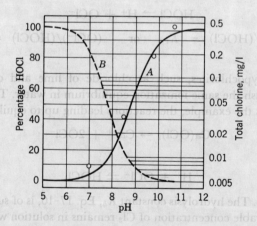

**Figure 17-5.**
Observed concentration of free available chlorine required for 99% kill of *Esch. coli* in 30 min at 2 to 5°C (curve A and right-hand scale), and percentage of HOCl in the total chlorine (curve B and left-hand scale). (See Fig. 7 in the Appendix.)

80:1 for the conditions of test. If the efficiencies of the two species of chlorine are additive, the total amount of chlorine, $R$, required to produce a given percentage of kill in a specified time at various pH values becomes $R = $ (HOCl) + (OCl$^-$) = (HOCl) [1 + (OCl$^-$)/(HOCl)]. By Eq. 17-12, therefore, $R$ = HOCl [1 + $K_i$/(H$^+$)]. If $c$ is the killing concentration of HOCl,

$$c = [\text{HOCl}] + r[\text{OCl}^-] = \frac{R}{1 + K_i/(\text{H}^+)} + r \frac{K_i}{(\text{H}^+)} \frac{R}{1 + K_i/(\text{H}^+)}$$

and

$$R = c \frac{1 + K_i/(\text{H}^+)}{1 + rK_i/(\text{H}^+)} \qquad (17\text{-}15)$$

Here brackets stand for the required concentration of chlorine and $r$ is the proportionate efficiency of OCl$^-$ ions relative to HOCl.

## Example 17-1.

From the data presented in Table 17-1 and Figs. 17-5 and 6a, draw a comparison between the disinfecting characteristics of (1) HOCl and (2) OCl$^-$ ion based on the results for pH values of 7.0 and 10.7.

1. At pH 7.0 about 80% of the disinfecting chlorine is present as HOCl; the coefficient of dilution, $n$, is 1.5, indicating that disinfectant concentration is more im-

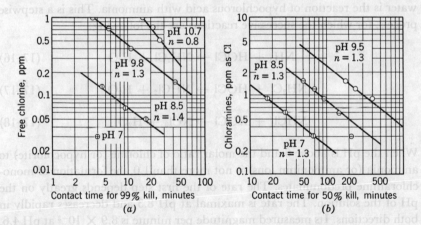

**Figure 17-6.**

Time-concentration relationships in disinfection. (a) Concentration of free available chlorine required for 99% kill of *Esch. coli* at 2 to 5°C; (b) concentration of combined available chlorine required for 50% kill of *Esch. coli* at 2 to 5°C.

portant than contact time; the energy of activation, $E = 8200$ cal, falls within the range of activation energies for diffusion processes; and the residual chlorine concentration required for a decontamination factor of 100 or 99% kill of *Esch. coli* in 10 min at 2 to 5°C is 0.017 mg/l, or $c^n t = 0.017^{1.5} \times 10 = 2.2 \times 10^{-2}$.

2. At pH 10.7 almost 100% of the disinfecting chlorine is present as $OCl^-$ ion; the coefficient of dilution, $n$, is 0.8, indicating that contact time is more important than disinfectant concentration; the energy of activation, $E = 15,000$ cal, falls within the range of activation energies for chemical reactions; and the residual chlorine concentration for 99% kill of *Esch. coli* in 10 min at 2 to 5°C is 1.7 mg/l, or $c^n t = 1.7^{0.8} \times 10 = 15.3$.

Therefore it appears that the rate-determining processes are different for HOCl and $OCl^-$ ions. The proportionate disinfecting efficiency, $r$, of $OCl^-$ ion, for 99% kill of *Esch. coli* in 30 min, is about $0.005/0.042 = 0.12$ or 12%. Substitution of $r = 0.12$, $c = 0.005$, and $K_i = 1.6 \times 10^{-8}$ in Eq. 17-15 gives the following numerical expression for total chlorine required for 99% kill of *Esch. coli* in 30 min at 2 to 5°C at pH 8.0: $R = 0.005[1 + (1.6 \times 10^{-8})/10^{-8}]/[1 + (1.9 \times 10^{-9})/10^{-8}] = 0.011$ mg/l. This is the basis of the curve shown in Fig. 17-5.

## 17-11 Combined Available Chlorine

The most important reaction of chlorine with compounds of nitrogen in water is the reaction of hypochlorous acid with ammonia. This is a stepwise process, for which the successive reactions are the following:

$$NH_3 + HOCl \rightarrow NH_2Cl + H_2O \qquad (17-16)$$

$$NH_2Cl + HOCl \rightarrow NHCl_2 + H_2O \qquad (17-17)$$

$$NHCl_2 + HOCl \rightarrow NCl_3 + H_2O \qquad (17-18)$$

When the pH is above 6 and the molar ratio of chlorine (or hypochlorite) to ammonia (or ammonium ions) is not more than 1.0, the formation of monochloramine predominates. The rate of the first step depends greatly on the pH of the solution. The rate is maximal at pH 8.3 and decreases rapidly in both directions. Its measured magnitude per minute is $8.9 \times 10^{-3}$ at pH 4.6, $5.8 \times 10^{-1}$ at pH 6.5, and $7.4 \times 10^{-3}$ at pH 12.1. At pH 8.3, 25°C, 0.8 mg per l of chlorine, and 0.32 mg per l of ammonia nitrogen, the reaction is 99% complete in about 1 min; at pH 5 the corresponding time is 210 min; at pH 11 it is 50 min. The reaction rate varies greatly also with temperature. Depending on pH, $Q_{10}$ lies between 2.0 and 2.5.

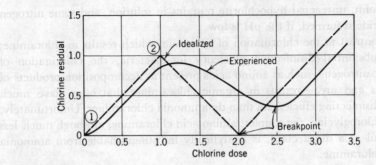

**Figure 17-7.**
**Schematic diagram of the breakpoint.**

Both Palin[10] and Morris[11] have shown that distribution is actually governed by the relative rates of formation of monochloramine and dichloramine. These change with the relative concentrations of chlorine and ammonia as well as with pH and temperature. Figure 8 of the Appendix traces the distribution at equimolar concentrations of chlorine and ammonia—therefore at a weight ratio of chlorine ($Cl_2$) to ammonia (N) of 5:1.

A comparison of Fig. 17-6a and b shows that combined available chlorine is a much less efficient colicidal agent than free available chlorine. The dilution coefficient $n = 1.3$ suggests that disinfectant concentration is somewhat more important than contact time, and the magnitudes of $E = 12,000$ to $20,000$ recorded in Table 17-1 lie within the range of chemical reactions.

## 17-12 Breakpoint Reactions of Ammonia

There is oxidation of ammonia and reduction of chlorine when the molar ratio of chlorine to ammonia is greater than 1.0. A substantially complete oxidation-reduction process occurs in the neighborhood of a 2:1 ratio and leads, in the course of time, to the disappearance from solution of all the ammonia and oxidizing chlorine. (This is the reason that the reaction in Eq. 17-18 and trichloramine is unimportant in disinfection.) This is called the *breakpoint* phenomenon. It is illustrated in Fig. 17-7. Between points 1 and 2, molar ratios of chlorine to ammonia are less than 1.0, and the residual oxidizing chlorine is essentially all monochloramine. Between point 2 and the breakpoint, oxidation of ammonia and reduction of chlorine increase until complete oxidation-reduction occurs at the breakpoint. In this region, again, the residual oxidizing chlorine is essentially all monochloramine. Beyond the

[10] A. T. Palin, The Estimation of Free Chlorine and Chloramine in Water, *J. Inst. Water Engrs.*, **3**, 100 (1949).

[11] J. C. Morris, Kinetics of Reactions Between Aqueous Chlorine and Nitrogen Compounds, in S. Faust and J. Hunter, Eds., *Principles and Applications of Water Chemistry*, John Wiley & Sons, New York, 1967.

breakpoint, unreacted hypochlorite remains in solution, and some nitrogen trichloride is formed, if the pH is low.

In contrast to the chlorination of ammonia, which results in chloramines with substantial disinfection efficiency for bacteria, the chlorination of organic nitrogen—such as amino acids present as decomposition products of proteins and urea—results in organic chloramines that often have much lower disinfecting efficiencies than do ammonia chloramines. Unfortunately, monochloroglycine, the simplest amino acid chloramine, although much less powerful as a disinfectant, is analytically indistinguishable from ammonia monochloramine.

The rate of the breakpoint reaction is strongly affected by pH. The maximum lies between pH 6.5 and 8.5. However, no clear-cut picture of the complex reactions involved can yet be presented. In practice, time requirements are determined by test; times of 30 min or more are common.

Important advantages of chlorinating to and beyond the breakpoint or otherwise obtaining free available chlorine residuals are that (1) most odors and tastes normal to water are destroyed and (2) rigorous disinfection is assured. However, in the presence of undecomposed urea, nitrogen trichloride is very likely to be formed and give rise to bad odors and tastes.

## 17-13 Dechlorination

When large amounts of chlorine have been added to water, for example to ensure disinfection before the water is to be consumed or to destroy odors and tastes, unwanted residuals can be removed by dechlorination. Intensive use of chlorine in this manner without the breakpoint reaction is called *superchlorination and dechlorination*. Some methods of dechlorination are the addition of reducing chemicals, passage through beds of granular activated carbon, and aeration. The reducing agents include sulfur dioxide, $SO_2$; sodium bisulfite, $NaHSO_3$; and sodium sulfite, $Na_2SO_3$. The bisulfite is ordinarily used in practice. It is cheaper and more stable than the sulfite. Samples of water collected for bacteriological analysis are usually dechlorinated by including sodium thiosulfate ($Na_2S_2O_3$) in the sampling bottles either in solution or as crystals. Granular activated carbon sorbs chlorine and is oxidized by it to carbon dioxide. In practice, contact with powdered activated carbon is generally too short to do this. Chlorine, hypochlorous acid, chlorine dioxide, and nitrogen trichloride are sufficiently volatile to be removed by aeration. Some species of chlorine are not.

The stoichiometric reactions of dechlorinating agents follow:

$$SO_2 + Cl_2 + 2 H_2O \rightarrow H_2SO_4 + 2 HCl \qquad (17\text{-}19)$$

$$NaHSO_3 + Cl_2 + H_2O \rightarrow NaHSO_4 + 2 HCl \qquad (17\text{-}20)$$

$$2\,Na_2S_2O_3 + Cl_2 \rightarrow Na_2S_4O_6 + 2\,NaCl \qquad (17\text{-}21)$$

$$C + 2\,Cl_2 + 2\,H_2O \rightarrow CO_2 + 4\,HCl \qquad (17\text{-}22)$$

## 17-14 Chemical Technology of Chlorine

For storage and shipment in steel cylinders or tanks, chlorine gas ($Cl_2$) can be liquefied at room temperatures, at pressures of 5 to 10 atm. One pound of the liquid produces 5 cu ft of gas. Under conditions of use, gas withdrawal lowers the temperature of the stored fluid. To keep the rate of withdrawal constant, heat loss must be balanced from without. For large rates of use, chlorine may be withdrawn as a liquid and vaporized in equipment similar to a hot-water heater. Direct application of heat at temperatures above 125°F is dangerous. Because reliquefaction of chlorine in measuring and dosing equipment produces erratic results, chlorine containers and gas lines must be kept cooler than the dispensing equipment.

The solubility of chlorine gas in water is about 7300 mg per l at 68°F and 1 atm. Below 49.2°F chlorine combines with water to form chlorine hydrate (usually $Cl_2 \cdot 8H_2O$), called *chlorine ice*. The hydrate may obstruct feeding equipment. Therefore, feed or sealing water coming into contact with the gas should be kept above 49.2°F.

Chlorine gas is a highly toxic irritant and must be handled with great care and under adequate safeguards. Its odor threshold in air is about 3.5 ppm by volume. Concentrations of 30 ppm or more induce coughing, and exposures for 30 min to concentrations of 40 to 60 ppm are dangerous. At 1000 ppm the gas is rapidly fatal.

For large uses liquid chlorine is cheapest. Yet the gas is so dangerous that its transport through crowded communities and its use in works within built-up areas must receive careful study and decision. Some species of chlorine other than liquid chlorine may be more satisfactory, among them the hypochlorites of calcium, $Ca(OCl)_2$, and sodium (eau de Javelle), ($NaOCl$); and chlorinated lime, $CaClOCl$. Purposeful combination of chlorine with ammonia and release of chlorine dioxide, $ClO_2$, from sodium chlorite, $NaClO_2$, create other useful chlorine disinfectants or deodorants. Chlorinated lime (a loose combination of chlorine with slaked lime), calcium hypochlorite, and sodium chlorite are generally marketed as solids; sodium hypochlorite, as a liquid (Appendix).

Of the substances employed in combination with chlorine, or as *antichlors* (chlorine-reducing substances), ammonia and sulfur dioxide are gases. They can be liquefied, stored, handled, and dispensed like chlorine. Ammonia is also available as ammonium hydroxide (aqua ammonia) and ammonium sulfate. Sodium bisulfite ($NaHSO_3$), a solid, may take the place of $SO_2$.

The strength of chlorine compounds, that is, their oxidizing capacity, is commonly expressed as *available chlorine*. This is analogous to reporting alkalinity as $CaCO_3$. *Chlorine equivalent* would be a more accurate designation. The percentage of *available chlorine* is based on (1) the moles of equivalent chlorine, or number of moles of chlorine that would have an oxidizing capacity equivalent to 1 mole of the compound and (2) the proportion by weight of the pure compound present in the commercial product. Chlorinated lime, for example, contains as its essential constituent about 62.5% of calcium oxychloride, $CaClOCl$ (mol w 127). Because each mole of $CaClOCl$ is equivalent to 1 mole of $Cl_2$ (mol w 71), the hypothetical weight of chlorine is $62.5 \times 71/127 = 35\%$ of the total, and the available chlorine is also 35%. The oxidizing capacity of nonchlorinous compounds is sometimes stated in terms of available chlorine, but this practice is not recommended. It is ambiguous for oxidants, such as $ClO_2$ and $KMnO_4$, that may undergo more than one oxidation reaction.

## 17-15 Operational Technology of Water Chlorination

Liquid chlorine, ammonia, and sulfur dioxide are generally added to water in controlled amounts through orifice flowmeters or dosimeters called chlorinators, ammoniators, and sulfonators, respectively. For given dosages, pressure drops across the orifice are kept constant. In dosimeters operated under pressure this is done by a pressure-reducing, pressure-compensating valve that keeps the influent pressure constant regardless of pressure changes in the container from which the gas is drawn. In dosimeters operated under a vacuum, the pressure drop across the orifice is regulated by a controlled vacuum on the inlet and outlet sides of the orifice. The purpose of vacuum feed is to lessen gas leakage. Some simplified pressure devices regulate the volumetric displacement of the gas (bubblers) rather than its rate of flow.

Chlorine gas may be fed directly into water through diffusers. However, because some gas may escape, chlorine gas is normally dissolved in a small flow of water that is passed through the gas-flow-regulating device to transport the dissolved gas to the point of application. A typical solution-feed chlorinator installation is shown in Fig. 17-8. Newly constructed or newly repaired water mains, wells, tanks, and masonry reservoirs can be chlorinated with the help of portable chlorinators. Initial chlorine concentrations are made high enough (about 50 mg per l) to overcome the chlorine demand of pollutants accumulated during construction. Chlorination is repeated, if necessary, until a residual of about 1 mg per l after 24 hours is reached. After this the structure is flushed out thoroughly before being placed in service.

Solutions of chlorinating, ammoniating, and sulfonating compounds are commonly added to water through chemical-reagent feeders (Fig. 16-14). For the chlorination of new mains, effective amounts of stabilized calcium

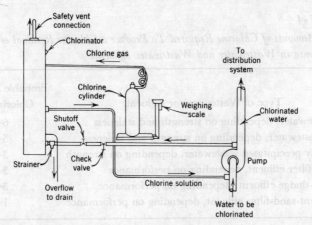

**Figure 17-8.**
**Typical chlorinator installation.**

hypochlorite may be spotted at suitable distances in the main during construction. Although chlorine can be manufactured at treatment works by the electrolysis of brine in electrolytic cells, it seldom pays to do this.

Rough estimates of chlorine dosage for *marginal* (minimal) treatment of water and sewage may be based on the values listed in Tables 17-3 and 4, if reasonable allowances are made for water temperature (Table 17-1) and water quality or wastewater strength. Dosimeter capacity must be flexible enough to meet both common and unusual chlorine demands. Examples are both sudden and longitudinal changes in water and wastewater quality by drought and flood or by accidental spills of industrial chemicals.

In drinking water, chlorine residuals of 0.2 to 1.0 mg per l after 15 to 30 min of contact will generally produce 99.9% destruction for *Esch. coli* and 37°C bacterial counts. A 15-min free-chlorine residual of 0.5 mg per l appears to be a safe average.

In treatment works, disinfecting chlorine may be added to the raw water (prechlorination), the partially treated water, or the finished water (post-

**Table 17-3**
*Minimum Chlorine Residuals for Drinking Water at 20°C*
(After Butterfield)

| pH value | 6-7 | 7-8 | 8-9 | 9-10 | 10-11 |
|---|---|---|---|---|---|
| Free available chlorine, mg/l after 10 min | 0.2 | 0.2 | 0.4 | 0.8 | 0.8 |
| Combined available chlorine, mg/l after 60 min | 1.0 | 1.5 | 1.8 | 1.8 | ... |

## Table 17-4

Probable Amounts of Chlorine Required To Produce a Chlorine Residual of 0.5 mg per l after 15 min in Wastewater and Wastewater Effluents

| Type of Wastewater or Effluent | Probable Amounts of Chlorine, mg/l |
|---|---|
| Raw wastewater, depending on strength and staleness | 6–24 |
| Settled wastewater, depending on strength and staleness | 3–18 |
| Chemically precipitated wastewater, depending on strength | 3–12 |
| Trickling-filter effluent, depending on performance | 3–9 |
| Activated-sludge effluent, depending on performance | 3–9 |
| Intermittent-sand-filter effluent, depending on performance | 1–6 |

chlorination). The water may be chlorinated more than once within the treatment plant and afterward. In-plant, pre- and postchlorination, and the rechlorination of the effluent from open distribution reservoirs are examples.

### 17-16 Other Uses of Chlorine

Chlorine is a useful disinfectant of bathing waters and shellfish as well as of drinking water and wastewaters. The heavy load of bathers imposed on swimming pools, the necessity of breathing through the mouth during speed swimming, and the plunger action exerted by air and water on the nasal passages and sinuses during diving combine to release large amounts of mucus and vast numbers of nasopharyngeal organisms into swimming pools. To keep contagion at minimum level, effective disinfectants must be present in all parts of swimming pools in sufficient concentration to ensure rapid disinfection of their waters without becoming irritants.

For the chemical disinfection of edible shellfish, the mollusks are transferred from their growing grounds to tanks to which enough chlorine is added to disinfect the water. The shellfish *drink* disinfected water whenever the concentration of chemicals drops to tolerable levels. Through successive exposures of this kind, the shellfish rid themselves of organisms ingested before they were harvested from their growing grounds.

Among uses of chlorine other than disinfection are the following: (1) the destruction or control of undesirable growths of algae and other organisms in water and wastewaters, examples being iron-fixing and slime-forming bacteria in pipelines and other water conduits, slime-forming bacteria in sewers and wastewater treatment works, freshwater mussels and clams in water conduits, and filter flies (*Psychoda*) and ponding slime growths on trickling filters; (2) the improvement of the coagulation of water and wastewaters and of the separation of grease from wastewaters; (3) the control of odors in water and wastewaters; (4) the stabilization of settling-tank sludges in water purification works and the control of odors associated with sludge treatment, including its drying; (5) the prevention of anaerobic conditions in sewerage systems and wastewater treatment works by delaying or reducing decomposition; (6) the conversion of cyanides to cyanates, such as NaOCN, in alkaline industrial wastes; (7) the destruction of hydrogen sulfide in water and wastewaters, and the protection of concrete, mortar, and paint against the corrosive action of this gas; (8) the reduction of the immediate oxygen requirements of returned activated sludge and of digester liquor within treatment plants; (9) the reduction or delay of the BOD of wastewaters discharged into receiving waters; and (10) the preparation at the plant of chlorinated copperas, a useful coagulant.

Of these purposes, the reduction of BOD by chlorination deserves amplification here. Four kinds of reactions are conceivably involved: (1) direct oxidation of BOD-exerting compounds; (2) formation with nitrogen compounds of bactericidal chloroamines by substitution of chlorine for hydrogen; (3) formation with carbon compounds of substances that are no longer decomposable, again by substitution of chlorine for hydrogen; and (4) addition of chlorine to unsaturated compounds to form nondecomposable substances. Often quoted is the observation that the application to municipal wastewaters of enough chlorine to produce a measurable residual after 15 min will reduce the five-day, 20°C BOD by 15 to 35%—or in the ratio of about 2 mg per l of BOD to 1 mg per l of chlorine. However, present knowledge is not sufficiently well founded to warrant the use of these figures. As Snow[12] has shown, BOD reduction depends not only on chlorine concentration, but also on the condition of the wastewater. Among his observations are the following. Chlorination of fresh wastewater to a trace of residual at 15 min gave a reduction of but 10%; in stale wastewater it was 25 to 40%. Doses of 100 to 300 mg per l were required for a reduction of 35% in fresh wastewater; breakpoint dosage eliminated 75%. Aeration prior to chlorination improved the BOD reduction of fresh wastewater. For all aerobic samples, a lowering of both the first-stage demand and its rate of BOD exertion was observed. In anaerobic, chlorinated wastewater, the BOD rate was stepped up.

Destruction of odors by oxidizing chemicals is successful when the reactions involved produce nonodorous substances. This is not always so. The reaction between chlorine and hydrogen sulfide is an example of a successful reaction. Elementary, nonodorous sulfur is precipitated. The stronger the oxidizing agent, the more certain is the elimination of the offending substances. That is why breakpoint chlorination is so effective. The production of chlorophenol

[12] W. B. Snow, Biochemical Oxygen Demand of Chlorinated Sewage, *Sewage Ind. Wastes*, **24**, 689 (1952).

by marginal chlorination and the intensification of tastes when water containing *Synura* or other algae is chlorinated speak against halfway measures.

## 17-17 Manageable Variables in Water Disinfection

As stated before (Sec. 17-7), degree of mixing or stirring, the length of detention, the pH, and the disinfectant concentration or intensity are the principal manageable variables in water disinfection. Although it is generally impractical to modify water quality or temperature for the specific purpose of improving disinfection, it is practical to choose a disinfecting process that is optimal for the water to be treated and to take advantage of in-plant conditions that may optimize disinfecting efficiency. Examples are (1) split treatment with chemical disinfectants before and after filtration in order to ensure lasting residuals in the product water; (2) adding of chlorine well in advance of water stabilization by lime in order to take advantage of a low pH value; (3) postponing chlorination until recarbonation has lowered the pH in lime-soda softening works; (4) chlorinating fresh urban wastewaters in preference to stale or septic ones; (5) preforming chloramines for the disinfection of urban and other wastewaters that have a large chlorine but low chloramine demand (useful for bacteria disinfection but not for virus disinfection); and (6) chlorinating stormwater overflows, main outflows from separate stormwater systems, and effluents from wastewater treatment plants in such fashion as to make use of available times of storage.

Disinfectant concentration is a matter of required dosimetry, that is, the selection of feeding equipment of sufficient capacity to cover not only normal requirements but also unusual demands that may occur when water and wastewater systems are placed under stress. Chlorine demands large enough to exceed the capacity of available equipment have been produced by (1) flash floods that pass rapidly through storage reservoirs by displacing the waters of small reservoirs or underrunning or overrunning even large reservoirs when the floodwaters are respectively colder or warmer than the stored waters; (2) overflow of swamps during spring freshets; (3) break-up of ice jams and release of pools of polluted water accumulating behind them; and (4) shifts in stream and lake currents that transport wastewater effluents to water intakes rapidly and in high concentration.

To ensure sufficient time of contact, holding or contact units should be given enough capacity, unless it is possible to capitalize on detention times otherwise available in the treatment works or in transmission and distribution systems. Quiescence and associated reduction in short-circuiting are not necessarily optimal conditions. In continuous treatment stirring or controlled power input may be more certain and more effective.

**eighteen**

# biological transfer processes

## 18-1 Origins and Functions of Biological Treatment

In the United States and in Britain, the historical sequence of engineering constructions for the transfer of suspended and dissolved nutrients from wastewaters to biomasses proceeds, in the main, from the operation of wastewater farms, through that of intermittent sand filters and contact beds, to the use of trickling filters and activated-sludge units (Figs. 18-1 and 19-4). That so-called biological treatment began as a farming practice is understandable. Wastewaters poured onto fields from urban drainage systems could irrigate soils thirsty for water and fertilize crops hungry for nutrients. Only as land became scarce and financial returns on wastewater-irrigated properties became unattractive was wastewater farming abandoned. On the natural sand deposits of New England, however, wastewater treatment by *intermittent sand filtration* was continued so long as dosages could be intensified and performance could be improved by pretreatment of the applied wastewaters in settling tanks, and eventually even in trickling filters or activated-sludge units (Sec. 14-3). On the tight soils of England this could not be done. There, *contact beds*—within tanks 4 to 6 ft deep and containing, at the beginning, layer upon layer of slate separated by supporting bricks but, later, filled more conveniently with slag or broken stone—were built instead. The contact surfaces of the beds, like those of the soil

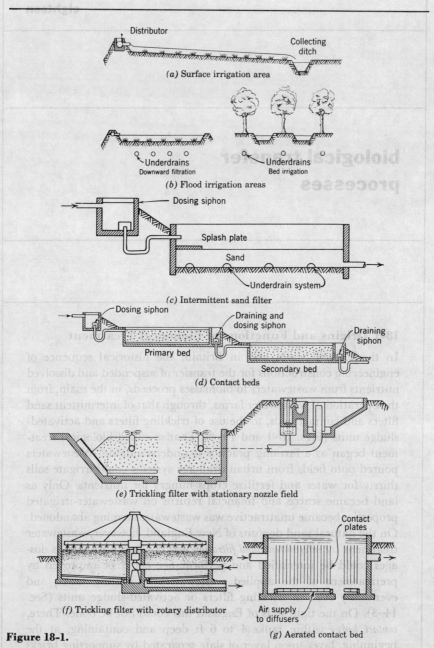

**Figure 18-1.**
Biological treatment on irrigation areas (a) and (b); on intermittent sand filters (c); in contact beds (d); in trickling filters (e) and (f); and in aerated contact beds (g). For activated-sludge aeration tanks, see Fig. 19-4.

and sand grains, were well suited for the colonization of large microbic populations, which, in their foraging, removed from the applied wastewaters nonsettling and dissolved solids, including putrescible organic compounds. In the course of time, the effective loading and performance of the beds, which were operated originally on a *fill-and-draw* and *resting cycle*, were raised appreciably by the addition of sprays that discharged the incoming wastewaters onto the bed surface in substantially continuous streams and more or less saturated with oxygen. No longer were the beds filled with wastewater. Instead they were left open to the air throughout their depth so that the wastewaters applied to them might trickle over the films, slime, or biomasses accumulating on the contact surfaces down to the underdrainage system while air swept through the pores and kept the beds aerobic. The much-improved beds—first called *bacteria beds* and, later, percolating filters in England—were named *trickling filters*[1] in the United States.

Somewhat later, continuing enquiry into the nature of treatment systems and imaginative laboratory research led to the construction of *activated-sludge units*, in which wastewaters and flocs or masses of microorganisms—not unlike the slime films of bacteria beds—were supplied with air for the dual purpose of keeping the units aerobic, in spite of heavy concentrations of living organisms, and the flocs in suspension in spite of the absence of fixed contact media.

Last in time, though least important in large-scale operations, was a linkage of solid contact materials and aeration (1) in *submerged-contact aerators* or *aerated contact beds* and (2) in partially submerged but rotating and, therefore, periodically emerging *nidus racks*.[2] In fully submerged units, stacks of eventually vertical plastic or asbestos-cement sheets were flushed by jets of air rising through submerged passageways between the contact materials colonized by surface growths of microorganisms. In the rotating, partially submerged, units, the emerging segment of the rotor was exposed to the atmosphere.

By the middle of the twentieth century, therefore, it had become possible to prescribe biological treatment for communities large enough and wealthy enough to afford the construction, operation, and technical supervision of necessary systems. Because small communities, situated principally in rural areas, were seldom able to afford works of this kind, however, they remained dependent principally on a combination of *septic* tanks with *subsurface irrigation* installed in individual properties. These systems were not suited to the diffusion of wastewaters into tight soils, and increasing availability of piped water and consequent greater water use often overtaxed the absorptive

---

[1] This term is by no means descriptive of the true nature of the system.
[2] Latin *nidus*, nest. See footnote 3.

capacity of even relatively permeable soils. Better treatment systems were needed for wastewater disposal from individual households and from small communities. *Package plants* incorporating settling and small-sized aerobic biological units were eventually marketed for both users. However, a less expensive treatment method was still badly needed for small rural communities. It was provided eventually in the form of shallow basins or ponds called *oxidation* or *stabilization ponds*. In a sense, these simple structures created a folkway of wastewater disposal that has since been explored and exploited with much success wherever local conditions have been suitable.

## 18-2 Transfer Mechanisms

As presently conceived and practiced, the biological treatment of wastewaters is not a single operation but a combination of interrelated operations that may differ in spatial distribution, proceed at different rates in time, and be accomplished by biomasses that are quite unlike in structure and support.

First in time and importance is the *transfer of nutrient substances* from the water to film, floc, or other forms of biomasses by interfacial contact and associated adsorptions and absorptions. This operation is fast and effective if the interface between the liquid and the biomass is large, if the concentration gradient of the substances to be removed from one phase to the other is steep, and if obstructive liquid films and concentrations of interfering substances do not build up on the interface (Sec. 11-18). Quality as well as extent of contact is therefore important.

Second in time but equally significant is the *preservation* of this *quality of contact*. It is accomplished primarily by the oxidation of organic matter and synthesis of new cells. As described by Buswell,[3] contact quality is preserved because of the tendency of dissolved matter to change in concentration in such fashion as to decrease the surface tension in the biotic film or floc. Substances concentrating at surfaces are adsorbed; adsorbed substances are decomposed by the accumulating enzymes of living cells; new cells are synthesized; and end products of decomposition are washed back into the waters whence they came or they are allowed to escape to the atmosphere. Examples are (1) the transfer of salts, such as nitrates, back to the wastewater because they increase the surface tension of the interface, and (2) the escape of gases, such as $CO_2$, because of their lower partial pressure in the contiguous atmosphere (Sec. 12-12). Third in time and of far-reaching importance is the conversion of the biomass into settleable or otherwise removable solids. This operation proceeds by *bioflocculation* (Sec. 18-6) in synchrony with the preservation of the quality of contact and determines the overall effectiveness of the process.

[3] A. M. Buswell, *Chemistry of Water and Sewage Treatment*, Chemical Catalog Co., New York, 1928.

The progress of biological purification is illustrated in Fig. 18–2. Interfacial transfer or adsorption is the rate-determining step. It can hold its lead, because slow operations—such as the preservation of contact quality and the settleability and stabilization of the biomass—are out of the time stream of happenings in the liquid phase and proceed at a more leisurely pace in the solid phase of the biomass.

If decomposition is designed to approach full stabilization of wasted flocs or sloughed films within the principal treatment unit itself, a fourth operation is added to overall treatment demands. Extended aeration in activated-sludge units and the protracted detention of biological films in trickling filters are examples. Stabilization of aerobic or essentially aerobic biomasses by anaerobic digestion is normally a fully independent operation (Sec. 20–13).

In bed irrigation, subsurface irrigation, and intermittent sand filtration, transferred substances must be almost completely mineralized or stabilized within the soil or sand. This cannot be done without a resting period. Otherwise the pores would clog, oxygen consumption would exceed oxygen absorption (reaeration), and the bed would become septic. Interception of substantial amounts of waste matters at the surface of the soil or sand reduces the subsurface load.

In trickling filters and activated-sludge units, films or flocs are removed respectively as settleable trickling-filter humus and waste (or excess) activated floc. These secondary or biological sludges differ in kind and vary in maturity. Depending on the duration of their retention, temperature, and factors such as film or floc biota and thickness, the sludges drop in energy content. However, they do retain more or less nitrogen and other chemical nutrients that impart some fertility to them and keep them putrescible.

The amount of film substance retained in trickling filters and the length of time during which transferred substances undergo decomposition, is a function of (1) temperature and through it of BOD, or rate of activity, (2) areal dimensions of the supporting surfaces and their exposure to moving liquid films or masses, (3) such scour as may be engendered by the moving fluids, and (4) the rate of diffusion of oxygen into the accumulating film. Diffusion is fast at the film or floc surface but drops off rapidly within the biomass proper (Sec. 11–16). Anaerobic decomposition of films normally proceeds at a slower rate than aerobic decomposition, releases gases, produces foul-smelling intermediates, and destroys the cohesive and adhesive gelatinous nature of the substrate. This, together with the formation and lifting effect of gas bubbles, causes the film to slough. Because transfer between liquid and film, floc, or biomass occurs only at the interface and because film, floc, and biomass can be captured as settleable secondary sludge, interfacial sloughing is a cardinal feature of trickling filters. In

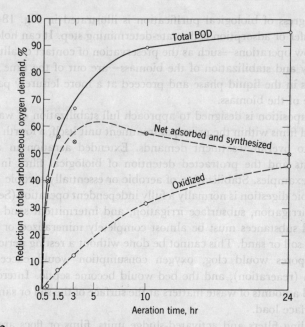

**Figure 18-2.**
**Removal of organic imbalances by biomasses in a batch operation.** [After C. C. Ruchhoft, Studies of Sewage Purification IX, *Public Health Repts.*, 54, 468 (1939).]

trickling filters sloughing reduces film thickness and, with it, fly breeding (Sec. 18-3).

In the activated-sludge process the amount of floc can be varied at will by regulating the volumes or weights of sludge wasted and returned. If flocs become too large and too heavy they will turn inactive and become anaerobic. It may then be difficult to keep them in suspension. If flocs become too small and too light they may remain highly active but they will foul the effluent and be lost to the treatment process because it may be impossible to remove them by sedimentation for return to the process.

## 18-3 Ecology of Biomasses

The populations of organisms responsible for the transfer of nutrients to biomasses are large and varied. Type genera are shown in Fig. 18-3. The principal and most numerous biological workmen are saprobic[4] microorganisms, including the autotrophic[5] bacteria. The gelatinous masses con-

---
[4] Greek, *sapros*, decaying and *bios*, life.
[5] Greek, *autos*, self and *trophicos*, nourishing.

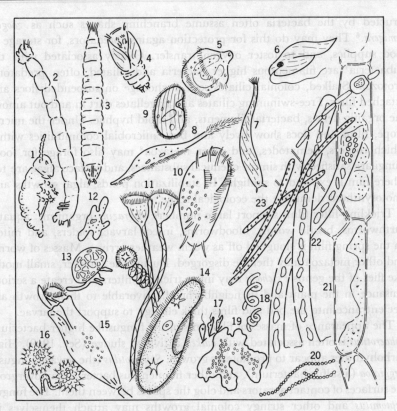

**Figure 18-3.**
**Organisms associated with the biological treatment of wastewater. (After Imhoff and Fair.)**

### Numbers 1 to 4, insects (×5)

1. Water springtail, *Podura*; the genus found on trickling filters is *Achorutes*.
2. Larva of bloodworm, *Chironomus*
3. Larva of filter fly, *Psychoda*
4. Pupa of filter fly, *Psychoda*

### Numbers 5 to 17, Protozoa (×150)*

5. *Didinium*
6. *Euglena*
7. *Chaenea*
8. *Lionotus*
9. *Colpidium*
10. *Stylonychia*
11. *Vorticella*
12. *Ameba*
13. *Arcella*
14. *Paramecium*
15. *Opercularia*
16. *Anthophysa*
17. *Oikomonas* ×1,500

### Numbers 18 to 23, Bacteria and fungi (×1,500)

18. *Thiospirillum*
19. *Zooglea ramigera*
20. *Streptococcus*
21. *Leptomitus*
22. *Sphaerotilus*
23. *Beggiatoa*

*Excepting No. 17, *Oikomonas*.

structed by the bacteria often assume branching shapes such as *zooglea ramigera*.[6] They may do this for protection against predators, for storage of food supplies, and to foster oxygen transfer. Closely associated with the eubacteria[7] are filamentous higher bacteria and ciliated, often predatory, protozoa. Stalked, colonial ciliates seek anchorage on suspended flocs and attached films. Free-swimming ciliates and flagellates dart in and out among the organic debris, bacterial filaments, and mold hyphae. Under the microscope, films and flocs show lively and busy microbial communities within which rotifers, nematodes, and other metazoa[8] may also forage for food. Fungi are satisfied with simple chemical substances, and surface algae are too where film is exposed to sunlight. In stabilization ponds, algal growths and photosynthesis dominate the ecological scene.

Trickling-filter films support large numbers of *grazing* organisms: aquatic earthworms, bristle worms, bloodworms, insect larvae, spiders, and mites. In the spring film is sloughed off as warm weather arrives. Masses of worms and other metazoa may then be disgorged. During the summer, small moth-like flies of the genus *Psychoda* may infest trickling filters and create a serious nuisance in the plant and its neighborhood. Favorable to their growth are free entrance into the bed and films thick enough to support the larvae.

The appearance of masses of filaments of the funguslike higher bacterium *Sphaerotilus* is often associated with *bulked* activated sludge (Sec. 19-9). High carbohydrates appear to promote its growth. *Sphaerotilus* (the sewage fungus), *Beggiatoa* (a sulfur bacterium), and other filamentous bacteria may overgrow the surfaces of contact aerators and clog the spaces between them. The fungus *Leptomitus* and other stringy colonial growths may attach themselves in unsightly masses to walls, gates, and baffles of wastewater channels and tanks.

The cultivation of environments that foster the welfare and, with it, the activity of seed or adventitious organisms lies at the base of good plant design and management. Toxic wastes inhibit biological activity and may destroy it; so may unwanted predators. The destruction of activated sludge by the bloodworm *Chironomus* is an example.

Even though the various aerobic biological systems produce much the same biochemical changes in a wastewater, their ecology is often quite different. Hawkes[9] has shown, for example, that the ecosystem of trickling filters is far more varied than that of activated-sludge units.

As pointed out by him, trickling filters do not create a truly aquatic or uniform environment. Instead, they differentiate into a succession of communities at different bed depths principally because of accompanying

---

[6] Greek, *zoion*, animal; *gloia*, glue; and Latin, *ramus*, branch.
[7] Greek, *eu*, true.
[8] Greek, *meta*, middle.
[9] H. A. Hawkes, *The Ecology of Wastewater Treatment*, Macmillan, New York, 1963.

changes in available nutrients. Higher forms of life graze on the microbial film-covering of the contact material, and exposure of the top of the bed to light encourages surface growths of algae and other photosynthetic organisms. As a result, the purification accomplished in trickling filters changes markedly in amount and kind at different levels, and the humus sloughing from the stones may consist not of fresh biomasses but of the waste products of the grazing fauna. As such, the humus is at a relatively low energy level and quite stable. Film age varies with season, and film structure is not affected significantly by conditions of flow. The downward discharge of unloading film makes it necessary to keep contact materials relatively large at all depths and largest at full depth.

*Activated-sludge units*, by contrast, proffer a truly aquatic and relatively uniform environment. Flocs are carried along by flows and contain much the same varieties and numbers of microscopic organisms. Grazers are almost entirely absent. Unless the tank contents are mixed longitudinally, the floc communities are exposed at the beginning to influent wastewaters and at the end to effluent wastewaters. However, if there is longitudinal mixing, much the same kind of purification can go on in all parts of the tank before the activated sludge is removed as microbial floc.

Floc age, which is governed by the amount of floc in circulation relative to the amount of new floc formed in a given time, determines the activity of the circulating or wasting floc and its energy level. As floc ages, it contains increasing proportions of dead cells and inert matter. Although it may still be active enzymatically and adsorptively, its ability to oxidize the adsorbed substances may die away as waste products accumulate. As floc increases in size, its combined surface area is reduced relative to its volume, and diffusion of nutrients into the floc and of waste substances out of the floc is hindered. Floc size cannot be determined precisely. Light-scattering techniques and photographic measurements are possible means.

In the ecology of aerobic *stabilization ponds*, algal photosynthesis takes command. Aerobic populations (largely bacteria) supply carbon dioxide to algal growths, which release oxygen to the wastewaters and help to keep the ponds aerobic (Eq. 18–2). Most ponds are bound to receive some settleable organics and to generate some settleable biomasses composed of zoogleal bacteria and dead algal cells. If the accumulating sludge deposits become thick enough, they turn anaerobic, and their organic constituents digest in the way they would in sludge-digestion tanks (Sec. 20–13).

In the United States stabilization ponds are normally colonized by both green and blue-green algae. The number of cells is large, but the number of genera composing even thick growths is small. Type species are members of the genera *Chlorella*, *Chlamydomonas*, and *Euglena* among the green algae, and *Oscillatoria* and *Anabaena* among the blue-greens. Chlorella is by far the most prevalent organism, perhaps because of its resistance to destruction by anaerobiosis and its tolerance of extremes of temperature. The cells are small (10 $\mu$ in diameter) and are not easily removed from the suspending water. *Chlamydomonas* and *Euglena* are more sensitive to adversity. Their sudden death may make heavy inroads on the oxygen resources of receiving waters.

Diurnal variations in the ecology of stabilization ponds make it difficult to assess their efficiency and to optimize their operation. Dissolved-oxygen concentrations and pH rise during daylight and drop at night. Carbon dioxide concentrations follow the opposite course. Tests for BOD and suspended solids are obscured by the presence of algae in pond and effluent samples.

## 18-4 Nutrients for Biomasses

More than any other group of organisms, the *saprophytic bacteria* are responsible for the processes of decomposition whereby waste organic matter is biochemically degraded and eventually mineralized or stabilized. Mineralization can be schematized as follows:

$$\left.\begin{matrix} \text{Elements in organic matter} \\ \left.\begin{matrix} C \\ H \\ N \\ S \\ P \end{matrix}\right\} + O_2 \xrightarrow{\text{Microorganisms}} \end{matrix}\right. \left.\begin{matrix} \text{Products of mineralization} \\ \left\{\begin{matrix} CO_2 \\ H_2O \\ NO_3^- \\ SO_4^= \\ PO_4^= \end{matrix}\right. \end{matrix}\right. \quad (18\text{-}1)$$

By contrast, the different families of algae contain chlorophyll and similar pigments that enable lower as well as higher plants to utilize radiant energy for converting carbon dioxide into organic compounds. Light energy from the sun changes water and carbon dioxide (as the sole source of carbon) into a simple sugar (glucose), oxygen being released as a byproduct.

$$6\ CO_2 + 6\ H_2O \xrightarrow[\text{photosynthesis}]{\text{algae}} C_6H_{12}O_6\ (\text{glucose}) + 6\ O_2 \quad (18\text{-}2)$$

The process is called *photosynthesis*. Accordingly the algae are both photosynthetic and autotrophic.

*Heterotrophic*[10] *bacteria* decompose carbonaceous and nitrogenous substances. Some also reduce nitrates and sulfates in the absence of free oxygen. Among autotrophic bacteria, pigmented forms derive energy by photosynthesis as shown in Eq. 18-3, and unpigmented forms by chemosynthesis as shown in Eqs. 18-4 and 5.

$$3\ H_2S + 6\ H_2O + 6\ CO_2 \xrightarrow[\text{photosynthesis}]{\text{pigmented bacteria}} C_6H_{12}O_6\ (\text{glucose}) + 3\ H_2SO_4 \quad (18\text{-}3)$$

---

[10] Greek, *heteros*, varied and *trophicos*, nourishing.

$$H_2S + \tfrac{1}{2} O_2 \xrightarrow[\text{chemosynthesis}]{Beggiatoa} S + H_2O + \text{energy} \quad (18\text{-}4a)$$

$$S + 1\tfrac{1}{2} O_2 + H_2O \xrightarrow[\text{chemosynthesis}]{Beggiatoa} H_2SO_4 + \text{energy} \quad (18\text{-}4b)$$

$$NH_3 + 1\tfrac{1}{2} O_2 \xrightarrow[\text{chemosynthesis}]{Nitrosomonas} HNO_2 + H_2O + \text{energy} \quad (18\text{-}5a)$$

$$NO_2 + \tfrac{1}{2} O_2 \xrightarrow[\text{chemosynthesis}]{Nitrobacter} NO_3 + \text{energy} \quad (18\text{-}5b)$$

**Essential Minerals.** With the exception of nitrates and phosphates, most of the minerals entering into the construction of active biomasses are normally available in the public water supplies that are drawn upon for the transport of waste substances. Nitrogen and phosphorus are added by domestic wastes and by some industrial wastes. Minimum mineral requirements are placed by Sawyer[11] at ratios of $BOD:N:P = 150:5:1$ and requirements for maximum N and P content of sludge at $90:5:1$. The conversion of ammonia to nitrate in a trickling filter is illustrated in Fig. 18-4.

Before the BOD concept was introduced into the management of wastewater-treatment works, plant effluents were generally considered satisfactory when they were rich in nitrates. Values of 10 to 15 mg per l were accepted as signs of good performance. This viewpoint is no longer held in the United States. Present-day targets are either high BOD removal without much nitrification or high nitrification followed by denitrification. The removal of phosphates is another target. Neither phosphates nor nitrates are changed significantly during normal biological treatment. In the presence of energy-supplying carbon, nitrogen gas is released from nitrates by denitrifying bacteria, a facultative group of heterotrophic organisms. Evolution of gaseous nitrogen in secondary settling tanks may then buoy up otherwise settleable solids and carry sludge to the tank surface where the nitrogen is lost to the atmosphere. Phosphorus can be transferred to the sludge by chemical coagulation (Sec. 16-10).

## 18-5 The Growth Curve of Biomasses

The growth of specific microorganisms or of compatible communities of organisms lies within one or more sequent portions of the growth curve[12] sketched in Fig. 18-5. Salient sections are encompassed by the zone of logistic growth, which can be treated as such (Sec. 2-6) or an exponential (logarithmic), nutritionally unrestricted growth followed by first-order,

---

[11] C. N. Sawyer, Bacterial Nutrition and Synthesis, in J. McCabe and W. W. Eckenfelder, Jr., Eds., *Biological Treatment of Sewage and Industrial Wastes*, Vol. 1, Reinhold, New York, 1956.

[12] J. Monod, The Growth of Bacterial Cultures, *Ann. Rev. Microbiol.*, **3**, 37 (1949). Figure 2-1, which idealizes the growth of human populations, offers useful comparisons.

536/*Biological Transfer Processes*

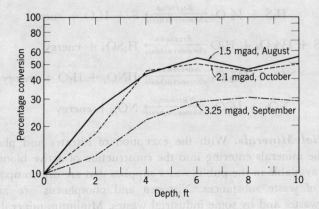

**Figure 18-4.**
**Conversion of ammonia nitrogen to nitrate nitrogen in a trickling filter.**
[After Buswell, *Illinois State Water Survey*, Bull. 26 (1928).]

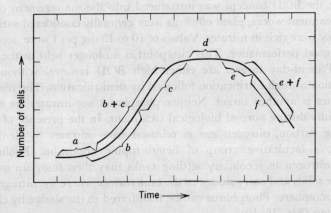

**Figure 18-5.**
**Idealized growth curve of bacterial cultures. Phases of bacterial growth: $a$, lag; $b$, logarithmic or exponential growth; $c$, first-order, or retardant, growth; $d$, stationary endogeny; $e$, mounting endogeny; $f$, complete endogeny and death; $b + c$, logistic or autocatalytic growth; $d + e + f$, phases of increasing endogeny.**

nutritionally restricted growth at or about the half-life of the logistic system. Environmental constraints eventually produce endogeny.[13] Possible causes are (1) reduction of the food supply or substrate below the trophic support levels required by the mounting population and (2) accumulation of toxic or

[13] Greek, *endo*, within; and *genesis*, growth, that is, growth on the metabolic products of growth.

interfering waste products. Cell multiplication usually continues, even though the concentration of the primary culture medium has been reduced below the minimum requirements of the saturation population (basal metabolism[14]). Cells facing starvation may draw upon stored metabolites for energy and reproduction or upon the protoplasm[15] of dead cells that have undergone lysis.[16] Enzymatic reactions (fermentations), too, are continued with the help of accumulated enzyme systems. Parts of Fig. 18-5 evolve in special circumstances alone. Acclimation, for example, is necessary only when the initial or *seed* population is not immediately adapted to the available food substances (substrates) or to some other environmental factor such as temperature. Endogeny implies continuing physiological activity in a stationary phase, which may eventually enter a declining phase.

For useful mathematical formulations of the growth curve, see Secs. 2–6 and 11–14.

**Substrate Utilization.** An understanding of growth as a metabolic response of microorganisms to food under specific environmental conditions is the key to the management of water quality in most circumstances. Growth may be wanted or unwanted. Examples of requisite management are (1) the avoidance, on the one hand, of nuisance growths of algae in reservoirs storing drinking water and (2) the promotion, on the other hand, of the algal colonization of wastewater stabilization or oxidation ponds. Where removal of nutrients or their conversion into removable form is the objective of wastewater treatment, the ecological system created must be optimal for the objectives to be reached. In forced biotic treatment procedures, for instance, transfer of nutrients from the liquid phase is generally effected in a matter of minutes by adsorption (or sorption) of organics and minerals onto biomasses, whereas the degradation of the nutrients within the biomasses is a matter of hours and days. Trickling and suspended contact[17] are prominent examples.

The rate of nutritional demands or substrate utilization, $dc_s$, of microbic cultures finds expression in the rate of population growth, $dy$, as follows:

$$dy/dc_s = y' \qquad (18\text{-}6)$$

where $y'$ is the rate of yield. By combining Eq. 18-6 with specific growth equations, useful equations for substrate utilization are obtained (Eqs. 11-15 and 16).

---

[14] Greek, *meta*, beyond and *ballein*, to throw; that is, the sum of the processes by which protoplasm is created (anabolism) and destroyed (catabolism).

[15] Greek, *protos*, first and *plasma*, form.

[16] Greek, *lysis*, loosening, that is, disintegration.

[17] Commonly referred to, for historical reasons, as trickling filtration and activated-sludge treatment.

## 18-6 Bioflocculation

In a physicochemical sense, bacteria are hydrophilic colloids that carry a net negative charge within the prevailing pH range of common treatment processes. However, their agglomeration is not brought about merely by reduction in the charge density and decrease in electrostatic repulsion. Instead, agglomeration may be looked on as an interaction of polyelectrolytes of natural origin. Polymers, polysaccharides, and polyacids, for instance, are excreted at the surface of microorganisms. This happens under all physiological conditions. When organisms grow prolifically, for example, new surfaces can be created faster than they can become covered with polymers. This brings about an improvement in bioflocculation during the declining growth phase and becomes optimal in the endogenous phase (Sec. 18-5). That polyacids of bacterial origin have been used successfully to flocculate bacteria supports this hypothesis (Sec. 16-4).

Dispersed microorganisms in effluents from biological wastewater works are readily flocculated by multivalent ions such as Fe(III) and Al(III) and by synthetic cationic as well as anionic polyelectrolytes (Fig. 16-4). Both sedimentation and filtration are improved thereby. Wastewater sludges, too, respond better to filtration after they have been coagulated (Sec. 20-9).

## 18-7 Load Equalization

Like most water- and wastewater-treatment processes, biological transfer operates inherently on a basis of diminishing returns. Even when the rate of stabilization is substantially constant—in the normal exertion of BOD, for example—the amount of work remaining to be done in time or distance decreases in proportion to the diminishing concentration of removable substances. Moreover, because nutrients that best lend themselves to biological utilization are normally removed fastest and first, the degree of removability of the nutrients as well as their concentration may be expected to contribute to the diminution of the purification pressure with time or in distance of travel.

The load of nutrients impressed on biological treatment units and presumably the work done by them can be equalized in some measure in the following ways: (1) by subdivision of the treatment structure into two or more units, or stages, and alternation of the lead unit (applicable only when the contact surfaces are fixed in place—for example, irrigation, intermittent sand filtration, or trickling filtration—called alternating double filtration); (2) by subdivision of the applied wastewater into two or more portions and their progressive introduction along the line of treatment (applied so far only when contact surfaces move with flow—for example, step loading or step aeration in the activated-sludge process); (3) by recirculation of effluent to the influent (Sec. 11-20) and subjection of the resultant mixture to treatment (applicable in all biological treatment systems); (4) by longitudinal, possibly complete, mixing of wastewater undergoing treatment to equalize nutrients, contact opportunity, and purification pressure in a single treatment structure; (5) by subdivision of the treatment structure into multiple units small enough to simplify and promote multidirectional mixing (complete mixing); and (6) by combinations of effluent recirculation with serial subdivision of the treatment units (alternation of component divisions being optional) and with progressive dosing of the treatment units (step aeration, step loading, and complete mixing). In these terms, the return of sludge in activated-sludge units (about 20% by volume) is but a rudimentary form of recirculation, and the incidental and possibly unintended short-circuiting or longitudinal mixing taking place in such units is but a rudimentary approach to complete mixing.

## 18-8 Kinetics of Treatment Systems

Because biological treatment is normally accomplished in a series of interlocking and often concurrent operations, it is possible to view the kinetics of treatment either for individual operations in a given system or for the systems as a whole. A typical concurrent series is the transfer of organic substances to activated sludge by adsorption, accompanied by the utilization of transferred organics by the biomass for energy and cell synthesis, and advanced by stirring the activated sludge for aeration and enhancement of contact opportunity (Sec. 11-17). Possible, moreover, is the formulation of biological processes in terms of changes in substrate concentration—that is, the economics of living communities—on the one hand, or in terms of microbic growth—that is, population dynamics—on the other hand. An example is the removal of BOD from wastewater substrates accompanied by an increase in biomass.

The active purification pressure is a function of the concentration of removable substances as a whole or of individual constituent fractions. Where longitudinal mixing is complete, the concentration of removable substances or nutrients or of specific fractions of these wastewater components is spatially invariant. Where there is little or no longitudinal mixing, removal and removability are bound to drop off during treatment in conformance with either a simple first-order reaction or a retardant first-order reaction. Illustrative of a simple first-order reaction is the *performance-time* relationship plotted semilogarithmically for the removal of BOD and ammonia nitrogen in activated-sludge units in Fig. 18-6, and for the spring and fall removals of BOD and ammonia in trickling filters in Fig. 18-7. The other curves in Fig. 18-7, including those for the removal of turbidity, exemplify the retardation

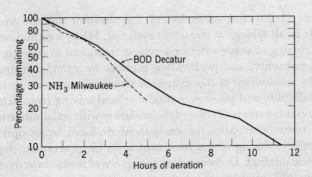

**Figure 18-6.**
**Performance-time relationship of activated-sludge units. (After Copeland, at Milwaukee, Wis., and Hatfield, at Decatur, Ill.)**

of first-order reactions. Organic nitrogen and COD would presumably exhibit behavior similar to that of BOD and ammonia.

If, in line with the bacterial growth curve (Fig. 18-4), the rate of purification in biological treatment units approximates a first-order reaction with or without retardation, the basic treatment relationship is given by the differential form of Eq. 11-13 as

$$dy/dt_d = k_0 \left(\frac{y_0 - y}{y_0}\right)^n (y_0 - y) \qquad (18\text{-}7)$$

Here $y_0$ is the initial or removable amount of substances present in the applied water; $y$ is the amount removed during passage through the treatment unit—vertically in trickling filters and in a general horizontal direction in most activated-sludge units; $t_d$ is the displacement or mean contact time within the unit; $k_0$ is a proportionality factor with the dimension (time)$^{-1}$; and $n$ is a measure of the degree of dropoff in treatment response or the retardation during the progress of purification (Sec. 11-14). The important contribution of biological systems to wastewater treatment is that their purification power is normally self-generated and self-maintained by their acceptance of waste substances as nutrients. Integration of Eq. 18-7 produces the relationships shown in Section 11-14.

Where a series of synchronous and interlocking actions and reactions can be characterized by an overall value of $k$, the contributions of the component actions or reactions are additive, that is, $k = \Sigma k$ or $kt = \Sigma kt$.

$$y/y_0 = 1 - (1 + n\Sigma kt)^{-1/n}, \text{ for } n > 0 \qquad (18\text{-}8)$$

$$y/y_0 = 1 - \exp(-\Sigma kt), \text{ for } n = 0 \qquad (18\text{-}9)$$

and

Kinetics of Treatment Systems/541

$$y/y_0 = 1 - (1 + \Sigma kt)^{-1}, \text{ for } n = 1 \qquad (18\text{-}10)$$

the value of $n$ being that of the overall reaction.

## Example 18-1.

Assuming that the semilogarithmic plot of BOD removal shown in Fig. 18-7 can be approximated by a straight line passing through the coordinates 100% at 0 hr and 10% at 11.4 hr, find (1) the overall value of the reaction rate and (2) the increase over a fundamental BOD-bottle rate of 0.23/day at 10°C that can be ascribed to the treatment process, that is, stepped-up BOD exertion by the biomass, nutrient transfer

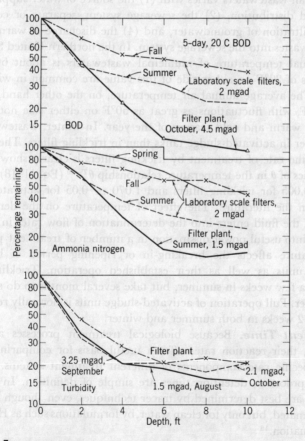

**Figure 18-7.**
**Performance-depth relationship of trickling filters operated at low rates.**
(After Buswell, *Illinois State Water Survey*, Bull. 26 (1928).]

to the floc by adsorption, and enhanced contact opportunity by useful power dissipation (aeration).

1. Because $n = 0$, as suggested by the goodness of fit of a straight line to the plotted information, Eq. 18-9 states that $e^{-\Sigma k t} = e^{-11.4 t} = 10/100 = 0.1$ or by Table 7 (Appendix), $11.4k = 2.3$, whence $\Sigma k = 2.3/11.4 = 0.2$ hr $= 4.8$/day.

2. The contribution of the treatment process to BOD removal beyond that of a bottle rate is $4.8 - 0.23 = 4.6$/day, or twentyfold.

***Temperature Effects.*** The effect of temperature on purification kinetics is given by the van't Hoff-Arrhenius relationship (Sec. 11-15). The temperature of urban wastewaters varies with (1) the source of water supply and its storage and distribution, (2) the sewerage system, separate or combined, (3) the infiltration of groundwater, and (4) the discharge of warm or cold industrial wastes into the sewerage system. In the northern United States the mean annual temperature of municipal wastewaters is about 60°F, and fluctuations of 20°F on either side of this value are common in winter and summer. The average annual air temperature, on the other hand, is more nearly 50°F, with fluctuations as great as 50°F on either side not unusual during the warm and cold seasons of the year. In winter, wastewaters are kept warmer in activated-sludge tanks than in trickling filters. The effect of season on the rate of treatment by trickling filters is clearly shown in Fig. 18-7. Values of $\theta$ in the temperature relationship $\theta^{T-20}$ (Eq. 11-18) equal to $1.040 \pm 0.005$ for trickling filters and $1.070 \pm 0.05$ for activated-sludge units are in the literature. The effect of temperature on the density and viscosity of the fluid enters into the determination of flow rates in trickling filters and into useful power dissipation in a number of treatment processes.

Temperature affects the breaking-in or ripening period of biological treatment units as well as their established operation. Trickling filters mature in a few weeks in summer, but take several months to do so during cold weather. Full operation of activated-sludge units is normally reached in 10 days to 2 weeks in both summer and winter.

***Treatment Time.*** Because biological treatment processes are time-dependent, their reaction rates offer a useful basis for comparing overall rates or specific rate components of different treatment systems. In most treatment processes, detention times are simple of definition. In trickling filters they are best determined by tracer techniques, even though they can be approximated, but only for clean water, by formulations such as Howland's general equation.[18]

[18] W. E. Howland, Flow over Porous Media as in a Trickling Filter, *Proc. 12th Ind. Waste Conf.*, Purdue Univ., **12**, 435 (1957). Also see M. D. Sinkoff, Ralph Porges, and J. H. McDermott, Mean Residence Time of a Liquid in Trickling Flters, *Proc. Am. Soc. Civil Eng.*, **85**, SA6, 51 (1959).

$$t_d = ch(\nu/g)^{1/3}\left(\frac{A/V}{Q}\right)^{2/3} \quad (18\text{-}11)$$

where $t_d$ is the detention time, $h$ is the filter depth, $\nu$ is the kinematic viscosity of the fluid, $g$ is the gravity constant, $A/V$ is the specific surface of the contact medium, $Q$ is the hydraulic load or flow per unit area and time, and $c$ is a coefficient that should reflect film buildup as well as contact-medium structure. Measured times range from 20 to 60 sec for common filter depths (6 ft) and low loadings (3 to 6 mgad). For high rates of flow (20 to 30 mgad), the detention times are presumably reduced in proportion to the two thirds power of $Q$, that is, to 6 to 20 sec.

The contact time in activated-sludge units is usually calculated as the quotient of the aeration-unit volume and the influent flow rate rather than the mixed-liquor flow rate. This is done because it represents the theoretical average time during which a constituent portion of the wastewater—which may find its way also into the sludge liquor from time to time—remains in the aeration unit.

*Example 18-2.*

The observed BOD removal illustrated in Fig. 18-7 for a trickling filter operated at a rate of 4.5 mgd was as follows:

| Depth, ft | 0 | 2 | 4 | 6 | 10 |
|---|---|---|---|---|---|
| BOD remaining, mg/l | 85 | 39 | 19 | 10 | 8 |

Find (1) the initial overall rate of reaction, assuming a retardation coefficient of 0.50, and (2) the calculated BOD remaining. Assume that the detention time is 20 min for a 10-ft filter.

1. For $n = 0.5$ and $y_0 = 85$ mg/l, Eq. 18-8 states that

$$\Sigma k = \frac{10}{(0.50 \times h \times 20)}\left\{\left[\frac{85}{(y_0-y)}\right]^{0.50} - 1\right\} = 0.24, 0.28, 0.32, 0.23$$

and averages 0.27/min or 390/day.

2. Substituting $\Sigma k$ in Eq. 18-8, $y_0 - y = 85/(1 + 0.50 \times 0.27 t_d)^2$, and the calculated BOD at depths of 2, 4, 6, and 10 ft are 37, 20, 13, and 8 mg/l respectively.

## 18-9 Nutrient Transfer

That the overall operation of nutrient transfer can be expressed mathematically as retardant or declining first-order reactions is in agreement with the observation that biomasses are normally within the declining phase of the growth curve (Fig. 18-5). Depending on the length of storage, that is, the age, of the floc or film within the system, the biomass may advance into the stationary and even into the endogenous phase. Pertinent decisions must be reached by the designer and operator. These they can make with more flexibility for activated-sludge units than for trickling filters.

Presently available information suggests that nutrient transfer is many times faster than nutrient oxidation. For the maintenance of treatment processes, however, conversion of adsorbed organics into settleable solids by oxidation may be the rate-determining step if a surrogate operation, such as chemical coagulation, is not introduced in its place (Sec. 18-6).

Katz and Rohlich[19] have shown that the BOD removed from wastewater by activated sludge during a contact period of 20 min can be fitted into a modified Freundlich adsorption isotherm (Eq. 11-30), namely

$$(X'/M)c_s' = K'c^{1/n'} \tag{18-12}$$

where $X'$ is the initial concentration of dissolved BOD in the wastewater, $M$ is the mass of the activated sludge ($c_s'$ being its concentration), $c$ is the residual dissolved BOD concentration, and $K'$ and $n'$ are constants. Calculated values of $K'$ varied with the origin of the sludge tested. They ranged approximately from $10^{-3}$ to $10^{-4}$ at sludge concentrations of $1 \times 10^3$ to $3 \times 10^3$ mg per l. The value of $n'$ was approximately unity.

## 18-10 Useful Power Dissipation

Power is dissipated usefully in biological treatment when it speeds up the transfer of nutrients to flocs and films. This it does by enhancing the contact opportunity. The pertinent power-dissipation function is $G^2 = P/(\mu C)$, and the associated mean temporal velocity gradient, $G$, has the dimension (time)$^{-1}$ (Sec. 11-18). Here $P$ is the *useful* power input, $\mu$ is the absolute viscosity of the liquid being treated, and $C$ is the volume of liquid in which the power is usefully dissipated. Because $G$ is a time-rate factor it can take its place with the other rate constants that govern treatment kinetics. To fit it typographically into the summation $\Sigma k$, the notation of $G$ can be changed to $k_G$. The possibility that not all of the actual power input, $P_a$, is usefully employed can be acknowledged at the same time by writing

$$k_G = \sqrt{pP_a/\mu C} \tag{18-13}$$

where $p$ is the proportion of the actual power input that enters directly and usefully into the treatment process proper.

Specifically, the important components of $k_G$ acquire the values listed in Table 18-1 for common wastewater-treatment processes.

[19] W. J. Katz and G. A. Rohlich, A Study of the Equilibria and Kinetics of Adsorption by Activated Sludge, in J. McCabe and W. W. Eckenfelder, Jr., Eds., *Biological Treatment of Sewage and Industrial Wastes*, Vol. 1, Reinhold, New York, 1956.

### Table 18-1
*Power Dissipation in Wastewater Treatment Processes*

| Treatment System | P | C | $k_G$ | $t_d$ |
|---|---|---|---|---|
| 1. Irrigation and intermittent filtration | $\rho g p h Q_d$ | $fhA$ | $\sqrt{gpQ_d/(\nu f A)}$ | $t_d = C/Q_d$ |
| 2. Trickling filters | $\rho g p h Q$ | $Q t_d$ | $\sqrt{gph/(\nu t_d)}$ | Eq. 18-11 |
| 3. Activated-sludge units | $\rho g p h V_a Q$ | $Q t_d$ | $\sqrt{gphV_a/(\nu t_d)}$ | $C/Q$ |

Here $Q_d$ is the actual rate of dosing the area or bed, not the average rate of flow, $Q$; $f$ is the porosity ratio of soils and sands; $\rho$ is the mass density and $\nu$ the kinematic viscosity of the fluid; and $V_a$ is the volume of air diffused into aeration units per unit volume of fluid treated.

### Example 18-3.

In a sand filter 3 ft deep with a porosity of 40% treating 200,000 gpad of clarified wastewater on four ¼-acre beds in single daily doses, the applied water occupies $(50,000/7.48)/[(43,500/4) \times 0.4] = 1.53$ ft of bed. Find the actual values of $G$, $t_d$, and $Gt_d$ for a kinematic viscosity $\nu = 1.27 \times 10^{-5}$ (sq ft)(sec) at 59°F.

In accordance with Item 1 of Table 18-1, $G = [32.2 \times 1.53/(1.27 \times 10^{-5} \times 0.4)]^{1/2} = 3.11 \times 10^3 \text{ sec}^{-1}$; $t_d = 0.4 \times 1.53 \times 43,560/(0.05 \times 1.548 \times 4) = 8.64 \times 10^4 \text{ sec} = 1 \text{ day}$; and $Gt_d = 3.11 \times 10^3 \times 8.65 \times 10^4 = 2.68 \times 10^8$.

### Example 18-4.

In a trickling filter 6 ft deep dosed at a rate of 25 mgad, the detention time is 10 min. Find the values of $G$, $t_d$, and $Gt_d$ at 59°F [$\nu = 1.27 \times 10^{-5}$ (sq ft)(sec)].

In accordance with Item 2 of Table 18-1, $G = [32.2 \times 6/(1.27 \times 10^{-5} \times 10 \times 60)]^{1/2} = 1.59 \times 10^2 \text{ sec}^{-1}$; $t_d = 10 \text{ min} = 6 \times 10^2 \text{ sec}$; and $Gt_d = 1.59 \times 10^2 \times 6 \times 10^2 = 9.54 \times 10^4$.

### Example 18-5.

An activated-sludge tank 10 ft deep treats 1.0 mgd of sewage during 5 hr with 0.75 cu ft of air per gal of wastewater at 59°F [$\nu = 1.27 \times 10^{-5}$ (sq ft)(sec)]. Find the values of $G$, $t_d$, and $Gt_d$.

In accordance with Item 3 of Table 18-1, $G = [32.2 \times 10 \times 0.75/(1.27 \times 10^{-5} \times 5 \times 3600)]^{1/2} = 3.24 \times 10 \text{ sec}^{-1}$; $t_d = 5 \text{ hr} = 1.8 \times 10^4 \text{ sec}$; and $Gt_d = 3.24 \times 10 \times 1.8 \times 10^4 = 5.84 \times 10^5$.

It is obvious from these examples that the proportionate usefulness of the applied power is quite small. Yet is is important. Comparative values of $G$ itself, not $k_G = G\sqrt{p}$, are largest for the long-time exposures generated in intermittent sand filters, smaller in trickling filters for greatly reduced periods of time, and smallest for activated-sludge units but for a relatively long time; the resulting products $Gt_d$ are much the same for trickling filters and activated-sludge units but considerably larger for intermittent sand filters. However, it is probable that $p$, the proportion of useful power dissipation, is least in intermittent sand filters and greatest in activated-sludge treatment because the unavailable power dissipation by frictional resistance to flow is presumably highest in intermittent sand filters.

## 18-11 Air Requirements

Because the oxygen introduced into aerobic, biological treatment systems must satisfy, within the treatment unit proper, the oxygen requirements of both the incoming wastewater and the active biomass,

$$y = ax + bz \quad \text{or} \quad y/z = (ax/z) + b \qquad (18\text{-}14)$$

where, in an activated-sludge unit, for example, $y$ is the rate at which oxygen is consumed in a given treatment unit (pounds per day), $x$ is the rate at which the 5-day, 20°C BOD or the COD is removed (pounds per day), and $z$ is the rate at which the returned sludge is introduced into the unit (pounds of mixed-liquor volatile suspended solids per day), $a$ and $b$ being coefficients respectively relating $O_2$ to BOD or COD oxidation and to biomass requirements. Expected values for the coefficient $a$ lie above 0.5 relative to BOD removal and somewhat lower relative to COD removal. In a given system experimental values of $y/z$ plotted against $x/z$ should yield a straight line[20] with slope $a$ and intercept $b$ at $x/z = 0$. However, about a quarter of the biomass synthetized is normally resistant to oxidation.

Most of the biomass is populated by microaerophilic organisms that, as their classification suggests, can thrive in the presence of less than 1 mg per l of DO. Therefore, if activated-sludge units are stirred mechanically at low concentrations of DO, the absorption of atmospheric oxygen at the air-water interface can usually satisfy the oxygen needs otherwise provided by compressed air. The rate of oxygen uptake must equal that of oxygen utilization by the biomass in the system or it will constitute the limiting operational factor.

[20] T. P. Quirk (Amenability of a Mixture of Sewage, Cereal, and Board-Mill Wastes to Biological Treatment, *Proc. 13th Ind. Waste Conf., Purdue Univ.*, 523 [1958]) presents a straight-line plot of $y/z$ versus $x/z$ from which the values of $a$ and $b$ can be read respectively as 0.52 and 0.15 relative to BOD removal and mixed-liquor suspended solids (MLSS); or $y = 0.52x + 0.15z$.

## 18-12 Floc and Film Volumes

Because many of the impurities of wastewaters are converted into voluminous and decomposable sludges, treatment processes must be examined not only by themselves but also in conjunction with possible sludge-conditioning methods. Estimates of the amount and nature of the sludge produced are therefore key matters of information. Terminal weights of the sludge solids settled from biotic effluents may be estimated roughly from the observation that between 50 and 60% of the 5-day, 20°C BOD is converted initially into floc as dry-weight volatile suspended solids in activated-sludge units and presumably also into the film of trickling-filter units. Amounts decrease thence with floc or film age, and in proportion to the amounts of floc in circulation. Observations by Wuhrmann[21] can be fitted by the equation

$$z_t/z_1 = \exp(-0.26t) \qquad (18\text{-}15)$$

where $z_t/z_1$ is the ratio of volatile solids (dry basis) in $t$-day-old floc to 1-day-old floc, $t$ being obtained by dividing the dry weight of suspended solids in the system by that in the daily inflow. Ratios of 0.59, 0.35, and 0.14 obtain, for example, when the floc is two, four, and eight days old. Observations by Heukelekian, Orford, and Manganelli[22] have been formulated by them as follows:

$$\Delta y = 0.5x - 0.055z \qquad (18\text{-}16)$$

where $\Delta y$ is the volatile-suspended-solids accumulation (pounds per day), $x$ is the BOD fed (pounds per day) and $z$ is the mixed-liquor volatile suspended solids (pounds), the coefficient 0.055 having the dimension (days)$^{-1}$.

If the weight of volatile suspended solids in the tank influent is $w$, it must be added to the mass balance, whence

$$\Delta y = a'x - b'z + c'w \qquad (18\text{-}17)$$

Here $a'$ and $b'$ are coefficients not unlike (in value) $a$ and $b$ in Eq. 18-14, and $c'$ is the fraction of the influent volatile solids that is not degraded during aeration.

## 18-13 Bench-Scale Testing

Information on the probable behavior of wastewaters can be obtained in the laboratory with specialized laboratory equipment or by relatively simple

[21] Karl Wuhrmann, Factors Affecting Efficiency and Solids Production in the Activated Sludge Process, in J. McCabe and W. W. Eckenfelder, Jr., Eds., *Biological Treatment of Sewage and Industrial Wastes*, Vol. 1, Reinhold, New York, 1956, p. 64.

[22] H. Heukelekian, H. E. Orford, and R. Manganelli, Factors Affecting the Quantity of Sludge Production in the Activated-Sludge Process, *Sew. Ind. Wastes*, **23**, 945 (1951).

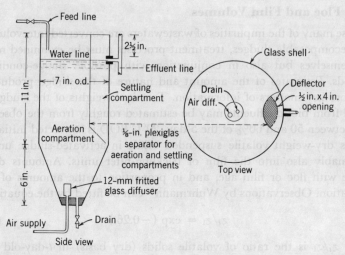

**Figure 18-8.**
Laboratory activated-sludge unit. Side view: Pyrex shell, $\frac{3}{8}$-in. wall drawn to a cone. Top view: separator made from a 10-in. section of 4-in. Plexiglas tube cut in half longitudinally. The bottom end is cut at a 45° angle. Part of the lower end is closed by a $\frac{1}{8}$-in. Plexiglas plate so that the two compartments are connected by a $\frac{1}{2} \times$ 4-in. opening. Waterproof rubber-to-metal cement joins Plexiglas to glass. (After Ludzack.[23])

adaptation of standard laboratory apparatus to the requirements of the investigation. Examples of such apparatus for the study of activated sludges are (1) cylindrical graduates or tubes large enough to hold 2.5 l of fluid and (2) fritted glass or porous ceramic diffusers. A more elaborate apparatus designed to operate as a general-purpose unit is illustrated in Fig. 18-8.[23] Why it is difficult to devise a bench model of a trickling filter is brought out in the discussion of trickling-filter versus activated-sludge ecology (Sec. 18-3). Lack of spatial uniformity of nutrients (*isotrophism*) and of associated biota does not allow significant departure from the vertical dimensions of full-sized units except by placing a large enough number of model beds in series to equal the full bed depth. Neither can there be significant reduction in the size and nature of contact media.

Basic purification behavior is generally identified in *batch* tests, and actual behavior of treatment plants in *continuous-flow* or *steady-state* tests, often referred to as *continuous-culture* techniques in laboratory studies of biological treatment. The purpose of most batch testing is to determine the magnitude of the coefficients included in treatment formulations, because the conversion

[23] F. W. Ludzack, Laboratory Model Activated-Sludge Unit, *J. Water Pollution Control Fed.*, **32**, 605 (1960).

of substrate into film, floc, or biomass can be assumed to proceed in step with the conversion of substrate into microbic populations (Sec. 18-6).

Laboratory studies are especially useful adjuncts to design and operation when industrial wastes are a large proportion of the wastewater or all of it. This is generally so because their composition is likely to be unique, whereas the composition of municipal wastewaters does not depart greatly from well-established normal values.

## 18-14 Loading Intensities

Two kinds of loads are impressed on biological treatment units: (1) hydraulic loads, which govern the hydraulic and, in conjunction with them, the pneumatic requirements of treatment units, and (2) process loads, in the form of degradable organic matter or nutrients contained in the applied wastewater, which govern the process requirements of the treatment system. To relate loadings to performance and make comparisons between them possible, they must be transformed into *loading intensities*. These are normally calculated for processes as a whole. Often, however, component details—adsorption, cell synthesis, oxidation of organic matter, endogeny, settleability, and film stabilization, for instance—are governing issues and must be considered individually.

*Hydraulic loads* are normally measured as rates of flow, such as *million gallons per day*. Their intensities then become velocities or times of passage and exposure within treatment units (Sec. 18-8). In flow-through systems the inverse of the flow rate is the *dilution* rate, because it tells how often in a given time interval the contents of a treatment unit are displaced. Examples of hydraulic load intensities are (1) the velocity factor, *million gallons per acre daily*, for irrigation areas, intermittent sand filters, stabilization ponds, and trickling filters;[24] (2) the reciprocal detention time, *million gallons daily per acre-ft, gallons per day per cubic yard*, or *gallons per day per 1000 cubic feet*, for trickling filters;[25] and (3) the detention time or dilution rate, *hours of aeration*, for activated-sludge units and submerged-contact aerators. The reciprocal detention time is also an indirect measure of the flushing action or hydraulic load per unit area of film surface because the volume of the bed—in acre-feet, for example—is implicitly also an approximate measure of the extent of its contact surfaces. Because the applied waters contain the nutrients to be

---

[24] Although the gallon as a unit volume of water does imply a length cubed, it does so unhappily as 231 cu in. Nevertheless, the term gallons per square foot daily does imply to the engineering practitioner a time-rate of directional displacement, that is, a velocity; and the term gallons per cubic foot (or per acre-foot) daily is recognized by him as expressing a rate of volumetric displacement with the dimension (time)$^{-1}$.

[25] 1 mgd/acre ft = 620 gpd/cu yd = 23,000 gpd/1000 cu ft = 3.061 day$^{-1}$. 1 lb BOD/(cu yd)(day) = 1613 lb BOD/(acre-ft)(day) = 37 lb BOD/(1000 cu ft)(day).

removed, the hydraulic-loading intensities are rough measures, too, of process-loading intensities.

Process loads are rationally expressed by the weights of removable nutrient impurities. Examples are the comprehensive measures BOD, COD, and suspended-solids loadings in *pounds daily*. In special circumstances and for certain purposes, other, more specific, determinants may be added or substituted, among them turbidity, organic nitrogen, ammonia, phosphorus, synthetics (including detergents), and bacteria or viruses. A *tributary population load* may be used instead by due expansion of storm runoff and industrial wastes into population equivalents.

To be converted into intensities, process loadings must be related to *contact opportunity*. A complex of design and operating factors may have to be introduced to do this. Examples are the power dissipation function $G^2 = P/\mu C$ and the detention period $t_d$, or the product $Gt_d$ as in Eqs. 11-25 and 27. Most parameters in common use are grossly inadequate measures of contact opportunity. Examples are *pounds of BOD daily per acre* for irrigation areas, intermittent sand filters, and trickling filters; for trickling filters, also, *pounds of BOD daily per acre-foot, cubic yard*, or *1000 cubic feet*; and for activated-sludge units and submerged contact aerators, (1) in reference to tank dimensions, *pounds of BOD daily per cubic foot* of tank volume, *square foot* of tank surface, or *foot* of tank length; (2) in reference to air use, *hours of aeration* or *cubic feet of free air per pound of BOD*; and (3) in reference to returned sludge (for the activated-sludge process only), *pounds of BOD daily per 1000 pounds daily of returned sludge*, or sludge returned relative to volume of entering wastewater flow during the same time period, *percentage of mixed-liquor suspended solids* (MLSS) or *volatile suspended solids* (MLVSS) with suitable inclusion of the sludge volume index (*milliliters of sludge per gram of dry weight* [Sec. 10-19]).

If area and time of contact by themselves or as part of the contact opportunity $Gt_d$ are considered controlling factors in biological treatment, a general parameter of loading intensity is prescribed as the weight of removable substance in the influent applied to a unit contact surface in a unit time. The weight of removable substance can normally be determined by analytical procedures and flow measurement. The areal extent of the contact surface can be measured directly for submerged contact aerators and trickling filters constructed of vertical plates, but only indirectly through evaluation of the void space and the specific surfaces of granular contact materials in trickling filters and of flocs in activated-sludge units.

## 18-15 Treatment Efficiency

Most advanced biological treatment systems are preceded by primary settling tanks. The biological or secondary component is then composed of the biological unit proper and its secondary settling tank. However, as

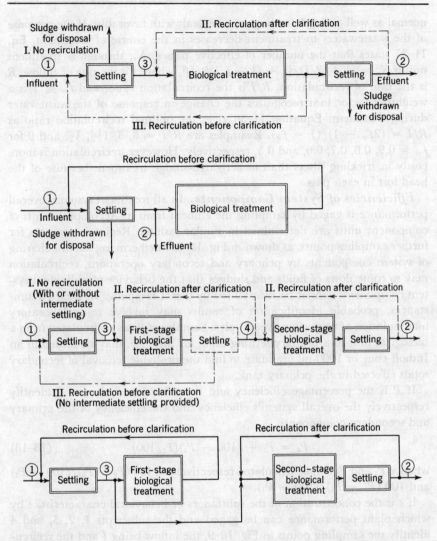

**Figure 18-9.**
**Flow charts for the recirculation and stage treatment of wastewaters.**

illustrated in Fig. 18-9, some partial or complete recirculating systems dispatch their solids to the primary tank, and some activated-sludge systems dispense with primary sedimentation or keep it to a minimum.

*Recirculation.* As suggested in Sec. 11-20, recirculation of wastewater flows through biological treatment units distributes the load of impurities imposed on the units and smoothes out the applied flow rates. In this way normal as well as shock loadings can be dealt with favorably. If the response of the wastewaters to treatment decreases in the course of treatment, Eq. 11-37 states that the number of effective passes, $N_e$, through a treatment unit is $N_e = (1 + R/I)/[1 + (1 - f_w) R/I]^2$, where $I$ is the rate of inflow, $R$ is the rate of recirculation, $R/I$ is the recirculation ratio, and $f_w \leqslant 1$ is a weighting factor that recognizes the change in response of the wastewater during treatment. Equation 11-38 gives the optimal recirculation ratio as $R/I = (2f_w - 1)/(1 - f_w)$. Examples are $R/I = 8, 3, 1\frac{1}{3}, \frac{1}{2}$, and 0 for $f_w = 0.9, 0.8, 0.7, 0.6$, and 0.5, respectively. However, recirculation is more costly in trickling filters than in activated-sludge treatment because of the head lost in each pass.

**Efficiencies of System Components.** In all treatment systems, overall performance is gaged by sampling the effluent from it. Accomplishments of component units are determined in similar fashion. Recirculation asks for further sampling points, as shown in Fig. 18-9. Furthermore, like the sharing of system components by primary and secondary operations, recirculation may so route flows of fluids and sludges that the behavior of individual system components cannot be ascertained by direct sampling. In these circumstances, probable identification of results may involve some laboratory manipulation of representative samples. Samples of flows recirculated from a biological unit to a primary settling tank, for instance, may be settled in an Imhoff cone or 1000-ml graduate to find the probable removal of secondary solids effected in the primary tank.

If $P$ is the percentage efficiency and the subscripts 0, 1, and 2 identify respectively the overall system's efficiency and the efficiency of the primary and secondary components,

$$P_0 = P_1 + (100 - P_1)(P_2/100) \tag{18-18}$$

whence $P_1$ and $P_2$ can be calculated respectively as $100 (P_0 - P_2)/(100 - P_2)$ and $100 (P_0 - P_1)/(100 - P_1)$.

If $c$ is the concentration of the substances or behavioral characteristics by which plant performance can be gaged and the subscripts 1, 2, 3, and 4 identify the sampling points in Fig. 18-9, the inflow being $I$ and the recirculated flow $R$, then the overall efficiency and the efficiencies of the primary and secondary components are the following:

1. For once-through flow,

$$P_{0,1,2} = 100(c_{1,1,3} - c_{2,3,2})/c_{1,1,3} \tag{18-19}$$

where the sequence of subscripts identifies the sequence of overall, primary, and secondary efficiencies, respectively.

2. For flow recirculated to the primary settling tank before clarification in the secondary settling tank and assuming that the removal efficiency of the

primary settling tank for recirculated flows is the same as that of the secondary tank, the overall removal $P_0$ being the same as that for $P_0$ in Eq.18-19,

$$P_1 = 100[I(c_1 - c_3) - R(c_3 - c_2)]/(c_1 I) \qquad (18\text{-}20)$$

and

$$P_2 = 100(I + R)(c_3 - c_2)/[Ic_3 + R(c_3 - c_2)] \qquad (18\text{-}21)$$

Similar relationships can be elaborated for the other flow diagrams of Fig. 18-9.

*Example 18-6.*

When a wastewater flow of 1 mgd containing 308 mg/l of BOD is passed through a settling tank and a filter (1) in straight-through flow, the primary-tank effluent contains 200 mg/l of BOD and the plant effluent 34 mg/l; (2) with recirculation of 1.5 mgd from the biological unit to the primary tank, its effluent contains 108 mg/l of BOD and the plant effluent 18 mg/l. Find the overall efficiencies and the efficiencies of the primary and secondary sections in these two operational arrangements.

1. By Eq. 18-19, $P_0 = 100(308 - 34)/308 = 89.0\%$; $P_1 = 100(308 - 200)/308 = 35\%$; and $P_2 = 100(200 - 34)/200 = 83\%$.

2. By Eq. 18-19, $P_0 = 100(308 - 18)/308 = 94.2\%$; by Eq. 18-20, $p_1 = 100[1.0(308 - 108) - 1.5(108 - 18)]/(1.0 \times 308) = 21.1\%$; and by Eq. 18-21, $P_2 = 100(1 + 1.5)(108 - 18)/[1.0 \times 108 + 1.5(108 - 18)] = 92.6\%$.

**Performance-Loading Relationships.** Based on such engineering concepts, the Committee on Sanitary Engineering of the National Research Council[26] formulated the performance-loading relationships of the biological portions of treatment plants in military installations, that is, communities producing predominantly domestic wastewaters, as follows:

$$P_2 = 100/(1 + mi^n) \qquad (18\text{-}22)$$

Here $P_2$ is the percentage efficiency of the secondary treatment components, $i$ is the loading intensity, and $m$ and $n$ are coefficients. The magnitude of $n$ reflects the variability of efficiency with load intensity; the numerical value of the coefficient $m$ depends on the units of measurement employed.[27]

[26] National Research Council, Sewage Treatment at Military Installations, *Sewage Works J.*, **18**, 796 (1946).

[27] The magnitudes of $m$ and $n$ can be read from a logarithmic plot of $[(100 - P_2)/P_2] = y$ against $i = x$, $m$ being the intercept of the straight line of best fit on the $y$-axis at $i = 1.0$, and $n$ being the slope of this line found from pairs of coordinates as $n = (\log y_2 - \log y_1)/(\log i_2 - \log i_1)$.

Summary plots of operational results analyzed by the National Research Council and by Fair and Thomas[28] are shown in Fig. 18-10. Numerical values for $n$ and $m$ and expressions of loading intensity are given in Table 18-2. Generally speaking, activated-sludge units, as suggested by the magnitude of $n$, appear to respond slightly better to high loading intensities than do trickling filters, and these in turn respond appreciably better than do plate contact aerators.

**Table 18-2**
*Values of Performance Coefficients for Biological Treatment Units*

| | $n$ | $m$ | $i$ | Units of $i$ for BOD[a] Removal |
|---|---|---|---|---|
| Plate contact aerators | 0.746 | $2.48 \times 10^{-1}$ | $y_0/(At)$ | lb/(1000 sq ft) (hr of contact) |
| Trickling filters First-stage[b] | 0.50 | $8.5 \times 10^{-3}$ | $y_0/(CF)$ | lb/(acre-ft) (recirculation factor) |
| Second-stage[c] | 0.50 | $8.5 \times 10^{-3}$ | $y_0/[W_2 t(100 - P_2)]$ | |
| Activated-sludge units | 0.42 | $3.0 \times 10^{-2}$ | $y_0/(W t_d)$ | lb/(1000 lb suspended solids) (hr of aeration) |

[a] Five-day, 20°C BOD.
[b] Also single-stage.
[c] $W_2$ and $P_2$ for second stage only.

*Example 18-7.*

A flow of 1 mgd of sewage containing 307 mg/l of BOD is passed through a primary settling tank, which removes 35% of the BOD, before being applied to a trickling filter and secondary settling tank; that is, the BOD load applied to the filter from the primary tank is $307 \times (1 - 0.35) \times 1 \times 8.34 = 1665$ lb/day. If the filter has a surface area of 0.185 acre and its depth is 3.0 ft, estimate the overall percentage reduction in BOD and the BOD remaining in the plant effluent (1) for once-through operation of the trickling filter, and (2) for recirculation of 1.5 mgd of plant effluent to the trickling filter.

1. The rate of dosage, or intensity of the hydraulic load, is $1.0/0.185 = 5.4$ mgd/acre, or $5.4/3 = 1.8$ mgd/acre-ft. The process-loading intensity is $y_0/(CN_e) = 1665/$

[28] G. M. Fair and H. A. Thomas, Jr., The Concept of Interface and Loading in Submerged, Aerobic, Biological Sewage-Treatment Systems, *J. and Proc. Inst. Sewage Purificat.*, Pt. 3, 235 (1950).

$(0.185 \times 3 \times 1) = 3000$ lb/acre-ft. In accordance with Eq. 18-22, the expected efficiency of the biological section of the plant is $P_2 = 100/[1 + 8.5 \times 10^{-3} (3000)^{1/2}]$ = 68.2. In accordance with Eq. 18-19, the expected overall plant efficiency is $P_0 = 35 + (100 - 35)68.2/100 = 79.5\%$, and the BOD remaining in the plant effluent is estimated at $(100 - 79.5)307/100 = 63$ mg/l.

2. By Eq. 11-37, recirculation of 1.5 mgd of sewage increases the rate of dosage of the filter $(1 + 1.5) = 2.5$-fold to 13.5 mgd/acre, or 4.5 mgd/acre-ft. If a weighting factor of 90% is assumed, the number of effective passes $N_e$ becomes $(1 + 1.5/1)/[1 + (1 - 0.9)(1.5/1)]^2 = 1.89$, in accordance with Eq. 11-37, and the process-loading intensity $y_0/(CN_e) = 1665/(0.185 \times 3 \times 1.89) = 1590$ lb per equivalent acre-ft.

In accordance with Eq. 18-21, therefore, the expected efficiency of the secondary section of the plant is $P_2 = 100/[1 + 8.5 \times 10^{-3} (1590)^{1/2}] = 74.6\%$. The resulting overall plant efficiency is $P_0 = 35 + (100 - 35)74.6/100 = 83.5\%$, and the BOD remaining in the plant effluent is estimated at $(100 - 83.5)307/100 = 51$ mg/l.

## 18-16 Manageable Variables

The performance-loading coefficients of Table 18-2 are founded on identifiable and determinable loading parameters. In each instance the loading of the treatment unit is stated as the weight of the five-day, 20°C BOD of the applied wastewater. Only in one instance, however—that of the plate contact aerator—can the magnitude of the *static contact opportunity* be computed from direct measurements of (1) the area of the liquid-solid or, proportionately, the liquid-film interface and (2) the time of contact or mean detention period, the magnitude of the *dynamic contact opportunity* being based on the useful power dissipation by the air blown into the system. For the activated-sludge process, the weight of the suspended solids rather than the active surface area of the floc, together with the time of exposure, is the commonly available information. The contact opportunity afforded in trickling filters is even less precisely categorized unless the use of vertical-plate contact media provides information on the areal extent of the contact surface and unless the time of contact is known from suitable flow tests. Otherwise the volume of bed, as reflecting in sequence—by assumption of reasonable proportionalities—film surface and time of exposure, is the parameter of choice. The magnitudes of the individual variables can be changed to reach wanted treatment efficiencies with due reference to the impact of the performance coefficients on the results.

Because the information on which the formulations of the National Research Council are based stems from unusual communities, the observational equations can only be assumed to offer background information. No specifically applicable design values can issue for their unqualified use.

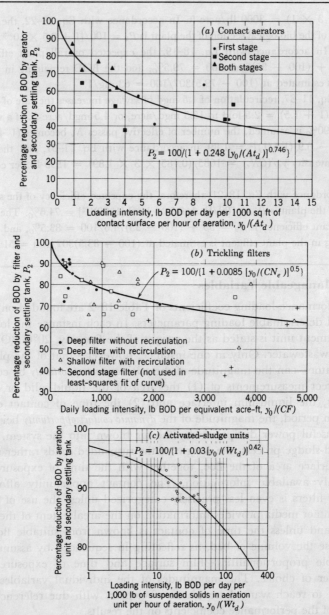

**Figure 18-10.**
**Performance-loading relationship for (a) contact aerators, (b) trickling filters, and (c) activated-sludge units. Plotted points in (a) and (b) show the results obtained at U.S. military posts.[26] Points in (c) are for diffused-air units in North America.[28]**

In summary, the manageable variables are the load itself, the area of interfacial contact, the time of exposure, and, wherever applicable, the conversion of static surfaces into dynamic interfaces by useful power dissipation (Sec. 18-10). In special circumstances, refinements may give attention to BOD:N:P relationships, temperature effects, and marginal oxygen requirements or air supply (Secs. 18-2, 8 and 11).

The theoretical counterparts of process performance have received attention in Secs. 18-7 to 9. As shown there, the only shared manageable variables are the applied load and the time of exposure. All other performance factors lie within the sum of the reaction coefficients $\Sigma k$, the individualized $k$ values being expected to account sequentially for the overall purification. An example is sequential interfacial transfer and adsorption of wastewater suspensoids and dissolved nutrients, the conversion of nutrients into energy or cell substance, and the promotion of overall activity by useful power dissipation.

The fact that as of today pertinent values of $n$ and $\Sigma k$ or $\Sigma kt$ can be ascertained only by bench-scale testing of the wastewaters actually to be purified (Sec. 18-13) is a valid reason for preserving observational data and empirical formulations such as Eq. 18-22 in the professional literature.

Neither the empirical nor the theoretical formulations give an account of the *built-in* side reactions by which (1) liquid-solid exchange surfaces are maintained, renewed, or reactivated on film or floc and (2) settleable, as well as partly or fully stable, sludges are produced.

## nineteen

# biological treatment systems

### 19-1 Purpose and Scope

The present chapter advances the discussion of biological transfer processes from the analysis of shared principles of biological treatment in Chapter 18 to the synthesis of successful treatment processes and operating features of specific treatment systems, that is, to process design and related hydraulic design.

Three present-day biological treatment systems are dealt with at length: (1) the trickling-contact or trickling-filter system, which, although it is a turn-of-the-century discovery, has proved itself adaptable to changing conditions and demands; (2) the suspended-contact or activated-sludge system, which was introduced after its timely and imaginative elaboration from bench-scale studies and has proved its great and growing potential as it has become better understood; and (3) the stabilization-pond system that was introduced in answer to the needs of small communities.

### 19-2 Irrigation Areas and Intermittent Sand Filters

Wastewaters are run onto grasslands, ploughed fields, and standing crops for two purposes: to improve the harvest and to dispose of applied waters. Agricultural benefits may derive from the water and fertilizing components individually or collectively. However, treatment is not necessarily proportional to agricultural benefits, and agricultural management of irrigation areas may come into conflict with their sanitary management.

Irrigation methods must be varied in conformity with the nature of the soil, rainfall magnitude and distribution, other climatic factors that affect crop yields, height of groundwater table, local topography and geology, and kinds of crops to be grown. In much of the world the annual *consumptive use* of water during the growing season is about 10 in., that is, from one half to one third the annual precipitation in well-watered regions and equivalent to about 750 gpd per acre over the year.

Of the wastewater reaching cropped lands, some evaporates, some seeps into the soil, and much is collected in ditches after but moderate contact with soil and crops. At best, this is only partial biological treatment. Moreover, action of the biological component of treatment ceases entirely in cold weather. In flood irrigation or land filtration the applied water must percolate through the soil to be collected by underdrains (Fig. 18-1). In open soils flows may join the natural groundwater and remain out of sight. The physical, chemical, and biological cleansing powers of the soil are mobilized in these circumstances, and treatment is more effective than in surface irrigation. Direct contact with crops is avoided when surface flows are confined to furrows between cropped beds. Organic residues are converted to *humus*, which functions as a soil builder and may improve the harvest. The power of the soil to bind all manner of chemicals, from simple to exotic in their structure, is remarkable.

The sanitary management of wastewater farms is beset by many difficulties. Unless degradable organics have been removed from the applied waters, there will be odors. Unless the standards and restrictions discussed in Sec. 10-12 are observed, there will be sanitary hazards. Spray irrigation is a particularly bad offender (Sec. 10-3), unless it is restricted to the disposal on waste and scrub land of industrial wastewaters (1) rich in organic matter but free from human excrement and (2) seasonal in production and thereby subjected to a long period of rest and recovery. When rainfalls are heavy, crops can do without supplemental irrigation; almost perversely, wastewater volumes to be disposed of are then swollen by storm runoff and groundwater infiltration into sewers. Hydrological difficulties may be overcome by holding about a quarter of the irrigation area in reserve, or by shunting-in stormwater detention basins or pretreatment devices. In sizing the areas, harvest times must be allowed for as well as rainy spells.

If they are to dry out, reaerate, and remain sweet, farmlands must be irrigated intermittently. Tight soils and overloaded fields become septic, sour, *sewage-sick*, and useless. Irrigation with raw wastewaters has become a rarity in industrialized and urbanized countries.

*Intermittent sand filters* are described in Sec. 14-3 and illustrated in Fig. 18-1. They are a logical modification of irrigation on farms in regions of thick sand deposits whenever direct agricultural exploitation of wastewaters can be dispensed with in favor of intensive wastewater treatment. Coarse solids left on the sand surface by downward-percolating raw wastewaters must be removed from time to time by raking the surfaces. Biological films within the sands undergo stabilization in place. To maintain aerobic conditions, beds are dosed intermittently for short periods of time between long resting periods.

Common loadings associated with removals of from 85 to 95% of the suspended solids, from 90 to 95% of the BOD, and from 95 to 98% of the bacteria are listed in the following schedule:

|  | gpd per acre |
|---|---|
| Irrigation of cultivated soils | 3,000 |
| Irrigation of grasslands | 25,000 |
| Intermittent filtration of raw wastewater | 80,000 |
| Intermittent filtration of settled wastewater | 200,000 |
| Intermittent filtration of biological effluent | 500,000 |

## 19-3 Contact Beds and Trickling Filters

The simple structures once employed for the transfer of dissolved organic matter and fine suspended solids from settled wastewaters to contact surfaces are illustrated in Fig. 18-1. The cycle of transient operations included filling, contact, emptying, and resting. Continuous operation of contact beds was eventually made possible by blowing air into the wastewaters flowing through them. The air was provided (1) in sufficient volume to keep the waters and slime surfaces aerobic, and (2) in sufficient intensity to tear away aging slime and solids accumulations for separation from the effluent in secondary settling tanks. The resulting *aerated-contact beds*—also called *submerged-contact-aerators*—are structurally similar to the prototype contact units. Hydraulically and pneumatically they are quite different. The settled wastewaters are displaced continuously through them and forced into circuitous paths as they are (1) lifted past the contact media by air introduced at the tank bottom and (2) returned to the bottom in counterflow. Microbic growths develop on the contact surfaces of the vertical sheets of durable plastics and asbestos cement hung into the tanks. Some of the growths are torn away mechanically by passing air and water; some are sloughed off when slimes become thick enough to become anaerobic, granular, and no longer adhesive as a result of the decomposition of the constituent organic matter and zoogleal (sticky) binding substances.

Structurally different from the prototype contact units but hydraulically and pneumatically similar are the trickling filters illustrated in Fig. 18-1.

In trickling contact, the wastewaters trickle more or less continuously in thin films over the biological growths covering broken stone or other contact material that composes the bed. Spray aeration and natural ventilation keep the applied waters and the slimes on the contact surfaces aerobic and sweet. Surface films slough off more or less continuously at some rates of filter dosage and intermittently at others. The films are normally intercepted by secondary sedimentation as trickling-filter *humus*.

Loadings associated with overall removals of 80 to 90% of the suspended solids, and from 65 to 85% of the BOD and COD, are 1 to 4.5 mgd per acre for trickling-filter beds with stationary nozzle fields and 3 to 45 mgd per acre for beds with traveling or rotary distributors, the higher loadings in the latter instance made possible by recirculation.

## 19-4 Design of Trickling Filters

Depending on plant size, trickling filters are currently designed (1) as circular structures enclosing contact media over which wastewaters are sprinkled from rotating distributors, rather than as the former rectangular units served by stationary spray nozzles; (2) as relatively shallow structures operated at rates 10 or more times as fast as the conventional deeper beds of the more distant past; (3) for more or less continuous operation and, in many cases, continuous or supplemental recirculation of effluent, instead of once-through operation alone; and (4) occasionally also for series operation of units in two or more sequent stages with possible alternation of the lead unit.

The schedule of information shown in Table 19-1 has been adapted from a joint manual of the American Society of Civil Engineers and the Water Pollution Control Federation.[1]

The contact material of trickling filters must be strong enough to support its own weight as well as reasonable live loads during construction, inspection, and repair; and not be subject to decay or disintregration under exposure to water, air, cold, biological growths, and flooding or chemical dosage for insect control. Crushed traprock, granite, and limestone are normally the contact media of choice, but hard coal, coke, cinders, blast-furnace slag, wood resistant to rotting, ceramic materials, and plastics have also been employed. The sodium sulfate soundness test[2] simulates the effects of alternate freezing and thawing on the disintegration of mineral contact media.

---

[1] American Society of Civil Engineers and Water Pollution Control Federation, Sewage Treatment Plant Design, *Manuals of Engineering Practice*, No. 36 and No. 8, respectively (1959).

[2] American Society of Civil Engineers, Filtering Materials for Sewage Treatment Plants, *Manual of Engineering Practice*, No. 13 (1937). The destructive force of repeated crystal formation within the pores of the rock forms lies at the root of the test.

### Table 19-1
Loadings and Depths of Low-Rate and High-Rate Trickling Filters

|  | Low-Rate Operation | High-Rate Operation |
|---|---|---|
| *Hydraulic loading* | | |
| gpd per sq ft | 25–100 | 200–1000 |
| mgad | 1.1–4.4 | 8.7–44 |
| *Process loading*—BOD | | |
| lb per 1000 cu ft daily | 5–25 | 25–300 |
| lb per acre-ft daily | 220–1,100 | 1,100–13,000 |
| *Depth*, ft | | |
| Single-stage | 5–8 | 3–8[a] |
| Multistage | 2.5–4 | 1.5–4 |
| *Relative recirculation in terms of influent flows* | | 0.5–10[b] |

[a] Commonly 3–6.
[b] Commonly 0.5–3.

To support a large surface of biological film, the contact material must be small, but not so small that its void space can be obstructed by biological growths and its passageways sealed by sloughing films. Normally specified is that individual pieces shall be substantially uniform in their principal dimensions, and that at least 95% of them (by weight) shall pass through a 4-in.-square wire screen but be retained on a 2½-in.-square wire screen. Rocks and fieldstones of this kind have a void space of about 40% and a total surface area of about 15 sq ft per cu ft. In comparison, polyvinyl chloride synthetic media[3] consisting of straight and corrugated plastic sheets, or cubical nodules, leave a void space of 97% and expose surface areas of 27 to 37 sq ft per cu ft.

### 19-5 Process Design of Trickling Filters

The observational performance-loading relationship noted in Fig. 18–10 can be rearranged as in Eq. 19–1 to identify the required equivalent acre-footage, $CN_e$, of a filter that is to remove $y$ pounds of BOD daily from an applied load of $y_0$ pounds daily, or effect a purification of $P_2$ per cent, that is,

$$CN_e = \frac{y_0}{13,800}\left(\frac{y}{y_0 - y}\right)^2 = \frac{y_0}{1.38 \times 10^4}\left(\frac{P_2}{100 - P_2}\right)^2 \quad (19\text{–}1)$$

[3] "Koroseal," manufactured by B. F. Goodrich Industrial Products Co.; "Surfpac," manufactured by Dow Chemical Co.; and "Flocor," made by Imperial Chemical Industries in Britain, are examples.

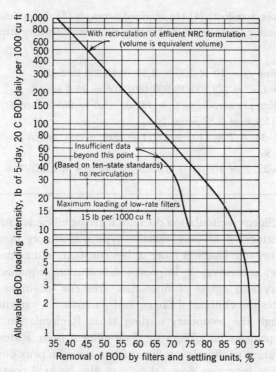

**Figure 19-1.**
**Allowable BOD loading intensities of trickling filters and settling tanks for given percentage reductions of the BOD (Sec. 19-5, footnote 4).**

Here effluent recirculation introduces the number of effective passes, $N_e$ of Eq. 11-37. Filter depth and the associated mean detention or residence time as well as the hydraulic load and specific surface of the contact material are tied into performance through Eq. 18-11 when the performance coefficients $k$ or $\Sigma k$ and $n$ are known. Displacement times should be measured rather than calculated. For single-stage operation without recirculation, the Ten-State Standards[4] trace the loading curves shown in Fig. 19-1.

*Example 19-1.*

A settled wastewater flow of 10 mgd containing 160 mg/l of 5-day, 20°C BOD is to be applied to a trickling filter 6 ft deep. Find (1) the BOD load in pounds daily; (2) the acre-feet of bed required to remove 80% of the BOD without recirculation; (3) the

[4] *Recommended Standards for Sewage Works*, Great Lakes-Upper Mississippi River Board of State Sanitary Engineers, Health Education Service, Albany, N.Y., 1968.

BOD loading per acre-foot; (4) the required acreage; (5) the optimal recirculation ratio for a weighting factor of 0.8; and (6) the increased efficiency attributable to this recirculation.

1. BOD loading $y_0 = 10 \times 160 \times 8.34 = 13,300$ lb daily.
2. By Eq. 19–1, $C = (13,300/13,800)(80/20)^2 = 15.4$ acre-ft.
3. BOD loading per acre-ft: $y_0/C = 13,300/15.4 = 865$ lb/acre-ft.
4. Acreage: $15.4/6 = 2.57$ acres.
5. By Eq. 11–38, $R/I = (1.6 - 1)/(1 - 0.8) = 3.0$, and $N_e = (1 + 3)/(1 + 0.2 \times 3)^2 = 1.56$.
6. By Eq. 18–22, $P = 100/[1 + 0.0085(13,300/15.4 \times 1.56)^{1/2}] = 83.5\%$.

---

Although effluent recirculation does not increase efficiency substantially, it has the advantage of keeping reaction-type distributors in motion, filter media moist, organic loadings more or less constant, and contact times with top film long; moreover, it improves distribution, equalizes unloading, obstructs entry and egress of filter flies, freshens incoming and applied sewage, reduces the chilling of filters, and narrows the variation in time of passage through the secondary settling tank. Recirculation of secondary sludge to the primary tank keeps the final effluent fresh. Rates of recirculation may be varied to offset low influent flows, dampen peak flows, or remain proportionate to incoming flows; they may also be held constant or made adjustable to more than a single value.

Because municipal wastewaters are generally warm, the temperature within trickling filters does not depart much from 20 to 25°F throughout the year and seldom varies by more than 5°F in any one day. High-rate filters remain warmer than low-rate filters, but recirculation pulls the temperature down in cold weather.

## 19-6 Hydraulic Design of Trickling Filters

To supply the saprobic organisms that colonize trickling filters with nutrients and oxygen in a continuous stream, the effluents of primary treatment units are distributed over the bed in droplets, thin sheets, or spreading jets from fixed or movable sprays (Fig. 18–1). Although rotating distributors have displaced fixed nozzle fields almost completely, a large acreage of fixed sprays remains in operation in the United States and abroad.[5]

[5] Traveling distributors that move up and down the length of a bed are not in favor because a relatively long time intervenes between successive passages of the distributor over a given portion of the bed. Dosage is concentrated (therefore at a high local rate) and widely intermittent.

**Distribution.** Rotating manifolds are composed of two or more horizontal, structurally balanced pipes or arms attached to a central supply and support column. The wastewaters undergoing treatment issue horizontally from ports on one side of the revolving arms. The ports may be orifices or detachable nozzles, and their jets may be fanned out by adjustable spreader plates to cover all parts of the bed. The pipes are mounted high enough above the bed surface to leave clearances of at least 3 in. after suitable allowance has been made for expected ice accumulations in cold climates. The arms are driven either by the reaction, that is, the jet action, of the spray or by an enclosed electric motor that turns the central column. Small distributors turn at approximately 2 rpm, large ones more nearly at ½ rpm.

Self-propelled distributors are normally built to cover beds from 20 to 200 ft in diameter. The piezometric head on the center line of their pipes is generally 12 to 24 in. for propulsion at minimal flows. Flow velocities are normally less than 4 fps. The revolving central column may be provided with a mercury, oil, or mechanical seal. Some distributors are equipped with overflow devices through which high flows are diverted into arms or sections of arms that are kept idle at low flows. A sampling of distributor capacities and dimensions is 0.16 to 1.25 to 3.30 mgd for filters 40 to 100 to 150 ft in diameter, respectively, feedlines being 6 to 10 to 14 in. in diameter while arms are 2½ to 4 to 8 in. in diameter.

Water-wheel drive is normally restricted to distributors covering beds 5 to 25 ft in diameter and operating under a minimum head of about 2 ft. Motor-driven distributors normally straddle beds 25 to 150 ft in diameter. They require little head. Hydraulic and motor-driven disks 12 to 30 in. in diameter and revolving at 300 rpm 20 in. above the bed surface under a head of 10 to 20 in. can cover beds 13 to 34 ft in diameter with spray. Stationary nozzle fields need heads of 5 ft or more.

The flow through a given port of a rotating manifold at a distance $l$ from the entrance is determined by the available head, $h$, which can be assumed to approximate the sum of the entrant static and velocity heads $(h_e + v_e^2/2g)$ reduced by the friction loss, $h_f$, as far as the port; and by the remaining velocity head $v^2/2g$, but increased by the head, $h_c$, engendered by the centrifugal force.[6] For a given manifold, these components of the general approximation

$$h \simeq h_e + v_e^2/2g - h_f + h_c - v^2/2g \qquad (19\text{-}2)$$

are determined by Eq. 12–9 for $h_f$, and by $h_c = (2\pi l N)^2/(2g \times 60^2)$ for the centrifugal head, where $N$ is the number of revolutions per min of the arm,

---

[6] C. J. Ordon, Manifolds, Rotating and Stationary, *J. Sanit. Eng. Div.*, Am. Soc. Civil Engrs., **92**, *SA*1, 269 (1966).

and $g$ is the gravity constant. It is assumed that the kinetic energy of the flowing water is fully recovered at each takeoff. Because this is not necessarily so, Eq. 19–2 is only approximately correct. Tapered arms suppress the recovery of velocity heads at takeoff and the difference term $(v_e^2 - v^2)/2g$.

**Dosing Tanks.** Unless the incoming flow is unusually even, or effluent is recirculated, an automatic dosing tank must be interpolated between the primary units and the filters. When flows drop below the rate required to turn the distributor, the feed is cut off until the tank has filled and can supply water at needed rates (Fig. 18–1). Dosing tanks add from 1 to 5 ft to hydraulic head requirements. Recirculation makes dosing tanks unnecessary.

When necessary, dosing tanks (Fig. 19–2) can be provided for intermittent sand filters, low-rate trickling filters, and other nonrecirculating filters. They are also introduced to keep hydraulically driven distributors from stopping at times of inadequate flows. Routing wastewaters through dosing tanks is quite the same thing as routing floods through reservoirs (Sec. 4–9). For a specified time element $\Delta t$, outflow $Q \Delta t$ equals inflow $I \Delta t$ plus change in tank storage $\Delta S$ (Eq. 4–7), because the outflow passes through the orifices of the distributor, $Q = K\sqrt{h}$, where $h$ is the head of the water surface above the center of the ports, and the coefficient $K$ includes pipe and distributor friction, the size of the port area, the orifice coefficient, and $\sqrt{2g}$; and $\Delta S = A \Delta h$, where $A$ is the cross-sectional area of a dosing tank with vertical sides (Fig. 19–2). In differential form, therefore,

$$A \, dh = (K\sqrt{h} - I) \, dt$$

or

$$dt = [A/(K\sqrt{h} - I)] \, dh \qquad (19-3)$$

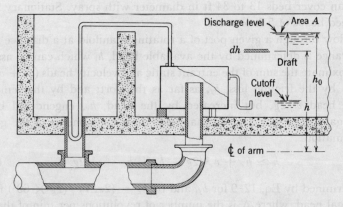

**Figure 19-2.**
**Dosing tank for rotary distributor.**

Integration between the limits 0 and $t$ and $h_0$ and $h$ (Fig. 19-2) then leads to the equation

$$t = \frac{2A}{K^2}\left[K(\sqrt{h_0} - \sqrt{h}) + I \log \frac{K\sqrt{h_0} - I}{K\sqrt{h} - I}\right] \quad (19\text{-}4)$$

A sampling of dosing-tank dimensions in feet is $(4 \times 4 \times 2\frac{1}{2})$ to $(10 \times 8 \times 2\frac{1}{2})$ to $(15 \times 12 \times 2\frac{1}{2})$, and static heads for 12-in. drawdowns of 18 and 18 to 21 in. for distributor capacities of 0.16 to 1.25 to 3.30 mgd, respectively.

**Underdrainage and Ventilation.** Filter floors are normally constructed of concrete with steel-mesh reinforcing. They are sloped at 0.5 to 5%. Diagonal or peripheral main drainage channels receive lateral flows from precast, perforated, vitrified-tile or concrete blocks. The blocks are laid at right angles to the main drains. Their perforations face upward and occupy 20% or more of their top surface. To promote ventilation, the channels within the blocks are designed to flow no more than half full. Velocities are normally set at 2 to 3 fps. Because sloughed film is carried along by the treated waters, it should be possible to clean all active channels. The laterals may be carried through the walls to permit flushing the underdrainage system from the outside and to assist in the ventilation of the filter. Hydraulically, the principal components of the underdrainage system of trickling filters function in much the same capacity as the washwater-collecting system of rapid sand filters. They are designed accordingly.

Because moist warm air is lighter than dry cold air, the air in trickling filters is normally displaced upward through them much of the time. According to Halvorson, Savage, and Piret,[7] air sinks through trickling filters when the outside air is warmer than the wastewater temperature by $+3.6°F$ or more; it rises through filters when the outside air is colder than the wastewater temperature or differs from it by less than $+3.6°F$.

Filters are generally encircled by reinforced concrete walls, which may rise well above bed level to act as windbreaks. Vitrified- and cement-block walls need to be reinforced. Other possibilities are precast silo construction and the use of corrugated-metal or chain-link fencing. If the beds are to be flooded for the control of trickling-filter flies, the walls must be built strong and tight, and there must be (1) gates or locks on the terminal drainage channels to close them down and (2) overflows at bed level to carry away possible spillage.

[7] H. O. Halvorson, G. M. Savage, and E. L. Piret, Some Fundamental Factors Concerned in the Operation of Trickling Filters, *Sewage Works J.*, **8**, 888 (1936).

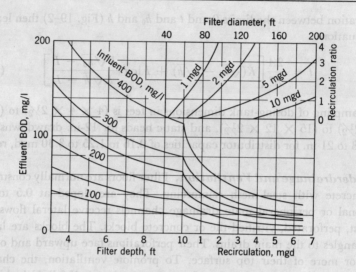

**Figure 19-3.**
**Optimal dimensions of trickling filters operated at 18°C (64°F) and maximum recirculation ratios. Hydraulic load 30 mgad, maximum depth 10 ft. (After Galler and Gotaas.)**

*Optimization.* A graphical determination of the optimal dimensions of trickling filters developed by Galler and Gotaas[8] is illustrated in Fig. 19-3. The curves shown were obtained by linear programming based, therefore, on a linear function expressing the objective of the program and a set of simultaneous linear inequalities expressing the constraints. Among the conclusions drawn in this study are the following: (1) that filter depths and recirculation ratios are functions of the process-loading and wanted efficiencies; (2) that hydraulic loading is maximal until the recirculation ratio is 4:1; (3) that at ratios less than this, the cost of increasing filter size is greater than the cost of increasing recirculation; (4) that where pumping the influent is not a factor, depth and adequate air circulation are rewarding; and (5) that except for the advantages of flexibility, single units are optimal.

*Example 19-2.*

Given a flow of 10 mgd and a BOD of 200 mg/l, find (1) the optimal depth ≤ 10 ft and the associated recirculation requirements of a trickling filter in conformance

[8] W. S. Galler and H. B. Gotaas, Optimization Analysis for Biological Filter Design, *J. Sanit. Eng. Div., Am. Soc. Civil Engrs.*, **92**, SA1, 163 (1966).

with assumptions and calculations of Galler and Gotaas;[8] (2) the optimal number of filters; and (3) the optimal hydraulic loading and diameter of the filters.

1. From Fig. 19-3, read for BOD values of 200 and 20 mg/l, respectively, the limiting depth as 10 ft and the recirculation ratio as 3.25.

2. Next find, for the hydraulic load of 10 mgd, a limiting diameter of 200 ft and a limiting recirculation ratio of 1:2. Accordingly, the number of filters must be more than 1—for example, 2.

3. For two equal filters and a hydraulic load of 5 mgd for each filter, find that the recirculation ratio of 3.25 can be obtained for two filters each 200 ft in diameter.

The National Research Council formulation would predict an effluent BOD of 18 mg/l for these filters.

## 19-7 Activated-Sludge Units

In activated-sludge units (Fig. 19-4), zoogleal microbic growths, matrices, or flocs produced within settled wastewaters and systematically returned to them are kept aerobic and in circulation and suspension by mechanical or pneumatic stirring. The structural units needed for this *suspended contact* are (1) tanks and terminating channels with appended settling tanks through which all of the flow passes, and (2) channels and ditches that close or return on themselves and may or may not discharge their effluents through separate settling units. The flocs teem with bacteria, fungi, and protozoa. For this reason they were called *activated-sludge flocs* by Ardern and Lockett, the originators of the process.[9]

The oxygen requirements of activated-sludge flocs are supplied by absorption from air originating either in the overlying atmosphere or in compressed air injected into the flowing wastewaters. Water-surface renewal by stirring the wastewaters, with or without the formation of droplets, promotes the absorption of air from the atmosphere. Air bubbles are introduced into the flowing waters as compressed air from diffusers or as air drawn into the wastewaters by hydraulic or mechanical devices. To reach needed floc concentrations and intensities of exposure, flocs formed in the course of many days are pumped back to the influent or flowing waters as returned sludge with or without intermediate settling, thickening, or reaeration (Sec. 19-9). After a steady state has been reached, some floc must be wasted as *excess sludge* instead of being recycled as *returned floc*.

[9] Edward Ardern and W. T. Lockett, Experiments on the Oxidation of Sewage without the Aid of Filters, *J. Soc. Chem. Ind.*, **33**, 523, 1122 (1914). This classical study should be known to every engineer and scientist in the field. Because the designation of wastewater solids as "sludge" normally connotes a waste product, it might be preferable to call the process biological flocculation or *bioflocculation* and to refer to the treatment units as *bioflocculation tanks* or *channels*.

# 570/Biological Treatment Systems

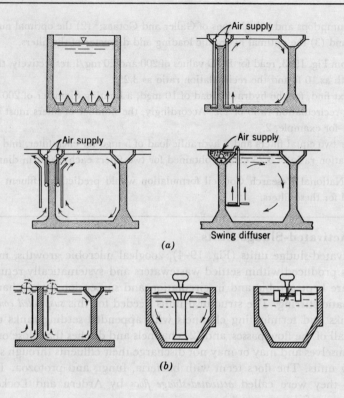

**Figure 19-4.**
Activated-sludge channels and tanks. (a) Diffused-air channels: longitudinal furrows, spiral flow with bottom diffusers, spiral flow with baffle and low-depth diffusers, and swing diffusers. (b) Mechanical aeration units: surface paddles, draft-tube and blade type.

## 19-8 Design of Activated-Sludge Units

A flow diagram for the prototype units constructed and tested at Manchester in 1916 and modified in the course of some years by Ardern and Lockett is shown in Fig. 19-5a. Because this system has been preserved with few changes in new as well as old treatment works, it is commonly referred to as the *conventional* design. Changing designs and operations can be classified usefully either in terms of the nutritional process being promoted, its hydraulics, and methods of accomplishment, or in terms of the constructions and appurtenances in which the process is conducted. Common examples are shown in Table 19-2. Table 19-3 illustrates typical loadings for various activated-sludge process formulations.

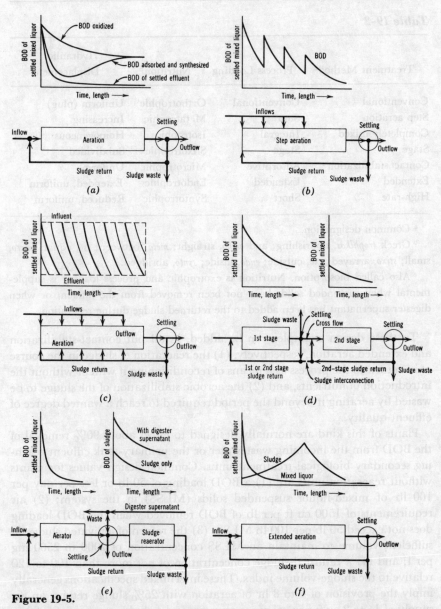

**Figure 19-5.**
Activated-sludge treatment methods. Salient flow and nutrient characteristics are identified in Sec. 19-8. (*a*) Conventional, orthotrophic; (*b*) step aeration, metatrophic; (*c*) completely mixed, isotrophic; (*d*) stage, stixotrophic; (*e*) contact stabilization, microtrophic; (*f*) extended aeration, endotrophic. (Not to scale.)

## Table 19-2

| Treatment Method[a] | Process Loading | Nutrition[b] | Hydraulic Displacement |
|---|---|---|---|
| Conventional | Conventional | Orthotrophic | Uniform (plug) |
| Step aeration | Step | Metatrophic | Increasing |
| Completely mixed | Integral | Isotrophic | Homogeneous |
| Stage | Stage | Stixotrophic | Subdivided |
| Contact stabilization[c] | Adsorptive | Microtrophic | Uniform |
| Extended | Extended | Endotrophic | Extended, uniform |
| High-rate | Short | Syntotrophic | Reduced, uniform |

[a] Common designation.

[b] Greek *trophikos*, nourishing; and *ortho*, straight; *meta*, between; *iso*, equal; *micro*, small; *stixo*, arrayed; *exo*, outside; *endo*, inside; *synto*, abridged.

[c] Also called biosorption. Nutrition is exotrophic and process loading is supplemental when suspended solids have not been removed from the influent or when digester supernatant has been added to the returned sludge during reaeration.

To these design examples can be added, in all but contact-stabilization and extended aeration, respectively, (1) the reaeration of sludge in the course of its return to the process as a means of reconditioning it with or without the introduction of nutrients, and (2) the aerobic stabilization of the sludge to be wasted by aerating it beyond the period required to reach a wanted degree of effluent quality.

Plants of this kind are normally designed to effect about 90% removal of the BOD from the incoming wastewater or the primary-tank effluent reaching secondary biological-treatment units. Common design values for plants without reaeration units are (1) a BOD loading of 50 lb or less per day per 100 lb of mixed-liquor suspended solids (MLSS) in the system; (2) air requirements of 1500 cu ft per lb of BOD removed when the BOD loading does not exceed 50 lb per 100 lb MLSS; (3) the return of activated sludge in sufficient volume to maintain an MLSS concentration of 2000 to 2500 mg per l; and (4) a returned-sludge concentration of not more than 100 to 120 relative to the sludge-volume index. These interlinked specifications generally imply the provision of 4 to 8 hr of aeration with 25% sludge return, an air supply of ½ to 2 cu ft per gal of wastewater, and a sludge wastage of 5000 to 10,000 gal per mg of wastewater.

Reaeration, although seldom included in the past in American treatment works, is recognized as being of value in (1) returning a well-aerated, thickened, and active sludge to the aeration units, (2) reducing the size of the

## Table 19-3
### Aeration Tank Capacities and Permissible Loadings

| Process | MGD-Plant Design Flow | Aeration Retention Period-Hours (Based on Design Flow) | Plant Design, lb BOD$_5$/Day | Aerator Loading, lb BOD$_5$/1000 cu ft | MLSS/lb BOD$_5$[a] |
|---|---|---|---|---|---|
| Conventional | To 0.5 | 7.5 | To 1000 | 30 | 2/1 to 4/1 |
|  | 0.5 to 1.5 | 7.5 to 6.0 | 1000 to 3000 | 30 to 40 |  |
|  | 1.5 up | 6.0 | 3000 up | 40 |  |
| Modified or "high rate" | All | 2.5 up | 2000 up | 100 | 1/1 (or less) |
| Step aeration | 0.5 to 1.5 | 7.5 to 5.0 | 1000 to 3000 | 30 to 50 | 2/1 to 5/1 |
|  | 1.5 up | 5.0 | 3000 up | 50 |  |
| Contact stabilization | To 0.5 | 3.0 (in contact zone)[b] | To 1000 | 30 | 2/1 to 5/1 |
|  | 0.5 to 1.5 | 3.0 to 2.0 (in contact zone)[b] | 1000 to 3000 | 30 to 50 |  |
|  | 1.5 up | 1.5 to 2.0 (in contact zone)[b] | 3000 up | 50 |  |
| Extended aeration | All | 24 |  | 12.5 | As low as 10/1 to As high as 20/1 |

[a] Normally recommended values at ratio of mixed liquor suspended solids under aeration to BOD loading.
[b] 30 percent to 35 percent of total aeration capacity. Reaeration zone comprises the balance of the aeration capacity.

From *Recommended Standards for Sewage Works*, footnote 4.

combined aeration units, (3) increasing the flexibility of the plant by holding in reserve sludge otherwise wasted, and (4) improving sludge-digester liquor that is returned to the plant for treatment.

The oxygen requirements of the process, the BOD removal (tested on settled samples), and the sludge buildup are all high at the beginning. They drop off as individual volumes of mixed liquor move through the aeration unit. Kinetically, overall progress is first-order or retarded first-order. The process becomes endogenous toward the end. Zooglcal, settleable flocs are formed as available nutrients are exhausted. Endogeny stabilizes the sludge during prolonged aeration or during reaeration.

Most modifications of the conventional process have looked to its improvement not so much in terms of attainable removals as in terms of reduced tank capacities, lowered air and power requirements, increased concentrations of returned and wasted sludge, and decreased process instability. Examples of these modifications follow.

*Load Distribution or Step Aeration* (Fig. 19-5b). By introducing the wastewater uniformly or in a number of increments into otherwise conventional aeration channels, the organic load is distributed over the operating range of the unit and the effects of fluctuations in influent and sludge quality are suppressed. Nutrients support the floc population more uniformly, oxygen demands and requirements drop within a narrower range, and steady-state operation as well as isotropic conditions or complete mixing are approached.

*Isotrophism or Nutrient Equalization by Complete Mixing* (Fig. 19-5c). This process introduces operational equality into large aeration channels by further elaboration of the load-distribution process. The performance curves of Fig. 19-5c approach horizontal straight lines but are conjoined in the sketch to indicate what is taking place within the unit.

*Two-Stage Treatment* (Fig. 19-5d). Flow is through a pair of aeration and sedimentation tanks in series. Sludge is either returned or wasted within each stage, or excess sludge from one stage is recycled to the other. Thence it is wasted along with the sludge of this stage. In this way the quality of both sludges can be exploited to best advantage. The process offers the physical option of contact stabilization.

*Contact Stabilization* (Fig. 19-5e). This process acknowledges the transfer and operational potential of the first 20 to 40 min of aeration. Accordingly, it separates the settleable sludge from the wastewater at the end of this period and continues the assimilation of soluble organics in a second aeration unit, to which digester supernatant may be added as a nutrient. Sludge from that unit alone is normally returned to the incoming wastewater.

*High-Rate Operation.* Partial treatment or an intermediate degree of treatment is the aim of this process. Short-period aeration is combined either with a thin mixed-liquor concentration of 650 mg per l, for example, or with a very thick mixed-liquor concentration. In the first instance the food to microorganism ratio is high; in the second, relatively low. Nevertheless, the rates of food assimilation can be essentially the same. Physical arrangements are foreshortened but otherwise much the same as those shown in Fig. 19-5a.

**Extended Aeration.** Because the principal service of extended aeration is the stabilization of the sludge to be wasted, Fig. 19-5f follows the BOD remaining in the sludge as well as that of the wastewater undergoing treatment. The dropoff in sludge BOD is brought about by endogeny.

## 19-9 Activated-Sludge Process Design

Activated-sludge aeration units can accept suspended as well as dissolved solids with impunity. Therefore some activated-sludge plants have confined primary or preparatory treatment to grit removal followed by screening; others, to brief sedimentation. A reason once advanced for eliminating normal primary settling is the production of as much marketable sludge as possible, but with disappearing markets this is no longer valid. Moreover, in accordance with the general principle that unit operations normally perform best when design accents their unique capacities, the great ability of activated sludge to remove organics from solution has generally persuaded engineers to relieve aeration units of the settleable load as much as possible. Similar in concept, but opposite in orientation, is the passage of wastewaters containing strong industrial wastewaters through *roughing filters*, that is, high-rate trickling filters, in advance of activated-sludge tanks. The purpose, in this instance, is to let the robust trickling filter protect the sensitive activated-sludge unit against shock loads of difficult and possibly toxic wastes.

Within activated-sludge aeration and sludge-separation units the stress in process design is on (1) the transfer of organics to dispersed bacterial cells and growths as well as to the active sludge, (2) agglomeration of the cells, biological flocs, and debris into settleable solids, (3) conditioning the floc for return to the process as *activated* sludge either in the normal course of aeration or by separate reaeration of the returned sludge, sometimes with the addition of sludge-digester liquor as a source of nutrients, including phosphate and nitrogen, and (4) the optional aerobic stabilization of excess solids during *extended aeration*, an option normally exercised by enlarging the principal aeration units or by constructing separate units for this specific purpose. Available reaeration units may also be diverted to this purpose. Further treatment of waste sludge by anaerobic digestion, for instance, may be avoided in this way.

In practice, the amount of returned sludge that activates the process is identified in three ways for design and operating purposes: (1) as the relative volume of activated sludge returned to the influent, (2) as the volume of suspended solids settling from the mixed liquor, and (3) as the dry-weight concentration of solids settling from the mixed liquor. Identified, too, is the ratio of (3) to (2) as the *sludge-volume index*.

*The percentage concentration of activated sludge by volume*, $P_s$, returned to the influent varies in practice from 10 to 30% with an average of 20% of the inflow. This figure is used in determining the capacities of pumps and conduits that transport the returned sludge. It is not a reliable measure of the amount of contact material or adsorptive surface in the aeration units, because the concentration of active solids in the returned sludge (and with it their relative surface area) varies widely in different plants and periodically in the same plant. A dry-solids concentration in the sludge of 0.2% by weight is undesirably low; one of 2% is desirably high.

*The volume concentration of suspended matter*, $P_v$, in the mixed liquor of the aeration unit varies normally between 10 and 25%. This ratio is generally found by measuring the volume of sludge settling from 1 l of mixed liquor in 30 min. The procedure is not analytically exact, because the settling and compacting properties of the suspended matter vary over wide ranges. But the test is inherently of the same order of magnitude as the percentage of returned sludge, because the influent settleable wastewater normally contains less than 0.05% of the settleable solids in the wastewater before primary settling.

*The dry-weight concentration of suspended solids* or volatile suspended solids in the mixed liquor can be determined with high analytical precision. In practice, the concentration of suspended solids by weight, $P_w$, varies between 0.06 and 0.4% (600 to 4000 mg per l) and averages 0.25%, the volatile portion being about three quarters of the total. Because no better means is currently available, this determination is used in the performance relationships discussed in this chapter, even though it is only an indirect measure of the area of contact surface provided.

*The sludge-volume index*, $I_v$, or ratio of the volume concentration to the dry-weight concentration of suspended solid, $P_v/P_w$, is specifically the volume in milliliters occupied by 1 g of sludge, dry weight, after 30 min of settling. Calculations are normally based on the ratio in the mixed liquor of the milliliters of settling sludge $\times 10^3$ to the milligrams of suspended solids per liter. The sludge-volume index of well-settling active sludge lies between 50 and 100. By contrast, a *bulked* sludge has an index of 200 or more.

The sludge-volume index is useful in treatment-process management. The amount of returned sludge that will maintain a given percentage of solids, $P_w$, in the mixed liquor is established by the recirculation ratio, $R/I$, and the sludge-volume index as

$$R/I = 1/\{[100/(P_w I_v)] - 1\} \tag{19-5}$$

Here $R$ is the recirculated flow and $I$ is the inflow, and it is assumed that the solids concentration of the sludge pumped from the secondary settlers is the same as that of the sludge-volume index test. Because settling conditions are different, there may be significant departures from this relationship. How sensitive operations are to the sludge volume index is shown in Example 19–3.

*Example 19–3.*

Find the percentage of activated sludge with a sludge-volume index of (1) 80 ml/g and (2) 200 ml/g that must be returned to an aeration tank in order to maintain a mixed-liquor suspended-solids concentration of 0.25% by weight.

1. By Eq. 19–5, $100 R/I = 100/\{[100/(0.25 \times 80)] - 1\} = 25\%$.
2. By Eq. 19–5, $100 R/I = 100/\{[100/(0.25 \times 200)] - 1\} = 100\%$, and operations would become marginal. They would be hampered further by the poor settling and compacting properties of the bulking sludge.

---

Because $W = 8.34 CP_w/(10^2 \times 10^3) = 8.34 \times 10^{-5} CP_w$, where $W$ is measured in thousands of pounds and $C$ is the tank volume in gallons, and because $t_d = 24C/I$, where $t_d$ is the detention time in hours and $I$ is the inflow in gallons daily, the capacity of a tank that will afford a contact opportunity $Wt_d$ is

$$C = 22.4\sqrt{I(Wt_d)/P_w} \tag{19-6}$$

and the associated detention time is

$$t_d = 538\sqrt{(Wt_d)/(IP_w)} \tag{19-7}$$

To these equations can be added the performance-loading relationship of Eq. 18–22, and the assessment of its coefficients shown in Table 18–2 and Fig. 18–10 in order to yield the contact opportunity as

$$Wt_d = \frac{y_0}{4200}\left(\frac{P_2}{100 - P_2}\right)^{2.38} = \frac{y_0}{4200}\left(\frac{y}{y_0 - y}\right)^{2.38} \tag{19-8}$$

in which the detention period $t_d$ establishes the volume $C$ of the aeration unit as $C = It_d$.

---

*Example 19–4.*

Find (1) the allowable loading intensity or required contact opportunity of a diffused-air, activated-sludge plant that is to effect 80% removal of BOD from 10 mgd of wastewater containing 160 mg/l of applied BOD, (2) the necessary hours of aeration to maintain a solids concentration of 0.25% in the aeration units, (3) the necessary solids concentration for a detention time of 6 hr, and (4) the approximate percentages of sludge to be returned to the influent under (2) and (3), if the sludge-volume index is 80.

1. Because the applied load is $160 \times 10 \times 8.34 = 13.300$ lb, Eq. 19–8 states that $Wt_d = (13,300/4200)(80/20)^{2.38} = 86,300$ lb-hr.

2. If the suspended-solids concentration in the aeration units is 0.25%, the detention time $t_d$ is given by Eq. 19–7 as $t_d = 538 \sqrt{86.3/(10^7 \times 0.25)} = 3.16$ hr.

3. If the time of aeration is 6 hr, Eq. 19–6 states that $P_w = (538)^2 \times 86.3/[10^7 \times (6)^2] = 0.0694\%$.

4. For $I_v = 80$, the percentage of returned sludge would have to be about as follows:
For $P_w = 0.25\%$, $100R/I = 100/\{[100/(0.25 \times 80)] - 1\} = 25\%$.
For $P_w = 0.0694\%$, $100R/I = 100/\{[100/(0.0694 \times 80)] - 1\} = 5.9\%$.

A commonly accepted value of the permissible loading of activated sludge is 50 lb of BOD per day per 100 lb of suspended solids in the aeration system. So long as this loading intensity is not exceeded, a reduction of 90% of the influent BOD can be expected. Higher loadings will result in lower relative removals.

### 19-10 Activated-Sludge Air Requirements

In diffused-air plants, needed oxygen is supplied by air diffused into the sewage (bubble aeration) and by the atmosphere in contact with the sewage surface (surface aeration). In mechanical-aeration plants it comes solely from the atmosphere. Common transfer agents are (1) water droplets formed by rotating brushlike or turbinelike devices that dip into the mixed liquor or pull it upward through draft tubes and shower droplets over the tank surface and (2) air bubbles generated by rotating turbinelike devices that pull air and mixed liquor downward through draft bubes. The transfer of oxygen to droplets is more rapid than the transfer of oxygen from bubbles of air, because the interfacial film is about a third as thick. However, the attainable time of exposure of droplets is ordinarily quite short.

The oxygen-transfer efficiency of bubble aeration lies normally between 5 and 15%. Because a liter of free air contains about 273 mg of oxygen, 14 to 41 mg of oxygen can be absorbed from each liter of air diffused into the tank.

The rate of oxygen demand of mixed liquor and the basic rate of returned sludge are idealized in Fig. 19–6 for the conventional treatment process. A maximum hourly demand of 50 to 80 mg per l per 1000 mg per l of volatile

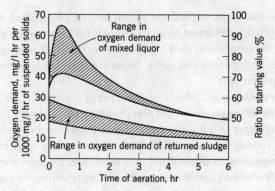

**Figure 19-6.**
**Oxygen demands in the activated-sludge process.**

suspended solids exerted near the beginning approaches the average hourly base rate of the returned sludge of about 20 mg per l per 1000 mg per l of volatile suspended solids in the course of 4 to 6 hr. Aeration of activated sludge by itself reduces its initial hourly base rate of 25 to 35 mg per l per 1000 mg per l of volatile suspended solids by about 25 to 50% during 4 to 6 hr. Therefore a liter of mixed liquor containing 0.25% (2500 mg per l) of suspended solids must be provided, during 5 hr of aeration (about 6 hr of detention), with $5 \times 2.5\{16 + \frac{1}{3}[(40 - 16) \text{ to } (64 - 16)]\}/[273 \times (0.05 \text{ to } 0.15)] = 7.3 \text{ to } 29.3$ l of air.[10] If it is assumed that half this amount, or 3.7 to 14.7 l, is derived from the atmosphere by surface aeration, diffused-air requirements may be estimated at 0.5 to 2.0 cu ft per gal of wastewater. Actual requirements range from 0.5 to 1.5 cu ft per gal of wastewater of average strength to much higher values for strong industrial wastes. High air supply and long aeration periods are generally required for strong wastewaters and the production of highly nitrified effluents. Low air supply and short aeration periods are adequate for weak wastewaters and the production of unnitrified effluents containing the minimum amount of dissolved oxygen ordinarily needed in aeration units (about 1 mg per l). Because the oxygen requirements of the conventional process decrease progressively in time (Fig. 19-6), air supply has been proportioned to them when wastewater and returned sludge are brought together at the influent to aeration units. To this purpose aeration has been tapered by providing at the beginning of the tank either more than the average number of diffusers per unit area of tank or diffusers of greater permeability. Air requirements are substantially uniform in unbaffled conventional units, in step-aeration or distributed-loading units, and in isotrophic units.

[10] Here the parabolic nature of the curve introduces the factor $\frac{1}{3}$ in the numerator.

Generally included in overall air requirements of diffused air plants are (1) the air supplied to returned-sludge channels for keeping the sludge in good condition and in suspension and (2) the air introduced into influent wastewater channels and mixing channels. Separate account is normally taken of reaeration requirements.

**Use of Pure Oxygen.** Much of the energy consumed in conventional aeration systems serves no useful purpose, because 80% of air is inert nitrogen. Oxygenation systems using pure oxygen gas began to be feasible with the availability of tonnage oxygen (95% pure), now widely used in steel manufacture. Tonnage oxygen is obtained and used directly from the liquefaction of air and its fractional distillation, without liquefaction or compression of the oxygen for storage. Studies using tonnage oxygen for biological wastewater treatment indicated that substantial savings in capital cost for aeration tanks are possible.[11] The use of pure oxygen, which has fivefold greater dissolved-oxygen saturation values than does air (Equation 12-2), permits much higher concentrations of biological floc to be maintained in the system, thereby reducing the required contact volume. An ancillary benefit of the high dissolved-oxygen levels is that filamentous microorganisms are suppressed and the biological floc has a low sludge volume index and is more easily thickened and dewatered. Pure oxygen offers promise in industrial areas, where the oxygen may be available as a utility, and in existing plants that need to be enlarged but where space is at a premium.

## 19-11 Hydraulic and Mechanical Design of Activated-Sludge Units

Aeration units are constructed as long aeration *channels* with much transverse and some longitudinal mixing, or as circular or square *tanks* with essentially complete mixing. Long channels have been used primarily for the treatment of large volumes of wastewater; circular and square tanks principally in small treatment works or package plants and for exploiting mechanical aeration devices.

Diffused-air and mechanical stirring perform three major functions: they supply needed oxygen, keep the active sludge in suspension, and foster contact between wastewater and sludge. Point velocities must be at least 0.5 fps and preferably 1.0 fps if sludge flocs are not to settle out and become septic. Displacement velocities in returned-sludge piping and channels are usually 2 fps. For this reason hydraulic circulation is an important element in tank design. To provide it, there are added to the longitudinal displacement

---

[11] Daniel A. Okun, System of Bio-precipitation of Organic Matter from Sewage, *Sewage Works Journal*, 21, 763 (1949) and W. E. Budd and G. F. Lambeth, High-Purity Oxygen in Biological Sewage Treatment, *Sewage and Industrial Wastes*, 29, 237 (1957). The advantages of using pure oxygen in covered conventional aeration tanks has been demonstrated by The Linde Division of Union Carbide Corp. in 1970.

velocity a relatively rapid upward vertical velocity, a complementary down ward velocity, and transverse links. Velocities well above 0.5 fps are induced If the aerator is placed on one side of a tank, spiral circulation is induced, which is well maintained if the channel width to be spanned is no more than three times its depth.

Circulation is helped by deflectors at the surface and fillets at the bottom of wide channels. A circular cross-section is then approximated in square channels and an elliptical cross-section in rectangular channels. Where air is supplied from tubular diffusers that are submerged only a few feet (Fig. 19-4), an air lift can be created by longitudinal baffles close to the diffusers. Channels are seldom less than 100 ft or more than 400 ft long. For structural economy, depths are generally 10 to 15 ft with up to 3 ft of freeboard. Expansion and coalescence of air bubbles also circumscribe the depth of diffused-air units. There is little evidence that fine bubbles are more efficient overall than coarse bubbles, provided that equal power is introduced.

The width of spiral-flow tanks is generally 1.5 to 2 times their depth, and the length of individual channels is normally from 10 to 20 times their depth. The diffuser area is from 2.5 to 15% of the maximum horizontal tank area.

Initially, mechanical surface aerators with vertical shafts were used in small "package" activated sludge plants, but their simplicity of design, installation, and operation has encouraged their adoption for large plants as well. Mechanical aerators are placed in the center in small tanks and are distributed uniformly over the area of large tanks, with each unit serving an area approximately square in shape. The units may be suspended from beams across the tank, or in large tanks from platforms supported by piers. Floating units may be used in very large tanks.

The draft tube aerator (Sec. 12-11) permits much deeper tanks to be used than is feasible for diffused air. Deep tanks are advantageous where space is limited or where the tanks are to be covered. Covered tanks may be desirable for aesthetic reasons or to contain oxygen-enriched air where pure oxygen is used for aeration.

Inasmuch as mechanical aerators are proprietary, their selection is based on their performance characteristics, much in the way that pumps are selected. The engineer will determine the type aerator to be used, but its size and the horsepower of the motor are chosen with the aid of the manufacturer's engineers.

## 19-12 Air Supply for Diffused-Air Units

Economic design of the air-distribution system commonly entails an overall distribution loss of about 25% of the wastewater depth. Therefore the required air pressure is $1.25 \times (10 \text{ to } 15)$ ft, or 5.4 to 8.1 psig. The resistance of the different portions of the system can be gaged from the following average component values: (1) air filters, $\frac{1}{8}$ to $\frac{3}{8}$ in. of water for viscous filters and $\frac{1}{2}$ in. for cloth filters; (2) air meters, 1 to 2 in. of water, or 18 to 24% of the differential head; (3) piping, $1.0\ v^2/2g$ ft head of air for 40 diameters of pipe[12] with $0.5\ v^2/2g$ ft head of air for elbows and $1.5\ v^2/2g$ ft head of air for globe valves, pipe velocities commonly being 2000 and 3000 fpm; (4) diffusers, 2 to 15 in. of water; and (5) wastewaters, 10 to 15 ft of water.

**Air Compression.** The theoretical power requirements of compressors are

$$P = 0.22Q[(p/14.7)^{0.283} - 1] \tag{19-9}$$

where $P$ is the theoretical horsepower for the compression of $Q$ cfm of free air from atmospheric pressure (14.7 psia) to a working pressure of $p$ psia, the ratio $n$ of the specific heats of air at constant pressure and constant volume being 1.40, or $(n - 1)/n = 0.283$. Compressor efficiencies are about 80%, and the power requirements of a compressor that will handle $10^6$ cu ft of free air daily (694 cfm) against a pressure of 7.0 psig are normally 23 hp. The corresponding energy requirements are 550 hp-hr and 410 kwhr. Mechanical stirring devices consume much the same amount of energy for identical rates of wastewater flow. This implies that substantially the same $G$ and $Gt$ values are produced in diffused-air and mechanical power dissipation. Rotary and centrifugal blowers are commonly employed as compressors.

**Air Piping.** For pipes less than 10 in. in diameter, Fritsche's formula for the Weisbach-Darcy friction factor $f$ is

$$f = 4.8 \times 10^{-2}\ D^{0.027}/Q^{0.148} \tag{19-10}$$

where $D$ is the diameter in inches and $Q$ is the rate of flow in cubic feet per minute. The value of $f$ varies otherwise for new piping from $2.5 \times 10^{-2}$ for 3-in. pipe to $1.6 \times 10^{-2}$ for 18-in. pipe, and for old piping from $4.9 \times 10^{-2}$ for 3-in. pipe to $2.8 \times 10^{-2}$ for 18-in. pipe. The pressure loss, $\Delta p$, is

$$\Delta p = flTQ^2/(38 \times 10^3\ pD^5) \tag{19-11}$$

where $l$ is the pipe length in feet, and $T = 520(p/14.7)^{0.283}$ deg F for adiabatic compression of air.

The length of pipe producing the same loss of head as elbows, tees, and globe valves is

$$l = kD/(1 + 3.6/D) \tag{19-12}$$

where $k = 7.6$ for elbows and tees, but 11.4 for the globe valves commonly controlling air flows.

[12] $fl/d = 40f = 1.0$, or $f = 0.025$ in the Weisbach-Darcy equation.

*Air Diffusers.* A number of air diffusers other than porous ceramic plates and tubes are illustrated in Fig. 19-7. Shown are (a) a sparger or nozzle, (b) an impinger that breaks up the air stream, (c) a water jet that draws in air and displaces it, and (d) a diaphragm of flexible and permeable cloth.

Plate diffusers are normally 12 in. square and 1 in. thick; tubular diffusers are 24 in. long and 3 in. in internal diameter, with a wall thickness of $5/8$ in. The air release of circular tubes is less uniform than for plates, because air bubbles issuing from the bottom half of the tube coalesce as they sweep around the walls of the diffuser. By convention, the *permeability* of diffusers is stated in terms of the face velocity (feet per minute) of air at 70°F and 10 to 25% relative humidity that will flow through the pores of the diffuser with a loss of head of 2 in. of water, when the diffuser is tested dry in a room kept at 70°F and with a relative humidity of 30 to 50%. Permeabilities of 30 to 60 fpm are commonly specified. When new, a square foot of submerged diffuser will then deliver 1.5 to 3 cfm of air with a pressure loss of 2 to 5 in. of water. Losses rise as diffusers become clogged in service. The diffusers are generally cleaned when the loss has trebled.

Diffusers clog on the air side when the air carries particulates originating in the atmosphere or in the air piping. They are clogged on the water side by suspended solids and growths of microorganisms. Clogging on the air side is kept to a minimum by cleaning the air before compression (which also prevents wear on the compressors) and by constructing the air-distribution system of noncorrosive materials or materials protected against erosion. Clogging on the water side is reduced if air pressures are maintained, without interruption, above the hydrostatic head of the overlying water. The nature of the clogging substances determines the method of cleaning.

---

*Example 19-5.*

A wastewater flow of 1 mgd is passed through an activated-sludge tank 15 ft deep. The air supplied to the tank proper is 0.9 cu ft per gal of wastewater treated. Find (1) the power consumption, (2) the mean temporal velocity gradient generated at 68°F, and (3) the contact opportunity during 6 hr of mixed-liquor flow.

1. $P = 62.4\, V_a h/t = 62.4 \times 0.9 \times 10^6 \times 15/(24 \times 60 \times 33 \times 10^3) = 116$ hp.
2. $G = \sqrt{P/C} = [16 \times 33 \times 10^3 \times 7.48/(2.1 \times 10^{-5} \times 10^6)]^{1/2} = 4.2 \times 10^2$ sec$^{-1}$.
3. $Gt = 9.1 \times 10^6$.

---

## 19-13 Oxidation Ditches

These variants of activated-sludge units were developed in the Netherlands as self-sufficient constructions for the extended aeration of wastewaters from

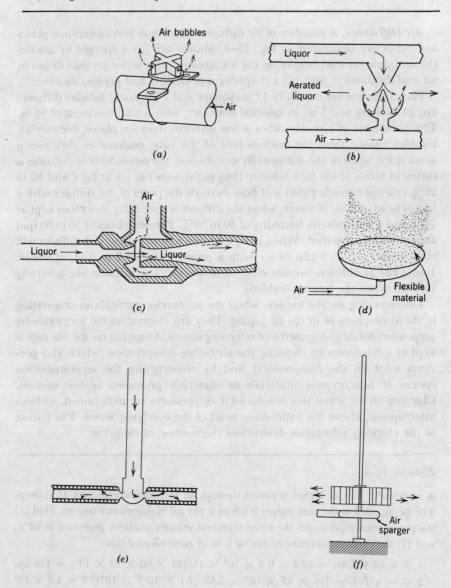

**Figure 19-7.**

**Submerged aeration devices. (a) Sparger; (b) impinger; (c) water jet; (d) flexible diaphragm; (e) porous ceramic diffuser; (f) rotating turbine aerator. [(a) through (d) after A. L. Downing, *J. Inst. Publ. Health Engrs.*, 59, 80 (1960); (e) and (f) after W. W. Eckenfelder and T. L. Moore, *Chem. Eng.* (Sept. 1955).]**

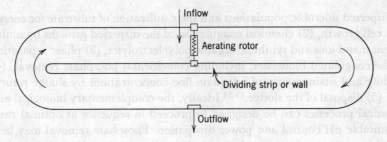

**Figure 19-8.**
**Basic oxidation ditch.**

small communities.[13] As shown in Fig. 19-8, the simplest oxidation ditch is a mechanically aerated continuous channel. Raw wastewaters are discharged into it directly. They are soon dispersed in the mixture of sewage and activated sludge and circulate with them through the ditch. A displacement velocity of about 1 fps keeps the floc in suspension. The rotor runs across the channel rather than along its side. In ditches designed for cyclical operation, depths lie between 3 and 5 ft and detention periods extend from 1 to 3 days. No effluent is withdrawn until the depth of water in the channel has built up to the highest operating level. The influent is then cut off, the rotor is stopped, an hour or two of quiescence is allowed for solids to settle, the clarified supernatant is withdrawn through an effluent launder, and excess sludge, if desired, is lifted from a section of the ditch to drying beds. The effluent is then cut off and the operating routine is repeated. Because the solids are well stabilized during the long aeration period, they are no longer putrescent and are readily dewatered. Cyclical operation can give way to continuous operation by the addition of an effluent settling tank from which useful amounts of sludge are returned to the ditch.

Good results are obtained even when much of the ditch is covered with ice. Depending on the climate, excess sludge can be dried on sand beds with an area of 1 to 3 sq ft per person. Ordinarily, areal requirements for oxidation ditches are smaller than for oxidation ponds (Sec. 19-17), but about the same as for package plants. Power needs are about the same.

## 19-14 Chemical Flocculation of Biological Masses

Chemical coagulation of biomasses in conjunction with their aerobic generation has much to offer. Essential components of the process are (1) creation

[13] J. K. Baars, The Use of Oxidation Ditches for Treatment of Sewage from Small Communities, *Bull. World Health Organ.*, **26**, 465 (1962).

of dispersed microbic populations and their utilization of substrate for energy and cell growth, (2) chemical coagulation of the dispersed growths by multivalent metal ions and synthetic organic polyelectrolytes, (3) phase separation of the coagulated biomasses, including coincidental phosphate removal, (4) seeding and maintenance of adequate floc concentrations by sludge return, and (5) disposal of the sludge.[14,15] Ideally, the complementary biological and chemical processes can be designed to proceed in sequence at optimal rates by suitable pH control and power dissipation. Phosphate removal may be a primary or a secondary consideration, depending on the relative importance of circumventing the eutrophication of receiving bodies of water along with a reduction of the general pollutional load.[16]

## 19-15 Anaerobic Flocculation and Digestion

Flocculation and digestion of wastewater impurities under anaerobic conditions are normally restricted to strong warm organic industrial wastewaters—some of the wastewaters from the meat- and poultry-packing industry, for instance. Essential components of the process are (1) creation of biological floc and the utilization of organic waste substances for energy and cell growth, (2) liberation of relatively stable end products either as methane, carbon dioxide, and other gases, or as liquefied substances that are either mineralized or humified, (3) seeding and maintenance of adequate floc concentrations by sludge return, and (4) disposal of any excess sludge. Subsequent treatment of the effluent by aerobic biological processes may also be necessary. The essentially new component of the process is the anaerobic biological floc. Because of this the process is referred to as anaerobic contact[17] and as anaerobic digestion.[18]

## 19-16 Stabilization Ponds

Stabilization ponds are simple earthwork structures open to the sun and air, which constitute the natural resources on which they can draw to accomplish

---

[14] For discussions of the Guggenheim process see G. H. Gleason and A. C. Loonam, Results of Six Months Operation of Chemical Sewage Purification; and E. B. Phelps and J. G. Bevan, A Laboratory Study of the Guggenheim Bio-Chemical Process, *Sewage Works J.*, **6**, 450 (1934), and **14**, 104 (1942), respectively.

[15] M. W. Tenney and Werner Stumm, Chemical Flocculation of Microorganisms in Biological Waste Treatment, *J. Water Pollution Control Fed.*, **37**, 1370 (1965).

[16] J. B. Nesbitt, Phosphorous Removal - The State of the Art, *J. Water Poll. Control Fed.*, **41**, 701 (1969).

[17] G. J. Schroepfer, W. J. Fullen, A. S. Johnson, N. R. Ziemke, and J. J. Anderson, The Anaerobic Contact Process as Applied to Packinghouse Wastes, *Sewage Ind. Wastes*, **27**, 460 (1955).

[18] A. E. J. Pettet, T. G. Tomlinson, and J. Hemens, The Treatment of Strong Organic Wastes by Anaerobic Digestion, *J. Inst. Publ. Health Engrs.*, **58**, 1–4, 170 (1959).

their mission. Apart from lagoons in which wastewaters—principally from industry—are stored until they can be safely discharged into available receiving waters, stabilization ponds take one of the following forms:

*Aerobic ponds*, in which suspended and dissolved degradable substances are stabilized by aerobic microbic populations supplied with needed oxygen by algal photosynthesis as well as by gas transfer at the pond surface, sometimes with support from mechanical or diffused-air aeration.

*Anaerobic ponds*, in which degradable substances are stabilized by anaerobic microbic populations in the continuous absence of DO.

*Hetero-aerobic* (or *amphi-aerobic*) *ponds*, in which degradable substances are stabilized by facultatively aerobic and anaerobic populations because of cyclical changes from aerobic to anaerobic conditions, such as the diurnal and seasonal changes in photosynthesis.

Plant and animal life are especially rich and varied in aerobic ponds with partly anaerobic benthal sludges. Settleable solids are precipitated close to the influent to be attacked first within the benthal environment by acid-forming bacteria. Organic acids produced by them are transferred not only to methane-forming bacteria within the same environment but also to algae in the supernatant waters. Coincidental anaerobic gasification releases $CO_2$ to the waters and the algae, and $CH_4$, via the supernatant, to the atmosphere. Dead algal cells settle to the pond bottom and are decomposed. At the same time, influent dissolved organic nutrients are being put to use by aerobic bacteria in the supernatant waters. The resulting biomasses draw on the oxygen supplied by the algae during photosynthesis and reciprocally provide $CO_2$ and $NH_3$ for further algal growths.

## 19-17 Design of Stabilization Ponds

Design procedures for stabilization ponds are at best imprecise. Involved in an engineering sense, at one and the same time, are operations such as sedimentation, oxidation, and digestion, gas exchange and photosynthesis, mechanical aeration, and evaporation and seepage. Rate-determining factors are (1) the detention time, $t_d$, which is ordinarily measured in days; (2) the pond depth, $h$, which is generally small, for example, 3 to 4 ft; (3) the pond loading, often expressed in pounds of five-day, 20°C BOD or as the underlying BOD concentration, $y_0$, of the influent wastewaters; (4) the pond temperature, $T$, in degrees Fahrenheit, but in degrees centigrade in connection with the van't Hoff-Arrhenius temperature factors $\theta = e^c = 1.072$ or $Q_{10} = 2.0$ (Sec. 11-15); (5) the visible-light energy, $S$, in langleys (calories daily per square centimeter) reaching the water surface; and (6) the efficiency of conversion of light energy into chemical energy, $p$, which ranges from 2 to 6% and averages 4%. Summer and winter values of visible-light energy, $S$, are shown for the United States in Fig. 19-9. The efficiency of its conversion into chemical energy depends not only on light intensity but also on the quality and concentration of nutrients, the relative duration of daylight, the water temperature, the degree of mixing, the detention period, and the pH. Including the numerical coefficient of Hermann and Gloyna,[19]

$$t_d = 0.49\ hy_0/(pS\theta^{T-20}) \qquad (19\text{-}13)$$

and for $C$ as the pond capacity in acre-feet serving a population $P$ discharging $q$ gpcd, and a critical pond temperature $T$,

$$C = 1.5 \times 10^{-7}\ Pqy_0/(pS\theta^{T-20}) \qquad (19\text{-}14)$$

To be noted is that the coefficient 0.49 in Eq. 19-13 includes a factor of safety in that the light energy $S$ is calculated to produce twice the amount of oxygen needed to satisfy 85% of the BOD.

*Example 19-6.*

A stabilization pond is to serve 1000 people in the eastern United States at latitude 40° north. The average rate of wastewater flow is 100 gpcd and its five-day, 20°C BOD is 200 mg/l. Assuming a pond depth of 3 ft, $\theta = 1.072$, and a summer temperature of 25°C, find (1) the requisite detention period in days, (2) the pond area in acres, and (3) the pond loading in persons/acre and lb of BOD daily/acre.

1. From Fig. 19-9, the insolation is 500 cal/(sq cm)(day), and by Eq. 19-13, $t_d = 0.49 \times 3 \times 200/(4 \times 10^{-2} \times 500 \times 1.072^5) = 10$ days.
2. The pond volume $C = 10^3 \times 10^2 \times 10/7.48 = 1.3 \times 10^5$ cu ft, and the pond area is $1.3 \times 10^5/(3 \times 43{,}560) = 1.0$ acre.
3. The population loading is 1000 persons/acre and the BOD loading is $10^5 \times 8.34 \times 10^{-6} \times 200/1.0 = 170$ lb of BOD daily/acre.

For the conditions of Example 19-6, but a critical mean temperature during the coldest month of 10°C and one fourth the insolation, the required pond area is 12 acres.

Marais and Shaw[20] have shown that construction of a chain of ponds in place of a single one improves the overall efficiency of BOD removal, and that

---

[19] E. R. Hermann and E. F. Gloyna, Waste Stabilization Ponds, III, Formulation of Design Equations, *J. Water Pollution Control Fed.*, **30**, 963 (1958).

[20] G. R. Marais and V. A. Shaw, A Rational Theory for the Design of Sewage Stabilization Ponds in Central and South Africa, *Trans. South African Inst. Civil Engrs.*, **3**, 205 (1961).

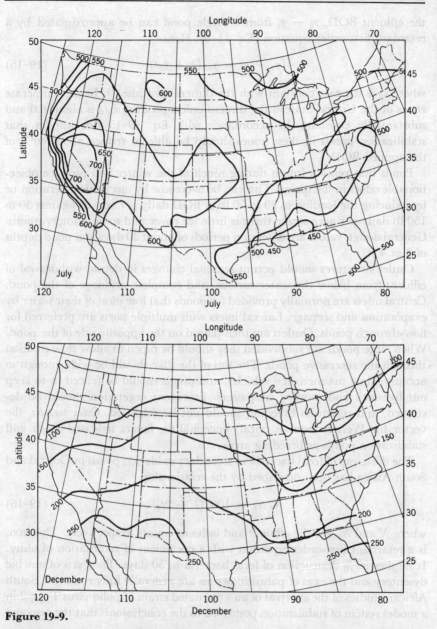

**Figure 19-9.**

Isoheliodynamic lines of average solar radiation (calories per square centimeter daily) received on a horizontal surface in the United States during days of average cloudiness in July and December, respectively. [After S. Fritz, Solar Energy on Clear and Cloudy Days, *Scientific Monthly*, 84, 55 (1957).]

the effluent BOD, $y_0 - y$, from a single pond can be approximated by a retardant first-order equation (Eq. 11-9). Hence

$$(y_0 - y) = y_0/(1 + k_0 t_d) \qquad (19\text{-}15)$$

where $k_0$ is the rate constant with the dimension (time)$^{-1}$. In the temperate zones of the United States[21] the dimensionless produce $k_0 t_d$ is about 6.0 and substantially constant. In accordance with Eq. 19-15 this implies that stabilization ponds in these zones should be able to reduce about 86% of the applied BOD.

Ponds that remain frozen during much of the winter will produce objectionable odors in the spring. This can be overcome by mechanical aeration or by reducing the loading to 20 to 50 lb of BOD daily per acre as against 50 to 150 lb daily per acre where there is little ice cover and some photosynthesis. General design values are detention periods of about 30 days and pond depths of 3 or 4 ft.

Outlet structures should permit seasonal changes in depth, withdrawal of effluent from below the water surface, and complete drainage of the pond. Central inlets are normally provided for ponds that lose most of their water by evaporation and seepage. Lateral inlets with multiple ports are preferred for flow-through ponds. Outlets are then placed on the opposite side of the pond. When large ponds are subdivided they should be piped to allow flow rotation through the successive ponds. The top of the dike should be wide enough to accommodate maintenance vehicles, and ponds should be fenced in to keep out livestock and unwary trespassers. Emergent vegetation should be destroyed, and mosquito breeding should not be tolerated. *Culex tarsalis*, the vector for Western equine (viral) encephalitis, favors irrigated lands and stabilization ponds as breeding areas.

The die-away of fecal bacteria in secondary oxidation ponds in Central and South Africa has been described by the relationship

$$N/N_0 = 100/(1 + 2t_d) \qquad (19\text{-}16)$$

where $N$ and $N_0$ are the effluent and influent counts, respectively. This, too, is a retardant first-order equation with a coefficient of retardation of unity. It implies 99% destruction of fecal bacteria in 50 days. The cysts of amebic dysentery and the eggs of parasitic worms are generally fully removed. South African studies of the survival of an attenuated strain of polio virus Type 2 in a model system of stabilization ponds led to the conclusion[22] that the response

---

[21] As shown by M. L. Granstrom for results obtained at Fayette, Mo., and Farmville, Va.

[22] H. H. Malherbe and O. J. Coetzee, The Survival of Type "2" Polio Virus in a Model System of Stabilization Ponds, *South African Council Sci. Ind. Res. Rept.*, 242 (1965).

of virus particles appears to be quite different from that of bacteria and that the removal of viruses probably depends on the opportunity for adsorption on static surfaces and exposure to the sun apart from natural die-away during detention periods of considerable length.

**Anaerobic Ponds.** Anaerobic ponds are normally 6 to 10 ft deep. Wanted performance is the same as in sludge digestion and is ensured by similar means (Sec. 20-14). To control odors, the ponds must be kept alkaline. Hydrogen sulfide then remains in solution as $HS^-$, the hydrosulfide ion. Pond effluents must normally be treated aerobically before discharge into receiving waters. Loadings and detention periods vary widely. Odors in secondary aerobic ponds are held in check by anaerobic holding periods of a day or two.

**Mechanically Aerated Ponds.** Mechanically aerated ponds phase into oxidation ditches and thence into extended-aeration activated-sludge units. Equipment includes (1) surface aerators like those on streams, reservoirs, and harbors, installed in fixed positions or placed on floats, (2) mechanical turbomixers with diffused-air spargers, and (3) compressed-air diffusion systems. Lead-weighted polyethylene tubing with nonclogging check valves can be laid on the bottom of oxidation ponds in gridiron fashion. Mixing is not so intense as in activated-sludge aeration tanks, and solids do settle here and there and decompose. However, detention periods are normally long enough to permit digestion of bottom deposits with little impact on the main aerobic process.

By aerating ponds only at night, a denitrification cycle can be established. During the day oxygen is produced in the photosynthetic zone while denitrifying bacteria release nitrogen gas from nitrates in the deeper parts of the pond. Nighttime aeration blows out the nitrogen gas and mixes oxygen downward to produce more nitrates.

Aeration is normally extended when industrial and municipal wastewaters are strong. Design information must be obtained by bench-scale testing and the operation of pilot plants (Sec. 18-13). Required information is the biodegradability of the wastewaters; their oxygen utilization; the performance of aeration units; the production, removal, and disposal of sludge; and the optimal loading of the proposed system.

twenty

# treatment and disposal of waste solids

## 20-1 Sources and Kinds of Waste Solids

The principal end products of water and wastewater treatment are (1) the finished water from a water-purification plant or the effluent from a wastewater-treatment plant and (2) byproduct slurries or sludges. Product water from a water-purification plant is sent to cities or industries; wastewater-treatment-plant effluent is usually discharged into receiving waters, more rarely onto receiving soils, or is reclaimed for industrial water supply. Slurries and sludges are not finished byproducts. Because of their origin, watery consistency, bulk, and putrescibility, most of them need to be processed prior to disposal. Processing is intended to reduce their volume and weight and to ensure their hygienic safety and sensory acceptability.

The dry weight of waste solids is that of the solids settleable at the time of solids separation or phase transfer from the suspending water. Included are (1) solids naturally present and settleable in waters and wastewaters, (2) additives—chemical coagulants and precipitants, for instance—converting unwanted nonsettleable solids into settleable solids, and (3) sloughed biological films and wasted biological flocs or other biomasses generated by living organisms from dispersed and dissolved nutrients during wastewater treatment.

Examples of waste solids from water and wastewaters are (1) leaves and other floating, largely organic, debris removed from surfacewaters by racks and screens on water intakes; (2) sand and other gritty mineral particles removed from groundwater by sand catchers; (3) dense, largely mineral, solids settled from turbid river waters with or without the benefit of coagulation; (4) flocculent mineral and organic solids settled from coagulated surfacewaters or contained in the washwaters of water filters; (5) flocculent, largely mineral, solids precipitated by aeration or by chemicals from waters containing iron or manganese in solution; (6) dense, largely mineral, *slurries* produced by chemical softening of hard waters or remaining after evaporation of saline waters; (7) sizable, largely organic, solids removed as *rakings* and *screenings* from wastewater racks and screens; (8) dense, largely mineral, solids removed as *grit* and *detritus* from grit chambers and detritus tanks; (9) rising, largely organic, solids removed as foams or *skimmings* from wastewaters by flotation, skimming, or settling tanks; and (10) settling, largely organic, solids removed as *sludges* or *underflows* from primary or secondary wastewater settling tanks and deriving their description from the treatment process in which they originate.

On reaching the bottom of settling units, most organic and mineral solids form loose, honeycombed structures of particulate and flocculent matter together with relatively large volumes of included water. As deposits build up they consolidate under their own weight, but water is not displaced from them with ease, and their moisture content often remains high. Thousands of gallons of putrescible and presumptively dangerous solids are removed similarly by sedimentation from each million gallons of municipal wastewater undergoing treatment. Industrial wastewaters contain settleable solids varying in amount from light pollutional loadings in spent cooling and rinse waters to heavy mineral and organic loadings in the process waters of some chemical and food industries. Satisfactory disposal of the resulting slurries and sludges creates economic problems of considerable magnitude.

Although individual classes of solids may contain, besides water, other substances of some economic value, they seldom do so in sufficient amount to make isolation or reclamation worthwhile. Examples of paying operations are few. The recovery of lime in water softening is one (Sec. 16–8). Conversion of wastes from the food and beverage industries into animal feed is another. Reclamation of fats and fibers from industrial wastes is a third. Radically different in concept is the utilization of energy (1) by burning combustible gases generated by digesting solids or (2) by incinerating dried highly organic solids.

Generally speaking, the management of slurries, sludges, and other treatment debris is a troublesome task in water purification and wastewater treatment. Solids management may indeed be a prime determinant in choosing from among specific water-purification and wastewater-treatment processes.

The unit operations of slurry and sludge treatment and disposal involve transport, concentration or thickening, digestion, dewatering, drying, incineration, and safe disposal (Sec. 11-9 and Fig. 20-1). The significance of most of the standard methods for examining waste solids should be clear from what has been said about their examination (Sec. 10-18). Common properties of sludges deserving of special consideration or determination include (1) moisture-weight-volume relationships, (2) density, viscosity, rigidity, and other flow characteristics, (3) response to concentration or thickening and to filtration, (3) fuel value, (5) digestibility, and (6) fertilizer value.

## 20-2 Moisture-Weight-Volume Relationships

The percentage moisture content, $p$, of a sludge containing a weight $w_s$ of dry solids and a weight $w$ of water is $p = 100w/(w + w_s)$, and the solids concentration is $(100 - p) = 100w_s/(w + w_s)$. The specific gravity of the dry

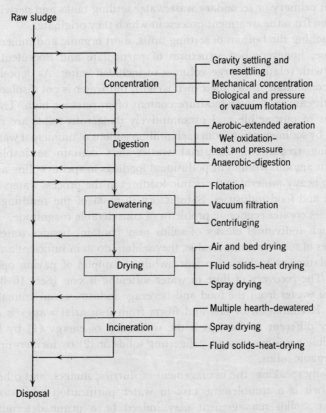

**Figure 20-1.**
**Flow chart for solids handling. Arrows indicate possible flow paths.**

sludge (or of its solids on a dry basis) is a function of the specific gravities of its components. If the two principal fractions (volatile and fixed), which are determined by evaporation of the moisture and ignition of the residue, possess specific gravities of $s_v$ and $s_f$, respectively, the specific gravity $s_s$ of the solids as a whole can be calculated from the percentage of volatile matter, $p_v$. Because $100/s_s = (p_v/s_v) + (100 - p_v)/s_f$,

$$s_s = 100 \, s_f s_v / [100 \, s_v + p_v(s_f - s_v)] \qquad (20\text{-}1)$$

and, for $s_v \simeq 1.0$ and $s_f \simeq 2.5$,

$$s_s \simeq 250/(100 + 1.5 \, p_v) \qquad (20\text{-}2)$$

The specific gravity of the wet sludge, $s$, is then the ratio of the combined weight of water and dry solids to the combined volume, that is,

$$s = \frac{p + (100 - p)}{(p/s_w) + (100 - p)/s_s} = \frac{100 \, s_s s_w}{p s_s + (100 - p) \, s_w} \qquad (20\text{-}3)$$

or, approximately,

$$s \simeq 25{,}000/[250 \, p + (100 - p)(100 + 1.5 \, p_v)] \qquad (20\text{-}4)$$

The volume $V$ of the wet sludge equals the combined volume of the water and the solids, or

$$V = \frac{w_s}{s_s \gamma} + \frac{w}{s_w \gamma} = \frac{w_s}{s_s \gamma} + \frac{p w_s}{(100 - p) \, s_w \gamma} = \frac{w_s}{\gamma} \frac{100 \, s_w + p(s_s - s_s)}{(100 - p) \, s_s s_w} \qquad (20\text{-}5)$$

Here $\gamma$ is the specific weight of water. It follows that removing water from a volume $V_0$ of sludge containing $p_0\%$ of water decreases the volume of the partially dewatered sludge to

$$V = V_0 \frac{[100 \, s_w + p(s_s - s_w)](100 - p_0)}{[100 \, s_w + p_0(s_s - s_w)](100 - p)} \qquad (20\text{-}6)$$

provided the voids between the solids remain full of water. Moreover, because $s_s$ closely equals $s_w$,

$$V \simeq V_0(100 - p_0)/(100 - p) \qquad (20\text{-}7)$$

Therefore the loss of water changes the volume approximately in the ratio of the solids concentration. For this reason solids concentration is, generally speaking, a more useful concept than sludge moisture.[1] As sludge loses water, it may acquire the plastic properties of a *paste*. Loss of so much moisture that

---

[1] A decrease in moisture content from 95% to 90% halves the sludge volume, a change better indicated by speaking of an increase in solids content from 5% to 10%.

the voids are no longer filled with water produces a *sludge cake*. Except for some consolidation, the volume of cake remains substantially constant at $V = w_s/[(1 - f) s_s \gamma]$, where $f$ is the porosity ratio (often 40 to 50%).

## Example 20-1.

A primary wastewater sludge, settling from domestic wastewater, has a moisture content of 95%, 72.2% of the dry solids being volatile. Find (1) the specific gravity of the dry solids on the assumption that the specific gravities of the volatile and fixed solids are respectively 1.0 and 2.5; (2) the specific gravity of the wet sludge on the assumption that the specific gravity of the included water is 1.0; and (3) the relative decrease in volume associated with a reduction in moisture of the wet sludge to 85%.

1. By Eq. 20-2, $s_s \simeq 250/(100 + 1.5 \times 72.2) = 1.20$.
2. By Eq. 20-3, $s \simeq 120/(95 \times 1.2 + 5) = 1.01$.
3. By Eq. 20-6, $V/V_0 = [(100 + 85 \times 0.20)5]/[(100 + 95 \times 0.20)15] = 0.328$.
4. By Eq. 20-7, $V/V_0 \simeq 5/15 = 0.33$.

## 20-3 Amounts of Water and Wastewater Sludges

The quantities of solids produced in different treatment processes are measured routinely in water-purification and wastewater-treatment works.

**Waterworks Sludges.**[2] The solids fractions in sludges withdrawn from settling and coagulation basins in municipal and industrial water-treatment works and in the washwater from rapid or slow filters vary with the nature of the water treated, the amounts and kinds of additives, and the reactions taking place during treatment. Some waterworks sludges are quite putrescible; coagulated, colored, or polluted waters present examples. The weight of settled solids drawn from coagulation tanks and washwater settling tanks may be as little as 0.1% of the weight of mixed liquor before thickening and as much as 2½% after thickening. Observed values vary with the nature of the raw water and the kind and concentration of chemicals employed.

**Wastewater Sludges.** The nature and characteristics of solids fractions in municipal wastewaters are briefly as follows:[3]

---

[2] *Disposal of Wastes from Water Treatment Plants*, Amer. Water Works Ass. Research Foundation Report, *J. Amer. Water Works Assn.*, **62**, 1, 63, 1970.

[3] R. S. Burd, *A Study of Sludge Handling and Disposal*, Federal Water Pollution Control Admin., Wash., D.C., 1968, offers a good summary of sludge-handling practice.

PLAIN SEDIMENTATION (Primary Treatment). The fresh solids contain most of the *settleable solids* in the raw wastewater. Anaerobic digestion destroys about 67% of the volatile matter, about a quarter of it being converted to fixed solids.

CHEMICAL COAGULATION (Primary Treatment). The fresh solids include the precipitated chemicals in more or less stoichiometric amounts, and from 70 to 90% of the solids suspended in the raw wastewater, depending on the effectiveness of chemical treatment. Digestion modifies the solids much as it does plain-sedimentation, primary solids.

TRICKLING FILTRATION (Secondary Treatment). Most of the dissolved organic matter and many of the otherwise nonsettleable solids in wastewaters applied to trickling filters are rendered settleable by adsorption and biological flocculation on the trickling-filter film. The film itself is modified by decomposition before it is sloughed off and picked up by the effluent. Destruction and loss vary with the length of storage in the bed. Common limits are 30% for filters operated at low rates against 10% for filters operated at high rates. Thus secondary sedimentation normally captures 50 to 60% and 80 to 90% of the nonsettleable suspended solids reaching low-rate and high-rate filters respectively. The *trickling-filter humus* is generally added to primary solids for subsequent handling. Changes in composition during anaerobic digestion are greater for high-rate humus than for low-rate humus.

ACTIVATION (Secondary Treatment). The controlling fraction of solids in fresh, excess activated sludge from conventional treatment works is normally richer in organic matter and higher in water content than trickling-filter humus. In conventional treatment plants, from 5 to 10% of the materials transferred are mineralized during formation and recirculation, depending on the proportion of solids return and the length of aeration, which, together, govern how long the activated sludge remains in circulation. Extended aeration raises the degree of mineralization appreciably. Excess activated sludge may be allowed to settle with primary solids, unless this practice is counterindicated because the combined sludge responds poorly to further treatment. The recovery of conventional excess sludge in secondary settling units is from 80 to 90% of the aerated, nonsettleable, suspended-solids load. Digestion brings about much the same changes in sludge composition, but not in concentration, as for plain-sedimentation, primary solids.

The accumulation of municipal wastewater solids depends on the composition of the wastewaters (Tables 10-5 and 6), the efficiencies of solids transfer (Table 11-2), and the degree of solids digestion incidental to treatment. Corresponding information must be obtained for other wastewaters. Common values for domestic wastewaters are summarized in Table 20-1.

## Table 20-1
*Proportion of Solids in Domestic Wastewater Sludges*[a]

| Treatment Process | Condition of Sludge | % Solids |
|---|---|---|
| 1. Plain sedimentation | Fresh—depending on method of removal from settling tanks | 2.5–5 |
| | Thickened—wet | 8–10 |
| | Digested—wet | 10–15 |
| 2. Chemical precipitation | Fresh—increasing in moisture with amount of chemical used | 2–5 |
| | Digested—wet | 10 |
| 3. Trickling filtration | Fresh—depending on length of storage in filter | 5–10 |
| | Thickened—settled | 7–10 |
| | Fresh—mixed with primary sludge | 3–6 |
| | Thickened | 7–9 |
| | Digested—wet | 10 |
| 4. Activation[b] | Fresh—depending on sludge age and length of aeration | 0.5–1 |
| | Thickened | 2.5–3 |
| | Digested—wet | 2–3 |
| | Fresh—settled with primary sludge | 4–5 |
| | Thickened | 5–10 |
| | Digested | 6–8 |

[a] Data for thickened sludges from footnote 12.
[b] If primary sedimentation is curtailed, suitable adjustments must be made.

*Example 20-2.*

Estimate for a conventional activated-sludge plant treating domestic wastewater and including primary and secondary sedimentation (1) the weight of dry solids in pounds daily/1000 persons and in pounds/million gallons of wastewater and (2) the volume of wet solids in cubic feet daily/1000 persons and in gallons/million gallons of wastewater. Find in each instance the separate and combined amounts of fresh solids and the amounts of digested solids if the primary solids are digested alone or in combination with the secondary solids. Assume a wastewater flow of 100 gpcd and specific gravities of 2.5 for fixed solids and 1.0 for volatile solids.

*Primary sludge*—fresh: From Table 10–5, the daily per capita production of fresh solids is 54 g, 39 g (or 72.2%) being volatile and 15 g (or 27.8%) fixed. Accordingly, the weight of dry solids is $54 \times 10^3/454 = 119$ lb daily/1000 persons, and $119 \times 10^6/(100 \times 10^3) = 1190$ lb/mg. By Eq. 20–2, the specific gravity of the dry solids is $250/(100 + 1.5 \times 72.2) = 1.20$. Assuming 5% solids in the sludge (Table 20–1), Eq. 20–3 gives its specific gravity as $120/(95 \times 1.20 + 5) = 1.01$. Therefore its volume is $119 \times 10^2/(5.0 \times 62.4 \times 1.01) = 37.8$ cu ft daily/1000 persons and $37.8 \times 10^6 \times 7.48/(100 \times 10^3) = 2830$ gal/mg. By similar calculations the following results are obtained for the remaining sludge components and for mixtures of the components.

*Excess activated sludge*—fresh: Dry solids 66 lb daily/1000 persons and 660 lb/mg, or 70.4 cu ft daily/1000 persons and 5620 gal/mg.

*Combined primary and excess activated sludge*—fresh: 185 lb daily/1000 persons and 1850 lb/mg, or 65.4 cu ft daily/1000 persons and 4890 gal/mg.

*Primary solids*—digested: 77 lb daily/1000 persons and 576 lb/mg, or 9.0 cu ft daily/1000 persons and 673 gal/mg.

*Combined primary solids and excess activated sludge*—digested: 119 lb daily/1000 persons and 890 lb/mg, or 26.5 cu ft daily/1000 persons and 1980 gal/mg.

## 20-4 Sludge Flow

All sludges and slurries are but pseudohomogeneous substances. Fresh plain-sedimentation solids are especially diverse in their composition; digested solids and activated sludges less so; and alum and iron flocs least of all. Because many wastewater sludges are non-Newtonian fluids with plastic rather than viscous properties, their flow resistance is a function of their concentration. Hydraulics of flow are further complicated because most sludges are *thixotropic*, implying, therefore, that their plastic properties change during stirring and turbulence. Gases or air released during flow add to the difficulty of identifying probable hydraulic behavior. Fundamental friction losses increase with solids content and decrease with temperature. In general, laminar or transitional flow persists at relatively high velocities, such as 1.5 to 4.5 fps for thick sludges flowing in pipes 5 to 12 in. in diameter. At turbulent velocities all sludges behave more nearly like water.

The following relationship[4] for laminar flow of plastic liquids in pipes is based on Poiseuille's equation for viscous liquids:

$$h/l = (32/g)(v/d^2)[(\eta/\rho) + \tfrac{1}{6}(\tau_y/\rho)(d/v)] \tag{20-8}$$

Here $h$ is the loss, in feet head of sludge; $l$ and $d$ are respectively the length and diameter of the pipe; $v$ is the velocity of the solids and their transporting

---

[4] For derivation of a similar equation, see H. E. Babbitt and D. H. Caldwell, Laminar Flow of Sludge in Pipes, *Univ. Ill. Bull.* 319 (1939); and V. C. Behn, Flow Equations for Sewage Sludges, *J. Water Pollution Control Fed.*, 32, 728 (1960).

liquid; $\rho$ is their mass density; and $\eta$ and $\tau_y$ are respectively their coefficient of rigidity and shearing stress at the yield point. The terms enclosed in brackets are analogous to the kinematic viscosity of a Newtonian liquid and possess the same dimensions. When the mixture is thin, its shearing stress at the yield point approaches zero and the coefficient of rigidity merges into the dynamic viscosity $\nu$ of the liquid. If the plastic-flow factors evaluated in the brackets are incorporated in a coefficient of kinematic rigidity $\nu_p$, the Reynolds number becomes $\mathbf{R} = vd/\nu_p$, and laminar flow should persist below $\mathbf{R} \sim 2000$ as it does in viscous flow.

Thixotropy makes the determination of $\eta/\rho$ and $\tau_y/\rho$ difficult. Neither flow tests nor viscometry are fully satisfactory. Reported magnitudes are few. Examples are $\eta = 0.03$ and $\tau_0 = 0.1$ for thick digested sludges. Fresh activated sludge, being low in solids (2% or less), has no measurable shearing resistance, and its $\eta/\rho$ value is substantially the same as the kinematic viscosity of water.

---

*Example 20-3.*

Estimate for digested, plain-sedimentation sludge and a velocity of 2 fps (1) the loss of head in 100 ft of 12-in. pipe and (2) the expected range of laminar flow. Assume a solids content of 10%.

1. If $\eta/\rho$ is $5 \times 10^{-4}$ sq ft/sec and $\tau_y/\rho$ is $1.5 \times 10^{-3}$ (ft/sec)$^2$, the loss of head is given by Eq. 20-8 as

$$h = 100(32/32.2)(2/1)[5 \times 10^{-4} + \tfrac{1}{6}(1.5 \times 10^{-3})(1/2)] = 0.13 \text{ ft}$$

2. At $\nu_p = 5 \times 10^{-4} + \tfrac{1}{6}(1.5 \times 10^{-3})/v$ and

$$\mathbf{R} = v \times 1/\nu_p = 2000, \quad v = 1.5 \text{ fps}$$

---

For turbulent flow, the loss of head of fairly homogeneous wastewater solids—digested primary solids, and activated sludge, for instance—is increased by not more than 1% for each 1% of solids. Fresh, plain-sedimentation solids are transported at losses 1.5 to 4 times those for water. Requisite self-cleaning velocities for the transportation of sand and gritty materials can be gaged from the formulations presented in Sec. 8-3. When the velocities of flow are small, the heavier and larger solids settle out and obstruct flow. Velocities in the reduced cross-sectional area of the conduit then rise until deposited material is scoured away. However, there may be pipe stoppages. Rates of sludge draw-off from tanks must not be so rapid that water breaks through the sludge blanket. Sludge conduits must be dimensioned accordingly.

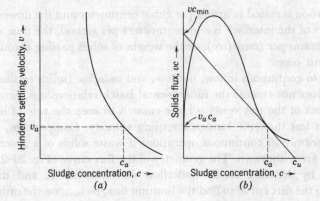

**Figure 20-2.**
**Graphical analysis of sludge settling curve. (a) Settling velocity of sludge and water interface at given sludge concentrations; (b) solids flux at given sludge concentrations.**[5]

Sludge pumps, like wastewater pumps in general, must contend with solids of all kinds. Plunger and diaphragm pumps and compressed-air ejectors are used at small rates of discharge. For higher rates of flow, centrifugal pumps (generally over 4 in. in size) provide sufficient clearance to prevent clogging. Air lifts are normally used only for activated sludges and for depths of submergence no more than twice the head.

## 20-5 Sludge Concentration

The concentration or thickening of sludge withdrawn from settling tanks can normally be increased by resettling it with or without stirring or chemical conditioning (coagulation). This is a simple operation. Necessary tank dimensions for resettling, effective rates of stirring, and chemical dosage can be derived from settling tests. Expected responses[5] are illustrated in Fig. 20-2. If the solids are uniformly dispersed and sufficiently concentrated, settling is *hindered*. The resulting hindered velocity equals the velocity $v$ of the sludge-water interface shown in Fig. 20-2$a$ in its relation to the solids concentration $c$. Sludges build up from the bottom and are compressed under their own mounting weight. So long as the solids remain in suspension, settling and buildup proceed at uniform rates. The product $vc$, called the solids flux (or flow), is plotted in Fig. 20-2$b$ against the solids concentration $c$. If the solids

---

[5] N. Yoshioka, Y. Hotta, S. Tanaka, S. Naito, and S. Tsugami, Continuous Thickening of Homogeneous Flocculated Slurries, *Kagaku Kogaku*, **26**, 66 (1957); also see H. S. Coe and C. H. Clevenger, Determining Thickener Unit Areas, *Trans. Am. Inst. Mining Engrs.*, **55**, 356 (1916).

concentration is stated in grams per cubic centimeter and the downward displacement of the interface is $v$ in centimeters per second, the flux, $vc$, is seen to equal grams per $(cm)^2(sec)$, or the weight of solids passing through a unit area in unit time.

A shift to continuous inflow, outflow, and *underflow* (solids or sludge withdrawal) does not change the fundamental batch relationships, provided the surface area of the test vessel is large enough to keep the rate of liquid displacement less than the hindered settling velocity. Compression, too, proceeds uniformly in continuous operations if waste solids of a given concentration $c_u$ are withdrawn. The geometry of the flux curve (Fig. 20-2b) can be exploited by deciding on an underflow concentration $c_u$ and drawing a tangent to the flux curve to find the limiting flux, $(vc)_{min}$, on the ordinate and the limiting concentration, $c_a$, at the point of tangency. Because $(vc)_{min} : c_u = v_a c_a : (c_u - c_a)$ by similar triangles, $(vc)_{min} = (v_a c_a) c_u/(c_u - c_a)$, and, because the rate of sludge withdrawal $Q/A$, where $Q$ is the flow in cubic centimeters per second and $A$ is the area in square centimeters, equals $(vc)_{min}/c_0$, where $c_0$ is the influent solids concentration,

$$A = Qc_0(c_u - c_a)/[(v_a c_a) c_u] \qquad (20\text{-}9)$$

*Example 20-4.*

An activated-sludge flow of 1 mgd (43 l/sec), with 5000 mg/l solids, is to be concentrated to 20,000 mg/l or 2% solids. The observed concentration and interface velocity at the point of tangency to the flux curve are found to be 14,000 mg/l and 0.018 × $10^{-2}$ cm/sec. Find the requisite thickener area

By Eq. 20-9, $A = 43 \times 10^3 (20 - 14)/(0.18 \times 10^{-2} \times 14 \times 20) = 26 \times 10^5$ cm$^2$ = 260m$^2$ = 2800 sq ft. Associated with it would be a surface loading of $10^6$/2800 = 360 gpd/sq ft, and a solids loading of $5 \times 10^3 \times 1 \times 8.34/2800$ = 15 lb/sq ft/day.

Solids surface loadings may be as high as 30 lb daily per sq ft for primary sludges and as low as 5 lb daily per sq ft for activated sludge yielding solids concentrations of 10% and 2.5%, respectively.

## 20-6 Sludge Dewatering

Processes of sludge dewatering include land methods, vacuum filtration, centrifugation, pressure filtration, and other methods occasionally used, such as gravity dewatering and freezing.[6] Land methods, suitable only for stable

[6] A comprehensive review of sludge dewatering technology is available in *Sludge Dewatering*, Water Pollution Control Federation Manual of Practice, No. 20, Washington, D.C., 1969.

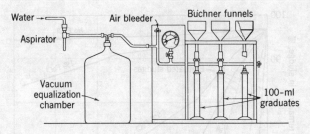

**Figure 20-3.**
**Apparatus for determining the release of sludge moisture by vacuum filtration.**

sludges, incorporate drying and are discussed in Section 20–13. Pressure filters are used widely in Europe, but have not been adopted in the United States, possibly because they are not readily adapted to continuous operation.

Mechanical gravity dewatering units are in the development stage. They operate in two steps, thickening followed by compression or cake-formation. Chemical conditioning is required while fine-mesh nylon screen cloth helps to produce a high-quality filtrate from the unit. Sludge freezing, to destroy its water-binding capacity, has found limited use in the treatment of waterworks sludges in Europe.[7]

## 20-7 Sludge Filtration

Concentrated sludge can be dewatered by pressing, aspirating, or draining moisture from it. This is done on filter presses, vacuum filters, and drying beds. In each instance water is withdrawn through a porous medium on which the sludge rests. In the air drying of sludge on sand beds, evaporation combines with drainage in the removal of moisture.

The *drainability* or *filterability* of waste solids is defined as the relative rate of moisture release. Its magnitude can be determined in the laboratory in an apparatus such as that shown in Fig. 20-3. The test is simple. A known weight of solids is placed on filter paper in a Büchner funnel, and the volume of filtrate collected in different time intervals is noted. The moisture content of the solids is calculated from a knowledge of their original moisture content. For example, if 50 ml of filtrate are collected in 10 min from 100 g of activated sludge containing 98% water at the outset, its moisture has been reduced to $(98 - 50)/(100 - 50) = 96\%$ in 10 min and its volume has been cut in half. Plots of observed values (Fig. 20–4) then identify the amenability of the solids to chemical or other conditioning agents. Correlation of test results with the performance of actual dewatering equipment becomes

---

[7] P. W. Doe, D. Benn and L. R. Bays. The Disposal of Washwater Sludge by Freezing, *J. Inst. Water Eng.*, **19**, 251 (1965).

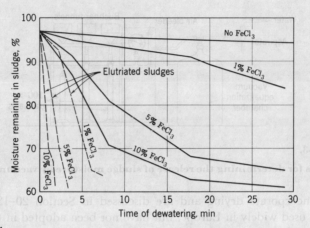

**Figure 20-4.**

**Rates of release of moisture by digested, plain-sedimentation wastewater sludge.**

meaningful if the laboratory work is conducted under standard conditions, namely, standard weights or thicknesses of waste solids, vacuum intensities, properties of filter paper—such as size, thickness, and permeability—and test temperatures. Measured weights must make allowances for added conditioning agents. Laboratory-scale vacuum filters, sand beds, and similar physical models can be employed instead.

**Hydraulics of Sludge Filtration.** In order to identify waste solids behavior during filtration, Coakley[8] has transformed the Blake-Kozeny equation (Eq. 14-6) to read as follows:

$$v = \frac{1}{K}\frac{1}{\mu}\frac{f^3}{(1-f)^2}\frac{\Delta p}{l}\frac{1}{S_s^2} \qquad (20\text{-}10)$$

Here $v = (dV_w/dt)/a = K_1 \Delta p/(\mu l)$ is the instantaneous velocity of flow measured as the volume of water $dV_w/dt$ flowing from an area $a$ of cake at time $t$; $K$ is a rate constant with the dimension (time)$^{-1}$; $\mu$ is the absolute viscosity of the displaced water; $f$ is the porosity of the cake; $\Delta p = \Delta h \rho g$ is the pressure difference in a cake of thickness $l$; $S_s$ is the specific surface $A/V$ of the cake solids; and $K_1 = 1/I = (1/K)(A/V)^{-2}f^3/(1-f)^2$, $K_1$ being the permeability of the cake and $I = 1/K_1$ its impedance.

Because $l = cQ/a - cV_w/a$ as cake with a total volume of $Q$ per unit time and a volume concentration of $c$ builds up on the filter, $dV_w/dt = aK_1\Delta p/(\mu l) = a\Delta p/(\mu I l) = a^2 \Delta p/(\mu I c V_w)$. Moreover, because the specific resistance to

[8] Peter Coakley, Theory and Practice of Sludge Dewatering, *J. Inst. Publ. Health Engrs.*, **64**, 34 (1965).

flow encountered by the sludge water equals the initial resistance of the filter medium plus the resistance of the cake,[9]

$$\frac{dV_w}{dt} = \frac{a^2 \Delta p}{\mu(icV_w + I_m a)} \tag{20-11}$$

Here $i$ is the resistance of a unit volume of cake and $I_m$ is the initial resistance of a unit area of filtering surface, $I$ being a length for dimensional consistency. For an initial dry-weight concentration of solids of $c_s$, and a specific weight of solids $\gamma_s$, $i_s = i/\gamma_s$ is the specific resistance, and

$$\frac{dV_w}{dt} = \frac{a^2 \Delta p}{\mu(i_s c_s V_w + I_m a)} \tag{20-12}$$

by integration,

$$\frac{t}{V_w} = \frac{\mu i_s c_s}{2a^2 \Delta p} V_w + \frac{\mu I_m}{a \Delta p} = nV_w + b \tag{20-13}$$

A straight line is obtained when log $t/V_w$ is plotted against log $V_w$, $V_w$ being the volume of filtrate collected in time $t$ after the cake has been formed. The magnitude of the intercept on the ordinate at $V_w = 1$ is $b = \mu I_m/a\Delta p$) and the slope of the line is $n = \mu i_s c_s/(2a^2 \Delta p)$.

It follows that the specific resistance to filtration of waste solids can be found experimentally by determining in a Büchner funnel test correlative values of $t$ and $V_w$ after measuring $c_s$, $p$, and $a$ as well as the temperature.

The resistance of compressible cake has been observed[10] to rise in proportion to the applied pressure, that is, as $i_{s_1}/i_{s_2} = (\Delta p_1/\Delta p_2)^q$, where $q$ is a coefficient of compressibility straddling unity according to Trubnick and Mueller.[11]

*Example 20–5.*

A Buchner funnel test of waste solids yields a straight line in a logarithmic plot of $t/V_w$ against $V_w$, the slope of the line being 1.66 sec/cm² and its intercept at $V_w = 1$ being 0.32 sec/cm³, and the vacuum applied is 28 in. Hg = 910g/cm² at 20°C. The

[9] P. C. Carman, A Study of the Mechanism of Filtration, Part I, *J. Soc. Chem. Ind. (Brit.)*, **52**, 280 t (1933).

[10] An acceptable theoretical relationship is $i_{s_1}/i_{s_2} = [1 + k(\Delta p_1)^q]/[1 + k(\Delta p_2)^q]$, where $k$ is a coefficient.

[11] E. H. Trubnick and P. K. Mueller, Vacuum Filtration Principles and Their Application to Sewage and Sludge Dewatering, in Joseph McCabe and W. W. Eckenfelder, Jr., Eds., *Biological Treatment of Sewage and Industrial Wastes*, Vol. 2, Reinhold, New York, Chaps. 3 and 4, 1958.

temperature is 20°C ($\mu = 1.01 \times 10^{-2}$ poise), the solids concentration is 0.10 g/ml, and the filter area is 40 cm². Find the specific resistance of the solids and the initial impedance of the filter.

Because $n = 1.66 = 1.01 \times 10^{-2} i_s \times 0.10/(2 \times 40^2 \times 910)$, $i_s = 4.8 \times 10^9$ cm/g, and because $b = 0.32 = 1.01 \times 10^{-2} I_m/(40 \times 910)$, $I_m = 115 \times 10^5$ cm.

The observed percentage[12] moisture of filter cake, $y$, relative to the percentage sludge solids concentration, is approximately $y = 88/(x^{1/8})$. For $x = 10\%$, for example, $y = 88/1.33 = 66\%$.

**Vacuum Filtration.** The dewatering of sludge by filtration through cloth and other filtering media originated in the chemical industries. The art of dewatering solids has advanced from batch or discontinuous operation of chamber, or leaf, filter presses to continuous, automatic operation of vacuum filters.

Among vacuum filters, the revolving-drum filter is most widely used (Fig. 20-5). In filters of this kind, a series of cells running the length of the drum can be placed under vacuum or pressure. The drum normally turns at a peripheral speed of 5 to 15 fpm. As it revolves it passes through a reservoir containing the sludge to be dewatered. A vacuum of 12 to 26 in. of mercury is applied to the submerged cells (from 15 to 40% of the filter surface) and attaches a mat of sludge of suitable thickness to the filter medium overlying the cells. The emerging mat is placed under a drying vacuum of 20 to 26 in. of mercury, and the sludge liquor is drawn into the vacuum cells and drained from them for treatment and disposal with other sludge liquors. The cake is removed from the drum by a scraper or by strings or coiled springs and carried away for heat drying, incineration, or disposal. In advance of scrapers, a slight pressure may be applied to the cake to facilitate its removal from the filter medium.

*Sludge cake* is usually about ¼ in. thick. Its solids content is as high as 32% for plain-sedimentation sludge (raw or digested) and as low as 20% for raw, activated sludge. Digested, activated sludge may be dewatered to about 25% solids. Trickling-filter sludge and mixtures of different sludges span the gap. The solids content of dewatered, trickling-filter humus depends on the length of storage of the sludge in the biological filters. The cake-producing capacity of vacuum filters varies ordinarily from 2 to 6 lb hourly per sq ft on a dry basis. Low yields are obtained with fresh, chemical sludge and activated sludge; high yields with digested, plain-sedimentation sludge. Filter speeds and vacuums are varied to suit operating conditions. Power requirements of drum filters and ancillary pneumatic and hydraulic equipment lie in the vicinity of ⅛ hp per sq ft of filter area.

[12] Derived from data by P. L. McCarthy, Sludge Concentration—Needs, Accomplishments, and Future Goals, *J. Water Pollution Control Fed.*, **38**, 493 (1966).

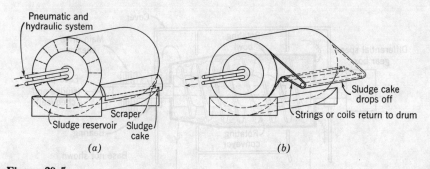

**Figure 20-5.**
**Sketches of drum vacuum filters. (a) Drum filter with sludge scraper; (b) drum filter with strings or coils.**

The filter medium is either (1) cotton, wool, or synthetic-fiber cloth stretched over a supporting layer of copper screening that covers the sides of the drum (Fig. 20-5a), or (2) a series of tightly spaced, parallel, metallic spring coils or plastic-covered wires that may be conducted over an idling drum of small diameter paralleling the filter drum (Fig. 20-5b) in order to lift the cake from the drum.

## 20-8 Centrifugation

Centrifuges were introduced during the earliest days of wastewater treatment, but they did not find a firm place in practice until the 1950s, when many of the mechanical problems associated with their use were resolved. A modern centrifuge, illustrated in Fig. 20-6, consists of a solid rotating bowl, with a screw conveyor inside the bowl that rotates at a different speed and moves the sludge cake to discharge ports. A modified design provides for concurrent flow, which allows feed and sludge cake to move in the same direction. To improve the economy of operation of centrifuges, the sludge to be dewatered may be chemically conditioned and thickened.

Auxiliary equipment for centrifuges include sludge feed pumps, centrate pump and piping where centrate cannot flow by gravity, a washwater system, and sludge cake handling facilities.

Bowl and conveyor design are a matter of centrifuge selection, while bowl and conveyor speeds and pool volume in the centrifuge may be operational variables. The efficiency of centrifugation is increased with increased bowl length and diameter. However, larger-diameter bowls have to operate at slower speeds. Optimum ratios of length to diameter of the bowl range from 2.5 to 3.5.[13]

---

[13] D. E. Albertson and E. J. Guidi, Jr. Centrifugation of Waste Sludges, *J. Water Pollution Control Federation*, **41**, 4, 607 (1969). Also see P. A. Vesilind, Estimation of Sludge Centrifuge Performance, *J. San. Eng. Div., Am. Soc. Civil Eng. Proc. Paper 7378*, **96**, SA3, 805 (1970).

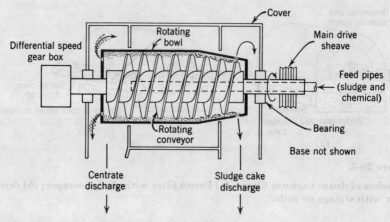

**Figure 20-6.**
**The solid bowl-conveyor sludge dewatering centrifuge assembly consists of a rotating unit comprising a bowl and conveyor. (Courtesy Bird Machine Co.)**

Centrifuge selection and operation serve two conflicting objectives: increased cake dryness and increased solids recovery. The factors that influence these objectives are shown below.[3]

|  | Increased Cake Dryness | Increased Solids Recovery |
| --- | --- | --- |
| Bowl speed | Increase | Increase |
| Pool volume | Decrease | Increase |
| Conveyor speed | Decrease | Decrease |
| Feed rate | Increase | Decrease |
| Feed consistency | Decrease | Increase |
| Flocculents | Do not use | Use |

Primary sludges, either raw or digested, are readily dewatered without chemical conditioning to cakes with 28 to 35% solids, and with solids recovery of 70 to 90%. With secondary sludges, sludge cakes of 15 to 30% solids are possible, but chemical conditioning is necessary if reasonable solids recovery, 80 to 95%, is to be obtained. Polymer dosages required range from 5 to 8 lbs per ton of solids treated.

## 20-9 Chemical Sludge Conditioning

Coagulation of the solids dispersed in sludge—called chemical conditioning—increases the rate of water removal by filtration or by air drying.

Common conditioning chemicals for wastewater sludge are listed in Table 20-2. Before coagulants can combine with the solids fraction of the sludge, they must satisfy the coagulant demand of its liquid fraction. This is exerted by the alkalinity or bicarbonates. The alkalinities of digested sludges are quite high—in some instances 1000 times those of fresh sludges. As a precipitant of bicarbonates (Sec. 16–7), lime may be substituted for the portion of the coagulant that combines with the liquid fraction. Lime does not form floc with this fraction, but only a precipitate.

**Table 20-2**
Common Conditioning Chemicals for Wastewater Sludge[a]

| Conditioning Chemical | Chemical Symbol | Molecular Weight |
|---|---|---|
| Ferric chloride | $FeCl_3$ | 162.2 |
| Chlorinated copperas | $FeSO_4Cl$ | 187.4 |
| Ferric sulfate | $Fe_2(SO_4)_3$ | 399.9 |
| Aluminum sulfate | $Al_2(SO_4)_3 \cdot 18\ H_2O$ | 666.4 |
| Lime | $CaO$ | 56.1 |

[a] Some synthetic polyelectrolytes are also good conditioning agents.

Coagulant or conditioner requirements are generally expressed as percentage ratios of the pure chemical to the weight of the solids fraction on a dry basis. Component parts are (1) the liquid-fraction requirement, approximated closely by the stoichiometry of the idealized chemical reactions, and (2) the solids-fraction requirement, which is a matter of experience. For ferric chloride, Genter[14] has suggested that

$$p_c = [1.08 \times 10^{-4} Ap/(1-p)] + 1.6\ p_v/p_f \qquad (20\text{-}14)$$

where $A$ is the alkalinity of the sludge moisture in milligrams per liter of $CaCO_3$, and $p_c$, $p$, $p_v$, and $p_f$ are respectively the percentages of chemical ($FeCl_3$), moisture, volatile matter, and fixed solids in the sludge, all on a dry basis. In accordance with the equation $2FeCl_3 + 3Ca(HCO_3)_2 \rightleftharpoons 2Fe(OH)_3 + 3CaCl_3 + 6CO_2$, 1 mg/l $CaCO_3(100)$ combines with $(2/3)(162.2/100) = 1.08$ mg/l of $FeCl_3(162.2)$. The term for the solids fraction $(1.6\ p_v/p_f)$ is derived from operating results for the vacuum filtration of ferric-chloride-treated wastewater. Because this term is a function of the volatile-matter content of the sludge, coagulant requirements can be reduced by digesting the sludge prior to coagulating it for dewatering. By contrast, the magnitude of the term for the liquid fraction $[1.08 \times 10^{-4} Ap/(1-p)]$ is greatly magnified by digestion. It can be reduced either by adding lime as a

[14] A. L. Genter, Computing Coagulant Requirements in Sludge Conditioning, *Trans. Am. Soc. Civil Engrs.*, **111**, 641 (1946).

precipitant or by washing out a share of the alkalinity with water of low alkalinity. This is called *elutriation*.

**Elutriation.** Sludge elutriation is carried out in single or multiple tanks by single or repeated washings, the washwater being used serially if desired. During washing, the solids are kept in suspension by air or mechanical agitation. Serial use of washwater is called *countercurrent elutriation*.

If a total of $R$ volumes of elutriating water with an alkalinity of $W$ milligrams per liter is added for each volume of moisture in the sludge, and the alkalinity of the moisture before elutriation is $A$ milligrams per liter, the alkalinity of the elutriated moisture, $E$, in milligrams per liter can be found by striking a mass balance between the alkalinities entering and leaving the elutriating tank or tanks. The following examples, in which subscripts denote the sequence of operations, show the development of general formulations:

MULTIPLE ELUTRIATION OF SLUDGE IN A SINGLE TANK. In this scheme $1/n$th of the elutriating water is used in each washing (Fig. 20–7a). Hence $(1/n) RW + A = (1/n) RE_1 + E_1$; $(1/n) RW + E_1 = (1/n) RE_2 + E_2$; and eventually $(1/n) RW + E_{n-2} = (1/n) RE_{n-1} + E_{n-1}$ and $(1/n) RW + E_{n-1} = (1/n) RE_n + E_n$.

COUNTERCURRENT ELUTRIATION IN MULTIPLE TANKS. In this scheme the clean washwater is introduced into the last tank, which receives the partially elutriated sludge from the next-to-the-last tank, the water from the last tank being used to wash the sludge coming into the next-to-the-last tank. Pro-

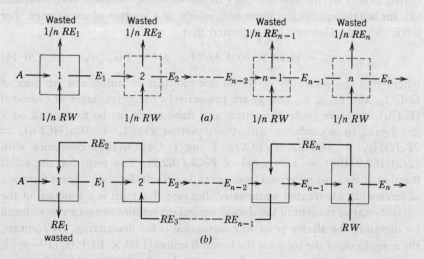

**Figure 20-7.**

**Sludge elutriation. (a) Multiple elutriation in a single tank; (b) countercurrent elutriation in multiple tanks.**

ceeding serially in this manner (Fig. 20-7b) to the first tank, $RW + E_{n-1} = RE_n + E_n$; $RE_n + E_{n-2} = RE_{n-1} + E_{n-1}$; and eventually $RE_3 + E_1 = RE_2 + E_2$ and $RE_2 + A = RE_1 + E_1$.

Solving respectively for the alkalinity $E_n$ of the final sludge and for the ratio $R$ of washwater to sludge, the following general equations are obtained for (1) single tank, multiple elutriation ($n$ stages):

$$E_n = \frac{A + W[(R/n + 1)^n - 1]}{(R/n + 1)^n} \qquad (20\text{-}15)$$

and

$$R = n\left[\left(\frac{A - W}{E - W}\right)^{1/n} - 1\right] \qquad (20\text{-}16)$$

and (2) multiple tank, countercurrent elutriation ($n$ tanks):

$$E_n = \frac{A(R - 1) + WR(R^n - 1)}{R^{n+1} - 1} \qquad (20\text{-}17)$$

and

$$(R^{n+1} - 1)/(R - 1) = (A - W)/(E - W) \qquad (20\text{-}18)$$

The ratio of elutriating water to wet sludge is usually close to 2:1. Reduction in the necessary amounts of conditioning chemicals is reflected in increased heat values of the dried cake and reduced heat requirements for drying and incineration. The washwater is treated or disposed of along with digester and filter liquor.

---

*Example 20-6.*

A digested sludge with 45% volatile solids on a dry basis and 90% moisture has an alkalinity of 3000 mg/l. Find the percentage of ferric chloride that must be added prior to vacuum filtration (1) if the sludge is not elutriated and (2) if the sludge is elutriated in two tanks by countercurrent operation in a ratio of 3:1 with water of 20 mg/l alkalinity. Find also (3) the elutriation ratio that will reduce the alkalinity of the sludge to 300 mg/l by countercurrent and two-stage elutriation.

1. Unelutriated sludge: By Eq. 20-14, $p_c = (1.08 \times 10^{-4}) 3000(90/10) + 1.6 (45/55) = 4.22\%$ of $FeCl_3$ on a dry-weight basis.

2. Elutriated sludge, countercurrent operation: By Eq. 20-17, $E_2 = (3000 \times 2 + 20 \times 3 \times 8)/26 = 250$ mg/l. By Eq. 20-14, $p_c = (1.08 \times 10^{-4}) 250(90/10) + 1.6 (45/55) = 1.55\%$ of $FeCl_3$ on a dry-weight basis.

3. Elutriation ratio: By Eq. 20-18, $(R^3 - 1)/(R - 1) = (3000 - 20)/(300 - 20) = 10.64$, or $(R^2 + R + 1) = 10.64$, whence $R = 2.64$ by countercurrent elutriation.

By Eq. 20-16, $R = 2[\sqrt{(3000 - 20)/(300 - 20)} - 1]$, or $R = 4.54$ for two washings by multiple elutriation. This is 70% more washwater than for countercurrent operation.

## 20-10 Fuel Value of Sludge

The fuel value of wastewater solids is normally determined in a bomb calorimeter. Statistical correlation between pairs of observed fuel values and volatile-solids values is good and can be expressed as follows for municipal waste solids:

$$Q = a\{[(100\, p_v)/(100 - p_c)] - b\}[(100 - p_c)/100] \quad (20\text{-}19)$$

Here $Q$ is the fuel value of the solids in British thermal units per pound, dry weight; $p_v$ and $p_c$ are respectively the proportions of volatile matter and chemical, precipitating or conditioning, reagent present in percent; and $a$ and $b$ are coefficients for different classes of waste solids. Examples are $a = 131$ and $b = 10$ for plain-sedimentation municipal wastewater solids (fresh and digested), and $a = 107$ and $b = 5$ for fresh activated sludge. Results compare favorably with the fuel values of fossil fuels, which range from 10,000 to 14,000 Btu per lb for coal and from 19,000 to 20,000 Btu per lb for crude petroleum and petroleum products.

*Example 20-7.*

Chemically precipitated municipal wastewater solids contain 68.0% volatile matter on a dry basis, and the precipitated chemical is 8.3% of the total weight. Estimate (1) the fuel value of the solids and (2) the percentage recovery of fuel value in gas by digestion of these solids, assuming that 67% of the volatile matter is destroyed by digestion and 72% of the gas by volume is $CH_4$, the remainder being $CO_2$.

1. By Eq. 20-19, using the values of $a$ and $b$ for plain-sedimentation solids, $Q = 131\{[(100 \times 68)/(100 - 8.3)] - 10\}[(100 - 8.3)/100] = 7700$ Btu per lb.

2. Because the weight of volatile matter destroyed is $0.67 \times 0.68 = 0.46$/lb of dry solids and gas production is 1.25 lb/lb of volatile matter destroyed (Example 20-10), the weight of gas produced is $0.46 \times 1.25 = 0.57$ lb/lb of dry solids. The volume of methane, therefore, is $(0.72 \times 0.57)/(0.72 \times 0.0446 + 0.28 \times 0.1225) = 6.2$ cu ft/lb of dry solids, and its fuel value is $958 \times 6.2 = 5900$ Btu. This is $100 \times 5900/7700 = 77\%$ of the fuel value of the dry solids.

## 20-11 Heat Drying and Incineration

When it is to be sold as a commercial fertilizer, wastewater sludge is commonly heat-dried to less than 10% moisture. Heat drying also generally precedes sludge incineration. Incineration itself may then provide the heat required for sludge drying as well as other plant purposes. Two types of rotary driers are in use: the kiln (direct-indirect) drier, common to the lime and cement industry, and the cage-mill flash drier, or fluid-solids system, illustrated in Fig. 20–8. A sufficient amount of previously dried sludge is

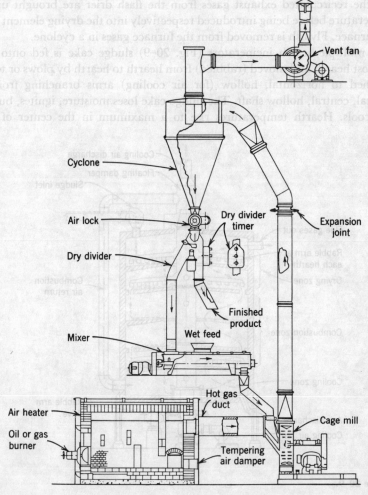

**Figure 20-8.**
**Fluid-solids heat drying. (Combustion Engineering Co.).**

added to the incoming cake to reduce the moisture of the mixture to about 50%. Drying gases are passed through the tumbling mixture in the cage mill and carry the dried dust to a cyclone, in which it is separated from the transporting gases. The temperature of the drying gases is reduced from 1000°F or higher to about 225°F. Volatilized-sludge odors are destroyed by incineration at about 1250°F. Heat is conserved by countercurrent flow. In drying kilns, sludge movement is also countercurrent.

Sludge-burning furnaces are operated at temperatures of about 2500°F. The furnace gases are passed through a regenerative preheater, in which air and the recirculated exhaust gases from the flash drier are brought up to temperature before being introduced respectively into the drying element and the furnace. Fly ash is removed from the furnace gases in a cyclone.

In multiple-hearth incinerators (Fig. 20–9) sludge cake is fed onto the topmost hearth and moved (rabbled) from hearth to hearth by plows or teeth attached to horizontal, hollow (for air cooling) arms branching from a vertical, central, hollow shaft. The sludge cake loses moisture, ignites, burns, and cools. Hearth temperatures rise to a maximum in the center of the

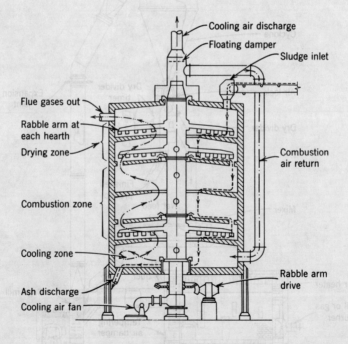

**Figure 20-9.**

**Multiple-hearth furnace for incinerating filter cake (Reprinted by permission of Black and Veatch, Consulting Engineers).**

incinerator. The exhaust gases are passed through a preheater or recuperator and heat the air blown into the furnace to support combustion. Cooling air is taken in through the central shaft.

Heat requirements are dictated by the temperature of the sludge and its moisture and by the efficiency of the furnace. If the temperature of the sludge is 60°F, the heat requirements are 1124 Btu per lb of moisture. Furnace efficiency usually varies between 45 and 70% in terms of total heat recovery (including credits on stack gas and latent heat of evaporation) and total heat input. Multiple-hearth furnaces have a combined efficiency of evaporation and incineration of about 55%. Sludge cake produced by the vacuum filtration of chemically conditioned, raw, plain-sedimentation sludge and its mixtures with fresh, trickling-filter, or activated sludge will ordinarily supply sufficient heat for self-incineration. Most digested sludges will not. Auxiliary fuel, possibly sludge-digester gas, is required for them. Incinerators for solid municipal wastes are a possible source of heat.

## Example 20–8.

An activated-sludge plant produces 7000 gpd of excess activated sludge from 1 mgd of sewage. The sludge contains 1.5% solids, of which 70% are volatile on a dry basis. In conditioning the sludge for dewatering on a vacuum filter, 6% of $FeCl_3$ on a dry basis is added. Find (1) the required vacuum filter area and (2) the auxiliary heat required to incinerate the filter cake. The specific gravity of the wet sludge may be taken as 1.002.

1. Area of vacuum filters: The daily weight of dry solids is $7000 \times 8.34 \times 1.002 \times 1.5 \times 10^{-2}$ = 880 lb. Assuming that each 1% of $FeCl_3$ increases the dry solids in the ratio of the molecular weight of $Fe(OH)_3$ to that of $FeCl_3$, or 106.8:162.2, the added chemical increases the weight of sludge by 0.66% for each 1% of ferric chloride. Hence the additional weight is $6 \times 0.66 \times 880/100$ = 35 lb. Assuming an allowable filter loading of 2.5 lb hourly/sq ft, the required filter area is $(880 + 35)/(24 \times 2.5)$ = 15.2 sq ft.

2. Auxiliary heat for incineration: By Eq. 20–19, the fuel value of the filter cake is $Q = 107\{[(100 \times 70)/(100 - 6 \times 0.66)] - 5\}[(100 - 6 \times 0.66)/100]$ = 7000 Btu/lb.

If the cake contains 20% solids and the efficiency of evaporation and incineration is 55%, the heat requirements are $1124(100 - 20)/(20 \times 0.55)$ = 8100 Btu per lb. Auxiliary heat must therefore be provided in the amount of $(880 + 35)(8100 - 7000)$ = 1.0 million Btu/day.

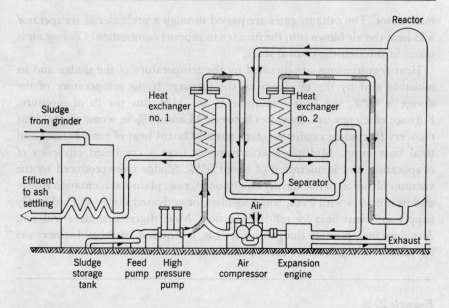

**Figure 20-10.**
**Wet air oxidation (Zimpro Division, Sterling Drug Co.)**

## 20-12 Wet-Air Oxidation of Sludge

In the wet-air oxidation or wet combustion of sludges, air pressures are raised to 1200 to 1800 psig and the accompanying temperature becomes 540°F or more. Under these conditions there is no vaporization of the sludge liquid. Equipment normally includes sludge pumps and air compressors, influent heat exchangers, main reactors, effluent coolers, and terminal vapor separators. External heat is applied to start oxidation of the organic matter in the sludge. Thereafter, oxidation increases until the rate reaches an equilibrium value. The final concentration of organic material in the reactor is substantially the same above 540°F. Effluent sludge is recycled from the reactor to the heat exchangers in countercurrent flow. To make the process thermally self-sufficient, initial concentration of organic matter in the system must lie above 2 or 3%. Up to 3000 Btu of heat are then supplied to a gallon of sludge. The heat produced ranges from 1200 to 1400 Btu per lb of required air. Fig. 20–10 illustrates the process.

In the course of the operation the COD is normally reduced by about 80%, for example, to less than 20 mg per l, and 90% of the volatile solids are removed. Control of corrosion is an inherent difficulty of the process.[15]

[15] E. Hurwitz and W. A. Dunbar, Wet Oxidation of Sewage Sludge, *J. Water Pollution Control Fed.*, 32, 918 (1960).

## 20-13 Air Drying

Under favorable climatic conditions, well-digested sludge run onto a porous bed to a depth of 8 to 12 in. dries in a week or two, generally without odor. By contrast, fresh sludge gives off bad odors while it dries and does not lose moisture satisfactorily in layers of reasonable thickness. Because of this, air drying is more or less confined to well-digested sludges.

Drying beds usually consist of graded layers of gravel or crushed stone beneath 8 to 18 in. of filter sand, $E = 0.3$ to $0.7$ mm. Agricultural tile or vitrified-tile sewer pipes laid with open joints serve as underdrains in much the same way as in intermittent sand filters. The beds are subdivided to meet plant-operating conditions. Their width is so chosen that the vehicle removing the dried sludge can be loaded conveniently. In small plants the width is normally 20 ft or less, and the length is generally held below 100 ft. These are the distances sludge will ordinarily spread and flow from a single outlet when the surface slope of the bed is 0.5% or less. Concrete posts and reinforced-concrete slabs or cypress planks rising about 12 in. above sand level confine the sludge to the bed and its subdivisions. Dried sludge is loaded into vehicles by hand forking or by mechanical loaders. Where used in large plants, as in England, bed dimensions are very great. Sludge is flowed onto them from multiple outlets along their sides, and dried sludge is stripped from them by mechanical devices that can be moved from bed to bed, where they straddle the width and travel the length of the bed.

Glass enclosures of the greenhouse variety, or sheds not unlike the platform sheds of transport stations, will protect the sludge from rain. If enclosures are properly ventilated, the number of dryings per year can be increased by 33 to 100%. In the northern part of the United States, about 5 dryings of 8 in. of wet digested sludge are feasible per year on open beds. Bed drying halves the volume of digested sludge and raises its solids content to about 40%.

In air drying, moisture is lost to the atmosphere by evaporation and to the bed by percolation. Both losses are important.

---

*Example 20–9.*

If the production of digested plain-sedimentation solids is 9 cu ft/1000 persons daily (Example 20–2), find (1) the area of open sludge-drying bed, in sq ft per capita, that must be provided in the northern United States, and (2) the volume of sludge cake, in cubic yards per capita, that must be removed annually.

1. Area of bed: assuming 5 dryings of sludge 8 in. in depth annually, the bed area is $(9 \times 365)/5 \times 1000 \times 8/12) = 1.0$ sq ft per capita. This is a common figure for open beds in the northern United States.

2. Volume of sludge cake: assuming 50% shrinkage during drying, the volume of the dried cake becomes $0.5 \times 9 \times 365/(27 \times 10^3) = 6 \times 10^{-2}$ cu yd annually per capita.

## 20-14 Anaerobic Sludge Digestion

The driving force in sludge digestion is the consumption of foodstuffs in the sludge by living organisms. Because the sludge is a concentrate of, by and large, the organic, decomposable substances removed from the suspending wastewaters, such oxygen as may have been dissolved in the water component of sludge is quickly exhausted. Unless the sludge is aerated vigorously, decomposition proceeds anaerobically. In batch operations, two sequent groups of bacteria are the principal food consumers: (1) facultatively anaerobic acid producers that convert carbohydrates, proteins, and fats into organic acids and alcohols, and (2) anaerobic methane fermenters that convert the acids and alcohols into methane and carbon dioxide. Organisms other than bacteria—protozoa, for example—appear to be of little account in sludge digestion. Not all of the organic matter in wastewaters is quickly decomposed, putrified, or digested. Lignin and other cellulosic substances are examples of resistant components that occur relatively frequently in domestic wastewaters and natural runoff. They remain substantially unaltered even during protracted digestion. Therefore an equational relationship of sludge digestion will read as follows:

$$\text{Bacteria} + \text{organic matter} = \text{bacteria} + \text{residual, resistant,} \\ \text{organic matter} + CH_4 + CO_2 + H_2O \qquad (20\text{-}20)$$

The kinetics of sludge digestion is governed by the fitness of the environment provided by the engineer. In the course of time it has been recognized that the digestion process is optimized under the following conditions.

(1) Digestion is a continuous rather than a transient, cyclical, or discontinuous operation. Fresh sludge enters and digested sludge leaves the process in continuous streams, one helping to displace the other. Because the organic matter is converted in part into gases that are withdrawn from the process, the streams are not identical in weight or volume. The organic matter remaining at the end of the digestion period is resistant to further decomposition, and the effluent solids are relatively stable. The colloidal, water-binding structure of the sludge has been destroyed, and the effluent solids are more or less granular and easily dewatered. However, the sludge liquor composing most of the effluent normally contains dissolved and nonsettleable substances that possess an immediate oxygen demand of substantial magnitude. In continuous operation, old sludge provides (1) the

seeding organisms for the incoming sludge and (2) the buffering capacity necessary to keep organic acids from lowering the pH of the digesting sludge to unfavorable levels.

(2) Digestion is a uniform operation in which incoming solids are quickly and uniformly dispersed in the digesting sludge and there is neither vertical nor horizontal differentiation between old and new sludge or between solids and liquids. This makes it necessary to mix or stir the digesting sludge effectively. Requisite energy is generally derived from external sources, but it may be provided by normal gas production in very deep tanks, in which the vertical rise of gas from thickened sludge becomes sufficiently fast and voluminous. The immediate objective is the provision of optimal contact opportunity between food and microorganisms, and optimal dispersion of accumulating enzymes. An incidental objective may be prevention of the formation of concentration pockets of unwanted biological intermediates such as floating scum blankets and end products, including organic acids.

(3) Digestion is expedited by thermal optimization of the process, heat being provided either from an outside source or reciprocally from the burning of methane and other combustible gases released during digestion, and

(4) Digestion is continued long enough to produce an end product that can be disposed of economically and without nuisance either immediately or after further treatment by dewatering, drying, or incineration.

Historically, however, sludge has been digested in most existing plants in a more or less discontinuous operation. Sludge has usually been drawn from settling tanks to digesters once or twice in each working shift. Digester liquor has been discharged from an acceptably stratified layer of liquid formed between (1) sludge solids settling toward the digester outlet and (2) top scum created by sludge particles light enough either to rise or to be buoyed up by evolving gas bubbles. Drawoff of this so-called supernatant liquor can be minimized, or even eliminated, by controlling the water content of the influent sludge. This is done by thickening the sludge or by careful monitoring of the sludge withdrawn from processing.

Controlling the water content of the influent sludge (1) reduces the supernatant, which imposes a burden on the treatment plant when it is returned to process, (2) reduces the required size of the digester, and (3) reduces the heat required to maintain desired digestion temperatures.

**Digestibility.** Some digestion characteristics of sludges can be determined conveniently in the laboratory on a batch basis. If expected operating conditions are well simulated, the results obtained can furnish information of use in the design of digestion units. Probable performance of continuous-digestion units may be assumed to equal that of a single daily charge of sludge during the period of required sludge storage (Fig. 20-11).

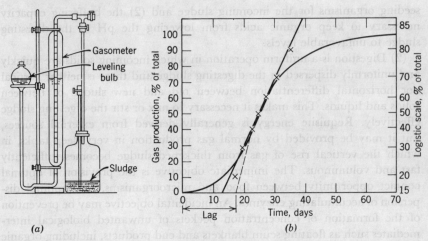

**Figure 20-11.**

**Progress of gas production by organic matter undergoing anaerobic decomposition.** (a) Functional testing equipment; (b) observed arithmetic and logistic course of gas production at 20°C.

Comparative tests may include experiments (1) with and without seeding, (2) at different temperatures, (3) with or without chemical treatment, such as liming, and (4) with or without stirring or circulation at different effective energy inputs. A suitable testing apparatus, shown in Fig. 20-11a, includes a gas collector attached to a bottle partly filled with sludge and standing in a temperature-controlled water bath. The amount of gas released in given time intervals and its composition are measures of digestibility (Fig. 20-11b). A mixture of two parts of fresh to one part of digested solids, both on a basis of dry, volatile solids, is a useful first charge for municipal wastewater solids. In continuous-digestion (steady-state) experiments, daily increments of fresh solids are added. Their amounts, after digestion becomes normal, must be such as to maintain a steady production of gas of even composition. For experimental purposes, the yield of total gas and methane (the difference being substantially all carbon dioxide) is conveniently reported in terms of (1) gas volume (total and $CH_4$) per unit weight of fresh volatile solids or volatile solids destroyed, or (2) gas weight (total and $CH_4$) per unit weight of volatile solids destroyed. Gas volumes may be related to tributary population on a per capita basis or to solids volume on a cubic-foot basis.

Digester gas generally contains 72% methane and is saturated with water vapor. Under standard conditions (0°C or 32°F, 1 atm or 29.9 in. Hg, and dryness), gas occupies a volume of 359 cu ft per lb-mole. Hence a cubic foot of $CH_4$ weighs $16/359 = 0.0446$ lb and a cubic foot of $CO_2$ weighs $44/359 =$

0.1225 lb. The net (or low-heat) fuel value of methane in 963 Btu per cu ft under standard conditions. The value for natural gas is 1080 Btu per cu ft. The net fuel value is the heat liberated in combustion minus the heat of condensation of water. The volume of gas recorded at a given temperature and pressure is reduced to standard conditions by Eq. 12–3. Digester gas is assumed to be saturated with moisture. The vapor pressure of water must therefore be subtracted from the observed pressure. For $p$ in inches of mercury and $T$ in degrees Fahrenheit,

$$V_0 = 16.4V(p - p_w)/(459.7 + T) \qquad (20\text{–}21)$$

where $V$ is the volume of gas, the subscript zero denotes standard conditions, and $p_w$ is the vapor pressure of water (Table 4, Appendix).

Digestibility can be measured in terms other than rate of gas production. Destruction of volatile matter and loss of calorific valve are examples.

*Example 20–10.*

Waste solids produced by primary (plain) sedimentation of municipal wastewater contain 72.2% volatile matter on a dry basis and 95% water. During experimental digestion at 30°C (86°F) of 384 g of fresh wet solids, the gas yield is 783 ml/g of volatile solids, 72% by volume being methane. The wet digested solids, exclusive of the seeding material, weigh 93 g and contain 87% water, the volatile matter being 38.2% on a dry basis. Find (1) cubic feet and pounds of gas per pound of volatile matter destroyed; (2) the net fuel value of the gas in British thermal units per cubic foot and per pound of volatile matter destroyed; (3) the cubic feet of gas per cubic foot of wet solids daily and per capita daily on the assumption that 37.8 cu ft of wet solids per 1000 persons are added daily to the digestion unit, and that the daily charge of the digestion unit is proportional to the batch charge of the experimental unit; and (4) the relative volume of gas measured at a temperature of 68°F and a barometric pressure of 29.6 in.

1. The weights of volatile matter in the fresh and digested solids are respectively $384[(100 - 95)/100](72.2/100) = 13.9$ g and $93[(100 - 87)/100](38.2/100) = 4.6$ g. Hence the weight of volatile matter destroyed is $(13.9 - 4.6) = 9.3$ g and the volume of gas per gram of volatile matter destroyed becomes $783 \times 13.9/9.3 = 1170$ ml. Because 1 cu ft = 28.3 l and 1 lb = 454 g, the volume of gas is $1.17 \times 454/28.3 = 18.8$ cu ft/lb of volatile matter destroyed. The weight of this gas is $(0.0446 \times 0.72 + 0.1225 \times 0.28)18.8 = 1.25$ lb/lb of volatile matter destroyed. This is a well-established value for the digestion of plain-sedimentation wastewater solids.

2. The net fuel value of the gas is $963 \times 0.72 = 690$ Btu/cu ft, or $18.8 \times 690 = 13{,}000$ Btu/lb of volatile matter destroyed.

3. The gas yield of 783 ml/g of volatile solids equals 783 × 0.05 × 0 722 = 28.3 ml/g of wet solids. In accordance with Example 20-1, the specific gravity of the fresh wet solids is 1.01. Therefore the gas yield is 28.3 × 1.01 = 28.5 ml/ml or cu ft/cu ft, and the daily per capita output of gas is 28.5 × 37.8/1000 = 1.08 cu ft. A value of about 1 cu ft of gas per capita daily is a useful background figure for the gas yield of digesting, plain-sedimentation solids.

4. By Eq. 20-21, $V/V_0 = (459.7 + 68.0)/[16.4(29.6 - 0.7)] = 1.11$. The vapor pressure of water at 68°F being 0.7 in. Hg (Table 4, Appendix), 1.11 cu ft of gas will be recorded for each cubic foot of gas under standard conditions.

**Kinetics of Sludge Digestion.** The digestion of wastewater sludges can be measured in terms of the reduction in their volatile matter content or their generation of gas. As shown in Fig. 20-11b, gas production from a single batch of organic material traces an S-shaped curve not unlike that described in Secs. 2-4 and 18-5 for the growth of human populations and of microorganisms within confined spaces, respectively. This means that, from the beginning of the process to a point in time near the halfway mark of total gas evolution, the rate of gasification becomes progressively greater. Thereafter it drops off as a limiting yield of gas is approached.

During continuous digestion, the enzymatic products of the reaction accumulate and promote the process. Provided that digestion is continuous and otherwise optimal, the process kinetics is expected to bear much the same marks as short-term aerobic decomposition. The broken line extended downward from the curve in Fig. 20-11b suggests this. It cuts off a *lag* period not unlike that of unseeded BOD samples or bacterial growth. Seeding will add digesting flora and fauna and their enzymatic systems. It also buffers the medium at a suitable pH value.

Nevertheless, anaerobic decomposition of organic matter is slower than its aerobic counterpart. At 20°C, for example, the daily rate of gasification of seeded sludge solids is seldom more than 8.4%, whereas the daily BOD rate is more than 30%.

**Formulation of Gasification.** Several meaningful equations will fit observed batch-gasification curves. Of interest, beside the equation of a simple first-order reaction (Eq. 11-8), are (1) the equation of a first-order reaction catalyzed either by products inherent in the reacting medium or generated within the medium, and (2) the equation of logistic growth (Sec. 18-5). The two autocatalytic equations are

$$dy/dt = k_1(L - y) + k_2(L - y)^2 \tag{20-22}$$

and

$$dy/dt = k_1(L - y) + k_2 y(L - y) \tag{20-23}$$

The logistic curve is characterized by the relationship $100 y/L = 100/(1 + m \exp nt)$. In each instance $y$ is the amount of gas produced in time $t$, $L$ is the saturation value to which gasification is asymptotic, and $k_1$ and $k_2$, or $m$ and $n$, are coefficients. For $L = 100$ and $y = 20, 50,$ and $80$ in Fig. 20–11, at $t_0 = 15$, 20, and 25 days, respectively, and conveniently replaced, in turn, by time units 0, 1, and 2, $y = 100/[1 + 4 \exp(-1.386t)]$.

**Temperature Effects.** As might be anticipated, the reaction velocity constants of anaerobic digestion increase with temperature in accordance with the van't Hoff-Arrhenius relationship (Sec. 11-15). Even when the constants of the autocatalytic first-order reaction or of the logistic curve have not been determined, it is possible to establish an overall evaluation of the temperature effect by identifying through inspection of the digestion results, the times required for gasification to reach a useful degree of completion, such as 90%.

First estimates of the effect of temperature may be had from observations such as those plotted in Fig. 20-12 for plain-sedimentation, primary sewage

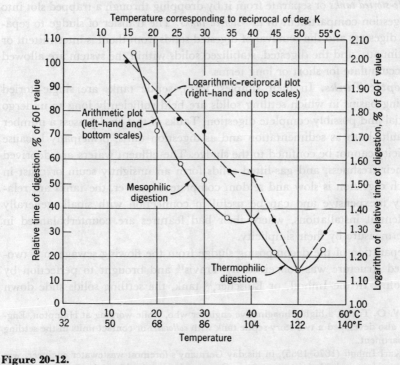

**Figure 20-12.**
Relative time required for 90% digestion of plain-sedimentation, primary sludge at different temperatures. The reference point is gas production at 60°F. [Data from G. M. Fair and E. W. Moore, *Sewage Works J.*, 9, 3 (1937).]

sludge. Digestion is seen to exhibit two significant response ranges to temperatures: (1) a range of responses to moderate temperatures, in which the common moderate-temperature-loving (mesophilic) saprophytes and methane formers are active, and (2) a range of responses to high temperatures, in which heat-loving (thermophilic) organisms are responsible for digestion. The intermediate upswing and dropoff of the mesophilic curve appears to be related to the *thermal death point* of normal saprophytes.

## 20-15 Sludge Digesters

The anaerobic decomposition or digestion of putrescible solids is made either a *concurrent function* of settling tanks or a *sequent function* of separate units to which the settled solids are transferred as sludge from sedimentation units (Fig. 20–13). *Septic tanks* perform the first function; *separate sludge digestion units*—also called *digesters*—perform the second. In septic tanks, the settling and digesting solids either keep in contact with the flowing wastewater in *single-storied tanks* or separate from it by dropping through a trapped slot into a digestion compartment in *two-storied tanks*. The transfer of sludge to separate digesters and the removal of digested solids from them is intermittent or continuous, and the digested, stabilized solids within the system are allowed to accumulate for short or long terms.

**Septic Tanks.** In their simplest form, septic tanks are single-storied settling basins in which settling solids are held sufficiently long to undergo partial and possibly complete digestion. Tanks of this kind possess a number of faults both as sedimentation and as digestion units, principally because septicity cannot be confined to the sludge. The effluent waters are deprived of their freshness, and gas-lifted solids form an unsightly scum or crust in which digestion is slow and seldom complete. However, the tanks are relatively inexpensive and can be useful in connection with small, generally residential installations, where their bad features are counterbalanced in some measure by their simplicity.

Separation of the decomposing sludge from the flowing sewage in a two-storied structure was conceived by Travis[16] and brought to perfection by Imhoff.[17] In the Imhoff, or Emscher,[18] tank, the settling solids slide down

---

[16] W. O. Travis, a highly imaginative engineer who, while working at Hampton, England, also developed a two-story septic tank with *colloiders* or contact units in the settling compartment.

[17] Karl Imhoff (1876–1965), in his day Germany's foremost wastewater engineer, was chief engineer and administrator of two German river authorities—the Emscher Genossenschaft and the Ruhrverband—both within the industrial complex of the Ruhr District. See G. M. Fair, Pollution Abatement in the Ruhr District, *J. Water Pollution Control Fed.*, **34**, 742 (1962).

[18] Named after the Emscher District, in which the Travis tank was modified.

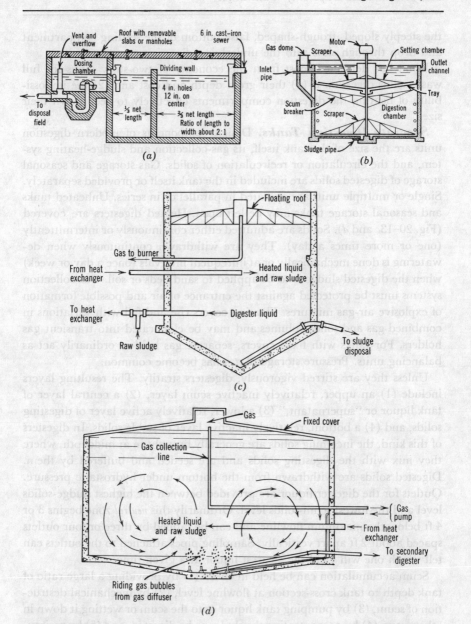

**Figure 20-13.**

**Sludge digesters.** (a) Two-compartment septic tank with dosing chamber; (b) two-story settling and digestion tank with scum breaker and sludge scraper (Dorr Co.); (c) separate digestion tank with floating cover; (d) separate digestion tank with fixed cover and gas-lift mixing system (Chicago Pump, FMC Corp.).

the steeply sloped, trough-shaped, false bottom of the settling compartment and drop through slots into the underlying digestion compartment.

What has detracted most from the economy of two-story tanks for full waste-solids digestion is (1) their great depth and cost and (2) the impossibility of heating the digestion compartments effectively to keep down their size.

**Separate Digestion Tanks.** Design components of modern digestion units are the size of the tank itself, its gas-collection and sludge-heating system, and the circulation or recirculation of solids. Gas storage and seasonal storage of digested solids are included in the tank itself or provided separately. Single or multiple units are operated in parallel or in series. Unheated tanks and seasonal storage tanks may be left open. Heated digesters are covered (Fig. 20-13c and d). Solids are admitted either continuously or intermittently (one or more times a day). They are withdrawn continuously when dewatering is done mechanically or at infrequent intervals (once a day or week) when the digested sludge is to be applied to sand beds or soil. Gas-collection systems must be protected against the entrance of air and possible formation of explosive air-gas mixtures. Floating covers rise and fall with variations in combined gas and solids volumes and may be elaborated into transient gas holders. For tanks with fixed covers, separate gas holders ordinarily act as balancing units. Pressure storage of gas has become common.

Unless they are stirred vigorously, digesters stratify. The resulting layers include (1) an upper, relatively inactive scum layer, (2) a central layer of tank liquor or "supernatant," (3) a lower, relatively active layer of digesting solids, and (4) a bottom, relatively inactive layer of stable solids. In digesters of this kind, the incoming solids are generally brought in at middepth, where they mix with the digesting solids and are seeded and buffered by them. Digested solids are withdrawn from the bottom under hydrostatic pressure. Outlets for the digester liquor are provided between the highest sludge-solids level and the lowest scum-solids level. Ordinarily this *neutral* zone begins 3 or 4 ft below the maximum flowline and can be tapped by three or four outlets spaced about 2 ft apart vertically. Sampling pipes attached to the outlets can tell which one will yield the clearest liquor at a given time.

Scum accumulation can be held in check (1) by providing a large ratio of tank depth to tank cross-section at flowline level, (2) by mechanical destruction of scum, (3) by pumping tank liquor onto the scum or wetting it down in other ways, (4) by concentrating the sludge to be digested, and (5) by recirculating gas into the digester bottom. A large depth-area ratio and a concentrated sludge release large volumes of gas per unit area, keep the scum in motion, mix the solids throughout the tank, make for good seeding, and ensure rapid heat transfer in heated tanks.

Besides occupying valuable space in single digesters, digester liquor is an often troublesome byproduct of noncontinuous digestion. It is usually high in immediate BOD and suspended solids. For this reason it is normally returned to the plant influent, in some instances by addition to activated sludge in reaeration units (Sec. 19-7). Otherwise it is added to the plant effluent after chemical coagulation, chlorination, and perhaps sand filtration. Waste liquor from sludge filters and elutriators is much like digestion-tank liquor and is disposed of or treated along with it or in much the same way.

Conversion of sludge digestion to a continuous process and its associated acceleration were long delayed because the treatment works in many places were unable to dry and dispose of digested solids during cold or wet weather. Not until the introduction of mechanical sludge dewatering was there a justifiable incentive to speed up digesters, with less emphasis on the "complete" digestion that is required where sludge is to be applied to drying beds.

A better grade of sludge liquor can be produced, and the design of high-temperature units can be functionalized, by placing two or more tanks in series (multistage digestion). The first or leading tank is designed to operate continuously and hold the sludge during its period of most active gas production. The unit is heated and stirred and its gas is collected. Sludge liquor is not withdrawn. The second tank is designed to hold the solids for the further length of time necessary to carry digestion to technical completion. The unit may be left unheated but may be equipped for gas collection. Because the rate of digestion is relatively low, clear liquor separates from the solids and can be withdrawn with little difficulty. Winter or operation storage may be incorporated into the second tank or into a third unit, which may be left uncovered.

## 20-16 Digester Capacities

The basic capacity of digestion tanks is a function of the sludge load, the time required for digestion, and the loss of sludge moisture (sludge liquor). Operational storage of gas and sludge (winter or seasonal storage) may be added elements. Capacities are expressed as cubic feet of tank volume per capita; cubic feet per pound of solids (on a dry basis) added daily; and cubic feet per pound of volatile solids (on a dry basis) added daily. Operational or loading parameters are the reciprocals of these capacities. The basic per capita capacity is determined in accordance with the general rules laid down in Sec. 20-3. The assumption is generally made that the rate of loss of sludge moisture or reduction in sludge volume, like the rate of gasification (Fig. 20-11), is substantially constant. It follows that progress of decomposition must be exponential and that the average difference in the volume of the fresh and digested solids is about two thirds the final difference. Formulated,

$$C = [V_f - \tfrac{2}{3}(V_f - V_d)]\, t \qquad (20\text{-}24)$$

where $C$ is the basic tank capacity in cubic feet per capita; $V$ is the daily per capita volume of solids in cubic feet, the subscripts $f$ and $d$ denoting the fresh and digested volumes, respectively; and $t$ is the time in days required for digestion. Required capacity is also a function of tank temperature and can be estimated from Fig. 20–12 for plain-sedimentation sludge submitted to noncontinuous (low-rate) digestion. Characteristic values are summarized in Table 20–3.

**Table 20-3**
*Noncontinuous Digestion of Plain-Sedimentation Sludge at Different Temperatures*

| Temperature, °F | 50 | 60 | 70 | 80 | 90 | 100 | 110 | 120 | 130 | 140 |
|---|---|---|---|---|---|---|---|---|---|---|
| Digestion period, days | 75 | 56 | 42 | 30 | 25 | 24 | 26 | 16 | 14 | 18 |
| Type of digestion | ←——————Mesophilic——————→ | | | | | | ←——Thermophilic——→ | | | |

The design capacity includes, in addition to the basic space requirement, allowances for sludge liquor, scum, and gas. A factor of safety, generally less than 2, is applied to the total.

*Example 20–11.*

Find the basic capacity of (1) the sludge compartment of an Imhoff tank in which plain-sedimentation primary sludge is to be digested at an average temperature of 60°F and (2) a heated, noncontinuous digestion tank in which (a) the same sludge and (b) a mixture of primary and activated sludge is to be digested at 90°F. Assume the per capita volumes of sludge calculated for domestic sewage in Example 20–2, namely, 37.8 cu ft of fresh and 9.0 cu ft of digested primary sludge, and 65.4 cu ft of fresh and 26.5 cu ft of digested combined sludge, all per 1000 persons daily.

1. The digestion period at 60°F is about 56 days. Hence, by Eq. 20–24, the required capacity of the Imhoff tank is $C = [37.8 - {}^2/_3(37.8 - 9.0)]56/1000 = 1.04$ cu ft per capita.

2. The digestion period at 90°F being about 25 days, the required per capita digester capacity for primary sludge is $C = 1.04 \times 25/56 = 0.46$ cu ft per capita; and for combined primary and activated sludge, $C = [65.4 - {}^2/_3(65.4 - 26.5)]25/1000 = 0.99$ cu ft per capita. The assumption has been made in this connection that the mixed sludge digests as rapidly as the primary sludge.

*Continuous operation*, including thickening the incoming sludge, allows the organic loadings on digesters operating at optimal mesophilic temperatures to be increased to three or four times the loadings on noncontinuous digesters. About half the rise can be accomplished by sludge thickening alone. Implied in these loading figures for continuous operation are detention times of 15 and 10 days, respectively, in comparison with 30 and 35 days for discontinuous operation. A factor limiting sludge thickening is loss of the sludge fluidity required for sludge transport and successful tank stirring. Factors limiting continuous operation at high rates are (1) increases in alkalinity and ammonia as biochemical end products and (2) decreases in the destruction of grease and volatile matter. As a general rule it is impractical to concentrate sludges to more than 6% total solids before digestion.[19] The digestibility of volatile organic matter in wastewater sludges is a function of the volatile to fixed solids ratio. The higher the content of volatile matter, the more effective is its reduction in a given time.

Winter storage can be calculated from the volume occupied by the fully digested sludge.

**Heat Requirements.** The heat supplied to heated sludge-digestion tanks must be sufficient to (1) raise the temperature of the incoming sludge to that of the tank, (2) offset the heat lost from the tank through its walls, bottom, and cover, and (3) compensate for heat lost in piping and other structures between the source of heat and the tank. The amount of heat entering into the digestion reactions and used up in the evaporation of water into the sludge gases is so small that it can be left out of engineering computations.

The specific heat of most sludges is substantially the same as that of water. The heat loss $Q$ through the walls, bottom, and cover of a tank is a function of the temperature difference $\Delta T$, the tank area $A$, and the coefficient of heat flow $C$, or

$$Q = CA(\Delta T) \tag{20-25}$$

The value of $C$ depends on (1) motion of the ambient fluids (sludge inside and air or water outside the tank), (2) thickness of specific portions of the tank and their relative conductance, and (3) opportunities for radiation. For concrete tanks, rough overall values for $C$ in British thermal units per square foot and hour are 0.10, 0.15, and 0.30 for exposures to dry earth, air, and wet earth, respectively. For equal exposure of all parts of a digester, heat is conserved best if tank geometry conforms as closely to a sphere as structural economy will allow. The largest ratio of sludge volume to tank surface is thereby obtained. For unequal exposures, the portions through which the greatest losses per unit occur should be kept at a minimum.

The heat requirements of digesters are generally supplied in the following ways: (1) the incoming sludge is heated outside the tank in countercurrent

[19] C. N. Sawyer and J. S. Grumbling, Fundamental Considerations in High Rate Digestion, *Proc. Am. Soc. Civil Engrs.*, 86, SA2, 49 (1960).

heat exchangers; (2) hot water is circulated through fixed or moving coils inside the tank; (3) gas is burned under water or in a heater submerged in the sludge; and (4) steam is introduced into the sludge. Heat transfer is analogous to heat loss. For stationary heating surfaces, the coefficient of heat flow in British thermal units per square foot and hour has approximate values of 8 to 12 for sludge of ordinary thickness and 35 to 45 for thin sludge or water. Moving surfaces increase the heat flow to about 60 Btu per sq ft and hr. From submerged burners, there is direct absorption of the heat of combustion. Sludge caking on the surface of heating units is held down by keeping the temperature of the heating surface at less than 140°F and by moving the unit through the sludge or by moving the sludge, either mechanically or thermally, past the unit.

*Example 20–12.*

A cylindrical digester 20 ft in diameter is built of concrete and is surrounded by dry earth or provided with equivalent insulation. The side walls are 16 ft high, and the roof, which is exposed to the air, rises 1 ft to a central, insulated gas dome. The bottom slopes 4 ft to a central sludge-withdrawal pipe. The daily weight of sludge added to the tank is 13,000 lb. Find (1) the daily heat requirements of the tank when the temperature of the incoming sludge, digesting sludge, earth, and air are 50, 90, 40, and 20°F, respectively, and the coefficient of heat flow is 0.20 Btu/(sq ft)(hr); (2) the required area of stationary heating coils through which water is circulated at an incoming temperature of 130°F and with a temperature drop of 10°F if the coefficient of heat flow is 10 Btu/(sq ft)(hr); and (3) the daily volume of heating water that must be circulated through the coils.

1. Heat requirements: area of the tank exposed to earth = $\pi \times 20(16 + \frac{1}{2}\sqrt{10^2 + 4^2})$ = 1344 sq ft; area of roof exposed to air = $\pi \times 10\sqrt{10^2 + 1^2}$ = 316 sq ft; heat requirements of the incoming sludge = $13{,}000 \times (90 - 50)$ = 520,000 Btu daily; exposure loss = $0.20 \times 24[1344(90 - 40) + 316(90 - 20)]$ = 430,000 Btu daily; and total heat requirements = $(520{,}000 + 430{,}000)$ = 950,000 Btu daily.

2. Area of heating coils: average temperature of heating water = $(130 - 5)$ = 125°F; and area of coils = $950{,}000/[24 \times 10(125 - 90)]$ = 113 sq ft.

3. Recirculating water: temperature drop = 10°F = 10 Btu/lb; weight of water = $950{,}000/10$ = 95,000 lb daily; volume of water = $95{,}000/8.34$ = 11,400 gpd = 8 gpm; and the calculated heat requirements do not take into account possible losses between the heat source and the digester.

As a rule of thumb, a 1°F-per-day drop in temperature would occur in the entire tank contents in the North, unless heat is added. This amounts to

$1 \times 62.5/24 = 2.6$ Btu per hour per cu ft. This figure is halved for the temperate zone, and reduced to one-third in the South.

**Digester Gas.** In addition to methane (combustible) and carbon dioxide (noncombustible), sludge gas always contains water vapor, occasionally hydrogen sulfide, and, more rarely, hydrogen (combustible) and nitrogen (inert). As stated before, hydrogen sulfide is an explosive as well as a toxic gas (Sec. 12-12). When hydrogen sulfide is burned, sulfur dioxide, a very corrosive gas, may be formed. For use in gas engines with exhaust gas heaters, not more than 10 grains of hydrogen sulfide[20] should be present in 100 cu ft of gas (0.015% by volume).

The fuel value of sludge gas can be put to use in various ways both within treatment plants and for nonplant purposes. The following are common: (1) plant heating—digesters, incinerators, buildings, and hot-water supply; (2) plant power production—pumping, air and gas compression, and operation of other mechanical equipment; (3) minor plant uses—gas supply to the plant laboratory for gas burners and refrigerators; and (4) motor fuel for municipal cars and trucks. Plants are heated and power is produced by burning the gas in a furnace, under a gas-fired hot-water or steam boiler, or in a gas engine equipped with a water jacket and an exhaust-gas heat exchanger. Minor plant uses may be served by direct combustion of the gas under the available plant pressure or by bottling the gas. Gas for motor fuel must be bottled under high pressure in steel containers.

Collection, storage, and utilization of sludge gas are economically justified only when the treatment works are large enough to warrant skilled attendance. Gas storage may be included in the design of digesters, but separate holders are normally preferred. The collection, storage, and distribution system must be kept under pressure in order to avoid the formation of explosive mixtures of gas and air.

Methane is burned as follows: $CH_4 + 2 O_2 = CO_2 + 2 H_2O$. Because air contains about 21% oxygen by volume, at least $2/0.21 = 9.5$ cu ft of air are required to burn 1.0 cu ft of methane. Explosive mixtures of methane and air are formed over a range of 5.6 to 13.5% of methane by volume. Flame speed is maximal at 9.6%. Above 13.5% the mixture burns quietly after ignition. Violent explosions with loss of life have occurred in wastewater treatment works.

The operating, protecting, and regulating devices of gas collection and distribution systems include (1) condensate traps and drains for water vapor, (2) flame traps that prevent flashbacks from gas burners and engines, (3) pressure-regulating valves, and (4) waste-gas burners. Gas lines and their appurtenances must be protected against freezing; yet all vents must termi-

---

[20] One pound = 7000 grains, and 1 cu ft of $H_2S$ weighs 0.095 lb.

nate in the open. Gas may be stored under a head of 3 to 6 in. of water without prior compression. The economical pressure for cylindrical and spherical pressure tanks is about 40 psig. Bottled gas is placed under a pressure of about 5000 psig. Before being compressed, the gas may be passed through scrubbers to remove unwanted constituents: carbon dioxide, hydrogen sulfide, and water vapor.

Hot-water boilers are neither as efficient (about 60%) nor as trouble free (sulfur dioxide corrosion) as steam boilers with heat exchangers for hot-water heating (about 80%). Water-jacketed gas engines equipped with exhaust-gas boilers have a water-heating efficiency of about 50% and a direct power efficiency of 22 to 27%, depending on the engine load (half load to full load, respectively). Conversion of gas-engine power into electrical power and use of electric motor-driven equipment entail a loss of about 25% of the engine power. For equal performance in automotive engines, about 160 cu ft of sludge gas containing 72% of methane (110,000 Btu) may be substituted for a gallon of gasoline (122,000 Btu).

*Example 20-13.*

If the 13,000 lb of sewage sludge added daily to the digester in Example 20-12 are assumed to contain 5% solids, 72.2% of which are volatile, and if 67% of the volatile matter is assumed to be destroyed in producing gas containing 72% methane and 28% carbon dioxide by volume, estimate (1) the volume of gas produced and its calorific power; (2) the power made available from a gas engine equipped with a waste-gas boiler for heating the digester and from an electrical generator driven by the engine; (3) the volume of gas available for plant purposes other than digester heating ($9.5 \times 10^5$ Btu daily) or required to be burned in a waste-gas burner; and (4) the volume of gas needed to deliver 1 million cu ft of air under a pressure of 7 psig to a conventional activated-sludge plant, and the adequacy of gas for both power production and digester heating.

1. Gas production: the volatile matter destroyed is $13 \times 10^3 \times 0.05 \times 0.722 \times 0.67 = 315$ lb; and at 1.25 lb of gas per lb of volatile matter destroyed and the given gas composition, the gas yield is $1.25 \times 315/(0.72 \times 0.0446 + 0.28 \times 0.1225) = 5930$ cu ft daily, or $5930 \times 963 \times 0.72 = 4.1 \times 10^6$ Btu daily.

2. Power of gas engine and generator: if the gas engine has a water-heating efficiency of 50%, the daily heat input will be $9.5 \times 10^5/0.5 = 1.9 \times 10^6$ Btu; at 2545 Btu/hp-hr or 0.7457 kwhr/hp-hr, and at an engine efficiency of 25% and a generator and motor efficiency of 75%, the heat and gas consumed are $2545/0.25 = 10^4$ Btu/brake hp-hr, and $10^4/(963 \times 0.72) = 14.5$ cu ft of gas/brake hp-hr; the engine power is then $(1.9 \times 10^6)/(24 \times 10^4) = 8.0$ hp, and the electrical power $8.0 \times 0.75 \times$

$0.7457 = 4.4$ kw. Incidentally, if all the available gas were supplied to the engine, the engine and generator power would be raised $(4.1 \times 10^6)/(1.9 \times 10^6) = 2.2$ times.

3. Volume of excess gas: with $(4.1 \times 10^6 - 1.9 \times 10^6) = 2.2 \times 10^6$ Btu not used by the engine, the volume of excess gas is $(2.2 \times 10^6)/(0.72 \times 963) = 3.2 \times 10^3$ cu ft daily.

4. Power requirements of a conventional activated-sludge plant: the energy requirements of conventional diffused-air, activated-sludge units being about 550 hp-hr or 410 kwhr/million cu ft of free air daily when the air pressure is 7 psig (Sec. 19-12), $550 \times 14.5 = 8.1 \times 10^3$ cu ft of gas are needed to deliver 2 million cu ft of air at a pressure of 7 psig by means of an engine-driven compressor, or $8.1 \times 10^3/0.75 = 10.7 \times 10^3$ cu ft of gas when electric drive is employed.

Because $10^6$ cu ft of air daily will treat about 1 mgd of sewage from $10^4$ people in conventional activated-sludge units, and because the gas production from primary solids and activated sludge is about 1.25 cu ft per capita, the available gas supply of $1.25 \times 10^4$ cu ft daily is normally sufficient to provide the necessary engine (and, where wanted, electrical) power and to keep the sludge-digestion tanks at optimal mesophilic temperatures. In primary-treatment plants with sludge digestion, incidentally, the gas yield of about 1 cu ft per capita is normally more than adequate for the principal plant purposes.

The flexibility of electrical operation in treatment works, including the possibility of using purchased electricity when the gas supply fails, often justifies the installation of electrical equipment. Standby fuel must otherwise be provided in the form of gasoline, oil, or municipal or natural gas. Standby equipment varies in accordance with the selected sources of energy.

## 20-17 Agricultural Value of Sludge

Most important among the chemical elements essential to the growth of agricultural crops are: (1) carbon, oxygen, and hydrogen secured freely from air and water; (2) nitrogen, phosphorus, potassium, calcium, magnesium, sulfur, and iron obtained in substantial quantities from the soil; and (3) boron, manganese, zinc, copper, and other *trace elements* leached in minute quantities from the soil.

Other elements, although they may not be beneficial to plants, are valuable components of the food of man and the higher animals. Examples are iodine, fluorine, chlorine, and sodium. Vitamins and hormones also play a part in plant growth.

Waste solids produced by wastewaters contain many of the fertilizing elements. In competition with commercial fertilizers, they are rated principally on their content of three substances: nitrogen as N, phosphorus as P or as phosphoric acid ($P_2O_5$), and potassium as K or as potash ($K_2O$). Like the humus they resemble, some waste solids are also good soil builders or soil conditioners.

The concentration of fertilizing chemicals is generally expressed as a percentage of the dry weight of the solids. Nitrogen varies in fresh municipal wastewater solids from 0.8 to 5% for plain-sedimentation solids (and proportionately less if they include chemical precipitating agents) to 3 to 10% for activated sludges. Only 1.5 to 5% nitrogen is left in trickling-filter humus, depending on the length of its storage in the filter. Anaerobic digestion reduces the nitrogen content of sludge by 40 to 50%. The phosphates in most wastewater sludges are small in amount (1 to 3%), and potash is even smaller (0.1 to 0.3%). In comparison, animal manures contain 1 to 4% nitrogen and about the same percentage of phosphate and potash. Animal tankage, blood, and fish scraps include 5 to 13% nitrogen and 0.5 to 14% phosphate. Cottonseed meal and castor pomace are about as rich in nitrogen and phosphate as animal tankage, blood, and fish scraps. In addition they include 1 to 2% potash.

The agricultural and horticultural use of wastewater solids is circumscribed by the hygienic hazards they may create. Pathogenic bacteria, viruses, protozoa (cysts), and worms (eggs) can survive wastewater treatment and be included in the waste solids. They are not fully destroyed during the normal course of digestion and air drying. Although the numbers of survivors decrease appreciably, only heat-dried solids can be considered fully safe where food crops that are to be eaten raw are grown.

## 20-18 Sludge Handling and Disposal

Wet sludge is normally transported in free-flow and pressure conduits (Sec. 20-4). Partially dewatered sludge (sludge paste or filter cake) is carried on belt conveyors. Granular (heat-dried) sludge is moved pneumatically or by belt or screw conveyors. All types of sludge can be handled in industrial cars and suitably constructed automotive equipment.

Sludge may be disposed of in any one of the states in which it is produced, that is, as wet sludge (both raw and digested), filter cake, sludge cake from air-drying beds, and heat-dried sludge. Alum and iron precipitates from water-purification plants may be discharged into sewers. If they are to be discharged into small streams, they should preferably be lagooned. Lagoons are natural depressions in the ground, or earth basins excavated for that purpose and possibly provided with overflows and underdrains. Heavy precipitates of calcium and magnesium from water-softening plants may be partially dewatered and used as fill for low-lying lands. Wet wastewater sludge may be pumped onto land and plowed under. Digested sludge is more suitable for this purpose than raw sludge. Wastewater sludges are no longer discharged into inland waters. Seacoast communities may transport either wet or partially dewatered fresh or digested sludge to dumping grounds at sea. Specially designed sludge tankers and scows or barges are used for this purpose. Wet, digested sludge may be pumped to deep-lying, hydraulically active portions of tidal estuaries. Sludge lagoons may be used both for the digestion of raw sludge and for the storage and consolidation of digested sludge. Odors rising from raw-sludge lagoons must be taken into consideration. Low-lying land and areas denuded by strip mining can be filled by lagooning or by dumping sludge cake or air-dried sludge. It may be economical to pump thickened or partially dewatered sludge appreciable distances to enrich poor agricultural soils with humus and wet them. Inclusion of partially dewatered sludge or sludge cake in sanitary land fills is a possibility. Air-dried sludge and filter cake may be hauled away by farmers to serve as soil builders. The sludge may be disintegrated for this purpose. Dried sludge is often used as a fertilizer or fertilizer base. The ash from incinerators can be dumped at sea or disposed of as fill.

**twenty-one**

# ecology and management of natural and receiving waters

**21-1 Scope of Inquiry**

To cure the ills of water to which it is subjected during its natural passage over and through the ground from the clouds to the sea and by the manifold uses to which it is put by man, the engineer—like the physician who treats a patient—must often put his trust in the healing power of nature.[1] Given a chance to exert themselves, these natural forces rid polluted water of corruption, even as the mobilization of the natural defenses of the body cleanses it of infection. Because this is nature's way, it behooves the engineer to become familiar not only with the syndrome of pollution and methods for its prevention or cure but also with the forces of natural or self-purification that of themselves cause a remission of the symptoms of pollution.

The natural forces of purification are many and varied. As a group they are closely interrelated and mutually dependent. Individually, they are also given attention in other chapters of this book, because they enter in one way or another into the treatment methods devised for waters and wastewaters. There they are purposely intensified in order to accomplish, in a brief time and in small space, the changes normally brought about in nature only in extended periods of time

---

[1] The *vis medicatrix naturae* recognized by early medical philosophers.

and over long distances of travel or wide areas of dispersion, with an accompanying deterioration in the quality of the aquatic environment.

The present chapter evaluates the significant forces of self-purification and demonstrates the application of existing knowledge to the solution of some of the engineering problems encountered in the quality management of natural and receiving waters.

## 21-2 Patterns of Pollution and Natural Purification

When pollutants are discharged into water, a succession of changes in water quality takes place. If the pollutants are emptied into a lake in which the currents about the outfall are sluggish and shift their direction with the wind, the changes occur in close proximity to each other, move their location sporadically, and cause much overlap. Moreover the resulting pattern of change is modified sharply by changes in season and hydrography. If, on the other hand, the water moves steadily away from the outfall, as in a stream, the successive changes occupy different river reaches and establish a profile of pollution and natural purification so well defined that it can be formulated mathematically with satisfying success. Yet there is no set pattern in most streams. Instead, the pattern shifts longitudinally up and down the water course and is modified in intensity seasonally and with changes in streamflow. The intensity of pollution rises during warm weather and at low river stages. It is suppressed during cold weather and when the stream is in flood. Ice cover imposes a pattern of its own.

When a single, heavy charge of putrescible matter is poured into a clean body of water, the water becomes turbid, sunlight is shut out of the depths, and green plants, which by photosynthesis remove carbon dioxide from the water and release oxygen to it, die off. Scavenging organisms increase in number until they match the food supply. The intensity of their life is mirrored in the intensity of the biochemical oxygen demand (BOD). The oxygen resources of the water are drawn upon heavily. In an overloaded receiving water the supply of dissolved oxygen may become exhausted. Nitrogen, carbon, sulfur, phosphorus, and other important nutritional elements run through their natural cycles, and sequences of microbic populations break down (1) the waste matters that have been added, (2) the natural polluting substances already within or otherwise entering the water, and (3) the food made available by the destruction of green plants and other organisms intolerant to pollution. The links of a food chain are forged from available nutrients by the growth and environmental adaptiveness of sequences of organisms.

Depending on the hydrography of the receiving water, suspended matter is carried along or removed to the bottom by sedimentation. Bottom (benthal or benthic) deposits are laid down in thicknesses varying from a thin pollutional carpet to heavy sludge banks. Their decomposition differs from that in the supernatant waters. In the presence of oxygen dissolved in the overlying waters, benthal decomposition varies with depth of deposit from largely aerobic to largely anaerobic conditions. The influence of the benthal factor on the stream varies accordingly.

The initial effect of pollution is to degrade the physical quality of the water. As decomposition intensifies, a shift to chemical degradation is biologically induced. At the same time biological degradation becomes evident in terms of the number, variety, and organization of the living things that persist or make their appearance. In the course of time and distance, the energy values of a single charge of polluting substances are used up. The biochemical oxygen demand is decreased, and the rate of absorption of oxygen from the atmosphere, which at first has lagged behind the rate of oxygen utilization, falls into step with it and eventually outruns it. The water becomes clear. Green plants flourish in the sunlight and release oxygen to the water during photosynthesis. Other higher aquatic organisms, including game fish, which are notably sensitive to pollution, reappear and thrive as in a balanced aquarium. The waters are returned to normal purity. Self-purification is gradually completed. Recovery has taken place.

The natural purification of polluted waters is never fast, and heavily polluted streams may traverse long distances during many days of flow before a significant degree of purification is accomplished.

If pollution is kept within bounds it contributes to the fertility of the receiving water and resultant aquatic populations. Fish then browse in the aquatic meadows that derive the elements for their growth from the nitrogen, phosphorus, and other fertilizing constituents of the waste matters. The use of settled wastewater for the fertilizing of fish ponds is indeed an example of controlled pollution; so is the possibility of harvesting proteinaceous plankton from oxidation ponds. In the fertilization of sources of water by domestic wastewaters, however, the danger of spreading disease through plant or animal foods must never be lost from sight; neither must the overfertilization and consequent *eutrophication* of receiving waters, especially lakes and estuaries. Natural pollution results in *natural eutrophication*, for which runoff from uncultivated or otherwise undeveloped catchment areas is responsible. Cultural pollution produces *cultural eutrophication*, for which man's water and land management or use is responsible.

The self-purification of polluted groundwaters differs appreciably from that of surfacewaters. The variety of living organisms that seize upon the pollutional substances for food is greatly restricted in the confinement and darkness of the pore space of the soil. But this reduction in biological purification is more offset by physical purification through filtration. In general, the rate of purification is stepped up greatly, and time and distance of pollutional travel shrink in dimension.

## 21-3 Parameters of Pollution and Natural Purification

Degrees of pollution and natural purification can be measured physically, chemically, and biologically. No single yardstick will do. Depending on the nature of the polluting substances and the uses the receiving body of water (or water taken from it) is to serve, measurements may be made of turbidity, color, odor, nitrogen in its various forms, phosphorus, BOD, organic matter, dissolved oxygen and other gases, mineral substances of many kinds, bacteria and other microorganisms, and the composition of the larger aquatic flora and fauna including that of the bottom.

When polluted waters are to be used for municipal purposes or as bathing waters, they are usually examined for the prevalence of the coliform group of organisms. Longitudinal changes in coliform concentrations establish (1) the progress of bacterial self-purification, (2) the relative hazard of infection incurred by ingesting the water, and (3) the degree of purification to which the water must be subjected before it can presumably be used with safety and satisfaction. When pollutional nuisance of receiving waters, such as unsightliness and odor, is the criterion, the DO and BOD, taken together, are relied on to trace the profile of pollution and natural purification on which engineering calculations of permissible pollutional loadings can be based. The BOD identifies in a comprehensive manner the degradable load added to the receiving water or remaining in it at a given time and place; the DO identifies the capacity of the body of water to assimilate the imposed load by itself or with the help of reaeration through oxygen absorbed mainly from the atmosphere, but possibly released to the water by green plants. Requisite standards of water quality will serve as guides to the tests that are meaningful in given circumstances; COD, for example, rather than BOD, may be the sentinel of choice when toxic wastes destroy saprophytes and other living things.

A single example[2] of observed changes in terms of some of the generally useful parameters of pollution and natural purification is illustrated in Fig. 21-1. The samples on which all but the benthal results are based were collected during midsummer. Because bottom muds are deposited over long periods of time, the bottom-dwelling organisms are generally characteristic of average conditions during the period of accumulation. That time of flow, rather than distance of flow, is the controlling factor is strikingly shown by the improvement in water quality during the relatively slow passage of the

---

[2] From a survey by the U.S. Public Health Service of the Mississippi River at Minneapolis and St. Paul before treatment of the metropolitan wastewaters.

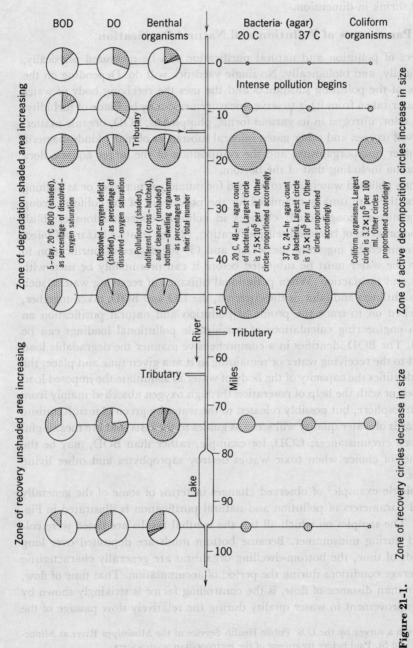

Figure 21-1.
Pollution and self-purification of a large stream.[2]

stream through the lake that occupies the lower river reaches in Fig. 21-1. However, this type of improvement may be transient and costly if eutrophication becomes intensive enough to be detrimental to the lake itself. The tributaries entering from the west (left) in Fig. 21-1 were themselves heavily polluted. The eastern tributary was significantly better than the mainstream.

## 21-4 Bacterial Self-Purification

The discharge into a receiving water of wastewaters rich in degradable organic matter vastly increases the number and genera of saprophytic bacteria that help to bring about self-purification. The multiplying organisms are derived in part from the wastewaters, and in part from the receiving waters. Other saprophytes enter with the runoff from agricultural and other soils. Only after they have come into balance with the food supply under prevailing environmental conditions do the number and variety of saprophytes begin to decline. The density of intestinal bacteria also rises appreciably below a wastewater outfall during the first 10 to 12 hr of flow. But this is so largely because clumps of fecal solids break up and disperse contained organisms in the water. Below the points of modal concentration of the different bacterial populations, their numbers drop off at varying rates. Among the reasons are (1) the presence, in polluted natural bodies of water, of predators such as the ciliated protozoa that feed on bacteria and (2) the operation of biophysical forces (such as sedimentation and biological flocculation and precipitation) that ally the processes of natural purification with those of a wastewater-treatment plant. Conversely, it has been suggested that a treatment plant is like a river wound up in small space.

Because the bacteria die not merely for lack of food, it follows that their die-away in receiving waters is only approximated by Chick's law (Sec. 11-14). A better fit of observed values is obtained by the general purification equation (Eq. 11-13). Generally speaking, the destruction of enteric bacteria is more rapid in (1) heavily polluted waters than in clean waters, (2) warm weather than in cold weather, (3) shallow, turbulent waters than in deep, sluggish water bodies, and (4) in salt water than in fresh water.

The time required for the bacterial self-purification of natural waters is long, and the associated distance of travel of stream waters may be great. The die-away of coliforms inoculated into natural seawater under laboratory conditions is reported to be about 25 times as rapid as in autoclaved seawater.[3] The lethal factors involved are apparently not the sea salts but

[3] B. H. Ketchum, C. L. Carey, and Margaret Briggs, Preliminary Studies on the Viability and Dispersal of Coliform Bacteria in the Sea, in *Limnological Aspects of Water Supply and Waste Disposal*, American Association for the Advancement of Science, Washington, D.C., 1949, p. 64.

organic, heat-labile substances. Activities of marine organisms may explain the accelerated die-away. A dilution factor of 200 to 250 for municipal wastewaters in seawater normally reduces the coliform organisms to 10 per ml or less (Sec. 21-12).

**Die-Away of Enteric Pathogens.** Dilution and die-away are important safeguards against waterborne enteric pathogens. Pertinent information on die-away is by no means complete. That assembled in Table 21-1 is useful only for similar conditions of test.

*Table 21-1*
*Average Time in Days for 99.9% Reduction in Original Titer of Enteric Organisms*

| Organism | Source of Water | | |
|---|---|---|---|
| | River,[a] 4°C | Sewage, 4°C | Clean Water, 4°C |
| Salmonella typhosa | ... | ... | 6.4 |
| Entameba histolytica | ... | ... | 40 |
| Aerobacter aerogenes | 15, 44 | 56 | ... |
| Streptococcus fecalis | 17, 57 | 48 | ... |
| Escherichia coli | 15, 18 | 48 | ... |
| Echo 7 virus | 26, 15 | 130 | ... |
| Echo 12 virus | 33, 19 | 60 | ... |
| Coxsackie A9 virus | 10, 20 | 12 | ... |
| Poliomyelitis I virus | 27, 19 | 110 | ... |

[a] Results for Little Miami and Ohio River water, respectively. See N. A. Clarke, G. Berg, P. W. Kabler, and S. L. Chang, Human Enteric Viruses in Water: Source, Survival, and Removability, *Proc. First Intern. Conf. Water Pollution Res.*, London, 1962.

## 21-5 The Oxygen Economy of Natural and Receiving Waters

In nature, clean waters are saturated with dissolved oxygen, or nearly so. Normally, therefore, waste matters discharged into natural waters undergo aerobic decomposition. Only when the supply of oxygen in solution or taken into solution—principally from the atmosphere—cannot keep pace with the BOD exerted by entrant natural or cultural nutrients does the receiving water, and with it the type of decomposition, become anaerobic. Although the ultimate result of both kinds of decomposition is a purified water, their manifestations are as different as the conditions obtaining in an activated-sludge unit on the one hand and a septic tank on the other. Aerobic receiving waters look reasonably clean and are free from odor. Within limits, they continue to support normal animal and plant populations. By contrast, anaerobic waters become black, unsightly, and malodorous, and their normal fauna and flora are destroyed.

It follows that the oxygen economy of receiving waters is of paramount esthetic and economic importance. In order to maintain a balanced ledger, biochemical oxygen demand on the debit side must not exceed available oxygen on the credit side. Exertion of the BOD results in *deoxygenation* of receiving waters. Absorption of oxygen from the atmosphere and from green plants during photosynthesis results in *reoxygenation* or reaeration. In streams, the interplay between deoxygenation and reaeration produces a dissolved-oxygen profile called the *oxygen sag* (Sec. 21-9).

## 21-6 Kinetics of Aerobic Decomposition

Terrestrial organisms draw oxygen from the atmosphere; aquatic organisms obtain theirs from the oxygen dissolved in water. Because water contains only about 0.8% oxygen by volume at normal temperatures (about 50°F), whereas the atmosphere holds about 21% by volume, the aquatic environment is inherently and critically sensitive to the oxygen demands of the organisms that populate it. Determination of the amount of oxygen dissolved in water (DO) relative to its saturation value and of the amount and rate of oxygen utilization (BOD), therefore, furnishes a ready and useful means for identifying the pollutional status of water and, by indirection, also the amount of decomposable or organic matter contained in it at a given time. As shown in Fig. 21-2, the progressive exertion of the BOD of freshly polluted water generally breaks down into two stages: a first stage, in which it is largely the carbonaceous matter that is oxidized; and a second stage, in which nitrogenous substances are attacked in significant amounts and nitrification takes place. If the temperature of freshly polluted water is 20°C, for example, the first stage extends about to the 10th day. During this period the amount of BOD exerted in a unit of time relative to the BOD remaining to be exerted during the first stage is substantially constant. In the succeeding second stage the BOD rises sharply as nitrification becomes dominant. Oxygen is then put to use at a fairly uniform rate.

A knowledge of the progressive utilization of oxygen by polluting substances is important for at least three reasons: (1) as a generalized measure of the amount of oxidizable matter contained in water, or the pollutional load placed on it, (2) as a means for predicting or following the progress of aerobic decomposition in polluted waters and the degree of self-purification accomplished in given intervals of time, and (3) as a yardstick of the removal of putrescible matter accompanying different treatment processes.

*644/Ecology and Management of Natural and Receiving Waters*

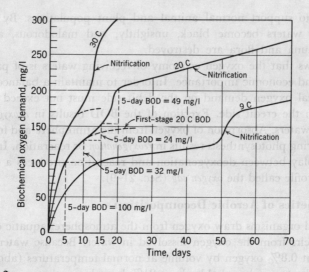

**Figure 21-2.**
Progress of biochemical oxygen demand (BOD) at 9, 20, and 30°C. (After Theriault.)

***The First-Stage BOD Curve.*** The first-stage BOD has generally been formulated as a first-order reaction. The concentration of oxidizable organic material present is the rate-determining factor, provided the oxygen concentration is adequate above about 2 mg per l at 20°C. The reactions involved are enzymatic.

The first-order equation may be written[4] $y = L[1 - \exp(-kt)]$, in which $L$ is the initial or first-stage BOD of the water, $y$ is the oxygen demand exerted in time $t$, and $k$ is the rate constant. The BOD remaining at time $t$ equals $(L - y)$ and the proportion of BOD exerted in time $t$ is $y/L = 1 - \exp(-kt)$.

Practical evaluation of this equation is complicated by the fact that $L$, as well as $k$, is usually unknown. Of the methods for finding the magnitudes of $L$ and $k$ from a series of observations of $y$ and $t$ that have been proposed, the *method of moments* developed by Moore, Thomas, and Snow,[5] appears to offer the most convenient analytical solution.

Studies of the behavior of wastewater samples have shown that $k$ may range from 0.16 to 0.70 per day at 20°C, and that the mean value is 0.39. It

---

[4] The engineering literature includes this equation also in the following form: $y = L(1 - 10^{-k't})$ where $k' = 0.4343k$.

[5] E. W. Moore, H. A. Thomas, Jr., and W. B. Snow, Simplified Method for Analysis of BOD Data, *Sewage Ind. Wastes*, **22**, 1343 (1950). See also H. A. Thomas, Jr., Graphical Determination of BOD Curve Constants, *Water and Sewage Works*, **97**, 123 (1950).

follows that comparisons should be drawn between the five-day, 20°C BOD values of different waters only if their reaction-velocity constants are identical. In the absence of actual tests the following values can be used in first estimates of expected BOD behavior of raw and treated municipal wastewaters. Values for tap water are added for comparison.

|  | $k$ | $L$ |
|---|---|---|
| Weak wastewater | 0.35 | 150 |
| Strong wastewater | 0.39 | 250 |
| Primary effluent | 0.35 | 75–150 |
| Secondary effluent | 0.12–0.23 | 15–75 |
| Tap water | <0.12 | 0–1 |

For a given sample of water, the reaction rate increases with temperature. The observed effect can be formulated in terms of the van't Hoff-Arrhenius relationship (Sec. 11–15) and expressed as (1) the energy of activation $E$, (2) the temperature characteristic $C$ and its power function, the temperature coefficient $\theta = e^C$, or (3) the temperature quotient $Q_{10} = e^{10C} = \theta^{10}$.

From about 15 to 30°C the activation energy $E$ of the BOD reaction is 7900 cal, corresponding to values of $C_k = 0.046$ per deg C, $\theta_k = 1.05$, and $Q_{10} = 1.6$. At lower temperatures $E$ increases to as much as 20,000 cal near 0°C ($\theta_k \simeq 1.11$ at 10°C and 1.15 at 5°C). Above 30°C a decrease in rate with increasing temperature is observed, probably because of thermal inactivation of the enzymes responsible for oxidation ($\theta_k \simeq 0.97$ at 35°C).

There are other complications in BOD testing. In unseeded samples (that is, samples with an inadequate flora and fauna to activate the BOD reactions), there are *lag* periods before the reaction proceeds normally. By contrast, wastewaters that have undergone partial anaerobic decomposition or contain reducing chemical substances may exert an *immediate* demand (sometimes called a chemical demand) at the beginning of the BOD run. This is not part of the normal BOD. In some instances, furthermore, $k$ values diminish as the percentage of reaction increases. Probably this results from differences in ease of oxidation of the materials present, the rate decreasing as more easily oxidized substances are used up. The onset of nitrification, finally, may increase the rate in the later stages of the reaction, particularly in highly diluted samples.

## 21-7 Deoxygenation of Polluted Waters

It is generally assumed that the demands made on the oxygen resources of polluted waters by living organisms are the same as those observed in the laboratory when samples of wastewater are mixed with convenient amounts of synthetic dilution water and incubated for the BOD tests. This is not necessarily so. Overlooked is the fact that the biophysical as well as the biochemical environment of BOD bottles does not simulate that of every kind of receiving water. Fortunately for engineering predictions of the deoxygenation of large and passive streams, the correlation between (1) laboratory observations of the BOD of polluted waters and (2) field investigations of the reaction of such streams to pollution is usually high.

When, by contrast, streams are shallow and turbulent, the rate of deoxygenation is stepped up appreciably. Deoxygenation may indeed approach that experienced in submerged biological treatment systems. At the same time the rate of BOD drops off along the course of the stream below the point of maximum pollution. If laboratory results are to be applied to actual situations, estimates must be in keeping with (1) the nature of the stream channel, flow, and flow variations; (2) the progress in BOD of the waste matters and the stream water itself; and (3) the transfer of pollutional load to the stream bottom and its benthal decomposition.

The five-day, 20°C BOD of domestic wastewaters on which engineering calculations of deoxygenation are generally based is shown in Table 10-6. If the wastewaters are derived from combined drainage systems or if they contain industrial wastes or other oxygen-demanding substances, the pollutional load can be accounted for in terms of the equivalent domestic population. Such wastewaters are generally much stronger than domestic wastes waters. The relations of organic industrial wastewaters to urban wastewater are not direct and simple. Wastewaters containing bactericidal substances, for instance, may not respond to the BOD test in a normal fashion, if at all. On the other hand, industrial wastewaters rich in organic matter may behave nearly like domestic wastewaters. If they do, it is possible to calculate the *population equivalent* of these industrial effluents by relating their total weight of BOD to the daily per capita BOD of 0.17 lb or 77 g daily of domestic wastewaters, for instance. Thus a cannery releasing 0.5 mgd of wastewater per day, containing 2500 mg per l of five-day, 20°C BOD, has a population equivalent of $0.5 \times 8.34 \times 2500/0.17$, or 61,000 people.

The oxygen demand of wastewaters emptied into small, relatively shallow and rapid streams can be calculated from the general purification equation (Eq. 11-13) when the pertinent coefficient of retardance is known or can be estimated with reasonable accuracy.

The BOD rate of polluted seawater appears to vary with the concentration of the sea salts. Up to about 25%, $k$ is larger than in fresh water. In straight seawater $k$ is less. The magnitude of the first-stage demand appears to remain unchanged. A second stage is observed, but it proceeds at a reduced pace.[6] The nitrification stage has been fitted by a logistic equation by Tsivoglou.[7]

[6] H. B. Gotaas, The Effect of Seawater on the Biochemical Oxygen Demand of Sewage, *Sewage Works J.*, **21**, 818 (1949).

[7] E. C. Tsivoglou, et al., Tracer Measurement of Stream Reaeration, *J. Water Pollution Control Fed.*, **37**, 10 (1965) and **40**, 285 (1968).

**Deoxygenation by Bottom Deposits.** Mud and sludge deposits are composites of settleable solids laid down and impounded, generally over long periods of time during which currents were neither sufficiently fast to prevent sedimentation of suspended matter nor sufficiently erosive to encourage bottom scour (Sec. 8-3). If the overriding waters contain dissolved oxygen, aerobic conditions are maintained at the surface of accumulating organic debris. However, downward diffusion of oxygen into the deposits is normally too slow to carry it into the deeper strata and keep them from becoming anaerobic.

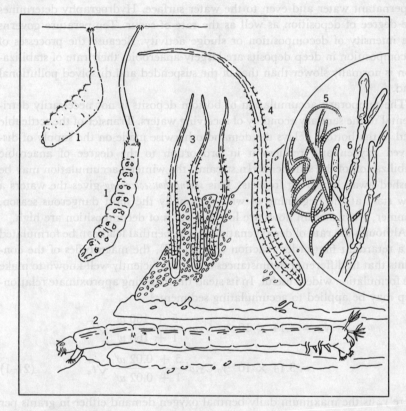

**Figure 21-3.**
**Bottom-dwelling organisms visible with the naked eye in polluted waters.**
1. Rat-tail maggot, *Eristalis*, the larva of the drone fly. ×5.
2. Bloodworm, *Chironomus*, the larva of the midge fly. ×5.
3. Sludge worm, *Tubifex*, a bristle worm. ×5.
4. Sludge worm, *Limnodrilus*, a bristle worm. ×5.
5. Sewage fungus, *Sphaerotilus*, a higher bacterium. ×500.
6. Sewage fungus, *Leptomitus*, a mold. ×500.

The sludge-water interface is by no means static. During periods of sedimentation, settling solids form new surface layers. During periods of scour, deposits are churned up. Indeed, the entire sludge load may be resuspended and rolled away. Some bottom-dwelling organisms, the sludge worms and insect larvae (Fig. 21-3), for instance, ingest subsurface debris and cast their fecal pellets on the mud surface; other organisms burrow into the deposits and expose the spoil to the flowing water. Gases of decomposition, principally $CO_2$, $CH_4$, and $H_2S$, are produced within the sludge. If they bubble up in sufficient volume, they may buoy some of the sludge into the supernatant water and even to the water surface. Hydrography determines the degree of deposition as well as the rate of scour. Temperature governs the intensity of decomposition or sludge activity. Because the processes of decomposition in deep deposits are largely anaerobic, their rate of stabilization is normally slower than that of the suspended and dissolved pollutional load.

The temporary accumulation of bottom deposits is not necessarily detrimental to the sanitary economy of receiving waters. Transfer of the settleable load to the bottom delays the demand otherwise made on the supply of dissolved oxygen and reduces it in proportion to the degree of anaerobic stabilization of the sediments. In streams, the winter's accumulation may be washed away by spring freshets. This *spring housecleaning* gives the waters a new start at the beginning of what is usually the most dangerous season, summer, when stream flows are low and rates of decomposition are high.

Although the rate of deoxygenation by the benthal load can be formulated as a retardant first-order reaction (Eq. 11-13), the magnitudes of the constants that fit different circumstances are not sufficiently well known to make the formulation widely useful. In its stead the following approximate relationship may be applied to accumulating sediment:[8]

$$y_m = 3.14 \times 10^{-2} y_0 \, C_T w \, \frac{5 + 160\,w}{1 + 160\,w} \sqrt{t_a}$$

$$= 3.14 \times 10^{-2} y_0 \, C_T w' \, \frac{5 + 0.02\,w'}{1 + 0.02\,w'} \sqrt{t_a} \qquad (21\text{-}1)$$

Here $y_m$ is the maximum daily benthal oxygen demand either in grams per square meter (for $w$) or in pounds per acre (for $w'$); $y_0$ is the five-day, 20°C BOD in grams per kilogram of volatile matter; $C_T = y/y_0 = [1 - \exp(-5k)]/[1 - \exp(-5k_0)]$ for $t = t_0 = 5$ days and $T_0 = 20°C$ in Eq. 11-18; $w$ and $w'$ are respectively the daily rates of deposition of volatile solids in kilograms per square meter and pounds per acre; and $t_a$ is the time in days, up to 365, during which accumulation takes place.

[8] G. M. Fair, E. W. Moore, and H. A. Thomas, Jr., The Natural Purification of River Muds and Pollutional Sediments, *Sewage Works J.*, **13**, 270, 756 (1941).

*Example 21-1.*

On a daily per capita basis, the five-day, 20°C BOD of 39 g of volatile settleable solids of a domestic wastewater is 19 g (Table 10-5), 10 g of volatile solids being deposited daily per m² of stream bottom during a period of 100 days. Find the maximum daily benthal oxygen demand of the accumulating sediment if the water temperature remains constant at 20°C.

Because a BOD of 19 g produced by 39 g of volatile solids equals $19 \times 1000/39 = 500$ g/kg of volatile matter, and because the temperature factor $C_T$ is unity at 20°C, Eq. 21-1 states that

$$y_m = 3.14 \times 10^{-2} \times 500 \times 1 \times 10 \times 10^{-3} \frac{5 + 160 \times 10 \times 10^{-3}}{1 + 160 \times 10 \times 10^{-3}} \sqrt{100} = 4.0$$

g/m² daily.

If deposits are laid down in equal increments and remain sufficiently thin to be decomposed aerobically, their maximum daily rate of deoxygenation will equal the BOD exerted by a single day's addition of settleable solids during a period of time equal to the period of sludge accumulation. In these circumstances the solids accumulating in five days at a daily rate of 10 g of volatile matter per m² of stream bottom and possessing a five-day, 20°C BOD of 500 g per kg of volatile matter, for example, may reach a maximum rate of deoxygenation of $500 \times 10 \times 10^{-3} = 5$ g/m² daily if the temperature is 20°C.

Scour may lift masses of sludge into supernatant waters and exert sudden sometimes overpowering, oxygen demands. However, even in the normal course of events, some products of anaerobic decomposition (from the nitrogen-containing substances, for example) leach into the supernatant water where they are oxidized. This is a complicating factor in the oxygen economy of streams. The seasonal destruction of bottom-dwelling organisms and aquatic weeds may also exert abnormal oxygen demands. If they are washed into ponds, lakes, reservoirs, and backwaters, they become secondary sources of BOD.

## 21-8 Atmospheric Reoxygenation of Polluted Waters

Aside from the oxygen released by green plants during photosynthesis, the oxygen dissolved in streams and other bodies of water is derived in nature from the overlying atmosphere. Although photosynthesis may make considerable amounts of oxygen available, oxygenation by green plants is confined to (1) waters that are calm enough to encourage plant growth and (a) either not so heavily polluted that green plants die off or (b) sufficiently recovered to reestablish the growth of green plants; (2) the hours of daylight; and (3) the warmer (growing) seasons of the year. During the night, aquatic

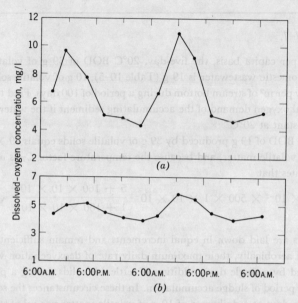

**Figure 21-4.**

**Diurnal variation in the dissolved-oxygen content of a stream.** (a) Above outfall; (b) below treatment-plant outfall. [After J. J. Gannon, River and Laboratory BOD Rate Considerations, *Proc. Am. Soc. Civil Engrs.* 92, SA7, 140 (1966).]

plants abstract oxygen from the water and release carbon dioxide to it. The result is a diurnal cycle of dissolved oxygen and carbon dioxide within waters rich in vegetation (Fig. 21–4). The amplitude of the cycle varies with the intensity of sunlight and the density of the plant population. Net effects can be ascertained by exposing pairs of opaque (black) bottles and clear bottles of the water within the water course and measuring the DO differences in the bottle pairs.

Important as this source of oxygen may be in the total oxygen economy of natural waters, it cannot generally be included in engineering calculations of their oxygen balance. Only the oxygen absorbed from the atmosphere at the air-water interface can be relied on to do its bit at all times except when ice cover eliminates the air contact. Cold winters may produce worse oxygen deficits than warm summers in spite of lowered rates of deoxygenation and higher oxygen-saturation values.

The rate at which water absorbs oxygen from the atmosphere is discussed in Sec. 12–4. If, in Eq. 12–5, $(c_s - c_0) = D_a$, $(c_s - c_t) = D$, and $K_g = r$, where $D_a$ is the initial dissolved-oxygen deficit, $D$ is the deficit after time $t$, and $r$ is the rate of reoxygenation of the body of water.

$$D = D_a \exp(-rt) \tag{21-2}$$

and

$$dD/dt = -rD \qquad (21\text{-}3)$$

Here the magnitude of $r$ is a function not only of water temperature but also of (1) the area of the air-water interface in relation to the volume of water and (2) the renewal of this interface by the film-reducing movements of the water and the air above it. The variation of $r$ with temperature can be formulated in accordance with the van't Hoff-Arrhenius equation as

$$r/r_0 = e^{C_r(T-T_0)} = \theta_r{}^{(T-T_0)} \qquad (21\text{-}4)$$

Here $C_r$ is the temperature characteristic of the rate of reoxygenation $r$ and $T$ is the temperature of the water, the subscript zero designating the reference values. Within the range of normal water temperatures, the magnitude of $C_r$ derived from Becker's observations (Sec. 12–4) is about 0.024 when the water temperature is measured in degrees centigrade.

The rate of reaeration, $r$, varies with surface exposure and volume of water and the rate at which the water is mixed by vertical and horizontal currents that distribute absorbed oxygen and bring fresh volumes of undersaturated water into contact with the atmosphere. Churchill, Elmore, and Buckingham[9] have concluded that the magnitude of $r$ at 20°C can be predicted with reasonable success by the observational relationship

$$r \simeq 5v/R^{5/3} = 7.5\, S^{1/2}/(Rn) \qquad (21\text{-}5)$$

where $v$ is the mean velocity of flow in a given river stretch, $R$ is its mean hydraulic radius, $n$ is Kutter's coefficient of roughness, and $S$ is the loss of head or drop in water surface in a river stretch of known length. Departures from predicted magnitudes are caused by photosynthesis, changes in the hydrological regimen of the stream, and structural shifts in the river channel.

---

*Example 21–2.*

A large stream about 400 ft wide with a mean hydraulic radius $R = 5.66$ ft flows at a velocity $v = 2.78$ fps. Its energy gradient $S = 3.70 \times 10^{-4}$ and its coefficient of roughness $n = 0.033$. The stream contains 2.2 mg/l of DO at 11.9°C. Predict (1) the rate of reaeration, (2) the amount of oxygen added in $l = 2650$ ft, and (3) the maximum rate of reoxygenation if decomposition is sufficiently active to keep the DO content of the stream at 2.2 mg/l. Assume a temperature characteristic $C_r = 0.024$.

[9] M. A. Churchill, H. L. Elmore, and R. A. Buckingham, The Prediction of Stream Reaeration Rates, in *Advances in Water Pollution Research*, Vol. 1, Pergamon Press, London, 1964, p. 89.

The DO saturation value of freshwater at 11.9°C is 10.8 mg/l (Table 20, Appendix). Hence the DO deficit is $D_a = (10.8 - 2.2) = 8.6$ mg/l.

1. By Eq. 21-5, $r = 5 \times 2.78/(5.66)^{5/3} = 0.773$/day at 20°C. By Eq. 21-4, $r = 0.773 \exp[0.024(11.9 - 20)] = 0.634$/day.

2. Time of flow = $2650/2.78 = 950$ sec = $1.10 \times 10^{-2}$ day. By Eq. 21-2, $D = 8.6 \exp(-0.634 \times 1.10 \times 10^{-2}) = 8.5$ mg/l.

3. By Eq. 21-3, $dD/dt = -0.634 \times 8.6 = 5.45$ mg/l daily. This is also the rate of deoxygenation.

Like other phenomena of hydraulic mixing and stirring, the rate of reaeration can also be stated in terms of the mean temporal velocity gradient $G$.

## 21-9 The Dissolved-Oxygen Sag of Polluted Streams

The interplay of the deoxygenation of polluted waters (BOD) and their reoxygenation, or reaeration, from the atmosphere creates a spoon-shaped profile of the dissolved-oxygen (DO) deficit along the path of water movement. The genesis of the resulting *dissolved-oxygen sag* is portrayed in Fig. 21-5. The general mathematical properties of the sag curve, which underlie engi-

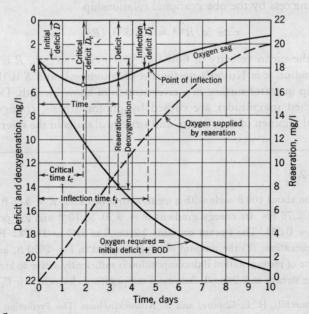

**Figure 21-5.**
**The dissolved-oxygen sag and its components: deoxygenation and reaeration.**

neering calculations of the permissible pollutional loading of receiving waters, have been formulated in the classical studies of Streeter and Phelps.[10]

The basic differential equation for the combined action of deoxygenation and reaeration states that the net rate of change in the DO deficit $(dD/dt)$ equals the difference between (1) the rate of oxygen utilization by BOD in the absence of reaeration $[dD/dt = k(L_a - y)]$ and (2) the rate of oxygen absorption by reaeration in the absence of BOD $(dD/dt = rD)$, or

$$dD/dt = k(L_a - y) - rD \qquad (21\text{-}6)$$

Integration between the limits $D_a$ at the point of pollution, or reference point $[t = 0, (L_a - y) = L_a]$, and $D$ at any point distant a time of flow $t$ from the reference point yields the equation

$$D = \frac{kL_a}{r - k} [\exp(-kt) - \exp(-rt)] + D_a \exp(-rt) \qquad (21\text{-}7)$$

This relationship may be used to find any point on the oxygen-sag curve. If the ratio between the rates of reaeration and deoxygenation $r/k$, which may be termed the oxygen-recovery or self-purification ratio $f$ of the particular body of water, is introduced insofar as possible, Eq. 21-7 becomes

$$D = \frac{L_a}{f - 1} \exp(-kt) \left\{ 1 - \exp[-(f - 1)kt] \left[ 1 - (f - 1)\frac{D_a}{L_a} \right] \right\} \qquad (21\text{-}8)$$

As an engineering concept, the sag curve possesses two characteristic points: (1) a point of maximum deficit, the critical point, with coordinates $D_c$ and $t_c$, and (2) a point of inflection, the point of maximum rate of recovery, with coordinates $D_i$ and $t_i$. The critical point is defined by the mathematical requirement $dD/dt = 0$ and $d^2D/dt^2 < 0$. Differentiation of Eq. 21-8 creates the following simplified expressions for the times $t_c$ and $t_i$, and the associated deficits $D_c$ and $D_i$:[11]

$$t_c = \frac{2.3}{k(f - 1)} \log \left\{ f \left[ 1 - (f - 1)\frac{D_a}{L_a} \right] \right\} \qquad (21\text{-}9)$$

or

$$D_c = \frac{L_a \exp(-kt_c)}{f} = \frac{L_a}{f\{f[1 - (f - 1)(D_a/L_a)]\}^{1/(f-1)}} \qquad (21\text{-}10)$$

and

---

[10] H. W. Streeter and E. B. Phelps, *U.S. Publ. Health Bull.*, 146 (1925).

[11] G. M. Fair, The Dissolved Oxygen Sag—An Analysis, *Sewage Works J.*, 11, 445 (1939). For the special case $r = k$ or $f = 1$, $D = (ktL_a + D_a)\exp(-kt)$; $t_c = [1 - (D_a/L_a)]/k$; $t_i = [2 - (D_a/L_a)]/k$; and $t_i - t_c = 1/k$, or $t_i/t_c = [2 - (D_a/L_a)]/[1 - (D_a/L_a)]$.

$$t_i = \frac{2.3}{k(f-1)} \log\left\{f^2\left[1-(f-1)\frac{D_a}{L_a}\right]\right\} \quad (21\text{-}11)$$

or

$$D_i = \frac{f+1}{f^2} L_a \exp(-kt_i) = \frac{(f+1)L_a}{f^2\{f^2[1-(f-1)(D_a/L_a)]\}^{1/(f-1)}} \quad (21\text{-}12)$$

The coordinates of these two points are related to each other as follows:

$$t_i - t_c = 2.3 (\log f)/[k(f-1)] \quad (21\text{-}13)$$

and

$$D_i/D_c = \frac{f+1}{f} \exp[-k(t_i - t_c)] = \frac{f+1}{f^{f/(f-1)}} \quad (21\text{-}14)$$

*Example 21-3.*

A large stream has a rate of self-purification $f = 2.4$ and a rate of deoxygenation $k = 0.23$/day. The DO deficit of the mixture of stream water and wastewater at the point of reference, $D_a$, is 3.2 mg/l, and its first-stage BOD, $L_a$, is 20.0 mg/l. Find (1) the DO deficit at a point one day distant from the point of reference; (2) the magnitudes of the critical time and critical deficit; and (3) the magnitudes of the inflection time and inflection deficit.

1. By Eq. 21-8, $D = (20.0/1.4)\exp(-0.23)\{1 - \exp(-1.4 \times 0.23)[1 -(1.4 \times 3.2/20.0)]\} = 5.0$ mg/l.

2. By Eq. 21-9, $t_c = [2.3/(0.23 \times 1.4)] \log\{2.4[1 -(1.4 \times 3.2/20.0)]\} = 1.9$ days, and by Eq. 21-10, $D_c = [20.0 \exp(-0.23 \times 1.9)]/2.4 = 5.3$ mg/l or $D_c = 20.0/[2.4(2.4 \times 0.776)^{1/1.4}] = 5.3$ mg/l.

3. By Eq. 21-11, $t_i = 1.9 + 2.3 \log[2.4/(0.23 \times 1.4)] = 4.6$ days, and by Eq. 21-12, $D_i = 5.3 \times 3.4/(2.4)^{2.4/1.4} = 4.0$ mg/l.

If part of the pollutional load settles in the immediate vicinity of the point of pollution, the benthal oxygen demand may be calculated by Eq. 21-10 and assumed to be an added DO deficit at this point. If the deposited load is dispersed over a long river stretch, Eqs. 21-6 and 7 may be expanded to include the rate of removal of BOD by benthal decomposition.[12] The resulting relationships are

$$dD/dt = kL_a \exp[-(k+d)t] - rD \quad (21\text{-}15)$$

and

[12] H. A. Thomas, Jr., Pollution Load Capacity of Streams, *Water and Sewage Works*, **95**, 409 (1948). Note that $k$ and $d$ are additive as exponents (Eq. 18-9).

$$D = \frac{kL_a}{r - (k + d)} \{\exp[-(k + d)t] - \exp(-rt)\} + D_a \exp(-rt)$$
(21-16)

where $d$ is the coefficient of deposition and must reflect (1) the composition of the wastewaters and the receiving water and (2) the relative quiescence of the receiving water. In times of considerable turbulence, scour of deposited sludge may render $d$ negative. The stream then receives an additional load. However, turbulence will increase reaeration at the same time, and this may more than offset the influence of the greater load.

If photosynthesis adds measurable amounts of oxygen to the stream, Eqs. 21-6, 7, 15 and 16 may be further expanded to include the rate of photosynthetic oxygenation $p$, the product $rD$ becoming $(r + p)D$, and $\exp(-rt)$ similarly $\exp[-(r + p)t]$.

## 21-10 Allowable BOD Loading of Receiving Streams

Inspection of Eqs. 21-6 to 8 shows that the allowable pollutional load $L_a$ for a given receiving water is determined by the magnitudes of the following parameters: (1) the deoxygenation constant $k$, (2) the self-purification constant $f = r/k$, (3) the critical deficit $D_c$, and (4) the initial deficit $D_a$.

As pointed out in Sec. 21-7, the value of the deoxygenation constant $k$ is expected to vary widely in different receiving waters and along their course. Only the magnitude $k = 0.23$ per day for large streams of normal velocity appears to be well founded. Departure from this value is expected to be noticeably great in shallow, swift streams filled with boulders and debris.

Present information on the self-purification constant $f = r/k$ supports the values at 20°C listed in Table 21-2.

The classification of different bodies of water is not sharply defined. Each class merges into its adjoining classes, and there is appreciable variation

*Table 21-2*
*Values of the Self-Purification Ratio $r/k = f$*

| Nature of Receiving Water | Magnitude of $f$ at 20°C |
|---|---|
| Small ponds and backwaters | 0.5–1.0 |
| Sluggish streams and large lakes or impoundments | 1.0–1.5 |
| Large streams of low velocity | 1.5–2.0 |
| Large streams of moderate velocity | 2.0–3.0 |
| Swift streams | 3.0–5.0 |
| Rapids and waterfalls | >5.0 |

within the types described, as well as within different reaches of the same body of water.

It is possible to determine the magnitude of $f$ for a given stream by identifying its critical point or point of maximum oxygen depletion, $D_c$, and finding the first-stage BOD, $L_c$, of the flowing waters. Equations 21–8 and 10 then state that $L_c = L_a \exp(-kt_c')$ and that $D_c = L_c/f$. Hence[13]

$$f = L_c/D_c \qquad (21\text{–}17)$$

This equation will assist in locating the critical point. Field tests for DO not only should confirm the position of the critical point but also account for changes in stream properties. Field measurements of $f$ at different river stages together with determinations of associated stream gradients make it possible to find congruous values of $r$ and $k$.

In accordance with Eq. 21–4, the variation of the self-purification constant $f = r/k$ with temperature is given by the relationship

$$f = f_0 \exp[(C_r - C_k)(T - T_0)] = f_0 \exp[C_f(T - T_0)] \quad (21\text{–}18)$$

For values of $C_r = 0.024$ and $C_k = 0.046$ within the range of normal water temperatures, $C_f = -0.022$; that is, the magnitude of $f$ decreases with rising temperatures and increases with falling temperatures by about 2% compounded per degree centigrade.

To avoid septic conditions, the maximum critical deficit $D_c$ must not exceed the DO saturation value, $S$, of the receiving water, for example, 9.2 mg per l at 20°C in freshwater. For the support of game fish, such as trout, the DO content must not fall below about 5 mg per l. The allowable critical deficit then becomes 4.2 mg per l at 20°C.

For given values of $k$, $f$, and $D_c$, the initial deficit $D_a$ establishes two boundary values for the maximum allowable loading on a receiving water: an upper limit associated with zero initial deficit or full DO saturation ($D_a = 0$), and a lower limit associated with an initial deficit equal to the critical deficit ($D_a = D_c$). These restrictions identify the boundary relationships for the maximum loading of receiving waters and for the coordinates of the characteristic points of the oxygen sag.[11]

Inspection of the boundary equations shows that the allowable loading and the coordinates of the characteristic points of the oxygen sag are simple functions of the coefficient of self-purification when the loading is expressed in terms of the critical deficit and the critical time is expressed in terms of the product $kt_c$. It follows that the important parameters of the sag curve can be

---

[13] C. H. J. Hull, *Report No. VI of the Low-Flow Augmentation Project*, Department of Sanitary Engineering and Water Resources, The Johns Hopkins University, Baltimore, 1961. Other simplifications of this kind are possible. See M. Le Bosquet, Jr. and E. C. Tsivoglou, Simplified Dissolved Oxygen Computations, *Sewage Ind. Wastes*, **22**, 1054 (1950).

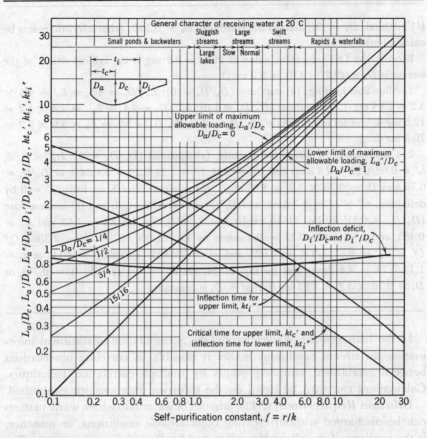

**Figure 21-6.**

**Allowable loading of receiving waters and associated coordinates of the critical point and the point of inflection of the dissolved-oxygen sag.** (See footnote 11.)

generalized in such fashion that their values need to be computed but once. They can then be recorded graphically as in Fig. 21-6 or in tabular form. Loading curves between the upper and lower boundary values make Fig. 21-6 widely useful.

*Example 21-4.*

Find from Fig. 21-6, for a water temperature of 20°C and a minimum DO content of 4.0 mg/l, (1) the allowable loading and (2) the coordinates of the characteristic points of the oxygen sag of a stream with a rate of self-purification $f = 2.4$ and a rate of deoxygenation $k = 0.23$/day under the following conditions: (a) maximal, $D_a = 0$;

(b) minimal, $D_a = D_c$; and (c) intermediate, $D_a = 0.50 D_c$ (characteristic points by calculation).

Because the DO saturation value at 20°C is 9.2 mg/l, the critical deficit of the stream is $D_c = (9.2 - 4.0) = 5.2$ mg/l.

1. Allowable loading: (a) maximal $(D_a/D_c = 0)$: $L_a'/D_c = 4.5$, or $L_a' = 4.5 \times 5.2 = 23.4$ mg/l; (b) minimal $(D_a/D_c = 1.0)$: $L_a''/D_c = 2.4$, or $L_a'' = 2.4 \times 5.2 = 12.5$ mg/l; and (c) intermediate $(D_a/D_c = 0.5)$: $L_a/D_c = 3.9$, or $L_a = 3.9 \times 5.2 = 20.3$ mg/l.

2. Coordinates of characteristic points: (a) maximal $(D_a/D_c = 0)$: $kt_c' = 0.62$, $t_c' = 0.62/0.23 = 2.7$ days and $kt_i' = 1.24$, $t_i' = 1.24/0.23 = 5.4$ days, or $D_i/D_c = 0.76$ and $D_i = 0.76 \times 5.2 = 3.95$ mg/l; (b) minimal $(D_a/D_c = 1.0)$: $t_c'' = 0$ by definition, $t_i'' = t_c' = 2.7$ days, and $D_i'' = D_i' = 3.95$ mg/l; and (c) intermediate $(D_a/D_c = 0.5)$: By Eq. 21-10, $\exp(kt_c) = L_a/(fD_c) = 3.9/2.4 = 1.63$ and $kt_c = 0.489$, and $t_c = 0.489/0.23 = 2.1$ days; by Eq. 21-13, $k(t_i - t_c) = 2.3(\log f)/(f - 1) = 0.8755/1.4 = 0.625$ and $kt_i = (0.625 + 0.489) = 1.11$, or $t_i = 1.11/0.23 = 4.8$ days; and by Eq. 21-14, $D_i/D_c = 3.4/(2.4^{2.4/1.4}) = 0.758$, or $D_i = 0.76 \times 5.2 = 3.95$ mg/l, that is, $D_i$ is constant.

---

In the management of river basins, the accumulation of statistical information on stream conditions makes it possible to correlate observations between established sampling points by multiple-correlation techniques. Calculations can then be based on the statistical relationships established.

**Dilution Requirements.** The amount of water into which waste matters can be discharged without creating objectionable conditions, or nuisance, is the converse of the allowable pollutional loading of receiving waters. The dilution parameter for combined systems of sewers is commonly expressed as stream flow, $Q$, in cubic feet per second per 1000 population required to avoid odor and related nuisances. Recommended values of $Q$ include (1) Hazen's estimate of 1898 that a sluggish stream, already partially depleted of oxygen at the point of combined wastewater discharge, may require a diluting runoff of as much as 10 cfs per 1000 population; (2) Stearns' estimate of 1890 that the lowest required dilution of normal streams is 2.5 cfs per 1000 population; and (3) a commonly quoted value of 4 cfs per 1000 population.

If waste matter with a first-stage BOD of $L$ lb per capita daily is discharged into a stream carrying $Q$ cfs per 1000 population, the BOD loading is 185.5 $L/Q$ mg per l. For a permissible loading of $L_a$, therefore, the required stream flow becomes

$$Q = 185.5 \, L/L_a \tag{21-19}$$

## Example 21–5.

If the first-stage, 20°C BOD of combined sewage is assumed to be 0.25 lb per capita daily, find the needed dilution corresponding to the allowable loadings of Example 21–4, namely, 23.4 mg/l (maximal), 20.3 mg/l (intermediate), and 12.5 mg/l (minimal) for a DO residual of 4.0 mg/l.

By Eq. 21–19, $Q = 185.5 \times 0.25/L_a = 46.4/L_a$ cfs or, respectively, 2.0, 2.3, and 3.7 cfs/1000 persons daily depending on the magnitude of the initial deficit ($D_a = 0$, 2.6, and 5.2 mg/l, respectively).

## 21.11 Pollution of Lakes and Impoundments

The introduction of nutrients into lakes and impoundments sets in motion a unique chain of happenings because standing waters are essentially closed communities in which foodstuffs are maintained or accumulated by circulating through the various trophic levels. Especially when inflows and outflows are small relative to the size of the lake or impoundment does the body of water become dependent on the photosynthesis of organic matter by its own stock of green plants. Although the concept of a lake or reservoir as a closed community must be applied with care, it does lend itself with some success to the classification of these bodies of water according to their total productivity. Associated physical, chemical, and biological conditions determine the various nutrient levels.

*Limnological Considerations.* Thermal factors more or less govern the response of lakes and impoundments to pollution. Most of the energy absorbed from the sun by a natural body of water is converted into heat. If the sun's rays were monochromatic and warming took place only by radiation, while the absorptive capacity itself remained uniform, the temperature of the water would decrease logarithmically from the surface to the bottom. Selective absorption, however, steepens the gradient within the upper layers, and even more radical shifts are induced by conduction and convection. Radiant energy absorbed by the bottom is released in the form of longer wavelengths that are trapped by the overlying water. Bottom sediments themselves are cooler than the overlying water in summer and warmer in winter; the deeper the deposit, the greater the difference.

Because the specific heat of water is about 4 times that of air and the mass of water is also relatively much greater, the temperature at the air-water interface would be the water temperature, were it not for wind and water movements. Of the two, air motion is normally the more vigorous, and heat is distributed through bodies of water, both horizontally and vertically, by wind action. Exclusion of wind-induced currents by ice cover explains why winter heat losses are small in spite of the lower specific heat and higher conductivity of ice.

Equilibrium is also upset by temperature changes caused by heat transfer from one fluid to another, including changes in temperature produced by evaporation. If surfacewater is cooled, either by contact with cold air or by evaporation, it becomes denser and sinks. Vertical (convective) currents are set in motion, and heat is transferred within the mass of water itself.

If the water temperature lies below the wet-bulb temperature of the overlying air, the water is warmed by the air; there is no thermal convection in either fluid; and little exchange of heat takes place. If, on the other hand, the water temperature is higher than the wet-bulb temperature of the air, the water is cooled by evaporation, and thermal convection currents are set in motion in both fluids.

Water driven to the windward shore of a lake or impoundment builds up a head that generates return currents. Depending on shore topography, these currents travel at or below the water surface. The returning water is displaced downward in coves and laterally where points of land jut out from shore. The vertical distribution of wind-induced currents in an idealized cross-section of lake or reservoir is illustrated in Fig. 21-7. The depth of the return currents depends on the water temperature. The greater the decrease in temperature with depth, the greater is the thermal resistance to mixing and the more will the return currents be confined to the upper layer of water. By thermal resistance is meant the resistance of colder and, therefore, denser and lower-lying water to be displaced by warmer and, therefore, lighter and higher-lying water. A shearing plane divides the surface currents that follow the wind from the return currents that run counter to the wind.

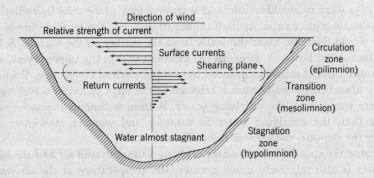

**Figure 21.7.**
**Direction and relative horizontal velocity of wind-induced currents in a lake or reservoir (idealized). (After G. C. Whipple, G. M. Fair, and M. C. Whipple, *Microscopy of Drinking Water*, 4th ed., Wiley, New York, 1948.)**

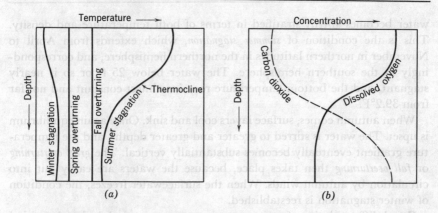

**Figure 21-8.**
**Vertical gradients of temperature and water quality in lakes, reservoirs, and other bodies of water (idealized). (a) Characteristic thermal gradients; (b) oxygen and carbon dioxide gradients during summer stagnation.**

The interplay of temperature, density, and wind during the different seasons of the year produces a sequence of characteristic patterns of thermal stratification in lakes and reservoirs. Figure 21–8a shows such a series for the temperature gradients of waters in lakes and reservoirs that are ice-covered during the winter months. Water immediately below the ice stands substantially at 32°F, although the ice is itself often much colder at and near its surface. At the same time the temperature at the bottom of the lake or reservoir is not far from that of maximum density (39.2°F). The water is in comparatively stable equilibrium and is stratified *inversely* in terms of temperature but *directly* in terms of density. This is the condition of *winter stagnation*. Ice cover shuts out wind disturbances of the underlying waters, and vertical or horizontal movements are suppressed.

When the ice breaks up in the spring, the waters near the surface begin to warm up. Until the temperature of maximum density is reached, they also become denser and tend to sink. Equilibrium is upset by diurnal fluctuations in temperature, and vertical circulation is aided by wind. When the water temperature is practically uniform at all depths and close to the temperature of maximum density, circulation becomes especially pronounced. This is the *spring circulation* or *spring overturning*. It may last several weeks, and it varies in length during different years.

As spring turns into summer, the surfacewater becomes progressively warmer. Soon lighter water overlies denser water once again and, as temperature differences increase, circulation is confined more and more to the upper waters. A second period of stable equilibrium is established, and the water becomes *directly* stratified in terms of both temperature and density. This is the condition of *summer stagnation*, which extends from April to November in northern latitudes in the northern hemisphere, and correspondingly in the southern hemisphere. The water below 25 ft or so is nearly stagnant, and the bottom temperature remains almost constant and not far from 39.2°F.

When autumn comes, surface layers cool and sink. Once again equilibrium is upset. The water is stirred to greater and greater depths, and the temperature gradient eventually becomes substantially vertical. The *great overturning* or *fall overturning* then takes place, because the waters are easily put into circulation by autumn winds. When the surfacewater freezes, the condition of winter stagnation is reestablished.

Zonal differentiation of water strata is most pronounced during the period of summer stagnation. Figure 21-7 identifies three zones: the circulation, stagnation, and transition zones. The *transition zone* is defined by convention as comprising a layer in which water temperatures fall off by 0.5°F or more in each vertical foot (Fig. 21-8a). Other words for the zones are the *epilimnion*, *thermocline* or *mesolimnion*, and *hypolimnion*, respectively.[14] Diurnal temperature changes in the surface strata establish a variable microenvironment.

The small range in density below 39.2°F and the large range above 39.2°F explain why the spring overturn is relatively short and the fall overturn much more protracted. The rapid drop in temperature within the thermocline during the summer produces a barrier not easily disturbed by the wind. Indeed, a lake, as suggested by Birge, is normally two lakes, one superimposed upon the other.

Because molecular diffusion is relatively slow, the thermal gradients of lakes and similar bodies of water are likewise gradients in the concentration of dissolved gases. The water surface is both a window through which radiant energy is received and a lung through which oxygen is taken in and carbon dioxide and other dissolved gases are released. The oxygen absorbed at the surface is distributed by the water circulating within the epilimnion. The gases of decomposition are released by contact with the air overlying the surface. Within the thermocline, there is a sharp drop in dissolved oxygen and a rise in the concentration of gases of decomposition (Fig. 21-8b). Below the thermocline, the concentration of dissolved oxygen reaches a minimum (often zero); that of gases of decomposition, a maximum. The degree of undersaturation of dissolved oxygen and the degree of supersaturation of gases of decomposition in the bottom water depend on the intensity and duration of the processes of decomposition. Decomposition is obviously a

---

[14] The prefixes of these words have their origin in the Greek *epi*, upon (above); *meso*, middle; and *hypo*, under (below). The word limnion comes from the Greek *limne*, a lake.

function of inherent water quality and the organic matter content of the bottom deposits.

Thermal stratification controls the mass seasonal movements in otherwise quiescent bodies of water. As a result it produces gradients in water quality that are images (for example, dissolved oxygen) or mirror images (for example, carbon dioxide) of the thermal gradient itself. There is, therefore, a vertical, seasonal variation in water quality within a reservoir or similar bodies of water as well as a seasonal variation in water temperature. This is why the depth of draft can be shifted to select water of the best quality available for the purposes it is to serve (Sec. 4-16).

The quality gradient, which like the thermal gradient is pronounced during summer stagnation and less apparent during winter stagnation, is suppressed during the overturns. The quality of the water at all depths then becomes substantially the same, and the selection of water of better than average quality becomes impossible. For this reason, the periods of overturning, in particular the fall or long overturning, are often periods of noticeably deteriorated water quality in terms of low oxygen and increased iron and manganese. During the overturns, moreover, the vertical mixing of waters may carry resting cells of light-loving organisms from the dark depths into which they have settled, upward to the surface, where they can flourish. At the same time food substances, too, are made available within the upper strata and the overturns may then be accompanied or followed by sudden, heavy growths or *pulses* of algae (particularly diatoms) and other microorganisms.

The productivity levels of lakes and reservoirs are generally classified as (1) *oligotrophic*, (2) *eutrophic*, and (3) *dystrophic*.[15]

*Oligotrophic* lakes are normally deep. Their hypolimnion[14] is large and cold, and they contain little organic matter in suspension or on the bottom. There is oxygen at all depths throughout the year. The plankton are quantitatively restricted. Algal blooms are rare, although algal species are many.

*Eutrophic* lakes are often shallow, sometimes deep. Shallow lakes contain little, if any, cold water. There is much organic matter in suspension and on the bottom. Deep stratified lakes that have become eutrophied contain little or no dissolved oxygen in the hypolimnion. The plankton is quantitatively abundant but varies in quality. Algal blooms are common.

*Dystrophic* lakes are usually shallow with abundant organic matter in suspension and on the bottom. In stratified dystrophic lakes dissolved oxygen is generally absent from deep waters. The plankton varies in composition and is usually low in numbers of species and in biomasses. Concentrations of

[15] Greek, *oligos*, small; *eu*, well; *dys*, badly; and *trophein*, to nourish.

calcium, phosphorus, and nitrogen are small; concentrations of humic materials are large. Algal blooms are infrequent.

In the course of time natural and cultural pollution change oligotrophic lakes into eutrophic lakes and eutrophic lakes eventually into dystrophic bodies of water. Impoundments begin with a level of biological productivity that depends on the nature of the area flooded and the degree of preparation in advance of its flooding.

**Lake Eutrophication.**[16] Sawyer[17] has concluded that any stratifying lake with more than 0.3 ppm of inorganic nitrogen and 0.01 to 0.05 ppm of inorganic phosphorus at the time of the spring overturn can be expected to produce nuisance blooms of algae. Some algae are extraordinarily efficient in utilizing phosphorus. According to Hutchinson,[18] they can do so from water containing only 0.001 ppm of this element.

An inductive derivation of tolerable amounts of phosphorus can be based on the fact that 1 mg of phosphorus requires, in one pass of the phosphorus cycle in lakes, about 130 mg of oxygen. Accordingly the concentration of phosphorus in the stagnant bottom waters should not exceed $1/130 = 0.08$ ppm if traces of oxygen are to remain in them. At the same time, this represents another type of nuisance, namely, an anaerobic bottom water that may be objectionable in itself. In back of this calculation is the mean stoichiometric relation between $C:N:P$ in planktonic material of $41:7:1$ by weight. This implies further that for every milligram of phosphorus, 7 mg of nitrogen and 41 mg of carbon enter into plankton protoplasm. The average ratio of $N:P$ is $14:1$ in urine, less than $4:1$ in fecal matter, and $(20\text{ to }50):(1\text{ to }13)$ in municipal wastewaters. Engelbrecht and Morgan[19] have reported that agricultural runoff contains a mean concentration of 0.05 ppm of phosphorus. Accordingly phosphorus presumably determines the degree of lake eutrophication. The rate of phosphate uptake is given by the Michelis-Menten equation.[20]

[16] National Academy of Sciences, *Eutrophication: Causes, Consequences, Correctives*, Proceedings of a Symposium, National Academy of Sciences, Washington, D.C., 1969.

[17] C. N. Sawyer, Fertilization of Lakes by Agricultural and Urban Drainage, *J. New England Water Works Assoc.*, **41**, 109 (1947) and Phosphates in Lake Fertilization, *Sewage Ind. Wastes*, **24**, 768 (1952). See also J. A. Borchardt, Eutrophication-Causes and Effects, *J. Am. Water Works Assn.*, **61**, 272 (1969) and *Eutrophication-A Review*, Calif. State Water Quality Control Board, Publ. No. 34, 1967.

[18] G. E. Hutchinson, *A Treatise on Limnology*, John Wiley & Sons, New York, 1957, p. 734.

[19] R. S. Engelbrecht and J. M. Morgan, Phosphates in Surface Waters, *J. Water Pollution Control Fed.*, **31**, 458 (1959).

[20] $k = ck_{max}/(c + c')$ where $k$ is the rate of uptake, $k_{max}$ the maximum rate, $c$ the concentration of phosphate, and $c'$ the concentration at which the uptake rate is half maximal. L. Michaelis and M. Menten, *Biochem. Z.*, **49**, 333 (1913).

The tie-in between the nutrient cycle in lakes and Sawyer's critical values is as follows: (1) algae take up and store phosphorus and nitrogen, (2) on dying, they settle to the lake bottom, (3) septic bottom deposits release phosphorus and nitrogen back to the water, and (4) the spring overturn carries phosphorus and nitrogen to the surface layers, where they promote the growth of new crops of algae.

Other pertinent observations are that (1) in the absence of available nitrogen some blue-green algae resort to *nitrogen fixation*, that is, the abstraction of needed nitrogen from the atmosphere, and (2) that lake waters can usually muster enough carbon dioxide from the atmosphere, decaying organic matter, and bicarbonate alkalinity to supply required amounts of carbon. To be remembered, in particular, is the operation of Liebig's law of the limiting or growth-determining nutrient.[21]

The conditions under which the progress of lake eutrophication can be halted and damage reversed remain under study. Experience with impoundages for water supply (Sec. 21-13) is probably not pertinent, because incoming flows are presumably quite clean and annual displacements of water are implicitly high enough to wash out accumulating nutrients.

## 21-12 Disposal of Wastewater Effluents into Lakes and the Sea

The natural dispersion of wastewater effluents in lakes and coastal waters is often poor. In the absence of wind and tide there is little turbulent mixing. Wastewaters and receiving waters normally differ in density. Both fluids follow paths of least resistance. Lake waters are normally colder and, therefore, heavier than wastewater effluents, and seawater is still heavier because of its salt content. Effluents discharged at or near the surface of denser receiving waters are likely to overrun them. Discharged at some depth below the surface, they rise like smoke plumes and, on reaching the surface, fan out radially. Because chemical diffusion is slow, natural dispersion or mixing of the unlike waters is mainly a function of wind, currents, and tides. Mixing requires the expenditure of energy (Sec. 11-16). To provide requisite energy the engineer will put natural forces to use wherever this can be done.

Hydrographic exploration of receiving waters will tell how effective natural dispersion or dilution will probably be and whether water intakes, bathing beaches, shellfish layings, and shore properties will be polluted or not. In lakes and ponds, displacement- and wind-induced currents as well as temperature and other density effects govern the degree of mixing. In the sea and its estuaries, tidal currents and the volume of the tidal prism are added variables.

[21] Baron Justus von Liebig (1803-73), eminent German chemist, father of agricultural chemistry and originator of the concept of the limiting or growth-determining nutrient, often referred to as Liebig's law of the minimum.

**Loading of Dispersal Areas.** Observations by the California State Board of Health[22] suggest that the area of the *sleek* field created by submerged discharge of raw wastewater into the sea is a function of tributary population as follows:

$$A = P(11.5 - 3.5 \log P) \tag{21-20}$$

where $A$ is the area in acres and $P$ ($\leq 1000$) is the population in thousands. Equation 21–20 gives a rough estimate of the length of outfall required to keep beach waters free from sleek and associated bacterial contamination. As a treatment process, dilution calls for a large area. For a per capita discharge of $Q$ gpd, the approximate areal rate of treatment $Q'$ in gallons per day and acre is

$$Q' = 1000 \, Q/(11.5 - 3.5 \log P) \tag{21-21}$$

At best this is an areal loading of the order of magnitude of intermittent sand filters (Sec. 19–2).

More exactly, the area and configuration of wastewater fields are functions of the following physical factors: rate of discharge; diameter, direction, and submergence of outlet nozzles; and speed of water currents. These control the thickness and horizontal spread of the field and the associated mixing or dilution. The use of multiple outlets normally produces some overlap but remains advantageous.

**Effluent Disposal in Tidal Estuaries.** At a given point in space and time, the tidal prism in an estuary is equal in volume to the upstream body of water lying between high and low tide. Rising and falling for 6.2 hr from ebb slack to flood slack, a tidal cycle is completed in 12.4 hr. Assuming that the change in water level is traced by a sine curve, the rates of displacement per hour range from zero at low and high tide to a maximum of $(\pi/12.4)$ times the volume of the tidal prism. Both fresh and salt water compose the prism. Mean values become important elements in necessary calculations.

Longitudinal mixing in tidal estuaries is kept in check by the failure of component freshwater and salt water flows to mix vertically. Engineering studies may be directed to salt-water *creep*, namely, the upstream progress of a tongue of salt water moving inland while overriding freshwater may still pour toward the ocean. Town and industrial water supplies may be affected, and the information sought may be fully as important as a concern for the dispersal of pollutants.[23]

---

[22] *Eng. News Rec.*, **123**, 690 (1939); see also E. A. Pearson, Ed., *Waste Disposal in the Marine Environment*, Pergamon Press, New York, 1960.

[23] D. J. O'Connor, Oxygen Balance of an Estuary, *Proc. Am. Soc. Civil Engrs.*, **86**, SA3, 35 (1960). Also see T. R. Camp, *Water and Its Impurities*, Reinhold, New York, 1963, pp. 310–315.

## 21-13 Natural Purification or Recovery of Impounded Reservoirs

The natural purification taking place in the course of many years after impounded reservoirs have been filled is illustrated in Fig. 21–9 for algal blooms and natural color. It can be formulated in much the same way as the self-purification of streams. Taking the first year after filling as the starting point of the time axis $t$, and the steady-state condition $y = L$ as the reference axis for water quality, reservoir improvement is described by the equation

$$dy/dt = -k(y - L) \tag{21-22}$$

whence, by integration between limits $y = y_1$ and $y = y$ and $t = 1$ and $t = t$,

$$y = L + (y_1 - L) \exp[-k(t - 1)] \tag{21-23}$$

Here $y$ is the color intensity or microscopic count of the reservoir (with a 1-year value $y = y_1$ and a terminal, or steady-state, value $y = L$), $t$ is the time after filling, and $k$ is the rate constant of improvement.

Observed stabilization of this kind suggests that lakes, ponds, and impoundments may rid themselves of organic matter in the course of about two decades, provided that the waters they hold are displaced by incoming flows or withdrawn for use in sufficient volume. However, the information presented in Fig. 21–9 does not imply that the eutrophication of large, deep, and unregulated lakes with relatively small catchment areas—the Great Lakes of America, for instance—will be reversed as soon as further pollution is stopped. Whether a eutrophic lake can be reclaimed appears to be a matter of lake hydrography and regional hydrology.

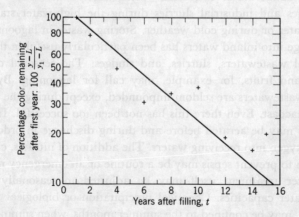

**Figure 21-9.**
**Natural rate of color reduction in large deep reservoirs flooding extensive swamps.**

## 21-14 Hydraulic Control of Receiving Waters

The volume of wastewater safely emptied into a natural receiving body of water can be increased not only by the proper location and design of outfalls and by suitable treatment of wastewater before discharge but also by (1) regulating waste discharge, (2) controlling the flow of the receiving water (low-water regulation), (3) supplementing, strengthening, or preserving the self-purifying power of the receiving water structurally, hydraulically, or by air injection, and (4) limiting sludge deposition. Properly conceived engineering works will husband the forces of self-purification inherent in natural bodies of water. Design and execution of necessary works are important engineering responsibilities.

**Regulation of Wastewater Discharge.** Variations in flow or volume and in other hydrographic conditions of the receiving water and changes in seasonal requirements may justify short or long storage of wastewaters in basins or lagoons. The release of waste matters can thereby be adjusted to the loading capacity of the receiving water. Fluctuations in rate of flow and strength of wastewaters can be ironed out at the same time. Impounded wastes can be released in quantities and concentrations that stand in optimal ratio to the flow, volume, and self-purifying power of the receiving water. Examples are (1) safeguarding water intakes by confining the discharge of wastewaters to certain periods of the year and withdrawing water only in the remaining periods, (2) protecting tidal waters by releasing wastewaters during certain stages of tide (normally the falling tide), and (3) discharging particularly dangerous or obnoxious wastes, wastes difficult to treat, wastewater sludges and industrial slurries during the high-water stages of the receiving water or during cold weather. Storing wastes in lagoons for selective discharge into inland waters has been particularly useful in the disposal of industrial wastewaters, slurries, and sludges. The seasonal canning of vegetables and fruits, for example, may call for lagooning. By contrast, municipal wastewaters are seldom impounded, except during the tidal cycle along the seacoast. Even there this has not been too successful. Impounded wastewaters may be aerated before and during discharge in order to carry necessary oxygen into receiving waters. The addition of nitrates, chlorine, or pure oxygen to prevent sepsis may be a routine or an emergency measure.

Wastewater treatment itself may be adjusted to seasonally changing receiving-water capacities. Chemical precipitation or biological treatment, for example, may be confined to the summer months, when summer populations are high, receiving waters are low, and rates of BOD are rapid. Complete treatment may be scheduled for the bathing season only, disinfection of treatment-plant effluents being stepped up at the same time or during the harvesting of shellfish from receiving waters. At other times treatment may stop with plain sedimentation.

**Low-Water Regulation.** It is usually immaterial whether the pollutional load imposed on a unit volume of water is reduced by (1) modifying the concentration or rate of discharge of the wastewater or (2) regulating the flow of diluting water. In terms of the oxygen sag, clean-water flows not only supply dissolved oxygen, they also step up reaeration rates by increasing turbulence or enlarging water surfaces—generally by doing both. More immediate effects of dilution on the concentration of conservative waste matters are self-evident.

Diluting waters may be supplied (1) from upstream impoundments of floodwaters, (2) by diversion works for waters from nearby catchment areas, (3) by pumping plants delivering water from lower reaches of the stream or from a lake or other body of water back into critical river stretches, and (4) by a combination of pumped storage and release of water for power generation during peak loading. The DO concentration of the diluting waters should be high. If it is not, the waters should be aerated before or during discharge from storage.

If low-water regulation is made a part of the regional development of available water resources, upstream reservoirs and pumped storage can be designed to serve the multiple purposes that optimize the regional water economy, namely, water supply, power development, flood control, navigation, irrigation, recreation, and the conservation of useful aquatic life.

**Regulation of Natural Purification.** It is possible to increase the self-purification accomplished in a receiving water (1) by lengthening the time of passage through a given stretch of water, (2) by stepping up the rate of reaeration, (3) by introducing mechanical aeration, (4) by controlling the stratification of lakes, ponds, and impoundments, and (5) by normalizing the accumulation of sludge deposits.

Impounding reservoirs lying below the point of pollution stretches out the time of flow, expands the water surface, and causes the deposition of suspended solids. The combined effect is well illustrated in Fig. 21–1. However, the load of nutrients that stimulate the production of algae and other aquatic organisms must be held in check along with the load of settleable solids. Impoundments must maintain a favorable oxygen balance. If the pollutional load becomes too heavy relative to stream flow and reservoir surface, the reservoir may be eutrophied. Yet oxidation ponds are an interesting and, in a sense, unique example of controlled self-purification by impoundage of undiluted wastewaters (Sec. 19–17).

Generally more economical and versatile than flow augmentation is the direct addition of oxygen to receiving waters by mechanical or pneumatic aeration. Air-injection systems may be stationary or mobile. It can be shown that the permissible loading of a stream is increased by the factor $f^{1/(f-1)}$ when the receiving water is saturated with oxygen at the point of pollution.[24]

In deep lakes, ponds, and impoundages, mechanical or pneumatic aeration may be assigned the additional task of preventing or destroying summer and winter stagnation.[25] The power needed to lift bottom water into the zone of circulation is a function of the difference in the specific weight of the coldest and warmest water, friction loss in the pump discharge and suction lines, and the rate of pumping.

**Regulation of Sludge Deposition.** In the absence of scour, streams are forced to digest large amounts of sludge in the course of a year. The permeability of river channels that furnish groundwater to the valley by bottom and lateral seepage from the river bed may be reduced, and the quality of the seepage water is certain to be adversely affected. When large storage works are constructed in river valleys or when floodwaters are diverted into other watersheds, scour may be lost.

In the hydrological management of a river system, scour can be regulated by controlling the discharge from impounding reservoirs and by canalizing receiving streams. The mechanical removal of sludge deposits by dredging is hardly ever justified in the interest of either water quality management or restoration of reservoir capacity.

## 21-15 Control of Algae

The control of the plankton or, more specifically, the algae of lakes, ponds, and reservoirs, and to a lesser extent of streams, is an important and often vexing problem in water-quality management. Among the nuisances created by the often sudden *blooming* of one or more algal genera are nauseous odors and tastes; fish kills; interference with stock watering; poisoned water fowl and cattle; poisoned mussels; shortened filter runs in water-purification plants; growths in pipes and other water conduits; and interference with industrial water uses. The odors and tastes associated with different organisms vary widely (Table 21-3). Ordinarily consumers complain when the concentration of odor-producing organisms lies above 500 to 1000 areal standard units of 400 $\mu^2$. However, some organisms become objectionable in much smaller concentrations. The bitter taste of *Synura*, for instance, becomes objectionable when the mere presence of this organism can be detected. Marginal amounts of chlorine, enough for disinfection but not enough for oxidation, may intensify plankton odors and tastes in much the same way as

[24] This factor is largest when the coefficient of self-purification is smallest; for example, it has magnitudes of 4.0, 2.7, 2.0, and 1.6 for values of $f$ = 0.5, 1.0, 2.0, and 4.0 respectively.
[25] James M. Symons, *Water Quality Behavior in Reservoirs*, U.S. Public Health Service, Bureau of Water Hygiene, Washington, D.C., 1969.

**Table 21-3**
Concentration of Copper and Chlorine Required to Kill Troublesome Growths of Organisms (After Hale)

| | Organism | Trouble | Copper sulfate, mg/l | Chlorine, mg/l |
|---|---|---|---|---|
| **Algae** | | | | |
| Diatoms | Asterionella, Synedra, Tabellaria | Odor: aromatic to fishy | 0.1–0.5 | 0.5–1.0 |
| | Fragillaria, Navicula | Turbidity | 0.1–0.3 | 0.3 |
| | Melosira | Turbidity | 0.2 | 2.0 |
| Grass-green | Eudorina,[a] Pandorina[a] | Odor: fishy | 2–10 | |
| | Volvox[a] | Odor: fishy | 0.25 | 0.3–1.0 |
| | Chara, Cladophora | Turbidity, scum | 0.1–0.5 | |
| | Coelastrum, Spirogyra | Turbidity, scum | 0.1–0.3 | 1.0–1.5 |
| Blue-green | Anabaena, Aphanizomenon | Odor: moldy, grassy, vile | 0.1–0.5 | 0.5–1.0 |
| | Clathrocystis, Coelosphaerium | Odor: grassy, vile | 0.1–0.3 | 0.5–1.0 |
| | Oscillatoria | Turbidity | 0.2–0.5 | 1.1 |
| Golden or | Cryptomonas[b] | Odor: aromatic | 0.2–0.5 | |
| yellow-brown | Dinobryon | Odor: aromatic to fishy | 0.2 | 0.3–1.0 |
| | Mallomonas | Odor: aromatic | 0.2–0.5 | |
| | Synura | Taste: cucumber | 0.1–0.3 | 0.3–1.0 |
| | Uroglenopsis | Odor: fishy. Taste: oily | 0.1–0.2 | 0.3–1.0 |
| Dinoflagellates | Ceratium | Odor: fishy, vile | 0.2–0.3 | 0.3–1.0 |
| | Glenodinium | Odor: fishy | 0.2–0.5 | |
| | Peridinium | Odor: fishy | 0.5–2.0 | |
| Filamentous bacteria | Beggiatoa (sulfur) | Odor: decayed. Pipe growths | 5.0 | 0.5 |
| | Crenothrix (iron) | Odor: decayed. Pipe growths | 0.3–0.5 | |
| Crustacea | Cyclops | [c] | 1.0–3.0 | |
| | Daphnia | [c] | 2.0 | 1.0–3.0 |
| Miscellaneous | Chironomus (bloodworm) | [c] | | 15–50 |
| | Craspedacusta (jellyfish) | [c] | 0.3 | |

[a] These organisms are classified also as flagellate protozoa. [b] Classification uncertain.
[c] These organisms are individually visible and cause consumer complaints.

they intensify phenolic tastes, whereas high concentrations of chlorine will destroy both the organisms and their odorous oils and cell matter.

**Chemical Destruction of Algae.** Biologists and engineers often differ with respect to the control or destruction of the plankton with copper sulfate, chlorine, activated carbon, and lime in lakes and reservoirs. Biologists hesitate to upset biological equilibria and may prefer to let nature take its course. Engineers may prefer to meet their immediate responsibilities of providing palatable water even though in the long run introduction of heroic measures of chemical control may involve biological inbalances.

The algicidal properties of copper sulfate were reported by Moore and Kellerman in 1904.[26] Since then copper salts have been applied to lakes and reservoirs more widely than any other chemical. Chlorine is a supplemental rather than a competitive agent. It destroys organisms more sensitive to its hydrolysis products than to copper ions (protozoa are examples), and oxidizes odors released when plankton are killed by copper and undergo decay. Activated carbon has been added with some success to small reservoirs and settling basins to shut out sunlight essential to the growth of photosynthetic organisms. It is also useful in removing algal odors and tastes. Reported *blackout dosage* is of the order of 0.2 to 0.5 lb per 1000 sq ft of water surface. Lime in amounts sufficient to produce caustic alkalinity deprives algae of needed carbon dioxide (Sec. 16-7).

The mechanism and kinetics of destruction of algae by copper and chlorine and of bacteria by disinfection are essentially alike. However, the rate of algal kills at the concentrations normally employed is relatively slow. For copper sulfate it is a matter of days rather than minutes. Destruction of large algal blooms by copper sulfate and by small concentrations of chlorine is usually accompanied by an intensification of odors and a rise in number of saprophytic bacteria feeding on the dead algal cells. The resulting depletion of dissolved oxygen may become serious enough to kill fish. Because destruction of one plankton genus may be followed by the rapid rise of another, treated bodies of water must be watched carefully. Remedial, follow-up treatment may be by chlorine or activated carbon. Lethal concentrations of copper sulfate and of chlorine for some of the most troublesome organisms are shown in Table 21-3. Stated values hold only for relatively soft, warm (60°F) waters. Because copper is precipitated as a carbonate in high-alkalinity waters, copper sulfate dosages may have to be raised by as much as 5% for each 10 mg per l of alkalinity (as $CaCO_3$). Carbon dioxide, on the other hand, reduces precipitation by decreasing the pH and $OH^-$. In the ab-

---

[26] G. T. Moore and K. F. Kellerman, A Method of Destroying or Preventing the Growth of Algae and Certain Pathogenic Bacteria in Water Supplies, *U.S. Bur. Plant Ind.*, Bull. 64 (1904).

sence of normal amounts of $CO_2$, copper sulfate dosage must be increased by about 5%. The $Q_{10}$ value of algicidal copper sulfate is about 1.3. The stage of development reached by the organisms to be destroyed is also a determining factor. Because organic matter competes for copper as well as for chlorine, copper sulfate dosage may have to be increased by 2% for each 10 mg per l of organic matter present.

The tolerance of fish to copper sulfate lies between 0.14 mg per l for trout and 2.1 mg per l for black bass, that is, within the range of concentrations required to destroy plankton growths. However, fish kills can be avoided because chemical dosage is based, not on the total amount of water in the body to be treated, but only on the volume contained within the limited (usually uppermost) strata in which the algae are concentrated. Fish can and do seek refuge in the untreated waters. The likely cause of massive fish kills following the destruction of heavy algal blooms may not be the copper sulfate, but the depletion of oxygen accompanying the decay of the algae on the one hand and the clogging of the gills of fish by dead algal cells on the other. Massive death of algae by natural causes, including their own comsumption of available foodstuffs, may also result in fish kills.

The copper needed for plankton control (Table 21-3) normally lies well below the allowable concentration of 1 mg per l in the drinking-water standards of the U.S. Public Health Service. Common odor- and taste-producing algae are shown in Figs. 21-10 and 11.

**Management of Chemical Algicides.** Large *pulses* of algal growths should not be allowed to develop in storage and distributing reservoirs or other bodies of water serving as only sources of drinking water. Fast propagation should be stopped as soon as limnologic or microscopic evidence sounds a warning. Implicit are adequate sampling and an understanding of growth response to season (heat and light) and to water movement. Except for highly objectionable organisms like *Synura*, drinking-water supplies need, as a rule, not be treated until algal concentrations exceed 500 to 1000 areal standard units. Whether to apply algicides to multiple and multipurpose sources of water is a more complicated decision. Even more complicated is a decision as to whether to allow the use of persisting poisons at all.

Algicides can be fed into the influent to small reservoirs, such as distribution reservoirs and settling basins. Algal seeding from upland storages can often be prevented by this means. Large reservoirs, or their bays and areas of shallow flowage, must be treated from boats or airplanes. Burlap bags filled with copper sulfate crystals and dragged through the water by boats will provide necessary concentrations. To cover moderately extensive areas, the boats may take a zigzag course over the surface, follow a mirror-image zigzag on the return run, and reach shore waters more intensively in a

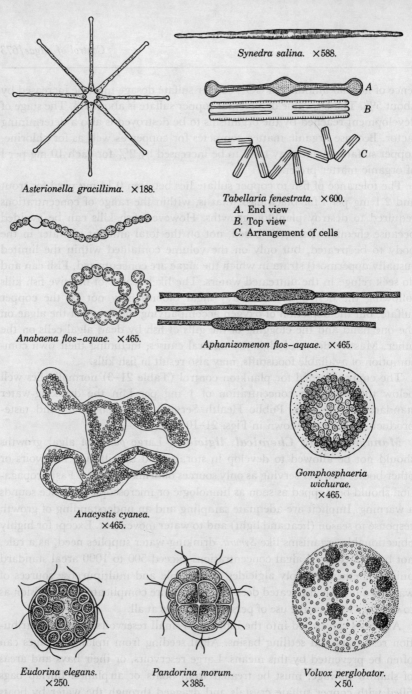

**Figure 21-10.**
**Common taste- and odor-producing diatoms and green and blue-green algae.** (Individual drawings from H. B. Ward and G. C. Whipple, *Fresh-Water Biology*, Wiley, New York, 1918.)

*Ceratium hirundinella.*
×325.

*Glenodinium pulvisculus.*
×500.

*Peridinium tabulatum.*
×320.

*Cryptomonas ovata.*
×350.

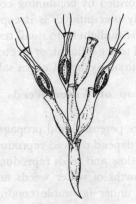
*Dinobryon sertularia.*
×750.

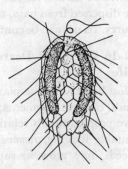
*Mallomonas species.*
×500.

*Synura uvella.* ×600.

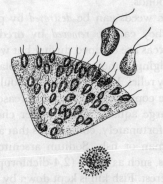
*Uroglenopsis americana.*
Individual cells. ×1,500.

**Figure 21-11.**
Common taste- and odor-producing golden or yellow-brown algae. (Individual drawings from H. B. Ward and G. C. Whipple, *Fresh-Water Biology*, Wiley, New York, 1918.)

perimetral run within about 20 ft of the shore line. Large areas may be covered by patterns of parallel paths 20 to 100 ft apart, first in one direction and then at right angles to it. The solubility of copper sulfate as $CuSO_4 \cdot 5H_2O$ varies from 19.5% by weight at 32°F to 31.3% at 86°F. However, the rate of solution of the crystals is sufficiently slow to make the bag method effective.

Dry-feed or solution-feed machines may be installed on boats as well as at fixed installations (Sec. 16-13). Dusters of the orchard-spray variety have been employed successfully from both boats and airplanes. Liquid chlorine, hypochlorites, and chloramines are applied as in water disinfection (Sec. 17-15). Cupric chloramine, formed by combining copper sulfate, chlorine, and ammonia, is not so easily precipitated as is copper sulfate. Powdered activated carbon for *blacking out* sunlight from small reservoirs and basins can be dispersed from bags, added to the influent, or ejected onto the surface as a slurry. Its use can be confined to sunny days. Copper salts are highly corrosive.

## 21-16 Chemical Destruction of Water Weeds and Other Aquatic Life

Most higher aquatic plants are perennial and propagate by runners, tubers, buds, or stem fragments; few depend on seed reproduction. Because vegetative propagation is relatively slow and seeds reproduce only when they find suitable lodging, *runaway* growths of water weeds are relatively rare. Yet weeds may grow so rapidly under favorable conditions that they create serious nuisances, especially in shallow bodies of water. The limiting depth for attached weeds is about 40 ft. Wastewaters rich in fertilizing elements promote heavy weed infestation of receiving waters and of stabilization ponds and fish ponds.

Aquatic weeds can be *destroyed* by poisons or by draining reservoirs and ponds. They can be *removed* by dredging, cutting, and dragging. When weeded areas are drained during hot weather, exposed plants and their roots die. Dredging removes the entire plant. Cutting and dragging offer but temporary relief. Flowering weeds should be cut before they have gone to seed.

Because copper and chlorine in reasonable concentration are not effective against large aquatic plants, other chemicals must be employed to poison them. Unfortunately, compounds that are effective against weeds may also be toxic to man or fish. Sodium arsenite and growth-promoting (hormonal) substances, such as 2,4-D (2,4-dichlorophenoxyacetic acid and higher esters), are of interest. Fish kill is kept down by proceeding from the shore outward to allow the fish to migrate into deep water. Arsenical compounds such as sodium arsenite will destroy a variety of water weeds; their use is limited to waters that are not sources of drinking water for man or cattle.

Proper treatment of the shore line of reservoirs will keep weeds from establishing themselves. Control of pollution will hold down their growth.

The destruction of a number of other aquatic plants and animals enters into water-quality management. A few examples will illustrate their range. Copper sulfate in concentrations of 0.5 to 2.0 mg per l will destroy the snail hosts of flukes that cause swimmer's itch (Sec. 10-3). Caustic alkalinity as well as chlorine will kill *Cyclops*, the intermediate host of the guinea worm (Sec. 10-3). Chlorine and iodine as well as storage of water for a day or two at temperatures normally prevailing in the tropics have destroyed the cercariae of the pathogenic schistosomes (blood flukes). Chironomid larvae (bloodworms) have been screened from returned activated sludge which they were consuming.

## 21-17 Biological Control of Aquatic Growths

There is a growing awareness of the dangers of indiscriminate and large-scale use of chemicals for the destruction of specific populations of living things. Dislocations of biological equilibria should be avoided where possible. Useful or otherwise wanted members of normal plant and animal communities should not be killed. Thinking has turned more and more to the promotion of biological self-regulating mechanisms that will right the imbalance—and not create a new imbalance—in abnormal ecological systems. The introduction of specific parasites that keep unwanted components of the plant or animal community in check is one example. Breaking the food chain or otherwise modifying the existing environment to the disadvantage of the organisms to be controlled is another. *Biological control*, in comparison with *chemical destruction*, is more likely to be a reversible or an incomplete reaction. Equilibrium reestablishes itself of its own accord. Damage does not become irrevocable.

At least one virus has been isolated that will destroy algae selectively.[27] The host-parasite relationship remains to be fully explored. The search for other viruses against algae as well as against higher aquatic plants is being continued.

---

[27] R. S. Saffernan and M.-E. Morris, Control of Algae with Viruses, *J. Am. Water Works Assoc.*, 56, 1217 (1964).

# twenty-two

# engineering projects

## 22-1 Role of Engineers

The planning, design, and construction of water and wastewater systems bring together a sizable and varied group of engineering practitioners for protracted periods of time. Studies, plans, specifications, and contracts for the construction of water and wastewater works are prepared by engineers engaged directly or as consultants by municipalities, water or wastewater districts, private water companies and industrial organizations. Engineers are employed directly as members of the professional staffs of the government or industry, but for larger projects, particularly those that come up only infrequently, firms of consulting engineers are used. Engineers for manufacturers of water and wastewater equipment as well as engineers of the contractors and construction companies have important roles to play in the engineering enterprise. In addition, engineers may be employed by the regulatory agencies that have the responsibility of approving the reports and the plans prior to construction, or they may be employed by administrative or research organizations.

Inasmuch as most projects fall within the purview of consulting engineering organizations, the specific role that they play in the initiation and creation of a project is summarized as follows:[1]

1. Engineers are invited by the prospective owner or client to submit proposals for preparing an engineering report that elucidates the project and offers alternative methods for meeting the specific need for the project.

2. The consultant is engaged, based on his qualifications. Competitive bidding is not appropriate, nor should the engineer be associated with either a construction company or the manufacturer of equipment.

3. The engineer prepares the report, which may take any of several forms depending on the specific requirement.

4. The report of the consulting engineer is examined by the client, and then may be accepted or rejected.

5. If the report is accepted, the consultant who prepared the report may be authorized to prepare plans and specifications, or proposals may be invited from other consultants.

6. The engineer submits plans and specifications, together with appropriate contract documents, including an estimate of construction costs.

7. After the plans and specifications have been accepted and approved by the regulatory agency, the engineer assists in the advertisement for construction contracts.

8. The bids from the contractors are examined and the lowest bid received from a qualified and financially responsible construction company is generally accepted.

9. Construction is then authorized to proceed under the supervision of a resident engineer generally employed by the consulting engineer. In larger municipalities or in industry, the resident engineer may be employed by the owner. This supervision is important, because it assures that the project as constructed represents the intention of the consulting engineer. Many changes need to be made on the site, and the resident engineer is responsible for initiating these changes. He is also responsible for approving necessary shop drawings supplied by the contractor for equipment and its installation. The resident engineer also authorizes payment periodically during construction and on completion of the construction, and sees to the preparation of a set of *as built* plans.

10. The completed works will be accepted on the recommendation of the consultant, who may then be retained for a month, a year, or even longer to assist in placing the system into operation. The consultant may be requested to prepare an operating manual and assist in the training of operating personnel.

---

[1] A guide for engaging engineering services has been prepared by the American Society of Civil Engineers as *Manual of Engineering Practice*, Number 45, 1964; role of the consulting engineer and the organization of a consulting office are described in *The Consulting Engineer* by C. M. Stanley, John Wiley & Sons, New York, 1961.

11. At an appropriate time, plans are made for financing the project. For smaller communities, the assistance of the engineering consultant is used, and in larger projects the additional assistance of a financial consultant may be useful.

## 22-2 The Engineering Report

Engineers are known to the public by their works and to their engineering colleagues by the quality of their reports. The report is the instrument by which the engineer communicates evaluation of the problem and his recommendations for solution to those who will be obligated to act on his recommendations. A poorly written or poorly organized report, although its technology may be sound, is likely to find its way to a bookshelf or storage rather than to council chambers or board rooms for implementation.

Engineering reports commissioned for the purpose of identifying the need for new or expanded water or wastewater projects and offering acceptable proposals for their development are expected to state their mission clearly; analyze and summarize available and needed data; assess the technical, economic, legal and political feasibility of the projected works; offer alternative answers to the questions asked; point out the one or two most suitable solutions, estimate the costs, and investigate methods of financing; and by these studies, lay a firm base for the recommendations made and the execution of a feasible scheme. If the report is well written its purpose will be understood. Its findings, conclusions, and recommendations will be quoted verbatim in news releases and give the public the information to which it is entitled. The bonds necessary for funding the proposed works can then receive public support. An imaginative and exhaustive report will become a document of reference for subsequent studies and further planning and development.

The young engineer is encouraged to inquire for the reports on which a project that he visits is based. The examination of reports of high quality can do much to stimulate the novice to emulation.

The engineering report generally presents alternative solutions, with the benefits and costs of each solution elaborated. Even where completely objective optimization based on costs is possible, the best solution may not be that which is cheapest. Community and regional history and tradition need to be taken into account. The most immediately economical system may not be the most acceptable. A recurrent example is the common preference, for a municipal water supply, for naturally clean water rather than for polluted water made clean by treatment. In many instances, upland supplies reaching the city by gravity meet the specification of naturally clean waters; supplies pumped from polluted rivers flowing past the city and purified in treatment works before delivery to its distribution system are characteristic of the alternative. Of the two, upland water supplies are usually more costly to develop, but their cost of operation and maintenance may be smaller. Aesthetic considerations cannot be ignored, nor can the potential hazards involved in utilizing a polluted source where there can be no certainty that the water purification is always entirely adequate to remove small concentrations of contaminants, some of which are not yet identified as important in the Public Health Service Drinking Water Standards. Mercury is a case in point.

## 22-3 Planning Water and Wastewater Projects

Engineering decisions are required to fix the area and the population to be served, the design period, the nature of the facilities to be provided, their location, the utilization of centralized treatment facilities or multiple points for treatment, and points of water-supply intake or wastewater disposal. Few projects are so clearly fixed and so straightforward in their possible development as to justify the adoption of a single design period. Optimization may call for the staging of plant capacity and/or for the degree of treatment to be provided. To be resolved for each stage are the capacities, interest charges, and funding; economies of scale; treatment capacities and degrees of treatment; investment of funds; and service charges. Uncertainties in studies of this kind are the difficulties in anticipating new technology and changes in the financing picture, the latter being characterized by increasing financing costs.

The waters to be handled may vary both in quantity and quality and in the degree of treatment required seasonally, monthly, daily, and even hourly. The engineer may use his ingenuity to mitigate these variations by provision of storage which, in the case of water supply, may be drawn on in periods of peak demand and, in the case of wastewaters, may hold discharges to be released during periods when the receiving stream may be more tolerant of them. Variations in quality may also be managed by the provision for the introduction of chemicals for water treatment such as coagulant filter aid polymers or activated carbon, or, in the case of wastewater treatment, by the use of coagulants of one type or another.

Plant siting is an important design decision, although not entirely independent of other considerations such as distribution or collection system layout, or the point of water pumping or wastewater discharge. Topography, foundations, and physical hazards are key siting determinants. Hillside construction may have an advantage in accommodating to head loss in the plant without excessive excavation, and may permit ground-level entrance to several floors in the service building. Foundation conditions are particularly important for construction. Wet sites must be dewatered and structures may have to be weighted down or otherwise structurally fitted to the hydrostatic uplift. On poor foundations, structures may need to be placed on piles or mattresses. Rocky sites may require costly excavation.

Flooding is a common hazard, as treatment plants are often located near rivers that are subject to flood. Plants are generally protected against 1000 or 10,000-year floods by building them above the high-water mark expected, by surrounding them by dikes, by making structures watertight, or by locating equipment that may be damaged above flood level.

Mechanization, instrumentation, and automation are becoming increasingly appropriate for water and wastewater treatment works in the United States and other industrialized countries. Mechanization replaces manual operations and serves functions that cannot be performed adequately by hand. One of the best examples is the removal of sludge from sedimentation tanks. Instrumentation involves the monitoring and recording of plant flows and performance. Automation combines instrumentation and mechanization to effect specific controls. Common examples are the controls exerted in response to level, pressure or flow signals from floats, electrodes, diaphragms, and bubblers; and other more sophisticated electronic sensors such as those that determine pH or chlorine residual.

Economy of scale is pronounced in the case of mechanization, instrumentation, and automation, as it permits attendants, instead of tending individual machines or operations, to control and repair the devices that eliminate such routine duties.

Considerable attention is given to the service buildings required at treatment works and pumping stations. They house offices and laboratories, washing and dressing rooms, shops and storerooms, utility rooms and garages, as well as specific treatment functions such as the addition of chemicals, the housing of pumps, and the handling of sludge. In mild climates, operating structures only need to be protected against rain and sun, while in harsher climates complete housing of all operating activities is advisable.

Provision needs to be made for such utilities as electricity for powering equipment and instruments as well as for indoor and outdoor lighting, plant water supply and drainage, roadways, parking areas and walkways, fencing and other services that make a facility safe for workers and pleasant for visitors. Often it is highly desirable to afford special facilities such as conference rooms or assembly rooms for groups of visitors who may be conducted through the facility.

## 22-4 Plans and Specifications

The preparation of detailed plans and specifications of the works to be built is expensive and time-consuming, rivalling the time required for construction. Plans, which comprise the drawings for the contractor, and specifications, which describe the quality of the materials to be used and in some instances the methods of construction, are both necessary in order for the owner, the designer, and the contractor to understand what the project is to be. The plans and specifications are intimately related, neither being comprehensible alone. They should be comprehensive and precise. Savings in preparing plans and specifications inevitably result in increased bids from the contractor or unsatisfactory construction.

Specifications may be by performance or by description. Where performance is specified, the contractor becomes responsible for both the design and the construction, which is often the case in so-called *turnkey* projects. This is common practice in industry and often in developing countries. Where descriptive specifications underlie design, the owner's engineer specifies what the work is to be, and he takes the responsibility for making certain that, if the construction is according to his plans and specifications, the plant will perform properly. As equipment becomes more complex, there is a tendency to shift to performance specifications. In such instances, the engineer becomes a coordinator of equipment and relinquishes some of his responsibility for design.

Performance specifications have a place when mass production is of benefit to the purchaser. They are justified also when performance can be pretested or when it is possible to test equipment after installation and to replace it if it does not meet specified performance. Pumps are examples of equipment selected on the basis of performance. Such performance specifications are usually supplemented by descriptive specifications that provide protection against overload and insure compatibility with other elements of the system, such as the quality of the materials to be handled. Performance specifications for an entire treatment plant or for individual units of treatment are rarely appropriate, because they can seldom be tested satisfactorily. The claim that performance specifications save money because the engineer need not prepare detailed designs is valid only when standardized *shelf items* can be incorporated in the projected system. When this is not so, the owner bears not only the cost of design but also the hidden cost of turnkey or other projects prepared for bids that were not successful. When a descriptive specification interferes with competitive bidding by suppliers of material or equipment that will serve the design purpose equally well, the engineers should prepare alternative designs and invite competitive bids. A common example is in selecting material for long pipe lines. Comparable costs for the same capacity include both material and construction cost.

## 22-5 Sources of Information

The engineer has available to him many sources of information provided by professional organizations, trade associations, and the government, as well as the catalogues of the manufacturers who provide equipment and materials for water and wastewater facilities. The standards and specifications of the American Water Works Association and the American Standards Association are often referred to in lieu of the presentation of detail. Manuals of engineering practice are prepared by the American Society of Civil Engineers, the American Water Works Association, and the Water Pollution Control Federation. Trade associations, such as the Cast Iron Pipe Research Association, the American Concrete Pipe Association, the Clay Sewer Pipe Association, the Portland Cement Association, and the Chlorine Institute, and commercial publications, such as *Public Works* and *The American City*, publish useful reference manuals. The bibliography at the end of this book provides a reasonably complete list of useful publications.

The availability of standards has done much to improve the quality and decrease the cost of water and wastewater facilities. Standardization of pipes and equipment reduces their cost and makes their maintenance much simpler. The standardization of materials and construction simplifies design and assures that construction with these materials will meet minimum standards of safety and compatibility. Standards of water quality establish a goal to be attained in the protection of waters. Standards of design by regulatory agencies protect small communities that may be served by engineers with limited experience in the field.

However, standards applied indiscriminately may also serve to stultify progress. They may impede the development and adoption of new processes and the fruitful introduction of new ideas. The engineer has a responsibility to ignore standards of practice when he can demonstrate that, as a result of new technology or imaginative design, violation of the standards promises improved performance or greater economy.

Health and safety standards occupy a special place, as they offer the regulatory agencies the opportunity to assure that the public welfare is met. However, the standards are usually minimum requirements, and the engineer may demonstrate that his project will usefully meet more rigorous standards.

## 22-6 Project Financing

Funds for the construction of major water and wastewater systems are usually borrowed. For relatively simple and straightforward projects, the engineer advises on the most suitable methods for borrowing the needed funds. For large projects, where funds are to be derived from several sources, special financial advice may be required. The methods of borrowing depend on the resources of the borrower, the sources from which funds can be borrowed, the regulations of appropriate local and national government agencies, and the nature of the repayment arrangements. The loans normally stipulate how the funds will be obtained for their repayment and for meeting the continuing obligations of the enterprise, including interest payments and operation, maintenance, and replacement costs.

Borrowing for public agencies is arranged through the sale of bonds, generally of three types:

1. *General obligation bonds* that are backed by the full faith and credit of the community with funds for repayment derived from *ad valorem* taxes on property. Such bonds generally carry the lowest interest rates.
2. *Revenue bonds* are based on repayments earned from the sale of water or from sewer service charges. The revenue bonds of an enterprise with a history of good management may carry as low an interest rate as general obligation bonds. New projects, or projects in which the quality of management is uncertain, may have to pay high interest rates. One advantage of revenue bonds is that they are generally not included in the legal limits on the bonded indebtedness of communities.
3. *Special assessment bonds*, like general obligation bonds, are backed by the value of the property they serve. They are generally short-term bonds that are designed to permit borrowing for a specific project serving only part of a community.

In a comprehensive system, borrowing will generally be from all three sources so that charges for service can bear a relation to the benefits received. If all borrowing is with tax-obligation bonds, large users of water are subsidized. On the other hand, if all income is produced by charges for water used, the owners of property in the community, whose property values rise because of the availability of water or wastewater service, and who pay no service charges, are subsidized. Such matters of equity are incorporated in the process of *rate making*. Rates must bring in sufficient income to cover fixed charges, that is, interest and amortization of the capital construction, normal operation, maintenance and replacement costs, and the cost of reasonable improvements. In some instances, the rates may be designed to provide a modest reserve for normal expansion of the system. Too large a reserve places an unfair burden on current users; too small a reserve entails frequent and expensive bond issues.

Water rates are normally structured to accord with the classes of consumer served and their water uses—wholesale, commercial, and residential—each category being subdivided according to the rate of draft. A minimum block of charges covers the cost of metering and meter reading and the billing and collecting, and is generally independent of the quantity of water drawn. The additional blocks represent charges in proportion to the cost of supplying water. In terms of cost, unit price is generally decreased for large users. However, where water is in short supply and each incremental use adds a higher cost for developing additional supplies, unit prices may be increased for larger users.

Water service for fire protection bears no relationship to the amount of water used, and charges for fire service must be handled separately. The added cost of providing transmission mains, pumps, and service storage for fire protection—as contrasted with water-supply needs alone—as well as the cost of hydrants, should be charged to the property owners protected, generally through property taxation.

Peak-flow costs are exemplified by the requirements for larger capacities of pipelines necessary for meeting seasonal peak demands, such as for lawn sprinkling. It is customary in the electrical industry to base charges in some measure on peak requirements, though this has not yet been adopted in the water industry for residential customers. Demand rates are beginning to be used as a method of charging wholesale customers, encouraging them to install their own storage to equalize system demand and reduce peak requirements.

Sewerage systems were historically paid for from general taxation. However, the current trend in financing is very much similar to that for water supply, often by the addition of a fixed percentage to the water bill, although flat rates are more common for wastewater handling than for water supply.

Surcharges may be imposed for industrial wastewaters discharged to municipal sewerage systems, such charges being computed on the basis of the quantity and strength of the wastewaters to be handled. In addition to meeting the cost of treating these wastewaters, the surcharges are an inducement to industries to reduce the volume and strength of their waste discharges by the recirculation of process water, the modification of manufacturing processes, the complete treatment of wastewaters, or their storage prior to discharge to the system.

## 22-7 Systems Management

To meet their responsibility to society in full measure, engineers should see to it that the systems they have designed accomplish their mission. Accordingly, they must be prepared to assist the community or the industry in the operation of projects as effectively as they did in their design and construction. To this purpose, consultants may be engaged by communities for introductory or continuing surveillance of systems operations and management. The engineer may prepare a manual describing the purpose and operations of each unit and the required sequence of operations for the works as a whole. Schematic diagrams that outline available methods of control, as well as record forms, data-collection sheets, equipment and maintenance cards, and computer programs, are often prepared by engineers for their clients.

Engineers frequently design flexibility into a system to meet varying contingencies. If the operators and managers of the works are not fully qualified, the built-in flexibility may never be utilized. State regulatory agencies, professional organizations, and educational institutions often assist in training plant personnel and other officials in the management of water and wastewater facilities. Undergraduate courses in engineering and postgraduate courses in environmental engineering are placing ever greater emphasis on the management of public and private enterprises.

# appendix

## Table 1
*Abbreviations and Symbols*

The abbreviations common to this volume are shown in the following schedule. Unless noted, there is no differentiation between the singular and plural.

| | | | |
|---|---|---|---|
| A | angstrom unit | csm | cubic foot per second per square mile |
| ac | alternating current | | |
| AM | before noon | cu | cubic |
| atm | atmosphere | dc | direct current |
| ave | average | deg | degree |
| bbl | barrel | DO | dissolved oxygen |
| Bé | Beaumé degree | DWF | dry-weather flow |
| BOD | biochemical oxygen demand | emf | electromotive force |
| Btu | British thermal unit | Eq. | equation |
| c | curie | Eqs. | equations |
| °C | centigrade (Celsius) degrees | °F | Fahrenheit degrees |
| cal | calorie | F | Froude number |
| cc | cubic centimeter | Fig. | figure |
| ccc | critical coagulation concentration | Figs. | figures |
| | | fpm | foot per minute |
| CCE | carbon-chloroform extract | fps | foot per second |
| cfm | cubic foot per minute | ft | foot |
| cfs | cubic foot per second | g | gram |
| cgs | centimeter-gram-second | gal | gallon |
| Chap. | chapter | gpad | gallon per acre daily |
| Chaps. | chapters | gpcd | gallon per capita daily |
| cm | centimeter | gpd | gallon per day |
| COD | chemical oxygen demand | gpm | gallon per minute |
| Col. | column | hp | horsepower |
| Cols. | columns | hp-hr | horsepower-hour |

| | | | |
|---|---|---|---|
| hr | hour | $N$ | normal |
| in. | inch | No. | number |
| kg | kilogram | p. | page |
| kw | kilowatt | pp. | pages |
| kwhr | kilowatthour | ppb | part per billion |
| l | liter | ppm | part per million |
| lb | pound | PM | afternoon |
| m | meter | pH | potential hydrogen or hydrogen-ion index (negative logarithm of the hydrogen-ion concentration) |
| $M$ | molar (mole per liter) | | |
| MAF | mean annual flow | | |
| me | milliequivalent | | |
| mg | million gallons, also milligram | psf | pound per square foot |
| mgad | million gallons daily per acre | psi | pound per square inch |
| mgd | million gallons daily | psia | pound per square inch, absolute |
| min | minute | | |
| ml | milliliter | psig | pound per square inch, gage |
| MLSS | mixed-liquor suspended solids | **R** | Reynolds number |
| MLVSS | mixed-liquor volatile suspended solids | rpm | revolution per minute |
| | | SDI | sludge density index |
| mm | millimeter | sec | second |
| mole | gram-molecular weight | Sec. | section |
| moles | gram-molecular weights | Secs. | sections |
| mol wt | molecular weight | sq | square |
| mph | mile per hour | SS | suspended solids |
| MPN | most probable number | SVI | sludge volume index |
| m$\mu$ | millimicron | TL$_m$ | median tolerance limit |
| $\mu$ | micron | U.S. | United States |
| $\mu$c | microcurie | Vol. | volume |
| $\mu\mu$c | picocurie | wt | weight |
| $\mu$g | microgram | yd | yard |

## Table 2
*Weights and Measures*

The American and English weights and measures referred to in this book are alike except for the gallon. The United States gallon is employed. The United States billion, which equals 1000 million, is also employed.

| Miles | Yards | Length Feet | Inches | Centimeters |
|---|---|---|---|---|
| 1 | 1760 | 5280 | ... | ... |
| ... | 1 | 3 | 36 | 91.44 |
| ... | ... | 1 | 12 | 30.48 |
| ... | ... | ... | 1 | 2.540 |

1 m = 100 cm = 3.281 ft = 39.37 in.

| Square Miles | Acres | Area Square Feet | Square Inches | Square Centimeters |
|---|---|---|---|---|
| 1 | 640 | ... | ... | ... |
| ... | 1 | 43,560 | ... | ... |
| ... | ... | 1 | 144 | 929.0 |
| ... | ... | ... | 1 | 6.452 |

1 sq m = 10.76 sq ft

| Cubic Feet | Imperial Gallons | Volume U.S. Gallons | Cubic Inches | Liters |
|---|---|---|---|---|
| 1 | 6.23 | 7.481 | 1728 | 28.32 |
| ... | 1 | 1.2 | 277.4 | 4.536 |
| ... | ... | 1 | 231 | 3.785 |
| ... | ... | ... | 57.75 | 0.946 |
| ... | ... | ... | 61.02 | 1 |

1 cu m = 35.31 cu ft = 264.2 gal

1 Imperial (UK) gal weighs 10 lb  
1 cu ft of water weighs 62.43 lb  
1 cu m = $10^3$ l and weighs 1000 kg

1 U.S. gal weighs 8.34 lb  
1 cu m weighs 2283 lb

## Velocity

| Miles per Hour | Feet per Second | Inches per Minute | Centimeters per Second | Kilometers per Hour |
|---|---|---|---|---|
| 1 | 1.467 | 1056 | ... | 1.609 |
| ... | 1 | 720 | 30.48 | ... |
| ... | ... | 1 | 0.423 | ... |

## Weight

| Tons | Pounds | Grams | Grains | Metric Tons |
|---|---|---|---|---|
| 1 | 2000 | ... | ... | 0.9078 |
| ... | 1 | 454 | 7000 | ... |
| ... | ... | 1 | 15.43 | ... |

1 long ton = 2240 lb

1 ppm = 1 mg/l = 8.34 lb per mg

## Discharge

| Cubic Feet per Second | Million Gallons Daily | Gallons per Minute |
|---|---|---|
| 1 | 0.6463 | 448.8 |
| 1.547 | 1 | 694.4 |

1 in. per hour per acre = 1.008 cfs

1 cu m/sec = 22.83 mgd = 35.32 cfs

## Pressure

| Pounds per Square Inch | Feet of Water | Inches of Mercury |
|---|---|---|
| 1 | 2.307 | 2.036 |
| 0.4335 | 1 | 0.8825 |
| 0.4912 | 1.133 | 1 |

1 atm = 14.70 psia = 29.92 in. Hg = 33.93 ft water = 76.0 cm Hg

## Power

| Kilowatts | Horsepower | Foot-Pounds per Second | Kilogram-Meters per Second |
|---|---|---|---|
| 1 | 1.341 | 737.6 | 102.0 |
| 0.7457 | 1 | 550 | 76.04 |

|  | Work, Energy, and Heat | | |
|---|---|---|---|
| Kilowatt-Hours | Horsepower-Hours | British Thermal Units | Calories |
| 1 | 1.341 | 3412 | $8.6 \times 10^5$ |
| 0.7457 | 1 | 2544 | $6.4 \times 10^5$ |

Temperature

Degrees Fahrenheit $= 32 + \frac{9}{5} \times$ degrees Centigrade (Celsius)

| 0 | 5 | 10 | 15 | 20 | 25 | 30 | 35 | 40 | 45 | 50 | 55 | 60 | °C |
|---|---|---|---|---|---|---|---|---|---|---|---|---|---|
| 32 | 41 | 50 | 59 | 68 | 77 | 86 | 95 | 104 | 113 | 122 | 131 | 140 | °F |

## Table 3

*Viscosity and Density of Water*

Calculated from *International Critical Tables*, 1928 and 1929

| Temperature, °C | Density $\rho, \gamma$ (grams/cm$^3$), also $s$ | Absolute Viscosity $\mu$, centipoises[a] | Kinematic Viscosity $\nu$, centistokes[b] | Temperature, °F |
|---|---|---|---|---|
| 0 | 0.99987 | 1.7921 | 1.7923 | 32.0 |
| 2 | 0.99997 | 1.6740 | 1.6741 | 35.6 |
| 4 | 1.00000 | 1.5676 | 1.5676 | 39.2 |
| 6 | 0.99997 | 1.4726 | 1.4726 | 42.8 |
| 8 | 0.99988 | 1.3872 | 1.3874 | 46.4 |
| 10 | 0.99973 | 1.3097 | 1.3101 | 50.0 |
| 12 | 0.99952 | 1.2390 | 1.2396 | 53.6 |
| 14 | 0.99927 | 1.1748 | 1.1756 | 57.2 |
| 16 | 0.99897 | 1.1156 | 1.1168 | 60.8 |
| 18 | 0.99862 | 1.0603 | 1.0618 | 64.4 |
| 20 | 0.99823 | 1.0087 | 1.0105 | 68.0 |
| 22 | 0.99780 | 0.9608 | 0.9629 | 71.6 |
| 24 | 0.99733 | 0.9161 | 0.9186 | 75.2 |
| 26 | 0.99681 | 0.8746 | 0.8774 | 78.8 |
| 28 | 0.99626 | 0.8363 | 0.8394 | 82.4 |
| 30 | 0.99568 | 0.8004 | 0.8039 | 86.0 |

[a] 1 centipoise $= 10^{-2}$ (gram mass)/(cm)(sec). To convert to (lb force)(sec)/(sq ft) multiply centipoise by $2.088 \times 10^{-5}$.

[b] 1 centistoke $= 10^{-2}$ cm$^2$/sec. To convert to (sq ft)/(sec) multiply centistoke by $1.075 \times 10^{-5}$.

$$1 \text{ gram/cm}^3 = 62.43 \text{ lb/cu ft}$$

## Table 4
*Vapor Pressure of Water and Surface Tension of Water in Contact with Air*

| Temperature, °C | 0 | 5 | 10 | 15 | 20 | 25 | 30 |
|---|---|---|---|---|---|---|---|
| Vapor pressure ($p_w$), mm Hg[a] | 4.58 | 6.54 | 9.21 | 12.8 | 17.5 | 23.8 | 31.8 |
| Surface tension ($\sigma$), dyne/cm[b] | 75.6 | 74.9 | 74.2 | 73.5 | 72.8 | 72.0 | 71.2 |

[a] To convert to in. Hg divide by 25.4.
[b] To convert to (lb force)/ft divide by 14.9.

## Table 5
*Areas under the Normal Probability Curve*

Fractional parts of the total area (1.0000) corresponding to distances between the mean ($\mu$) and given values ($x$) in terms of the standard deviation ($\sigma$), i.e., $(x - \mu)/\sigma = t$.

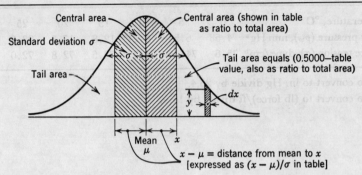

| t | .00 | .01 | .02 | .03 | .04 | .05 | .06 | .07 | .08 | .09 |
|---|---|---|---|---|---|---|---|---|---|---|
| 0.0 | .0000 | .0040 | .0080 | .0120 | .0160 | .0199 | .0239 | .0279 | .0319 | .0359 |
| 0.1 | .0398 | .0438 | .0478 | .0517 | .0557 | .0596 | .0636 | .0675 | .0714 | .0753 |
| 0.2 | .0793 | .0832 | .0871 | .0910 | .0948 | .0987 | .1026 | .1064 | .1103 | .1141 |
| 0.3 | .1179 | .1217 | .1255 | .1293 | .1331 | .1368 | .1406 | .1443 | .1480 | .1517 |
| 0.4 | .1554 | .1591 | .1628 | .1664 | .1700 | .1736 | .1772 | .1808 | .1844 | .1879 |
| 0.5 | .1915 | .1950 | .1985 | .2019 | .2054 | .2088 | .2123 | .2157 | .2190 | .2224 |
| 0.6 | .2257 | .2291 | .2324 | .2357 | .2389 | .2422 | .2454 | .2486 | .2517 | .2549 |
| 0.7 | .2580 | .2611 | .2642 | .2673 | .2704 | .2734 | .2764 | .2794 | .2823 | .2852 |
| 0.8 | .2881 | .2910 | .2939 | .2967 | .2995 | .3023 | .3051 | .3078 | .3106 | .3133 |
| 0.9 | .3159 | .3186 | .3212 | .3238 | .3264 | .3289 | .3315 | .3340 | .3365 | .3389 |
| 1.0 | .3413 | .3438 | .3461 | .3485 | .3508 | .3531 | .3554 | .3577 | .3599 | .3621 |
| 1.1 | .3643 | .3665 | .3686 | .3708 | .3729 | .3749 | .3770 | .3790 | .3810 | .3830 |
| 1.2 | .3849 | .3869 | .3888 | .3907 | .3925 | .3944 | .3962 | .3980 | .3997 | .4015 |
| 1.3 | .4032 | .4049 | .4066 | .4083 | .4099 | .4115 | .4131 | .4147 | .4162 | .4177 |
| 1.4 | .4192 | .4207 | .4222 | .4236 | .4251 | .4265 | .4279 | .4292 | .4306 | .4319 |
| 1.5 | .4332 | .4345 | .4357 | .4370 | .4382 | .4394 | .4406 | .4418 | .4429 | .4441 |
| 1.6 | .4452 | .4463 | .4474 | .4484 | .4495 | .4505 | .4515 | .4525 | .4535 | .4545 |
| 1.7 | .4554 | .4564 | .4573 | .4582 | .4591 | .4599 | .4608 | .4616 | .4625 | .4633 |
| 1.8 | .4641 | .4649 | .4656 | .4664 | .4671 | .4678 | .4686 | .4693 | .4699 | .4706 |
| 1.9 | .4713 | .4719 | .4726 | .4732 | .4738 | .4744 | .4750 | .4758 | .4761 | .4767 |
| 2.0 | .4772 | .4778 | .4783 | .4788 | .4793 | .4798 | .4803 | .4808 | .4812 | .4817 |
| 2.1 | .4821 | .4826 | .4830 | .4834 | .4838 | .4842 | .4846 | .4850 | .4854 | .4857 |
| 2.2 | .4861 | .4864 | .4868 | .4871 | .4875 | .4878 | .4881 | .4884 | .4887 | .4890 |
| 2.3 | .4893 | .4896 | .4898 | .4901 | .4904 | .4906 | .4909 | .4911 | .4913 | .4916 |
| 2.4 | .4918 | .4920 | .4922 | .4925 | .4927 | .4929 | .4931 | .4932 | .4934 | .4936 |
| 2.5 | .4938 | .4940 | .4941 | .4943 | .4945 | .4946 | .4948 | .4949 | .4951 | .4952 |
| 2.6 | .4953 | .4955 | .4956 | .4957 | .4959 | .4960 | .4961 | .4962 | .4963 | .4964 |
| 2.7 | .4965 | .4966 | .4967 | .4968 | .4969 | .4970 | .4971 | .4972 | .4973 | .4974 |
| 2.8 | .4974 | .4975 | .4976 | .4977 | .4977 | .4978 | .4979 | .4979 | .4980 | .4981 |
| 2.9 | .4981 | .4982 | .4982 | .4983 | .4984 | .4984 | .4985 | .4985 | .4986 | .4986 |
| 3.0 | .4987 | .4987 | .4987 | .4988 | .4988 | .4989 | .4989 | .4989 | .4990 | .4990 |
| 3.5 | .499367 | | 4.0 | .499968 | | 4.5 | .499997 | | 5.0 | .4999997 |

*Example:* For $x = 10.8$, $\mu = 9.0$, and $\sigma = 2.0$, $(x - \mu)/\sigma = 0.90$, and $0.3159 = 31.59\%$ of the area is included between $x = 10.8$ and $\mu = 9.0$.

## Table 6

*Values of the Exponential $e^{-x}$ for x Ranging from 0.00 to 10.00*

| x   | 0     | 1     | 2     | 3     | 4     | 5     | 6     | 7     | 8     | 9     |
|-----|-------|-------|-------|-------|-------|-------|-------|-------|-------|-------|
| 0.0 | 1.000 | 0.990 | 0.980 | 0.970 | 0.961 | 0.951 | 0.942 | 0.932 | 0.932 | 0.914 |
| 0.1 | 0.905 | 0.896 | 0.887 | 0.878 | 0.869 | 0.861 | 0.852 | 0.844 | 0.835 | 0.827 |
| 0.2 | 0.819 | 0.811 | 0.803 | 0.794 | 0.787 | 0.779 | 0.771 | 0.763 | 0.756 | 0.748 |
| 0.3 | 0.741 | 0.733 | 0.726 | 0.719 | 0.712 | 0.705 | 0.698 | 0.691 | 0.684 | 0.677 |
| 0.4 | 0.670 | 0.664 | 0.657 | 0.651 | 0.644 | 0.638 | 0.631 | 0.625 | 0.619 | 0.613 |
| 0.5 | 0.607 | 0.600 | 0.595 | 0.589 | 0.583 | 0.577 | 0.571 | 0.566 | 0.560 | 0.554 |
| 0.6 | 0.549 | 0.543 | 0.538 | 0.533 | 0.527 | 0.522 | 0.517 | 0.512 | 0.507 | 0.502 |
| 0.7 | 0.497 | 0.492 | 0.487 | 0.482 | 0.477 | 0.472 | 0.468 | 0.463 | 0.458 | 0.454 |
| 0.8 | 0.449 | 0.445 | 0.440 | 0.436 | 0.432 | 0.427 | 0.423 | 0.419 | 0.415 | 0.411 |
| 0.9 | 0.407 | 0.403 | 0.399 | 0.395 | 0.391 | 0.387 | 0.383 | 0.379 | 0.375 | 0.372 |
| 1.0 | 0.368 | 0.364 | 0.361 | 0.357 | 0.353 | 0.350 | 0.347 | 0.343 | 0.340 | 0.336 |
| 1.1 | 0.333 | 0.330 | 0.326 | 0.323 | 0.320 | 0.317 | 0.313 | 0.310 | 0.307 | 0.304 |
| 1.2 | 0.301 | 0.298 | 0.295 | 0.292 | 0.289 | 0.287 | 0.284 | 0.281 | 0.278 | 0.275 |
| 1.3 | 0.273 | 0.270 | 0.267 | 0.264 | 0.262 | 0.259 | 0.257 | 0.254 | 0.252 | 0.249 |
| 1.4 | 0.247 | 0.244 | 0.242 | 0.239 | 0.237 | 0.235 | 0.232 | 0.230 | 0.228 | 0.225 |
| 1.5 | 0.223 | 0.221 | 0.219 | 0.217 | 0.214 | 0.212 | 0.210 | 0.208 | 0.206 | 0.204 |
| 1.6 | 0.202 | 0.200 | 0.198 | 0.196 | 0.194 | 0.192 | 0.190 | 0.188 | 0.186 | 0.185 |
| 1.7 | 0.183 | 0.181 | 0.179 | 0.177 | 0.176 | 0.173 | 0.172 | 0.170 | 0.169 | 0.167 |
| 1.8 | 0.165 | 0.164 | 0.162 | 0.160 | 0.159 | 0.157 | 0.156 | 0.154 | 0.153 | 0.151 |
| 1.9 | 0.150 | 0.148 | 0.147 | 0.145 | 0.144 | 0.142 | 0.141 | 0.139 | 0.138 | 0.137 |

| x    | $e^{-x}$ | x    | $e^{-x}$ | x    | $e^{-x}$ | x    | $e^{-x}$ | x    | $e^{-x}$  |
|------|----------|------|----------|------|----------|------|----------|------|-----------|
| 2.00 | 0.135    | 3.00 | 0.0498   | 4.00 | 0.0183   | 5.00 | 0.00674  | 6.0  | 0.00248   |
| 2.05 | 0.129    | 3.05 | 0.0474   | 4.05 | 0.0174   | 5.05 | 0.00641  | 6.2  | 0.00203   |
| 2.10 | 0.122    | 3.10 | 0.0450   | 4.10 | 0.0166   | 5.10 | 0.00610  | 6.4  | 0.00166   |
| 2.15 | 0.116    | 3.15 | 0.0429   | 4.15 | 0.0158   | 5.15 | 0.00580  | 6.6  | 0.00136   |
| 2.20 | 0.111    | 3.20 | 0.0408   | 4.20 | 0.0150   | 5.20 | 0.00552  | 6.8  | 0.00111   |
| 2.25 | 0.105    | 3.25 | 0.0388   | 4.25 | 0.0143   | 5.25 | 0.00525  | 7.0  | 0.000912  |
| 2.30 | 0.100    | 3.30 | 0.0369   | 4.30 | 0.0136   | 5.30 | 0.00499  | 7.2  | 0.000747  |
| 2.35 | 0 0954   | 3.35 | 0.0351   | 4.35 | 0.0129   | 5.35 | 0.00475  | 7.4  | 0.000611  |
| 2.40 | 0.0907   | 3.40 | 0.0334   | 4.40 | 0.0123   | 5.40 | 0.00452  | 7.6  | 0.000500  |
| 2.45 | 0.0863   | 3.45 | 0.0317   | 4.45 | 0.0117   | 5.45 | 0.00430  | 7.8  | 0.000410  |
| 2.50 | 0.0821   | 3.50 | 0.0302   | 4.50 | 0.0111   | 5.50 | 0.00409  | 8.0  | 0.000335  |
| 2.55 | 0.0781   | 3.55 | 0.0287   | 4.55 | 0.0106   | 5.55 | 0.00389  | 8.2  | 0.000275  |
| 2.60 | 0.0743   | 3.60 | 0.0273   | 4.60 | 0.0101   | 5.60 | 0.00370  | 8.4  | 0.000225  |
| 2.65 | 0.0707   | 3.65 | 0.0260   | 4.65 | 0.00956  | 5.65 | 0.00352  | 8.6  | 0.000184  |
| 2.70 | 0.0672   | 3.70 | 0.0247   | 4.70 | 0.00910  | 5.70 | 0.00335  | 8.8  | 0.000151  |
| 2.75 | 0.0639   | 3.75 | 0.0235   | 4.75 | 0.00865  | 5.75 | 0.00319  | 9.0  | 0.000123  |
| 2.80 | 0.0608   | 3.80 | 0.0224   | 4.80 | 0.00823  | 5.80 | 0.00303  | 9.2  | 0.000101  |
| 2.85 | 0.0578   | 3.85 | 0.0213   | 4.85 | 0.00783  | 5.85 | 0.00288  | 9.4  | 0.000083  |
| 2.90 | 0.0550   | 3.90 | 0.0202   | 4.90 | 0.00745  | 5.90 | 0.00274  | 9.6  | 0.000068  |
| 2.95 | 0.0523   | 3.95 | 0.0193   | 4.95 | 0.00708  | 5.95 | 0.00261  | 9.8  | 0.000055  |
| 3.00 | 0.0498   | 4.00 | 0.0183   | 5.00 | 0.00674  | 6.00 | 0.00248  | 10.0 | 0.000045  |

$e = 2.7183$; $\log e = 0.43429$; $\ln 10 = 2.30258$.
$\pi = 3.1415$; $\sqrt{\pi} = 1.7724$; $\sqrt{2\pi} = 2.5066$.

## Table 7

*Values of the Well Function $W(u)$ for Various Values of $u$ From U.S. Geological Survey Water Supply Paper 887*

| $N$ | $N \times 10^{-15}$ | $N \times 10^{-14}$ | $N \times 10^{-13}$ | $N \times 10^{-12}$ | $N \times 10^{-11}$ | $N \times 10^{-10}$ | $N \times 10^{-9}$ | $N \times 10^{-8}$ |
|---|---|---|---|---|---|---|---|---|
| 1.0 | 33.96 | 31.66 | 29.36 | 27.05 | 24.75 | 22.45 | 20.15 | 17.84 |
| 1.5 | 33.56 | 31.25 | 28.95 | 26.65 | 24.35 | 22.04 | 19.74 | 17.44 |
| 2.0 | 33.27 | 30.97 | 28.66 | 26.36 | 24.06 | 21.67 | 19.45 | 17.15 |
| 2.5 | 33.05 | 30.74 | 28.44 | 26.14 | 23.83 | 21.53 | 19.23 | 16.93 |
| 3.0 | 32.86 | 30.56 | 28.26 | 25.96 | 23.65 | 21.35 | 19.05 | 16.75 |
| 3.5 | 32.71 | 30.41 | 28.10 | 25.80 | 23.50 | 21.20 | 18.89 | 16.59 |
| 4.0 | 32.56 | 30.27 | 27.97 | 25.67 | 23.36 | 21.06 | 18.76 | 16.46 |
| 4.5 | 32.47 | 30.15 | 27.85 | 25.55 | 23.25 | 20.94 | 18.64 | 16.34 |
| 5.0 | 32.35 | 30.05 | 27.75 | 25.44 | 23.14 | 20.84 | 18.54 | 16.23 |
| 5.5 | 32.26 | 29.95 | 27.65 | 25.35 | 23.05 | 20.74 | 18.44 | 16.14 |
| 6.0 | 32.17 | 29.87 | 27.56 | 25.26 | 22.96 | 20.66 | 18.35 | 16.05 |
| 6.5 | 32.09 | 29.79 | 27.48 | 25.18 | 22.88 | 20.58 | 18.27 | 15.97 |
| 7.0 | 32.02 | 29.71 | 27.41 | 25.11 | 22.81 | 20.50 | 18.20 | 15.90 |
| 7.5 | 31.95 | 29.64 | 27.34 | 25.04 | 22.74 | 20.43 | 18.13 | 15.83 |
| 8.0 | 31.88 | 29.58 | 27.28 | 24.97 | 22.67 | 20.37 | 18.07 | 15.76 |
| 8.5 | 31.82 | 29.52 | 27.22 | 24.91 | 22.61 | 20.31 | 18.01 | 15.70 |
| 9.0 | 31.76 | 29.46 | 27.16 | 24.86 | 22.55 | 20.25 | 17.95 | 15.65 |
| 9.5 | 31.71 | 29.41 | 27.11 | 24.80 | 22.50 | 20.20 | 17.89 | 15.59 |

| $N$ | $N \times 10^{-7}$ | $N \times 10^{-6}$ | $N \times 10^{-5}$ | $N \times 10^{-4}$ | $N \times 10^{-3}$ | $N \times 10^{-2}$ | $N \times 10^{-1}$ | $N$ |
|---|---|---|---|---|---|---|---|---|
| 1.0 | 15.54 | 13.24 | 10.94 | 8.633 | 6.332 | 4.038 | 1.823 | $2.194 \times 10^{-1}$ |
| 1.5 | 15.14 | 12.83 | 10.53 | 8.228 | 5.927 | 3.637 | 1.465 | $1.000 \times 10^{-1}$ |
| 2.0 | 14.85 | 12.55 | 10.24 | 7.940 | 5.639 | 3.355 | 1.223 | $4.890 \times 10^{-2}$ |
| 2.5 | 14.62 | 12.32 | 10.02 | 7.717 | 5.417 | 3.137 | 1.044 | $2.491 \times 10^{-2}$ |
| 3.0 | 14.44 | 12.14 | 9.837 | 7.535 | 5.235 | 2.959 | 0.9057 | $1.305 \times 10^{-2}$ |
| 3.5 | 14.29 | 11.99 | 9.683 | 7.381 | 5.081 | 2.810 | 0.7942 | $6.970 \times 10^{-3}$ |
| 4.0 | 14.15 | 11.85 | 9.550 | 7.247 | 4.948 | 2.681 | 0.7024 | $3.779 \times 10^{-3}$ |
| 4.5 | 14.04 | 11.73 | 9.432 | 7.130 | 4.831 | 2.568 | 0.6253 | $2.073 \times 10^{-3}$ |
| 5.0 | 13.93 | 11.63 | 9.326 | 7.024 | 4.726 | 2.468 | 0.5598 | $1.148 \times 10^{-3}$ |
| 5.5 | 13.84 | 11.53 | 9.231 | 6.929 | 4.631 | 2.378 | 0.5034 | $6.409 \times 10^{-4}$ |
| 6.0 | 13.75 | 11.45 | 9.144 | 6.842 | 4.545 | 2.295 | 0.4544 | $3.601 \times 10^{-4}$ |
| 6.5 | 13.67 | 11.37 | 9.064 | 6.762 | 4.465 | 2.220 | 0.4115 | $2.034 \times 10^{-4}$ |
| 7.0 | 13.60 | 11.29 | 8.990 | 6.688 | 4.392 | 2.151 | 0.3738 | $1.155 \times 10^{-4}$ |
| 7.5 | 13.53 | 11.22 | 8.921 | 6.619 | 4.323 | 2.087 | 0.3403 | $6.583 \times 10^{-6}$ |
| 8.0 | 13.46 | 11.16 | 8.856 | 6.555 | 4.259 | 2.027 | 0.3106 | $3.767 \times 10^{-5}$ |
| 8.5 | 13.40 | 11.10 | 8.796 | 6.494 | 4.199 | 1.971 | 0.2840 | $2.162 \times 10^{-5}$ |
| 9.0 | 13.34 | 11.04 | 8.739 | 6.437 | 4.142 | 1.919 | 0.2602 | $1.245 \times 10^{-5}$ |
| 9.5 | 13.29 | 10.99 | 8.685 | 6.383 | 4.089 | 1.870 | 0.2387 | $7.185 \times 10^{-6}$ |

## Table 8

*Velocity of Flow and Rate of Discharge for Pipes Flowing Full When Frictional Resistance is 2 ft per 1000 (2‰) and C is 100 in Hazen-Williams Formula*

| Diameter, $d$, in. | Area, $A$, sq ft | Velocity, $v$, fps | Discharge $Q$, 1000 gpd |
|---|---|---|---|
| (1) | (2) | (3) | (4) |
| 4  | 0.0873 | 0.96 | 54.3 |
| 5  | 0.137  | 1.11 | 97.5 |
| 6  | 0.196  | 1.24 | 157 |
| 8  | 0.349  | 1.49 | 336 |
| 10 | 0.546  | 1.71 | 602 |
| 12 | 0.785  | 1.92 | 971 |
| 14 | 1.07   | 2.12 | 1,380 |
| 16 | 1.40   | 2.29 | 2,080 |
| 18 | 1.77   | 2.48 | 2,830 |
| 20 | 2.18   | 2.64 | 3,760 |
| 24 | 3.14   | 2.97 | 6,060 |
| 30 | 4.91   | 3.42 | 10,800 |
| 36 | 7.07   | 3.83 | 17,500 |
| 42 | 9.62   | 4.32 | 26,200 |
| 48 | 12.57  | 4.60 | 37,300 |
| 54 | 15.90  | 4.93 | 50,900 |
| 60 | 19.64  | 5.29 | 67,200 |

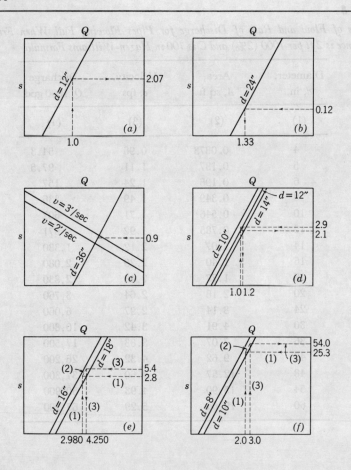

**Figure 1.**

**Use of Hazen-Williams pipe-flow diagram (Fig. 2).**
(a) Given $Q$ and $d$; to find $s$. (b) Given $d$ and $s$; to find $Q$. (c) Given $d$ and $s$; to find $v$. (d) Given $Q$ and $s$; to find $d$. (e) Given $Q$ and $h$; to find $Q$ for different $h$. (f) Given $Q$ and $h$; to find $h$ for different $Q$.

For $C$ other than 100: (1) multiply given $Q$ or $v$ by $(100/C)$ to find $s$; or multiply found value of $Q$ or $v$ by $(C/100)$ for given $s$.

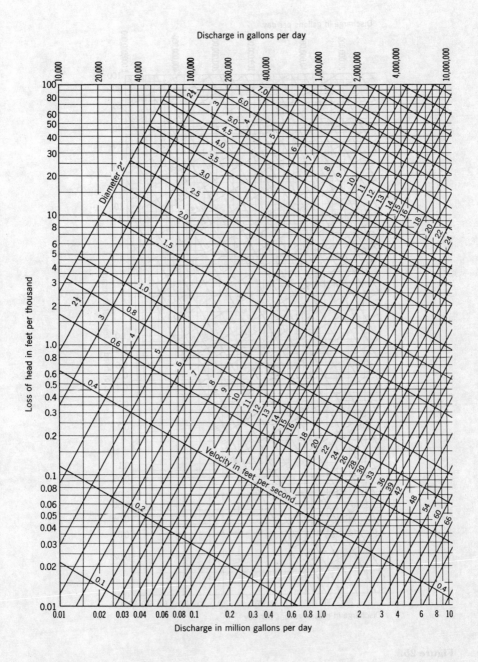

**Figure 2a.**
**Hazen-Williams pipe-flow diagram for discharges of 10,000 to 10,000,000 gpd.**

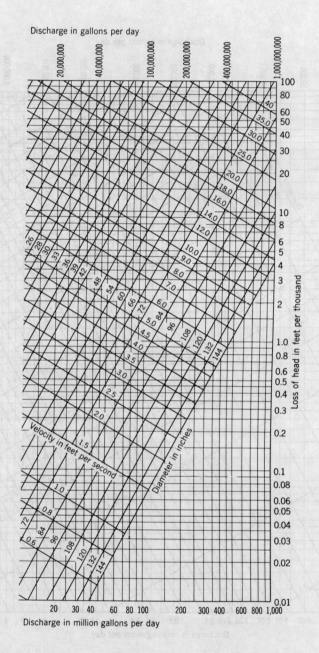

**Figure 2b.**

**Hazen-Williams pipe-flow diagram for discharges of 1 to 100 mgd for C-100.**

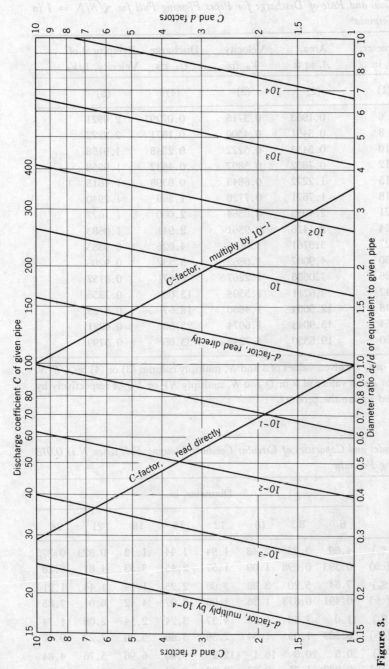

**Figure 3.**

Length, diameter, and coefficient ($l_e$, $d_e$, and $C_e = 100$) hydraulically equivalent to an existing pipe of given length, diameter, and coefficient ($l$, $d$, and $C$).

*Example:* Find the length of a 24-in. pipe, $C_e = 100$, equivalent to a 12-in. pipe, $C = 130$, $l = 2000$ ft. At $C = 130$ read the C-factor $6.2 \times 10^{-1}$. At $d_e/d = 2.0$, read the d-factor $2.9 \times 10$. Hence $l_e/l = (6.2 \times 10^{-1})(2.9 \times 10) = 18$ or $l_e = 2000 \times 18 = 36{,}000$ ft.

701

## Table 9

Velocity of Flow and Rate of Discharge for Pipes Flowing Full for $\sqrt{S}/N = 1$ in Manning's Formula[a]

| Diameter, $D$, in. | Area, $A$, sq ft | Velocity, $V_0$, fps | Discharge, $Q_0$, cfs | Reciprocal of Velocity, $1/V_0$ |
|---|---|---|---|---|
| (1) | (2) | (3) | (4) | (5) |
| 6  | 0.1963  | 0.3715 | 0.07293 | 2.6921 |
| 8  | 0.3491  | 0.4500 | 0.1571  | 2.2222 |
| 10 | 0.5455  | 0.5222 | 0.2848  | 1.9158 |
| 12 | 0.7852  | 0.5897 | 0.4632  | 1.6958 |
| 15 | 1.2272  | 0.6843 | 0.8398  | 1.4613 |
| 18 | 1.7671  | 0.7728 | 1.366   | 1.2940 |
| 21 | 2.4053  | 0.8564 | 2.060   | 1.1677 |
| 24 | 3.1416  | 0.9361 | 2.941   | 1.0683 |
| 27 | 3.9761  | 1.0116 | 4.026   | 0.9885 |
| 30 | 4.9087  | 1.0863 | 5.332   | 0.9206 |
| 36 | 7.0686  | 1.2267 | 8.671   | 0.8152 |
| 42 | 9.6211  | 1.3594 | 13.08   | 0.7356 |
| 48 | 12.5664 | 1.4860 | 18.67   | 0.6729 |
| 54 | 15.9043 | 1.6074 | 25.56   | 0.6221 |
| 60 | 19.6350 | 1.7244 | 33.86   | 0.5799 |

[a] To find $V$ or $Q$ for given values of $S$ and $N$, multiply column (3) or (4) by $\sqrt{S}/N$. To find $S$ for given values of $V$ or $Q$ and $N$, multiply $NV$ or $NQ/A$ respectively by column (5) and square the product.

## Table 10

Minimum Grades and Capacities of Circular Conduits Flowing Full when $N$ is 0.013 in the Manning Formula

| Velocity, fps | | Diameter, in. | | | | | | | |
|---|---|---|---|---|---|---|---|---|---|
| | | 6 | 8 | 10 | 12 | 15 | 18 | 21 | 24 |
| 2.0 | $S$(‰) | 4.89 | 3.33 | 2.48 | 1.94 | 1.44 | 1.13 | 0.923 | 0.775 |
|     | $Q$(cfs) | 0.393 | 0.698 | 1.09 | 1.57 | 2.45 | 3.53 | 4.81 | 6.28 |
| 2.5 | $S$(‰) | 7.64 | 5.20 | 3.88 | 3.06 | 2.25 | 1.74 | 1.44 | 1.21 |
|     | $Q$(cfs) | 0.491 | 0.873 | 1.36 | 1.96 | 3.07 | 4.42 | 6.01 | 7.85 |
| 3.0 | $S$(‰) | 11.0 | 7.50 | 5.58 | 4.37 | 3.24 | 2.54 | 2.08 | 1.74 |
|     | $Q$(cfs) | 0.589 | 1.05 | 1.64 | 2.36 | 3.68 | 5.30 | 7.22 | 9.42 |
| 5.0 | $S$(‰) | 30.5 | 20.8 | 16.1 | 12.2 | 9.00 | 6.96 | 5.76 | 4.84 |
|     | $Q$(cfs) | 0.982 | 1.75 | 2.73 | 3.93 | 6.14 | 8.84 | 12.0 | 15.7 |

## Table 11
*Hydraulic Elements of Circular Conduits*

Central angle: $\cos \frac{1}{2}\theta = 1 - 2d/D$

Area: $\dfrac{D^2}{4}\left(\dfrac{\pi\theta}{360} - \dfrac{\sin\theta}{2}\right)$

Wetted perimeter: $\pi D\theta/360$

Hydraulic radius:
$$\frac{D}{4}\left(1 - \frac{360\sin\theta}{2\pi\theta}\right)$$

Velocity: $\dfrac{1.49}{n} r^{2/3} s^{1/2}$

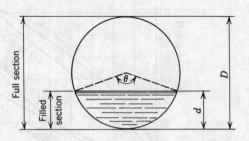

| Depth | Area | Hydraulic Radius | | | Velocity | Discharge | Rough- |
|---|---|---|---|---|---|---|---|
| $d/D$ | $a/A$ | $r/R$ | $R/r$ | $(r/R)^{1/6}$ | $v/V$ for $N/n = 1.0$ | $q/Q$ | ness $N/n$ |
| (1) | (2) | (3) | (4) | (5) | (6)[a] | (7)[a] | (8) |
| 1.000 | 1.000 | 1.000 | 1.000 | 1.000 | 1.000 | 1.000 | 1.00 |
| 0.900 | 0.949 | 1.192 | 0.839 | 1.030 | 1.124 | 1.066 | 0.94 |
| 0.800 | 0.858 | 1.217 | 0.822 | 1.033 | 1.140 | 0.988 | 0.88 |
| 0.700 | 0.748 | 1.185 | 0.843 | 1.029 | 1.120 | 0.838 | 0.85 |
| 0.600 | 0.626 | 1.110 | 0.900 | 1.018 | 1.072 | 0.671 | 0.83 |
| 0.500 | 0.500 | 1.000 | 1.00 | 1.000 | 1.000 | 0.500 | 0.81 |
| 0.400 | 0.373 | 0.857 | 1.17 | 0.975 | 0.902 | 0.337 | 0.79 |
| 0.300 | 0.252 | 0.684 | 1.46 | 0.939 | 0.776 | 0.196 | 0.78 |
| 0.200 | 0.143 | 0.482 | 2.07 | 0.886 | 0.615 | 0.088 | 0.79 |
| 0.100 | 0.052 | 0.254 | 3.94 | 0.796 | 0.401 | 0.021 | 0.82 |
| 0.000 | 0.000 | ... | ... | ... | ... | 0.000 | ... |

[a] For values corrected for variations in roughness with depth multiply by roughness ratio $N/n$ in column 8.

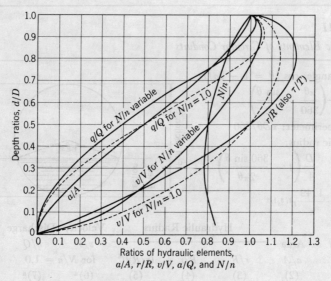

**Figure 4.**

**Basic hydraulic elements of circular conduits for all values of roughness and slope.**

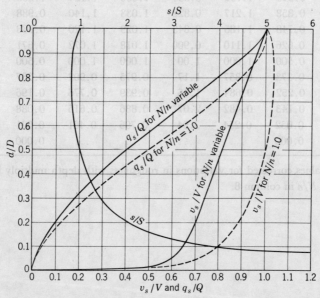

**Figure 5.**

**Hydraulic elements of circular conduits with equal self-cleansing properties at all depths.**

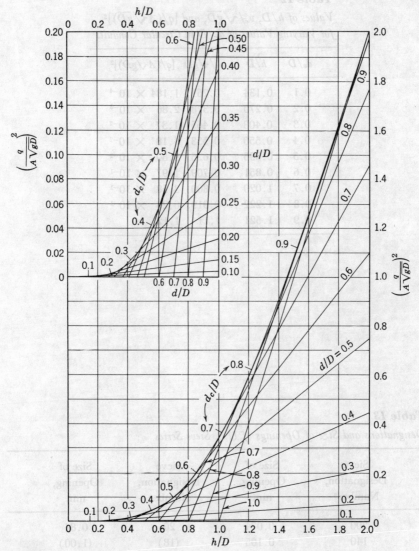

**Figure 6.**
Alternate stages and critical depths of flow in circular conduits.

## Table 12

Values of $h/D$, $v_c/\sqrt{gD}$, and $[q/(A\sqrt{gD})]^2$ for Varying Values of $d_c/D$ in Circular Conduits

| $d_c/D$ | $h/D$ | $v_c/\sqrt{gD}$ | $[q/(A\sqrt{gD})]^2$ |
|---|---|---|---|
| 0.1 | 0.134 | 0.261 | $1.184 \times 10^{-4}$ |
| 0.2 | 0.270 | 0.378 | $2.86 \times 10^{-3}$ |
| 0.3 | 0.408 | 0.465 | $1.37 \times 10^{-2}$ |
| 0.4 | 0.550 | 0.553 | $4.18 \times 10^{-2}$ |
| 0.5 | 0.696 | 0.626 | $9.80 \times 10^{-2}$ |
| 0.6 | 0.851 | 0.709 | $1.97 \times 10^{-1}$ |
| 0.7 | 1.020 | 0.800 | $3.58 \times 10^{-1}$ |
| 0.8 | 1.222 | 0.919 | $6.23 \times 10^{-1}$ |
| 0.9 | 1.521 | 1.11 | 1.12 |

## Table 13

Designations and Size of Openings of U.S. Sieve Series

| Sieve Designation, Number[a] | Size of Opening, mm | Sieve Designation, Number | Size of Opening, mm |
|---|---|---|---|
| 200 | 0.074 | 20 | 0.84 |
| 140 | 0.105 | (18) | (1.00) |
| 100 | 0.149 | 16 | 1.19 |
| 70 | 0.210 | 12 | 1.68 |
| 50 | 0.297 | 8 | 2.38 |
| 40 | 0.42 | 6 | 3.36 |
| 30 | 0.59 | 4 | 4.76 |

[a] Approximately the number of meshes per inch.

## Table 14

Values of $1/(1 - f_e)$ Corresponding to Values of $f_e^3/(1 - f_e)$ Ranging from 0.1 to 9.9.

| $\dfrac{f_e^3}{1 - f_e}$ | 0.0 | 0.1 | 0.2 | 0.3 | 0.4 | 0.5 | 0.6 | 0.7 | 0.8 | 0.9 |
|---|---|---|---|---|---|---|---|---|---|---|
| 0 | 0.00 | 1.62 | 1.89 | 2.10 | 2.28 | 2.44 | 2.59 | 2.74 | 2.88 | 3.01 |
| 1 | 3.14 | 3.27 | 3.40 | 3.52 | 3.65 | 3.78 | 3.89 | 4.01 | 4.13 | 4.24 |
| 2 | 4.35 | 4.47 | 4.58 | 4.70 | 4.81 | 4.93 | 5.05 | 5.16 | 5.27 | 5.38 |
| 3 | 5.49 | 5.60 | 5.71 | 5.82 | 5.92 | 6.03 | 6.14 | 6.24 | 6.35 | 6.46 |
| 4 | 6.57 | 6.68 | 6.78 | 6.88 | 6.99 | 7.10 | 7.20 | 7.31 | 7.41 | 7.52 |
| 5 | 7.62 | 7.73 | 7.83 | 7.94 | 8.04 | 8.15 | 8.25 | 8.35 | 8.46 | 8.56 |
| 6 | 8.67 | 8.77 | 8.88 | 8.98 | 9.08 | 9.18 | 9.29 | 9.39 | 9.49 | 9.60 |
| 7 | 9.70 | 9.81 | 9.91 | 10.01 | 10.11 | 10.21 | 10.32 | 10.42 | 10.52 | 10.62 |
| 8 | 10.72 | 10.83 | 10.93 | 11.03 | 11.14 | 11.24 | 11.35 | 11.45 | 11.56 | 11.66 |
| 9 | 11.76 | 11.86 | 11.96 | 12.06 | 12.16 | 12.27 | 12.37 | 12.47 | 12.58 | 12.68 |

## Table 15
Atomic Numbers, Weights, and Valences of Chemical Elements[a]

| Element | Symbol | Atomic Number | International Atomic Weight (1952) | Valence |
|---|---|---|---|---|
| Aluminum | Al | 13 | 26.98 | 3 |
| Arsenic | As | 33 | 74.91 | 3, 5 |
| Barium | Ba | 56 | 137.36 | 2 |
| Boron | B | 5 | 10.82 | 3 |
| Bromine | Br | 35 | 79.92 | 1, 3, 5, 7 |
| Cadmium | Cd | 48 | 112.41 | 2 |
| Calcium | Ca | 20 | 40.08 | 2 |
| Carbon | C | 6 | 12.01 | 2, 4 |
| Chlorine | Cl | 17 | 35.46 | 1, 3, 5, 7 |
| Chromium | Cr | 24 | 52.01 | 2, 3, 6 |
| Cobalt | Co | 27 | 58.94 | 2, 3 |
| Copper | Cu | 29 | 63.54 | 1, 2 |
| Fluorine | F | 9 | 19.00 | 1 |
| Gold (*aurum*) | Au | 79 | 197.2 | 1, 3 |
| Hydrogen | H | 1 | 1.008 | 1 |
| Iodine | I | 53 | 126.92 | 1, 3, 5, 7 |
| Iron (*ferrum*) | Fe | 26 | 55.85 | 2, 3 |
| Lead (*plumbum*) | Pb | 82 | 207.21 | 2, 4 |
| Magnesium | Mg | 12 | 24.32 | 2 |
| Manganese | Mn | 25 | 54.93 | 2, 3, 4, 6, 7 |
| Mercury (*hydrargyrum*) | Hg | 80 | 200.61 | 1, 2 |
| Nickel | Ni | 28 | 58.69 | 2, 3 |
| Nitrogen | N | 7 | 14.01 | 3, 5 |
| Oxygen | O | 8 | 16.00 | 2 |
| Phosphorus | P | 15 | 30.98 | 3, 5 |
| Platinum | Pt | 78 | 195.23 | 2, 4 |
| Potassium (*kalium*) | K | 19 | 39.10 | 1 |
| Selenium | Se | 34 | 78.96 | 2, 4, 6 |
| Silicon | Si | 14 | 28.09 | 4 |
| Silver (*argentum*) | Ag | 47 | 107.88 | 1 |
| Sodium (*natrium*) | Na | 11 | 23.00 | 1 |
| Strontium | Sr | 38 | 87.63 | 2 |
| Sulfur | S | 16 | 32.07 | 2, 4, 6 |
| Tin (*stannum*) | Sn | 50 | 118.70 | 2, 4 |
| Zinc | Zn | 30 | 65.38 | 2 |

[a] Elements encountered in radioactive wastes are not included. For a complete list, see *Handbook of Chemistry and Physics*, Chemical Rubber Publishing Company, Cleveland, Ohio.

## Table 16

Coefficients of Diffusion of Aqueous Solutions into Water

| Substance | Molarity, gram-molecules/l | Temperature, °C | $k_d$ cm²/hr | cm²/day |
|---|---|---|---|---|
| Acetic acid | 0.2 | 13.5 | $3.2 \times 10^{-2}$ | 0.77 |
| Ammonia | 1.0 | 15.23 | $6.4 \times 10^{-2}$ | 1.54 |
| Bromine | 0.1 | 12.0 | $3 \times 10^{-2}$ | 0.8 |
| Calcium chloride | 2.0 | 10.0 | $2.8 \times 10^{-2}$ | 0.68 |
| Chlorine | 0.1 | 12.0 | $5.1 \times 10^{-2}$ | 1.22 |
| Copper sulfate | 0.1 | 17.0 | $1.6 \times 10^{-2}$ | 0.39 |
| Glucose | ... | 18.0 | $2.0 \times 10^{-2}$ | 0.49 |
| Glycerine | 0.1 | 10.14 | $1.5 \times 10^{-2}$ | 0.357 |
| Hydrochloric acid | 0.1 | 19.2 | $9.2 \times 10^{-2}$ | 2.21 |
| Iodine | 0.1 | 12.0 | $2 \times 10^{-2}$ | 0.5 |
| Magnesium sulfate | 1.0 | 7.0 | $1.3 \times 10^{-2}$ | 0.30 |
| Nitric acid | 0.1 | 19.5 | $8.6 \times 10^{-2}$ | 2.07 |
| Sodium chloride | 0.1 | 15.0 | $3.9 \times 10^{-2}$ | 0.94 |
| Sugar | 1.0 | 12.0 | $1.1 \times 10^{-2}$ | 0.254 |
| Sulfuric acid | 1.0 | 12.0 | $4.7 \times 10^{-2}$ | 1.12 |
| Urea | 0.1 | 14.8 | $4.0 \times 10^{-2}$ | 0.97 |

## Table 17

*Absorption Coefficients of Common Gases in Water*

(Milliliters of gas, reduced to 0°C and 760 mm Hg, per liter of water when partial pressure of gas is 760 mm Hg)

| Gas | Mol. wt | Weight at 0°C and 760 mm Hg, g/l | Absorption Coefficients | | | | Boiling Point, °C |
|---|---|---|---|---|---|---|---|
| | | | 0°C | 10°C | 20°C | 30°C | |
| Hydrogen, $H_2$ | 2.016 | 0.08988 | 21.4 | 19.6 | 18.2 | 17.0 | −253 |
| Methane, $CH_4$ | 16.014 | 0.7168 | 55.6 | 41.8 | 33.1 | 27.6 | −162 |
| Nitrogen, $N_2$ | 28.01 | 1.251 | 23.0 | 18.5 | 15.5 | 13.6 | −196 |
| Oxygen, $O_2$ | 32.00 | 1.429 | 49.3 | 38.4 | 31.4 | 26.7 | −183 |
| Ammonia, $NH_3$[a] | 17.03 | 0.7710 | 1,300 | 910 | 711 | — | −33.4 |
| Hydrogen sulfide, $H_2S$[a] | 34.08 | 1.539 | 4,690 | 3,520 | 2,670 | ... | −61.8 |
| Carbon dioxide, $CO_2$[a] | 44.01 | 1.977 | 1,710 | 1,190 | 878 | 665 | −78.5 |
| Ozone, $O_3$ | 48.00 | 2.144 | 641 | 520 | 368 | 233 | −112 |
| Sulfur dioxide, $SO_2$[a] | 64.07 | 2.927 | 79,800 | 56,600 | 39,700 | 27,200 | −10.0 |
| Chlorine, $Cl_2$[a] | 70.91 | 3.214 | 4,610 | 3,100 | 2,260 | 1,770 | −34.6 |
| Air[b] | ... | 1.2928 | 28.8 | 22.6 | 18.7 | 16.1 | ... |

[a] Total solubility.

[b] At sea level dry air contains 78.08% $N_2$, 20.95% $O_2$, 0.93% A, 0.03% $CO_2$, and 0.01% other gases by volume. For ordinary purposes it is assumed to be composed of 79% $N_2$ and 21% $O_2$.

## Table 18
*Transfer Coefficients and Liquid Film Thicknesses for Gases of Low Solubility in Water*

| Gas | Temperature, °C | Test Condition | Transfer Coefficient, $k_{d(l)}$, cm/hr | Diffusion Coefficient, $k_d$, cm²/hr | Film Thickness, $k_d/k_{d(l)}$, cm | Observer |
|---|---|---|---|---|---|---|
| \multicolumn{7}{c}{*Plane surfaces in contact with 50 ml of water at stated rate of stirring*} |
| $O_2$ | 25 | 0 rpm | 0.41 | $9.4 \times 10^{-2}$ | $2.3 \times 10^{-1}$ | Hutchinson[a] |
|  |  | 76 rpm | 1.20 |  | $7.8 \times 10^{-2}$ |  |
|  |  | 171 rpm | 3.00 |  | $3.1 \times 10^{-2}$ |  |
|  |  | 486 rpm | 5.43 |  | $1.7 \times 10^{-2}$ |  |
|  |  | 1025 rpm | 7.64 |  | $1.2 \times 10^{-2}$ |  |
| $CO_2$ | 25 | 0 rpm | 1.46 | $7.4 \times 10^{-2}$ | $5.1 \times 10^{-2}$ |  |
|  |  | 93 rpm | 1.65 |  | $4.5 \times 10^{-2}$ |  |
|  |  | 171 rpm | 2.73 |  | $2.7 \times 10^{-2}$ |  |
|  |  | 486 rpm | 4.84 |  | $1.5 \times 10^{-2}$ |  |
|  |  | 1025 rpm | 8.35 |  | $8.9 \times 10^{-2}$ |  |
| \multicolumn{7}{c}{*Gas bubbles of stated size in water*} |
| $O_2$ | 25 | 10 ml | 20 | $9.4 \times 10^{-2}$ | $4.7 \times 10^{-3}$ | Schwab[b] |
|  | 20–21 | 1.51–2.13 ml | 88 |  | $1.1 \times 10^{-3}$ | Ippen[c] |
| \multicolumn{7}{c}{*Water droplets of stated size*} |
| $CO_2$ | 24 | 0.1 ml | 260 | $7.4 \times 10^{-2}$ | $2.8 \times 10^{-3}$ | Whitman[d] |

[a] M. M. Hutchinson and T. K. Sherwood, Liquid Film in Gas Absorption, *Ind. Eng. Chem.*, **29**, 836 (1937). The rate of absorption varies approximately as the ⅔ power of the rate of stirring. [b] Quoted by Hutchinson and Sherwood.

[c] A. T. Ippen, L. G. Campbell, and C. E. Carver, Jr., The Determination of Oxygen Absorption in Aeration Processes, *Hydrodynamics Lab., Mass. Inst. Technol., Tech. Rept.* No. 7 (1952).

[d] W. G. Whitman, The Two-Film Theory of Gas Absorption, *Chem. Metal. Eng.*, **29**, 146 (1923).

## Table 19

*Solubility Constants of Important Chemical Substances in Water at 25°C*

| Substance | Ion Product | $pK_s = -\log K_s$ | Application |
|---|---|---|---|
| $Al(OH)_3$ | $(Al^{+++})(OH^-)^3$ | 32.9 | Coagulation |
| $AlPO_4$ | $(Al^{+++})(PO_4^=)$ | 22.0 | Phosphate removal |
| $CaCO_3$ | $(Ca^{++})(CO_3^-)$ | 8.32 | Softening; corrosion control |
| $CaF_2$ | $(Ca^{++})(F^-)^2$ | 10.41 | Fluoridation |
| $Ca(OH)_2$ | $(Ca^{++})(OH^-)^2$ | 5.26 | Softening |
| $Ca_3(PO_4)_2$ | $(Ca^{++})^3(PO_4^=)^2$ | 26.0 | Softening; phosphate cycle |
| $CaHPO_4$ | $(Ca^{++})(HPO_4^-)$ | 7.0 | Phosphate cycle |
| $CaSO_4 \cdot 2H_2O$ | $(Ca^{++})(SO_4^-)$ | 4.62 | Hardness; scale formation |
| $Cu(OH)_2$ | $(Cu^{++})(OH^-)^2$ | 19.25 | Algae control |
| $Fe(OH)_2$ | $(Fe^{++})(OH^-)^2$ | 15.1 | Corrosion; deferrization |
| $FeCO_3$ | $(Fe^{++})(CO_3^-)$ | 10.68 | Iron cycle |
| $FeS$ | $(Fe^{++})(S^-)$ | 17.3 | Anaerobic corrosion |
| $Fe(OH)_3$ | $(Fe^{+++})(OH^-)^3$ | 38.04 | Coagulation; deferrization |
| $FePO_4$ | $(Fe^{+++})(PO_4^=)$ | 21.9 | Phosphate cycle |
| $MgCO_3 \cdot 3H_2O$ | $(Mg^{++})(CO_3^-)$ | 4.25 | Hardness |
| $MgF_2$ | $(Mg^{++})(F^-)^2$ | 8.18 | Defluoridation |
| $Mg(OH)_2$ | $(Mg^{++})(OH^-)^2$ | 10.74 | Softening |
| $MnCO_3$ | $(Mn^{++})(CO_3^-)$ | 10.06 | Manganese cycle |
| $Mn(OH)_2$ | $(Mn^{++})(OH^-)^2$ | 12.8 | Demanganization |
| $Zn(OH)_2$ | $(Zn^{++})(OH^-)^2$ | 17.15 | Corrosion |

## Table 20

Saturation Values of Dissolved Oxygen in Fresh and Sea Water Exposed to an Atmosphere Containing 20.9% Oxygen under a Pressure of 760 mm of Mercury[a]

| Temperature, °C | Dissolved Oxygen (mg/l) for Stated Concentrations of Chloride (mg/l) | | | | | Difference per 1000 mg/l Chloride |
|---|---|---|---|---|---|---|
| | 0 | 5000 | 10000 | 15000 | 20000 | |
| 0 | 14.7 | 13.8 | 13.0 | 12.1 | 11.3 | 0.165 |
| 1 | 14.3 | 13.5 | 12.7 | 11.9 | 11.1 | 0.160 |
| 2 | 13.9 | 13.1 | 12.3 | 11.6 | 10.8 | 0.154 |
| 3 | 13.5 | 12.8 | 12.0 | 11.3 | 10.5 | 0.149 |
| 4 | 13.1 | 12.4 | 11.7 | 11.0 | 10.3 | 0.144 |
| 5 | 12.8 | 12.1 | 11.4 | 10.7 | 10.0 | 0.140 |
| 6 | 12.5 | 11.8 | 11.0 | 10.4 | 9.8 | 0.135 |
| 7 | 12.1 | 11.5 | 10.8 | 10.2 | 9.6 | 0.130 |
| 8 | 11.8 | 11.2 | 10.6 | 10.0 | 9.4 | 0.125 |
| 9 | 11.6 | 11.0 | 10.4 | 9.7 | 9.1 | 0.121 |
| 10 | 11.3 | 10.7 | 10.1 | 9.5 | 8.9 | 0.118 |
| 11 | 11.0 | 10.4 | 9.9 | 9.3 | 8.7 | 0.114 |
| 12 | 10.8 | 10.2 | 9.7 | 9.1 | 8.6 | 0.110 |
| 13 | 10.5 | 10.0 | 9.4 | 8.9 | 8.4 | 0.107 |
| 14 | 10.3 | 9.7 | 9.2 | 8.7 | 8.2 | 0.104 |
| 15 | 10.0 | 9.5 | 9.0 | 8.5 | 8.0 | 0.100 |
| 16 | 9.8 | 9.3 | 8.8 | 8.4 | 7.9 | 0.098 |
| 17 | 9.6 | 9.1 | 8.7 | 8.2 | 7.7 | 0.095 |
| 18 | 9.4 | 9.0 | 8.5 | 8.0 | 7.6 | 0.092 |
| 19 | 9.2 | 8.8 | 8.3 | 7.9 | 7.4 | 0.089 |
| 20 | 9.0 | 8.6 | 8.1 | 7.7 | 7.3 | 0.088 |
| 21 | 8.8 | 8.4 | 8.0 | 7.6 | 7.1 | 0.086 |
| 22 | 8.7 | 8.3 | 7.8 | 7.4 | 7.0 | 0.084 |
| 23 | 8.5 | 8.1 | 7.7 | 7.3 | 6.8 | 0.083 |
| 24 | 8.3 | 7.9 | 7.5 | 7.1 | 6.7 | 0.083 |
| 25 | 8.2 | 7.8 | 7.4 | 7.0 | 6.5 | 0.082 |
| 26 | 8.0 | 7.6 | 7.2 | 6.8 | 6.4 | 0.080 |
| 27 | 7.9 | 7.5 | 7.1 | 6.7 | 6.3 | 0.079 |
| 28 | 7.7 | 7.3 | 6.9 | 6.6 | 6.2 | 0.078 |
| 29 | 7.6 | 7.2 | 6.8 | 6.5 | 6.1 | 0.076 |
| 30 | 7.4 | 7.1 | 6.7 | 6.3 | 6.0 | 0.075 |

[a] For derivation of information see Sec. 12–3. For barometric pressures other than 760 mm of Hg, the solubilities vary approximately in proportion to the ratio of the actual pressure to the standard pressure.

## Table 21

*Electrode Reactions and Standard Potentials at 25° C*

| Electrode Reaction | $E^0$, v | Remarks (Direction) |
|---|---|---|
| *Metallic corrosion and cathodic protection* | | |
| $Mg^{++} + 2e = Mg$ | −2.37 | Sacrificial anode (←) |
| $Al^{+++} + 3e = Al$ | −1.67 | Sacrificial anode (←) |
| $Zn^{++} + 2e = Zn$ | −0.762 | Sacrificial anode (←) |
| $Fe^{++} + 2e = Fe$ | −0.409 | Initial corrosion (←) |
| $Pb^{++} + 2e = Pb$ | −0.126 | Initial corrosion (←) |
| $Cu^{++} + 2e = Cu$ | 0.345 | Initial corrosion (←) |
| $Fe^{+++} + e = Fe^{++}$ | 0.771 | Ferrous oxidation (←) |
| $Fe(OH)_3 + e = Fe(OH)_2 + OH^-$ | −0.56 | Rust formation (←) |
| *Reduction of water and dissolved oxygen* | | |
| $2\,H_2O + 2e = H_2 + 2\,OH^-(10^{-7}M)$ or $2\,H^+(10^{-7}\,M) + 2e = H_2$ | −0.414 | Cathodic reduction of water at pH 7 (→) |
| $O_2 + 2\,H_2O + 4e = 4\,OH^-$ | 0.401 | $O_2$ reduction at pH 14 (→) |
| $O_2 + 4\,H^+(10^{-7}\,M) + 4e = 2\,H_2O$ | 0.815 | $O_2$ reduction at pH 7 (→) |
| $O_2 + 4\,H^+ + 4e = 2\,H_2O$ | 1.229 | $O_2$ reduction at pH 0 (→) |
| *Oxidations related to disinfection and odor control* | | |
| $O_3 + 2\,H^+ + 2e = O_2 + H_2O$ | 2.07 | Most powerful; (→) 1.68 at pH 7 |
| $H_2O_2 + 2\,H^+ + 2e = 2\,H_2O$ | 1.77 | 1.26 at pH 7 (→) |
| $MnO_4^- + 4\,H^+ + 3e = MnO_2 + 2\,H_2O$ | 1.68 | Odor oxidation (→); 1.13 at pH 7 |
| $HOCl + H^+ + 2e = Cl^- + H_2O$ | 1.49 | 1.28 at pH 7 (→) |
| $Cl_2 + 2e = 2\,Cl^-$ | 1.358 | $Cl_2 > Br_2 > I_2$ (→) |
| $ClO_2 + e = ClO_2^-$ | 1.15 | $ClO_2 <$ HOCl (→) |
| $Br_2 + 2e = 2\,Br^-$ | 1.087 | |
| $OCl^- + H_2O + 2e = Cl^- + 2\,OH^-$ | 0.88 | $OCl^-$ oxidation at pH 14 (→) |
| *Dechlorination, deoxygenation, inorganic biochemical metabolism* | | |
| $SO_4^- + 4\,H^+ + 2e = H_2SO_3 + H_2O$ | 0.172 | Dechlorination, deoxygenation (←) |
| $SO_4^- + 8\,H^+ + 6e = S + 4\,H_2O$ | 0.357 | Bacterial reduction (→); −0.21 at pH 7 |
| $S + 2\,H^+ + 2e = H_2S$ | 0.142 | Sulfur bacteria (→); −0.33 at pH 7 |
| $HNO_2 + 7\,H^+ + 6e = NH_4^+ + 2\,H_2O$ | 0.86 | Nitrification (←); 0.38 at pH 7 |
| $NO_3^- + H_2O + 2e = NO_2^- + 2\,OH^-$ | 0.01 | Nitrification (←); 0.42 at pH 7 |

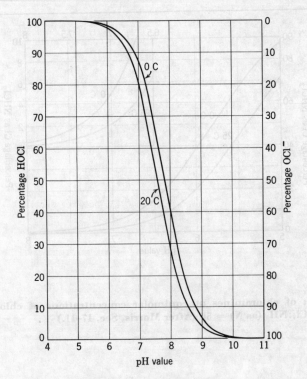

**Figure 7.**
**Distribution of hypochlorous acid and hypochlorite ion in water at different pH values and temperature. (After Morris, Sec. 17-10.)**

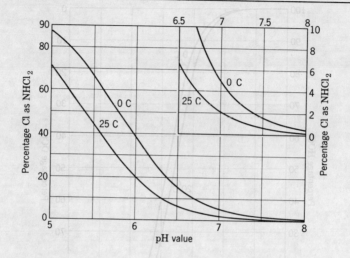

**Figure 8.**

Distribution of chloramines at equimolar concentrations of chlorine and ammonia. [$Cl_2:NH_3$ (as N) = 5]. (After Morris, Sec. 17-11.)

## Table 22

*Water and Wastewater Treatment Chemicals*

Information compiled from table prepared by B.I.F. Industries.

---

1. *Activated carbon*, C; Aqua Nuchar, Hydrodarco, Norit; available as powder and granules in 35-lb bags, 5- and 25-lb drums, and carloads; dusty, arches in hoppers; weighs 8 to 28 lb/cu ft, averaging 12; fed dry or as slurry of 1 lb/gal maximum; handled dry in iron or steel, wet in stainless steel, rubber, Duriron, and bronze.

2. *Activated silica*, —$SiO_2$—; Silica sol, produced at site from sodium silicate, $Na_2SiO_3$, and alum, ammonium sulfate, chlorine, carbon dioxide, sodium bicarbonate, or sulfuric acid; 41° Bé sodium silicate is diluted to 1.5% $SiO_2$ before activation; insoluble, will gel at high concentrations; fed as 0.6% solution to prevent gel formation; handled in iron, steel, rubber, and stainless steel.

3. *Aluminum sulfate*, $Al_2(SO_4)_3 \cdot 14\ H_2O$; alum, filter alum, sulfate of alumina; available in ground, rice, powder, and lump form in 100- and 200-lb bags, 325- and 400-lb barrels, 25-, 100-, and 250-lb drums, and carloads; dusty, astringent, only slightly hygroscopic; weighs 60 to 75 lb/cu ft; should contain at least 17% $Al_2O_3$; fed dry in ground and rice form; maximum concentration 0.5 lb/gal; handled dry in iron, steel, and concrete, wet in lead, rubber, Duriron, asphalt, cypress, and stainless steel 316; available also as 50% solution.

4. *Anhydrous ammonia*, $NH_3$; available in steel cylinders containing 50-, 100-, and 150-lb, also in 50,000-lb tank cars; pungent, irritating odor, liquid causes burns; 99 to 100% $NH_3$; fed as dry gas and aqueous solution through gas feeder or ammoniator; handled in iron, steel, glass, nickel, and Monel.

5. *Calcium oxide*, $CaO$; quicklime, burnt lime, chemical lime, unslaked lime; available as lumps, pebbles, crushed or ground, in 100-lb, moisture-proof bags, wooden barrels, and carloads; unstable, caustic, and irritating; slakes to calcium hydroxide with evolution of heat when water is added; weighs 55 to 70 lb/cu ft; should contain 70 to 90% $CaO$; best fed dry as ¾-in. pebbles or crushed to pass 1-in. ring; requires from 0.4 to 0.7 gal of water for continuous solution; final dilution should be 10%; should not be stored for more than 60 days even in tight container; handled wet in iron, steel, rubber hose, and concrete; clogs pipes.

6. *Calcium hydroxide*, $Ca(OH)_2$; hydrated lime, slaked lime; available as powder in 50-lb bags, 100-lb barrels, and carloads; must be stored in dry place; caustic, dusty, and irritant; weighs 35 to 50 lb/cu ft; should contain 62 to 74% $CaO$: fed dry, 0.5 lb/gal maximum, and as slurry 0.93 lb/gal maximum; handled in rubber hose, iron, steel, asphalt, and concrete; clogs pipes.

7. *Calcium hypochlorite*, $Ca(OCl)_2 \cdot 4 H_2O$; HTH, Perchloron, Pittchlor; available as powder, granules, and pellets in 115-lb barrels, 5-, 15-, 100-, and 300-lb cans, and 800-lb drums; corrosive and odorous; must be stored dry; weighs 50 to 55 lb/cu ft; should contain 70% available $Cl_2$; fed as solution up to 2% strength (0.25 lb/gal); handled in ceramics, glass, plastics, and rubber-lined tanks.

8. *Chlorine*, $Cl_2$; chlorine gas, liquid chlorine; available as liquified gas under pressure in 100- and 150-lb steel cylinders, ton containers, cars with 15-ton containers, and tank cars of 16-, 30-, and 55-ton capacity; corrosive, poisonous gas; should contain 99.8% $Cl_2$; fed as gas vaporized from liquid and as aqueous solution through gas feeder or chlorinator; dry liquid or gas handled in black iron, copper, and steel, wet gas in glass, silver, hard rubber, and tantalum.

9. *Copper sulfate*, $CuSO_4 \cdot 5 H_2O$; available ground and as powder or lumps in 100-lb bags and 450-lb barrels or drums; poisonous; weighs 75 to 90 lb/cu ft ground, 73 to 80 lb/cu ft as powder, and 60 to 64 lb/cu ft as lumps; should be 99% pure; best fed ground and as powder; maximum concentration 0.25 lb/gal; handled in stainless steel, asphalt, Duriron, rubber, plastics, and ceramics.

10. *Ferric chloride*, $FeCl_3$ (anhydrous and as solution), $FeCl_3 \cdot 6 H_2O$ (crystal); chloride of iron, ferrichlor; available as solution, lumps, and granules in 5- and 13-gal carboys and in tank trucks; hygroscopic; solution weighs 11.2 to 12.4 lb/gal, crystals 60 to 64 lb/cu ft, anhydrous chemical 85 to 90 lb/cu ft; solution should contain 35 to 45%, crystals 60%, and anhydrous chemical 96 to 97% $FeCl_3$; fed as solution containing up to 45% $FeCl_3$; handled in rubber, glass, ceramics, and plastics.

11. *Ferric sulfate*, $Fe_2(SO_4)_3 \cdot 3 H_2O$ and $Fe_2(SO_4)_3 \cdot 2 H_2O$; Ferrifloc, Ferriclear, iron sulfate; available as granules in 100-lb bags, 400- and 425-lb drums, and carloads; hygroscopic, must be stored in tight containers; weighs 70 to 72 lb/cu ft;

$Fe_2(SO_4)_3 \cdot 3\ H_2O$ should contain 18.5% Fe, $Fe_2(SO_4)_3 \cdot 2\ H_2O$ should contain 21% Fe; best fed dry, 1.4 to 2.4 lb/gal, detention time 20 min; handled in stainless steel 316, rubber, lead, ceramics, and Duriron.

12. *Ferrous sulfate*, $FeSO_4 \cdot 7\ H_2O$; copperas, iron sulfate, sugar sulfate, green vitriol; available as granules, crystals, powder, and lumps in 100-lb bags, 400-lb barrels, and bulk; weighs 63 to 66 lb/cu ft; should contain 20% Fe; best fed as dry granules, 0.5 lb/gal, detention time 5 min; handled dry in iron, steel, and concrete, wet in lead, rubber, iron, asphalt, cypress, and stainless steel.

13. *Sodium carbonate*, $Na_2CO_3$; soda ash; available as crystals and powder in 100-lb bags, 100-lb barrels; 25- and 100-lb drums, and carloads; weighs 30 to 65 lb/cu ft extra light to dense; should contain 58% $Na_2O$; best fed as dense crystals 0.25 lb/gal, detention time 10 min, more for higher concentrations; handled in iron, steel, and rubber hose.

14. *Sodium chloride*, NaCl; common salt, salt; available as rock, powder, crystals, and granules in 100-lb bags, barrels, 25-lb drums, and carloads; rock weighs 50 to 60 lb/cu ft, fine weighs 58 to 70 lb/cu ft; should contain 98% NaCl; fed as saturated brine; handled in galvanized iron, rubber, and Monel.

15. *Sodium fluoride*, NaF; fluoride; available as granules (crystals) and powder in 100-lb bags, 25-, 125-, 375-lb drums; powder weighs 66 to 100 lb/cu ft, granules weigh 90 to 106 lb/cu ft; should contain 43 to 44% F; best fed as granules 1 lb to 12 gal; handled dry in iron and steel, wet in rubber, plastics, stainless steel, asphalt, and cypress.

16. *Sodium silicate*, $Na_2O(SiO_2)_{3.25\ approx}$; water glass; available as liquid in 1-, 5-, and 55-gal drums, in ton trucks and tank cars; weighs about 11.7 lb/gal; should contain about 9% $Na_2O$ and 29% $SiO_2$; fed in solution as received; handled in cast iron, steel, and rubber.

17. *Sulfur dioxide*, $SO_2$; available as liquefied gas under pressure in 100-, 150-, and 200-lb steel cylinders; 100% $SO_2$; fed as gas, handled dry in steel, wet in glass, rubber, and ceramics.

# bibliography

**REFERENCE WORKS**

**Water Resource Management**

American Chemical Society, *Cleaning our environment. The chemical basis for action*, Washington, D.C., 1969.

American Society of Civil Engineers, *Manual on groundwater basin management*, Manual of Engineering Practice No. 40, New York, 1961.

American Society of Civil Engineers, *Water and metropolitan man* (Report on the Second Conference on Urban Water Resources Research), New York, 1969.

Blake, N. M., *Water for the cities*, Syracuse University, Syracuse, N.Y., 1956.

Camp, T. R., *Water and its impurities*, Reinhold, New York, 1963.

Cleary, Edward J., *The Orsanco story: water quality management in the Ohio Valley under an interstate compact*, Resources for the Future, Washington, D.C., Johns Hopkins Press, Baltimore, Md., 1967.

Craine, Lyle Eggleston, *Water management innovations in England*, Resources for the Future, Washington, D.C.; distributed by Johns Hopkins Press, Baltimore, 1969.

Daley, Robert, *The world beneath the city*, J. B. Lippincott, Philadelphia, 1959.

Dieterich, B. H., and J. M. Henderson, *Urban water supply conditions and needs in seventy-five developing countries*, World Health Organization, Geneva, Switzerland, Public Health Papers No. 23, 1963.

Carson, R. L., *The sea around us*, Oxford University Press, New York, 1951.

Dorsey, N. E., *Properties of ordinary water substance in all its phases: water-vapor, water, and all ices*, Reinhold, New York, 1940.

Dunno, E. S., *Sir John Harrington's a new discourse of a stale subject, called the metamorphosis of Ajax*, Routledge & Kegan Paul, London, 1962.

Durfor, C. N., and Edith Becker, *Public water supplies of the hundred largest cities in the United States, 1962*, U.S. Geological Survey, Water Supply Paper 1812, 1964.

Eisenberg, D., and Kauzmann, *The structure and properties of water*, Clarendon, Oxford, 1969.

Federal Water Pollution Control Administration, *Water quality criteria*, Government Printing Office, washington, D.C., 1968.

Furon, Raymond, *The problem of water: a world study*, American Elsevier, New York, 1967.

Hirshleifer, J. J., J. C. De Haven, and J. W. Milliman, *Water supply—economics, technology, and policy*, University of Chicago Press, Chicago, 1960.

Hirshleifer, Jack, *Water supply. Economic, technology and policy* (with recent postscript on New York and Southern California), U. of Chicago Press, 1969.

Kneese, Allen V., and Bower, Blair T., *Managing water quality: economics, technology, insitutions*, Resources for the Future, Inc.; distributed by Johns Hopkins Press, Baltimore, 1968.

Kneese, A. V., *The economies of regional water quality management*, Johns Hopkins University Press, Baltimore, 1964.

Linsley, R. K., and J. B. Franzini, *Water resource engineering*, McGraw-Hill, New York, 1964.

Maass, Arthur, M. M. Hufschmidt, Robert Dorfman, H. A. Thomas, Jr., S. A. Marglin, and G. M. Fair, *Design of water resource systems*, Harvard University Press, Cambridge, 1962. Mass.,

McGauhey, P. H., *Engineering meanagement of water quality*, McGraw-Hill, 1968.

McJunkin, Frederick, E., *Community water supply in developing countries*, Agency for International Development, United States Department of State, Washington, D.C., 1970.

McKee, J. E., and H. W. Wolf, *Water quality criteria*, State Water Quality Control Broad, Resources Agency of California, Sacramento, 1963.

Office of Saline Water, United States Department of the Interior, *1969–70 Saline water conversion report*, U.S. Government Printing Office, Washington, D.C., 1969.

Reynolds, Reginald, *Cleanliness and godliness*, Doubleday and Co., Garden City, New York, 1946.

Rouse, Hunter, and Simon Ince, *History of hydraulics*, Iowa Institute of Hydraulic Research, State University of Iowa, Ames, 1957.

United States Resource Council, *The nation's water resources, summary report*, Washington, D.C., 1968.

University of Michigan Law School, *Water resources and the law*, Ann Arbor, Mich., 1958.

Whipple, G. C., *State sanitation*, 2 vols. Harvard University Press, Cambridge, Mass., 1917.

White, Gilbert F., *Strategies of American water management*, U. of Michigan Press, Ann Arbor, Michigan, 1969.

Williams, Virginia R., and Hulun Williams, *Basic physical chemistry for the life sciences*, U. H. Freeman, San Francisco, 1967.

World Health Organization, *International standards for drinking water*, 2nd ed., Geneva, Switzerland, 1963.

Wright, Lawrence, *Clean and decent*, Viking, New York, 1962.

Zimmerman, Josef D., *Irrigation*, Wiley, New York, 1966.

## Water and Health

Berg, Gerald, Ed., *Transmission of viruses by the water route*, Wiley, New York, 1967.

Environmental Control Administration, United States Department of Health, Education and Welfare, *Public Health Service drinking water standards—1962*, Public Health Service Publication No. 956, United States Government Printing Office, Washington, D.C., 1969.

Faber, H. A., and L. J. Bryson, Eds., *Proceedings, conference on physiological aspects of water quality*, U.S. Public Health Service, Washington, D.C., 1960.

Longmate, Norman, *King cholera: the biography of a disease*, Hamish Hamilton, London, 1966.

McCartney, William, *Olfaction and odours*, Spring, Berlin, 1968.
Miller, A. P., *Water and man's health*, Agency for International Development, Washington, D.C., 1962.
Pollitzer, R., *Cholera*, WHO Monogram #43, World Health Organization, Geneva, Switzerland, 1959.
Sartwell, P. E., Maxcy-Rosenau, *Preventive medicine and public health*, 9th ed., Appleton-Century-Crofts, New York, 1965.
World Health Organization, *Endemic Goitre*, Geneva, Switzerland, 1960.
U.S. National Bureau of Standards, *Handbooks of radioactivity control*, Government Printing Office, Washington, D.C., 1949.

### Water and Wastewater Engineering

American Society for Testing Materials, *Manual on industrial water*, 2nd ed., Philadelphia, 1960.
American Society of Civil Engineers and Water Pollution Control Federation, *Design and construction of sanitary and storm sewers*, Manual of Engineering Practice (ASCE) No. 37, New York, Manual of Practice (WPCF) No. 9, Washington, D.C., 1969.
Babbitt, H. E., J. J. Doland, and J. L. Cleasby, *Water supply engineering*, 6th ed., McGraw-Hill, New York, 1962.
Bosick, Joseph F., *Corrosion prevention for practicing engineers*, Barnes and Noble, New York, 1970.
Davis, C. V., and Sorenson, K. E., Eds., *Handbook of applied hydraulics*, 3rd. ed., McGraw-Hill, New York, 1969, Secs., 35–37 by T. R. Camp and J. C. Lawler; Sec. 38 by T. R. Camp; Sec. 40 by S. A. Greeley, W. E. Stanley, and Donald Newton; and Sec. 41 by S. A. Greeley, W. E. Stanley, and K. V. Hill.
Fair, G. M., J. C. Geyer, and D. A. Okun, *Water and wastewater engineering*, Vol. 1, *Water supply and wastewater removal* and Vol. 2, *Water purification and wastewater treatment*, Wiley, New York, 1966 and 1968 respectively.
Hardenbergh, W. A., and E. R. Rodie, *Water supply and waste disposal*, International Textbook, Scranton, Pa., 1961.
International Union of Pure and Applied Chemistry, *Re-use of water in industry*, Butterworths, London, 1963.
Isaac, P. C. G., *Public health engineering*, Spon, London, 1953.
Krenkel, Peter A., and Frank L. Parker, *Biological aspects of thermal pollution* (proceedings of the National Symposia on Thermal Pollution, June 3–5, 1968) Vanderbilt University Press, Nashville, Tennessee, 1968.
Myers, John J., Ed., *Handbook of ocean and underwater engineering*, McGraw-Hill, New York, 1970.
McGuiness, W. J., Benjamin Skin, C. M. Gay, and Charles De van Fawcett, *Mechanical and electrical equipment for buildings*, 4th ed., Wiley, New York, 1964.
National Academy of Sciences—National Research Council, *Waste management & control*, Washington, D.C., 1966.
Parker, Frank L., and Peter A. Krenkel, *Engineering aspects of thermal pollution* (proceedings of the National Symposia on Thermal Pollution) Vanderbilt University Press, Nashville, Tennessee, 1969.
Pearson, E. A., Ed., *Waste disposal in the marine environment*, Pergamon, New York, 1960.

Riehl, M. L., *Hoover's water supply and treatment*, 9th ed., National Lime Association, Washington, D.C., 1962.

Salvato, J. A., Jr., *Environmental sanitation*, 2nd ed., Wiley, New York, 1969.

Skeat, W. O., Ed., *Manual of British water engineering practice*, 4th ed., Institution of Water Engineers, Heffer, Cambridge, England, 1969.

Steel, E. W., *Water supply and sewerage*, 4th ed., McGraw-Hill, New York, 1960.

Wagner, E. G., and J. N. Lanoix, *Water supply for rural areas and small communities* and *Excreta disposal in rural areas and small communities*, World Health Organization, Geneva, Switzerland, 1959.

### Water and Wastewater Treatment

Besselievre, E. G., *Industrial waste treatment*, 2nd ed., McGraw-Hill, New York, 1969.

Degremont, *Water treatment handbook*, 3rd ed., H. K. Elliott, London, 1965.

Eckenfelder, W. W., Jr., *Industrial water pollution control*, McGraw-Hill, New York, 1966.

Eckenfelder, W. W., Jr., and D. J. O'Connor, *Biological waste treatment*, Pergamon, New York, 1961.

Gloyna, Ernest F., Ed., *Advances in water quality improvement* (Water Resources Symposium No. I) University of Texas Free Press, Austin, Texas, 1968.

Gurnham, C. F., Ed., *Industrial wastewater control*, Vol. 2, Academic Press, New York, 1965.

Hardenbergh, William A., and Edward R. Rodie, *Water supply and waste disposal*, International Textbook Co., Scranton, Pa., 1970.

Imhoff, Carl, and Gordon Maskew Fair, *Sewage treatment*, 2nd edition, John Wiley & Sons, New York, 1956.

Isaac, P. C. G., Ed., *Proceedings, symposium on the treatment of waste waters*, Pergamon, New York, 1960.

Maier, F. J., *Manual of water fluoridation practice*, McGraw-Hill, New York, 1963.

Nemerow, N. L., *Theories and practices of industrial waste treatment*, Addison-Wesley, Reading, Mass., 1963.

Nordell, Eskel, *Water treatment for industrial and other uses*, Reinhold, New York, 1961.

Poole, J. D., and D. Doyle, *Solid-liquid separation*, H. M. Stationery Office, London, 1966.

Rich, L. G., *Unit operations of sanitary engineering*, Wiley, New York, 1961.

Rich, L. G., *Unit processes of sanitary engineering*, Wiley, New York, 1963.

Spiegler, K. S., *Principles of desalination*, Academic Press, New York, 1966.

Water Pollution Control Federation, *Anaerobic sludge digestion*, Manual of Practice, No. 16, Washington, D.C., 1968.

Water Pollution Control Federation, *Sewage treatment plant design*, Manual of Practice, No. 8, American Society of Civil Engineers, Manuals of engineering practice, No. 36, Washington, D.C., 1959.

### Hydrology and Hydraulics

American Society of Civil Engineers, *Nomenclature for hydraulics*, Manual and Report on Engineering Practice, No. 43, New York, 1962.

Bureau of Reclamation, *Design of small dams*, Government Printing Office, Washington, D.C., 1960.

Chow, Ven T., Ed., *Handbook of applied hydrology*, McGraw-Hill, New York, 1964.

De Wiest, R. J. M., *Geohydrology*, Wiley, New York, 1965.
Davis, S. N., and R. J. M. De Wiest, *Hydrogeology*, Wiley, New York, 1966.
Fiering, M. B., *Streamflow synthesis*, Harvard University Press, Cambridge, Mass., 1967.
Hufschmidt, M. M., and M. B. Fiering, *Simulation techniques for design of water resource systems*, Harvard University Press, Cambridge, Mass., 1966.
Karassik, Igor, and Roy Carter, *Centrifugal pumps*, F. W. Dodge, New York, 1960.
Langbein, W. B., *Salinity and hydrology of closed lakes*, Government Printing Office, Washington, D.C., 1961.

### Examination of Water and Wastewater

American Public Health Association, *Standard methods for the examination of water and wastewater including bottom sediments and sludges*, 12th ed., New York, 1965.
American Society for Testing and Materials, *Manual on water*, 3rd ed., Philadelphia, Pennsylvania, 1969.
Federal Water Pollution Control Administration, *Water quality criteria*, Government Printing Office, Washington, D.C., 1968.
*Groundwater and wells*, Edward E. Johnson, Inc., St. Paul, Minnesota, 1966.
McCoy, James W., *Chemical analysis of industrial water*, Chemical Pub. Co., New York, 1969.
Taylor, E. W., Thresh, *The examination of waters and water supplies*, 7th ed., Little Brown, Boston, 1958.
U.S. National Bureau of Standards, *Handbooks of radioactivity control*, Government Printing Office, Washington, D.C., 1949.
Water Pollution Control Federation, *Simplified laboratory procedures for wastewater examination*, Manual of Practice, No. 18, Washington, D.C., 1970.

### Chemistry

Coulson, J. M., and J. F. Richardson, *Chemical engineering*, 2nd ed., Pergamon Press, Oxford, 1968.
Foust, A. S., L. A. Wenzel, C. W. Clumb, L. Maus, and L. B. Andersen, *Principles of unit operations*, Wiley, New York, 1960.
Kunin, Robert, *Elements of ion exchange*, Reinhold, New York, 1960.
Marshall, C. E., *The physical chemistry and mineralogy of soils*, Wiley, New York, 1964.
Millar, E. C., L. M. Turk, and H. D. Froth, *Fundamentals of soil science*, 4th ed., Wiley, New York, 1965.
Sawyer, C. N. and McCarty, P. L., *Chemistry for sanitary engineers*, McGraw-Hill, New York 2nd ed., 1967.
Shreir, L. L., Ed., *Corrosion*, 2 vols., Wiley, New York, 1963.
Uhlig, H. H., *Corrosion and corrosion control*, Wiley, New York, 1963.
van Olphen, H., *An introduction to clay colloid chemistry*, Interscience, New York, 1963.
Weast, Robert C., Ed., *Handbook of chemistry and physics*, Chemical Rubber Co., Cleveland, latest edition.

### Biology

Alexander, Martin, *Introduction to soil microbiology*, Wiley, New York, 1961.
Brock, T. D., *Biology of micro-organisms*, Prentice-Hall, Englewood Cliffs, New Jersey, 1970.

Burges, Alan, *Micro-organisms in the soil*, Hutchinson University Library, London, 1964.

Casey, E. J., *Biophysics, concepts and mechanisms*, Reinhold, New York, 1962.

Chichester, E. O., Ed., *Research in pesticides, Conference on research needs and approaches to the use of agricultural chemicals from a public health point of view*, Academic Press, New York, 1965.

Dean, A. C. R., and Cyril Hinshelwood, *Growth, function and regulation in bacterial cells*, Clarendon, Oxford, 1960.

Gunsalus, I. C., and Stanier, R. Y., Eds., *The bacteria*, Academic Press, New York, 5 vols., 1960–1964.

Jackson, Daniel J., Ed., *Algae, man and the environment* (proceedings of a symposium), Syracuse University Press, Syracuse, New York, 1968.

Lamanna, Carl, and M. Frank, *Basic bacteriology, its biological and chemical background*, Williams & Wilkins, Baltimore, Md., 1965.

Mascoli, C. C., and R. G. Burrell, *Experimental virology*, Burgess, Minneapolis, 1965.

McElroy, W. D., *Cellular physiology and biochemistry*, Prentice-Hall, Englewood Cliffs, N.J., 1961.

Setlow, R. B., and E. C. Pollard, *Molecular biophysics*, Addison-Wesley, Reading, Mass., 1962.

Sistrom, W. R., *Microbial life*, Holt, Rinehart and Winston, New York, 1962.

Smith, C. A. B., *Biomathematics; the principles of mathematics for students of biological and general science*, 4th ed., Hafner, New York, 1966.

Society of American Bacteriologists, *Bergey's manual of determinative bacteriology*, 7th ed., Williams & Wilkins, Baltimore, 1957.

Stanier, R. Y., Michael Duodoroff, and E. A. Adelberg, *The microbial world*, Prentice-Hall, Englewood Cliffs, N.J., 3rd ed., 1970.

Thimann, K. V., *The life of bacteria: their growth, metabolisms, and relationships*, Macmillan, New York, 1965.

Umbreit, W. W., *Modern microbiology*, Freeman, San Francisco, 1962.

United States Department of Health, Education and Welfare, *Report of the Secretary's Commission on pesticides and their relationship to environmental health*, Government Printing Office, Washington, D.C., 1969.

Zajic, James E., *Properties and products of algae*, Plenum Press, New York, 1970.

**Ecology and Limnology**

Brock, T. P., *Principles of microbial ecology*, Prentice-Hall, Englewood Cliffs, N.J., 1966.

Edmondson, W. T., Ed., Ward and Whipple's *Fresh water biology*, 2nd ed., Wiley, New York, 1959.

Hawkes, H. A., *The ecology of wastewater treatment*, Macmillan, New York, 1963.

Heukelekian, H., and N. C. Dondero, Eds., *Principles and applications in aquatic microbiology*, Wiley, New York, 1954.

Hutchinson, G. E., *A treatise on limnology:* Vol. 1, *Geography, physics, and chemistry;* Vol. 2, *Introduction to lake biology and the limnoplankton*, Wiley, New York, 1957 and 1966 respectively.

Hynes, H. B. N., *The biology of polluted waters*, Liverpool University, England, 1960.

International Symposium on Eutrophication, *Eutrophication: causes, consequences, correctives; proceedings of a symposium*, National Academy of Sciences, Washington, D.C., 1969.
Jackson, D. F., Ed., *Algae and man*, Plenum Press, New York, 1964.
Klein, Louis, *River pollution*, 3 vols., Butterworths, London, 1959–1966.
Lowe-McConnell, R. H., Ed., *Man-made lakes*, Academic Press, London, 1966.
MacKenthun, Kenneth M., *The practice of water pollution biology*, Federal Water Pollution Control Administration, United States Department of the Interior, Washington, D.C., 1969.
McKinney, R. E., *Microbiology for sanitary engineers*, McGraw-Hill, New York, 1962.
Odum, E. P., *Ecology*, Holt, Rinehart and Winston, New York, 1963.
Odum, E. P., and H. T. Odum, *Fundamentals of ecology*, 2nd ed., Saunders, Philadelphia, 1959.
Phelps, E. B., *Stream sanitation*, Wiley, New York, 1944.
Reid, G. K., *Ecology of inland waters and estuaries*, Reinhold, New York, 1961.
Ruttner, Franz, *Fundamentals of limnology*, University of Toronto, 1963.
U.S. Public Health Service, *Limnological aspects of recreational lakes*, Government Printing Office, Washington, D.C., 1964.
Wilber, Charles G., *The biological aspects of water pollution*, Charles C Thomas, Springfield, Illinois, 1969.
Wood, E. J. F., *Marine microbial ecology*, Reinhold, New York, 1965.

## SERIAL PUBLICATIONS

### Professional Societies

American Geophysical Union, *Transactions*, quarterly; *Water Resources Research*, quarterly; and *Journal of Geophysical Research*, monthly.
American Society of Civil Engineers, *Transactions*, annual; *Proceedings*, bimonthly, printed as Journals of the Divisions of the Society, for example, *Sanitary Engineering, Hydraulics, Pipeline, Waterways and Harbors*, formerly monthly and as separates; *Civil Engineering*, monthly; and *Manuals of Engineering Practice*.
American Water Works Association, *Journal*, monthly; *Manuals; and Standards*.
American Water Resources Association, *Hydata, International Review of Periodical Contents*, monthly; and *Journal*, monthly.
British Waterworks Association, London, *Journal*, several times a year.
Institute of Water Pollution Control, London, *Journal*, alternate months.
Institution of Civil Engineers, London, *Proceedings*, monthly.
Institution of Public Health Engineers, London, *Journal*, several times a year.
Institution of Water Engineers, London, *Journal*, 6 to 8 times a year.
International Water Pollution Research Association, *Proceedings* (biennial), 1962 (London), 1964 (Tokyo), and 1969 (Prague), Pergamon, New York, 1966 (Munich), Water Pollution Control Federation, Washington, D.C.
New England Water Works Association, *Journal*, quarterly.
Water Pollution Control Federation, *Journal* (formerly *Sewage and Industrial Wastes* and, before that, *Sewage Works Journal*), monthly; and *Manuals of Practice*.

## Governmental Agencies

U.S. Department of the Interior, Geological Survey, *Water Supply Papers: Office of Saline Water*, series.

Water Quality Office, Environmental Protection Agency, Washington, D.C., *Summary Reports, AdvancedWaste Treatment Program*, irregularly issued.

Office of Water Resources Research, United States Department of the Interior, Washington, D.C., *Selected Water Resource Abstracts*, semimonthly.

California State Water Quality Control Board, Sacramento, series.

Illinois State Water Survey, Champaign-Urbana, series.

New York State Department of Health, Albany, research series.

H. M. Stationery Office, London, *Water Pollution Research*, annual reports, and *Water Pollution Abstracts*, monthly.

## University Conferences

Purdue University, Lafayette, Ind., *Industrial Waste Conference*.

University of Kansas, Lawrence, Kans., *Sanitary Engineering Conference*.

University of North Carolina, Duke University, and North Carolina State University, *Southern Water Resources and Pollution Control Conference*.

Vanderbilt University, Nashville, Tenn., *Sanitary and Water Resources Engineering Conference*.

## Magazines

Buttenheim Publishing Corp., Pittsfield, Mass. *The American City*, monthly.

Case-Sheppard-Mann Publishing Co., New York, *Wastes Engineering*, monthly; and *Water Works Engineering*, monthly.

McGraw-Hill Publishing Co., New York, *Engineering News-Record*, weekly.

Public Works Journal Corp., Ridgewood, N. J., *Public Works*, monthly.

Reuben H. Donnelly Corp., Lancaster, Pa., *Water and Wastes Engineering*, monthly.

Scranton Publishing Co., Chicago, *Water and Sewage Works*, monthly.

Select Publications, New York, *Industrial Water Engineering*, monthly.

# Index

Abbreviations and symbols, table, 688
Absorption, of oxygen, rate of, 638
Absorption coefficients, of common gases, in water, table, 710
Acclimation of organisms, 537
Activated carbon, 445
  beds, 448
  blackout dosage, 672
  control by algae, 672
  phenol value, 446
  process technology, 448
Activated sludge, 677
  bulked, 532
  ecology of, 533
  permissable loading of, 578
  by volume, percentage concentration of, 576
Activated-sludge floc, 569
Activated-sludge process, 530, 556
  air requirements, 578
  design, 575
  floc control, 530
  systems, 551
  treatment methods, 571
Activated-sludge units, 527, 533, 546, 554, 569
  contact time, 543
  design of, 570
  hydraulic design of, 578
  mechanical design of, 578
  performance-loading relationship for, 555
  performance-time relationship, 540
Activation, 597
  energy, 645
Adams, Julius W., 14

Adeney, W. E., 318
Adenovirus, 285
  disinfection of, 507
  resistance to disinfection, 505
Adsorbates, 313, 336
Adsorbents, 313, 336, 444
Adsorption, 313
  equilibria, 336
  isotherms, 337
  kinetics, 336
  of odors and tastes, 446
  rates, 445
  selectivity of, 444
  solid-liquid, 443
  transfer mechanisms, 444
Adsorptive, capacity, 445
  properties, colloids, 465
Advanced waste treatment, 322
Aerated-contact beds, 560
Aeration, extended, 575
  for gas exchange, 344
  mechanical, 669
  step, 574
  tank design, 357
Aerators, 351
  draft tube, 581
  mechanical, 581
*Aerobacter aerogenes*, 642
Aerobic, 533
  bonds, 587
  decomposition, kinetics of, 643
  stabilization ponds, ecology of, 533
Agricultural, value of sludge, 633
  runoff, 664
Air, compression, 582

727

diffusers, 353, 583
drying of sledges, 317, 617
lifts, 184
piping, 582
requirements, 546
scouring, rate of, 426
supply for diffused-air units, 581
use, 550
Albertson, D. E., 607n
Algae, 288, 664
chemical destruction of, 672
control, 316, 670
control, using chlorine, 522
green and blue green, 533
Algal genera, 670
growths, 532
Algicides, 673
Alkalinity, total, 477
Allowable BOD loading of receiving streams, 655
Allowable pollutional load, 655
Alternate stages, 234–236
Aluminum, coagulation with, 468–471
Amebiasis, 284
Amebic cysts, 284, 435, 501
Amebic dysentary, 284
cysts of, 590
*The American City*, 684
American Concrete Pipe Association, 684
American Insurance Association (AIA), 34
fire fighting requirements, 195
fire reserve storage, 216
American Society of Civil Engineers, 561n, 684
American Standards Association, 684
American Water Works Association, 684
Ammonia, breakpoint reactions of, 517–518
chlorination of, 518
removal of, in trickling filters, 539
Ammonia nitrogen, removal of in activated sludge units, 539
*Amphi-aerobic* ponds, 587
*Anabaena*, 533
Anaerobic, contact, 586
decomposition, 649
digestion, 575, 586
floccalation, 586
ponds, 587, 591
sludge digestion, 618
Anchor ice: intakes, 109

Anderson, A. G., 243n
Anderson, J. J., 586n
Anderson, L. B., 311
Analog model of corrosion, 490–491
Anion exchangers, 455
Anisotropy, definition, 116
Antichlors, 519
Apportionment method, population, 24
Aquatic earthworms, 532
Aquatic growth, biological control, 677
Aquatic life, chamical destruction of, 676
Aqueducts, 163
Aquiclude, definition, 116
Aquifer, boundaries, 144
characteristics, 123
definition, 116
evaluation, 160
location, 146
types, 118
Aquifers, confined, 118, 123
nonsteady radial flow, 127
recharge of, 397
Aquitard, definition, 116
Archibald, R. S., 375
Ardern, Edward, 569
Areas under the normal probability curve, table, 694
Artesian aquifers, 118
Atmosphere reoxygenation of polluted waters, 649
Atomic numbers, weights, and valences of chemical elements, table, 708
Auerbach, V. H., 504n
Available chlorine, 520
Average-end-area method, 84
Averages, types, 44
Avogadro's hypothesis, 345, 346
Axial flow pumps, 183

Babbitt, H. E., 599n
Bacillary dysentery, 283
Backwater curve, 237
calculation, 239
Bacteria, 283, 501
autotrophic, 530, 534
beds, 527
fecal, die away of, 590
to flocculate, 538
heterotrophic, 534
methane forming, 587

## Index

pathogenic, 634
pigmented, 534
saprophytic, 534
Bacterial cultures, idealized growth curve of, 536
Bacterial infections, 283
Bacterial self-purification, 641
Bacteriological analysis units, 304
Baffle boards in settling tanks, 386
Baffles, 439
Baker, M. N., 11n
Bakhmeteff, B. A., 240n
Batch tests, 548
Bathing waters, 295
Baylis, J. R., 422n
Bays, L. R., 605n
Bazalgette, John, 14
Becker, Edith, 27n
Becker, H. G., 328, 348
Becker's observations, 651
Beggiatoa, 532, 535
Behn, V. C., 599n
Bench scale testing, 547
Benn, D., 603n
Benthal, oxygen demand, 654
 sediment examination, 303
 sludges, 587
Benton, G. S., 245n
Bentonite, 473
Berg, G., 642
Berg, Gerald, 285n, 507n
Bernoulli's theorem, 243
Bevan, J. G., 586n
Bioassay of toxic wastes, 306–307
Biochemical oxygen demand, 637
Bioflocculation, 528, 538, 569
 ions in, 538
 pH in, 538
 polyelectrolytes in, 538
 polymers in, 538
Biological, control of aquatic growths, 677
 flocculation, 641
 flotation of sludges, 317
 masses, chemical flocculation of, 585
 purification, 530
Biological treatment, 544
 associated organisms, 531
 fill and draw, 527
 history of, 525
 resting cycle, 527

systems, 550
Biological treatment units, 538, 549
 breaking in of, 542
 ripening period of, 542
Biomass, idealized growth curve, 536
Biomasses, chemical coagulation of, 585
 ecology of, 530
 essential minerals for, 535
 growth curve of, 535
 nutrients for, 534
Biomass requirements, 546
Black, H. P., 359n
Black, W. M., 333
Blackout dosage of activated carbon, 672
Blair, A. H., 153n
Blake, N. M., 12n
Blake-Kozeny equation, 408, 604
Blood flukes, 677
Bloodworms, 532, 677
Blooming, 670
Blue-green algae, nitrogen fixation, 665
BOD, 550, 639, 642, 643, 652
 allowable loading of receiving streams, 655
 converted, 547
 curve, first stage, 644
 oxidation, 546
 rate of polluted seawater, 646
 reaction, activation energy, 645
 reduction by chlorination, 523
 values, for municipal wastewaters, 307
 testing, 301
  municipal wastewater, 307
 units, 304
 unseeded samples, 645
BOD removal, 535, 539, 654
 in activated sludge units, 539
 in settling tanks, 390
 in trickling filters, 539
 from wastewaters, 322
Boil-water orders, 502
Boltzmann's constant, 345
Bomb calorimeter, 612
Borchardt, J. A., 644n
Borda loss, 168
Borda's loss function, 421
Bottom, deposits, 637
 deoxygination by, 647
 muds, 639
Bouches, P. L., 363n
Boylton, N. S., 134n

Boyle's law, 345
Bradley, J. N., 106n
Branches, 247, 249
Breakpoint reactions of ammonia, 517–518
Briggs, Margaret, 641n
Bristle worms, 532
Bromine as disinfectant, 503
Brooks, N. H., 377
Brown, R. H., 127n, 141
Buckingham, R. A., 651
Büchner funnel, 603, 605
Bulked activated sludge, 532
Bulked sludge, 576
Burd, R. S., 596n
Burdon, R. S., 394n
Bureau of the Census, U. S., 19, 24
Burr, Aaron, 12
Bushwell, A. M., 528
Butterfield, C. T., 508

Caldwell, D. H., 599n
California State Board of Health, 666
Camp, T. R., 214, 233, 297n, 335, 391, 427n, 666n
Cancer, 289
Capacity design in storm drainage, 271–274
Carbonate precipitation, 471–483
Carbon dioxide, water-quality surveys, 299
Carbon dioxide removal, 358–359
Cardiovascular disease, 288
Carey, C. L., 641n
Carman, P. C., 405, 605n
Cascades, 351
Cassidy, J. J., 245n
Cast Iron Pipe Research Association, 684
Cast-iron pipes, corrosion of, 493
Catch basins, 255, 256
Catches, 4
Catchment areas, 90
Catchment quality control, 91
Cation exchangers, 455
Cavitation, 185
Celerity of propagation, 242
Centrifugation, 607
Centrifuging of sludges, 317
Chadwick, Edwin, 13
Chambers, C. W., 508n
Chang, S. L., 642
Charles law, 345
Chelating reagents, 462

Chemical, algicides, management of, 673
  aquatic life, 676
  coagulation, 312, 597
  conditioning of sludges, 317, 608
  destruction, of algae, 672
    of water weeds, 676
  precipitation, 313, 368
  treatment wastes, disposal of, 498
Chemicals, handling, storing, and feeding of, 496
Chemosynthesis, 534
Chézy, Antoine, 166n
Chézy formula, 166, 225, 355
Chick's law, 328, 506, 641
Chironimus, 532
Chironomid larvae, 677
Chlamydomonas, 533, 534
Chloramines, 676
Chlorella, 533
Chlorination, marginal, 524
  operational, technology of, 520
  reduction of BOD, 523
  residuals, 520
Chlorinators, 520, 521
Chlorine, 670, 677
  available, 520
  breakpoint phenomenon, 517
  chemical technology of, 519
  colicidal efficiency, 514
  combined available, 516
  control of algae, 672
  demand, 512
  disinfection, 512
  equivalent, 520
  free available, 512, 513
  hydrolysis constant of, 513
  hydrolysis equation for, 513
  ice, 519
  ionization equation for, 513
  liquid, 676
  sludge stabilization, 523
  toxicity of, 519
  use as a disinfectant, 504
  uses of, 522
Chlorine Institute, 684
Chlorophyll, 534
Cholera, 283
Churchill, M. A., 651
Ciliates, 532
Cisterns, 4

# Index

Clark, N. A., 642
Clays, groundwater, 118
Clay Sewer Pipe Association, 684
Cleasby, J. L., 329n
Clevenger, C. H., 601n
Clevenger, W. A., 102n
Cloaca maxima, 2n
Clump, C. W., 311
COD, 550, 613, 639
   oxidation, 546
   removal from wastewaters, 322
   testing, 301
   values for municipal wastewaters, 308
Coagulation, 312, 467
   with aluminum, 468
   chemical, 597
   coagulant aids, 473
   of colloids, 467
   energy hill, 467
   with ferric iron, 468
   hydrolized Fe(III) and Al(III), 469
   jar tests, 473
   with polyelectrolytes, 471
   of solids, 608
   systems, 473
Coakley, Peter, 604n
Coe, H. S., 601n
Coefficient, of attenuation or extinction, 502
   of drag, 368
   of deposition, 655
   of molecular diffusion, 333
   of nonuniformity, 404
   of permeability, 121
   of retardance, 21
   of self-purification, 404
   of storage, 123
   of transmissivity, 123
   of variation, 47
Coefficients of diffusion of aqueous solutions into water, table, 709
Coetzee, O. J., 590n
Coincident draft, 194, 199
Colas, R., 41n
Coliform, organisms, 290
   die away of, 641
   test, 301
   test units, 304
Collecting systems, choice of, 260
   design information, 261

Collection works, definition, 4
Colloidal state, 464–468
Colloids, coagulation of, 467
   critical mass of, 473
   electrokinetic properties, 465
   hydrophilic, 465
   hydrophobic, 465
   stability of, 465
   surface properties of, 465
Color removal, 434
Color units, 304
Combined, sewer systems, 2, 6
   sewerage systems, 255
   sewers, 261, 264
      cleansing velocity, 255
   systems, 307
Combined collection of wastewaters and stormwaters, 255
Comminutors, 366–367
Common design values, 572
Common loadings, 560
Compressor efficiencies, 582
Computor programs, 686
Concentration, sludge, 601
Conditioning, chemical sludge, 608
Conjunction kinetics, 325, 334
Continuous draft, 69
Continuous operation of digesters, 628
Continuous culture techniques, 548
Continuous-flow tests, *see* Continuous-culture techniques
Contract, aerators, 532
   performance loading relationship for, 555
   submerged, 527
   anaerobic, 586
   beds, 525, 560
   aerated, 527
   freezing, 454
   interfacial, 528
   opportunity, 550
   measures of, 550
   dynamic, 556
   static, 556
   quality of, 528
   stabilization, 574
   suspended, 569
   time, activated sludge units, 543
Control, biological, of aquatic growth, 677
   of corrosion, 616
   of algae, 670

Cooling waters, 291
Cooper, H. H., 131
Copper sulfate, control of algae, 672
  solubility of, 676
  tolerance of fish to, 673
Copper sulfating of water, 316
Corrosion, 490
  control of, 494, 616
  electromotive forces, 491
  galvanic series, 493
  inhibitors, 496
  losses, 490
  protection against, 495
  reaction, 491
  tuberculation, 490
Corrosion-resistant materials, 494–495
Costs, peak flow, 685
Coxsackie, A and B viruses, 285
  virus, 642
    disinfection of, 506
    resistance to chlorination, 505
Cranston, R. I., 269n
Creep, saltwater, 666
Cribs, intakes, 107
Critical deficit, 655
Crocker, E. C., 289n
Cross-connections, 6
  protected, 193, 220
Cross correlation, 49
Crustacean, 284
Culex tarsalis, 590
Cunettes, 232, 256, 257
Cunningham, J. W., 363n
Cyclops, 284, 504, 677
Cysts, 435, 504

Dalton's law, 345, 346
Dams, 95
  berm, 96
  common dimensions, 97
  dimensions, 99
  diversions, 107
  embankment, 95
  riprap, 96, 102
Darcy, unit, 115
  friction formula, 146
Darcy's Law, 121–123, 408
Darcy-Weisbach formula, 164
Davis, S. N., 117n
Dechlorinating agents, 518

Dechlorination, 518–519
Decomposition, aerobic, kinetics of, 643
  anaerobic, 649
  gases, 648
  tests, 302
Degree day, 176
Demineralization, 449, 460
Dentrification, 535
Deoxygenation, 643
  by bottom deposits, 647
  of polluted waters, 645, 652
  rate of, 648
Depth of draft, 663
Dermatitis, cercarial, 284
Desalination of water, 317
Desalting, 449
  distillation, 450
  freezing, 453
  reverse osmosis, 451
  solvent extraction, 451
Design, flexibility, 687
  storage, 77
  storm, 274
Designation and size of openings of U.S.
    sieve series, table, 706
Desorption of gases, 345–347
Destruction, chemical, of algae, 672
  of water weeds and other aquatic life, 676
Detention tanks, 326
Detention time, 549
Dewatering, gravity, 603
Dewatering sludge, 602
DeWiest, R-J. M., 117n
Dewpoint, definition, 430
Dialysis, 462–463
Dialytic processes, 317
Diatomaceous earth, body feed, 435
  filters, 435
  precoating, 435
  septum, 435
Dichloramine formation rate, 517
Die-away, of coliforms, 641
  of enteric pathogens, 642
  of pathogens, 325, 500
Diffused-air units, air supply for, 581
Diffusers, air, 583
  permeability of, 583
Diffusion, kinetics, 325
  processes, 317
Digester gas, 620, 631

# Index

Digesters, 624
  capacity, 627
  continuous operation of, 628
  scum accumulation of, 626
  of sludge, 624
  stratification, 626
Digestibility of sludge, 619
Digestion, anaerobic, 586
  anaerobic of sludge, 618
  kinetics of, 618, 622
  seperate sludge units, 624, 626
Dikes, 95
Dillingham, J. H., 214n
Diluting waters, 669
Dilution, 8, 666
  rate, 549
  requirements, 658
Dimeric species, 468
Discharge, waste water, regulation of, 668
Dishpan hands, 288
Disinfectants, concentration of, 501
  properties of, 501
Disinfecting efficiency, 501
Disinfection, by alkalis and acids, 503
  of bathing waters, 522
  breakpoint phenomenon, 517
  by chemicals, 503
  by chlorine, 512
  concentration of, 506
  concentration of organisms, 507
  and disinfectants, 500–501
  by heat, 502
  kinetics of chemical, 505
  by light, 502
  management of variables in, 524
  by metal ions, 503
  by oxidizing chemicals, 503
  by ozone, 509
  by rate of kill, 506
  residual protection, 501
  of shellfish, 522
  by surface-active chemicals, 503
  surviving fraction, 506
  temperature of, 507
  theory of chemical, 504
  time of contact, 506
  by ultraviolet light, 502
  of water, 316
  of water, importance of, 500
  ways of, 502

Dispersal areas, loading of wastewaters, 666
Disposal, effluent, in tidal estuaries, 666
Disposal of waste water affluents into lakes
    and seas, 665
Dissolved oxygen, 642
  measurement, 299
  sag, 652
  of polluted streams, 652
Distillation, 450
  multiple-effect evaporators, 451
  multistage flash evaporation, 451
  vapor compression process, 451
Distribution, 565
Distribution of chloramines at equimolar concentrations of chlorine and ammonia, 916
Distribution graph, *see* Unit hydrograph
Distribution of hypochlorous acid and hypochlorite ion in water, table, 715
Distribution system capacity, 194
  components, 197
  patterns, 191
  pressure, 195
Distribution systems, 190
  area served criterion, 267
  one-and-two-directional flow, 190
  red, brown or black waters, 483
Distribution works, definition, 4
Distributors, rotating, 564
Diurnal variations, 534
  in the dissolved-oxygen content of a stream, 650
Diversion conduits, 110
Diversion works, 110
DO, 639, 652
  content for fish, 656
  saturation value, 347
  testing, 301
Doe, P. W., 603
Domestic waste water, 6, 8, 9
Domestic waste water composition, 309
Domestic water analysis, 300–302
  consumption, 27
  demand variations, 33
Donaldson, Wellington, 305n
Donovan, D. C., 509n
Draft, coincident, 194, 199
  continuous, 69
Drag coefficient, 368
Drain, building, 247, 249
Drainage system maintenance, 279

Drainage system operation, 279
Drains, 247, 249
Drawdown, 126
　calculation, 240
　constant discharge, 136
　curve, 237
　intermittent discharge, 140
　prediction, 136
　variable discharge, 140
Dredging, 670
Drinking water, 282–285
　bacteriological characteristics, 290
　chemical characteristics, 291
　physical characteristics, 290
　quality standards, 290
Drinking water standards, U.S. Public Health Service, 673, 681
Dry intakes, 109
Dry-weather flow, 249
Dual supply, 220
Dufor, C. N., 27n
Dunbar, W. A., 616n
Dyes, use of, 326
Dystrophic lakes, 663

Eakin, H. M., 83n
Earth dams, see Embankment dams
Echo virus, 285, 642
Ecology, of natural waters, 636
　of receiving waters, 636
Economic usefulness of water, 301
Eddying resistance, 369
Edward E. Johnson, Inc., 148n
Effective passes through treatment units, 342
Effective rain, see Unit hydrograph
Effective velocity, groundwater, 121
Efficiencies, of sewage treatment processes, 321
　of system components, 552
Effluent, acceptability, 311
　characteristics, 300
　disposal in tidal estuaries, 666
　recirculation, 564
　sampling, 552
　weirs, 386
Electric analyzers, 214
Electrode reactions and standard potentials at 25°C, table, 714
Electrodialysis, 6, 462
Electrokinetic properties, colloids, 465

Electroneutrality, 457
Electroneutrality balance, 480–481
Electro-osmosis, 317
Electroosmosis, 465
Electrophoresis, 465
Ellis, M. M., 296
Elmore, H. L., 651
Elutriation, 610
　countercurrent, 610
　of digested sludge, 340
　multiple of sludge, 610
　sludge, 610
　of sludges, 317
Eminent domain, 93
Emscher tank, 624
Endogeny, 536
Engelbrecht, R. S., 664
Engineer, sources of information for, 683
*Engineering News-Record*, cost index, 107
Engineering report, 680
Engineers, role of, 678
*Entamba histolytica*, 642
*Entamoeba histolytica*, resistance to disinfection, 504
Enteric organisms, 283–286
Enzymatic reactions, 537
Enzyme destruction in disinfection, 504
Epilimnion zone, 662
Equilibrium constant, 477
*Escherichia coli*, 504, 642
　destruction concentrations, 509
　disinfection of, 507
　disinfection concentrations of chlorine and chloramines, 509
　irradiation of, 502
Esthetic acceptability of water, tests, 301
Eubacteria, 532
Euglena, 533, 534
Eutectic freezing, 454
Eutrophication, 638, 667
　lake, 664
　prevention of, 488
Eutrophic lakes, 663
Evans, G. R., 363n
Evaporation, 42, 44
Evapotranspiration, 69
Exchange, anion, 455
　cation, 455
Exchange capacity, 457
Expected frequency, 45

# Index

Expression of analytical results, 304–305
Ey, L. F., 482n

Fair, G. M., 16n, 26n, 41n, 75n, 324n, 328, 329, 409, 554, 648n, 653n
Fall overturning, 662
Feeders, chemical, 499
Ferric iron, coagulation with, 468–471
Ferris, J. G., 127n
Fick's law of diffusion, 333
Field sampling, 299
Field surveys, 298
Fiering, M. B., 74n
Fill, critical density, 99
  dams, 96
  maximum density, 99
Filter, appurtenances, 429
  backwashing, 412, 433
  beds, 405
  breakthrough index, 419
  cracks, 433
  design, 418
  dimensions, 429
  floors, 424
  grain shape and shape variation, 404
  grain size and size distribution, 403
  jetting and sand boils, 433
  maintenance, 398
  materials, coefficient of nonuniformity, 404
  effective size, 404
  mud boils, 433
  operation, 398
  performance, 434
  revolving drum, 606
  runs, shortened, 670
  run length, 430
  sand leakage, 433
  sand preparation, 406
  spreading rates, 402
Filtering materials, granular, 402–406
Filters, clogging of, 432
  conduit dimensions, 429
  controlling, head loss, 421
  diatomaceous-earth, 435
  early water, 397
  English, 398
  filter-waste connection, 430
  floc breakthrough, 432
  granular wastewater, 402
  granular water, 398

intermittent sand, 402, 558, 560
hydraulic radius, 409
mixed media, 419
perforated laterials, 421
pressure, 603
pipe grid, 421
rapid sand, 398
rewash connection, 430
roughing, 418, 575
slow sand, 398
underdrainage system, 420, 421, 424
Filtration, 315
  air-bound, 432
  equipment, 431
  head loss, 408
  hydraulics, 408
  kinetics, 417
  mechanical, 425
  natural, 397
  negative head, 432
  sludge, 603
    hydraulics of, 604
  theory, 417
  trickling, 597
  vacuum, 606
  viscosity effects, 419
Finer, S. E., 13n
Fire, damp, 358
  demands, 34
  protection, water for, 686
  reserve storage, 216
Fish, effects of pH change, 296
  kills, 670, 673, 676
  waters management, 296
Fishing and shellfish waters, water-quality management, 296–397
Fish-life hazards, 296
Fixed-spray pressure aerator design, 355
Flagellates, 532
Flash mixes, 441
Flawn, P. T., 162n
Flentje, M. E., 449n
Floc, age, 533
  buildup, 379
  growth, 438
  returned, 569
  shear, 438
Flocculate bacteria, 538
Flocculation, 312, 437
  agitation, 438

allowable loading, 440
anaerobic, 586
baffled channels, 438
biological, 641
bridging mechanism, 472
jar tests, 473
mechanical mixing and stirring, 440
mixing devices, 438
paddle velocity, 442
paddles, 440
by polyelectrolytes, 472
pneumatic mixing and stirring, 439
practice, 442
propellers, 440
stirring, 438
stirring devices, 438
theory, 335, 372, 380
turbines, 440
useful power input, 438, 440
Flocculators, paddle, 440
performance, 443
Flood flows, analysis, 61–62
formulas, 66–68
Flood routing, 85–90
Flotation, 315
agents, 315
air, 393
natural, 393
reagents, 394
tanks, 395
Flow, augmentation, 669
in partially filled sewers, 229
in sewer bends, 242
in sewer overfalls, 242
in sewer junctions, 242
in sewer transitions, 232
turbulent, 600
Fluidized bed, 399
Fluidized beds, activated carbon, 448
hydraulics of, 412
porosity ratio, 412
removal of impurities, 416
scour intensification, 424
Fluid transport, 164–170
Flume, standing-wave, 391
Fluorapatites, 288
Fluoride, 288
in drinking-water, 291
Flux curve, geometry of, 602
Forchheimer, Philip, 244

Form resistance, to flow, 164, 168
Foster, H. A., 62
Fourier series, 333
Foust, A. S., 311n
Fox, D. M., 329n
Franklin, Benjamin, 12
Frazil ice, intakes, 109
Freeboard, definition, 104
Free chlorine, 504
Free fall, particle, 371
Free flow, 6
Freeman, J. R., 195n
Freezing, contact, 454
eutectic, 454
Freundlich adsorption isotherm, 544
Freundlich equation, 337, 446, 447
Frictional resistance, 164
Fritsche's formula, 582
Frontinus, Julius Sextus, Roman water
commissioner, 12
Froude number, 241
Fuel value of sludge, 612
Fullen, W. J., 586n
Fuller, George W., 399
Fuller, W. E., flood flow formula, 68, 90
Functional tests, for sludges and sediments, 304
for water, 302
Fundulus, 306
Fungi, 283, 532
parasitic, 285

Galler, W. S., 568
Galton, Sir Francis, 49n
Ganguillet, E., 225
Gas, absorption, 345
absorption rates, 348
desorption rates, 348
diffusion, 332
dissolution, 350
digester, 620, 631
precipitation, 350
release, 350
removal, 358
sludge, 631
fuel value of, 631
transfer, 312
coefficient of, 348
transfer equation, 354
transfer objectives, 344

# Index

Gas law, general, 345
Gases or decomposition, 648, 662
Gasification, formulation of, 622
Gaussian curve, 45
Gay-Lussac's law, 345
Generalized storage values, 80–81
Genter, A. L., 609
Geometric mean, 47
Geyer, J. C., 11n, 16n, 26n, 27n, 39n, 41n, 63n, 75n, 245n, 264n
Gibb, John, 397
Gifft, H. M., 39
Gizienski, S. F., 102n
Gleason, G. H., 586
Gloyna, E. F., 588
Glucose, 534
Goethe, Faust, quotation, 15
Goiter, 288
Golden or yellow brown algae, 675
Goodson, J. P., Jr., 359n
Grab sampling, errors of, 299
Graham's law, 345
Granstrom, M. L., 590n
Graphical fitting, 23, 58
Gravels, groundwater, 117
Gravity, aerator design, 354–355
  aerators, 351
  dams, design, 102
  dewatering, 603
Grease interceptors, 248
Grease traps, 396
Great overturning, 662
Green, D. E., 504n
Green and blue-green algae, 533
Grit chambers, 390–393
Grotaas, H. B., 568, 646n
Groundwater, 8
  biological contamination, 161
  criteria, 113
  development, 112
  formation constants, 125, 147
  geology, 116
  movement, 120
  quality management, 161
  seepage, flow, 37
  sources, 5
  usage, 112
Groundwaters, composition, 282
  self-purification, 638
Growth curve, mathematics of, 537

Grumbling, J. S., 629n
Guidi, E. J., 607n
Guinea worm, 284, 504, 677
Gumbel, E. J., 62
Gumbel's distribution, 62
Gutter flow equations, 245–246

Hagen, G. F., 121
Hager, D. G., 449n
Halogens as disinfectants, 503
Halvorson, H. O., 567
Hamilton, Alexander, 12
Haney, P. D., 354n
Hantush, M. S., 127, 135
Hardness, noncarbonate, 481
Hardy-Gross method, 204–210
Hartung, H. O., 107n, 363n
Hatch, L. P., 372, 409
Hawkes, H. A., 532
Hazen, Allen, 47, 78n, 80, 166n, 202, 332n, 374, 375, 404
Hazen, H. L., 214
Hazen, Richard, 189
Hazen-Williams coefficient values, 167
Hazen-Williams formula, 166, 225
Hazen-Williams pipe-flow diagram, 698, 699, 700
Head loss, controlling, 421
  definition, 164
Health and safety standards, 684
Healy, T., 472n
Heat, requirements of sludge digestion tanks, 629
Heat drying of sludges, 318, 613
Heath, R. C., 124n
Hemena, J., 586n
Hemorrhagic jaundice, 283
Henry's law, 345, 346
Hepatitis virus, 285
Hermann, E. K., 588n
Hetero-aerobic ponds, 587
Heterotrophic, bacteria, 534
  organisms, 535
Heukelekian, H., 547n
High-service works, 192
High-value district supplies, 192
Hindered settling of discrete particles, 371
Hinds, Julian, 232
HOCl, killing concentration of, 515
Holmes, Justice Oliver Wendell, quotation, 281

Homogeneity, definition, 116
Hoop tension, 173
Hoover process, 483
Horn, H. L., van, 482n
Horseshoe sewer sections, 257
Hotta, N., 601n
House sewer, 247, 249, 250
House sewers, 226
Howe, H. E., 392n
Howland, W. E., 542
Howland's equation, 542
Hudson, H. E., Jr., 419, 443n
Hull, C. H. J., 656n
Humus, 561
Hurwitz, E., 616n
Hutchinson, G. E., 664
Hydrant-flow calculations, 200
Hydrant-flow tests, 199
Hydrant pressures and spacing, 196
Hydraulic, conductivity, 121
  conductivity, measurements, 125
  control of receiving waters, 668
  diffusivity, definition, 124
  design, combined sewers, 279
    sanitary sewerage, 268
    storm drainage, 276
  elements of conduits, 229
  jumps, 241
  loads, 549
  rams, 183
  transients, 169
Hydraulic elements of circular conduits, table, 703
Hydraulics, of filtration, 408
Hydrogen sulfide removal, 359-360
Hydrological, cycle, 343
  data, collection, 43
  frequency functions, 44
Hydrologic equation, 160
Hydrology, definition, 41
Hydrolysis of chlorine, 513
Hyetograph, 274, 276
Hypochlorites, 676
  use as disinfectants, 504
Hypochlorous acid, ionization of, 513-514
  reaction with ammonia, 516
Hypolimnion, 663
Hypolimnion zone, 662

Ice, chlorine, 519

Ideal settling basin efficiency, 373-374
Igneous rocks, 116
Imhoff, Karl, 359n, 624n
Imhoff tank, 624
Impoundage, 71-73
Impounded reservoirs, natural purification of, 667
Impounded supplies, treatment, 71
Impounding reservoirs, 73, 669
Impoundments, pollution of, 659
Incineration of sludges, 318, 613
Industrial, organic chemicals, 288
  purposes water tests, 301
  wastes, definition, 6
  waste waters surcharge, 686
  waste waters flow, 39, 40
  water consumption, 30
  water systems, 220
  water, use, 1, 30
Industrial-waste surveys, 298
Infection, chains of, 285
Infections from water-related sources, 285-286
Infectious hepatitis, 284, 296
Infest trickling filters, 532
Ingersoll, A. C., 377
Injection aerator design, 356-357
Inlet time, 271
Insect larvae, 532
Insufficient mineralization, 288
Intake, conduits, 109
Intakes, 106
  exposed, 107
Intensities, high loading, 554
  loading, 549
Intensity-duration-frequency relationships, 58-60
Intensity of rainfall, 274
Interceptors, 257
  U.K. design, 258
  U.S. design, 257
Interfaces between phases, 325
Interfacial contact of organisms, 316
Intermittent sand filters, 546
Intermittant sand filtration, 525
Interpolation, arithmetic progression, 23
  geometric progression, 23
Intraparticle diffusion, 337, 445
  transport, 337
Intrusive rocks, 116

# Index

Invert drop, 232, 233
Inverted siphons, 256, 257
Iodide, 288
Iodine, 677
  as disinfectant, 503
Ion, concentration, 461
  diffusion, 462
  transfer, 312
Ion exchange, 313
  and exchangers, 454
  kinetics, 325
  selectivity coefficients, 457
Ion-exchange equations, 457
  equilibria, 457
  membranes, 462
  technology, 459
  water softening, 460
Ion-exchanger breakthrough capacity, 460
Ion exchangers, exchange capacities, 456
Ionization of chlorine, 513
Ions, multivalent metal, 469
Ippen, A. T., 104n
Iron, precipitation of, 486
  redox reactions of, 484
  removal, 434, 483
    engineering management of, 487
  solubility of, 483
Iron ration, storage value, 78
Irrigation, 8
  areas, 558
  methods, 559
  subsurface, 527
  water quality, 297
  waters management, 297
Isotopes, use of, 326
Isotrophism, 548, 594
Isotropy, definition, 116
Ives, K. J., 329, 417
Iwasaki, T., 417
Iwasaki, V. T., 328
Izzard, C. F., 275
Izzard's surface detention equation, 275

Jacob, C. E., 123n, 124, 127, 130, 131, 135, 145n
Jenkins, J. E., 83n
Jet pumps, 184
Johnson, A. S., 586n
Johnson, C. F., 230
Johnson, George A., 512

Kahler, P. W., 642
Kaltenbach, A. B., 63n
Kammerer, L. C., 112n
Karassik, I. J., 186n
Katz, W. J., 544
Keifer, C. J., 275n
Kellerman, K. F., 672
Kerby, W. S., 272
Kessenes brush aerator, 357
Ketcham, B. H., 641n
Kimberley, A. E., 482n
Kinematic viscosity, 600
Kinetics, 325
  of aerobic decomposition, 643
  of chemical disinfection, 509
  diffusion, 332
  of filtration, 417
  of oxygenation, 486
  purification, 542
  reaction, 325
  of sludge digestion, 619
  of treatment systems, 539
Kirkwood, James P., 14, 398
Kirschmer, O., 366
Knapp, J. W., 63n
Knowles, D. B., 127n
Knox, W. E., 504n
Kolles, Lewis R., 503n
Kozeny formula, 148
Kurtosis, 46
Kutter, W. R., 225
Kutter-Ganguillet formula, 225
Kutter's coefficient of roughness, 651

Laboratory-activated sludge unit, 548
Laboratory analysis, 299
Lagrangian optimization, 171
Lake intakes, 107
Lakes, disposal of waste-water effluents into, 655
  dystrophic, 663
  eutrophic, 663
  eutrophication of, 664
  limnological considerations, 659
  oligotrophic, 663
  overturning, 661
  pollution of, 659
  thermal stratification in, 663
  vertical gradients of temperature and water quality, 661

La Mer, V. K., 472n
Laminar flow, groundwater, 122
Lang, S. M., 147n
Langelier's saturation index, 475
Langmuir, Irving, 94
Langmuir equation, 338, 339, 446, 447
Layout design, sanitary sewerage, 268
  storm drainage, 276
Leach, C. N., 286
Leakage, controllable, 30
Leaky aquifers, 118, 120
  nonsteady radial flow, 135
Leal, John L., 512
Leaping weirs, 259
Least-squares method, 26, 48–49, 59
Le Bosquet, M. Jr., 656
Lentz, J. J., 39n, 264n
Leptomitus, 532
Leptospirae, 295
Leptospirosis, 283
Lewis, W. K., 328
Lewis and Whitman theory, 348
Li, W. H., 245n
Licensed drain layers, 38
Leibig's law, 665
Lime, control of algae, 672
Limestone formations, groundwater, 117
Limiting velocities of flow, sewers, 226
Limnology, 73
Linaweaver, F. P. Jr., 27n
Lischer, V. C., 107n, 363n
Little, A. D., 311
Load, distribution, 574
  equalization, 538
Loading, of dispersal areas, 666
  intensities, 549
  solids, 602
Lockett, W. T., 569
Logistic growth, 535
  curve, 25
  of population, 331
  zone of, 535
Logistic scale, 25
Longitudinal mixing in tidal estuaries, 666
Longitudinal stress in pipes, 173
Loonam, A. C., 586n
Lower alternate stage flow, sewers, 232, 235
Lowland catchments, 91
Low-service works, 192
Low-water regulation, 669

Ludzack, F. W., 548
Lykken-Estabrook process, 483

McCarthy, P. L., 606n
McDermott, J. H., 542
McGuiness, C. L., 118
McIlroy, M. S., 214
McIlroy network analyzer, 214
McKee, J. E., 297, 377
Mackichan, K. A., 112n
McLean, J. C., 25n
McNown, John S., 275n, 370n
McMath's runoff formula, 275
MPN, use of, 304
Mains, dead-end, 122, 192
  dual, 192
Maintenance of distribution systems, 221
Malaika, Jamil, 370n
Malherbe, H. H., 590n
Manageable variables, 556
Management of distribution systems, 221
Manganelli, R., 547
Manganese, precipitation of, 486
Manhole, flushing, 254
Manning, formula, 225
  gutter, 245
Manning, Robert, 225
Manufacturing rate, definition, 29
Marais, G. K., 588n
Masonry dams, 97
  dimensions, 102
  practical profiles, 102
Mathematical modelling, 311
Mavis, F. T., 392
Maus, L., 311n
Maystre, Y., 11n
Maxcy, K. F., 286
Maximum available drawdown, 151
Maximum well yield, 151
Mean, arithmetic, 45, 46, 47
Mechanical aeration, 669
Mechanical aerator design, 357–358
Mechanical aerators, 353
Median, 44, 47
Median tolerance limit, 306
Megregian, Stephen, 508n
Meinzer, O. E., 65n
Meinzer unit, 122
Membranes, use of, 462
Metal ions, 503

# Index

Metamorphic rocks, 116, 117
Mesolimnion zone, 662
Mesophilic saprophytes, 624
Metazoa, 532
Meterage, 29
Methane, explosive mixtures of, 631
  formers, 624
  removal, 358
Method of images, 144, 146
Method of moments, 644
Michaelis, L., 664n
Michaelis-Menten equation, 664
Microaerphilic organisms, 546
Mills, Hiram F., 14
Mineralization, 534
  excessive, 288
  products of, 534
Minerals, essential, 535
Minimum grades and capacities of circular conduits flowing full, table, 702
Mitchell, R. D., 363n
Mites, 532
Mixed-liquor, volatile suspended solids, 547
Mixing, 665
  complete nutrient equalization by, 574
  in tanks and basins, 327
Moak, C. E., 83n
Mode, 44
Modelling, 311
"Modified" characteristic curves, 188
Mohlman, F. W., 305n
Moisture-weight-volume relationships, 594
Mold hyphae, 532
Molecular diffusion, coefficient of, 333
  of dissolved substances, 332
Molecular transfer, 316
Moments of distributions, 46
Monochloramine formation rate, 517
Monod, J., 535n
Moore, E. W., 328, 329, 644, 648n
Moore, G. T., 672
Moore, W. A., 347n
Morgan, J. J., 485n
Morgan, J. M., 664
Morning-glory spillway, 106
Morris, J. C., 445, 517
Morris, M. E., 677n
Movable-spray aerator design, 356
Mueller, P. K., 605n
Multiple, sampling, 299

transmission lines, 171
well systems, 142
Multiple transmission lines, 171–172
Multistage pumps, 183
Municipal wastewater, definition, 6
Municipal water analysis, 300–302

Naito, S., 601n
National Research Council, 449
Natural bathing water standards, 295
Natural increases, population, 24
Natural purification, 325, 636, 637, 639
  of impounded reservoirs, 667
  regulation of, 669
Natural waters, 281–282
  oxygen economy of, 642
Nematrode, 284, 532
Nesbitt, J. B., 586n
Network circuit laws, 214
Networks, computer programming for, 214
New Delhi infectious hepatitis outbreak, 285
Nidus racks, 527
Nitrification, 535
Nitrobacter, 535
Nitrogen, 535, 633, 634, 637, 663, 665
Nitrogen fixation, by blue-green algae, 665
Nitrosomonas, 535
Nonsteady radial flow, recovery method, 133
  semi-logarithmic approximation, 129
  wells, 127
Nonuniform flow, sewers, 232
Nonuniformity, coefficient of, 404
Normal distribution, 44
Normal dynamic pressure of a distribution system, 197
Normal probability curve, 45
Nozzle design, 355
Nutrient equilization, 574
Nutrient transfer, 316, 543
Nutrients, load of, 538
Nutritional demands, 537

O'Connor, D. J., 666n
Odor, removal, 360, 443, 446, 523
  removal by dechlorination, 518
  threshold, 447
  units, 304
Odors, 289
Off-peak-hour draft, 29
Ogea spillway, 105

Oil interceptors, 248
Okun, D. A., 16n, 26n, 41n, 75n, 83n, 289n
Oligotrophic lakes, 663
Once-through cooling, 293
Open channels, 7, 39
Operating storage, see Equalizing storage
Operation, high rate, 574
Operation of distribution systems, 221–223
Ordon, C. J., 565
Orford, H. E., 547
Organic acids, 587
Organic matter, oxidation, 435
Organisms, troublesome growth of, concentration of copper and chlorine required to kill, 671
Organism removal, 434–435
Orthokinetic motion, 334
Oscillatoria, 533
Osmosis, reverse, 6, 317, 452
Overflow capacity, storage, 218
Overflows, 259
Oxidation, ditches, 583
  ponds, 528, 669
Oxidative deferrization and demanganization, engineering management of, 487
Oxidizing chemicals, 503
Oxygen, demand, benthal, 654
  demand of wastewaters, 646
  dissolved, 642
  economy of natural waters, 642
  economy of receiving waters, 642
  sag, 643, 656
  uptake, rate of, 546
Oxygenation, kinetics of, 486–487
Oxygenation reactions, 485
Oxygen-transfer efficiency, 578
Ozonation plants, modern, 511
Ozone, decontamination factor of, 510
  disinfection, 509
  distribution, coefficient of, 510
  production of, 509
  toxicity of, 509
Ozonizing towers, 340

Package plants, 528
Palatability, 289
Palatability tests, 301
Palin, A. T., 517
Parasitic organisms infective to man, 283
Paratyphoid fever, 283

Pardoe, W. S., 202
Pardoe method of sectioning, 202
Parmakian, John, 181
Partial penetration, wells, 148
Particle contact, theory, 335
Pathogen destruction, 500
Pathogenic bacteria, 634
Pathogens, human euteric, 501
Peak flow costs, 686
Pearl, Raymond, 25n
Pearson, E. A., 666n
Pearson's type-III distribution, 62n
Perched water table, 120
Performance loading relationship, 553
  for activated-sludge units, 555
  for contact aerators, 555
  for trickling filters, 555
Performance-time relationship, 539
Perikenetic motion, 334
Permeability, coefficient of, 121
  definition, 115
  of river channels, 670
Perviousness, definition, 115
Pesticides, 288
Peterka, A. J., 106n
Pettet, A. E. J., 546n
pH, effect on breakpoint reaction, 518
  range, 538
  testing, 301
  units, 304
Phase boundaries, 325
Phase transfer, 311
Phelps, E. B., 328, 333, 586n, 653
Phenolic tastes, 672
Phenol removal, 360
Phosphate, precipitation of, 488
  removal, 535
  uptake, rate of, 664
Phosphates, 634
  removal of, 535
Phosphorus, 535, 633, 637, 663, 665
Photosynthesis, 344, 435, 532, 534, 637, 643, 649, 655, 659
Photosynthetic oxygenation, rate of, 655
Phreatic surface, definition, 118
Phytotoxins, 288
Piezometric surface, definition, 120
Pilot-plant tests, 449
Pipe, crown of, 173
  equivalence method, 210

## Index

network analysis, 201
standard lengths, 177
Pipeline leakage units, 178
Pipelines, protection of, 495
Pipes, flexural stress in, 174
growth in, 670
Piret, E. L., 567
Plain sedimentation, 597
Planck's constant, 502
Plankton, 663
Planning water and wastewater projects, 681
Plans and specification, water projects, 682
Plant siting, 681
Plastic waterpipes, 288
Point of concentration, 60
Point rainfalls, 52
Poiseuille, J. L., 121
Poiseuille's equation, 599
Poliomyelitis virus, 285, 590, 642
disinfection of, 507
resistance to chlorination, 505
Polluted streams, dissolved oxygen sag of, 652
Polluted waters, atmospheric reoxygenation of, 649
deoxygenation of, 645, 652
reaeration of, 652
reoxygenation of, 652
Pollution, by backflow, 284
by cross-connections, 284
parameters of, 639
patterns of, 637
shellfish waters, 296
tests, 301
Pollutional surveys, 298
Polyacids, 538
Polyacrylate, 471
Polyamino acids, 471
Polyelectrolytes, 288, 471, 538
Polymerization, 468
Polymers, 471, 538
Polyprotonic acids, (Ion exchange), 457
Polysaccharides, 538
Polyvinylpyridinium salts, 471
Ponds, aerobic, 587
anaerobic, 587
amphi-aerobic, 587
hetero-aerobic, 587
mechanically aerated, 591
stabilization, 586
design of, 587

Population, densities, 27
equivalent, 304, 646
growth, rate of, 537
Porges, Ralph, 542
Porosity, definition, 113
effective, definition, 115
Portland Cement Association, 684
Postchlorination, 521
Potassium, 634
Power dissipation function, 335, 550
Prechlorination, 521
Precipitation, 641
chemical, 474
of Fe and Mn, 486
Pressure, filters, 603
Pressure surveys, 199
Primary treatment, 322, 597
Prior, J. C., 174n
Prior's observational equation, 174
Probability paper, 47
Process loads, 550
Process waters, 293
quality standards, 294
Project financing, 684
borrowing, 684
general obligation bonds, 684
revenue bonds, 684
special assessment bonds, 684
Projects, planning, water and waste water, 681
Propeller pumps, 183
Property drains, 249, 255
Proportionality factor, sewers, 232, 233
Protozoa, 283, 634
ciliated, 641
Protozoal infections, 284
*Psychoda,* 522, 532
*Public Works,* 684
Pump characteristics, 184
curves, 187
Pumping stations, 183
intakes, 109
Pumping system head relationships, 185
Pump priming, 186
Pumps, centrifugal, 183
diagonal flow, 183
displacement, 183
displacement ejectors, 184
sludge feed, 607
types, 183
uses, 183

Pure oxygen, use of, 580
Purification, efficiency, 324
 kinetics, 328
  effects of temperature on, 542
  natural, of impounded reservoirs, 667
  regulation of, 669
 works, definition, 4
 time rates, 330
Purification kinetics, 328
Purification works, definition, 4

Quiescent settling, 372
Quirk, T. P., 546

Rth moment about origin, 46
Racks, 314
 on water intakes, 362–363
Radial flow pumps, 183
Radioactive substances, 288
Radioactivity in drinking water, 291
Rainfall, and runoff analysis, 50
 annual, 50
Rainwater, collection, 111
 sources, 4
 yield, 111
Rapid filters, materials, 402
Rate, of deoxygenation, 648
 of gas absorption, 348
 of gas desorption, 350
 of phosphate uptake, 664
 of photosynthetic oxygenation, 655
 of reaeration, 651
 of treatment response, 328
Rates, water, 684
Ratio, self-purification, 653
Rational method, 62–63
Raw data, 41
Rawlinson, Robert, 14
Reaction orders, 331
Reactions, enzymatic, 537, 644
Ready-to-serve requirements, 194
Reaeration, 572, 639
 of polluted waters, 652
 rate of, 651
Reagents, activating, 394
 collecting, 394
 depressing, 394
 flotation, 394
 foaming, 394
 frothing, 394

Recarbonation, 345
Receiving streams, allowable BOD loading of, 655
Receiving waters, hydraulic control of, 669
 oxygen economy of, 642
Recharge boundaries, 144–145
Reciprocal detention time, 549
Recirculating systems, 551
Recirculation, 341, 551
 ratio, 576
 of wastewater, 551
Reclamation of wastewaters, 498
Recovery of impounded reservoirs, 667
Rectangular sewer sections, 257
Redox reactions of Fe and Mn, 483
Reduction of infections, 286
Red water, see Corrosion
Regression analysis, 49
Regulation, low-water, 669
 of natural purification, 669
 of sludge disposition, 670
 of wastewater discharge, 668
Rejected recharge, groundwater, 160
Relaxation method, Hardy Cross procedure, 204
Reliability, distribution, 47
Removal, of ammonia nitrogen, 539
 of BOD, 539
 of turbidity, 539
Renovation of water, 322–324
Reo virus, 285
Reoxygenation, 643
 of polluted waters, 652
Reservoir, evaporation control, 94
 intakes, 107
 management, 94
 quality control, 94
 siting, 92
 storage as a function of draft and runoff, 74
Reservoirs, area and volume, 84
 distributing, types, 218
 impounded natural purification of, 667
 site preparation, 93
 thermal stratification in, 661
 vertical gradients of temperature and water quality, 661
Residential water use, 1
Respiration, 344
Response factor of reaction velocity, 329
Retardance, 60

# Index

Retarding basins, 259
Reverse osmosis, 6, 317, 452
Revolving, drum filter, 606
Reynolds, Reginald, 11n
Reynolds number, 368, 369
Riffle plates, 352
Rippl mass-method, see Rippl method
Rippl method, 74, 75, 77
Rippl, W., 74n
Riprap dams, 96, 102
River intakes, 107
Rocks, extrusive, 116
Rohlich, G. A., 544
Romero, J. C., 162
Rorabough, M. I., 150n
Rose, H. E., 409n
Rotary, distributor design, 356
  pumps, 183
Rotating distributors, 564
Rotifers, 532
Rotodynamic pumps, 183
Roundworm, 284
Rouse, Hunter, 242
Rules of thumb, sewer transition flow, 233
Run-of-bank sands, 398, 406
Runoff, agricultural, 664
  annual, 51
  coefficients, 272
  empirical, formulations, 275
  hydrograph, 274
Rural wastewaters, flow, 40
Rural water consumption, 32

Safe yield, of an Aquifer, 160
  streams, 74
Safferman, R. S., 677n
Sag curve, critical point, 653
Sag pipes, 256
Saline-water conversion processes, 450
Salinity of water, classification, 449
*Salmonella typhosa,* 642
Salmonellosis, 283
Salt water conversion, 449
Salt water creep, 666
Sample analysis, 299
Sampling, 299
Sand filters, intermittent, 546, 558, 560
Sand formations, groundwater, 117
Sand interceptors, 248
Sanitary sewer average flow, U.S., 249

Sanitary sewerage, capacity design, 267
Sanitary sewers, 2, 6, 261
  design, 262
Sanitary surveys, 298
Sanitation history, 11–14
Saprobic microorganisms, 530
Saprobic organisms, 316, 564
Saprophytes, 343, 639, 641
Saprophytic bacteria, 534
Saturation index, 475
Saturation values of dissolved oxygen in fresh
  and sea water, table, 713
Savage, G. M., 567
Sawyer, C. N., 535, 629n, 664
Schaake, J. C., 63n
Schistosome cercariae, 295
Schistosomes, 504, 611
Schistosomiasis, 284
Schmutzdecke, definition, 416
Schroeffer, G. J., 586n
Scott, R. D., 482n
Scour, 649
  air, filters, 399
  of bottom deposits, 377
  mechanical, filters, 399
  surface, filters, 399
Screened bellmouths, intakes, 107
Screening, 362
Screens, 314
Scum accumulation in digesters, 626
Sea, disposal of wastewater effluents into, 665
Secondary treatment, 322, 597
Sectioning, 201–204
Sedgwick-Rafter method, 305n
Sediment, composition tests, 303
  concentration tests, 303
  condition tests, 303
Sedimentary formations, 116, 117
Sedimentation, 314, 367, 641
  plain, 368, 597
Seed populations, 20
Seepage, 670
  loss, storage, 81, 82
Seggem, M. E. von, 237n
Selective draft, 70–71
Self-purification, bacterial, 641
  coefficient of, 656
  constant, 655
  of groundwater, 638
  ratio, 653

values of, 655
Semiconfined aquifers, 118
Separate sewer systems, 2, 6
Separate sludge digestion units, 624, 626
Separators, 248
Septic tanks, 161, 527, 624
Segment peak procedure storage, 74
Serial correlation, 49
Service headers, 192
Services to premises, 193
Settling, curves, 389
  of flocculent suspensions, 372
  hindered, 601
  paths of discrete particles, 373
  velocities of discrete particles, 368
Settling basin, dead spaces, 376
  efficiency, 374
    effect of currents, 374
  performance curves, 375
  short-circuiting, 375
  stability, 375
  zones, 373
Settling tank loadings, 388
  detention periods, 388
  performance, 389
Settling tanks, allowable BOD of, 563
  affluent weirs, 386
  general dimensions of, 382
  inflow and outflow structures, 387
  inlet hydraulics, 385
  outlet hydraulics, 386
  sludge removal, 384
Sewage, farms, 398
  treatment plants, typical layouts, 323
  treatment processes efficiencies, 321
Sewage-sick, 559
Sewer, building, 247, 249, 250
  construction details, 251
  cross sections, 256
  discontinuous surge fronts, 241
  flow, nature, 224
  flows, damping effects, 39
  gas, 255
  length criterion, 267
  profiles, common elements, 264
  street, 247, 249, 250
  transition lengths, 237
  transitions, 232
  use-charge, value received basis, 11
Sewers, bed loads, 226

critical depth, 234, 240
critical stage flow, 255
damaging velocities, 229
depressed, 256
egg-shaped, 232, 256, 257
systems, 686
Shaft spillway, 106
Shannon, W. L., frost depth equation, 175
Shape factor, definition, 405
Shattuck, Lemuel, 13
Shaw, V. A., 588n
Shelf items, standardized, 683
Shellfish, disinfection of, 522
Shellfish waters, disease hazards, 296
  management, 296
Sherard, J. L., 102n
Shields, A., 228
Shigellosis, 283
Sholji, I., 417
Shortened filter runs, 670
Shutoff head, 186
Side-channel spillway, 106
Siemens, Werner Von, 509ı
Sieves, American (U.S.) standard, 403
  manufacturer's rating, 404
Silica, activated, 473
Sitting, storage, 81, 83–84
Silts, groundwater, 118
Simpson, James, 14, 397
Single mains, 192
Single-stage pumps, 183
Sinkoff, M. D., 542
Siphons, 174
Siphon spillways, 259, 260
Sizing of conduits, 170
Skewness, distribution, 44, 46
Skimming tanks, 395
Sleek field, 666
Slope, embankments, 99
Sludge, activated, 602
  agricultural value of, 633
  air drying, 617
  benthal, 587
  blanket filtration, 380
  cake, 606
  coagulation of, 608
  centrifuging, 317
  chemical, conditioning, 608
  composition tests, 303
  concentration, 601

# Index

concentration tests, 303
condition tests, 303
density indes (SDI), 305
disposition, regulation of, 670
dewatering, 317, 602
digested, elutration of, 340
digesters, 624
digestion, 318
  anaerobic, 618
  kinetics of, 619, 622
digestion tanks, heat requirements of, 629
digestion units, 624, 626
disposal, 634
drying beds, 617
elutration, 317
excess, 569
feed pumps, 607
filtration, 603
  hydraulics of, 604
flow, 599
freezing, 603
fuel value of, 612
handling, 634
heat drying, 613
incineration of, 613
lagoons, 634
liquor, 321, 627
pipes, 385
returned, 550
specific gravity, 305
stabilization using chlorine, 523
thickening, 317
treatment, 322
volume index (SVI), 305, 576
wet, specific gravity, 595
wet-air oxidation, 616
Sludge, dry combustion of, 318
  primary, 602
  proportion of solids, 598
  wastewater, 596
  water, amounts of, 596
  waterworks, 596
Sludge-burning furnaces, 614
Sludge gas, 631
  collection devices, 631
  fuel value of, 631
Slurries, 593
Smith, Robert, 11n
Smith, Stephen, 14
Smoluchowski, Miron von, 334, 335

Smoluchowski's orthokinetic flocculation, 378
Snow, W. B., 523, 644
Sodium, 288
Softening, by ion-exchange, 460
  stoichiometry of, 479
  water, 477
  stabilization by $CO_2$, 481
Softness, 288
Soil stripping, 94
Solids, loading, 602
  management, 593
  stabilization, 317
  transfer, 314
  waste, 592
Solids, concentration, 317-318
Solubility constants of important chemical substances in water at $25°C$, table, 712
Solubility of gas, 346
Solute stabilization, 313-314
Solvent extraction, 451
Sorption kinetics, 325
Sorteberg, K. K., 245n
Soucek, E., 392n
Sources, of gases in water, 343-344
  of information for engineer, 683
Southgate, B. A., 296n
Spalding, G., 445
Spargers, 353
Specific capacity-drawdown curve, 151
Specific capacity, wells, actual, 148
  theoretical, 147
Specific speed, pumps, 184
Specifications, water projects, 682
Specific yield, definition, 115
Spent water, 2, 27, 37, 247
  collection, 249
Sphaerotilus, 532
Sphencity, definition, 405
Spiders, 532
Spillway, capacity, design, 45
  channel, 105
  drop-inlet, 106
  types, 104, 106
Spillways, 104
Spirochetes, 283
Spray aeritors, 352
Spring circulation, 661
Springing line, 173
Spring overturning, 661
Stabilization, chemical, 475

Stabilization ponds, 528, 532, 534, 586
   diurnal variations, 534
   ecology of, 532, 533
   ecology of aerobic, 533
Stacks, 247, 249, 250
Stallman, R. W., 127n, 135
Standard deviation, 45, 46
   arithmetic, 47
   geometric, 47
Standard fire stream, 195
Standardization, 684
Standardized shelf items, 683
"Standard Methods," 299, 301
Standard tests, water and wastewater, 300–304
Stand-by requirements, fire service, 194
Standpipes, 218–220
Stanley, C. M., 679
Staphylococci, resistance to disinfection, 505
Statistical hydrology, definition, 41
Steady state tests, *see* Continuous culture techniques
Stein, P. C., 335
Sterilization of water, 500
Stirling's formula for approximating factorials, 327
Stoke's law, 370, 393
Storage, coefficient of, 123
   coefficient, groundwater, 124
   elevation, 218
   emergency reserve, 217
   equalizing, 215
   evaporation loss, 81
   location, 217
Storm, drains, 261, 264
   intensity, 53
      units, 53
   rainfall, 52
   runoff flows, 60
      estimation methods, 62
   sewers, 2, 6
Storm-pattern analysis, 274–275
Storms, frequency of intense, 55
Stormwater, collection, 254
   flow, 38
   sewers, 2, 6
Stormwaters, chlorination of, 524
Straight run maxima, 251
Straining of solids, 314
Stratification, thermal, 661, 663
   of digestors, 626

Straub, L. G., 243n
Street, inlets, 244, 255
   types, 245
   sewer, 247, 249, 250
Streeter, H. W., 328, 653
*Streptoccus fecalis,* 642
Stumm, Werner, 485n, 586n
Stumpf, P. K., 504n
Submerged aeration devices, 584
Submerged-contact-aerators, *see* Aerated contact beds
Submerged intakes, 107
Substrate utilization, 331, 537
   management of, 537
   rate of, 537
Subsurface disposal, 162
Suction specific speed, pumps, 186
Summer stagnation, 662, 670
Sumps, 248
Superchlorination, 518
Supernatant water, 649
Supply, dual, 220
Surface resistance, exponential equation, 164
   rational equation, 164
Surface tension of water, table, 693
Surface waters, composition, 281
   sources, 4, 69
Surfactants, 503
Surge front profile, 242
Surge tanks, 170
Suspended matter, volume concentration of, 576
Suspended solids, dry-weight concentration of, 576
Swamp drainage, 92
Swamp waters, composition, 282
Swift, Jonathan, 2n
Swimmer's itch, 284
Swimming pool bacteriology, 295
Swimming pool standards, 295
Symbols, table, 688
Symons, James M., 670n
*Synura,* 360, 524, 670, 673
System head, definition, 184
Systems management, 686

Tank, design, elements of, 378
   dimensions, 550
   Emscher, 624
   Imhoff, 624

secondary settling, 550
Tanka, S., 601n
Tanks, dosing, 566
  elevated, 218
  primary setting, 550
  separate digestion, 626
  septic, 624
Taste, removal, 360, 443, 446
  removal by dechlorination, 518
  threshold, 447
Tastes, 289
Technical equilibrium, 324
Temperature, effects, 542, 623
  on chemical and biological processes, 332
  importance in surveys, 298
Tenney, M. W., 586n
Ten-State Standards, 563
Tertiary treatment, 322, 402
Testing, bench scale, 547
Theis, C. V., 123n, 127, 145n
Theis equation, 128
Theis type-curve, 130
Thermal death point of saprophytes, 624
Thermal stratification, 661, 663
  in lakes and reservoirs, 661
Thermocline zone, 662
Thermophilic organisms, 624
Thickening of sludges, 317
Thirumurthi, D., 377n
Thixotropic, 599, 600
Tholin, A. L., 275
Thomas, H. A., 237n, 375
Thomas, H. A., Jr., 74n, 329, 554, 644, 468, 654
Thomas, H. E., 118
Tidal estuaries, effluent disposal in, 666
  longitudinal mixing, 666
Time, of concentration, 60, 271
  of exposure, 326, 329
  of passage, 326
Time-concentration curve of a basin, 377
Time-rate factor, 329
Tomlinson, T. G., 586n
Total storage, 217
Toxicity measurement, 306–307
Toxic metal tests, 301
Tracers, use of, 326, 376
Trace elements, 633
Trainer, F. W., 124n
Transfer, gradient, 325

kinetics, 325
mechanisms, biological treatment, 528
nutrient, 528, 543
processes, concurrent operation of, 339
  countercurrent operation of, 339
Transfer coefficient and liquid film thicknesses for gases of low solubility in water, table, 711
Transition zone, 662
Transmission line, air valves, 180
  altitude control valves, 183
  anchorages, 182
  appurtenances, 178
  blowoff valves, 180
  check valves, 182
  expansion joints, 182
  insulation joints, 182
  kick blocks, 182
  location, 174
  manholes, 181
  pressure-reducing values, 183
  shutoff valves, 183
  surge tanks, 182
  valves, 180
  Venturi meters, 183
Transmission lines, carrying capacity, 176
  cross-section, 172
  depth of cover, 175
  durability, 177
  leakage, 178
  line and grade, 174
  maintenance, 178
  materials of construction, 177
  materials transportation, 177
  safety, 177
  strength, 177
  structural requirements, 173
  vertical and horizontal curves, 175
Transmission systems, 163
Transmission works, definition, 4
Transmissivity, coefficient of, 123
  definition, 123
Transpiration, 42, 44
Transporting velocities, sewers, 227
Trap efficiency, reservoirs, 83
Traps, 248
Trash racks, 362
Travis, W. O., 624
Treatment, efficiency, 550
  kinetics, 324

primary, 597
processes tests, 303
secondary, 597
systems, kinetics of, 539
time, 542
two stage, 574
works, definition, 4, see also Purification works
system design, 318
Treatment systems, kinetics of, 539
Tributary population load, 550
Truckling filters, 527, 554, 556, 560
allowable BOD of, 563
contact material of, 561
design of, 561
ecology of, 532
film control, 529
hydraulic design of, 564
loading of, 562
optimal dimensions of, 568
performance-depth relationship, 541
performance-loading relationship for, 555
process design of, 562
removal of BOD, 539
Trickling-filter, films, 532
humus, 597
Trubnick, E. H., 605n
Tsivoglou, E. C., 646, 656
Tsugami, S., 601n
Turbidity, 289
removal of, 434, 539
units, 304
Turbulent flow, groundwater, 120
Turcan, A. N., 148n
Turnkey projects, 683
Turre, G. W., 363n
Typhoid fever, 283, 296
Typhoid organism secretion, 285

Uhlig, H. H., 495n
Ultraviolet light irradiation, 502–503
Ultimate adsorptive capacity, 445
Unconfined aquifers, 118, 124
nonsteady radial flow, 134
Underdrainage, 567
Underdrains, 559
Underflow, 602
Unit hydrograph, 63–66
Unit operations, concept, 311
coordination, 318

effects on water characteristics, 319
optimization, 401
Unsteady flow, sewers, 232
Upflow clarification, 379–381
Upland catchments, 91
Upper alternate stage flow, sewers, 232, 235
Urban water consumption, 1, 29
Use cycle of water, 280
Useful power dissipation, 544
Useful power input, 544
Useful storage, 84
U.S. Census dates, 19
U.S. Geological Survey, groundwater reports, 118
maps, 175
U.S. Public Health Service, 290, 434
U.S. Public Health Service Drinking Water Standards, 290, 673, 681
Utilities, 223, 279
Utilization of oxygen by polluting substances, 643

Values of the exponential $e^{-x}$ for x, table, 695
Values of the well function $W(u)$ for various values of u, table, 696
van der Waals force, 467
van der Waal forces, 416
van't Hoff-Archenius equation, 332, 651
van't Hoff-Arrhenius relationship, 508, 542, 623, 645
Vapor pressure of water, 347
table, 683
Variability, distribution, 44
Variance, 46
Variation, coefficient of, 47
in water demand, 33–36
Velocities, self cleaning, 600
Velocity, factor, 549
head, 243
Velocity of flow and rate of discharge for pipes flowing full, table, 697, 702
Vents, 248
Verhulst, P. F., 25
Vertical flow tank sections, 379
Vesilund, P. A., 607n
Viral infections, 284–285
Viruses, 283, 501, 634
Viscosity, 122
kinematic, 122, 600
Viscosity and density of water, table, 692

# Index

Volatile-suspended-solids accumulation, 547
von Liebing, Baron, 665

W(u, r/B) function values, 137
Wagner, W. E., 106n
Wall effect in settling, 371
Walton, W. C., 127n
Waring, F. W., Jr., 482n
Wash, high-velocity, filters, 399
Washwater troughs, 426–428
Waste cycle of water, 280
Waste products, disposal of, 498
Waste solids, drainability, 603
    filterability, 603
    sources, 592
    treatment and disposal of, 592
    kinds, 592
        detritus, 593
        grit, 593
        rakings, 593
        screenings, 593
        skimmings, 593
        sludges, 593
        slurries, 593
        underflow, 593
Waste treatment biology, 316
Wastewater, analysis, 302
    anticipated volumes, 37
    composition tests, 302
    concentration tests, 302
    condition tests, 302
    cost of municipal systems, 10
    discharge, regulation of, 668
    effluents, disposal into lake and spas, 665
    examination, 302
    examination objectives, 297
    farms, management of, 559
    flocculation, 586
    flows, variations, 39
    functional tests, 303
    municipal, composition, 307
    projects, planning, 681
    properties, 8
    racks, 363
    reclamation, 297
    sludge, examination, 303
    sludges, 596
    solids, hazards of, 634
    surveys, 298
    treatment, 319
    purpose of, 310
    tertiary, 402
Wastewaters, chlorination of, 524
    oxygen demand of, 646
    probable behavior of, 547
Water, analysis, 300
    interpretation, 305
    methodology, 299
    color of, 289
    consumption, 27, 28
    consumptive use, 559
    cycle, 42
    deionization, 460
    examination objectives, 297
    for fire protection, 686
    hammer, 169
    phases, 332
    projects, planning, 681
    purification, 319
    quality criteria, 15
    rates, 684
    resources, global, 41
    reuse, irrigation, 297
    sludges, amounts of, 596
    supernatant, 649
    supplies, cost of municipal, 10
    table, aquifers, 118
        definition, 118
    treatment, effects on water characteristics, 319
    use, 1
    and wastewater systems, design period, 18
        performance and required capacities, 17
        population data, 19
    and wastewater treatment chemicals, 716
Waterborne poisons, 287–288
Water Pollution Control Federation, 561, 684
Water quality, 639
    analysis, 299
    chemical management, 299
    sampling, 299
    standards, industrial, 291
    surveys, 298
Water-quality management, 670, 677
    irrigation waters, 297
    objectives, 280
    reduction of infections, 286
Water-surface response, 81–82
Water treatment, plants, typical layouts, 320
    purpose of, 310

Water-treatment plant performance, 434
  removal, of bacteria efficiency, 434
    of color, 434
    of iron, 434
    of turbidity, 434
Water weeds, chemical destruction of, 676
Waterworks sludges, 596
Watson, R. M., 186n
Wattie, Elsie, 508n
Weber-Fechner law, 446
Weber, W. J., Jr., 445n
Wegmann, Edward, 102n
Wegmann's Practical Profile Types for dams, 102
Weights and measures, table, 690
Weil's disease, 283
Weir, proportional-flow, 391
Weir, Sutro, 392
Weirs, diverting, 259
Weisbach-Darcy, 582
Weisbach-Darcy equation, 377
Well, characteristics, 147
  construction, 154
  design, 152
  development, 158
  effective radius, definition, 148
  hydraulics, 126
  loss, 147
  maintenance, 159
  measurements, 149
  protection, 158
  pumps, 157
  testing, 158
  types, 155
  yield, 150
Wells, area of influence, 126
  bored, 156
  collector, 157
  cone of depression, 126
  drilled, 157
  driven and jetted, 156

dug, 155
formation, 105, 147
Wenzel, L. A., 311n
Wet-air oxydation of sludge, 116
Wet combustion of sludges, 318
Wet intakes, 109
Wet-weather flow, 249
Whipple, G. C., 13n, 305n
Whitlock, W. E., 363n
Whitman, W. G., 328, 348n
Willcox, E. R., 230
Williams, Gardner S., 166
Winter chapping, 288
Winter stagnation, 661, 670
Wislicensus, G. F., 186n
Wolf, H. W., 297
Wolfenden, Sir John, 311n
Wolff, J. B., 27n
Woodward, R. J., 102n
Woodward, S. M., 230
Worm, guinea, 677
Worm infections, 284
Worms, 283, 435, 634
  parasitic, 590
Wuhrmann, Karl, 547n

Yarnell, D. L., 230
Yoshioka, N., 601n
Yu, Y. S., 275n

Zeolites, 454
Zeta potential, 465
Ziemke, N. R., 586n
Zone, transition, 662
  epilimnion, 662
  thermocline, 662
  mesolimnion, 662
  hypolimnion, 662
*Zooglea ramigera*, 532